TEXTBOOK OF WORK PHYSIOLOGY

PHYSIOLOGICAL BASES OF EXERCISE

McGraw-Hill Series in Health Education, Physical Education, and Recreation

Deobold B. Van Dalen, *Consulting Editor*

Åstrand and Rodahl: Textbook of Work Physiology: Physiological Bases of Exercise
Jensen, Schultz, and Bangerter: Applied Kinesiology and Biomechanics

TEXTBOOK OF WORK PHYSIOLOGY

PHYSIOLOGICAL BASES OF EXERCISE

THIRD EDITION

Per-Olof Åstrand, M.D.

Swedish College of Physical Education
Stockholm, Sweden

Kaare Rodahl, M.D.

Institute of Work Physiology
Norwegian College of Physical Education
Oslo, Norway

McGRAW-HILL BOOK COMPANY

New York   St. Louis   San Francisco   Auckland   Bogotá
Hamburg   London   Madrid   Mexico   Montreal   New Delhi
Panama   Paris   São Paulo   Singapore   Sydney   Tokyo   Toronto

To
PROFESSOR ERIK HOHWÜ CHRISTENSEN,
who first introduced us to the field of work physiology.
It is to a large measure due to his encouragement
and continuous and active interest
that the writing of this book was undertaken.

TEXTBOOK OF WORK PHYSIOLOGY
Physiological Bases of Exercise
INTERNATIONAL EDITION 1986

Exclusive rights by McGraw-Hill Book Co. - Singapore
for manufacture and export. This book cannot be re-exported
from the country to which it is consigned by McGraw-Hill.

9 0 KKP 9 4

This book was set in Times Roman by J. M. Post Graphics, Corp.
The editor was Marian D. Provenzano;
The production supervisor was Marietta Breitwieser.

Library of Congress Cataloging-in-Publication Data

Astrand, Per-Olof.
  Textbook of work physiology.

  (McGraw-Hill series in health education, physical education,
and recreation)
  Includes bibliographies and index.
  1. Work-Physiological aspects. 2. Exercise-
Physiological aspects. I. Rodahl, Kåre, 1917-
II. Title. III. Series. (DNLM: 1. Exertion.
2. Physical Education and Training. WE 103 A115t)
QP301.A67   1986   612'.042   85-18225
ISBN 0-07-002416-2

When ordering this title use ISBN 0-07-100114-X

Printed in Singapore

CONTENTS

PREFACE   xi

1   Our Biologic Heritage   1

References   10

2   The Muscle and Its Contraction   12

The Cell   13
Architecture of the Skeletal Muscle   19
Myofibrillar Fine Structure   23
Interaction of Actin and Myosin   28
Different Fiber Types   33
Types of Muscle Activation   40
Mechanical Efficiency of Muscle Contraction   45
Storage and Utilization of Elastic Energy   46
References   50

3   Neuromuscular Function   54

The Motoneuron and Transmission of Nerve Impulses   56
Transmission of Impulses from Nerve Axon to Skeletal Muscle   70
Reflex Activity and Some of the Basic Functions of Proprioceptors   73
Properties of the Muscle and Muscular Contraction   100

Muscle Fatigue 115
References 122

4 Body Fluids, Blood and Circulation **127**

Body Fluids 129
Blood 130
Heart 139
Hemodynamics 144
Blood Vessels 149
Regulation of Circulation at Rest 158
Regulation of Circulation during Exercise 168
Cardiac Output and the Transportation of Oxygen 174
Summary 201
References 202

5 Respiration **209**

Main Function 210
Anatomy and Histology 211
"Air Condition" 218
Filtration and Cleansing Mechanisms 218
Mechanics of Breathing 219
Volume Changes 222
Compliance 227
Airway Resistance 227
Pulmonary Ventilation at Rest and During Exercise 228
Diffusion in Lung Tissues, Gas Pressures 240
Ventilation and Perfusion 244
Oxygen Pressure and Oxygen-Binding Capacity of the Blood 247
Regulation of Breathing 249
Breathlessness (Dyspnea) 262
Second Wind 262
High Air Pressures, Breath Holding, Diving 263
References 267

6 Skeletal System **273**

The Many Functions of Bone 273
Evolution of the Skeletal System 274
Structure 274
Function 275
Growth 279

Repair–Adaptation 280
Joints 281
Ligaments and Tendons 286
Pathophysiology of the Back 286
References 293

7   Physical Performance   295

Demand-Capability 296
Aerobic Processes 299
Anaerobic Processes 314
Lactate Production, Distribution, and Disappearance 320
Interaction Between Aerobic and Anaerobic Energy Yield 325
Maximal Aerobic Power—Age and Sex 330
Anaerobic Power, Sex, and Age 341
Muscle Strength, Sex, and Age 342
Performance in Sports—Women Versus Men 346
Concluding Remarks 347
References 348

**8   Evaluation of Physical Performance on the
      Basis of Tests   354**

Physical Fitness Tests 355
Tests of Maximal Aerobic Power 355
Evaluation of the Anaerobic Power 384
Measurements of Muscular Strength 386
Exercise Electrocardiogram 386
References 387

9   Body Dimensions and Muscular Exercise   391

Statics 392
Dynamics 392
Summary 408
References 410

10   Physical Training   412

Introduction 413
Training Principles 420
Training Methods and Biological Long-Term Effects of Training 427

Miscellaneous   468
Ischemic Heart Disease   471
Contradictions for Physical Training   474
References   476

11   Applied Work Physiology   486

Introduction   487
Factors Affecting the Ability to Perform Sustained Physical Work   488
Assessment of Work Load in Relation to Work Capacity   490
Energy Expenditure of Work, Rest, and Leisure   501
Fatigue   512
Vibration   515
Circadian Rhythms and Performance   515
Effects of Menstruation   518
References   518

12   Nutrition and Physical Performance   523

Energy Liberation and Transfer   524
Energy Transformation   524
The Oxidation of Fuel   530
Control Systems   532
Interplay Between Anaerobic and Aerobic Energy Yield   534
Relative Importance of the Different Energy Stores   535
Summary   537
Nutrition and Physical Performance   538
Nutrition in General   540
Digestion   541
Energy Metabolism and the Factors Governing the Selection of Fuel for
     Muscular Exercise   543
Regulatory Mechanisms   557
Food for the Athlete   562
Physical Activity, Food Intake, and Body Weight   566
References   576

13   Temperature Regulation   583

Heat Balance   584
Methods of Assessing Heat Balance   585
Magnitude of Metabolic Rate   588
Effect of Climate   589
Effect of Exercise   595

Temperature Regulation 600
Acclimatization 605
Limits of Tolerance 610
Coordinated Movements 618
Mental Work Capacity 619
Water Balance 619
Practical Application 625
References 636

14   **Applied Sports Physiology** 646

Introduction 646
Analysis of Specific Athletic Events 649
Sex Differences 678
References 679

15   **Factors Affecting Performance** 683

Introduction 684
High Altitude 684
High Gas Pressures 706
Tobacco Smoking 709
Alcohol and Exercise 711
Caffeine 713
Doping 713
The Will to Win 715
References 717

APPENDIX 725

Definition of Terms and Units 725
Prefixes, Symbols, Unit Abbreviations 726
Conversion Tables 727
List of Symbols 733
GLOSSARY 735
INDEX 739

PREFACE

The purpose of this new, revised third edition of the *Textbook of Work Physiology* is the same as that of the original text: to bring together into one volume the various factors affecting human physical performance in a manner that is comprehensible to the physiologist, the physical educator, and the clinician. Contrary to most conventional textbooks of physiology, in which the emphasis is on the regulation of the various functions of the body at rest, the regulatory mechanisms studied during physical activity have been especially emphasized in this book. It is assumed that the reader has some knowledge of elementary physics and chemistry, as well as human anatomy and physiology. However, to facilitate the understanding of some of the physiological and biochemical events encountered during work stress and physical exercise, a certain amount of basic physiology and biochemistry has been included.

In this edition we have systematically tried to use the term exercise for physical activity of all kinds, instead of work. However, work, which is defined as the application of force over a distance, has been used in some cases in keeping with common practice. The same applies to work rate as work performed per unit time, and work load meaning the load placed upon the individual who performs the work.

In the selection of the material, an attempt has been made to meet the modern needs of the student of physical education at both the undergraduate and the postgraduate levels. More references have been included than is customary in most textbooks. Inevitably, since the submission of our revised manuscript, new developments have taken place which we were unable to include in this edition.

We are aware that the curriculum in many physical education programs may not permit as comprehensive a study of physiology as this book may entail. For this reason, each chapter has been written as a fairly complete entity, relatively independent from the rest of the book. With this arrangement, the book may also be useful to those students who wish to penetrate more deeply into a particular field or a limited area of study.

It is our hope that this text may be useful not only in the teaching of physical education, but also in the teaching of clinical and applied physiology and that it may serve to stimulate the appreciation of the role of physical education for young and old, in health and disease.

Much of the unpublished material in this book has been gathered in collaboration with our colleagues at the College of Physical Education in Stockholm, and at the Institute of Work Physiology and the College of Physical Education in Oslo. Their kind cooperation is gratefully acknowledged. We have also benefited greatly from our personal associations and frequent discussions with our many colleagues in these and other institutions. We are also very grateful for the technical assistance provided by Joan Rodahl, Maria Hellström, and Svanhild Kurseth in the preparation of the manuscript.

Previous editions of this textbook have appeared in the following translations: Italian, Japanese, Portugese, Spanish and French.

Per-Olof Åstrand

Kaare Rodahl

CHAPTER 1

OUR BIOLOGIC HERITAGE

To remind ourselves that it has taken mankind so long to develop into the human beings we are today, let us start by presenting a brief sketch of our evolutionary history. In so doing, we shall also be reminded of the fact that many of the biologic processes, which are the basis of our present physical performance that will be dealt with in detail in this book, are in fact thousands of millions of years old. Thus, a comprehensive textbook of biochemistry written some 1,500 million years ago, would no doubt still be up-to-date in its treatment of the functions of the cell.

It is assumed that our solar system was created some 4,600 million years ago (see Dickerson, 1978). At that time the atmosphere surrounding our planet did not contain oxygen; this was a prerequisite for the evolution of life from nonliving organic matter, for without atmospheric oxygen, the ultraviolet radiation from the sun, in the absence of high altitude ozone, could reach the surface of the earth. This radiation could then provide the energy for the photosynthesis of organic compounds from molecules such as: water, carbon dioxide, and ammonia. The photosynthetic process that enabled living organisms to capture solar energy for the synthesis of organic molecules like glucose, can be clearly traced in fossils dated about 3,500 million years before present. Anaerobic fermentation, or glycolysis is probably the oldest energy-extracting pathway found in life on earth.*

Ancient organisms split water by photosynthesis and gradually released free oxygen into the atmosphere. The development of a biologic facility for this oxygen-generating process was indeed a momentous event. Yet it may have taken 2,000 million years

*For further reading on this subject, the following sources are recommended: The September issue of Scientific American 1978 (*Evolution*), and the British Museum publication, "Man's Place in Evolution," Cambridge University Press, 1980.

to create an atmosphere where one out of every five molecules was oxygen. Oxygen became toxic to many of the original oxygen producers and new metabolic patterns were developed; i.e., aerobic energy yield, which utilizes oxygen as a hydrogen acceptor. Another result of the production of oxygen was the formation of an ozone layer in the upper atmosphere. Nonbiologic synthesis of organic matter ceased, because ultraviolet radiation now had to pass through the ozone layer in order to reach the surface of our planet and became markedly reduced. However, solar energy still managed to reach the earth in the visible wave lengths, and this energy could then be utilized for biologic photosynthesis.

A new milestone in biologic evolution occurred about 1,500 million years ago, when the unicellular organism *with a nucleus* (the eukaryote) was developed (see Vidal, 1984). In these primitive organisms, representing the simplest form of aerobic life, fundamental functions such as: metabolism, excitability, locomotion, and reproduction were incorporated, based on very elaborate and complicated biologic phenomena. From this point of view, it is hardly fair to call them primitive. Thus, the energy-absorbing and energy-yielding processes typical of present cellular activity are merely repetitions of events occurring thousands of millions of years ago, such as the ATP-ADP system; the primary mode of transportation for chemical energy in every cellular reaction (see Schopf, 1978).

Actually, ATP is the principal medium for the storage and exchange of energy in almost all living organisms. A large amount of energy is released when ATP is hydrolyzed by water into ADP and a phosphate ion. However, because it is a heavy fuel, the supply of ATP is very limited. Depending on how physically active she/he is, within 24 hours an individual could expend the energy equivalent to the amount of energy stored in ATP weighing 50 to 100 percent more than his or her own body weight. Therefore, a very rapid resynthesis of ATP is essential; the *anaerobic* processes, several millions of years old, are supplemented by the *aerobic* energy yield taking place inside the mitochondria.

Calcium ions (Ca^{2+}) play a key role in the regulation of many processes in the body, including the activation of heart and skeletal muscles. A special protein (calmodulin) serves as an intracellular Ca^{2+}-receptor and mediates the Ca^{2+}-regulatory functions. This protein is structurally conserved and functionally preserved throughout the plant and animal kingdoms. This is yet another example of how a mechanism, that developed thousands of millions of years ago, has proved to be very efficient, and has survived the test of time.

A cell, microscopic in size from a few micrometers (μm) up to a small number of millimeters (mm), using simple physical forces is able to transport nutrients, waste products, electrolytes, and dissolved gases both intra- and extracellularly. Diffusion and osmosis (differences in cellular concentrations) are the main driving forces. In addition, energy-consuming biologic processes assist in this exchange of matter and molecules.

Thus, it may be summarized that during thousands of millions of years of evolution, a unicellular living organism evolved. Through trial and error, the fundamental biologic principles for maintaining life were developed; these processes are still in efficient operation.

Evolution was now ready for the next major step, the development of larger animals; probably beginning some 700 million years ago (see Valentine, 1978). In the evolution of larger organisms it was impossible simply for the cell to increase in size, which might jeopardize its supply of oxygen and fuel. The living cell needs oxygen for its metabolism, which reduces the oxygen tension inside the cell. As long as oxygen is available in the environment, it will diffuse toward the place of metabolism: the mitochondria. The distance that the oxygen molecules have to travel and the difference in oxygen tension between the solution extra- and intracellularly determine the rate of diffusion. It has been calculated that a hypothetic cell with a 10 mm radius and a reasonable metabolic rate would need an external oxygen pressure of 25 times the barometric pressure at sea level in order to secure the oxygen supply to the center of the cell by diffusion alone (Krogh, 1941). This, of course, is out of the question. Furthermore, the available oxygen pressure in ambient air is only about one fifth of the barometric pressure.

Similarly, the transport of fuel into the cell by diffusion also limits the size of the individual cell. Suppose one million sugar molecules were placed on the bottom of a cylinder filled with water. After one hour, half of them would have traveled 1 mm or more by diffusion but only 20 molecules would have covered a distance of 7 mm. It would take about 100 years to obtain an even concentration of the sugar molecules up to a level of 1 m. In other words, diffusion is an efficient transport mechanism over short distances, but inefficient over long distances.

Consequently, in the evolution of larger animals, the individual cell retained its original size, i.e., the same size as the unicellular organism living more than one thousand million years ago. However, more of these single cells were piled together as a means of increasing the size of the organism. Special structures in the cell, the genes, were encoded with detailed instructions about the proliferation of the cell mass, guiding a specialization in shape, structure, and function. Human beings have about 200 different types of cells. Certain cells aggregate to form tissues and organs with a relatively homogeneous composition. Some tissues became specialized for support (bone, cartilage, connective tissue), others developed the potential for motion (muscles), while other cells specialized in dealing with excitability and conduction of information (sensory cells, nerve cells), etc. All living cells have a certain metabolic capability, but certain groups of cells are capable of handling specific metabolic tasks, as in the case of liver cells and gastrointestinal tract cells. In order for the organism to survive as a whole, it is essential that its individual cells collaborate according to the principle of one for all, and all for one.

As an inevitable consequence of piling thousands of millions of some 200 different types of cells together in one organism, the individual cell lost intimate contact with the external environment. Furthermore, throughout the course of evolution, this environment changed, in the case of some organisms, from water to air. Transporting adequate supplies of building material, fuel, and oxygen to each cell and removing waste products are two of the greatest challenges facing an organism, as the number of cells increase. Both problems are solved by bathing each cell in water, i.e., the interstitial fluid. Like the amoeba, each cell in our body (with some exceptions) is surrounded by fluid, basically similar in composition to that of the ancient oceans.

The organism brought the sea water with it, so to speak, in a bag made of skin. The distance between the cell interior and its external environment is so small that gases and substances are easily transferred. For optimal function, the cell needs an environment that is as stable as possible. Thus, the composition of the cell-bathing fluid must be kept fairly constant, and prevented from large fluctuations. Its content of organic compounds such as: fatty acids, glucose, hormones, and enzymes and of inorganic substances such as: sodium, potassium, and calcium exerts a vital influence on the cell. The continuous supply of oxygen and removal of carbon dioxide is crucial. The tolerance for an increase in hydrogen ions, i.e., a decrease in pH, during heavy exercise is remarkable (pH less than 7 in the arterial blood) but there is a limit. The tolerance for changes in the body temperature is also limited. All warm-blooded animals live their lives only a few degrees away from their thermal death points.

In the course of diversification of multicellular organisms, taking place over the last 700 million years, new types of organisms have appeared and radiations occurred within already established groups. The earliest fossil traces of animal life are burrows that begin to appear in rocks younger than 700 million years (see Valentine, 1978). Fig. 1-1 depicts the major events in the evolution of multicellular organisms. It should be noted that the history of the mammals covers the last 220 million years, if not more. The first primates (the order including human beings) can be traced 60 to 70 million years ago, to a period when the dinosaurs still dominated the scene. With the extinction of the dinosaurs, there was a mammalian radiation into newly available niches. Sixty to 70 million years ago an evolutionary explosion occurred with a radiation of flowering plants, birds, and mammals.

What then, are the mechanisms that underlie the origin of species and the evolutionary relationships between them, i.e., Darwinism? Lewin (1980) has summarized the current views held by different researchers in this field. According to the modern synthesis, evolution is a consequence of the gradual accumulation of genetic differences due to point mutations and rearrangements in the chromosomes. The direction that an evolutionary change takes is then determined by natural selection, promoting those variants that are best adapted to their environment. However, the fact remains that on the whole, the fossils do not document a smooth transition from old morphologies to new ones; this was also discussed by Darwin. For millions of years species remain unchanged in the fossil record, suddenly to be replaced by something that is substantially different but clearly related (Lewin, 1980). Because the accumulation of small genetic changes cannot solely account for the development of new species, a new theory called punctuated equilibrium has been advanced. According to this theory, individual species may remain virtually unchanged for longer periods of time. Then suddenly they are "punctuated" by abrupt events in the environment and a new species arises from the original stock. Incidentally, these "abrupt" events may take 50,000 years, but this is, of course, a very short period of time in evolutionary terms. It is indeed conceivable, however, that future fossil discoveries may fill many of the gaps and provide some of the missing links.

It may be only 10 or as much as 20 million years ago that the family tree of primates developed a branch called the hominids, which finally resulted in *Homo sapiens*

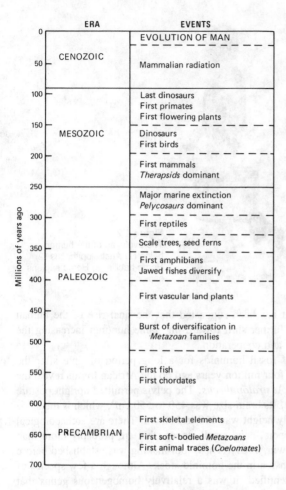

Figure 1-1
Major events in the evolution of multicellular organisms. (*Modified from J.W. Valentine, Sci. Am., 239(3):68, 1978.*)

sapiens, the only surviving hominid (Fig. 1-2). Another branch led to the development of the anthropoid apes; the orangutan, gorilla, and chimpanzee. According to evolutionary theory, the variants that survive are those best adapted to their particular environment. These individuals reached adulthood and produced viable offspring. The weaker variants failed to survive. There were periods of tropical climate in large areas, as well as glacial periods of which there were four in the past million years.

Somewhere along the line a prototypic anthropoid ape abandoned life in the trees and started to forage and hunt on the ground. A species related to these creatures may have been *Ramapithecus*. According to some experts this transition occurred some 14 million years ago; others believe it happened several million years later. Mankind thereby initiated a bipedal adaptation to terrestrial life, first at the margins of forests and then gradually out on the savanna, forming bands of hunters and gatherers. Ac-

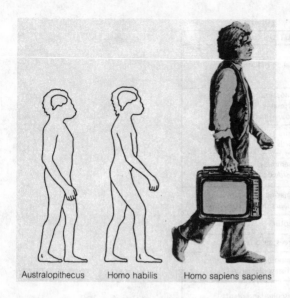

Australopithecus   Homo habilis   Homo sapiens sapiens

Figure 1-2
The evolution of the human
being (1) Australopithecus (2)
Homo habilis (3) Homo sapiens
sapiens

cording to Valentine (1978), it has been suggested that the final rise of the human species was associated with a further shift toward big game hunting; increasing the value of cunning, intelligence, and cooperation.

There is a puzzling lack of fossil hominids for a long period of time after the *Ramapithecus*. Not until about four million years ago do the African fossils reveal the presence of the hominid genus *Australopithecus*. The pelvis permitted upright posture with bipedal gait and free arms. Its brain size was 450 to 550 cm^3, which is the same size as a gorilla brain. The body height was 110 to 120 cm. There are archaeological records of tools, pebble "choppers," and small stones, that are probably more than three million years old (see Lewin, 1981). Thus, tool-making was established before there was a marked brain expansion in the hominid stock. Although a few species of *Australopithecus* have been identified, it was a relatively homogeneous genus that survived for more than two million years. The last representatives of the Australopithecines became extinct less than two million years ago. Another member of the hominid family tree has been named, by some authorities, *Homo habilis* existing from 2.3 to 1.5 million years ago, with a brain volume ranging from 600 to 800 cm^3. The first well-identified true man in the fossil record is *Homo erectus* (see British Museum, 1980). By this time, the brain had nearly doubled in size, to an average of 1050 cm^3. "In the struggle for survival through technology, selection for bigger and more efficient brains would seem to have been at work. For tools to be greatly improved and diversified, the appropriate brain capacity had to evolve" (Weiner, 1971). *Homo erectus* had a truly modern pelvis, and moved with a striding gait. They lived as hunters and gatherers with a wide geograhic range. Their body height was probably 150 to 160 cm. They made use of fire, as evidenced by a hominid occupation site 1.4 million years old (Gowlett et al., 1981).

The most familiar human ancestor to the general public is probably the Neanderthal man. According to archaeologic findings *Homo sapiens neanderthalensis* appears to be well-established more than 100,000 years ago (see Trinkaus and Howells, 1979). They were skilled hunters of both large and small game, forming bands similar to those of more recent hunting people, probably linked into tribal groupings, or at least into groups sharing a common language. They formed a human population complex extending from Gibraltar across Europe into East Asia. The Neanderthal population was as homogeneous as the human population is today. The brain encased in the Neanderthal skull, however, was slightly larger on average compared with the brain of modern man. According to Trinkhaus and Howells (1979), this anatomical feature is undoubtedly related to the fact that the musculature of the Neanderthals was more substantial than that of modern man. Neanderthals had apparently the same postural abilities, manual dexterity, and range and character of movement that is typical for modern man. But they had more massive limb bones, and as previously mentioned, a larger muscular mass and power. The departure of the Neanderthals occurred some 35,000 years ago; when they disappeared from the scene anatomically modern man, *Homo sapiens sapiens*, was already in existence. No one knows why man of the modern type, came to the fore and the Neanderthals disappeared. Quoting Washburn (1978): "Increasing in brain size does seem to be correlated with the increasing complexity of stone tools over hundreds of thousands of years in a way that is not evident during the past 100,000 years of evolution." The size of modern man's brain is, on the average, 1,300 cm^3.

Most likely, a human being living 50,000 years ago had the same potential for physical and intellectual performance, such as playing a piano or constructing a computer, as anyone living today. As mentioned, the Neanderthals had a larger brain and greater muscle mass than modern man, but these findings do not suggest any difference in intellectual or behavioral capacities (Trinkhaus and Howells, 1979). From all indications, *Homo sapiens sapiens* has remained biologically unchanged at least during the last 50,000 years. By 30,000 years ago, modern man had spread to nearly all parts of the world. It was not until some 10,000 years ago that the transition from a roaming hunter and gatherer to a stationary farmer began.

To illustrate the evolutionary time scale in a graphic manner let us compare the 4,600 million years our planet has existed with a 460 km journey (Fig. 1-3). Life began after the first 100 km of the trip had been covered. It took another 200 km before the unicellular organism with a nucleus was born. Multicellular animals were living when we arrived at the 400 km mark. An evolutionary radiation of the mammalian stock started somewhere around the 453 km milepost. The first hominid probably appeared approximately 6 km further on. *Australopithecus* joined the journey about 450 to 200 meters from the end and the Neanderthals disappeared just about 3.5 meters from the finishing line, where they were replaced by modern man. The cultivation of land and keeping of live stock occurred one meter from our present position. A person, one hundred years old today, has merely covered a distance of ten millimeters of the entire 460 km journey.

The purpose of this very brief summary of an ongoing evolution, is to provide an outline of our genetic background. Many structures and functions are common to different species in the animal kingdom. For instance, there appears to be no funda-

OUR BIOLOGICAL EVOLUTION

Start

0	km	The earth is created	4,600,000,000	years ago
110	km	Biochemical processes developed that can trap solar energy	3,500,000,000	years ago
310	km	Single-celled organisms with a nucleus appear	1,500,000,000	years ago
390	km	Multicellular organisms appear	700,000,000	years ago
453	km	Modern mammals appear	70,000,000	years ago
459	km	Hominid branch being developed	10,000,000	years ago
459.6	km	*Australopithecus* appears	4,000,000	years ago
459.99	km	Neanderthal man appears	100,000	years ago
459.999	km	Agriculture introduced	10,000	years ago
459.99999	km	Present 100-year-old man born	100	years ago

Figure 1-3
In an attempt to illustrate our evolutionary time scale, we may compare the 4,600 million years our planet has existed with a 460 km journey. Life began after the first 100 km of the trip had been covered. A person, one hundred years old today, has merely covered a distance of ten mm of the 460 km journey.

mental difference in structure, chemistry, or function between the neurons and synapses in man and those of a squid, a snail, or a leech (Kandel, 1979). We can, therefore, learn a great deal from studying different species. It is remarkable that all living organisms have a genetic code based on the same principle. Data indicate, for instance, that man and chimpanzees share more than 99 percent of their genetic material (Washburn, 1978). However, minimal genetic changes can affect major morphologic modification. Consequently, one should be careful when extrapolating findings from one species to another, including man. Over millions of years many species undergo minor or major modifications in physical and other characteristics. In general, however, evolution is a very conservative process. All vertebrates, including the hominids, have backbones. Backbones are complicated in design, yet they are quite similar in all animals that have them. This finding supports the hypothesis that backbones have evolved only once. In other words, it appears that all vertebrates share a common ancestor with a backbone.

Other examples of structures and functions shared by many species can be taken from the mammals: all mammals have three separate bones in the middle ear; the females have milk-producing glands, although the composition of the milk varies markedly among different species. As mentioned, the cradle of the mammals can be traced back some 220 million years. One reason for the conservatism of evolution must be its functional power and the fact that no other invention has proved to be better.

Among vertebrates, locomotion is generally genetically programmed. Fish are able to swim as soon as they are born; birds may walk as soon as they are hatched. Many species of mammals are well developed at birth, some are able to walk or even run as soon as they are born, and some of these are able to attain a speed of up to 35 km per hour when they are only a few days old. Ultimately their survival may depend on their ability to get away. In the case of human beings, that are utterly helpless at birth and entirely dependent on parental care, it may be to his or her advantage not to be able to remove itself very far from his or her parents, until he or she is mature enough to stand on his or her own two feet.

The evolutionary process continues, and more recent mammalian history has seen a wave of extinction, particularly severe for large mammals, including the hominids. Extinctions are a measure of the success of evolution in adapting organisms for particular environmental conditions. New forms have their chances when their adaptations provide entry into a relatively empty niche. In the balance between existance and extinction, the odds are not too favorable: It has been estimated that two thousand million species have appeared on earth during the last 700 million years, but the number of multicellular species now living are in the order of two million, that is, only 0.1 percent have survived.

The cortex of the human brain mirrors man's evolutionary success, just as the proportions of the human hand, with its large opposable and muscular thumb, reflects a successful arboreal adaptation, and later on for the use of tools, so does the anatomy of the human brain reflect a successful adaptation for manual and intellectual skills. Adults of most vertebrate species allocate 2 to 8 percent of their basal metabolism for maintenance of the central nervous system (CNS) (Mink et al., 1981). These authors hypothesize that " . . . an optimal functional relationship between the energy requirements of an animal's executor system (muscle metabolism) and its control system (CNS metabolism) was established early in vertebrate evolution." One important exception is *Homo sapiens* whose CNS consumes 20 percent of its basal metabolism.

Just as upright walking and toolmaking were the unique adaptations of early phases of human evolution, the physiologic capacity for speech was the biological basis for the later stages. Indeed, it is through language that human social systems are mediated. Speech is the form of behavior that differentiates man from other animals more than any other behavior. It is the passing down of knowledge and experience by language from one generation to the next that has enabled man, biologically unchanged for tens of thousands of years, to accelerate progress so dramatically, and to apply its endowed intellectual resource in a technical revolution leading to entirely new and complex tools, weapons, shelters, boats, wheeled locomotion, exploratory voyages, and the attainment of the impossible: landing on the moon. And yet, in the midst of these splendid achievements, there are those who now wonder whether the evolution of the human brain has gone too far. While its ability to conceive, invent, create, and construct is astonishing, it remains to be seen whether or not man's brain has retained or developed equally well its capability of ethical conduct or the responsible application of its endowed potential. At the time when our ancestors roamed around in small bands, any destructive consequence of their activity was quite limited. But today, because of social developments and technical innovations, basically the same brain is capable of turning man into a self-destructive monster.

Man, like all higher animals, is basically designed for mobility. Consequently, our locomotive apparatus and service organs constitute the majority of our total body mass. The shape and dimensions of the human skeleton and musculature are such that the human body cannot compete with a gazelle in speed or with an elephant in sturdiness, but in diversity human beings are indeed outstanding.

The basic instrument of mobility is the muscle. It is a very old tissue. As already mentioned, the earliest animal fossils were the burrowers living some 700 million years ago. Evidently, using muscle force, these animals could dig in the sea bed. Muscles have retained the metabolic pathways developed when the air had no oxygen, i.e., the anaerobic energy yield. The pyruvic acid formed in our muscles under anaerobic conditions is removed by the formation of lactic acid. One old-fashioned alternative could have been the transformation of the pyruvate into ethyl alcohol. There may be those among us who regret that skeletal muscles did not select this alternative route; if so, producing pyruvate by exercising to exhaustion or running uphill, might have been a very popular endeavor!

At any rate the skeletal muscle is unique in that it can vary its metabolic rate to a greater degree than any other tissue. In fact, active skeletal muscles may increase their oxidative processes more than 50 times the resting level. Such an enormous variation in metabolic rate must necessarily create serious problems for the muscle cell, because while the consumption of fuel and oxygen increases 50-fold, the rate of removal of heat, carbon dioxide, water, and waste products must be similarly increased. To maintain the chemical and physical equilibrium of the cell, there must also be a tremendous increase in the exchange of molecules between intra- and extracellular fluid; "fresh" fluid must continuously flush the exercising cell. When muscles are thrown into vigorous activity, the ability to maintain the internal equilibria necessary to continue the exercise is entirely dependent on those organs that service the muscles. This dependence is especially true in the case of respiration and circulation, but certainly food intake, digestion and handling of substrates, kidney function, and water balance are also affected by variations in the metabolic rate.

As a summary of this brief outline of the evolution, close to 100 percent of the biologic existance of our species has been dominated by outdoor activity. Hunting and foraging for food and other necessities in the wilds have been a condition of human life for millions of years. We are adapted to that style of life; this applies to our emotional and social lives, and intellectual skills. After a brief spell in an agrarian culture we have ended up in an urbanized, highly technologic society. There is obviously no way to revert to our natural way of life, which by the way, was not without problems. But with insight into our biologic heritage we may yet be able to modify our current life style. Knowledge of the function of the body at rest, as well as during exercise under various conditions is important as a basis for an optimization of our existence.

REFERENCES

British Museum: "Man's Place in Evolution," Cambridge University Press, Cambridge, 1980.

Dickerson, R. E.: "Chemical Evolution and the Origin of Life," *Sc. Am.*, **239**(3):62, 1978.

Gowlett, J. A. J., J. W. K. Harris, D. Walton, and B. A. Wood: Early Archeological Sites, Hominid Remains and Traces of Fire from Chesowanja, Kenya, *Nature*, **294**(12 Nov.):125, 1981.

Kandel, E. R.: Small systems of neurons, *Sc. Am.,* **241(3):**67, 1979.

Krogh, A.: "The Comparative Physiology of Respiratory Mechanisms," University of Pennsylvania Press, Philadelphia, 1941.

Lewin, R.: Evolutionary Theory Under Fire, *Science,* **210:**883, 1980.

Lewin, R.: Ethiopian Stone Tools are World's Oldest, *Science,* **211:**806, 1981.

Mink, J. W., R. J. Blumenschine, and D. B. Adams: Ratio of Central Nervous System to Body Metabolism in Vertebrates: Its Constancy and Functional Bases, *Am. J. Physiol.,* **241:**R203–212, 1981.

Schopf, J. W.: "The Evolution of the Earliest Cells," *Sc. Am.,* **239(3):**85, 1978.

Scientific American: "Evolution," W. H. Freedman and Company, San Francisco, 1978.

Wickramasinghe C.: "Evolution from Space," Dent, London, 1981.

Trinkaus, E., and W. W. Howells: The Neanderthals, *Sc. Am.,* **241(6):**94, 1979.

Valentine, J. W.: The Evolution of Multicellular Plants and Animals, *Sc. Am.,* **239(3):**67, 1978.

Vidal, G.: The Oldest Eukaryotic Cells, *Sc. Am.,* **250(2):**32, 1984.

Washburn, S. L.: The Evolution of Man, *Sc. Am.,* **239(3):**146, 1978.

Weiner, J. S.: "Man's Natural History," Weidenfeld and Nicolsen, London, 1971.

CHAPTER 2

THE MUSCLE AND ITS CONTRACTION

CONTENTS

THE CELL
What is Known About the Transport Mechanism Through
the Cell Membrane
The Cell Interior
ARCHITECTURE OF THE SKELETAL MUSCLE
MYOFIBRILLAR FINE STRUCTURE
INTERACTION OF ACTIN AND MYOSIN
At Rest
Excitation
Contraction
Relaxation
Mechanism of Calcium Transport
"The Heart of the Whole Problem"
DIFFERENT FIBER TYPES
Contractile Properties, Histochemical and Ultrastructural Bases
for Fiber Typing
Metabolic Characteristics
Training Can Modify the Metabolic Profile of the Different
Fiber Types
What Factors Determine Whether a Muscle Fiber Becomes Type I
or Type II

There are Great Individual Variations in Fiber Types
Muscle Biopsies

TYPES OF MUSCLE ACTIVATION
Force in Relation to Muscle Length and Speed of Contraction;
Force-length
The Optimal Position to Produce Force is Not Only Dependent on the
    Initial Length of the Engaged Muscles
Force-velocity

MECHANICAL EFFICIENCY OF MUSCLE
    CONTRACTION
STORAGE AND UTILIZATION OF ELASTIC ENERGY

THE CELL

First, let us begin by summarizing the general design of a cell (Fig. 2-1), with particular emphasis on the *cell membrane,* which plays a crucial role in almost all cellular activity. The cell membrane actively regulates the internal environment of the cell and the transport of substances in and out of it. The structural framework of the membrane, approximately 5 nm thick, is a double layer of lipid molecules. The individual lipid molecule has a head and two tails. Hydrophilic heads (soluble in water) form the outer and inner membrane surfaces, and hydrophobic tails (with poor affinity to water) meet in the membrane interior. This structure serves as an anchor for other components of the membrane, such as proteins and glycoproteins. Actually, what makes one cell membrane different from another are the various specific proteins attached to the membrane in one way or another. There are five different classes of membrane proteins: pumps, channels, receptors, enzymes, and structural proteins. The five classes of membrane proteins are not necessarily specialized for one of these tasks, a particular protein might simultaneously be a receptor, an enzyme, and a pump. They can transfer information and instructions from the environment into the cell, in some cases, using hormones as triggers. Many molecules, such as O_2 and CO_2, can diffuse through the membrane, but in the case of other molecules, including water-soluble compounds, the lipid core of the membrane forms a barrier to molecular diffusion. Accordingly, most biologically important molecules and ions have specific transport mechanisms available for their fluxes between the cell environment and its interior. Studies indicate that the lipid cell membrane derive some permeability for water from the aqueous pores in the membrane brought about by the incorporation of pore-forming molecules (Fettiplace and Haydon, 1980).

Rothman (1980) also supports the assumption that proteins can pass through cell membrane by "simple" diffusion, moving through membrane pores or channels, driven by the permeable molecules' kinetic energy and along a concentration gradient. The cell membrane maintains a difference in ion concentrations between the extra- and intracellular fluid. This concentration gradient is of vital importance for the cell's activity, including the activation of nerve and muscle cells. Energy-consuming "active"

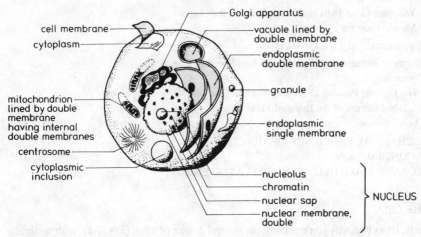

Figure 2-1
A typical cell. The cell membrane plays a crucial role in almost all cellular activity. It can actively regulate the internal environment of the cell and transport substances in and out of it. The structural framework of the membrane is a double layer of lipid molecules. The individual lipid molecule has a head and two tails. Hydrophilic heads (soluble in water) form the outer and inner membrane surface, and hydrophobic tails (with poor affinity to water) meet in the membrane interior. This structure is then the anchorage for other components of the membrane, like proteins and glycoproteins. These molecules can also provide additional support, i.e., they can act as enzymes or function as "pumps" and carriers of material across the cell membrane. Many of the lipids in mammalian cell membranes are unsaturated and are liquid at body temperature. Therefore, the membrane has the consistency of a light oil within the sheetlike structure, and both lipid and protein molecules are relatively free to move about within the membrane. Protein molecules may lie close to either membrane surface; they may penetrate the membrane surfaces and eventually bridge them completely. The inner membranes of mitochondria have a similar structure and many of the enzyme proteins combine to form supramolecular aggregates organized in an orderly array throughout the membrane (Capaldi, 1974). Encapsulated within the boundaries of the cell membrane is the cytoplasm, which contains a number of formed and dissolved elements, including enzymes which support the anaerobic metabolic processes of the cell. The mitochondria, which take up oxygen, are rod-shaped bodies, surrounded by a double-walled membrane and represent the "powerhouse" of the cell. Here fuel and oxygen enter into energy-yielding processes resulting in the formation of ATP. In the endoplasmic reticulum, a network of canaliculi formed by a system of membranes may extend all the way from the outer surface of the cell to the membrane surrounding the nucleus. Through these canaliculi, substances may move from the outer membrane of the cell to the membrane of the nucleus. The dots that line the endoplasmic reticulum are ribosomes. These are the sites of protein synthesis. The nucleus contains the chromosomes, which contain genes and deoxyribonucleic acid and are the carriers of the hereditary factors. In cell division, the pair of chromosomes, shown in longitudinal section (rods) and in cross section (circles), parts to form two poles of an apparatus that separates two duplicate sets of chromosomes.

processes are engaged in the transport of substances across the cell membrane. The energy cost of the transport of ions and molecules is paid for by ATP. Several ATPases essential to transport are located in the cell membrane. A common characteristic of these enzymes is that they catalyze the same overall reaction, i.e., the hydrolysis of ATP to ADP and inorganic phosphate. The free energy yielded by this reaction is then used for the transport (Gunn, 1980; Schuurmans Stekhoven and Bonting, 1981).

What Is Known About the Transport Mechanisms Through the Cell Membrane?

Many of the lipids in mammalian cell membranes are unsaturated and are liquid at body temperature. Therefore, the membrane has the consistency of a light oil within a sheetlike structure; both lipid and protein molecules are relatively free to move about within the membrane. Protein molecules may lie close to either membrane surface; they may penetrate the membrane surfaces and eventually bridge them completely; parts may stick out above and below the membrane like floating icebergs. It was mentioned that some of the proteins are receptors and that they combine selectively with molecules, e.g., other proteins, with a configuration specific for a particular receptor. Actually some receptors are less specific but often they are exquisitely specific; they are tailored to fit the shape of a particular molecule like a lock accepting one particular key. "The receptor in effect plucks one ingredient from the extracellular soup—even if the molecule is present in a low concentration or with a vast excess of unrelated molecules—and holds it fast" (Dautry-Varsat and Lodish, 1984). In special regions of the membrane the ligand-receptor complex is engulfed in a "coated pit," i.e., an invagination of the membrane (Fig. 2-2). These pits may account for some 2 percent of the cell surface. This binding triggers the formation of a membrane-bounded vesicle enclosing the ligand; this *coated vesicle* is transported to the lysosomes inside the cell (a *receptor-mediated endocystosis*). When the ligand is dispatched to its intracellular destination the receptor is recycled to the surface ready to bind more of the ligand. In some cases receptors seem to cycle continuously, like a shuttlebus, in other cases the cycling is induced by the binding of an available ligand (more like a taxicab finding a passenger).

What then is the mechanism behind the attachment and dissociation between the

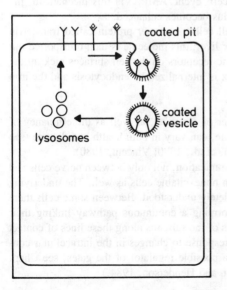

Figure 2-2
A protein or a peptide may be engulfed in an invagination of the membrane (a coated pit), which is then cut off and transported as a coated vesicle to the lysosomes inside the cell. (*Modified from K. Wiman, 1981.*)

ligand and the receptor? It is known that the pH inside a cell is lower than that of its surrounding aqueous medium. There are data supporting the assumption that the receptors bind their ligands tightly at the pH level of the extracellular fluid, but the lower pH in the intracellular fluid promotes a dissociation (Dautry-Varsat and Lodish, 1984).

A few examples may serve to illustrate how the receptor-mediated endocystosis works:

1 In a region of the motor endplate of the skeletal muscle fiber there are receptors that bind acetylcholine, the transmitter substance that initiates the muscle contraction. This acetylcholine-receptor complex leads to a conformational change associated with an opening of an ion channel, thereby permitting an influx of sodium ions into the cell and a depolarization of the cell membrane. This is the second step in the activation of the muscle (for details see Fambrough, 1979; Vincent, 1980).

2 Other types of receptors combine with insulin, forming a functional unit that stimulates an increased influx of glucose and amino acids into the cell's cytoplasm. This is true for muscle cells as well as for liver and fat cells. How, then, is the message passed from the hormone to the cytoplasm? One method, used by many hormones, including insulin, is the activation of an enzyme (adenyl cyclase) attached to the inside of the cell membrane. This enzyme immediately causes the conversion of cytoplasmic ATP into cyclic AMP (cyclic $3',5'$-adenosine monophosphate). The cyclic AMP will then initiate cellular functions that are determined by the character of the target cell itself. As mentioned, the membrane permeability for glucose and amino acids can be modified. Enzyme activities can be altered, synthesis or degradation of intracellular chemical compounds can be initiated, and so forth. This is an example of a transducer, e.g., insulin, that can activate an enzyme inside the membrane, which catalyzes the synthesis of a second messenger in the cell, cyclic AMP. Via this mechanism, the effect of a relatively weak primary stimulus becomes enhanced.

3 Iron is an essential constituent of all cells. A carrier protein called transferrin binds ferric ions (Fe^{3+}) in the intestine or liver; this molecule is then transported out to the tissues by the blood stream. Specific receptors on the cell surfaces pick up the molecule; the receptor-transferrin complex is internalized by endocytosis and the iron is delivered.

It should be mentioned that the number of receptors, as well as the efficiency of the intermediate steps that they may initiate, can vary in both health and disease (see examples in Flier et al., 1979; Kolternam, et al., 1980; Vincent, 1980).

There are systems for cell-to-cell communication, not only between nerve cells and between nerves and muscles, but between nonexcitable cells as well. The underlying mechanism in the latter case is not completely understood. Between some cells there seem to be short intercellular "pipes" forming a continuous pathway linking their interiors. Small tilting and sliding motions of the subunits along these lines of contact can open and close the gap junctions in response to changes in the intracellular conditions. Calcium has been suggested as a possible regulator of the gates, see Chap. 3. (For details, see Gergely, 1981; Unwin and Henderson, 1984.)

Summary The cell membrane is an extremely thin but tenaciously stable film of lipid and protein molecules. It is like a very rich soup that protects the integrity of the cell, which is surrounded by extracellular fluid that is derived from the blood. The cell membrane controls the trafficking of ingredients from this medium, and from the body as a whole. Special proteins play key roles in the transport of cellular building materials, nutrients, and hormones by delivering specific intracellular signals, like cyclic AMP. They also play a part in the transportation of waste products and toxic materials to certain cells that are able to break them down. One important (although not the only) mechanism is receptor-mediated endocytosis.

The Cell Interior

Encapsulated within the boundary of the cell membrane is the *cytoplasm*. It contains a number of formed and dissolved elements, including enzymes that support the anaerobic (without oxygen) metabolic processes for the regeneration of ATP in the cell. There is an extensive distribution of membrane organelles within the cell similar in design to the membrane structure bounding the cell. In part, they form a highly complicated arrangement of tubules and vesicles, the *endoplasmic reticulum*. Its network of canaliculi may extend all the way from the outer surface of the cell to the membrane surrounding the nucleus. It may have the potential to provide an "intercellular circulatory system," but it is very unclear how proteins and other substances are transported between the organelles and how such transportation is regulated. Many membranes are studded with granules on their cytoplastic surface. These granules are rich in ribonucleic acid (RNA) and are called *ribosomes;* they are the factories for protein synthesis. A cell that produces large amounts of protein, like a liver cell, may be packed with as many as 100 million ribosomes.

The *mitochondria* are rod-shaped bodies variable in size from 0.5 μm to 12 μm in length, separated from the cytoplasm by a double-walled membrane. Many of the enzyme proteins combine to form supramolecular aggregates that are organized in an orderly array throughout the mitochondrial membranes (Capaldi, 1974). The mitochondria are the "power-houses" of the cells and their energy yield is aerobic for the regeneration of ATP, which means that they consume oxygen and produce carbon dioxide. A skeletal muscle cell may contain several thousand mitochondria indicating the potential for a high metabolic rate. In Chapter 10, we shall present data indicating that regular training may increase the number of mitochondria in the skeletal muscle cell and that prolonged inactivity has the opposite effect. Cells with a modest energy requirement for their functional activities have merely a few hundred mitochondria.

The *nucleus* contains the primary genetic material of the cell (RNA and deoxyribonucleic acid or DNA), and its chromosomes are the carriers of the hereditary factors. The enormous quantity of information and instructions for development and function of the cell, stored in a single cell, has been expressed in a variety of ways. For example: It would take 3,500 volumes, each 500 pages long in order to spell out all of the genetic material in clear text. Furthermore, the first pages of this material were formulated thousands of millions of years ago; most likely nothing new has been added to the material dealing with *Homo sapiens* during the past 10,000 years.

Cell activity can include hundreds of simultaneous chemical reactions, each catalyzed by highly specific *enzymes*. Most enzymes will catalyze only one step of a chemical transformation. As mentioned, the enzymes required for the anaerobic regeneration of ATP are accommodated in the cell's cytoplasm; enzymes operating within the Krebs cycle and the electron transport chain are located in the mitochondria.

Assuming that the reader has a basic background in biochemistry, we may at this

Figure 2-3
Organic compounds (fats, carbohydrates, proteins) containing carbon and hydrogen may be totally combusted to CO_2 and H_2O by oxygen, that is, by the process of a catalytic oxidation with the aid of enzymes in the living cell. For some of the steps the number of carbon atoms involved are given.

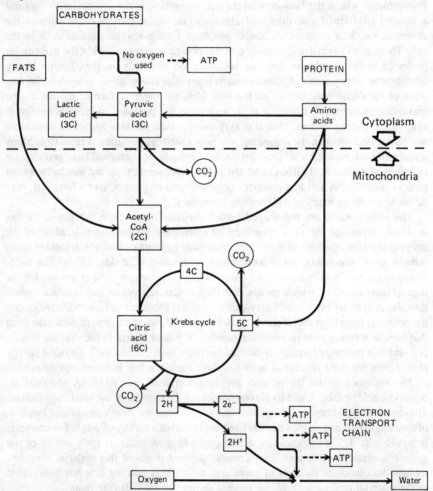

stage limit ourselves to the presentation of a simple summary of the cell respiration in Fig. 2-3. Further details will be included in Chapter 12.

ARCHITECTURE OF THE SKELETAL MUSCLE

Now we shall turn to the question of how chemical energy is transformed into mechanical energy and describe the anatomy, histology, and mechanical design of skeletal muscle. In this chapter, we will concentrate on the properties of the skeletal muscle as such. The muscle function is greatly dependent on muscle innervation and activation from the central nervous system. Therefore some aspects of the function of the muscle will be discussed in the next chapter where this system is tied together with the skeletal muscles.

Standard textbooks of anatomy and histology (e.g., Bloom and Fawcett, 1975), or more specialized texts, should be consulted for detailed descriptions of the muscle (Carlson and Wilkie, 1974; Peachey et al., 1983). In this review, we shall simply present a summary of the architecture of the muscle tissue, with some details of the "working" unit within the muscle. Of the three types of mammalian muscles—smooth muscles, heart muscle (striated involuntary muscle), and skeletal muscles (striated voluntary muscles)—only the skeletal muscle tissue will be discussed here.

The *muscle groups,* known by their Latin names, such as brachialis, are groups of muscle bundles that join into a tendon at each end. On the outside the muscle is covered by a fascia of fibrous connective tissue known as the *epimysium.* Each bundle is separately wrapped in a sheath of connective tissue called *perimysium.* The bundle is made up of thousands of muscle fibers, each embedded in a fine layer of connective tissue (*endomysium*) (Fig. 2-4). The various sheaths of connective tissue blend with the tendon in a way that is determined by function and space.

The functional unit within a muscle is the group of muscle fibers innervated by a single motor nerve fiber, the *motor unit.* The individual muscle fiber is, however, anatomically separated from the neighboring fibers by the endomysium.

The amount of connective tissue (collagen fibers, elastic fibers, and other cells) varies in different muscles and in different species of animals. In human muscle the number of fibers in a muscle group is probably finally established after the embryo has reached the age of four to five months (MacCallum, 1898; Gollnick et al., 1981). However, the thickness of a fiber can vary. At birth the fiber is about twice as thick as in the fourth fetal month, but has only one-fifth the adult thickness (Lockhart, 1973). (See also chapter on training.)

The skeletal muscle fiber is a cylindrical, elongated cell. Its thickness varies in different muscles or even in the same muscle, and may be from 10 μm to 100 μm. In many muscles, the length of the individual cell extends all the way from the tendon of origin to the tendon of insertion. Whether this is so in the longest muscles, such as the sartorius, is not clear. However, cells more than 30 cm long have been traced in this muscle. The muscle cell is multinucleated, with sometimes as many as several hundred nuclei in a single fiber.

Other elements in the cell are the sarcolemma, myofibrils, and sarcoplasm. The *sarcolemma* is a thin, elastic noncellular membrane, less than 10 nm thick, enveloping

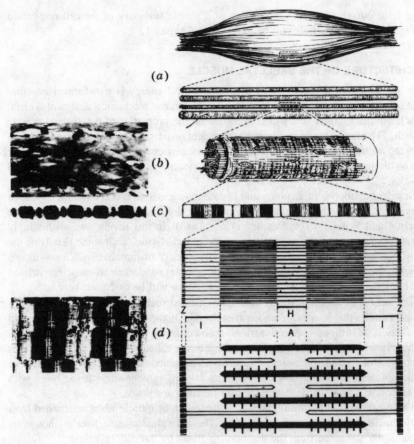

Figure 2-4
Schematic drawing of striated muscle (right) with corresponding photomicrographs (left). The striated muscle (a) is made up of muscle fibers (b), which appear striated in the light microscope. The small branching structures at the surface of the fibers are the "endplates" of motor nerves, which signal the fibers to contract. Each single muscle fiber (c) is made up of myofibrils, beside which lie cell nuclei and mitochondria. In a single myofibril (d), the striations are resolved into a repeating pattern of light and dark bands. A single unit of this pattern consists of a Z line, then an I band, then an A band which is interrupted by an H zone, then the next I band, and finally, the next Z line. This repeating band pattern is due to the overlapping of thick and thin filaments (bottom part of diagram). (*Redrawn from H. E. Huxley, 1958.*)

the striated muscle fiber. Its structure is very similar to the internal membranes of other cells (nerve cells, Schwann cells, etc.). In some places the sarcolemma has tunnels, cavelike invaginations, or open vesicles. These structures are morphological manifestations of active transport mechanisms. The sarcolemma also has remarkable electrical properties.

The contractile element of the cell consists of the *myofibrils*. The structure and function of these fibrils have been extensively studied and described by A. F. Huxley and H. E. Huxley and coworkers (reference papers in "Cold Spring Harbor Symposia on Quantitative Biology," 1972; Eisenberg and Greene, 1980; A. F. Huxley, 1974; Wray and Holmes, 1981). Methods that have been used in the study of these structures, range from electron microscopy and small-angle x-ray diffraction to light microscopy and optic techniques for measuring changes in fluorescence or absorbance using voltage-sensitive dyes. Furthermore, biochemical studies of the protein components of skeletal muscles have contributed to the present understanding of muscular function.

In each muscle fiber (or cell) there are many myofibrils, each 1 to 3 μm thick, arranged parallel to one another. The individual myofibrils are aligned within the sarcolemma so that points with the same density lie at the same level, giving the appearance of disks crossing the whole thickness of the muscle fiber. This arrangement is illustrated in Figs. 2-4 and 2-5. Each repeat is called a *sarcomere* and is bordered by a narrow membrane called the Z line, which, like a disk, crosscuts the myofibril in units. In the middle region of the sarcomere, there is a dark band, the A band (detected by using the deep-focusing position on the microscope). A stands for anisotropic, which refers to the optical property of the tissue. The alternate light bands

Figure 2-5
(a) Schematic drawing of the protein rods or filaments of the myofibril. The thick filaments consist of myosin; the thin filaments consist of actin.
(b) Diagram indicating the sliding of thick and thin filaments that occurs when a muscle is stretched. Note the constancy of the lengths of thick and thin filaments. (*From Carlson and Wilkie, 1974.*)

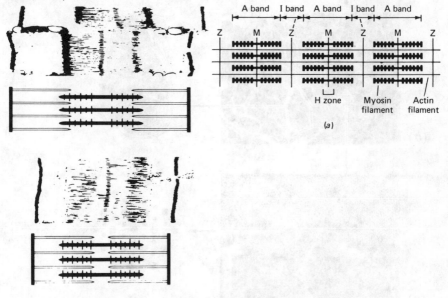

are isotropic or *I* bands. The *Z* line (*Z* stands for the German, *zwischen,* which means between) is located in the middle of the *I* band. The region in the center of the *A* band is called the *H* zone, and is less refractile (lighter) than the rest of the *A* band (*H* is named after Hensen, who first described it in 1868; see Needham, 1971, p. 129). In the center of the *H* zone there is a darker structure, called the *M* line (from the German, *mittelinie,* meaning midline).

Figure 2-6
Cross section of (*a*) striated muscle. The larger magnification shows (*b*) the individual fibers and (*c*) the fibrils with the very regular pattern of thick and thin filaments.

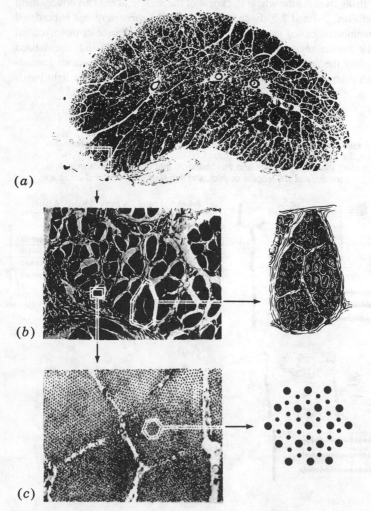

(*a*)

(*b*)

(*c*)

This terminology may be summarized as follows (Fig. 2-5): Two bands of protein rods or filaments are distributed in the myofibrils in parallel order. Filaments of *myosin* (about 1.6 μm in length and 15 nm thick), arranged lengthwise, fill the *A* band. The other band of filaments consists of the protein *actin*. They are thinner (about 6 nm) and run from the area of the *Z* line into the *A* band as far as the beginning of the *H* zone. The length of the actin filament is about 1.0 μm on each side of the *Z* line. Evidently, the *I* band is occupied only by thin filaments, the *H* zone only by the thicker filaments, and the outer parts of the *A* band by both filament types. A cross section of muscle fiber in the overlapping zone reveals a regular pattern. The thick filaments lie about 45 nm apart with thin filaments in between so that each of the thin ones is "shared" by three thick filaments (Fig. 2-5, 2-6).

The Sarcoplasm   The filaments just described are embedded in a fluid in which there are soluble proteins (such as myoglobin), glycogen droplets, fat droplets, phosphate compounds, other small molecules, and ions. This aqueous phase is called the *sarcoplasmic matrix*.

A fraction of the sarcoplasm is called the *sarcoplasmic reticulum*. It consists of elaborate anastomosing tiny tubular sacs, vesicles, and channels of varying caliber extending in the spaces between the myofibrils, resembling a lace cuff or sleeve (Fig. 2-7). The membranes bounding the channels and vesicles are similar in structure to the sarcolemma. Adjacent to the *Z* lines are transversely oriented compartments or cisternae (*terminal cisternae*), which are parts of the sarcoplasmic reticulum that are thickened to form continuous sacs. Located across the myofibrils, they run in close contact with a tubular system, the *transverse tubular system (T system)*. However, these two membrane systems, the sarcoplasmic reticulum and the *T* system, are separated from each other but are in close association in a triad structure (Fig. 2-7). The excitation of a muscle moving along the sarcolemma of the fibril can penetrate into the fibril via this tubular system. Actually, the *T* system starts as an inward extension of the sarcolemma. In mammalian muscle, the penetration of the *T*-tubules occurs at the A-1 interface, so that there is one on each side of the *A* band for a single sarcomere (Eisenberg, 1983). The *T*-tubules are invaginations that do not open into the cytoplasm. Ca^{2+} ions, which play a key role as trigger of the actin-myosin interaction, are released from the sarcoplasmic reticulum. We shall now present a more detailed description of the contractile units.

MYOFIBRILLAR FINE STRUCTURE

The basic contractile components of the muscle fiber are assembled by four proteins into the two multimolecular aggregates, the previously mentioned thick *myosin* and thinner *actin* filaments. *Tropomyosin* and *troponin* are the other two proteins. Neither protein, by itself, is contractile. In vitro, however, myosin and actin can, under certain conditions, form a complex protein, actomyosin, that can contract. The individual *myosin* molecule (about 50 percent of the muscle protein) resembles a golf club with a "head" (of about one-sixth the total length), and a long shaft or "tail." The site responsible for its enzymatic activity (see below), and its affinity for actin, is located

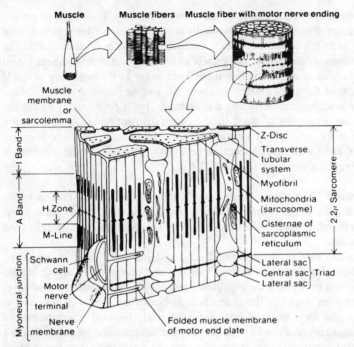

Figure 2-7
Enlarged section of a single striated muscle fiber showing the myofibrillar system, the transverse tubular system, the sarcoplasmic reticulum, and the myoneural junction. (*From Carlson and Wilkie, 1974.*)

Figure 2-8
(a) Structure of actin is represented by two chains of beads twisted into a double helix (top). The contact of actin and myosin might be made in the manner schematically illustrated in the middle part of the figure. The thin actin filaments at top and bottom are so shaped that certain sites are closest to the thick myosin filament in the middle. The heads of individual myosin molecules (zigzag lines) extend as cross-bridges to the actin filament at these close sites. At the bottom of the figure is shown double overlap of thin filaments from each side of the sarcomere, which would result if the sliding-filament hypothesis is correct. The tension would fall when thin filaments cross the center and interact with improperly oriented cross-bridges. (*From H. E. Huxley, 1965.*)
(b) Diagrammatic representation of the two filaments of the myofibril. On the myosin filament the heads, with the potential to form cross-bridges with the actin filaments, are arranged in pairs at regular intervals, each being rotated by about 120° from the preceding pair. The troponin complex is assumed to consist of equimolar amounts of the calcium-binding protein (black), the inhibitory protein (cross-hatched) and troponin-T (stippled). The tropomyosin (dark line) is represented as lying in each groove of the actin filament. (*From Ebashi, 1980.*)

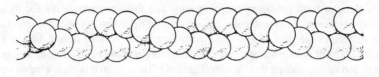

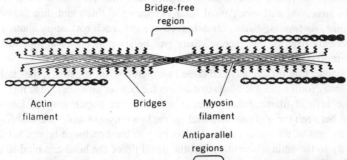

Bridge-free
region

Actin
filament

Bridges

Myosin
filament

Antiparallel
regions

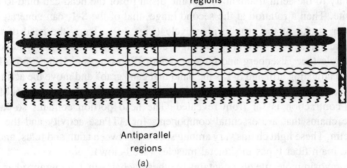

Antiparallel
regions

(a)

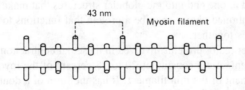

43 nm

Myosin filament

(b)

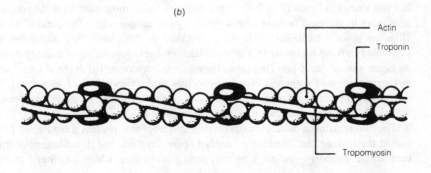

Actin

Troponin

Tropomyosin

in its globular head (of heavy meromyosin), and the sites responsible for its affinity to other adjacent myosin molecules are in its tail (of light meromyosin). Several hundred such molecules are packed in a sheaf with their heads pointed in one direction along half the filament, and in the opposite direction along the other half, leaving a projection-free region midway along their length (Fig. 2-8). Thus, from this cigar-shaped structure (backbone), the heads project out toward the thin filaments; these cross-bridges are the only structural and mechanical links between the thick and thin filaments. The projections, resembling barbs, are arranged in pairs, each rotated by about 120° from the preceding pair. In this "screw" there will be six projections in a period of about 43 nm, representing one turn (Fig. 2-8).

The connecting region between these two meromyosin segments, the "tail" and the "head," may form a flexible hinge that allows the head or cross-bridge to rotate outward from the myosin filament backbone. A second hinge region may be located at the junction between the globular terminal on the head myosin molecule (subfragment S-1) and the rest of the molecule (S-2). Rotation in the first hinge brings S-1 into close proximity to the actin filament so that the distal tip of the head can bind to a specific actin site. Then a rotation at the second hinge, that of the S-1, can generate the force or power stroke of the cross-bridge cycle (Fig. 2-9). Thus, the head of the cross-bridge may attach to actin at a 90° angle and then change its angle of attachment to a 45° angle (see Eisenberg and Greene, 1980; Eisenberg and Hill, 1985). In a way, S-2 is an adjusting arm and S-1 is the hand that can "grab" and move the actin filament. The exact degree of movement and rotation of the S-1 and S-2 myosin subfragments during contraction is far from revealed. The head portion contains so-called light peptide chains that are essential components for ATPase activity and the ability to bind actin. These light chains vary among species, between adult and fetus, and between the two main fiber types in skeletal muscles (see below).

The myosin molecule also contains two heavy chains that are wrapped around each other, and folded at one end into the globular structures that make up part of the head. The M line, mentioned above, may be a protein that functions to hold the bundles of myosin molecules together.

The three other major proteins involved in muscular contraction are all incorporated in the thin filament. *Actin* constitutes 20 to 25 percent of the myofibril protein and is the main component of the thin filament. It has the form of a double helix, consisting of two chains of roughly globular subunits (monomers) twisted around each other like two strings of beads (Fig. 2-8). Apparently, the actin monomers have the potential to interact in identical fashion with a given myosin cross-bridge. The actin filaments (like the myosin filaments) indicate a structural polarization—their molecules are assembled into the filaments in a front-to-back manner. A reversal of polarity occurs on either side of the Z line (the thin filaments are interconnected at the Z line); with this arrangement of actin and myosin molecules in the two halves of an A band, we can expect the actin filaments on either side of the sarcomere to move in opposite directions, that is, toward one another in the middle of the sarcomere.

The two remaining proteins, tropomyosin and troponin, possess a regulatory function in the making and breaking of contacts between thick and thin filaments during contraction. *Tropomyosins* are long polypeptide molecules, which attach end to end,

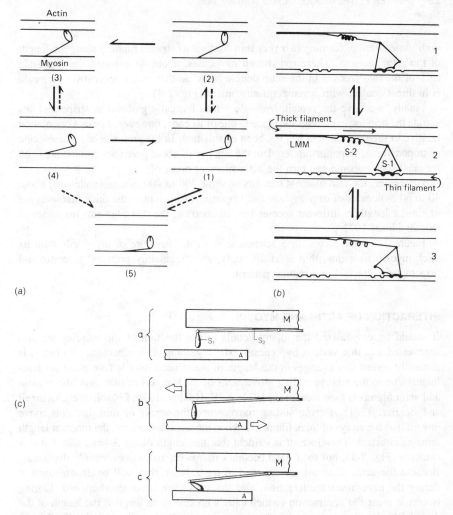

Figure 2-9

(a) A schematic diagram of the physical events thought to take place during the contraction of a striated muscle. In step 1 the myosin cross-bridges are separated from the actin; step 2 is symbolic of attachment at a specific orientation, which then changes to that indicated by step 3, generating tension, and for a submaximal load, shortening. Step 4 represents the cross-bridge redissociated from actin, presumably by the binding of ATP. Step 5 represents the extremely well-ordered cross-bridge array that gives rise to the live, resting myosin x-ray pattern. This last may be only a more ordered form of step 1. (*From Lymn and Huxley, 1972.*)

(b) Development from (a) proposed by A. F. Huxley and Simmons to incorporate the elastic and stepwise-shortening elements for which they gave evidence derived from tension transients. The strength of binding of the attached sites is higher in position 2 than position 1, and in position 3 than position 2. During isometric contraction the myosin head oscillates rapidly between its three stable positions. The myosin head can be detached from position 3 with the utilization of a molecule of ATP; this is the predominant process during shortening. During stretch, the myosin head can dissociate from position 1 without utilization of ATP. (*From A. F. Huxley, 1974.*)

(c) Schematic diagram of the kinematic cycle proposed to be performed by an individual myosin molecule, showing the forces operating during the displacements of the actin (*A*) and the myosin (*M*) filaments. (*From Borejdo et al., 1981.*)

with some overlap, forming two very thin continuous strands running along the length of the actin filament. These rod-shaped molecules, about 40 nm long, lie alongside each of the two grooves in the actin double helix, so that each tropomyosin molecule is in direct contact with seven actin monomers (Fig. 2-8).

Finally, we have the protein *troponin*.* It is basically globular in shape and sits astride the tropomyosin molecule close to one of its ends, however, its exact orientation relative to the thin filament has not been established. In the presence of one molecule of tropomyosin, troponin can regulate the activity of about seven actin monomers. As we shall see, troponin can turn the actin filament on or off. .

Altogether, the thin filament consists of some 300 to 400 thin molecules and about 40 to 60 molecules of tropomyosin and troponin. (Apparently, the thick filaments are of similar length in different species but, in contrast, the thin filaments may vary in length; Close, 1972.)

Figure 2-8 summarizes, in a schematic way, the structure of myofibrils with its thick myosin filaments, thin actin filaments, and "regulatory proteins" troponin and tropomyosin attached in a regular pattern.

INTERACTION OF ACTIN AND MYOSIN

It should be emphasized that many details in the function of the muscles are still unrevealed and that various hypotheses exist regarding its mechanisms. So far, it is generally agreed that changes in the length of the striated muscle take place predominantly due to the relative sliding movements of the two sets of filaments, the myosin and actin filaments (see Needham, 1971; A. F. Huxley, 1974; Ebashi, 1980; Grinnell and Brazier, 1981). During sliding movements, the arrays of thin filaments move inward into the arrays of thick filaments; when the muscle shortens, the filament length remains constant. Therefore, it is evident that the length of the A band also remains constant (Fig. 2-5), but the I band becomes narrower, and will eventually disappear. Because the actin filaments are attached to the Z lines, they will be drawn together during the myosin-actin interaction, and the sarcomere becomes shortened. During isometric muscular contraction (which causes no change in length), the length of the A and I bands remains constant, however, when extending the muscle, the I band broadens.

At Rest

None of the cross-bridges are attached to the actin filaments; the "bards" are probably located very near to the myosin "backbone," tilted in one of their hinges. It seems as if, in the resting muscle, the tropomyosin rods lie toward the edge of the thin filament groove; in this position they can directly or indirectly block the actin sites, which would otherwise react with the cross-bridges. With this organization of the contractile

---

*The troponin molecule contains three subunits; one is a calcium-binding protein, TN-C; one is an inhibitory protein, TN-1, that is capable of inhibiting the interaction of actin with myosin cross-bridges; the third subunit, TN-T, is a protein that binds very strongly to tropomyosin.

proteins into separate actin and myosin filaments, the resistance to passive extensibility is very modest.

Excitation

(1) The motor nerve stimulates the muscle, and the propagated action potential depolarizes the muscle cell membrane (see Chap. 3). Next, there is an inward spread of the action potential along the T system, which, as mentioned, invaginates from the surface sarcolemma.[†]

(2) This event will, in one way or another (see below), result in a release of Ca^{2+} ions from the terminal cisternae of the sarcoplasmic reticulum into the fluid surrounding the myofibrils. (At rest the sarcoplasm is almost free from Ca^{2+}.) The Ca^{2+} ions bind to troponin on the actin filament (actually, two Ca^{2+} ions attach themselves to "specific regulatory sites" on the troponin molecule, troponin C).

Contraction

(3) This binding of calcium to troponin causes a change in the troponin-tropomyosin-actin complex that will remove the inhibition for an interaction between the myosin head and actin. There is evidence to suggest that the tropomyosin rods are now moved from their blocking positions. In other words, with the binding of Ca^{2+} to troponin, the tropomyosin strands are drawn from the periphery toward the center of the groove of the actin filaments, thus allowing actin to react with myosin; the actin monomers are now released from the preexisting inhibitory influence of the troponin-tropomyosin complex.[‡]

(4) The heads of myosin molecules move out perpendicularly from the thick filament core toward the actin filament and attach (the S-I subunits) to the actin molecules within reach. The cross-bridges are "energized" by Mg-ATP (or its intermediate), which is bound to the myosin head.

(5) The heads undergo an energy-yielding conformational change, so that the cross-bridges change their angular relationship to the axis of the myosin core; the actin filaments are moved or pulled along the myosin filaments. This is termed the power stroke of the myosin heads as they flip into a different conformational state (Fig. 2-9a).

The actual torque is probably applied at the S-I-actin interface. "If the S-I impeller is likened to an oar one must conclude that the torque is not, as in conventional rowing, applied at the oar lock, but rather at the point where the paddle dips in the water." (Borejdo, Putnam and Morales, 1981).

[†]Actually, the electric activity, in connection with the activation of the muscle cells, can be recorded with electrodes introduced into the muscle on a needle or a wire, alternatively with electrodes placed on the skin over the muscle groups of interest to study; electromyography, EMG.

[‡]According to one hypothesis, Ca^{2+} free troponin forces tropomyosin to cover peripheral sites of all actin molecules, favorably competing with the myosin heads with their potential to react with the same sites (see Murray and Weber, 1981). Seymour and O'Brien (1980), on the other hand, assume that tropomyosin binding at one place, will affect the binding of myosin at a different site on the actin filament.

(6) ATP (or its intermediate) becomes hydrolyzed because the rotations of the cross-bridges are energy-demanding and driven or coupled to ATP hydrolysis. Actin will activate the myosin ATPase residues in the S-I. ADP and free phosphate will leave the binding sites on the myosin head. (For details about myofibrillar protein-phosphorylation, see Bárány and Bárány, 1980.)

Relaxation

(7) Fresh ATP is taken up by the myosin head, which will promptly dissociate actin from myosin. Ca^{2+} is released from the troponin and transported across the membrane back into the cisternae of the sarcoplasmic reticulum. This transport is energy-consuming, requiring hydrolysis of ATP (one ATP for transport of two Ca^{2+}). The tropomyosin again changes its position relative to the actin subunits and inhibits actin from interaction with the cross-bridges. The cross-bridges then switch back to their original conformation.

Mechanism of Calcium Transport

It is still a mystery how the depolarization of the T tubule causes a release of Ca^{2+} from the sarcoplasmic reticulum (SR) into the myofibrillar space. It was mentioned that the T system is separate from the SR membrane. Contiguity, not continuity characterizes the "junction" (triad). The nature of communication between the T tubule, SR, and also the Ca^{2+} release process itself is still unknown (Tada et al., 1978; Grinnell and Brazier, 1981). "This is the least understood part among the entire series of events through which spread of an action potential in a muscle cell leads to contraction" (Endo, 1977). In a resting muscle, most of the calcium is stored in a large-capacity but low-affinity calcium-binding protein, probably calsequestrin located in the terminal cisternae, not in the longitudinal tubules of the SR. Apparently, the triad junction between the T tubules and the SR is the site of the so-called excitation-contraction coupling. According to one hypothesis there is an electrically coupled transfer of a "message" between the two compartments. Others emphasize that the T system is, in a way, electrically insulated from the SR, but it can also function as a chemical bridge or a "voltage-dependent charge movement" which will start a quick transport of Ca^{2+} from the terminal cisternae via the longitudinal tubules of the SR system out to the troponin (see Grinnell and Brazier, 1981). One problem is that the SR membrane is, at rest, extremely impermeable to Ca^{2+}, and SR membrane charges initiated by stimulation have, so far, not been established. The stimulation must involve a breakdown of a diffusion barrier, perhaps by opening Ca^{2+} channels and/or activating a carrier mechanism for a Ca^{2+} translocation out to the myofibrillar space (see Tada et al., 1978). There is a discussion of the role of Ca^{2+} itself in the transmission step from the T system because the SR membrane clearly contains calcium receptors capable of initiating or regulating the Ca^{2+} ion release from the cisternae, at least in the heart muscle (see Endo, 1977; Stephenson, 1981). Questions remain concerning the mechanisms behind the SR activity, the calcium pump, which removes Ca^{2+} from the troponin and the myofibrillar space back to the sink in the cisternae, essential for the breaking down of the cross-bridges (see Tada et al., 1978).

Summary Ca^{2+} is the link between the excitation and contraction of a muscle. The sarcoplasmic reticulum is specialized for the storage and release of calcium on the command of the T system. When a nerve signal arrives at the muscle cell, it initiates a release of Ca^{2+} into the fluid surrounding the filaments from special storage vesicles in the sarcoplasmic reticulum. By combining with troponin, it initiates the removal of a hindrance for a potential interaction between actin and myosin filaments, a steric hindrance, or perhaps another effect on actin conformation. In any case, tropomyosin, which in the resting muscle has turned off the active sites of actin, will now switch them on; actin can now activate an ATP hydrolysis of the myosin heads. This allows myosin heads to attach to sites on the actin filament. An attached bridge, during its action, exerts a longitudinal force for a certain distance in which probably one molecule of ATP is split, and that pulls the actin filament along toward the center of the A band. The system is switched off when Ca^{2+} is rebound by the sarcoplasmic reticulum. In the absence of calcium, the troponin-tropomyosin complex again prevents the interaction between actin and myosin filaments.

These events are repeated as long as the muscle is stimulated and the cross-bridges attach, swivel, and detach cyclically, thus propelling the thin filaments past the thick ones shortening the muscle. The active actin site may successively react with several linearly arranged myosin groups, and the actin filament travels along-side the myosin filament. If the muscle is contracting isometrically, the same molecular groups may react with one another repeatedly, but eventually the actin filament will slip back if enough tension cannot be generated. It is easy to understand why troponin and tropomyosin are called regulatory proteins.

Murray and Weber (1974) present an analogy:

The process of liberating the energy of ATP hydrolysis can be likened to the firing of a gun. The gun must first be loaded by placing an appropriate cartridge (ATP) in a specific chamber (the myosin head). This combination (the myosin-ATP) is converted to a special metastable form by cocking the gun (the second step in the hydrolysis). If the cocked gun (or charged form) is left alone, it is more or less stable. If the trigger is squeezed (or if an actin molecule is available), however, the stored energy is rapidly released and work is done on a bullet (or a cross-bridge). The process is completed by ejecting the spent cartridge (the products of hydrolysis, ADP and phosphate) and reloading.

"The Heart of the Whole Problem"

In the closing remark at the symposium on "The Mechanism of Muscle Contraction" at Cold Spring Harbor in 1972, H. E. Huxley stated: "However, there is a very large gap in our present knowledge, unfortunately right at the heart of the whole problem. It concerns the nature of the structural changes in the myosin head accompanying the different stages of ATP-splitting which are thought to produce the changes in the effective angle of attachment to actin during the operation of the cross-bridge." The problem is still with us (see Wray and Holmes, 1981; Pollack, 1983, has in his review article presented some critical thoughts concerning many details of the cross-bridge theory).

According to A. F. Huxley (1974), the filaments themselves are stiff, but within each cross-bridge there is an elastic element in series with a nonlinear element. The attachment of a cross-bridge to actin is followed by a movement in a small number of steps (Fig. 2-9). The total range of movement may be 10 to 12 nm, but part of this movement (some 5 nm) will stretch the elastic element even in an isometric contraction. By this mechanism, some of the yielded energy is stored in a spring-like fashion, and then the extended spring may shorten and release its stored energy.

It is assumed that the tension generated by an attached cross-bridge can vary in a series of steps depending on the effective angle of attachment of the S-I subunit to actin and the degree of extension of the elastic S-2 linkage (connecting S-I to the light-meromyosin backbone of the thick ligament). The attachment of bridges may be the rate-limiting step that controls the speed of shortening at moderate loads. A consequence is that the number of attached cross-bridges at any particular moment decreases as the velocity of shortening increases. Even during maximal activation only part of the cross-bridges are effectively linked to sites on actin (perhaps less than 50 percent), while the remaining move randomly about in the interfilament space.

Apparently, the equilibrium position of a bridge is sensitive to changes in its environment, and there may be "messages" passed between neighboring bridges affecting the position and eventual movements, even if the heads are not in a position to interact with actin (H. E. Huxley, 1972).

It is not known which structure is the elastic element in the cross-bridge, what structure undergoes the stepwise change, what kind of bonds hold the myosin head to the actin filament, or how the binding of ATP causes myosin to dissociate from actin (A. F. Huxley, 1974). It was mentioned that one ATP is probably yielding energy for one working cycle of a cross-bridge. The energy demand of the transport of Ca^{2+} is an additional expense requiring ATP as an energy source. In the resting muscle cell, ATP undergoes a very slow hydrolysis when bound to myosin. It may be partly reacting with myosin (M), forming a "charged" $M \cdot ADP \cdot P$ complex. It can then react with actin (A), which will markedly accelerate the ATP hydrolysis by combining with the $M \cdot ADP \cdot P$ intermediate. (Tropomyosin also exerts an activating effect on actomyosin ATPase.) The resting rate of ATP hydrolysis is probably less than 1 percent of the maximal rate during contraction. The sequence may be as follows:

$$M + ATP \rightarrow M \cdot ATP \rightarrow M \cdot ADP \cdot P$$
$$A + M \cdot ADP \cdot P \rightarrow A\text{-}M \cdot ADP \cdot P \rightarrow A\text{-}M + ADP + P_i$$

Binding of ATP (plus Mg^{++}) to the actomyosin complex is very rapid and the dissociation of the A-M complex follows also very rapidly:

$$A\text{-}M + ATP \rightarrow A + M \cdot ATP$$

and again we obtain a M · ADP · P intermediate. (For a detailed discussion see Eisenberg and Hill, 1985.)

Normally, there is plenty of ATP available to keep the system going, but if the concentration of Mg · ATP is very low, the cross-bridges will remain attached to actin, and the muscle is brought into sustained contraction (rigor). In this situation, the cross-bridge is apparently locked at the. end of the active stroke, the point at which the next ATP molecule normally comes along to dissociate the bridge.

When the muscle shortens, the ends of the thin filaments will slide toward one another in the center of the A band, and they may even overlap (see the bottom illustration in Fig. 2-8). (A dense zone appears in the center of the A band, and in a transverse section of this zone, an electron micrograph shows twice as many thin filaments as in a relaxed muscle, proving that actin filaments may overlap during contraction.) This may explain the observed decrease in maximal force generated by a muscle as it shortens. The projections are absent at the center of the thick filaments, and as the thin filaments from one Z line continue into the "wrong" part of the A band, the orientation of the molecules becomes abnormal from a functional viewpoint. In this region, the filaments would not be expected to contribute to the development of force by the muscle, and they may even interfere in a negative way with the interaction of the correctly oriented actin and myosin molecules.

The maximal force a muscle fiber can develop is dependent on the length of the muscle, i.e., the sarcomere lengths, which will be discussed in greater detail below.

DIFFERENT FIBER TYPES

Contractile Properties, Histochemical, and Ultrastructural Bases for Fiber Typing

From a functional point of view, the muscle cells do not constitute a homogeneous tissue. Most muscles are built up of muscle fibers with different mechanical and contractile properties. Based on the time it takes for the fibers to reach their peak tension, which is related to the relaxation time, one can identify two main classes of fiber types: Those with a relatively long time-to-peak tension are the *slow-twitch fibers* or using a more "neutral" nomenclature; *type I fibers*. The fibers with a short time-to-peak tension are the *fast-twitch fibers* or type II fibers. (In the case of human skeletal muscle, experts have not agreed on the use of a single nomenclature when describing the fiber types.) For human skeletal muscles, there are studies indicating that the time-to-peak tension in a maximal isometric contraction is 80 to 100 ms for type I fibers and about 40 ms for type II fibers (see Saltin and Gollnick, 1983).

It is a complicated matter to study the contractile properties of the subunits in an intact skeletal muscle. For this reason, the classification of fibers is usually based on the histochemical differentiation of myofibrillar ATPase. There is strong evidence indicating that the quantity of ATPase bound to myosin is actually rate-limiting in the shortening process, and that by exposing a muscle sample to buffers with different pHs and histochemical staining, one can identify the two main groups of muscle fibers just mentioned (see Brooke and Kaiser, 1970). Pre-incubation of muscle sections at pH 10.3 causes the myosin of the type I fibers to lose its ATP activity. Therefore, there is a loss of the histochemically demonstrable stain for myofibrillar ATPase, whereas type II fibers will stain intensively. At a pH of 4.3, the picture is reversed. An example is given in Fig. 2-10.

The type II fibers can be further subdivided into IIa, IIb, and IIc fibers based on their resistance to loss of histochemically identified myofibrillar ATPase at the lower pH range (Brooke and Kaiser, 1970). The type IIa and IIb fibers have similar kinetic properties. Less is known with regard to the type IIc fibers, but in rat soleus muscle they have contractile characteristics intermediate to type I and type II (Kugelberg, 1976). Not only is there a difference in the myosin ATPase activity between type I

Figure 2-10
Top: Serial cross sections of human skeletal muscle samples. "A" (stained for myofibrillar ATPase) showing fast twitch fibers, Type II (dark), and slow twitch fibers, Type I (light); "B" (stained for glycogen with PAS reaction staining glycogen, dark). By comparison with "A," the fibers can be identified in "B." Note the selective depletion of glycogen in the slow twitch fibers.
Bottom: Sections of skeletal muscle (stained for myofibrillar ATPase as in "A"). "C" was obtained from a male sprinter who ran 100 m in 10 s; "D" is from an elite long-distance runner. Note the dominance of fast twitch fibers in the sprinter's muscle and the reversed picture for the endurance athlete. (*By courtesy of Gollnick and Saltin.*)

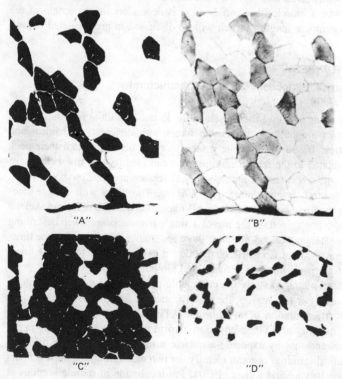

"A"   "B"

"C"   "D"

and type II fibers, but there are also differences in the myosin subunit composition. As previously mentioned, the myosin molecule contains heavy-chain peptides as well as light-chain peptides. Using immunohistochemical procedures and electrophoretic techniques the peptide pattern of the different fiber types can be analyzed. (For discussion see Billeter et al., 1981; Pierobon-Bormioli et al., 1981.) Type I fibers have so-called *slow heavy chains*, a predominance of *slow light chains*, and a variable amount of *fast light chains peptides*. (In the human skeletal muscle there are apparently five different light chains.) A variation in the light chain pattern may modify the actin-activated ATPase of myosin which, as mentioned, is highly correlated with the contraction speed, which may, at least partly, explain the wide spectrum of contraction times noted for type I fibers.

Type IIa and type IIb fibers contain identical fast light chain patterns but biochemical differences are evident in the *fast heavy chain* structure (although they have no slow heavy chain peptides). Intermediate properties between type IIa and IIb fibers may appear. Perhaps they are fibers in transition from one type to another. Thus, in conclusion, it is evident that different fiber types can be identified by distinct myosin properties. In addition, the troponin structure is different in type I fibers as compared with type II fibers. There is approximately twice as much sarcoplasmic reticulum in the fast (type II) compared with the slow fibers (type I). The consequence is hypothesized to be that the rate of uptake of Ca^{2+} by the sarcoplasmic reticulum and also by troponin is faster in the fast muscle fiber, which should promote a quick reactivation of the fiber.

Differences exist in the structures of the Z line and M line between the type I, type IIa, and type IIb fibers (Sjöström et al., 1982). The myosin/actin ratio is not different in the two types of muscle fibers. This may explain why, in the human skeletal muscle, there is no apparent difference in the force generating potential between type I and type II fibers. The maximal force depends on the cross-sectional area of the muscle, irrespective of the fiber distribution, possibly with the exception of very fast contraction, where the type II fibers may be superior (see Saltin and Gollnick, 1983; Schantz et al., 1983; Maughan and Nimmo, 1984).

Metabolic Characteristics

As mentioned, a low activity of myosin ATPase in the muscle fiber is usually associated with a relatively slow contraction. This is typical for the type I fibers. Their glycolytic enzyme system is less developed. On the other hand, type I fibers have a high potential for aerobic metabolism. They are richer in mitochondria and myoglobin, and the density of the capillary network is more developed around the slow fibers compared with the fast, type II fibers. The slow-twitch fibers are well adapted for prolonged activity; they are much more resistant to fatigue than the fast-twitch fibers.

From this description, it is evident that the "average" fast fibers with their high activity of myosin ATPase and short contraction time have a relatively well-developed glycolytic enzyme system, the mitochondrial content and oxidative activity are lower, and the fibers fatigue rapidly. However, the subgroups within the type II fibers have different metabolic profiles. Type IIa fibers have a high oxidative potential and a high glycolytic power. They are relatively resistant to fatigue. Type IIb is the "typical" fast-twitch fiber with a low aerobic potential. Type IIc is considered a relatively little differentiated fiber. Normally there are few type IIc fibers in the human skeletal muscle (approximately 1 percent). Schantz and Henriksson (1983) reported a marked increase in the number of type IIc fibers in the triceps brachii muscle in subjects who had been skiing and had covered a distance of 800 km in 36 days (from 2 to 15 percent). They discussed the possibility that the type IIc fiber may be an intermediate fiber due to a transformation of type IIa and IIb fibers into type I fibers. Actually, type IIc fibers contain a mixture of slow (type I) and fast (type II), heavy and light chain peptides. Table 2-1 summarizes some of the functional properties of the fiber types.

TABLE 2-1
CLASSIFICATION OF HUMAN SKELETAL MUSCLE FIBERS*

Property	Type I Slow twich fiber	Type II Fast twitch fiber a	b
Motoneuron	Small	Large	Large
Discharge	Low	High	High
Contractile speed	Low	High	High
Endurance	High	Medium	Low
Capillary density	High	Medium	Low
Myoglobin content	High	Medium	Low
Glycogen content	ND		
Glycogenolytic enzyme activity	Low	High	High
Mitochondrial enzyme activity	High	Medium	Low
Myofibrillar ATPase activity	Low	High	High
ATPase activity after preincubation at pH 10.3	0	High	High
ATPase activity after preincubation at pH 4.6 − 4.8 and 10.3	0	0	High

ND = no difference.
*BASED ON HISTOCHEMICAL STAINING AND SOME FUNCTIONAL PROPER-
TIES. TRAINING CAN MODIFY SOME OF THE PROPERTIES, AND THE DIFFER-
ENCE BETWEEN "HIGH" AND "LOW" IS NOT ALWAYS LARGE. A STAINING FOR
THE MYOFIBRILLAR ATPᴀsᴇ ACTIVITY IS MADE AT A GIVEN pH AFTER PREIN-
CUBATION AT VARIOUS pHs. IT IS ASSUMED THAT THE DIFFERENCES IN THE
STAINING PATTERN ARE DUE TO THE PRESENCE OF DIFFERENT ISOENZYMES
OF THE MYOFIBRILLAR PROTEINS IN THE FIBER TYPES.

Training Can Modify the Metabolic Profile of the Different Fiber Types

It should be emphasized that in contrast to the myosin ATPase activity, the metabolic enzymes seem to be influenced by the level of habitual activity, use or disuse, and may thus change in response to endurance training. Endurance training causes an increase in the concentration of mitochondrial enzymes, as well as an increase in the number and volume of mitochondria *in all fiber types*. The capillary density will also increase, and since capillaries are usually shared by muscle fibers of all types, the potential for an exchange of gases and substrates between blood and mitochondria will also be improved by training in all fiber types. It should also be emphasized that, despite a superior potential for anaerobic energy yield in the fast-twitch fibers as compared with the slow-twitch fibers, the anaerobic energy yield is at a level sufficient to provide a high anaerobic power in the case of slow-twitch fibers as well. Consequently, the metabolic difference in the potential to perform aerobic and anaerobic exercise in the case of type I and of type II fibers, is more a question of habitual use or disuse than a consequence of genetic endowment. As an example, longitudinal studies have indicated that endurance training will reduce the number of type IIb fibers that apparently have been transformed into type IIa fibers. In other words, the relative number of fibers in different subgroups within the type II family may vary at different times for a single individual. The overall picture is that there is a relatively wide range of contraction times for a particular fiber type. In the case of enzyme activities, endurance, or capillary density, there is a plasticity that may result in a definite overlap in the characteristics of the different fiber types. It should also be emphasized that there is some confusion regarding terminology in the literature; a classification suitable for the skeletal muscle of one species may not be suitable for describing fiber types in another species. (For a more complete discussion, see Buchtal and Schmalbruch, 1980; Jolesz and Sreter, 1981; Saltin and Gollnick, 1983; Gollnick et al., 1983.)

What Factors Determine Whether a Muscle Fiber Becomes Type I or Type II?

Under "normal" conditions the proportions of type I and type II fibers seem to be a genetic matter. Identical twins, for example, have a very similar proportion of fiber types in a given muscle, in contrast to fraternal twins (Komi et al., 1977). The specialization of different fiber types seems to start in the 20-week-old human fetus with maturation occurring at various times in different muscles of the same species; fiber differentiation is nearly complete at approximately one year of age (see Colling-Saltin, 1980). There is strong evidence to suggest that neural influences determine the fundamental dynamic properties of the contractile material; in other words, the nerve can in one way or another exert influence on the contractile properties of the muscle fiber it innervates.

The motoneurons, contacting type I fibers, are relatively small and their firing rates are modest in comparison with the motoneurons innervating type II fibers (see Chap. 3). They may pass on different chemical compounds to the muscles. A motoneuron, via its axon, establishes functional endplates on denervated muscle fibers and can effectively excite them. Buller et al. (1960) describe experiments performed on kittens and cats, in which the nerves to a slow muscle (such as the soleus) and a fast muscle (such as the flexor digitorum longus) were divided and cross-sutured. When a nerve from the "fast" motoneurons had innervated the slow muscle, the muscle was gradually transformed into a fast muscle, even in the adult cat. Likewise, "slow" motoneurons converted fast muscles to slow muscles. More recent experiments have shown that a changed speed of muscle contraction after reinnervation is due to changes in the kinetic properties of myofibrillar ATPase and other enzymes. After cross-innervation, there are also changes in the myosin, troponin, and sarcoplasmic reticulum, as well as in the contractile material itself and the regulatory protein system.

If muscle fibers of type II are subjected to chronic electrical stimulation, e.g., by electrodes implanted on their motor nerves with a low frequency typical for the motoneurons innervating type I fibers, say 10 Hz, there will, over a period of a month, be a transformation of the fibers. The former fast-twitch fibers will, when subjected to this change in their normal pattern of use, become indistinguishable from slow-twitch fibers morphologically, physiologically, and biochemically (see Salmons and Henriksson, 1981). Following removal of the stimulus, the muscle fibers will gradually regain their former characteristics. It is more difficult to transform type I fibers into type II fibers, however, continuous electrical stimulation with a high frequency, say 100 Hz, in denervated type I fibers may give them characteristics typical of type II fibers. (See also Jolesz and Sreter, 1981.) These findings illustrate the plasticity of the skeletal muscle. The pattern of nervous activity is apparently of decisive importance for the characteristics of the muscle fibers, particularly with regard to their contractile properties. There may also be effects mediated by molecular transportation along the motor axons and across to the muscle fibers via the neuromuscular synapses, e.g., a neurotrophic effect. The fact that all muscle fibers in a motor unit have almost identical histochemical, biochemical, and physiological properties in all of their respective sarcomeres, supports the concept of the neural effect on the muscle fiber (see Nemeth et al., 1981). (The motor unit is a single motoneuron and all of the muscle fibers that it innervates.)

As mentioned previously, the occurrence of type IIc fibers, in response to hours of daily endurance training, *may* indicate a transformation of type II fibers into type I fibers. There is good evidence to support the hypothesis that all muscle fibers have in their genetic code the potential to synthesize all of the components associated with the molecular structures and functions typical for both fiber types. When some muscle cells becomes denervated, due to a trauma or other reasons, there is a good chance that another intact nerve axon or terminal by a so-called collateral sprouting will reinnervate the deprived muscle cells. There is no guarantee, however, that a type II fiber will form a synapse with the axons of a large motoneuron. The fiber may just as well become a "slave" to a small motoneuron with an activity pattern that will gradually transform the muscle fiber into a type I fiber. It is therefore conceivable that at least some of the type IIc fibers are in fact such intermediate or transitional fibers reinnervated by the "wrong" motoneurons.

There Are Great Individual Variations in Fiber Types

The proportion of slow-twitch fibers can be anywhere from approximately 10 up to 95 percent. On the average, it is approximately 50 percent in the vastus lateralis, deltoidus, and gastrocnecius muscle, but the percentage is as high as 70 percent in the soleus (see Saltin and Gollnick, 1983). There is no systematic difference in fiber distribution in the skeletal muscle between men and women. The fiber type distribution and ultrastructure of the skeletal muscle in the 6-year-old child is not different from normal adult tissue (Bell et al., 1980). The high percentage of type 1 fibers that occurs in top athletes, specializing in endurance events, is most likely a consequence of the necessity of this fiber composition for success, and the fact that natural endowment is more decisive than the amount or intensity of training. (Saltin and Gollnick, 1983, have summarized data from the literature on "typical" fiber composition in the limb muscles of humans in training for various events.)

It should be mentioned that the increase in muscle length during growth is brought about by an increase in the number of sarcomeres to the ends of the existing fibers without marked changes in the individual sarcomere length. The increase in the cross-sectional area of a muscle fiber during growth is the result of longitudinal splitting of myofibrils, and thereby an increase in the number of myofibrils when they reach a critical size (Goldspink, 1970).

The recruitment and function of the muscle, and its different fiber types are closely tied to the function of the nervous system. Further discussion of this matter will therefore be addressed in the next chapter. More details on the adaptive responses of the skeletal muscles will appear in the chapter on training (Chap. 10). The proportions of different fiber types within a muscle and within different muscle groups in the same individual, as well as the relationship between fiber types and muscle performance will be discussed in subsequent chapters.

Summary   Human skeletal muscle is composed of two main types of muscle fibers, each type with different contractile properties. Therefore, the time it takes for one fiber type, the fast-twitch, or type II fibers, to reach peak tension in a maximal contraction is shorter than that for the slow-twitch, or type I fibers. The fiber types also differ in some of their morphological and histochemical properties, and of particular significance is the higher activity pattern of the myofibrillar ATPase in the fast-twitch fibers. One might generalize and say that the type I fibers are specialized for an aerobic energy yield and are quite resistant to fatigue, which is in direct contrast to type II fibers, which possess a biochemical machinery developed more for anaerobic processes, and are therefore more fatiguable. However, there are subgroups in the type II family with type IIa fibers more closely related to type I fibers in their metabolic potential; type IIb fibers, which are less equipped for an aerobic metabolism; and type IIc fibers, which may actually be a transformational stage between type I and type II fibers. Actually, there is a wide spectrum of physiological properties and an ability to adapt to different demands in all fibers, but the functional difference in their respective kinetic properties is apparently genetically predetermined. (See also Pette, 1980.)

Muscle Biopsies

Most of our knowledge concerning the fiber composition and characteristics of human skeletal muscle is based on the analysis of samples obtained from muscle biopsy. The question is, "how representative is such a biopsy?" Blomstrand and Ekblom (1982) report that the percentage of type I fibers varied 6 percent between duplicate biopsies taken from one leg. When the biopsies were taken from both legs, the variation was 12 percent. The variation in the percentage of type IIa fibers was 4.4 percent and 7.3 percent, respectively. Lexell, et al. (1983), have analyzed the fiber type distribution throughout the vastus lateralis quadriceps (which is the most commonly used muscle) in terms of fiber composition and metabolic parameters. They report that the distribution of different fibers varied intramuscularly, primarily as a function of depth, with a predominance toward type II fibers at the surface and type I fibers in deeper regions within the muscle. They recommend that multiple biopsies should be performed and that the biopsy depth in the muscle should be defined. They note that the coefficient of intramuscular variability was + 15 to 20 percent. Evidently, data obtained from the analyses of single muscle biopsies should be evaluated quite critically.

TYPES OF MUSCLE ACTIVATION

In the activated muscle, the contractile components (i.e., the myofibrils) shorten and stretch the elastic elements (in the cross-bridges as well as in connective tissue, and in tendons). When both ends of the muscle are fixed and no movement occurs in the joint(s) involved, the contraction is called *isometric* ("same length"; *static*). If the muscle varies its length when activated to produce a given force, the contraction is *isotonic* ("same tension"; *dynamic*). It should be emphasized that there are probably no contractions involved in human performance in which the force is kept the same throughout a movement. Even if the external load is kept constant, the force developed by the muscle varies as the lever arms become shorter or longer. Therefore the use of the term isotonic should be discouraged and replaced by the term *dynamic* exercise. In this case, external work can be done, and the amount of work can be calculated from the product of the force multiplied by the distance through which it is expressed (and measured in Newton meter, Nm). Since the distance is zero in an isometric contraction, no mechanical work is done according to physical laws. However, isometric activity demands energy and can be very fatiguing. From a physiological viewpoint, the "work" performed is definitely related more to developed force times contraction time rather than to force times the displacement. In dynamic exercise, the muscle may shorten, in which case the work is labeled *positive* or *concentric,* or the muscle may lengthen, in which case the work is *negative* or *eccentric.*

Force in Relation to Muscle Length and Speed of Contraction; Force-Length

The force exerted by a muscle, stimulated by a given impulse traffic, depends on the initial muscle length. It was mentioned previously, that the length of the sarcomere, a small unit in the muscle fiber, is crucial in this respect. An unattached, unstimulated muscle is at its equilibrium length. If stretched, the passive elastic components will create a force that increases as an exponential function of length, actually over a range of up to 200 percent of the equilibrium length (isolated muscle). Normally, when attached by its tendons to the skeleton, the muscle is under slight tension, since it is moderately stretched (at resting length). Measurements of the force developed by an activated muscle show that the isometric force is maximal when the initial length of the muscle at the time of activation is approximately 20 percent above the equilibrium length (relative length 1.2:1). The force falls roughly linearly at lengths below this optimum and is zero when the muscle is maximally shortened (see Fig. 2-11). When stretched beyond the relative length 1.2:1, the "active" force produced by the stimulated muscle becomes progressively smaller and is zero when the muscle is elongated about twice its resting length. This failure to yield force when overstretched can be explained by the sliding hypothesis of muscular contraction. A stretch beyond the resting length does diminish the distance of overlap between the myosin and actin filaments. The more they slide apart, the fewer cross-bridges between the two types of filaments are possible. Actually, at a sarcomere length beyond 3.5 μm, there is no overlap at all. The explanation of why *less* force is exerted when the muscle shortens is not as evident. One may speculate as follows: During a variation in the sarcomere length, there is

Figure 2-11
Isometric force-length data for human triceps muscle. (The ordinate gives the maximal tension in arbitrary units.) To obtain the "net voluntary force curve," the values from the "passive force curve" were subtracted from the "total force curve." For further explanation see text. (*From Univ. Calif., "Fundamental Studies of Human Locomotion and Other Information Relating to Design of Artificial Limbs," 2, 1947.*) (*See Ruch and Fulton, 1960.*)

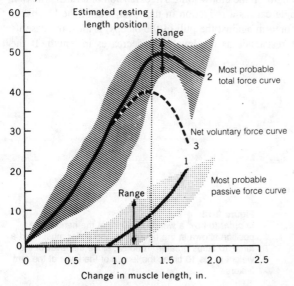

also a variation in the actin-to-myosin distance. For example, in toad muscle, it has been estimated that the surface-to-surface distance between actin and myosin is about 5 nm at a sarcomere length of 3.5 μm, but 13 nm at a sarcomere length of 1.8 μm (Elliot, 1967). It could be that there are mechanical disadvantages in generating force because of the increasing distance between the actin and myosin filaments as the muscle shortens. The flexible cross-bridges are able to extend from the myosin backbone, thanks to their hinges, and reach the actin filaments despite an increase in the interfilament distance, however, the leverage for the interaction must be different.

Fig. 2-11 summarizes this discussion by illustrating experimental data from studies in which measurements were made of (1) the force produced by passive stretching of the triceps muscle (lowest curve), and (2) the force developed during a maximal voluntary contraction (top curve).

The Optimal Position to Produce Force is not Only Dependent on the Initial Length of the Engaged Muscles

In the intact body, anatomic limitation of the joint usually restricts muscular lengthening and shortening; the range usually varies between 0.7 to 1.2:1 (for some muscles up to 1.4:1) of the equilibrium length. Therefore, in general, maximal force can be exerted when the muscle is maximally stretched, and as it shortens, the force produced decreases. Since the skeletal muscles exert their effect on external resistance via levers, the geometric arrangement of the bony levers must be included in an analysis of the optimal exercise positions and the most effective utilization of the forces of muscle contraction. Figure 2-12 illustrates a position in which the flexor of the arm holding a weight must produce a force that is about 10 times greater than the force of gravity acting on the load. The optimal lever arm for the biceps muscle is obtained with the arm flexed at an angle of about 90° in the elbow joint. An extension or a further flexion of the arm from the right angle causes a reduction in the lever arm of the biceps. If a constant force acts on the forearm against the pull of the biceps, it must increase its exerted force to balance the resistance as the lever arm decreases in length. If the

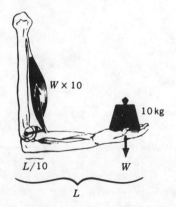

Figure 2-12
In order to hold a weight of 10 kg in the hand in the position shown in the figure, the arm flexor must exert a force 10 times greater (1000 N), since the lever of the weight (L) is 10 times the length of the lever of the arm flexors.

tendon of the biceps were closer to the hand, its ability to flex the forearm against resistance would certainly be favored—but at the expense of speed of movement.

Summary In any analysis of the most favorable position for maximal force to be exerted against an external resistance, consideration must be given to the following factors: (1) The maximal tension that any muscle fiber can develop depends on the relative length of the muscle fiber at the time of contraction. The tension reaches a maximum at a relative length of about 1.2:1 and decreases at lower and higher lengths. (2) The lever arms in the body, through which the muscle forces are transformed into pulls, pushes, etc., alter with the changing positions of movable joints.

In functional muscle activities, several anatomically different muscles collaborate. The parts of the muscle group that act in synergism may change with the position of the limb. Consequently, it is very difficult to predict, from theoretical considerations, the most efficient position that will produce the greatest strength.

Force-Velocity

Figure 2-13 illustrates the variation in maximal force developed by muscles during contraction with different speeds, showing the well-known S-shape relationship, i.e., the classical force-velocity curve. The highest force that can be developed is in a fast eccentric contraction; the maximal force then declines to a minimum when the muscle is activated in a concentric contraction at high speed. When measuring maximal dynamic strength, it is therefore extremely important to control the speed of contraction carefully. For example, in a strength test performed before and after a training program, a slight reduction in velocity of a concentric contraction in the second test can erroneously simulate an increase in muscle strength. An apparatus has been developed for the measurement of force potentials at given velocities, i.e., an isokinetic (same speed) loading dynamometer (for references, see Thorstensson, 1976). Figure 2-13 also shows the power, i.e., the force times distance per unit of time, that is developed or absorbed in maximal contraction at various speeds. The highest power is attained when velocity of contraction is 25 to 30 percent of the maximal value; the force is then about 30 percent of the maximal isometric strength.

Electromyographic (EMG) studies (Bigland and Lippold, 1954) show the following: (1) When a voluntary isometric contraction of a muscle is made, the electrical activity, measured by integrating the action potentials from surface electrodes, bears a linear relation to the force that is being exerted. (2) At constant velocity of shortening or lengthening, the electrical activity in the muscle is directly proportional to the force. However, the slope of this correlation falls off during lengthening (Fig. 2-14). This means that the degree of muscle excitation required to produce a given force of contraction is smaller when the active muscle is forcibly stretched than it is when the muscle shortens at the same velocity. (3) At constant force, the electrical activity increases linearly with velocity of shortening, but it decreases when the muscle is being lengthened. (See also Milner-Brown and Stein, 1975.) The integrated EMG activity is the same in maximal contraction of different speeds, indicating the same degree of muscle recruitment in maximal efforts; it does not matter whether the con-

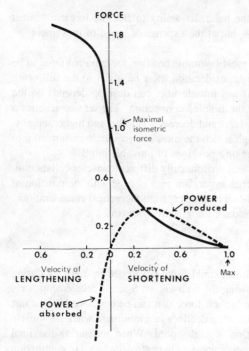

Figure 2-13
The classical force-velocity curve (full line) obtained on an isolated muscle, showing the maximal force that can be developed when a muscle is contracting at various speeds. Note that the maximal force in a concentric activity is less than in an isometric contraction. Highest force can be attained in a rapid eccentric contraction. The dotted line gives the maximal power, i.e., the force times the velocity of contraction. Curves differ according to the muscles studied. In in vivo experiments, the analysis is complicated because in movements, the muscles are rarely contracting isotonically; the leverage changes during a dynamic contraction and therefore the force demand on a muscle is not constant even when the external load is kept constant.

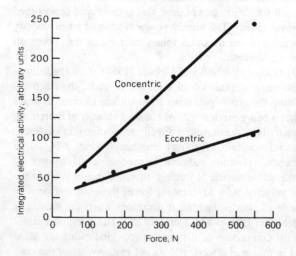

Figure 2-14
The relation between integrated electrical activity and force in the human calf muscles. Recordings were made from electrodes placed on the skin over the muscles. Shortening at constant velocity (above) and lengthening at the same velocity (below). Each point is the mean of the first 10 observations on one subject. Actually the force represents the weight lifted and is approximately one-tenth of the tension developed in the tendon. (*Bigland and Lippold, 1954.*)

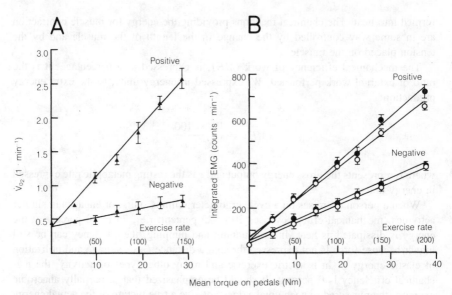

Figure 2-15
(A) the mean rates of oxygen uptake ($\dot{V}_{O_2}$), and (B) the integrated EMG recordings obtained in all experiments on the best trained subjects over the course of several months, plotted against the mean torque on the pedals for positive and negative exercise. Filled circles represent records from left leg; open circles, right leg. The work rates (watt) corresponding to each exerted torque are shown in parentheses. The pedal length was 17 cm. The slope of the $\dot{V}_{O_2}$/torque and the integrated EMG/torque relationships were obtained from the calculated regression lines (continuous lines). Vertical bars show the standard error of the mean. (*Modified from Bigland-Ritchie and Woods, 1976.*)

traction is concentric, isometric, or eccentric (Komi, 1973; Rodgers and Berger, 1974). Roughly, the oxygen consumption of active muscles should reflect the electrical activity displayed. As a matter of fact, the oxygen uptake increases linearly with the rate of exercise in many types of activities. A comparison of the oxygen uptake at a given rate of work, in exercise where the muscle groups contract concentrically ("positive exercise"), and eccentrically ("negative exercise") respectively, reveals that the negative exercise demands much less oxygen (Fig. 2-15) (Asmussen, 1953; Bigland and Lippold, 1954). These results fit in with the electromyographic studies, but a physiological explanation for the difference in efficiency in eccentric and concentric exercise is not available.

MECHANICAL EFFICIENCY OF MUSCLE CONTRACTION

An activation of muscle fibers is always associated with increased heat production (Hill, 1958). During isometric contraction, all extra energy output is converted into heat. In dynamic exercises, various fractions of chemically available energy are transformed into heat. The chemical reactions providing the energy for muscle contraction are in some way controlled by the change in the length of the muscle and by the tension placed on the muscle.

The mechanical efficiency of work (ME), is expressed as a percentage, it is the ratio of external work performed, W, expressed in energy units, to the extra energy production:

$$(ME) \ = \ \frac{W \times 100}{E - e}$$

where E represents the gross energy output and e is the resting metabolic rate expressed in energy units.

When a person exercises on a cycle ergometer, climbs stairs, or engages in similar activities, mechanical efficiency rises 20 to 25 percent; i.e., 75 to 80 percent of the energy is dissipated as heat. In fast running and jumping, the efficiency can be still higher because force is not expressed through a distance (see "storage and utilization of elastic energy"). In isometric exercise and many other types of activity, the mechanical efficiency is 0 percent. It should be emphasized that in partially anaerobic exercise, the measured oxygen uptake does *not* give a true picture of the actual energy demand. We have no methods available for an exact measurement of the mechanical efficiency of such types of exercise.

The heat produced increases the temperature of the muscle. Within limits, the elevated temperature improves the performance of the muscle. This can be explained on both a chemical and physical basis. The increased force produced by the muscle fibers upon repeated stimulation (Fig. 3-21) can be partially explained by the beneficial effect of elevated tissue temperature. To prevent overheating during prolonged activity of the muscle fibers, an increased local blood flow is essential, as well as an increased heat conductance of the skin.

STORAGE AND UTILIZATION OF ELASTIC ENERGY

It has already been mentioned that a contracted muscle has elastic properties. If an activated muscle is stretched, part of the added energy is degenerated into heat, but another part can be absorbed by the "series elastic elements" of the muscle. This stored energy can be made available during the contraction. In fact, under certain conditions the developed tension in concentric exercise can exceed the maximal strength. In many exercises kinetic and potential energies created by the muscles in one phase of a movement, are taken up again by the muscles in another phase of the movement. As pointed out by Cavagna (1977) the muscles "brake" as well as "propel" the body. When the activated muscles are forcibly stretched, they are doing negative or eccentric exercise. Cavagna et al. (1971) measured the power exerted during sprint running by means of a platform (force-platform) sensitive to the force impressed by the foot. They noticed that at speeds of 6 to 7 m · s^{-1}, the contractile component alone was responsible for the power output of the muscles. In other words, the muscle behavior followed the power-velocity curve (Fig. 2-13). It should be recalled that the maximal power is "normally" attained at about one-third the maximal speed of the shortening of a muscle. At higher velocities, the maximal force and power should decrease and this reduction must soon limit the running speed. However, at speeds greater than 6 to 7 m · s⁻¹, the good sprinter could provide an "abnormal" increase in power (Fig. 2-16). Cavagna and his coworkers (1971) explained this finding as follows: (1) The extensor muscles of the leg were activated before the foot reached the ground. (2) There was a phase of deceleration and kinetic energy that stretched these muscles. (3) Part of this energy was stored in the elastic elements during stretching of the contracted muscles, during negative exercise. (4) The energy was then utilized in the next phase of the step when the muscles did positive work. Thereby, the muscle can produce more power than when it merely contracts because part of the stored energy is released, supporting the power generated by the cross-bridge interactions.

The calculated average power developed during a push at each step reached 2,200 to 3,000 W at the highest speed. A technique factor is evidently involved, for not every subject could make good use of the energy absorbed by the muscles at the higher

Figure 2-16
The average power developed by the muscles during the push in sprinting, as a function of the speed of the run (abscissa). The continuous line is traced by hand through the experimental points. In the tracing the "contractile" curve (interrupted line partially overlapping the continuous line) indicates the mechanical power developed by the contractile component of the muscles. The curve "elastic" was obtained by subtracting the "contractile" curve from the continuous line. It indicates the part of the total power output due to the mechanical energy stored in the series elastic elements during the stretching of the contracted muscles. (*Cavagna et al., 1971.*)

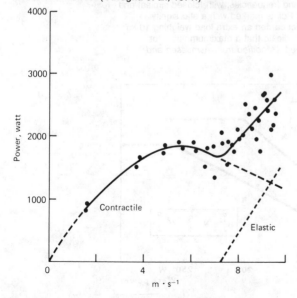

speeds. Apparently, the properties of the activated muscles have some of the qualities of a springboard, that can also be utilized in a more or less efficient way by the diver. Cavagna (1977) points out that the most striking example of improved efficiency is demonstrated in running, both in the case of humans and animals. The efficiency can increase with the speed of running up to 60 to 70 percent. Asmussen and Bonde-Petersen (1974a, b) have provided additional data on the storage of elastic energy in muscles. Maximal vertical jumps were performed with and without countermovement (bouncing) and after jumps down onto the force-platform from various heights. When the subjects bounced, an average of 23 percent of the developed power could be recovered during the jump. When they jumped down from 0.40 m before the upward jump, the stored energy was 10 percent of the totally developed energy. The actual gain in jumping height averaged 10 percent at the most. This was attained after the jump down from 0.40 m, (compared with the jump from a semisquatting position). In another series of experiments, the same authors (1974b) measured the energy cost of lowering and lifting the body by flexing and extending the legs from a standing or sitting position, that is, with or without a rebound in the deepest position. Fig. 2-17 illustrates the marked difference in efficiency in the two types of exercises. At a power of 170 watts the metabolic power was 870 watts in knee bending without rebound, but only 650 watts with rebound. The mechanical efficiency of one subject was around 25 percent and 40 percent, respectively. The authors conclude that the difference in efficiency must depend on the possibility of using the energy absorbed and stored in the muscles as elastic energy during a phase of negative exercise, when the positive phase follows immediately after the negative phase. In 1977, Komi and coworkers

Figure 2-17
Metabolic rate against the power developed in rhythmic, deep leg flexions and extensions at varying frequencies, with (closed circles) and without (open circles) rebound. Points marked with a star signify experiments in which the subject carried an extra load weighing 10 kg on the shoulders. Dotted lines indicate that a maximum value for oxygen uptake has been reached. (*Modified from Asmussen and Bonde-Petersen, 1974b.*)

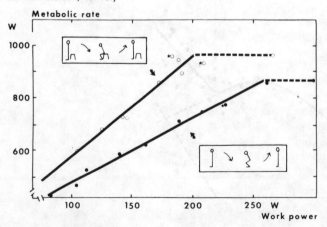

(see Bosco, 1982) reported similar findings with an increase in mechanical efficiency from 20 percent (no rebound jump) to 30 percent (rebound jump). It should again be emphasized that the energy cost of eccentric, negative exercise is relatively modest (Fig. 2-15). Secondly, when the muscle is stretched it can produce more power (moving up the length-tension curve). (Fig. 2-11). Thus, at least two advantages exist in the rebound technique so commonly employed in various activities. Cavagna et al. (1975) also suggest that there may be a greater or more efficient mobilization of chemical energy by the contractile component induced by previous stretching of the muscle.

The activation of extensor muscles before the foot reaches the ground in walking and running is an effect of a central pattern generator discussed in Chap. 3. When the body weight then stretches the muscles, there is eventually no need for an additional recruitment of muscles to take the load. There is a similar predetermined pattern of neuromuscular activity in a repetitive, rhythmic hopping movement, i.e., the gastrocnemius muscle is activated before contact with the ground is made (Melwill Jones and Watt, 1971). A stretch reflex would occur too late to avoid a "catastrophic landing." The preferred frequency of hopping is apparently chosen so that the takeoff phase of the muscular activity is timed to make maximal use of the stretch reflex, that is, maximal support from the muscle spindles (see Chap. 3)

Children have a lower efficiency than adults when running at a given speed (Åstrand, 1952). The explanation for this could be due to less economical coordination in children, or it may be caused by the fact that their lighter bodies do not provide enough kinetic and gravitational energy to be stored during the eccentric phase of the muscle contraction. Davies, (1980), who found that external loading equivalent to 5 percent of the body weight actually reduced the oxygen cost when calculated per kg of gross body weight, supports this hypothesis. The improved efficiency was particularly evident when the subjects (11 to 13 years of age) were running at higher speeds. Bosco (1982) reports that young children do not have the same potential to improve a squatting jump by a counter-movement as do adults. Here we have a similar situation to running, i.e., in such activities children do not have an optimal mechanical efficiency.

Evidence suggests that most of the elastic properties or compliance of a skeletal muscle are located in the contractile machinery, that is in the cross-bridges between the actin and myosin filaments (for references, see Cavagna, 1977). Since the slow-twitch fibers retain the cross-bridge attachment for a slightly longer period of time than the fast-twitch fibers, they may have an advantage in slow type ballistic movements (see Bosco and Rusko, 1983).

Summary In exercise, where activated muscles are forcibly stretched, kinetic and/or gravitational potential energy can be stored in the elastic components of the muscles for utilization in the chemical energy yield if the stretching is immediately followed by a shortening. Such a recoil of tense elastic elements, like a stretched rubber band, can (1) increase the mechanical efficiency, and (2) enhance the peak force and power of a contraction so that the mass affected by the muscles can be subjected to a greater acceleration. These positive effects are noticed particularly in "rebound" exercises such as running and jumping. The time interval between stretching of the active muscle and its subsequent shortening, is very critical, a factor that has to do with technique. There are events in which this mechanism can hardly be utilized, e.g., in swimming (except at the start).

REFERENCES

Asmussen, E.: Positive and Negative Work, *Acta Physiol. Scand.,* **28:**364, 1953.

Asmussen, E., and F. Bonde-Petersen: Storage of Elastic Energy in Skeletal Muscles in Man, *Acta Physiol. Scand.,* **91:**385, 1974a.

Asmussen, E., and F. Bonde-Petersen: Apparent Efficiency and Storage of Elastic Energy in Human Muscles during Exercise, *Acta Physiol. Scand.,* **92:**537, 1974b.

Åstrand, P-O.: "Experimental Studies of Physical Working Capacity in Relation to Sex and Age," Munksgaard, Copenhagen, 1952.

Bárány, M. and K. Bárány: Phosphorylation of the Myofibrillar Proteins, *Ann. Rev. Physiol.,* **42:**275, 1980.

Bell, R. D., J. D. MacDougall, R. Billeter, and H. Howald: Muscle Fiber Types and Morphometric Analysis of Skeletal Muscle in Six-year-old Children, *Med. Sci. Sports Exc.,* **12:**28, 1980.

Bigland, B., and O. C. J. Lippold: The Relation between Force. Velocity and Intergrated Electrical Activity in Human Muscles, *J. Physiol.,* **123:**214, 1954.

Bigland-Ritchie, B., and J. J. Woods: Integrated electromyogram and oxygen uptake during positive and negative work, *J. Physiol.,* **260:**267, 1976.

Billeter, R., C. W. Heizmann, H. Howald, and E. Jenny: Analysis of Myosin Light and Heavy Chain Types in Single Human Skeletal Muscle Fibers, *Eur. J. Biochem.,* **116:**389, 1981.

Blomstrand, E., and B. Ekblom: The Needle Biopsy Technique for Fibre Type Determination in Human Skeletal Muscle—a Methodological Study, *Acta Physiol. Scand.,* **116:**437, 1982.

Bloom, W., and D. W. Fawcett: "Textbook of Histology," 10th ed., Saunders, Philadelphia, 1975.

Borejdo, J., S. Putnam, and M. F. Morales: The Contractile Mechanism, in A. D. Brinnell, and M. A. B. Brazier (eds.), "The Regulation of Muscle Contraction: Excitation-Contraction Coupling," p. 321, Academic Press, New York, 1981.

Bosco, C.: Stretch-shortening Cycle in Skeletal Muscle Function, thesis, *Studies in Sport, Physical Education and Health,* **15,** U. Jyväskylä, Finland, 1982.

Bosco, C., and H. Rusko: The Effect of Prolonged Skeletal Muscle Stretch-shortening Cycle on Recoil of Elastic Energy and on Energy Expenditure, *Acta Physiol. Scand.,* **119:**219, 1983.

Brooke, M. H., and K. K. Kaiser: Muscle Fiber Types: How Many and What Kind, *Arch. Neurol.,* **23:**369, 1970.

Buchtal, F., and H. Schmalbruch: Motor Unit of Mammalian Muscle, *Physiol. Rev.,* **60:**91, 1980.

Buller, A. J., J. C. Eccles, and R. M. Eccles: Interactions Between Motoneurons and Muscles in Respect of the Characteristic Speeds of Their Responses, *J. Physiol.,* **150:**417, 1960.

Capaldi, R. A.: A Dynamic Model of Cell Membranes, *Sc. Am.,* **230:**26, 1974.

Carlson, F. D., and D. R. Wilkie: "Muscle Physiology," Prentice-Hall, Inc., Englewood Cliffs, N.J., 1974.

Cavagna, G. A.: Storage and Utilization of Elastic Energy in Skeletal Muscle, in R. S. Hutton (ed.), "Exercise and Sport Sciences Reviews," vol. 5, p. 89, J. Publ. Affiliates, Santa Barbara, Calif., 1977.

Cavagna, G. A., L. Komarek, and S. Mazzoleni: The Mechanics of Sprint Running, *J. Physiol.*, **217:**709, 1971.

Cavagna, G. A., G. Citterio, and P. Jacini: The Additional Mechanical Energy Delivered by the Contractile Component of the Previously Stretched Muscle, *J. Physiol.*, **251:**65P, 1975.

Close, R. J.: Dynamic Properties of Mammalian Skeletal Muscles, *Physiol. Rev.*, **52:**129, 1972.

"Cold Spring Harbor Symposia on Quantitative Biology," vol. 37, Cold Spring Harbor, N.Y., 1972.

Colling-Saltin, A. S.: Skeletal Muscle Development in the Human Foetus and During Childhood, in R. Berg, and B. Eriksson, (eds.), "Children and Exercise IX," p. 193, University Park Press, Baltimore, 1980.

Dautry-Varsat, A., and H. F. Lodish: How Receptors Bring Proteins and Particles into Cells, *Sc. Am.*, **250:**48, 1984.

Davies, C. T. M.: Metabolic Cost of Exercise and Physical Performance in Children with Some Observations on External Loading, *Eur. J. Appl. Physiol.*, **45:**95, 1980.

Ebashi, S.: Regulation of Muscle Contraction, *Proc. R. Soc. Lond. B.*, **207:**259, 1980. (errata in ibidem, **208:**483).

Eisenberg, B. R.: Quantitative Ultrastructure of Mammalian Skeletal Muscle, chap. 3, in L. D. Peachy, R. H. Adrian, and S. G. Geiger (eds.): "Handbook of Physiology: Skeletal Muscle," sect. 10, pp. 73–112, American Physiological Society, Distributed by Williams and Wilkins, Baltimore, 1983.

Eisenberg, E., and L. E. Greene: The Relation of Muscle Biochemistry to Muscle Physiology, *Ann. Rev. Physiol.*, **42:**293, 1980.

Eisenberg, E., and T. L. Hill: Muscle Contraction and Free Energy Transduction in Biological Systems, *Science*, **227:**999, 1985.

Elliot, G. F.: Variations of the Contractile Apparatus in Smooth and Striated Muscles. X-ray Diffraction Studies at Rest and in Contraction, *J. Gen. Physiol.*, **50:**171, 1967.

Endo, M.: Calcium Release from the Sarcoplasmic Reticulum, *Physiol. Rev.*, **57:**71, 1977.

Fambrough, D. M.: Control of Acetylcholine Receptors in Skeletal Muscle, *Physiol. Rev.*, **59:**165, 1979.

Fettiplace, R., and D. A. Haydon: Water Permeability of Lipid Membranes, *Physiol. Rev.*, **60:**510, 1980.

Flier, J. S., C. R. Kahn, and J. Roth: Receptors, Antireceptor Antibodies and Mechanisms of Insulin Resistance, *N. Engl. J. Med.* **300:**413, 1979.

Gergely, J. (section ed.): Cell and Membrane Physiology, *Ann. Rev. Physiol.*, **43:**477, 1981.

Goldspink, G.: The Proliferation of Myofibrils during Muscle Fiber Growth, *J. Cell. Sci.*, **6:**593, 1970.

Gollnick, P. D., B. F. Timson, R. L. Moore, and M. Riedy: Muscular Enlargement and Number of Fibers in Skeletal Muslces of Rat, *J. Appl. Physiol.: Resp. Env. Exer. Physl.*, **50:**936, 1981.

Gollnick, P. D., D. Parsons, and C. R. Oakley: Differentiation of Fiber Types in Skeletal Muscle from a Sequential Inactivation of Myofibrillar Actomyosin ATPase during Acid Preincubation, *Histochemistry,* **77:**543, 1983.

Grinnell, A. D., and M. A. B. Brazier (eds.): "The Regulation of Muscle Contraction: Excitation-Contraction Coupling," Academic Press, New York, 1981.

Gunn, R. B.: Co-and Counter-Transport Mechanisms in Cell Membranes, *Ann. Rev. Physiol.*, **42:**249, 1980.

Hill, A. V.: The Priority of the Heat Production in a Muscle Twitch, *Proc. Roy. Soc. (Biol.)* **148:**397, 1958.

Huxley, A. F.: Muscular Contraction, *J. Physiol.*, **243:**1, 1974.

Huxley, H. E.: The Contraction of Muscle, *Sc. Am.*, **19:**3, 1958.

Huxley, H. E.: The Mechanism of Muscular Contraction, *Sc. Am.*, **213:**18, 1965.

Huxley, H. E.: Structural Changes in the Actin- and Myosin-containing Filaments during Contraction, *Cold Spring Harbor Symposium in Quart. Biol.*, **37:**361, 1972.

Jolesz, F., and F. A. Sreter: Development, Innervation and Activity-pattern Induced Changes in Skeletal Muscle, *Ann. Rev. Physiol.*, **43:**531, 1981.

Kolternam, O. G., J. Insel, M. Sackow, and J. M. Okfsky: Mechanisms of Insulin Resistance in Human Obesity. Evidence for Receptor and Postreceptor Defects, *J. Clin. Invest.*, **65:**1272, 1980.

Komi, P. V.: Relationship between Muscle Tension, EMG and Velocity of Contraction under Concentric and Eccentric Work, in J. E. Desmedt (ed.), "New Developments in Electromyography and Clinical Neurophysiology," vol. 1, p. 596, Karger, Basel, 1973.

Komi, P. V., J. H. T. Viitasalo, M. Havu, A. Thorstensson, B. Sjödin, and J. Karlsson: Skeletal Muscle Fibers and Muscle Enzyme Activities in Monozygous and Dizygous Twins of Both Sexes, *Acta Physiol. Scand.*, **100:**385, 1977.

Kugelberg, E.: Adaptive Transformation of Rat Soleus Motor Units during Growth. Histochemistry and Contraction Speed, *J. Neurol. Sci.*, **27:**269, 1976.

Lexell, J., K. Henriksson-Larsén, and M. Sjöström: Distribution of Different Fibre Types in Human Skeletal Muscles, 2, *Acta Physiol. Scand.*, **117:**115, 1983.

Lockhart, R. D.: Anatomy of Muscles and Their Relation to Movement and Posture, in G. H. Bourne (ed.), "The Structure and Function of Muscle," vol. 1, p. 1, Academic Press, Inc., New York, 1973.

Lymn, R. W., and H. E. Huxley: X-ray Diagrams from Skeletal Muscle in the Presence of ATP Analogs, *Cold Spring Harbor Symposium Quart. Biol.*, **37:**449, 1972.

Maughan, R. J., and M. A. Nimmo: The Influence of Variations in Muscle Fibre Composition on Muscle Strength and Cross-Sectional Area in Untrained Males, *J. Physiol.*, **351:**229, 1984.

MacCallum, J. B.: On the Histogenesis of Striated Muscle Fiber, and the Growth of the Human Sartorius Muscle, *Bull. John Hopkins Hosp.*, **9:**208, 1898.

Melwill Jones, G., and D. G. D. Watt: Observations on the Control of Stepping and Hopping Movements in Man, *J. Physiol.*, **219:**709, 1971.

Milner-Brown, H. S., and R. B. Stein: The Relation between the Surface Electromyogram and Muscular Force, *J. Physiol.*, **246:**549, 1975.

Murray, J. M., and A. Weber: The Cooperative Action of Muscle Proteins, *Sc. Am.*, **230:**59, 1974.

Murray, J. M., and A. Weber: Competition Between Tropomyosin and Myosin and Cooperativity of the Tropomyosin-Actin Filament, in A. D. Grinnell, and M. A. B. Brazier (eds.): "The Regulation of Muscle Contraction: Excitation-Contraction Coupling," p. 261, Academic Press, N.Y., 1981.

Needham, D. M.: "Machina Carnis: The Biochemistry of Muscular Contraction in Its Historical Development," Cambridge University Press, Cambridge, 1971.

Nemeth, P. M., D. Pette, and G. Vrbová: Comparison of Enzyme Activities Among Single Muscle Fibres Within Defined Motor Units, *J. Physiol.*, **311:**489, 1981.

Peachey, L. D., R. H. Adrian, and S. G. Geiger (eds.): "Handbook of Physiology: Skeletal Muscle," sect 10, p.555, *American Physiological Society*, distributed by Williams and Wilkins, Baltimore, 1983.

Pette, D. (ed.): "Plasticity of Muscle," Walter de Gruyter, Berlin, New York, 1980.

Pierobon-Bormioli, S., S. Sartore, L. Dalla Libera, M. Vitadello, and S. Schiaffino: "Fast" Isomyosins and Fiber Types in Mammalian Skeletal Muscle, *J. Histochem. Cytochem.* **29:**1179, 1981.

Pollack, G. H.: The Cross-bridge Theory, *Physiol. Rev.*, **63:**1049, 1983.

Rodgers, K. L., and R. A. Berger: Motor-unit Involvement and Tension during Maximum, Voluntary Concentric, Eccentric, and Isometric Contractions of the Elbow Flexors, *Med. Sci. in Sports,* **6:**253, 1974.

Rothman, S. S.: Passages of Proteins Through Membranes—Old Assumptions and New Perspectives, *Am. J. Physiol.,* **238:**391, 1980.

Ruch, T. C., and J. F. Fulton: "Medical Physiology and Biophysics," 18th ed., W. B. Saunders Company, Philadelphia, 1960.

Salmons, S., and J. Henriksson: The Adaptive Response of Skeletal Muscle to Increased Use, *Muscle and Nerve,* **4:**94, 1981.

Saltin, B., and P. D. Gollnick: Skeletal Muscle Adaptability: Significance for Metabolism and Performance, in L. D. Peachey, R. H. Adrian, and S. R. Geiger (eds.): "Handbook of Physiology: Skeletal Muscle," sect 10, p. 555, *American Physiological Society,* distributed by Williams and Wilkins, Baltimore, 1983.

Schantz, P., and J. Henriksson: Increases in Myofibrillar ATPase Intermediate Human Skeletal Muscle Fibers in Response to Endurance Training, *Muscle and Nerve,* **6:**553, 1983.

Schantz, P., E. Randall-Fox, W. Hutchinson, A. Tydén, and P-O. Åstrand: Muscle Fibre Distribution, Muscle Cross-sectional Areas and Maximal Voluntary Strength in Humans, *Acta Physiol. Scand.,* **117:**219, 1983.

Schuurmans Stekhoven, F., and S. L. Bonting: Transport Adenosine Triphosphatases Properties and Functions, *Physiol. Rev.,* **61:**1, 1981.

Seymour, J., and E. J. O'Brien: The Position of Tropomyosin in Muscle Third Filaments, *Nature,* **283:**680, 1980.

Sjöström, M., S. Kidman, K. Larsen-Henriksson, and K. A. Ängqvist: Z-and M-band Appearance in Different Histochemically Defined Types of Human Skeletal Muscle Fibers, *J. Histochem. Cytochem.,* **30:**1, 1982.

Stephenson, E. W.: Activation of Fast Skeletal Muscle: Contributions of Studies on Skinned Fibers, *Am. J. Physiol.,* **240:**CI, 1981.

Tada, M., T. Yamamoto, and Y. Tonomura: Molecular Mechanism of Active Calcium Transport by Sarcoplasmic Reticulum, *Physiol. Rev.,* **58:**1, 1978.

Thorstensson, A.: Muscle Strength, Fibre Types and Enzyme Activities in Man, *Acta Physiol. Scand.,* (suppl. 443), 1976.

Unwin, N., and R. Henderson: The Structure of Proteins in Biological Membranes, *Sci. Am.,* **250:**56, 1984.

Vincent, A.: Immunology of Acetylcholine Receptors in Relation to Myasthenia Gravis, *Physiol. Rev.,* **60:**756, 1980.

Wiman, K.: Cell receptorer och deras sjukdomar, introduktion, *Läkartidningen,* **78:**1310, 1981.

Wray, J. S., and K. C. Holmes: X-ray Diffraction Studies of Muscle, *Ann. Rev. Physiol.,* **43:**553, 1981.

CHAPTER **3**

NEUROMUSCULAR FUNCTION

CONTENTS

THE MOTONEURON AND TRANSMISSION OF NERVE IMPULSES

    Anatomy

    Resting Membrane Potential

    Facilitation-Excitation

    Inhibition

    Inhibition and Excitation

    Neurotransmission

    Nerve Axon Conduction Velocity

TRANSMISSION OF IMPULSES FROM NERVE AXON TO SKELETAL
    MUSCLE

REFLEX ACTIVITY AND SOME OF THE BASIC FUNCTIONS OF
    PROPRIOCEPTORS

    Renshaw Cells

    The Muscle Spindle and the Gamma Motor System

    The Golgi Tendon Organs

    Reciprocal Inhibition (Innervation)

    Joint Receptors

    The Resting and Activated Muscle

    Lateral Inhibition

    Supraspinal Control of Motoneurons

    Cerebral Cortex

    Cerebellum

Various Nuclei in the Brain Involved in Movement
Integration of the Neuronal Activity in Movement
Learning and Memory

PROPERTIES OF THE MUSCLE AND MUSCULAR CONTRACTION
The Motor Unit
Regulation of Force
Muscular Strength
Coordination
Degeneration and Regeneration of Nerves
Posture
Motor Development

MUSCLE FATIGUE
Fatigue in Static Exercise
Blood Supply
What Then Causes Fatigue When Activating Skeletal Muscles in "All-Out"
    Performance Exercise Up to Some 10 Minutes?
Effect of Prolonged Exercise
"Fatigue" in Joints
Motivation
Cramp

The central nervous system (CNS) receives information concerning the outside world via *exteroceptors* as they react to light, sound, touch, temperature, or chemical agents, and *interoceptors* that are stimulated by changes within the body. The latter include *proprioceptors* (such as muscle spindles, tendon-end organs of Golgi, joint receptors, and vestibular receptors), *chemoreceptors,* and *visceroceptors* (Fig. 3-1). The CNS is equipped to receive, interpret, and handle information, and then eventually to transform the result into movements. Even the process of feeding information to the CNS may involve some muscular activity; for example, the eye muscles are almost continuously active, especially while the individual is awake. Reaction to various stimuli includes gestures, speech, writing, and sometimes very vigorous muscular contractions, as in the event of an approaching danger. There is scarcely any stimulus that does not, through reflexes, affect smooth muscle, heart muscle, or skeletal muscle.

A presentation of the physiology of work and exercise would justify a very detailed discussion of the function of the CNS. In fact, it is difficult to decide which aspects are more essential and which are less essential for the understanding of muscular movements, athletic techniques, and training. However, we shall limit our discussion mainly to the units more directly involved in the activation of the skeletal muscle. For a more comprehensive description of the CNS, the reader should consult basic physiological textbooks and reviews (Brooks, 1979; Eccles, 1973b; Kandel and Schwartz, 1981). It is assumed that the reader is familiar with the anatomy and histology of the nervous system.

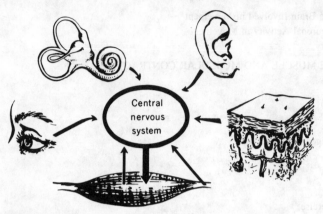

Figure 3-1
The central nervous system receives information via afferent nerves from the various receptors. Impulses are transmitted through the "final common path," which is the only available route to the skeletal muscle.

THE MOTONEURON AND TRANSMISSION OF NERVE IMPULSES

Anatomy

The brain is made up of individual units, the nerve cells, the communicating elements, and the supporting and nourishing glial cells. The number of nerve cells in our body is somewhere around a hundred billion (10^{11}), give or take a factor of 10; it is about the same number as the stars in our galaxy! There are approximately 1,000 different types of nerve cells, or neurons. Each has a diameter from 5 to 100 μm and consists of four morphological regions: (1) the cell body, or *soma,* the "heart" of the nerve cell, (2) the *dendrites,* a set of short, fine arborizing processes radiating from the cell body, (3) the *axon,* 10 to 20 μm in diameter and from 1 mm to 1 m long, and (4) the *terminals* of the axon. Each region has a distinctive function which will be described in more detail below (Fig. 3-3). Generally, the dendrites and the cell body receive signals, the cell body combines and integrates them and decides whether or not the signals shall be transmitted via the axon to the terminals. The axon may branch off near its beginning, but more often the branching takes place closer to the end. As we shall see, the signals are mediated electrically inside the cell. Communication with other cells is, in most cases, based on chemical processes in the nerve terminal, the synapsis. A typical nerve cell may have anywhere from 1,000 to 10,000 synapses on its surface which relay information from something like 1,000 other neurons. The task of the single nerve cell is to receive a message and then to pass it on to other cells; the message may go to a nerve cell, muscle cell, or cells of other types. Some nerve cells are specialized in "holding the message back" and preventing other cells from reacting to impulses (Fig. 3-2). A membrane, 5 nm thick, covers the soma and the

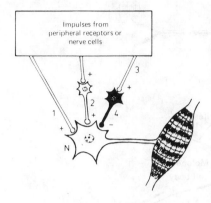

Figure 3-2
The motoneuron, N, in the ventral horn of the gray matter of the spinal cord can be excited (+) directly (1) or via an interneuron (2). Thereby, an impulse is propagated in the nerve fiber, and the muscle is stimulated, causing muscular activity. However, other nerve terminals can prevent the motoneuron from responding to the exciting impulse. Schematically, this is illustrated as follows: Nerve end (3) stimulates an interneuron nerve cell (4). But this cell inhibits (−) the motoneuron's reaction to the stimulation by nerve 1. The mechanisms involved are discussed in the text.

processes. The density of mitochondria in the cell indicates a high level of metabolic activity. In fact, the oxygen uptake of nerve cells per unit of weight is probably of the same order as that of maximal contraction of skeletal muscles. (If we take a power of 80W as being a reasonable figure for a resting adult person, 15 to 20W is the portion of this power used by the nervous system. This level is, however, relatively constant whether the person is sleeping or concentrating very hard on an intricate intellectual problem. This, in contrast to the metabolic rate of the skeletal muscles, which at rest may also expend some 15 to 20W but during maximal exercise, as in the case of top athletes, may exceed 2000 W.)

In the nerve branches, fine threads, known as neurofibrils, run parallel to the axis and are bathed in the axoplasm. The membrane, or *axolemma,* constitutes a barrier between the intra- and extracellular fluid. In the peripheral part, the axon of the motoneuron is covered by a *myelin sheath*. This myelin sheath is interrupted at intervals by the nodes of Ranvier. (Numerous nerves in the CNS, the thinner nerve fibers in the somatic nervous system, and postganglionic fibers of the autonomic nervous system are unmyelinated.) For the nutrition of the dendrite and axon, an uninterrupted connection with the cell body is essential. The axoplasm is, in a way, an artery for busy molecular traffic moving up and down between the cell body and its terminals. There is a slow outgoing flow at a rate of 0.5 to 3 mm per day, but also a fast transport running at approximately 400 mm per day. This axoplasmic flow is essential for the function of the nerve cell. As we shall see the nerve cells are secretory cells, and fast transport of the products is important. Just think of the phrenic nerve of a giraffe! In the axons, there is also a retrograde flow of substrates, which apparently are essential as some sort of an information carrier to the nerve cell body about events that occur at the distal ends of its axonal processes (see Grafstein and Forman, 1980).

Axonal terminals (*synaptic knobs*) of other nerves end at the surface of the soma (Fig. 3-3). Actually, the surface of the soma, the dendritic stumps, and even the axonal origin may be almost completely covered by synaptic knobs from thousands of other nerve cells. The space between the membranes of the synaptic knobs and the contacted nerve cell (postsynaptic membrane), known as the *synaptic cleft,* is about 20–30 nm.

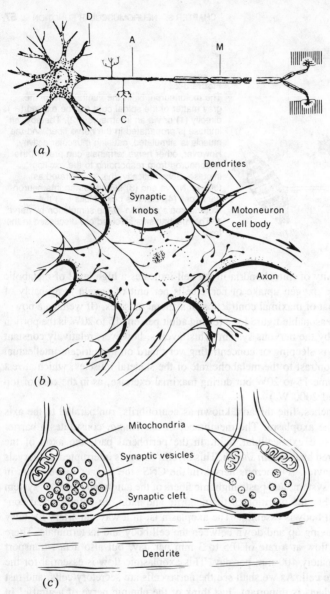

Figure 3-3
(a) Motoneuron: A, axon; C, collateral; D, dendrite; M, myelin, (K. E. Schreiner and A. Schreiner, 1964.)
(b) A motoneuron cell body and its dendrites covered with synaptic knobs of which only a few are included in the figure. These knobs are the terminals of the impulse-carrying nerve fibers (axons) from other nerve cells. A typical motoneuron in the human spinal cord has some 10,000 synaptic contacts on its surface of which one fifth may be on the cell body and the rest on its dendrites.
(c) When stimulated, the synaptic knobs deliver a chemical transmitter substance into the synaptic cleft, where it can act on the surface of the opposite nerve cell membrane. The chemical transmitter substance is stored in numerous vesicles. [(b) and (c) from Eccles, 1965.]

Resting Membrane Potential

In the nerve cells, as in other cells, the chemical composition of the fluid on either side of the semipermeable cell membranes is different. In its composition, the external fluid medium is an ultrafiltrate of blood; i.e., it has a concentration of various particles very similar to a protein-free filtrate of blood plasma. However, the internal fluid contains a lower concentration of sodium and chloride ions and a higher concentration of potassium. These differences in the concentration of ions would cause diffusion across the membrane, if perfusibility and electrochemical gradients could permit this. The membrane, however, does not permit a free passage of ions, it acts as a physical barrier to diffusion. As mentioned in Chap. 2, the cell membrane contains proteins with different functions. Some of the proteins form ionic channels, which can be passive, i.e., they are always open, or they are active, i.e., they have a gate across the ionic channel which can be either open or closed. Proteins often change their shape as they function. In Chap. 2, we discussed the dramatic conformational change in the arrangements of the protein molecules during the muscle contraction and relaxation. In the case of membrane channels, slight movement of critically placed parts of the protein molecule apparently plays a decisive role in the permeability of the membrane for ions. In one type of molecule, a change in the potential across the membrane can open or close the gate (*voltage gated*). Another type of channel is *chemically gated*, which means that the channels open when a particular molecule (transmitter) binds to a receptor region on the channel protein. A third type of channel is opened by *physical stimuli* and quite selectively by light (eye), vibrations (ear), pressure, temperature (skin), etc. The ions of interest in this connection are K^+, Na^+, Cl^-, and Ca^{2+}, and the interaction between ions and parts of the channel structure is decisive for the specificity. (The density of a particular type of channel on a cell ranges from zero up to some 10,000 per μm^2.)

Important characteristics of the membrane and its proteins account for the nerve impulse and other complex features of neuron function. We shall now discuss some of the electric and chemical events that take place on both sides of the nerve cell membrane, as well as inside the membrane at rest and when active.

The cell contains a high concentration of protein anions that cannot escape through the membrane into the fluid outside the cell where the concentration of protein is low. Electric charges of opposite sign attract each other, and the excess protein anions inside the cell therefore attract cations. The membrane is much less permeable to Na^+ than to K^+; the tendency for K^+ to diffuse out of the cell according to a difference in concentration gradient is counteracted by the attraction from the protein anions inside the cell. To maintain the difference in concentration of ions across the resting cell membrane, there must be a forced movement of Na^+ outward and K^+ inward, virtually equaling the diffusion in the opposite directions (Fig. 3-4). Actually, aerobic metabolic processes supply the energy to keep a sodium-potassium ATP pump (or more often simply called the sodium pump) going, whereby ions that diffuse or "leak" in or out of the cell are forced back.

In order to convey a rough idea of the capacity of the sodium pump, one might mention that when operating at its maximal rate each pump can transport some 200

Figure 3-4
Schematic diagram showing the K^+ and Na^+ fluxes through the surface membrane of the nerve cell in the resting state. The slopes in the flux channels across the membrane represent the respective electrochemical gradients. The voltage drop across the surface membrane is about 70 mV, with the inside being negative. (In different cells the voltage drop may range from 45 to 75 mV.) Under these conditions, chloride ions diffuse inward and outward at equal rates. But the concentration of sodium and potassium has to be maintained by some sort of a metabolic pump. The negative potential inside the nerve cell membrane is 20 mV short of the equilibrium potential for potassium ions. Therefore, the pump has to actively prevent an outward diffusion of potassium ions. For sodium ions, the potential across the membrane is 130 mV in the opposite direction. To keep the sodium out, a very active pump is needed.
(*Modified from Eccles, 1965.*)

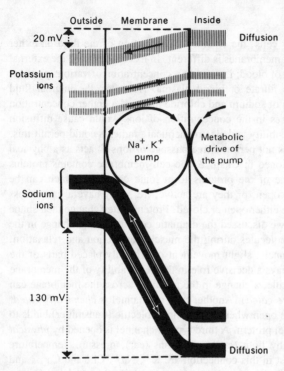

Na^+ and 130 K^+ ions per second across the membrane. One neuron may have one million pumps with a potential to move approximately 200 million Na^+ ions per second (Stevens, 1979).

The net effect of these membrane characteristics is that there are approximately 10 times more Na^+ ions in the external medium than inside the cell, but for the K^+ concentration the picture is reversed. Furthermore, there is a *higher concentration of anions in the interior of the soma and axon.* With a microelectrode inserted in the cell, the potential difference has been measured and found to be approximately 70 mV over the membrane, the interior being negative with respect to the exterior. (In some cells it may be as high as 75 mV, in others the resting membrane potential may be as low as 45 mV.) The cell contains nondiffusible protein; a disturbance of the resting potential difference would essentially occur if the sodium and potassium ions were allowed to move according to the diffusion gradients (Fig. 3-4). Such a flux of ions across the surface membrane would disturb the resting potential of 70 mV. Hence, an inward flow of positive ions, for example, sodium, would decrease the potential and cause a *depolarization;* an outward flow of positive ions, say potassium, would give rise to a *hyperpolarization.* Actually, such fluxes of ions occur normally when the nerves terminating at the neuron are stimulated.

Facilitation-Excitation

One of the best studied nerve cells is the *motoneuron,* the cell that ultimately controls the skeletal muscles. Let us see what happens when terminals of other nerve cells contacting a motoneuron are "signaling." The electron microscope has revealed the construction of the synaptic junction: the synaptic knob of approximately 2 μm in diameter (the terminal end), synaptic cleft, and the subsynaptic membrane covering the motoneuron. Within the synaptic knob there are numerous vesicles approximately 30 nm in diameter (Fig. 3-5). They contain the chemical substance that is responsible for the transmission of the impulse across the synaptic junction. The arrival of an impulse at the synaptic knob causes a synchronous ejection of many vesicles. A corresponding number of uniform packages, quanta, of transmitter substance is liberated. A trigger for this almost instantaneous chemical "injection" into the synaptic cleft is the entry of Ca^{2+} into the presynaptic terminal. In one way or another, the nerve impulse (causing a depolarization of the cell membrane) will open gates in the membrane channels, allowing Ca^{2+} to diffuse rapidly down a very large electrochemical gradient. The transmitter substance diffuses readily to receptors on the subsynaptic membrane which are specifically affected by this transmitter substance. With a microelectrode inserted in the cell, the effect of this event can be recorded: a slight depolarization can be assigned to a net inflow of sodium ions. (Actually excitory transmitters open up chemically-gated channels permitting a simultaneous movement of Na^+ ions into, and K^+ ions out of the neuron. Due to the steeper gradient for sodium (Fig. 3-4) the Na^+ current will dominate.) Eventually the active sodium pump will restore the resting membrane potential of -70 mV by extracting the Na^+ ions. However, if the membrane potential is reduced to a critical value, say -60 mV, there is suddenly a free passage through the membrane of sodium ions, and they move in the direction of their concentration gradient, causing a depolarization of the membrane. The equilibrium potential for Na^+ ions will actually be about $+60$ mV (it varies in different cells), with the interior of the motoneuron 60 mV positive to the extracellullar

Figure 3-5
Synaptic vesicles. Unfilled dots denote transmitter substance (e.g., acetylcholine), filled dots symbolize an inactivator (e.g., cholinesterase). (*Redrawn from Eccles, 1957.*)

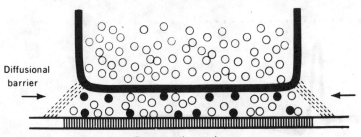

Synaptic knob

Diffusional barrier

Postsynaptic membrane

fluid. However, the potential charge or height of the recorded "spike" (*action potential*) will not be $70 + 60 = 130$ mV, but only about 100 mV, for, immediately after the beginning of the influx of sodium, there is an outward flow of potassium ions along their electrochemical gradient. Consequently, the number of positive ions within the cell will again decrease. The channel proteins are voltage-gated, i.e., a change in the potential over the cell membrane can trigger the opening of the gates. It is assumed that Ca^{2+} reaches critical sites close to the regions for the release of the transmitter inside the terminal. It causes, or facilitates, fusion of the vesicular membrane with the terminal membrane and the vesicle can empty its contents. As we shall see the Ca^{2+} influx can be graded and therefore there is also a graded release of quantal units of transmitter substance. Here again we have an example of voltage-gated channels, actually of two types: Na^+ channels that open first, and K^+ channels that are turned on slightly later. The efflux of potassium ions actually begins early during the "rising" limb of the spike and reduces the spike potential. Within 1 ms the resting membrane potential of -70 mV is almost restored by trading potassium ions for sodium ions across the cell membrane. The movement of ions back to where they belong is now affected by a "slow" process requiring metabolic energy in running the Na^+-K^+ pump. Following the action potential, there is a prolonged phase (15 to 100 ms) when the permeability of the neutral membrane for potassium is still increased, and during this phase there is an *after-hyperpolarization* adding 5 mV to the resting potential of -70 mV (Fig. 3-6).

The events just described have far-reaching effects: The electrochemical charges initiated by the transmitter mechanism, if exceeding the given threshold of about -60 mV (it varies in different cells) will cause similar sequential changes in the ionic permeability of the membrane of the whole nerve axon. This traffic is associated with the passage of a change in membrane potential (an action potential) during activity along the axon cell interior, changing from -70 mV to about $+30$ mV with respect to the exterior (Fig. 3-6). A propagated nerve impulse, traveling at a speed of up to 100 m $\cdot$ s^{-1}, will reach the muscle and stimulate it to contract.

Figure 3-6
Changes in membrane potential during activity along an axon. The interior voltage changes from -70 mV to about $+30$ mV with respect to the exterior. Note the hyperpolarization after the "spike." Time base, 2,000 cps. (*Modified from Kuffler et al., 1951.*)

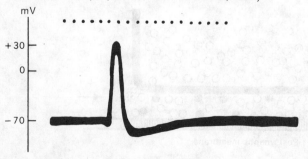

The depolarization, after a stimulus of the different nerves, may not be strong enough to pass the mentioned threshold. It is then *subliminal* and gives rise only to a local potential change under the activated synaptic knob. This condition is referred to as an *excitatory postsynaptic potential* (EPSP); it is also called *facilitation*. The EPSP propagates along the dendrites or cell body and it decreases progressively in amplitude with distance; its amplitude 1 mm down the axon may be only one-third of what it was at the site of generation. The primary function of these graded potentials is to convey signals for very short distances. After a few milliseconds, the sodium pump restores the normal membrane potential of -70 mV, and a new impulse of the same strength at an interval of 5 ms or more will merely repeat the local depolarization, say to -63 mV, the interior still being negative with respect to the exterior. If, however, the next stimulus arrives at the synaptic knob close to the preceding volley and initiates an additional ejection of transmitter substance while the subsynaptic membrane is still hypopolarized, say to -66 mV, the new sodium influx will be superimposed. This will decrease the potential to about -59 mV. This voltage change can evoke a depolarization and generate a propagating action or spike potential in the axon. If the single stimulus is below the minimal strength, successive stimuli can elicit a depolarization by a summation of the receptor potential. In this case, it is called a *temporal summation*. With several synaptic knobs terminating close together on a nerve cell, a simultaneous excitation may deliver the quanta of transmitter substance required to produce the sieve for sodium and potassium on the subsynaptic membrane, by opening enough channel gates. This is an example of *spatial summation*.

If the stimulus is adequate to elicit a propagated spike, the magnitude of the spike potential of a single nerve fiber is independent of the strength of the stimulus; the nerve responds to the utmost of its ability. The answer to a stimulus is an *all-or-none response*. There is no alternative. During the spike potential, neither excitability nor conductivity is present in the nerve; it is in an *absolute refractory period*. Apparently, for a new Na^+ flux to occur as a consequence of a nerve stimulation, the potential must become more negative than about 50 mV. For the next few milliseconds a *partial refractory* period exists, during which time a stronger stimulus is necessary to evoke an action potential. The normal excitability is gradually restored in the course of about 100 ms. Axons with large diameters recover 90 percent of the normal state within 1 ms, with the consequence that they can generate impulses at a frequency of 1,000 s^{-1}. The intermittent impulse traffic in nerves is due to the refractory period, and the recovery of the nerve cell is certainly enhanced by this mechanism. The maximal frequency by which a nerve cell can send its messages depends on the time of the absolute refractory period; if it takes 2 ms, this maximum will be 500 impulses per second.

Thus far, we have discussed the *synaptic potentials* by which electrical energy can be transformed into chemical energy with the release of a transmitter, which then will trigger electric changes. Eventually, these changes are of such magnitude, if depolarization of the membrane does occur, that they can produce a propagating *action potential*. There is also a *receptor* or *generator potential* which serves to transform sensory stimuli such as stretching, vibration, light affecting specific receptors of the sensory nerve endings, into electric energy. This potential has a function comparable to that of EPSP in the synaptic potentials. It is locally restricted and graded, and can produce an action potential.

Summary The synapse is a unique structure capable of transferring information from one nerve cell to another, or to other cell types. (1) When an action-potential spike reaches the terminal of a synaptic knob, the postsynaptic membrane is converted into a sieve by a chemical transmitter substance that is released from the knob. The action potential triggers an influx of Ca^{2+} into the terminal which is essential for this release. (2) By acting on receptors on the postsynaptic membrane, channels will open, permitting an inward flux of sodium ions and an outward flux of potassium. Thus, the resting voltage difference of 70 mV (it can vary from 45 to 75 mV) across the cell membrane is momentarily reduced. If the stimulus is not strong enough, i.e., the quanta of transmitter release is too small to change the membrane potential beyond a certain threshold, the sodium pump will restore the resting potential. (3) Repeated (temporal) or concurrent (spatial) impulses can add to this depolarization so that the threshold level for initiating an action potential in the affected neuron may be reached. This threshold is usually 10 to 18 mV above the resting potential in the spinal moto-neurons. Actually, an addition of a fraction of a millivolt may be the difference between a subthreshold depolarization and the "explosion" into a full-blown action potential. This voltage reversal propagates down the axon to its terminals. (4) At the synapse the transmitter is inactivated, and an energy-demanding Na^+-K^+ pump will extract the sodium ions from the intracellular fluid and bring back the potassium ions (Fig. 3-7). Protein-lined channels that span the nerve cell membrane may remain closed or may be opened by gates operated by voltage changes over the membrane or by specific chemical agents. These channels are selectively available for Na^+, K^+, Cl^-, and Ca^{2+}.

Inhibition

From a functional viewpoint, there is another type of neuron with short axons ending in synaptic connection with other nerve cells, forming interneurons. When stimulated, their synaptic knobs liberate a chemical transmitter, which in turn, permits a diffusion of K^+ out of the cell and Cl^- inward. The membrane retains its high degree of impermeability to Na^+ ions, however. The result is an outward current and an increase in the voltage difference between the interior of the cell membrane and its exterior, that is, a *hyperpolarization*. It rises to a summit in 1.5 to 2 ms after its onset and fades exponentially. During this phase, a stronger stimulus is needed to discharge an impulse in the motoneuron; i.e., an *inhibitory postsynaptic potential* (IPSP) has developed. The IPSP can bring the voltage to − 80 mV (which is the equilibrium potential of the inhibitory ionic mechanism). The further the membrane is removed from the resting potential, the larger is the IPSP. Recovery will occur as the intracellular potassium ion concentration is restored by the operation of a potassium pump.

In Fig. 3-2, nerve 3 represents such an inhibitory interneuron. Within a motoneuron there is a spatial as well as a temporal summation of inhibition, as is true for excitatory stimuli. The IPSP is virtually a mirror image of the EPSP, differing in its longer

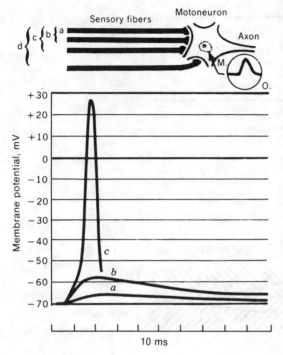

latency and shorter time constant of decay. The total duration of the inhibition by the IPSP is 2 to 3 ms at most.

Summary A specific transmitter substance causes an inhibitory synaptic response on areas of a neuron by changing the ionic flux of potassium and chloride ions through the subsynaptic membrane. During the hyperpolarization caused by this ion flux, a stronger excitatory action is necessary for initiation of a depolarization, evoking the discharge of an impulse. The terminology is summarized in Fig. 3-8.

Inhibition and Excitation

Figure 3-9 provides a quantitative evaluation of the effect of an excitatory impulse on the membrane potential of a motoneuron simultaneously subjected to a stimulus of an inhibitory nerve fiber.

Motoneurons are subjected to a continuous bombardment of impulses, which expose their membranes to both inhibitory and excitatory transmitter substances. They discharge impulses only when the excitatory synaptic activity is so great that the depolarization exceeds the threshold value, causing a short circuit of the membrane and the launching of an action potential.

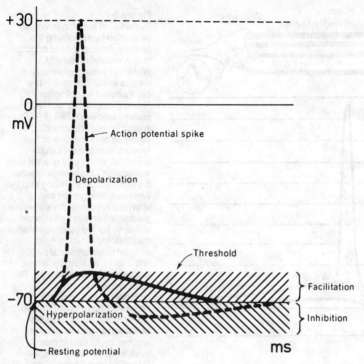

Figure 3-8
Summary of the terminology used when discussing the electrical activity in a
nerve cell under various conditions.

The excitatory synapses tend to be peripherally located on the dendrites, while the
inhibitory synapses are likely to be concentrated on the soma and at the origin of the
axon. Actually, the initial segment of the axon usually has the lowest threshold for
the initiation of an action potential; it is a trigger zone for action. This is an example
of good strategic design, because the inhibitory synapses located on the soma have
final control over whether certain impulses will be allowed to propagate down the
axon (Eccles, 1973b).

As a result of the longer latency for inhibitory impulses than for excitatory impulses
(about 1 ms), the former are most effective when they precede the excitatory volley.

Functionally, there are only two types of nerve cells, excitatory and inhibitory. All
inhibitory nerve cells are short axon neurons lying in the gray matter, whereas all
transmission pathways, including the peripheral afferent and efferent pathways, are
formed by the axons of excitatory neurons. Many short axon interneurons also belong
to the class of excitatory neurons.

If one inhibitory interneuron, preventing a contacted nerve cell from responding to
stimuli, becomes subjected to the influence of other inhibitory interneurons, it may be
forced to remain inactive. This means that its inhibitory effect has been neutralized

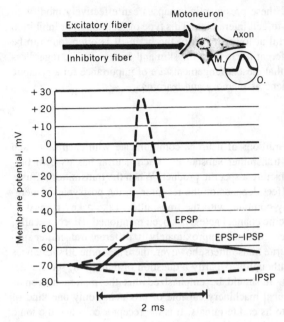

Figure 3-9
With a microelectrode (M.) implanted in a motoneuron, the changes in the cell's internal electric potential can be displayed on an oscilloscope (O.). When the excitatory nerve fiber is stimulated, an EPSP is evoked, and the threshold for producing a spike is eventually reached (dotted line at top). Stimulation of the inhibitory nerve fiber gives rise to an IPSP (dotted line at bottom). Inhibition of a spike discharge is an electrical subtraction process. The IPSP widens the gap between the cell's internal potential and the firing threshold (here about -55 mV). Thus, if the motoneuron is simultaneously subjected to both excitatory and inhibitory stimulation, the IPSP is subtracted from the EPSP, and no spike occurs (full line). (*Modified from Eccles, 1965.*)

and the nerve cell which it contacts synaptically, thus released from inhibition, can now respond. This mechanism is called *disinhibition*.

It has been emphasized that the action potential is an all or nothing response to effective stimuli, which means that the amplitude of this potential has a standard size, being stereotypic and similar in all axons (110 mV and with a duration of 1 msec). Whether or not a stimulus is effective is evidently determined by the algebraic summation of the impact of the EPSPs and IPSPs at the integrative, decision-making point, the trigger zone. How is the intensity and nature of the stimulus reported and interpreted? In the individual neuron, the intensity of the stimulus is conveyed by changes in the number of spikes in a "train" or volley and the interval between the spikes. It is a *frequency code*. But a *population code,* based on varying the number of participating cells in roughly the same type of operation, is also of major informational importance. The nature of the stimulus and its behavioral end results are determined by what part of the brain, and which sets of neurons receive the signals. Later in the text we shall point out that various sensory surfaces of our body are represented in a topographical manner in the brain, the same holds true for muscles and movements. These two types of maps, one for sensory perception and the other for motor commands are interconnected in various combinations.

Some axons send their terminals to the terminals of adjacent neurons forming *presynaptic synapses.* Apparently they can control the Ca^{2+} influx, and in some synapses the axons can enhance and in others reduce the Ca^{2+} diffusion. It should be recalled that Ca^{2+} plays a crucial role in the release of the transmitter substance from the presynaptic membrane, and that the transmitter substance is released in proportion to the Ca^{2+} influx. Therefore, these presynaptic synapses can effectively modify an otherwise given pattern of cell-to-cell interaction. This type of facilitation or inhibition is very widespread and powerful at lower levels of the brain. It is often found to be acting on terminals of large afferent fibers from the skin and from muscles, regardless of function. It is also believed that these arrangements are of importance for a synaptic plasticity behind such phenomena as memory and learning.

Neurotransmission

As pointed out, thousands of millions of neurons communicate with each other and with other cells by means of transmitter substances. Each neuron has its own biochemical competence to manufacture a specific product. When the transmitter diffuses across the synaptic cleft, its effect depends on the receptor on the postsynaptic membrane to recognize and quickly combine with the transmitter. Like a key it must fit into a lock. Its guidance of specific channel gates, has been mentioned. Biochemically, neurotransmitters are quite heterogeneous. Approximately 10 different transmitter substances have been identified as true transmitters, however, there are some 30 substances that are possible candidates (although it is partly a question of definition, see Kandel and Schwartz, 1981, p. 106). It should be emphasized that the mature neuron is specialized and has a biochemical machinery capable of producing only one kind of transmitter which is available to its end terminals. It is the receptor cells and the ionic channels that they interact with, which determine the synaptic effect of the transmitter, e.g., whether the synaptic event is going to be an EPSP or IPSP. Norepinephrine (which is the same as noradrenaline, and the nerve releasing it, sometimes called adrenergic) may have receptors mediating excitation on some postsynaptic membranes, while mediating inhibition on other membranes. There are many neuroactive peptides and amino acids. Glycine and gamma amino butyric-acid (GABA) can exert inhibitory effects. Dopamine exerts an excitatory effect and appears to play an important role in the control of complex movements.

There are other interesting substances with opiate-like properties, the endorphins, but there is at present a discussion as to whether these peptides should be regarded as neurotransmitters. They are widely distributed in the brain and may be involved in the perception of emotion, pain, pleasure, etc. (see Hökfelt et al., 1980). It has been reported that physical activity can enhance the release of endorphins (Allen, 1983; Harber and Sutton, 1984).

The endocrine system is, in a way, related to the synaptic strategy, but its transmitter substances, the hormones, are released far from their intended sites of action. The precision and the very restricted area of action are unique for the process of neurotransmission.

It is evident that when the transmitter has performed its job on the postsynaptic membrane, its action must be rapidly terminated; otherwise, its effect would last too long and the precision control of synaptic traffic would be lost. The removal of the transmitter can be achieved by diffusion, but in many cases a barrier of some sort keeps the transmitter fairly effectively within the synaptic cleft. It is assumed that the glial cells, surrounding the nerve cell, function as chemical insulators, thus preventing a diffusion of transmitters away from their sites of release. The glial cells may metabolize the neurotransmitters thereby keeping the extracellular spaces of the CNS free from disturbance by free-floating transmitters. In one way or other, inactivation may take place by enzymatic degradation. However, the most common design is to recapture the substance back into the presynaptic nerve terminal by a very fast and efficient transport process (see Krujevic, 1974).

As mentioned, the motoneuron has been intensively studied, as has the communication between the motoneuron and its "slave," the skeletal muscle. Acetylcholine (ACh) has been identified as the transmitter substance; it is the chemical messenger from the motoneuron to the skeletal muscle at the motor endplate. It would therefore seem reasonable to use ACh as an example to illustrate one type of strategy used in the communication from cell to cell. In the terminal ending there are a great number of minute bags or vesicles containing the transmitter substance, ACh in this case (some 2×10^4 ACh molecules in each vesicle). In the "resting" neuron, a so-called miniature endplate potential can be recorded. It is probably produced by the ejection of a few quanta of ACh from the prepacked vesicles that spontaneously burst on the surface of the membrane. The postsynaptic depolarization is, however, subliminal (see Miledi et al., 1983).

The arrival of an action potential at the terminal will initiate an inflow of Ca^{2+} ions causing an almost synchronous ejection of 100 to 200 quanta of ACh. The muscle fiber membrane has receptors for ACh. Within microseconds, transmitter molecules bind to these receptors. The chain reaction proceeds with an opening of Na^+ channels, and after a short delay, the opening of channels allowing K^+ to escape out of the muscle cell. (A single channel is open for about 1 msec, on the average; during this time, some 10,000 Na^+ and K^+ ions can pass through the channel. With a single nerve impulse, as many as 250,000 channels may open, see Peper et al., 1982). This is the birth of an action potential that will spread into and along the muscle fiber triggering its contraction (see Chap. 2). In the synapse, an enzyme, cholinesterase, present in the postsynaptic structure, will inactivate ACh by its enzymatic breakdown into choline and acetic acid. Much of the choline thus formed is recaptured from the synaptic cleft and brought back into the nerve terminal. In this way, it becomes available for the synthesis of ACh, which will be packed into vesicles and become available for reuse. Actually, a release of ACh will accelerate its synthesis, so that, under optimal conditions, the level of ACh remains relatively constant (Hebb, 1972). Even the vesicle membranes are retrieved and brought back to the terminal for reuse; thus, there is incentive to minimize waste built into the system. ACh is also a transmitter substance found in the excitatory synapses in the central nervous system, as well as in the parasympathetic system. Nerves that use ACh as a messenger are called cholinergic.

In primitive animals, a direct electric transmission exists between nerves. Channels run from the cytoplasm of the presynaptic to the postsynaptic neuron, forming a bridge capable of mediating electric currents. In higher vertebrates, however, the rule is that the gap between nerve cells is unbridged, and the synapse is chemically operated. This permits a more flexible function compared with the stereotyped bridged connection. Eccles (1982) points out that chemical transmission was an essential evolutionary advance. It made possible the design of brains with all of the infinite complexity required for higher nervous functions. The chemical transmission of information provided an opportunity for the evolution of the mechanism of inhibition, a process essential to the function of the CNS, and one that would not be possible with electrical transmission alone.

Nerve Axon Conduction Velocity

The conduction velocity in a nerve fiber would be slow if a myelin sheath was not wrapped around its "internode" sections. At the nodes, however, the sheathing is interrupted (Fig. 3-3), and it is apparently only at these zones that the bar nerve membrane has gated channels for N^+ and K^+ ions. In the propagation of an impulse in such a myelinated fiber, the current flow through the membrane is therefore restricted to the nodes. The impulse jumps from node to node (a saltatory transmission) with the long internodal sections remaining passive, which greatly enhances the velocity of the transmission. Furthermore, by ionic fluxes restricted to less than 1 percent of the surface area of the nerve fiber, there is an enormous metabolic advantage of the myelination (for details, see Eccles, 1973b).

The normal range of velocity in myelinated fibers is 12 to 120 m · s^{-1}. The fastest fibers are concerned with the control of movement. Unmyelated fibers have much slower conduction velocities, or 0.2 to 2 m · s^{-1}. Such nerves are typical pathways carrying less urgent, visceral information.

TRANSMISSION OF IMPULSES FROM NERVE AXON TO SKELETAL MUSCLE

As the axon of the motoneuron approaches the muscle fiber, it loses its myelin sheath. The axis cylinder branches and the terminals make close contact with the sarcoplasm of the muscle, forming *motor endplates* (Fig. 3-10). Each motoneuron supplies from about five (eye muscles) up to several thousand muscle fibers (limb muscles). The motoneuron and the muscle fibers it supplies function as a *motor unit,* since an impulse in the axon will activate all of the fibers almost simultaneously. The gain in muscular tension with the activation of one motor unit will depend on the number of muscle fibers included in the unit. The fewer there are, the finer the grading of the contraction can be. The fibers in a motor unit may be scattered and intermingled with fibers of other units, and they can be spread over an approximately circular region with an average diameter of 5 mm (Buchtal and Schmalbruch, 1980). Single muscle fibers may be innervated by more than one motoneuron, but the majority of the muscle fibers have only one endplate, which is supplied by a single nerve fiber. The terminal branches of the axon end near the middle of the muscle fiber. (The motoneurons in the human spinal cord and their dependent motor units number about 300,000 altogether.)

Figure 3-11 presents a schematic drawing of an endplate. The surface membranes of the nerve fiber and muscle fiber are comparable with the pre- and postsynaptic

Figure 3-10
Motor endplate. In the motor endplate, the knoblike terminations of the telodendron
of a motor axon fit into depressions of a multinucleated mass of sarcoplasm. (*After
Willy Schwartz, from Elias and Pauly, 1961.*)

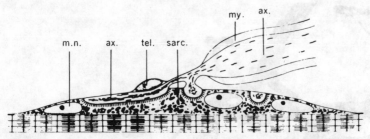

Figure 3-11
Schematic drawing of endplate: ax., axoplasm with its mitochondria; my., myelin
sheath; tel., teloglia; sarc., sarcoplasm with its mitochondria; m.n., muscle nuclei.
(*From R. Couteaux, 1960.*)

membranes, respectively, of a central synapse. It contains numerous vesicles filled
with ACh forming the vehicles for the neuromuscular transmitter.

It should be recalled that within the CNS, the algebraic sum of the electrochemical
effects of impulses arriving at excitatory and inhibitory synaptic knobs in contact with
a neuron determines whether or not it will become excited. In the neuromuscular
function there is, however, *only one transmitter,* and *its effect is excitatory.* The quanta
of ACh released is usually more than adequate to excite the muscle fiber. It should
be emphasized that the only way to promote relaxation of the muscle fibers is to
decrease or stop the discharge of their motoneurons.

The motor unit also obeys the all-or-none law: An effective stimulus of the motoneuron causes a maximal muscle action potential, and the muscle fibers react "to
the best of their ability." However, the strength developed will depend on the length
of the muscle fiber, on temperature and oxygen supply, and on the frequency of stimuli.
These factors will be discussed later. The action potentials can be recorded with needle
electrodes or surface electrodes attached to the skin over the active muscle, and the
obtained *electromyogram* (EMG) reveals the sum of the motor unit activity (for discussion and references, see Milner-Brown and Stein, 1975; Winter, 1979).

Summary An impulse started in a motoneuron propagates in the axon and is
transmitted to the motor endplate where ACh is released. The release of ACh affects
the muscle membrane, giving rise to an ion flux which reverses the resting membrane
potential. Similar to the impulse traffic between nerve cells, we have in the neuromuscular transmission a transducing mechanism, where electrical signals (nerve impulses) are transduced to chemical signals (ACh), and then back to electrical signals
(muscle action potentials). The muscle action potential propagates over the membrane
and initiates the mechanical-chemical mechanisms that cause the myosin and actin in
the sarcolemma of the muscle to react (Chap. 2). The developed muscle tension will
depend on (1) the number of motor units activated and (2) the frequency with which
they are stimulated.

REFLEX ACTIVITY AND SOME OF THE BASIC FUNCTIONS OF PROPRIOCEPTORS

Some of human behavior is innate and follows a stereotypic pattern, identical in all individuals. Examples of such behavior patterns are swallowing, coughing, breathing, coordination of the eye movements, blinking, vomiting, orgasm, and startling when taken by surprise. Central programs in the nervous system can coordinate the motoneurons involved in such activities, and once initiated the motoneurons do not require incoming (afferent) sensory feedback for the continuation of their essential pattern, even if a dozen or more muscle groups are involved. Even more complex movements like walking and running are basically genetically programmed. However, sensory inputs are essential for the adaptation and modification of these programs to changing environmental conditions, including the appearance of obstacles. Many universal emotional expressions like anger, fear, sorrow, disgust, or joy have innate rather fixed and rigid components, but they can be modified and controlled by will.

During childhood, we learn new movements, and there is an ongoing process seeking to modify behavior as a result of experience. For an understanding of some basic principles underlying the function of the central nervous system, it is useful to examine how relatively simple reflexes are brought about and how they are modified, for the reflex is an elementary model of behavior.

In this section, we shall try to describe some important components of behavior.

It has been pointed out that various nerve cells, either directly or through interneuronal relays, expose the motoneurons to a continuous bombardment of transmitter substances. Actually, every motoneuron may be exposed to as many as 3,000 to 5,000 neurons of the great intermediate net. They are intercalated between the true sensory side and the true motoric side as components of a computational network (see Nauta and Feirtag, 1979).

There is a minimum of two neurons in a reflex chain: the *afferent* (receptor) neuron and the *efferent* (effector) neuron. The nutrient cell of the afferent neuron is in the dorsal root ganglion (or cranial equivalent), and it conveys information from cutaneous, muscular, or special senses. The cell body of the efferent nerve of the motoneuron is located in the ventral horn (or motor cranial nucleus). The afferent fibers entering the dorsal root of the spinal cord may thus end monosynaptically around motoneurons, but in general several or many connecting interneurons intervene between the afferent and efferent neurons (Fig. 3-12).

Most neural control of skeletal muscles is reflex in nature. The excitability of the motoneurons is increased or decreased, depending on the algebraic sum of the excitatory and inhibitory activity in the synaptic knobs. If the muscles are the slaves of the motoneurons, the motoneurons are slaves of spinal and supraspinal mechanisms. It may be pointed out that most of the pyramidal tract fibers from the cerebral cortex area evidently terminate on interneurons. Two systems of interneurons are involved, one at the cortical level and one at the spinal stage of pyramidal tract innervation. "It can be stated that by establishing synaptic connections with interneurons rather than with motoneurons, the pyramidal tract and the other descending tracts (from the cerebellum, the red nucleus, the reticular formation) are able to operate through the

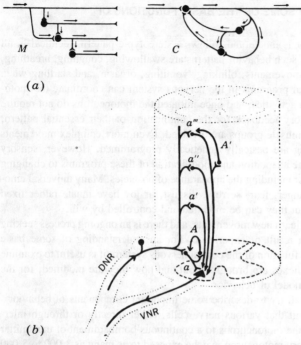

Figure 3-12
(a) Plans of the two fundamental types of neuron circuits. *M*,
multiple chain; *C*, closed chain. (After Lorente de No, 1938.) (b)
Role of internuncial neurons in reflex action. Diagrammatic section
of spinal cord. DNR = dorsal (posterior) nerve root; VNR = ventral
(anterior) nerve root; a, a', a'', a''', A, A' = internuncial neurons.
The shortest reflex arc is via DNR, a, VNR. Longer reflex arcs
involving delay paths are shown via a', a'', or a'''. A, A' are
"reverberators." Note how an impulse passing from DNR to VNR
along a may branch to excite A, which in turn reexcites a; similarly,
an impulse along a''' may branch to excite A', which in turn
reexcites a''. (*From Keele et al., 1982.*)

coordinative mechanisms at the segmental levels of the spinal cord" (Eccles, 1957).
Otherwise, the only alternative of an impulse traffic in the tracts would be an excitation
of the motoneuron. As it is, the intervening interneuron can give rise to excitation *or*
inhibition. The descending impulses from the higher levels of the brain, including
those voluntarily evoked, will necessarily be modified on the basis of coincident
information from all kinds of receptors. It also seems important that, via interneuronal
circuits, information based on past experiences will permit modification of responses.
Thus, the *impulse traffic in the final common path, the axons of the motoneurons, will
reflect an integration of synaptic activity based on both past and present experience*.

The synapses may be created in the same segment of the spinal cord as the afferent
fiber enters, or collaterals may pass up or down the spinal cord, eventually reaching

the brain (Fig. 3-12). Theoretically, there are innumerable pathways that an impulse can travel via nerve fibers, but some tracts are preferred, partially on an anatomic basis. One nerve may branch and terminate with several synaptic knobs on the same nerve cell and its dendrites.

In the synapse there is a delay of about 0.5 ms. "It used to be thought that one of the distinguishing properties of synapses was fatigue, but this is belied by the performance of synaptic mechanisms in the normal functioning of the brain" (Eccles, 1973b). Actually, synapses and neurons of the central nervous system are well adapted metabolically to respond continuously at frequencies as high as 100 s^{-1} with peaks of 500 s^{-1} or, for some species of neurons, as high as 1000 s^{-1}.

The most important features of the synaptic functions may be summarized as follows: If fibers from two afferent nerves ending on the same nerve cell are stimulated separately, neither causes a reflex response if the stimulus is subliminal (below threshold). However, when excited, each afferent nerve liberates sufficient chemical transmitter to cause an EPSP. When both nerves are stimulated simultaneously, the liberated quanta of transmitter substance may be enough to discharge a spike potential, and a reflex response is evoked (*spatial summation.*) The other type of summation is the *temporal summation:* Individual stimuli may be ineffective, but if repeated at a high frequency, a depolarization of the membrane of the connected nerve cell occurs.

Afterdischarge is another example of spatial summation. If the axon of a motoneuron is stimulated with an electrical shock of sufficient strength, the muscle supplied responds with a single twitch. If the stimulus is applied to the afferent nerve, the tension may remain for a much longer time because the muscle responds with repetitive twitches. Figure 3-12 explains how the motoneuron can be persistently stimulated, since volleys continuously arrive after having passed through the shorter or longer delay paths formed by the intricate maze of interneurons.

Renshaw Cells

Motoneurons give off collateral branches as they traverse the spinal cord to emerge in a ventral root. They form excitatory synaptic contacts with interneurons located in the ventromedial region of the ventral horn. These *Renshaw cells* send axons that affect inhibitory synaptic connections with the same and other motoneurons and interneurons of that segmental level in an overlapping and diffuse fashion (Fig. 3-13). Renshaw cells may also be excited by afferent impulses from muscles and skin. The Renshaw cells provide a feedback, and a single volley in the axon of the motoneuron may evoke a repetitive discharge of the Renshaw cell with the consequent tendency to dampen the motoneural activity. The effect is thus a reduction in the motoneural discharge frequency. This may protect against convulsive activity and overloading of the muscles.

In other words, the Renshaw cells and other inhibitory interneurons will keep down the level of excitation by suppressing discharges from all weakly excited neurons. Only the strongly excited neurons will survive this inhibitory barrier. In addition, as pointed out by Eccles (1973b), they participate very effectively in a neuronal integration.

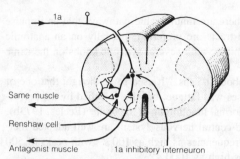

Figure 3-13
Some motoneurons give off branches, called recurrent collaterals, before their axons leave the gray matter of the spinal cord. These collaterals synapse with inhibitory interneurons called Renshaw cells. They synapse in turn with motoneurons, including the one that gave rise to its input. The figure also illustrates an example of inhibition of inhibition (disinhibition). The impulse in the afferent nerve (1a) also stimulates the inhibitory interneuron contacting the antagonist muscle. If the Renshaw cell is activated by the recurrent collateral it will inhibit the inhibitory interneuron and as a consequence the antagonist muscle will be released from the inhibition and may more easily respond to excitatory impulses (Inhibitory neurons are shown in black.) (*Modified from Kandel and Schwartz, 1981.*)

Figure 3-13 illustrates how an inhibition itself may be inhibited. Many motoneurons are normally subjected to a steady background inhibition by the activity of inhibitory nerve cells, eventually driven from higher levels of the CNS. Renshaw cells may form synapses with such inhibitory neurons. When these Renshaw cells are stimulated, they will inhibit or depress the activity of the inhibitory neurons and free the motoneurons from inhibition. The effect will eventually be a "recurrent facilitation" or *disinhibition* of the motoneurons.

The Muscle Spindle and the Gamma Motor System

From the skeletal muscles, afferent nerves report to the CNS about their tension, length, and position. They are activated by special receptors, one of which is the *muscle spindle*. The axons of the motoneurons so far discussed are of the so-called *alpha* (α) *type* (12- to 20-μm diameter), and they supply the skeletal muscles (*extrafusal* fibers). In the ventral horn of the spinal gray matter, there are also other types of motoneurons giving rise to axons, which are thinner, and they are of the *gamma* (γ) type. These gamma fibers supply the *intrafusal* muscle fibers of the muscle spindle. This proprioceptor sense organ is shown diagrammatically in Fig. 3-14a. It is an elongated fusiform capsule lying *parallel* to the extrafusal muscle fibers. Its central part consists of a noncontractile elastic "nuclear bag" or elastic "nuclear chain" (below, in Fig. 3-14a). On either side, some intrafusal muscle fibers are attached (consequently named nuclear bag fibers and nuclear chain fibers, respectively). The other ends of these fibers terminate either at a tendon or in the endomysium of extrafusal muscle fibers.

Activation of the gamma fibers causes the intrafusal muscle fibers to contract, and the equatorial parts between the polar regions become stretched. Receptors are located at these central parts, and if the stretch is strong enough, they will react with action potentials which, via the afferent nerve fibers, are propagating toward the spinal cord.

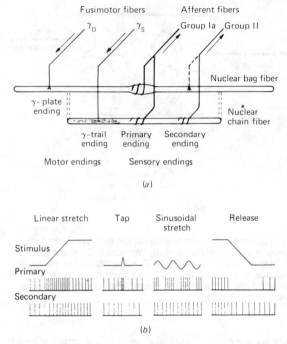

Figure 3-14
(a) Diagram of the mammalian muscle spindle mechano-receptors showing the main functional component involved in the transduction of muscular displacement into a receptor potential that is considered to be the direct governor of the afferent discharge pattern (γ_D = dynamic gamma axon; γ_S = static gamma axon). (*From Rudford, 1972.*)
(b) Diagrammatic comparison of the responses of "typical" primary and secondary endings to various stimuli. The responses are drawn as if the muscle were under moderate initial stretch and as if there were no fusimotor activity. (*From Matthews, 1964.*)

Some impulses are monosynaptically transmitted to alpha neurons of the same muscle (and its synergists), providing an excitation. When the muscle contracts, the ends of the muscle spindle come closer together, and the stretch on the nuclear bag and chain is reduced. The stimulus of the receptors will be decreased or will cease completely (Fig. 3-15). The afferent impulses, via an interneuron, can also cause an inhibitory effect on the alpha neuron of the antagonists.

Gamma motoneurons for a particular muscle are located within the alpha motor neuron pool for the same muscle. Actually, most of the nerves descending from the higher levels of the CNS which impinge upon the alpha motoneurons, will also stimulate the gamma motoneurons. This is called *coactivation*. The gamma motoneurons can adjust the length of the muscle spindle so that it can maintain its sensitivity to stretch over a wide range of change in length of the extrafusal musculature during voluntary, as well as reflex contraction. A passive stretch of the muscle, irrespective of its length, can therefore cause the muscle spindle to fire, and by reflex the same muscle will respond by contracting, thus counteracting the stretching. Tapping the quadriceps tendon, as is routinely performed in medical examinations, stretches the muscle and its contained muscle spindle, causing a jerk of the lower leg in the "knee reflex." The explanation for this is the one presented above.

The reflex can be evoked by (1) stretch of the muscle and (2) increased activity in the gamma fiber.

Regarding the latter, it should be emphasized that the activation of the gamma

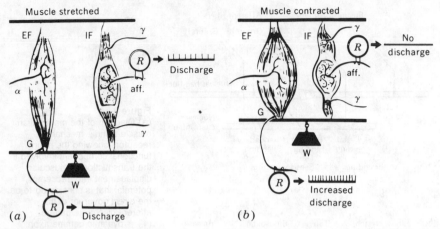

Figure 3-15
Schematic illustration of the activity in afferent fibers from muscle spindle (aff.) and Golgi tendon organ (G, lower record) when the muscle is (a) stretched and (b) contracted. EF = extrafusal muscle fibers, IF = intrafusal fibers innervated by gamma fibers. R = recording instrument with its electroneurogram. The muscle spindle is stimulated by stretch, but there is eventually a pause in its discharge during an extrafusal muscle contraction, during which activity there is an accelerated rate of discharge from the tendon organ.

fibers per se does not cause any increase in the muscular tension. An intact afferent nerve from the muscle spindle is necessary to induce a contraction as a response to gamma activity or stretch. Adaptation in the muscle spindle takes place very slowly; that is, the discharge in the afferent nerve fiber continues for as long as the muscle is stretched, though the frequency of the discharge gradually declines. A stimulus that increases slowly in intensity produces a lower frequency of impulses than a stimulus that rises very rapidly to the same level. Thus, a pull on a muscle of a cat with a given force attained within 1 s produces an afferent impulse frequency of more than $100 \cdot s^{-1}$. But a slower increase in stretch, continuing until the same force is applied, within 6 s will give a peak volley of about 40 impulses $\cdot s^{-1}$ (Granit, 1962).

It may thus be concluded that the muscle has an excellent instrument to measure its length (or more precisely, the difference in length between the parallel extrafusal and intrafusal muscle fibers), the extent of a mechanical stimulation, and the rate with which the stretch is applied. Furthermore, the spindle control enormously increases the number of functions that a slightly activated alpha pool of motoneurons can perform. (See Matthews 1972; Granit, 1975.)

The structure and function of the muscle spindles are much more complicated than revealed by the discussion presented. There are actually *two types of gamma efferents* and it is not settled exactly how the nuclear bag and nuclear chain are innervated (see Boyd, 1980; Burke, 1980). Most likely the so-called *dynamic gamma axons* are innervating nuclear bag fibers while axons of *static gamma* axons can innervate both nuclear bag and chain fibers. The intrafusal muscle fibers of the nuclear bag are slow fibers, they build up their state of contraction slowly, and they possess damping. The muscle fibers of the nuclear chain are fast-twitch fibers and they respond by twitches but with little viscous damping. It is only the static gamma fibers that adapt the length of the spindle muscle fibers during a contraction of the extrafusal muscle fibers thus preventing it from slackening. The *afferent* nerves from the spindles are also of two different kinds: the *primary*, fast conduction *nerve fibers* (group Ia) which are equatorically located in both bag and chain fibers; and the *secondary*, slower conducting *nerve fibers* (group II). The large majority of the secondaries are found on the chain fibers, with a minority on the bag fibers.

Fig. 3-14b illustrates the different properties of the responses of primary and secondary nerve endings to various stimuli. (1) The primaries are particularly sensitive to a stretch. They cease to fire when the stretch is released. The secondaries on the other hand are better adapted for signaling the actual length of the muscle and they do not show a period of silence after release of pull. (2) The primaries respond strongly to a "tap" on the tendon of the muscle which is also in contrast to the secondary nerve endings. (3) If the muscle or its tendon is subjected to sinusoidal stretches or high-frequency vibrations, it is only the primaries which, to a marked degree, vary their action potential frequency. (When recording from single afferent nerve fibers, this characteristic of the primaries can be used for identification.) It should be pointed out that γ activity during the release of pull can adapt the length of the muscle spindles, and the pause noticed in firing from the primary nerve endings in Fig. 3-14b may be filled out. As will be discussed, both static and dynamic γ fibers are activated or inhibited in voluntary muscle activities.

It is the primary nerve fibers that monosynaptically can activate the motoneurons supplying their own and synergistic muscles. In fact, a primary afferent fiber makes contact with virtually all homonymous motoneurons. They inhibit via interneurons those of the antagonistic muscles. This is the "classic" function of the muscle spindles. The effect of impulses from secondary endings is more complicated and partly unknown. They may polysynaptically excite flexor motoneurons irrespective of whether the endings themselves lie in flexor or in extensor muscles. The effect on extensors may be an inhibition. We can conclude by stating that in general muscle spindle primaries and secondaries have been found to behave similarly, but the primary nerve endings are more sensitive to the velocity at which a muscle is being stretched (dynamic conditions), and the secondary nerve endings more precisely report the muscle length (static conditions). On a subconscious level, they can so provide the CNS with essential information about the state of the muscle. The final effects on the α motoneurons from the ingoing impulse traffic of the different nerve endings may evidently be essentially similar or antagonistic. (See Boyd, 1980.) The role of the muscle spindles in locomotion will be discussed later on.

The Golgi Tendon Organs

These spindle-shaped corpuscles, a few millimeters in size, are connected in *series* with extrafusal muscle fibers and inserted between the muscles and their tendons. The relatively few muscle fibers attached to a tendon organ belong to different motor units. Its afferent nerve fiber has many branches twisted within the braids of collagen fiber bundles. A stretch of the Golgi tendon organ will give rise to action potentials that will be conducted centrally in relatively thick nerve fibers (the so-called Ib fibers). For mechanical reasons, the receptor will be particularly sensitive to "active" tension caused by a contraction of the extrafusal motor units that are attached in series to the portion of the tendon in which the Golgi tendon organ is located. A "passive" tension created by pulling the muscle is a much less effective stimulus on the tendon organ because the force is exerted on a much larger tissue area.

When stimulated, the afferent nerve fibers cause a reflex inhibition of their corresponding muscle which is elicited via interneurons. Excitatory interneurons are also stimulated, but they form synapses on the motoneurons to the antagonistic muscles. It should be emphasized that the reflex arc of the gamma loop is monosynaptic, but the Golgi tendon organ stimulates via interneurons, and this delays the impulse. The tendon receptors inhibit not only the alpha motoneuron of the same muscle, but also the gamma motoneuron. This is a wise arrangement if the elicited reflex is effectively to prevent the development of a dangerous tension in muscle and tendon. The Golgi tendon reflex also causes the smooth retardation of a movement. One may generalize and claim that the Golgi tendon organs measure the tension developed by the muscle, and the muscle spindles measure its length.

Reciprocal Inhibition (Innervation)

One very important design in locomotion is the principle of reciprocal innervation. If a muscle is stretched, impulses are evoked in the afferent fibers from the muscle spindles, as already mentioned. These fibers, by direct synaptical contact, i.e., monosynaptically, excite the motoneurons of both the same and synergistic muscles. Their collaterals, releasing the same transmitter, will excite adjacent interneurons, which in turn will inhibit the antagonist muscle (disynaptically). It is certainly functionally economic that a synergist team of muscles when activated are not faced with the resistance of their antagonists. However, in many movements, the muscle groups, on both sides, are activated simultaneously. Figure 3-16 illustrates in a schematic way how the reciprocal inhibition results in an integrated motor response.

Joint Receptors

The ligaments and the capsules of the joints contain different kinds of receptors (Ruffini end organs and receptors of the Pacinian and Golgi type). Some of these proprioceptors are specialized to respond to movement of the joint; others show an impulse discharge that varies with the exact position of the joint but are less sensitive to movement. The afferent nerve fiber's synapse in the spinal cord and other neurons can transmit impulses to the cerebellum, thalamus, and sensory cortex respectively. In other words, the information from joint receptors reaches neurons at all levels of the CNS as do the nerve activities originating in other peripheral receptors. These neurons usually also receive convergent signals from not only joints, but from skin and muscle receptors as well. There are indications that joint afferents do not play a crucial role in the sense of the static limb position, but most likely their reports are important in locomotion. They are part of the information that can modulate the activity of neurons in the motor cortex, cerebellum, and other areas of the brain.

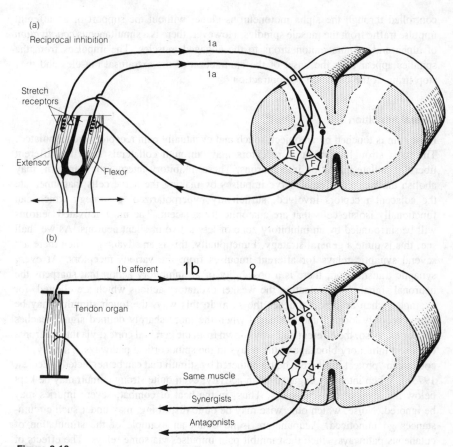

Figure 3-16
Reciprocal innervation-inhibition.
(a) Cross-sectional analysis of the monosynaptic excitation of the muscle spindle's "own" muscle and the disynaptic inhibition of its antagonist when the muscle spindle via its *1a* axon activates the nerve cells in the spinal cord, (E = extensor; F = flexor motoneuron)
(b) Central connections of the *1b* afferent fibers ensuring the contact between the Golgi tendon organ and the spinal cord. All connections to motoneurons are through interneurons. In the reflex mediated by the *1b* afferent system, inhibitory interneurons (black) inhibit motoneurons to the muscle of origin and its synergists, and excitatory interneurons excite the antagonists.
(*Adapted from Kandel and Schwartz, 1981.*)

The Resting and Activated Muscle

At rest, the skeletal muscles are relaxed and normally no EMG can be recorded. Vallbo (1973) reports that, when he was recording from afferent nerves of spindle primaries in finger muscles, only a small proportion of the nerves studied were continuously discharging in his conscious human subjects sitting in comfortable resting positions.

Vallbo (1973) concludes that in the human, the *onset* of voluntary movement is controlled through the alpha motoneurons alone, without the support of an afferent impulse traffic from the muscle spindles. However, there is a simultaneous *coactivation* of alpha and gamma motoneurons to the engaged muscles. The impulses from the spindles appear after the onset of the contraction of the extrafusal muscles and they stop firing at the cessation of contraction.

Lateral Inhibition

If the skin is touched by an object, touch and eventually pain receptors are stimulated. The most strongly stimulated receptors may, through collaterals from the afferent fibers, stimulate inhibition interneurons. These inhibitory interneurons, in turn, may abolish the weaker afferent nerve impulses by making the nerve cells that innervate the adjacent receptors involved, sufficiently hyperpolarized, a strategy which can functionally isolate cells that are anatomically adjacent. The most activated neurons will be surrounded by an inhibitory zone of less active or silent neurons. As we shall see, this is quite a general strategy. Functionally, this is an advantage when there are several synaptic relays for afferent impulses from the various receptors. At every synaptic transmission, there is a possibility of an inhibitory action that sharpens the neuronal signals by eliminating the weaker excitatory actions which are elicited, for example, when an object touches the skin. In this way, the touch stimulus may be more precisely located and evaluated when the more sharply defined signal reaches the cortex. Also, the efferent pathways down from the cerebral cortex, via interneurons, can exert inhibitory blockage at the relays in the spinocortical pathways. Thereby, the cortex can "protect itself from being bothered by stimuli that can be neglected" (Eccles, 1973b). When intensely engaged in an activity, even quite strong stimuli may be kept below the level of consciousness. Thus, in the heat of combat, severe injuries may be ignored. Noise, which otherwise may be quite disturbing, may under such circumstances go unnoticed. Acupuncture is no doubt an example of the stimulation of cutaneous pathways which then inhibit pain impulses via some relays. The effects of hypnosis and yoga are other examples of lateral inhibition. Similar mechanisms may explain the fact that only a small minority of the ongoing traffic in afferent axons will eventually reach key neurons in the sensory areas of the brain and consciousness.

Supraspinal Control of Motoneurons

In its biologic evolution, primarily a development of the CNS and, to a lesser extent, a reconstruction of existing parts have taken place. The basic neural machinery essential for the generation of reflex action is located in the spinal cord. However, these local spinal circuits can be continuously "manipulated" by descending axons from higher brain areas as well as by other regions within the spinal cord. The motoneurons can be facilitated/activated or inhibited by incoming electrically mediated messengers. Therefore, let us turn our attention to the higher levels of the nervous system and try to analyze what is going on there. The cerebellum has been developed, so to speak, parallel with other parts of the brain. In particular, the evolutionary building of a linked system between the cerebrum and the cerebellum gave humans their immense superiority for survival. The more recently developed parts of the brain have assumed a functional dominance, compared with the older parts. This is especially true in the case of the cerebral cortex.

Nauta and Feirtag (1979) describe some evolutionary aspects of the development of the human cerebral cortex. It is estimated to contain no less than 70% of all the neurons in the CNS. "This is the neocortex. It is the latest form of cortex to appear in evolution. We owe it to a branching; beyond the reptiles one strain of animals elaborated on the reptilian pattern and became the birds, while a more venturesome strain developed the neocortex as it became the mammals. From a strictly phylogenetic point of view birds are thus the logical end of the brain's traditional development and the mammals are deviants, since there are no birds in their ancestry. In one of the many radiations of mammalian evolution the primates appeared, an order in which the neocortex reaches its maximal development. We human beings are heir to all the consequences, perhaps including psychiatry."

In the introduction to his book *The Understanding of the Brain,* Eccles (1973b) claims that "this human brain that doesn't look too distinguished on the outside, weighing about 1.5 kilograms, is without any qualification the most highly organized and most complexly organized matter in the universe." He believes that attempts to understand the brain will occupy hundreds of years of our future.

Generally speaking, it appears that the nerve cells in the brain have a tendency to fire all the time; there is an incessant, irregular discharge from billions of nerve cells, giving a massive "background" noise. The number of synapses on a nerve cell may vary from a few hundred to more than 200,000. In other words, it is a matter of a pronounced convergence. Probably none of the neurons in the brain is exposed only to excitation, and certainly no nerve cells are affected solely by inhibitory reception. It is therefore often a question of a modification of a basic frequency more or less, when a nerve cell is exposed to a synaptic effect. Eccles (1973b) has pointed out that the "voice" of a single neuron in the brain would be lost in the "chorus" of all the neurons humming together. It appears that the same kinds of neurons are organized together, that they receive the same kinds of coded message, and that they will eventually forward the message to another cluster of neurons. "The neurons have to shout together, as it were, to get the message across and so make a reliable signal despite all the background noise" (Eccles, 1973b). It may be compared with the situation at a fully packed stadium during an exciting athletic event. The single shouts are drowned in the roar of the masses but many voices joining in a claque of yellers can be identified! The final common path, the motoneurons, are usually not bothered with raw data but are served synopses of information by neurons of the great intermediate net, which are continuously communicating with the periphery and the CNS (see Nauta and Feirtag, 1979).

Cerebral Cortex

The region just anterior to the central fissure has a key position in the voluntary control of muscular movements (precentral gyrus). It is not the prime initiator of movement, but it is the final relay station receiving instructions from widely dispersed areas in the brain and putting them into effect via the motoneurons. Most of the pyramidal cells are arranged in narrow columns (about 0.5 to 2.0 nm in diameter) which are vertically oriented to the cortical surface. One column (with a few hundred pyramidal cells and some 10,000 other types of cells) projects on a specific muscle group or to synergistic muscle groups. Thus, stimulation of a column can produce one single movement and other stimuli may give rise to several movements, depending on which part of the column is stimulated. The afferent input to such cell columns from skin, joint, and muscle spindle receptors is clearly related to the direction of the movement. Thus, all the various parts of the body are represented in the striplike map established in the motor cortex. The muscles controlling the tongue, lips, larynx, and thumb have a particularly large representation in such columns. In the postcentral gyrus of the cortex, the primary somatic sensory cortex, there is a similar representation of the different parts of the body.

Also in this part of the sensory system there are columnar arrays receiving inputs from a specific muscle receptor or from a region of skin specifically related to the function of that muscle. Thus, there is a topographic relationship between sensory input and motor output of importance for the control of movement. It is actually a basic organizational principle of the cerebral cortex that columns of neurons are activated by the same single submodality, e.g., found on a particular point on the skin. They are like computational modules that can receive information and then distribute a modified version of that information to the other regions of the brain. Often there is a parallel processing of given sensory information, providing security and adding "richness to our perception." (See Kandel and Schwartz, 1981).

Afferent fibers reaching the motor cortex can exert inhibitory or excitatory actions on the pyramidal cells (which are themselves excitatory). Interneurons can secure a fast, vertical spread of excitation through the whole depth of a column. They can, by a positive feedback, build up a strong, amplified excitation in the column. With multiple synapses, one afferent fiber can excite many cells. Inhibitory neurons in a column, which can probably be activated by recurrent collaterals of the pyramidal cells, often have their axons distributed to the immediately surrounding parts of the cortex. By this design, activation of a column may suppress the nerve activities in adjacent columns, often with antagonistic functions, in a reciprocal fashion. Thus, the activated column may cause a flexion of a finger and the inhibited surrounding columns may be projecting on extensor muscles that will now relax. It should be pointed out that the pyramidal cells are not tied together in rigid relationships. They can plastically operate reciprocally and simultaneously and can combine in different patterns of activity.

It is estimated that the pyramidal tract in the human contains about 1.2 million fibers (in a cat, less than 200,000). Of these fibers approximately 60 percent originate in neurons of the motor cortex. In the higher primates it is estimated that some 10 percent of the pyramidal endings in the spinal cord do form monosynaptic contacts with both α and γ motoneurons. Many such fast-conducting fibers connect with muscles involved in precise movements of the hands and in speech. Therefore, the motoneurons innervating the intercostal muscles also have some direct connection with the motor cortex. Those muscles certainly guide such skilled activities as speech and song. A destruction of these tracts does not interfere with the breathing but the intercostal muscles become useless for speech and song (see Evarts, 1979.) With the α motoneurons excited to discharge impulses, the innervated muscle fibers will contract. At approximately the same moment, the γ motoneurons will also discharge, thus exciting the muscle spindles in the same muscle, and the γ loop is put in operation. There is a *coactivation of the α and γ motoneurons* with the exception of very fast movement, which probably is carried out by pure α activity. The two pools of motoneurons are also *coinhibited*. The sequence is: (1) The skeletal muscle is activated by the α motoneuron discharge; (2) the thinner γ fibers conduct impulses more slowly than the α fibers, and therefore the muscle spindle contraction will begin somewhat later than the extrafusal contraction; and (3) the afferent discharge from the muscle spindles follows, with the triple role of directly supporting the α activity to the same muscle and inhibiting the antagonists as well as reporting to the brain about the progress of the initiated movement.

One important feature of the pathways from the cerebral motor cortex is their profound collateral branching to brain stem nuclei (particularly to relays in the pontine gray matter but also to those in the lateral reticular nucleus and the olive) (Fig. 3-17). Many of the relay nuclei project to the cerebellum. The result is that the cerebral cortex cannot keep any secrets with regard to its signal traffic to the motoneurons!

Pathways from the motor cortex can also, via 'an interneuron, powerfully *inhibit both α and γ motoneurons*. Clough et al. (1971) point out the importance of a control of γ activity to muscle spindles of "unwanted" synergists and of withdrawal of support from the spindles of antagonistic muscles that might otherwise undergo autogenic stiffening when lengthened passively by the action of the prime movers.

In summary, we have discussed an inhibition of muscles not needed for a special movement on two levels: by a lateral inhibition in the motor cortex, and by interneurons at the spinal level.

The motor area, i.e., the precentral cortex of the brain, is not only a source of motor output but also a target for sensory input. The afferent input is strongest to a particular subdivision of the motor cortex from the part of the body controlled by that subdivision (e.g., from receptors in one hand to the area of the motor cortex that can induce movements in that hand).

Figure 3-18 illustrates schematically how three neurons can relay impulses starting in different muscle spindles and eventually reach the sensory cortex. Interneurons inform the motor cortex about the peripheral event, and may, via the corticospinal tract give appropriate commands directly to the muscles, in this case moving a finger.

In experiments with monkeys and human subjects, the time has been measured from initiation of unexpected movement of a finger or a hand until the previously learned movement-response occurred. There was a difference in the reaction time noted for monkey and man, but this could be explained by differences in body size. The experiments showed that in less than 0.1 s a message originating from peripheral sense organs in the hand can reach the brain, be interpreted, and provoke a muscle action. (See Evarts, 1973; Kandel and Schwartz, 1981). There is a response with a shorter reflex time but it operates exclusively at the level of the spinal cord. Collaterals and other neurons also involve transcortical engagements. Evarts (1979) points out

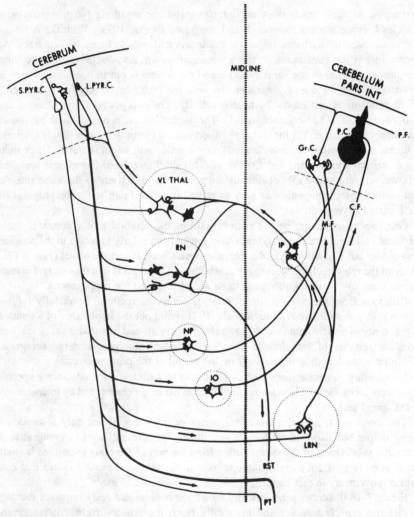

Figure 3-17
Pathways linking the sensorimotor areas of the cerebrum with the cerebellum. [C.F. = climbing fiber; Gr.C. = granule cell; IO = interior olive, IP = interpositus nucleus (intracerebellar nucleus); L.PYR.C. = large pyramidal cell; LRN = lateral reticular nucleus; M.F. = mossy fiber; NP = nuclei pontis; P.C. = Purkinje cell; P.F. = parallel fiber; PT = pyramidal tract; RN = red nucleus; RST = rubrospinal tract; S.PYR.C. = small pyramidal cell; VL THAL = ventrolateral thalamus.] (*From Eccles, 1973b.*)

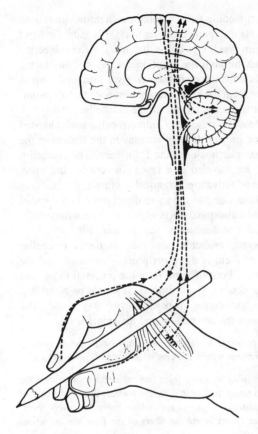

Figure 3-18
Illustrating schematically how three neurons may relay impulses starting in different muscle spindles and eventually reach the sensory cortex. Interneurons inform the motor cortex about the peripheral event, and may, via the corticospinal tract give appropriate commands directly to the muscles, in this case moving a finger.

that the mammalian motor cortex, which phylogenetically is a new part of the brain, is subject to the same laws of reflex action that characterize the brain's older components. In addition, the motor cortex can be driven by a second set of inputs. These inputs underlie the internally generated motor programs that are a product of activity in the basal ganglia and the cerebellum and reach the motor cortex by way of the thalamus.

Cerebellum

The cerebellum has a key function in the smooth and efficient control of movement. It integrates and organizes information flowing into it along various pathways, from peripheral receptors (in skin, joints, muscles, tendons, ears, and eyes), and from various nerve centers in the brain. (For further studies, see Eccles, 1973b; Allen and Tsukahara, 1974; Harvey, 1980.)

There are two distinct *afferent tracts* feeding the cerebellum with information from the periphery: (1) the direct pathways via the spinocerebellar tracts, (2) and an indirect tract which first projects to the cerebrum (via relays in the thalamus). After subjection to modifications, the information goes through the mentioned relay stations in the brain stem to the cerebellum. In addition, the cerebellum receives "original" inputs from many association areas as well as from the sensorimotor cortex of the cerebrum. Most of the end points of the afferent fibers are located in the cerebellar cortex.

The *efferent tracts* originate from the numerous "deep" intracerebellar nuclei located in the interior of the cerebellum. They project to relay stations in the thalamus, the red nucleus, and the vestibular and reticular nuclei of the brain stem. The cerebello-cerebral pathways (via the thalamus) are traveled in a few milliseconds, and using these routes, the cerebellum is capable of activating pyramidal neurons and modifying their impulse discharges. The cerebellum also has the more direct route to the spinal motoneurons via the rubospinal and vestibularspinal tracts which can monosynaptically or polysynaptically activate both α and γ motoneurons (see Grillner, 1981.)

We may conclude that there are several available loops that can tie the cerebellar nuclei to the locomotor organs. One of them is more peripherally oriented, and the other includes the cerebral motor cortex. Both of them have the potential to operate the motoneurons and participate in the control of movements. It should be noted that the pathways between the cerebellum and cerebrum permit a two-way traffic. The same is true for the connections between the cerebellum and the spinal level.

In order to illustrate an architectural principle in nerve connections, some details will be given about the organization of the elements within the cerebellar cortex (Fig. 3-19).

The cerebellar cortex is more uniform in its structure from one region to another than the cerebral cortex. Its construction is also rather similar in different animals than is the case of the cerebral cortex. It receives inputs with two fiber types, *climbing fibers* and *mossy fibers*. The only pathway out of the cerebellar cortex is via the fibers of the *Purkinje cell* axons that contact the intracerebellar nuclear cells.

The afferent fibers exert their effects on the Purkinje cell in different manners. One climbing fiber innervates one Purkinje cell by forming hundreds of synapses on its dendritic tree. The climbing fibers exert a very powerful excitatory action when discharging. Originating in the olive, they relay messages from muscles and skin as well as from the precentral and postcentral gyri of the cerebral cortex. The mossy fibers branch enormously and synapse on the *granule cells* (one mossy fiber innervates some 450 granule cells). Their axons bifurcate to form the densely packed *parallel fibers* that extend along the folium for about 1 mm in each direction from the bifurcating point. These fibers form excitatory synapses to Purkinje cells, and it is estimated that one Purkinje cell can receive as many as 80,000 parallel fibers, or, eventually, up to 200,000 fibers. (Some 400,000 parallel fibers may pass through the dendritic tree of one Purkinje cell. To give an idea of the richness of nerve cells: There are about 30,000 million granule cells and 30 million Purkinje cells in the human cerebellum; Eccles, 1973a. In addition, the parallel fibers can activate inhibitory cells: the *basket cells,* the *stellate cells,* and the *Golgi cells* which can also be stimulated by mossy fibers.) The Golgi cells have the potential to inhibit the granulate cells in a negative feedback fashion; they have a profuse branching, so that one Golgi cell can inhibit some 10,000 granule cells. *Functionally,* the Purkinje cells are arranged in columns, and, in some areas, there is a strict topographic representation of distinct movements similar to the arrangement of the pyramidal

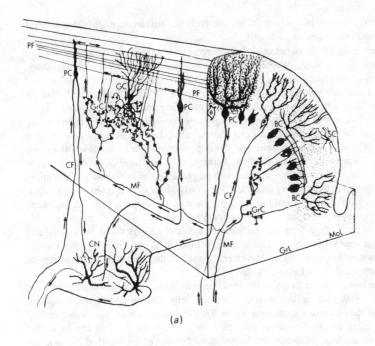

(a)

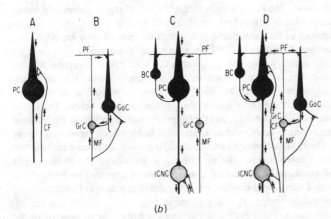

(b)

Figure 3-19
Top: Schematic drawing of a segment of a cerebellar folium. (*From C. A. Fox, 1962.*)
Below: The most significant cells and their synaptic connections in the cerebellar cortex. The component circuits of A, B, and C are assembled in D. Arrows show lines of operation. Inhibitory cells are shown in black. (*From Eccles, 1973b.*)
(BC = basket cell; CF = climbing fiber; CN or ICNC = intracerebellar nuclear cell; GoC or GC = Golgi cell; GrC = granule cell; GrL = granular layer; MF = mossy fiber; MoL = molecular layer; PC = Purkinje cell; PF = parallel fiber; SC = stellate cell.)

cells in the cerebral cortex. In other words, specific Purkinje cells can integrate both peripheral and cortical inputs that represent a limb.

The Purkinje cells are "spontaneously" discharging, driven by a constant bombardment via the numerous mossy fibers. With a more or less synchronous activation of mossy fibers, generated from (1) peripheral receptors (and initiated by movements) and/or (2) other parts of the brain, a compact granule cell channel will excite a cluster of Purkinje cells via the parallel fibers. They will cover a rectangular field extending about 2 mm. In addition, the climbing fibers transmit an excitation conveying much the same information as the mossy fibers. However, the *granule cell-basket cell channel will inhibit the Purkinje cells on each side of the field,* and an inhibitory state will develop on either side of the beam of excited Purkinje cells. Depending on their location, Purkinje cells will increase or decrease their discharge rate. They can respond to an input with all gradations of excitation and inhibition. A single climbing fiber has a strong influence. The mossy fibers work by their large number, each with a very weak, almost insignificant, influence on the Purkinje cell. An individual Purkinje cell performs an independent computation on a specific subset of inputs available to it. It can integrate inputs from several areas of the cerebral cortex and from peripheral receptors. There appear to be many centers for many kinds of integration, with much overlap and replication in an irregular manner. However, "No part of the cerebellum knows in general what other parts are doing" (Eccles, 1973b).

It was mentioned that the axons of the Purkinje cells form the sole connection between the cerebellar cortex and the intracerebral nuclear cells. (As an additional example of convergence in the neuronal connections, up to 200 Purkinje cells may contact one nuclear neuron.) The effect on these nuclei is *inhibitory* and the neurons therefore can be further inhibited or released from inhibition (by disinhibition) because of the modified Purkinje cell discharge. After about 10 ms, the Purkinje cells in the "on-beam" region become inhibited, probably by the negative feedback from basket cells and Golgi cells.

Because all the neurons of the cerebellar cortex except the granule cells are inhibitory, the inhibition has dominance. Eccles (1973b) points out that "there can be no prolonged chattering in chains of excitatory neurons . . . . within 0.1 s after some computation, that area of the cerebellar cortex is 'clean', ready for the next computation. This automatic 'cleansing' is very important in giving reliable performance during quick movements." Excitation actions have to fight their way against the opposite inhibitory action. The lateral inhibitions by the basket cells, stellate cells, and the Golgi cells can sharply focus the Purkinje cell response to limited cortical areas, and in a reciprocal fashion keep "unwanted" nerve cells in adjacent foci silent. With Purkinje cells inhibited, the synaptically contacted nuclear cells are relieved from a depression and the net effect is an excitation. The Purkinje cells that are activated will consequently reduce the discharge rate of projected nuclear cells or eventually silence them. During rapid movements, many different patterns of excitatory and inhibitory interactions will be completed within a second.

The continuous irregular activity, even under resting conditions, in all neurons of the cerebellar cortex and nuclei makes them primed to respond to changes in the afferent input by transitory variants in frequency that constitute signals for computation. A large number of parallel lines carrying similar information converge on the same target neuron. This onslaught provides an automatic averaging which will ensure an improved reliability and an elimination of "noise." The role of the climbing fibers is not clear (see the subsection "Learning and Memory"). The signal traffic mediated by these fibers is delayed, and the Purkinje cell response comes at least 10 ms later than the response elicited by the mossy fibers.

Slow-conducting mossy fibers give off collaterals to the intracerebellar cells (Fig. 3-19). These excitatory fibers can provide a strong background, a nonspecific drive onto the nuclear cells, which the Purkinje cells can modulate downward by inhibition. A specific stimulus of the nuclear cells by this route is delayed and arrives at approximately the same time as the Purkinje cell inhibitory control. Eccles points out that this simultaneity is important for an effective computational performance by the nuclear cells; there is a closely timed clash of the opposed influences of excitation and inhibition.

To give an idea of the time relations, it may be mentioned that a tap on a cat's forepaw pads can, via mossy fibers, stimulate Purkinje cells after a latency of about 12 ms, and after 16 to 18 ms, the response comes from nuclear cells (Eccles, 1973a).

Summary   The cerebellum receives nerve impulses from peripheral receptors and other parts of the brain via mossy fibers and climbing fibers. In the cortex there are five neuronal species, four of them being inhibitory in their effects. The only output from the cortex is via the axons of the Purkinje cells. These cells can be fired by either cortical or peripheral inputs alone, but in general, the two inputs are cooperative. The Purkinje cells "have the role of reading out the computation by the neuronal machinery of the cerebellar cortex and transmitting their coded message to the nuclear cells" (Eccles, 1973a). These intracerebellar nuclei then feed back "messages" to the cerebral cortex and to the periphery. There is strong evidence that the cerebellum plays a key role in the planning and initiation of movement after a general command has been given by higher brain centers. For successful muscular performance the cerebellum is also essential for it can adapt the central motor programs to the environmental conditions (see below).

Various Nuclei in the Brain Involved in Movement

The pyramidal tract originating in the cerebral motor cortex has been mentioned. The *extrapyramidal tracts* constitute other pathways contacting the motoneurons mono- or polysynaptically. They run from the basal ganglia, red nucleus, mesencephalic and pontine reticular formation, vestibular nucleus, and cerebellum. The *basal ganglia* (consisting of striatum and pallidum) form important relay stations for pathways between cerebral cortex, thalamus, and brain stem nuclei. The *red nucleus* is specially concerned in transmitting the cerebellar output down the spinal cord in the rubrospinal tract.

The *reticular formation* is a poorly defined part of the brain stem (mainly midbrain, pons, and the tegmentum of the medulla). It is formed by many nuclei scattered throughout the central part of the brain stem. It is also characterized by an interlacing network of nerve fibers giving a reticular appearance, hence the name: reticular formation. The axons of the nerve cells of the reticular formation have a far-reaching distribution in both rostral and caudal directions. The reticular formation sends and relays excitatory and facilitatory extrapyramidal reticulospinal fibers to the spinal neurons, both to α and γ neurons. A stimulus of the facilitatory reticular neurons increases the rate of discharge of the gamma efferents. Conversely, the activated inhibitory reticular neurons reduce or annul the gamma activity. Similar effects can be elicited from the hypothalamus. Movements induced either by stimuli in the motor cortex or by reflex mechanisms can be abolished by impulses from the inhibitory reticular nerves. The net effect of the descending impulse traffic normally is a facilitation of spinal neurons, especially those innervating the extensor (antigravity) muscles.

The *thalamus* is a large, ovoid mass of gray matter. It is important as an integrating and funneling relay station handling information coming in from the spinal cord, the cerebellum, and the basal ganglia, and it reports to the cerebral cortex including the primary cortical sensory area. It also receives numerous axons from the cerebral cortex and, as mentioned, it is involved in the transmission between the cerebellum and cerebrum.

Integration of the Neuronal Activity in Movement

What is the beginning of the neural events that lead to a muscular movement? It has been noted that there is a slowly rising negative electrical potential over a wide area of the cerebral cortex as early as about 0.8 s before the onset of a voluntary movement (a "readiness potential"). About 30 to 150 ms prior to the onset of muscular contraction, there appear sharper potentials, the so-called motor potentials, located more specifically over an area of the cerebral motor cortex concerned in the movement. An inhibition of the antagonist may precede the excitatory response of the agonists by some 15 to 20 ms. In other words, nerve cells in the motor cortex are important components in a circuit that initiates muscular contractions.

It has been pointed out that the cerebrum cannot begin instituting any movement without the cerebellum's immediately knowing about it. In fact, the cerebellum, many association areas in the cerebral cortex, and the basal ganglia are engaged in planning the movement and in translating the idea to move into a patterned activation of certain motor cortical columns. It should be recalled that the entire cerebral cortex sends fibers to both the cerebellum and the basal ganglia, and that these two structures in turn have massive connections back to the motor cortex via the thalamus. By recording the electrical activity in single nerve cells, it has been shown that both the cerebellum and the basal ganglia become active in advance of muscular contraction (Evarts, 1973).

There are apparently "memory stores" in the cerebellar nuclei related to specific innate as well as learned movements. The cerebellum continuously receives messages from peripheral receptors about joint positions, muscle length and tension, movements, environment, and so on. When a call for movement is reported to the cerebellum, it has all the necessary requirements for execution and control of the movement. At the time the motor command descends to the motoneurons, the cerebellum updates the intended movement on the basis of the somatosensory description of body position and velocity on which the movement is to be superimposed. Within a fraction of a second, the return circuit from the cerebellum can take over the output from the cerebrum and modify the discharges down the pyramidal tract. In addition, the cerebellum can affect the motoneurons via extrapyramidal tracts.

The muscle spindles and other proprioceptors are part of a mechanism for checking the execution of movement in relation to command. The cerebellum, in particular, has the potential to compute and integrate the total sum of information about a movement and to execute a follow-up correction. It has been emphasized that there is usually a coactivation of both α and γ motoneurons. In fact, one descending nerve axon can activate both α and γ motoneurons, at least those innervating the intercostal muscles (see Granit, 1975). If the muscle spindle's own contraction is set for shortening, which is suddenly prevented by an unexpected increase in the load, an accelerated afferent impulse traffic from the spindles driving the motoneuron pool automatically compensates for the extra load. This drive stops when the intrafusal and extrafusal lengths are functionally equal. The muscle spindles can serve as quick error detectors. In taking compensatory measures, the cerebellum has a key function.

We can see how a movement planned within the cortex in close cooperation with the cerebellum and basal ganglia is finally executed by the "upper motoneurons" in the cerebral motor cortex via the "lower motoneurons," most of them located in the spinal cord. There is a plan-ahead activity based on previous experience (memories) or memory fragments of fixed movement patterns, modified by a continuous updating of the motor command at its beginning and throughout its duration. There is always some kind of preprogram available. This is particularly important in very rapid movements, for they cannot be adequately updated once they begin. Therefore, in learning an exercise, one must first execute it slowly; the cerebral cortex continually intervenes, but the cerebellum does participate and it is capable of "learning." With practice, the movement can become preprogrammed and can eventually be executed more and more rapidly. This means that the feedback from the periphery becomes less essential for a successful execution. For learned movements, the cerebellum provides an internal substitute for the external world (for details, see Eccles, 1973a,b; Allen and Tsukahara, 1974). The cerebellum's way of functioning has been compared with the servomechanisms commonly used in modern technology (the automatic pilot, industrial control systems, antirobot weapons, and so on).

At various levels of the nervous system, there are "generators" that can initiate and coordinate movements. The reciprocal innervation is a basic feature of such movements and, as we have seen, it is established all the way down to the spinal level. A brief description of some reflexes will clarify this point. The stretch reflex has been discussed in connection with the muscle spindle presentation. If pain receptors in the skin are stimulated (by a needle prick in the sole of the foot, for example), the limb is withdrawn from the source of the pain even before any sensation of pain has been experienced. The response is executed by means of excitatory connections in the spinal cord between the afferent fibers whose nerve endings were stimulated and the motoneurons of the flexor muscles of the extremity. Several interneurons are included in the reflex arc. Some of the intercalated interneurons will have an inhibitory influence on the motoneurons of the extensors of the extremity. This is an example of an ipsilateral flexor reflex.

If the nocuous stimulation is strong enough, the extensor muscle of the opposite limb of an animal is contracted and its flexor muscles are reciprocally inhibited almost simultaneously with the development of flexion on the ipsilateral limb. This cross-extensor reaction (or contralateral extensor reflex) will further assist in removing the foot from the irritating needle. A stimulus of the hind limb may spread to the forelimbs via long nerve collaterals within the spinal cord. Examples of muscular activities that are normally executed and controlled by generators in the brain stem are breathing, swallowing, blinking, and coughing. At any moment one can voluntarily interfere with most such normally subconscious muscle actions. Motoneurons innervating the respiratory muscles can be activated by two parallel pathways, an "automatic" one serving the metabolic function of breathing, and a voluntary behavioral one originating in the cerebral cortex. All of these movements are based on central programs, many of them laid down in the spinal cord.

Grillner (1975) reports that cats with chronic spinal transections can generate the essential and basic features of the step cycle as well as coordinate the limbs and shift from one type of gait to another. This also holds true for the deafferented animal (with the dorsal roots cut). It means that locomotor patterns are generated in the spinal cord by central programs: central pattern generators. There are several structures in the brain stem from which locomotion can be initiated, and these patterns are exerted via pathways descending from the brain stem. There are interesting observations which indicate that a *continuous* electrical stimulation of an area called the mesencephalic locomotor region in decerebrated cats can elicit active walking movements, i.e., *rythmic* activity. By increasing the strength of the stimulation cats changed stride frequency and did change to trot, and eventually to gallop (Grillner, 1981). The afferent input is important when, for one reason or another, the locomotor movements are disturbed. Grillner and his group (see Grillner, 1981) noticed that a light touch on the dorsum of a spinal kitten's paw during the flexion phase in walking gave an additional flexion. However, the same stimulus during the extension phase produced no response in the flexor muscles, but enhanced the extensor activity. In other words, there was a reversal of the reflex. In one position, the afferent pathway to the flexor motoneurons was wide open; in a different position, it was shut off. Receptors influenced by the hip position, can apparently modify the response to a given stimulus. Anesthesia of the skin abolished these reflexes. Also, deafferented higher primates, like monkeys, can develop motor skill; they can walk and run, and the timing of the muscles can be normal so the afferent signals from a limb are not necessary for movement. The eyes can take over part of the information input to the CNS. However, in patients with nonfunctioning dorsal roots, one notices a retardation and prolongation of the execution of voluntary movements, with particularly the slower movements being affected. More rapid movements are least influenced. This can be explained by the basic importance of a central programming, especially of the fast movements. It should be emphasized that the engagement of specific muscle groups and the demand for force in the performance of a given movement in a joint (at a given speed) are dependent on whether gravity works for or against the movement, the heaviness of extra external loads, the degree of muscle fatigue, etc. Numerous sense organs of different kinds provide a feedback mechanism that can report from the periphery, and they are therefore essential for very efficient and skilled movements. Allen and Tsukahara (1974) point out that the cerebellum with its "computer" provides the dominant input to the pyramidal neurons, whereas the direct somatosensory and cortical association areas exert subsidiary influences on these neurons.

Summary Central pattern generators control stereotypic locomotor movements like walking and running. The neural networks are to a significant degree located within the spinal cord (Grillner, 1981). Actually each leg (e.g., in the cat) has an individualized central program, which is coordinated by interneurons with the pattern generators of the other three legs. Supraspinal regions can activate the relevant spinal programs as well as control and modify these programs. Similarly, a powerful signal traffic from peripheral receptors can control the central pattern generators, more directly on the spinal level or via loops passing higher levels of the CNS. The hip position, as well as the receptors reporting the load on a limb can control and modify a step cycle. The central program does not require an afferent feedback for its essential pattern or maintenance, but becomes functionally more efficient and adaptable to unexpected events when fed with afferent signals. Faster movements are less dependent on external cues than slower ones. (For a general discussion see Brooks, 1979.)

"It is already clear from a large amount of experimental work that diverse animals, both vertebrate and invertebrate, who locomote by such different means as swimming, flying, or walking, use the same general neural principles in locomotion. It appears that evolution may have found an optimal solution to accomplish locomotion: a central program as the basic motor framework, with afferent input providing reflex adjustment to the program to compensate for a changing environment. How could man pass up such an elegant solution?" (Carew, in Kandel and Schwartz, 1981, chap. 28). To continue the evolutionary aspect of locomotion: Vilensky and Gehlsen (1984) conclude that: ". . the volution of human bipedality did not require many changes in the neurological control of limb movements, as these movements relate to speed."

Nauta and Feirtag (1979) illustrate the same point with a modern analogy: In the spinal cord there are functional subunits of motoneurons and their guiding neuronal pools, each of them a local motor apparatus that corresponds to the various parts of the body. "Each local motor apparatus is, it seems, a kind of file room in which blueprints, each one representing a possible movement of a particular body part, are stored. The brain, with its descending fiber systems, reaches down and selects the appropriate blueprint."

The traditional point of view has been that the cerebral cortex reigns at the highest level in the hierarchical organization for the brain's motor function. However, Evarts (1973) emphasizes that the cerebral motor cortex is at a rather low level of the motor control system, close to the muscular apparatus itself. The cerebellum and basal ganglia are at a higher functional level in the neural chain of command that initiates and controls movement. The primary function of the cerebral motor cortex may not be volition, but rather, the refined control of motor activity. Naturally, evolution has created a specialization. The cerebellum is particularly involved in rapid ballistic movements, whereas the basal ganglia are preferentially active in slow movements. The motor cortex is engaged in both types of activities. In addition, the basal ganglia are important (1) to inhibit muscular tone and (2) to provide a "background activity" which is necessary for all movements. For example, if someone in a standing position performs a precision movement with the fingers, these movements are primarily guided by the motor cortex, but the leg, trunk, and arm muscles serve to stabilize the hand;

their engagement takes place subconsciously, probably largely through the basal ganglia. A malfunction of these ganglia causes postural disturbances, muscular tremor at rest, increased tonus, muscular rigidity, and difficulty in the initiation of movement. Such symptoms are typical in Parkinson's disease, which is a neurological disorder resulting from damage to the basal ganglia. With damage to the cerebellum, a muscular tremor is also one symptom, but it is most severe during voluntary movement. The patient's movements are clumsy, carried out in a more or less disorderly fashion, and slow. The person also suffers from poor equilibrium. (It should be recalled that the cerebral cortex in the left hemisphere is "connected" with the right part of the cerebellum and skeletal muscles and with peripheral sense organs also on the right side of the body, and vice versa.) Higher primates with the cerebral motor cortex destroyed can still perform many activities, but there is a loss of fine motor control, particularly of the fingers, and decreased voluntary movement. In most individuals the left cerebral hemisphere is dominant, in liaison with the conscious self, and analytic and sequential; the right hemisphere has no liaison with consciousness, and it is nonverbal, synthetic, and musical, to list a few characteristics. Apparently the two hemispheres are complementary (Eccles, 1973b). Thanks to the massive impulse traffic through the corpus callosum, the minor hemisphere achieves consciousness by its communication with the dominant hemisphere. (See Sperry, 1981.)

To summarize this section, one may quote Eccles (1973b):

Let us now try to visualize what would be happening in the cerebro-cerebellar circuits during some skilled action, for example, a golf stroke. In the first place we will assume a loop time of the cerebro-cerebellar circuit of a fiftieth of a second, so that a motor command to start the stroke will result in a "wise" comment from the cerebellum to modify the pyramidal tract (PT) discharge in accord with its learnt performance. The modified (PT) discharge is reported to the cerebellum and this in turn evokes a further corrective comment from the cerebellum. Thus there is this continuous on-going cerebellar modification of PT discharge. Furthermore, it has to be remembered that, before the initial PT discharge that started off the golf stroke, there was already the combination background discharge of the PT cells. Moreover, continuous background discharges would also be occurring from Purkinje cells and nuclear cells (in the cerebellum), and in fact for all the cells in the circuit. Thus, when the movement is about to start, the whole circuit is in dynamic operation. Every action that you make is already superimposed on background discharges in the various neuronal circuits. They are "ticking over" all the time, and the cell discharges at all stages of the loop are modulated up or down in the process of the continuous flow of information. I think this concept of dynamic loop operation gives the essence of the motor control by the cerebellum. It is very important to recognize that all pyramidal tract discharges are provisional and are unceasingly subject to revision by the feedback operation of this cerebellar loop. It should be mentioned that there is also a feedback loop from the periphery that is concerned in motor control, but the loop time is much longer, about a tenth of a second. Presumably the cerebellum carries an immense store of information coded in its specific neuronal connectivities so that, in response to any pattern of pyramidal tract input, computation by the integrational machinery of the cerebellum leads to an output to the cerebrum that appropriately corrects its PT discharge. These hypotheses of the manner of cerebellar action in the control of movement provide great challenges for future research.

Learning and Memory

When unfamiliar and complicated movements are performed, they are performed clumsily and with difficulty. With proper practice they become smooth and easy. However, we are far from understanding the reasons for this in physiological terms, how new connections are formed during learning, etc. The movements of the newborn are characterized by uncoordinated movements of many parts of the body, but gradually, coordinated reflexes develop (postural reflexes, tonic neck reflexes, writing reflexes, walking, and so on). The extrapyramidal and pyramidal motor centers and tracts develop, so that the movements become more complex. Nevertheless, when beginning school, the child's extrapyramidal centers dominate the movements, and specific practical tasks directed toward external objects constitute the activity in play, imitations, and other activities. The child usually fails in the performance of more complicated "artificial" movements. It is during puberty that the pyramidal system first attains full functional maturity and the child becomes capable of developing the fine coordinated movements necessary for writing and other precise actions, based on the integration of nervous activity from various levels of the CNS and the impact from all peripheral receptors. The changing body dimensions during adolescence necessitate a continuous modification of the intepretation of impulses exchanged between muscles and CNS to secure a correct innervation pattern for given tasks. Apparently there are definite anatomic and physiological limitations to the complex movements that can be performed during early adolescence.

The CNS, like other tissues, is characterized by a plasticity in its structural, biochemical, and functional properties (Eccles, 1973b; Kandel and Schwartz, 1981; Tsukahara, 1981; Squire, 1982; Thompson et al., 1983). The gross neuronal connectivities are established in their final form before birth. After the initial formation, no change occurs in the neuronal pathways or growth in the brain of mammals. A genetic instruction secures the development, and there is some kind of mechanical guidance and, at the end, a chemical sensing and recognition (Cowan, 1979; Stent, 1981; Lewi-Montalcini, 1982). The initial formation of the connection is primarily under genetic and developmental control; there appears to be no contribution from the effects of learning. However, in one way or another learning can make one combination of neural connections more efficient, so that it can be quickly repeated with remarkable precision, and it can be retrieved with similar precision, even if not practiced for years. One may see someone's face merely for a few seconds and yet one may never forget it. Another face may remain anonymous even after having been seen numerous times. When checking in at a hotel, the room number is well-stored somewhere in the brain for quick recollection at the hotel reception, but there is no problem forgetting it and remembering a new room number when moving to another hotel. In adults, as well as children, a structural plasticity is evident at the microlevel, that is, some synapses may develop while others regress. Frequent impulse traffic through synapses seems primarily to result in a maintenance of the dendritic spine synapses, but it can also modify the synapses by a branching of the spines to form secondary synapses (fiber sprouting). There may be a hypertrophy of the synaptic knobs. In other words, the process of learning seems to stimulate a protein synthesis of activated synapses which gives a selective synaptic hypertrophy. In one way or another, repetition of a movement can be coded and memory engrams can be stored and played back with great accuracy and precision. Available results suggest that the cerebellum is directly involved in the learning of motor actions. There are many unanswered questions concerning the role of several neuroactive substances, including peptides, amino acids, norepinephrine, and hormones with potential of influencing a variety of processes involved in learning and memory (Thompson et al., 1983). Generally speaking, recently acquired memories are more readily disrupted than older memories. When a strategy is made for a movement of the body by cooperation between many regions within the CNS, perhaps with the cerebellum as a controller, the result is a short-term memory of the motor output. From the periphery, reports arrive telling whether or not the movement was satisfactory. If the report is favorable, the pattern is suitable for storage in a more permanent long-term memory, (Fig. 3-20). This is a model with many similarities to a "black box." It is generally recognized that the combination of ambition with a will to learn definitely enhances the learning process!

In view of the very sharply defined projection of impulses within and between different areas of the brain, it is difficult to understand how a "general movement pattern" may function. The organization does not appear to favor a "transfer" effect, i.e., the learning by practice of a certain movement pattern does not in itself enhance the performance of another movement pattern, not even one that is relatively similar. However, the technique of learning new tasks can be improved. One can learn and memorize specific activities which can then be utilized and woven together in different combinations. The pianist, having practiced many hours at the keyboard, has the potential to learn new pieces of music, quickly. Whether it should be called a transfer effect is a matter of definition, however, the pianist can learn to play a piece of music slowly and softly, but once he or she has learned it, the pianist can play it fast and loud just as well. If someone is asked to write his or her name in letters 1 m high, his or her normal signature can easily be identified in the greatly exaggerated letters. Apparently, a central program for the shape and pattern of movements is developed,

Fig. 3-20
Model of memory storage system. (*From Kandel and Schwartz, 1981, Chap. 47.*)

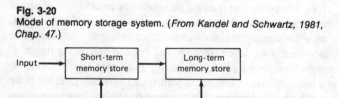

but there is no rigid and formal program, with a given number of motor units activated at a given frequency of contractions. Many of the movements learned must be conducted by central programs because they can be executed without peripheral feedback, whether it be visual, auditory or factual. For instance, the pianist can play and write his or her signature with closed eyes.

Habituation is in a way the most ubiquitous of all forms of learning. "Through habituation, animals and humans learn to ignore stimuli that have lost novelty or meaning, thereby freeing them to attend to stimuli that are rewarding or significant for survival" (Kandel and Schwartz, 1981). Kandel describes some interesting experiments that were performed on the marine snail *Apalysia california*. A certain stimulus initiates a brisk gill withdrawal. If this stimulus is repeated, say 10 times, without reward or noxiousness, the synaptic potentials generated by sensory neurons, in the motoneurons, become gradually reduced, and the behavior becomes less vigorous, eventually stopping altogether. A memory for habituation remains and can be objectively recorded as a persistent reduction in the effectiveness of the synaptic connections between the sensory neurons and the motoneurons. This effect can last for hours and repeated "training sessions" can produce long-term habituation that lasts for weeks. It was noticed that the repeated stimuli reduced the Ca^{2+} influx at the presynaptic terminals by a prolonged shutting off of the Ca^{2+} channels. As a consequence, there was a diminished release of the transmitter substance which will inevitably reduce the postsynaptic excitatory potential (PSEP). Somehow, the habituation was due to a plastic change in the configuration of the presynaptic membrane channel proteins. Such an inactivation of synaptic connections between sensory and motor neurons can be quickly reactivated, which suggests that the changes behind the habituation were more of a biochemical modification than an extensive morphologic rearrangement.

There are no similar observations reported that indicate a sensitization or facilitation of synaptic connections in learning. Within the CNS, however, if there is a potential for a prolonged reduction of Ca^{2+} influx into the axon terminals triggered by a given stimulus, the effect could in some regions conceivably be the reverse, i.e., an enhanced Ca^{2+} diffusion.

Learned movements can be easily performed subconsciously, but *the motor cortex is still engaged in the control of the movement regardless of whether the movement is innate or learned*. The same nerve cells seem to control the pattern of muscle contraction regardless of the circumstances or context of the movement (Evarts, 1973).

Little by little, a "sense" is developed, based on both endowed and acquired capabilities. Granit (1972) reports old experiments showing that a person, when lifting two equally heavy spheres, one 4 cm and the other 10 cm in diameter, feels the larger sphere to be lighter. Subjects who lifted equally heavy objects (112 g) with volumes ranging from 10 ml to 2,000 ml felt that the largest ones were also lighter; they were then asked to add weights to the apparently lighter object (2,000 ml in volume) until it felt as heavy as the 10 ml one. It turned out that the amount required averaged about 112 g!

Granit and Burke (1973) quoted Hagbarth's observation: "If, in a completely darkened room, a normal subject swings his arm in an arc around the elbow joint and the arm is stroboscopically illuminated only when it is at 90°, the subject has a strong sensation that he is not really moving his arm at all. On closing his eyes, this illusion immediately disappears." No doubt the visual input has a relatively dominant influence for the determination of position. It can also as mentioned partially compensate for defects in the afferent signal input.

To *summarize* in a slightly different way: we learn by experience; we memorize by trial and error. We acquire behaviors by supplementing innate, stereotyped behaviors with the goal of providing pleasure and satisfaction, but also of vital importance for survival. In our sophisticated, technological society, we can easily cheat our brain by exposing it to completely new inputs. There is now a wealth of evidence to suggest that one factor behind learning and memory is a synaptic plasticity in the CNS, including a sprouting of nerve endings. Thereby the synaptic connections can be modified.

PROPERTIES OF THE MUSCLE AND MUSCULAR CONTRACTION

In this section we shall consider aspects of the CNS and the muscular system as *one* cooperative system. A logical place to start is with the motor unit.

The Motor Unit

The motor unit consists of the alpha motoneuron and the muscle fibers which it innervates. As discussed earlier, the number of muscle fibers in a unit varies from about 5 to 2,000. A single muscle fiber rarely has a polyneuronal innervation. When a nerve impulse reaches the motor endplate, ACh is liberated and the membrane of the muscle fibers is locally depolarized. This evokes an action potential propagating along the muscle fiber at a speed of about 5 m $\cdot$ s^{-1}. The whole length of each muscle fiber and all the fibers of the motor unit are therefore brought into action almost instantly, even if the unit is acted upon by a nerve fiber at only one point in its length. The main feature of the muscle action potential is very similar to the action potential of the nerve. It is generated and propagated by basically the same mechanisms. A single, adequately strong, stimulus of the motor nerve gives rise to a twitch of the innervated muscle. After a short period the force developed by the muscle increases and thereafter it decreases again. The muscle action potential is completed during the early phase of the contraction (Fig. 3-21a). If the motor nerve of the muscle is stimulated repeatedly and the second impulse reaches the muscle before it has relaxed after the first stimulus, it contracts again. Since the second twitch starts from a higher force level, the force resulting from the two stimuli will be considerably greater than that from a single stimulus of the same strength (summation). At high rates of stimulation, the muscle does not relax before the next contraction, and the muscle fibers are in full tetanus; that is, there is a complete mechanical fusion of the contractions. The force developed by a muscle in tetanus may be 4 to 5 times greater than that exerted during a single twitch (see Taylor and Stephens, 1976). At lower rates of stimulation, the mechanical fusion is incomplete and the developed tension is submaximal (Fig. 3-21b). Even in a tetanus, the muscle action potential develops in response to each

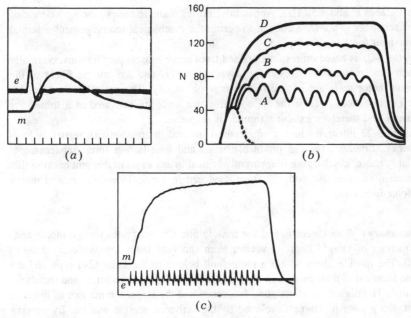

Fig. 3-21
Action potentials (e) and mechanical changes (m) in skeletal muscle in response to stimulation of its motor nerve (from mammalian nerve-muscle preparation). (a) A single stimulus gives rise to a single twitch. The diphasic action potential is completed in the early part of the contraction phase. Each mark on the abscissa denotes 0.05 s.
(b) Time (abscissa) is not marked; the ordinate gives tension in Newton (N). The lower curve, partly dotted, shows one response of the muscle to a single maximal stimulation to its nerve. Curves A to D show the responses to repetitive, maximal stimuli; A at 19, B at 24, C at 35, and D at 115 Hz. Note that the higher the frequency of stimulation, the greater is the tension which is developed, and it is maintained more steadily. From a state of partial or incomplete tetanus (curves A to C), the muscle goes into full tetanus (curve D).
(c) At a high frequency of stimulation (in this case 67 Hz), the muscle goes into tetanus, but the action potentials appear at the same frequency as the stimulation rate. [(a) and (c) *after Sherrington et al., 1932*, (b) *after Cooper and Eccles, 1930*.]

stimulus, revealing the rate at which the nerve is being stimulated (Fig. 3-21c). It was mentioned in Chap. 2 (and shown in Table 2-1) that the skeletal muscle is composed of two main types of muscle fibers with different mechanical properties: type 1 or *slow-twitch fiber*, and type II or *fast-twitch fiber*. The time it takes to reach peak isometric tension in human skeletal muscle is reported to be 80 to 100 ms for type 1 fibers and approximately 40 ms for the type II fibers (see Saltin and Gollnick, 1983). The speed of contraction is proportional to the activity of the myofibrillar (myosin) ATPase activity of the fibers. The fast-twitch fibers are innervated by larger motoneurons than the slow fibers. *Each motor unit is composed of only one kind of fiber.* Because of a relatively long afterhyperpolarization of the slow fibers, they will fuse their contraction into tetanus at a lower frequency (around 20 Hz) than do the fast fibers (tetanus at about 50 Hz). Apparently, during a single twitch, the active state of the muscle fiber is not long enough to permit the mechanical rearrangement essential for maximal force to develop.

Figure 3-21 is based on results obtained from nerve-muscle preparations. Normally the activated motoneurons discharge asynchronously, and the muscle fibers of the different motor units are in a different stage of activity. The net effect is a smooth muscle contraction, even if the individual motor units are activated at a subtetanus frequency and therefore exhibit a tremulous response.

Figure 3-22 illustrates how a single short interval in an otherwise relatively low-frequency stimulus train can produce marked and long-lasting force enhancement. Similar "double" discharging is apparently critical in any rapid movement because this mechanism can increase both the force produced in a muscle and the rate at which that force increases.

Summary A motoneuron and the muscle fibers it innervates form a motor unit. The number of muscle fibers in such a team can vary from approximately 5 up to 2,000. The muscle fibers within a motor unit belong to the same fiber type and are almost identical in their molecular structure, contractile characteristics, and metabolic potential. The higher the stimulation frequency and the larger the number of fibers in a unit, the greater the force developed by the activated muscle will be. By varying the number of active motor units the force can be effectively modified. The magnitude of this particular aspect of neuromuscular function is illustrated by the fact that in 10 mm^3 of skeletal muscle there is approximately 500,000 km of myofibrils that have to be coordinated.

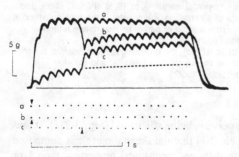

Figure 3-22
The "catch property" in a single muscle unit. Mechanical responses of a type I unit in cat MG (superimposed records) to three stimulation sequences, each with mean frequency of about 12.8 Hz (basic interstimulus interval of 78 ms). Sequence a included one extra stimulus following the first with an interval of 5 ms (arrow), which resulted in marked tension enhancement over what would have resulted from a constant frequency train without the extra impulse (dashed line). Sequence b also had the same short interval to start (arrow) plus a later, longer interval which caused tension to drop to a different level, at which it was "caught" by resumption of the 12.8 Hz stimulus rate. Sequence c included an interstimulus interval of about 20 ms (arrow), which resulted in less tension enhancement than the shorter intervals of sequences a and b. (*From Burke and Edgerton, 1975.*)

"Regulation of Force"

Muscle strength is a very complicated function, and the number of muscle fibers available is only part of the story. It is not surprising that even in well-standardized measurements of muscle strength, the standard deviation of the results obtained in repeated tests on the same subjects is of the order of ±10 percent or even higher (Table 3-1) (Tornvall, 1963; Simonson, 1971, p. 244).

The main objection against the method for the testing of muscle strength with the aid of the muscle dynamometer is its dependence upon the motivation of the subject to exert maximal effort. In fact, the practical application of this method may ultimately depend upon objective criteria for the evaluation of the state of motivation of the subject to be tested. In an effort to search for such criteria, use was made of the observation made by athletic performers that there is a relationship between the rise in heart rate and the level of endeavor or exertion immediately before the start of an athletic competition. Heart rates were recorded with the aid of a recording electrocardiograph for a 10-s period immediately before and again immediately after the muscle dynamometer test, in five normal subjects, with a total of 137 observations, and correlated with the measured muscle strength. A statistically significant correlation ($r = 0.738$, $P < 0.001$) was observed between the subject's heart rate before the test and his measured muscle strength; and also between his heart rate immediately after the test and his muscle strength ($r = 0.798, P < 0.001$). This finding suggests the possibility of using the heart rate as an indication of motivation in the selection of motivated subjects for experiments involving testing of muscle strength.

From studies on individuals of different body size, age, and sex, Asmussen et al. (1965) and Lambert (1965) concluded that the correlation r between symmetrical muscle groups (right and left) is quite high ($r = 0.8$). Between flexors and extensors of the same extremity, it is fairly high, but between muscles from different parts of the body, the correlation is rather low ($r =$ about 0.4 or less). Therefore, they con-

TABLE 3-1
TESTING OF MUSCLE STRENGTH IN FOUR
SUBJECTS AT HALF-HOURLY INTERVALS
FOR ONE DAY*

Subject	Right Arm		Left Arm	
F.A.	33.1 ±	1.0	33.8 ±	2.0
	(13)	3.6	(13)	6.1
G.B.	31.1 ±	1.1	28.9 ±	1.2
	(13)	3.9	(13)	4.3
E.D.	37.5 ±	1.5	38.6 ±	1.1
	(10)	4.6	(10)	3.5
J.S.	30.6 ±	1.0	29.9 ±	0.8
	(13)	3.5	(13)	2.9

*Figures (kg) denote mean, standard error of the
mean, number of observations, and standard deviation.
One kg is roughly 10 N.

clude, the general muscle strength should not be evaluated from measurements in one single muscle group, such as the finger flexors in a handgrip, but from the application of a battery of well-standardized, selected tests of muscular strength. The correlation between maximal dynamic and isometric strength recorded in these studies was 0.8. Thus, an extrapolation from one type of testing to another has its limitations, especially if the subject is well-trained in a particular type of exercise.

With this in mind, we may proceed to discuss the "regulation of strength." In a completely relaxed muscle, no motor units are active and the muscle is electrically silent. As a result of the elasticity of the myofibrils and fibrous tissues, there is, however, a certain tension (tonus) even in the relaxed muscle. Data have been accumulated illustrating the recruitment pattern of different motor units in human muscles activated in voluntary contraction (studied with EMG and electroneurogram). Hannerz (1974) and Grimby and Hannerz (1976) have reported experiments in which they mapped out the discharge pattern and recruitment order of single motor units in voluntary contraction of human anterior tibial muscle. Low-threshold units were recruited in mild, sustained contractions starting at a frequency of 5 to 10 Hz. During such activity, when the exerted tension reached a certain value, a particular motor unit started to contract and continued its activity until the tension again dropped below the threshold level. The same unit usually started its activity at about the same tension level and at a regular rate. With increasing contraction strength, the units increased their discharge rate up to a relatively low maximum (around 25 Hz). New motor units were recruited but they started at a higher frequency (in some cases exceeding 30 Hz), and they attained a higher maximum (up to 65 Hz). It was noted that new motor units were recruited at all levels of tension up to maximal effort. Units with high threshold, coming in later, tended to exhibit a discontinuous discharge, particularly when the force exceeded 80 percent of the maximum. As an example, a motor unit recruited at a sustained contraction with 15 percent of maximal isometric strength had a minimal frequency of about 10 Hz and at maximal voluntary tension it reached 30 Hz. A unit starting at 60 percent load at about 20 Hz increased its rate up to 45 Hz, and the "90 percent unit" began at 35 Hz and finished at about 60 Hz. In twitch contractions and rapid movements, the recruitment order of the motor units differed considerably from the pattern noticed in sustained contractions, and high-threshold units could be activated first with rapid firing in bursts. Slow-twitch fibers are activated by a motoneuron activity covering a frequency range from about 5 up to 20 to 30 Hz. The fast-twitch fibers are driven with a frequency range from approximately 30 up to 60–65 Hz.

During steady contractions, the firing range extends from approximately 8 to 30 Hz. It is only possible to produce high-frequency discharges (80 to 120 Hz) for very short periods of time (some 100 ms) and then mainly for the generation of ballistic contractions. Firing rates exceeding 30 Hz are used exclusively for variations in the speed of the muscle contraction. Due to the inherent time course of the muscle twitch contraction, the fastest possible alternating human finger or hand movements cannot exceed rates of about 7 to 8 Hz (see Freund, 1983).

One can assume that the low-threshold units are slow-twitch fibers innervated by small motoneurons, and that the high-threshold motor units are composed of fast-twitch fibers with nerve axons from large α motoneurons. From experiments, in which they manipulated with the proprioceptive afferent input, the authors conclude that the low frequency motor units have more support by the afferent feedback than the high-threshold units. The recruitment order can be modified or reversed by various procedures changing this feedback. Different muscles may exhibit different orders of motoneuron recruitment, but the pattern just described seems to be quite basic.

By recording the discharge properties of single motor units in the short extensors of the toes, Grimby (1984) observed that the single motor units were recruited in the same order during slowly increasing speed of locomotion as they were during slowly increasing voluntary tension. Low-threshold units fired five to ten times at 80 to 40 ms intervals in each step cycle during walking at normal speed. The number of units recruited, and their firing rates, increased with the speed of locomotion. At a given speed the variations in the pattern from one step cycle to the next were small. In rapid walking or running, high-threshold motor units were also recruited and they fired only in short high-frequency bursts. Some of them only fired during rapid acceleration, rapid changes of direction, or in other rapid corrective movements.

Figure 3-23 illustrates an important principle in the orderly manner in which the motor units are normally recruited. It is the size principle (see Winter, 1979; Henneman, 1980; Freund, 1983). The low-threshold motoneurons, activated when demand for muscular force is low, usually activate motor units with relatively few muscle fibers per unit. This provides a means for achieving precisely controlled and finely graded movements by the selective mobilization of varying numbers of small motor units. When additional force is demanded, high-threshold motor units are gradually recruited. These motor units contain many muscle fibers. The largest motor units in the human calf muscle can develop 200 times more tension than the smallest ones. Therefore,

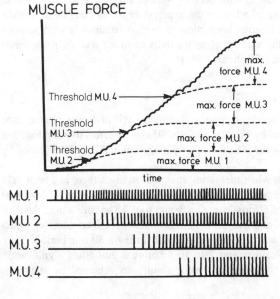

Figure 3-23
Illustration of size principle of recruitment of motor units. Smaller motor units are recruited first; successively larger units begin firing at increasing tension levels. In all cases the newly recruited unit fires at a base frequency, then increases to a maximum. M.U. = motor unit, (*Modified from Winter, 1979.*)

when such units are brought into activity, progressively greater increments of force are added. As illustrated in Fig. 3-21, increasing the firing rate of the motoneurons will also produce a greater muscular force. Then there is the strategy of "double" discharging, see Fig. 3-22, when there is a demand for fast and forceful muscle contractions.

Summary Slowly contracting small motor units are recruited first in activities with low demands on force (both in reflex and voluntary contractions). Their frequency is relatively low. With increasing demand for force, the "old" motor units increase their discharge rate and in addition new motor units are recruited. The fast-contracting motor units gradually start their activity and then at a relatively high frequency. The *gradation of a muscle contraction is brought about by varying the number of active motor units (recruitment) and their frequency of excitation (rate coding).* The recruitment of slow-twitch fibers in sustained contractions of low tension is a "wise" arrangement, for the fibers have the potential to work aerobically and are fatigue resistant in their activity at a relatively low frequency. By recruiting new motor units with many muscle fibers activated when larger tension is demanded, the metabolic load on the individual muscle fiber can be kept low. Stein (1974) points out that rate coding would be less useful than recruitment for maintaining a steady force, because there is only a limited range of rates at which motor units can produce a reasonable steady force without fatigue, i.e., at low force levels. Rate coding becomes more prominent when higher tension is demanded, however. (See also Milner-Brown et al., 1974, who give an example of subjects who were able to increase the force of their hand muscles from 15 to a maximum of 20 N without recruiting new motor units.)

The relatively stereotyped recruitment order of given motor units in specific tasks must be a consequence of central pattern generators at the level of the spinal cord and, certainly, at higher levels of the central nervous system (innate or learned). These centers are capable of selecting in advance the appropriate recruitment order for the task intended. There is a continuous interaction of central commands with sensory feedback which can modify the number of motor units recruited and their frequency of contraction, (see Garnett and Stephens, 1981).

Muscular Strength

There are many factors to be considered when evaluating what is of decisive importance for maximal strength developed by skeletal muscles. This is an effort to bring together in a condensed manner some of these factors, partly discussed in this and the preceding chapter.

It has been mentioned that, when measuring maximal strength there is a relatively large variation when measurements are repeated in the same individual. Maximal performance is the synthesis of efforts from "body and soul." The following, classical study is quite instructive: Ikai and Steinhaus (1961) conducted experiments in which the subjects made a maximal arm flexion every minute during a 30-min period. They found that the firing of a 22-caliber gun 2 to 10 s before a pull could significantly modify the exerted maximal strength. The same result was achieved by shouting, giving various drugs, or exposing the subject to hypnosis. Thus, the performance was distinctly higher after the gunshot than before. Shouting, hypnosis, epinephrine, and amphetamine also tend to improve performance as compared with the controls. The positive effect on strength was noticeable on untrained subjects, but slight or absent on well-trained athletes, such as weight lifters. Ikai and Steinhaus cite Pavlov's statement that "any unusual sensory experience or excitement may inhibit inhibitions" (with an up-dated terminology it is a disinhibition). They emphasize that their findings "support the thesis that in every voluntarily executed, all-out maximal effort, psychological rather than physiological factors determine the limits of performance. Because such psychological factors are readily modified, the implications of this position gravely challenge all estimates of fitness and training effects based on testing programs that involve measures of all-out or maximal performance."

There are many anecdotal reports indicating that individuals in a stress situation can perform better than otherwise; they become exceptionally powerful. In controlled experiments, it is established that epinephrine and norepinephrine increase both excitability and contractility above normal values, but the mechanism for such effects is not clear. During stress the production of these hormones is increased, and we have pointed out that norepinephrine is a transmitter substance important in the communication between many neurons involved in physical efforts.

Muscle strength is definitely a very complicated function. It depends on the number of motor units activated and their frequency of contraction. With increasing load, recruitment of more motor units with more muscle fibers per motor unit, is most important until the load becomes heavy; then an increase of the firing rate becomes the most prominent mechanism for the development of more force. The maximal tension can be produced when the muscle is lengthened, and it declines as the muscle shortens. The maximal strength in rapid eccentric movements exceeds the maximum in isometric contractions; it is less in concentric movements, and particularly in rapid concentric movements.

A given force can be produced at a lower energy yield in eccentric exercise than in concentric activities. When stretching an activated muscle, part of the mechanical energy can be stored because of the elastic properties of the muscle. In the immediately following contraction, this energy can be released and can support the contractile energy. In our opinion, there is overwhelming evidence to show that a voluntary maximal muscle effort, in most situations and with unconditioned subjects, does not engage all the motor units of the active muscle at tetanus frequency. Effective inhibitions of varying degree exist on some motoneurons, depending on supraspinal and proprioceptor activity. In a specific situation, say an emergency, and perhaps as an effect of training, inhibition decreases (or facilitation increases), and the muscle mass can become more completely utilized in the contraction. The reader is reminded of the rather massive inhibitory interaction between interneuronal paths in the spinal cord and the many inhibitory or excitatory impulses that may descend from higher levels of the CNS. Many athletes shout in the critical stage of an effort, and in the light of the experiments by Ikai and Steinhaus (1961), this action may increase the motoneuron activity. Training in technique may empirically teach the athlete how to take advantage of facilitating stimuli and reflex mechanisms and how to avoid inhibitions which reduce the performance. The bouncing maneuver is typical in many activities. A proper understanding of the underlying mechanism is very important and useful, not only for sport activities, but also in physical therapy.

The thesis that central factors are of decisive significance for the development of strength is also supported by the observation that the strength can increase without a proportional hypertrophy of the muscles. In one recent study, McDonagh et al. (1983), report that a 5-week training producing a 20 percent increase in the force of a maximal voluntary isometric contraction, was not combined with any modification of the electrically evoked twitch and tetanus maximal force. They concluded that the improvement was based on factors other than the force-generating potential of the muscle fibers themselves. There are other observations indicating that during the first weeks of strength training, the improvement is not associated with an increase in the cross-sectional area of the muscle groups involved. Then there is often a gradual increase in both strength and the muscle cross-section area, particularly of the type II fibers. These events will be further discussed in Chap. 10.

On the basis of the available data on the maximal force that can be exerted in standardized tests on, for instance, the hand grip strength or a knee extension, one may be tempted to see whether or not the maximal force produced with both left and right limbs involved simultaneously is simply the sum of the individually measured forces (i.e., left and right limb performance). The literature provides conflicting data on this point. Some studies indicate an inhibition on the left side of the body's performance when the right side is simultaneously involved and vice-versa (e.g., Ohtsuki, 1981; Coyle et al., 1981, in untrained subjects). Others have noticed that the two limb strength exceeded the sum of the right and left forces measured separately. One side of the body did enhance the performance on the other side (see Wawrzinoszek and Kramer, 1984). It would appear logical that when one arm is not enough to accomplish a task, it should not be a disadvantage for that arm when the other arm is brought into use to assist it. However, there may be other implications to be considered: with two arms (or legs) pushing at maximum simultaneously, the strain on the back, for instance, might be so great that an inhibition might be warranted.

As will be discussed in Chap. 10, there is also specificity in the effect of strength training; the cause of this specificity is more likely related to events in the central nervous system rather than in the muscles (see Coyle et al., 1981).

When striving for muscle strength for a particular activity, the best training is that activity. The gain in strength when one is engaged in "unfamiliar" procedures is comparatively modest even when they activate the muscles which are being trained.

The explanation of these results may be that a gain in strength after a training program is due not only to changes in the muscle tissue but also to a modification of the impulse traffic reaching the motoneurons. In a different procedure, receptors, nerves, and synapses which are not "trained" are engaged. Furthermore, a slight change in a movement may reduce the load on one muscle and increase it on another.

Basmajian (1979), in his studies of the elbow flexors (the biceps brachii, the brachialis, and the brachioradialis), found that these muscles show a wide range of individual patterns of activity during flexion and extension of the forearm. These muscles also differ in their flexor activity in the three positions of the forearm: prone,

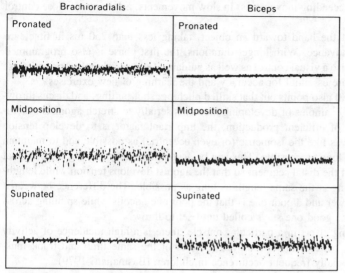

Figure 3-24
Electromyograms recorded from brachioradialis and biceps brachii during maximal isometric contraction (90° flexion in the elbow) in three different positions of the forearm. Note the high electrical activity in brachioradialis and the "silence" in biceps when the forearm is pronated. In the supinated position, the biceps is very active but the brachioradialis contributes much less than in the pronated position. A training of the flexor muscles of the arm with the forearm pronated cannot be so effective if the aim is to improve the ability to perform with the arm supinated, since this position partly engages different muscles. (*By courtesy of S. Carlsöö.*)

semiprone, and supine (Fig. 3-24). The speed of a movement also influences the activity pattern. Therefore, a training of the elbow flexors with the forearm in a prone position may be quite ineffective for a performance done with a supine forearm.

Coordination

In fast (ballistic) movements, at least a spurt of activity in the agonist produces momentum and kinetic energy in the segment, and then it relaxes as the limb proceeds by its own momentum. By reciprocal inhibition the antagonist relaxes completely except perhaps at the end of a movement or when the movement is stopped by the limits of the joint or an external force. Also, in slow movements in some activities, discrete bursts of neural activity are observed in agonists and antagonists to produce muscular impulses, which alternately act independently to accelerate and decelerate the segment (Hubbard, 1960). An integration of central programs and the feedback loops from the proprioceptors are probably responsible for a periodicity and modulation of the motoneural activity. In fast movements, proprioceptor and visual stimuli are probably relayed too late to correct a misdirected movement but in time to make adjustments in succeeding movements. In slow movements, a continuous close control is possible.

Movements of the hand toward an object, taking less than 250 ms in time, are programmed in advance. With longer durations, the first phase is also programmed but then there is also a visual control as well as guidance from peripheral proprioceptors. Such feedbacks are essential for accuracy and the learning of new exercises.

Hubbard (1960) also points out that skilled pitching, shot-putting, and discus throwing are excellent examples of developing moments serially to stretch agonists successively. In terms of efficient production, the important factor is to develop tension under conditions as like the isometric (or even eccentric) as possible and to maintain this condition as long as possible. This can be accomplished by moving the proximal segment ahead of the distal segment so that the agonist develops tension while lengthening or remaining at the same length as long as possible. The difference between a good discus thrower and a poor one is that the poor one uncoils while spinning across the circle, and the good one stays coiled until set to throw.

During a strong effort in a particular muscle, there is a high incidence of activity (in a predictable pattern) in far-removed muscles of the same limb and trunk musculature, an especially frequent occurrence in children (Basmajian, 1979).

Degeneration and Regeneration of Nerves

In adults, neurons in the CNS have lost their mitotic cycle. Therefore, they are no longer capable of cell division. After damage of an axon, the loss of the normal metabolic connection with the cell body causes a degeneration of the axis cylinder and myelin sheath distal to the site of injury. But there are also retrograde changes, which are often severe and may include presynaptic nerves. Glial cells and Schwann cells play an important role in healing. A proliferation of Glial cells can give rise to scars around the zone of the trauma which can, unfortunately, effectively block restoration of synaptic connections within the brain and spinal cord. The conditions for a regeneration of peripheral nerves is well developed. It is facilitated by the presence of uninterrupted endoneural tubes. The regenerating axons grow into the peripheral endoneural tubes at a rate of several millimeters per day. How the regenerating nerve finds its way is not exactly known. It is believed, however, that chemical factors released from the degenerated nerve fibers stimulate and guide the collateral growth of the axons (see Bennett, 1983).

After a partial denervation of a peripheral nerve the remaining axons will, after a few days, begin sprouting at the terminals and the last node of Ranvier (collateral sprouting). This sprouting can actually be prevented by direct stimulation of the muscle, which seems to inhibit the release of a diffusable growth factor stimulating the sprouting. The sprouts can reach nearby synaptic sites. Actually, there is a competitive reinnervation, and as in the developing embryo, the muscle fiber receives a polysynaptic innervation. During this phase, an intact motoneuron may increase up to five-fold. After the return of the cut axon to the synaptic sites, there is an elimination of all synapses to a muscle fiber but one, and the size of the sprouted motoneuron declines.

As mentioned, the synaptic sites on a muscle fiber cannot distinguish between axon terminals and sprouts that are the same family (e.g., fast-twitch type), but belong to a different motoneuron. Thus there are significant chances for a cross innervation under normal conditions as well (see Bennett, 1983; Brown, 1984).

Many of the sprouting central axons may get lost before reaching the peripheral sheath; other axons will establish faulty connections. A touch fiber originally supplying one receptor may connect with a touch receptor in a different region or may even connect with a temperature receptor. The effect will be reduced sensitivity, false localization, or complete misinterpretation of a stimulus of the receptor, such as touch giving a sensation of heat. Prolonged practice may be necessary for a return of the finer and discriminative sensibility, as well as appropriate neuromuscular control.

Motoneurons that fail to form appropriate synapses will ultimately die. Apparently, a factor within the muscle is necessary for the survival of the motoneuron. *With age, there is a loss in muscle mass* mainly due to a loss of muscle fibers. It is an open question as to whether the reason for this consequence of ageing is a primary death of motoneurons or a primary degeneration and gradual disappearance of muscle fibers (see Grimby and Saltin, 1983). It should be noted that there are normally no signs of degenerated fibers in the ageing muscle and the fiber composition is relatively constant. On the other hand, the muscle fibers per motor unit have been reported to increase, which may be interpreted as a reduction in motoneurons, partly compensated for by peripheral sprouting and reinnervation of the muscle fibers that lost their original motoneurons. To what extent changes in habitual physical activity can change this pattern is unknown (see also Chap. 10).

Summary   A functional regeneration of damaged nerves within the CNS is very limited, at least in the adult. (In young children neural function can be restored much better, even after severe damage, Kandel and Schwartz, 1981). The prognosis for a regeneration of peripheral nerves is good in the presence of uninterrupted endoneural tubes. Denervated skeletal muscles are reinnervated by nerve terminal sprouting and collateral sprouting from neighboring axons. Some trophic factor released from the denervated, inactive muscle fiber stimulates to a competitive multisynaptic formation. As time passes all but one of these synapses will normally disappear. Only those motoneurons that make successful contacts with their target muscle fibers will survive. However, new unfamiliar connections are quite common resulting in the reinnervation of muscle fibers and different receptors. Plasticity within the CNS can often restore normal function, but the potential is limited.

Forssberg and Svartengren (1983) studied the locomotor output of the gastrocnemius muscle in cats after a transposition giving the muscle a dorsiflexing action around the ankle. The old extensor pattern was retained in all six cats throughout the study period, more than one year after surgery. It was concluded that neither the spinal locomotor network controlling the gastrocnemius muscle nor the supraspinal circuits influencing the network exhibit a high degree of plasticity in response to locating the muscle in a position antagonistic to the original. They point out that in humans most transposed muscles can be voluntarily controlled in their new positions, while only some muscles seem to convert their activity pattern during locomotion.

Posture

Life began in water. In such an environment, it is much less of a problem to maintain a given posture than it is for an organism living in air. The density of animal tissues is close to that of water but quite different from that of air. In air, appropriate muscles have to be activated at sufficient force, and the body's center of gravity has to be kept within the area of support. In a two-legged posture, most of the corrective movements take place in the ankle joints, to a lesser degree in the knees and the hips. For control of posture (as well as in walking and many other exercises) three sets of information are essential: information from (1) the proprioceptive system (receptors in joints, muscles, tendons, skin); (2) the visual system; and (3) the vestibular system. The proprioceptive afferent inflow from the ankle is of special importance as well as nerve impulses from the sole of the foot.

The upright position is maintained by muscular activity against the force of gravity. The myotatic stretch reflex is one important factor in maintaining an adequate posture. The muscles antagonizing the pull of gravity are stretched, thereby the muscle spindles located in the muscles are also stretched. Afferent impulses are evoked, and the muscles contract so that the pull of gravity is counterbalanced. Since the intrafusal muscle fibers of the muscle spindle can be activated from higher centers, via the gamma fibers, its receptor may be more or less prone to respond to a stretch. A feeling of happiness, alertness, or attention may increase the gamma activity, whereas unhappiness, drowsiness, or lack of attention may reduce the activity. In this way, the very noticeable relationship between an individual's mood and posture may be explained.

A descending control of spinal mechanisms is, however, very important in the maintenance of posture, as it is, in the control of locomotion. The stretch reflex is a mechanism that helps an animal maintain a set position in the face of changing external forces such as gravity. However, there are central pattern generators within the CNS, driving a tonic activity in extensor muscles during standing.

In many joints it is the ligament, not the muscle, that normally maintains the integrity of the joint. For example, muscles that are able to support the arches in the foot are generally inactive when standing at rest. When sitting upright with the arms hanging in the relaxed neutral position with heavy downward pull applied to an arm, the muscles that cross the shoulder joint and the elbow joint are not active in preventing dislocation of these joints (Basmajian, 1979). It is an interesting observation that during activity with forward flexion of the spinal column, there is a marked muscular activity until flexion is extreme. At that point ligament structures assume the forces and the EMG recorded discharge from the trunk muscles ceases (Floyd and Silver, 1955; Basmajian, 1979). With the center of gravity of the head and trunk very close to the supporting column of bones, man has the most economical antigravity mechanism among the mammals, once the upright posture is attained. (The aerobic power in the erect position is only slightly elevated, say from 0.25 liter · min^{-1} in the supine to 0.30 to 0.35 liter · min^{-1}.) Many of the antigravity muscles are of the slow-twitch fiber type (often named tonic muscles); they are more affected by the gamma loop than the fast-twitch fibers. A stronger afferent impulse traffic recruits more motoneurons rather than increasing the rate of discharge in those units already active. EMG studies reveal that the antigravity muscles are activated at a frequency of 5 to 20 Hz unless they are under special stress.

As an illustration of the cooperation between the proprioceptors involved in the maintenance of a normal posture and the spinal and supraspinal networks, the following experiments may briefly be discussed. Nashner's (1976) subjects performed stance tasks requiring several different postural "sets." The stabilizing role of stretch reflexes acting upon the ankle musculature was studied. The principal tool used to impose disturbances was a servo-controlled moveable platform upon which the subject stood. (1) By sliding the platform backwards, the subject sways in the opposite direction, i.e., forward. (2) A tilting of the platform upward will cause the subject to sway backward (rotation about an axis colinear with the subject's ankle joints and with similar deflexion in both experiments). By these procedures, the movement in both situations stretches the ankle extensors with the potential to elicite a stretch reflex. In case (1) this reflex would be appropriate and reduce the sway, while in case (2) it would increase the sway and hence would be undesirable. What then does actually happen?

Three types of responses to stabilize posture and to prevent the subject's falling were observed: an immediate muscle stiffness; activation of a "postural synergy" acting at approximately 120 ms latency; and "vestibular" activation acting at a latency longer than 180 ms. The myotatic stretch reflex acting at 45 to 50 ms delay was never significant.

The long-latency reflex (about 120 ms, the postural synergy), involves many muscle groups, starting in the peripheral muscles. It seems to involve supraspinal as well as spinal circuits. In some of the subjects this reflex dominated. Interestingly, when the rotational stimulation of the ankle (platform sliding upwards) was unexpectedly repeated, the first EMG response from the gastrocnemius muscle was vigorous, but it was certainly inappropriate. However, the responses evoked by the succeeding stimuli became progressively weaker, and the sway, therefore, not so pronounced. With a similar protocol, but now with the platform sliding backward, the successive trials in the functional stretch reflexes gave an adaptive facilitation, which also decreased the extent of the swaying.

In other subjects, there was a delay in the compensatory responses, and apparently visual and/or vestibular inputs were involved (with 180 to 200 ms latency).

If the subject in similar experiments supported itself with the arm(s), the reflex response was switched off. These are examples illustrating how higher levels of the CNS evaluate whether a reflex response is appropriate or not, and whether it can be extensively modified to optimize the response; descending influences can gate basic reflex loops by opening them (facilitation) or closing them (inhibition).

The so-called antigravity muscles have the very important function of producing the powerful movements necessary for the changes from a supine to a sitting or standing position and of providing a firm, but flexible foundation for the variety of muscular activity of everyday life. Actually, posture is the basis of movement. All movements start from and end in a posture. In other words, not only do peripheral factors automatically evoke postural responses, but voluntary movements (e.g., of an arm) will also initiate an activation of the postural muscles. The pattern of sequences in such an activation is similar to the events typical of an automatic postural reaction. Certainly, any movement disturbs the position of the body to some degree. It has been observed that changes in the activity of postural muscles may preceed the actual start of voluntary movement, particularly if that movement disturbs the body's center of gravity. It appears that principal postural components are included in the composition of the central program of an impending movement, but the exact pattern is determined by the state of the locomotor organs at that particular time (see Kholmsgorova, 1982).

The free normal posture is characterized by a "postural sway," so that the center of gravity varies with respect to its projection on the ground with a frequency of 5 to 6 "cycles" · min^{-1}. With the eyes closed (or in the dark), the swaying is about 50 percent more pronounced.

Summary   As soon as one begins to fall, reflex compensatory muscular programs are activated, which restore the state of equilibrium. Any of the muscles of the trunk and legs act as antigravity muscles. For the maintenance of a balanced body position in standing, in locomotion, and in any kind of exercise, an integrated coordination of the proprioceptive, visual, and vestibular systems is essential. The proprioceptive and visual systems can jointly manage most of the neuromuscular interactions necessary to secure an optimal situation; the vestibular system seems to function more as a reference system controlling which adaptive modifications should be performed in the corrections elicited by the proprioceptive and visual systems. The myotatic stretch reflexes are important, but there seem to be central pattern generators also for posture. A reflex response is evaluated in its effect, and a readjustment can take place, probably with the cerebellum as an important relay station.

Motor Development

In the following we shall briefly consider the "maturation" of posture and locomotion in the child. (For further discussion and references see Forssberg and Nashner, 1982.)

Posture   A child only a few weeks old already possesses visual depth perception and early in life will avoid any steep edges or precipices. Over time this perception is further developed, which is essential for an optimal postural control. Children who have just learned to stand have started to form their supporting reflexes, and their proprioceptors are well developed, but they are far from perfect. When submitted to experiments using platforms that can be rotated or moved backward and forward (see above), the response in children below the age of 7 to 8 is much more variable when compared with older individuals. There is no adaptation response possible when young subjects are repeatedly exposed to situations that trigger inadequate reflex responses in order to maintain posture. When facing conflicting information from the visual and proprioceptive systems they usually fail to stand. Evidently, the three systems of key importance for postural balance are not integrated until the age of about 8 years.

Locomotion   The human bipedal walking pattern probably has many features in common with quadrupedal walking. But the technique is different, with the human heel hitting the ground first, while in quadrupeds, the heel never hits the ground in normal walking. Over the course of evolution, the locomotor activity generated in the CNS has, in important aspects, been modified to make bipedal walking possible. A newborn child can execute a walking behavior, if supported and moved across the floor. But the pattern of leg movement is not the typical pattern of *Homo sapiens,* but is more related to quadruped locomotion. The generators of this "primitive" locomotion are probably located in the spinal cord. Gradually, a modification in the impulse traffic to the muscles engaged in walking takes place from the higher brain areas. Most likely other generators within the CNS take over the coordination of muscles. To a large extent, walking is a question of balance control and of motivation (courage) when the child starts to walk without support. A child who cannot walk without support may stand and walk if offered a stick to grab. If noticing that no one holds on the other end of the stick, it immediately sits down. Walking is an innate behavior; one day the child can walk, with or without previous training or education, but at individually different times. With time, there is an optimizing of the sequence of muscle activation and, as pointed out, at the age of approximately 8 years the proprioceptive, the vestibular, and the visual systems are integrated. This will enhance the child's choice of exercises and improve the flexibility and modifiability of the locomotor system so that new demands, whether expected or unexpected, can be handled adequately.

MUSCLE FATIGUE

Fatigue is a very complex conception, especially since both psychological and physiological factors are involved. Feelings of fatigue without preceding exercise are not uncommon. A given power output may be perceived heavier today than it was yesterday. Exercise does load respiration and circulation, as well as the neuromuscular function (see Simonson, 1971). A more general discussion of physical fatigue, largely based on Christensen's definition as a state of disturbed homeostasis (Christensen, 1960), will be presented in Chặp. 11. In this section, we shall merely deal with local muscular fatigue, which has recently been defined by Edwards (1983) as: "failure to maintain the required expected power output." It should be kept in mind, however, that even muscular fatigue may include a very powerful psychological or mental element as is evident at any sports arena. Thus, at the end of an Olympic marathon race, the loser may collapse, while the winner, probably equally fatigued, has the stamina to run an extra round in jubilation.

Before discussing possible sites for muscular fatigue we will present some data on "fatigue curves" in sustained and repeated contractions, respectively.

Fatigue in static exercise produces a sensation of discomfort and sometimes even pain. The disposition to subdue the feeling of fatigue is very different among individuals. Cooperative and well-motivated subjects can maintain a muscular contraction to the point of muscular fatigue, whereas others terminate the activation before reaching that point. It should be emphasized that experiments on fatigue factors must be very carefully controlled with respect to initial length of the involved muscles and the position of the joints.

Figure 3-25 summarizes data on maximal holding time in sustained isometric contractions at various levels of forces related to the maximal force of contraction (MFC), or maximal voluntary contraction (MVC). The maximal force can be maintained for

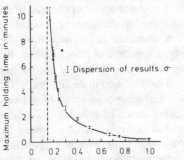

Figure 3-25
Maximal work time plotted versus force in isometric muscular contraction against a load expressed in percentage of maximal isometric strength in the same type of work. Average of results obtained in studies on different muscle groups. Note that a 50 percent load can be maintained for just 1 min, but as long as the muscular force is less than 15 percent of the maximal force (dashed line), the contraction may be maintained almost "indefinitely." The subjects experienced discomfort and aches in the working muscle some time before they were compelled to terminate the effort because of muscular exhaustion. (*From Rohmert, 1968.*)

only a few seconds, 50 percent of MFC for about 1 min, but at the 15 percent level and below, Rohmert and others found that the isometric contraction could be held for more than 10 min and even up to hours (Simonson and Lind, 1971). Some studies indicated, however, that the upper limit for isometric contraction maintained for "indefinite" time is below 10 percent of MFC (Björksten and Jonsson, 1977). One must expect a variation depending on the muscle groups studied, fiber types, and individual variations in maximal strength. In repeated isometric exercise, the combination of force and frequency of repetitions determine the length of time that the exercise can be endured. Figure 3-26 shows results from experiments in which the subjects performed rhythmic, maximal isometric contractions on a dynamometer in pace with a metronome (Molbech, 1963). The force gradually decreased because of fatigue, but they finally leveled off at a value that could be maintained for a long time. With ten contractions/min, about 80 percent of the maximal isometric strength could be applied without impairment. With 30 contractions/min, the force was reduced to 60 percent. The values seemed to be independent of the size of the activated muscle group. There appears to be an optimum capacity to exercise isometrically when the ratio between period of contraction time and period of rest is 1:2.

In exercises including frequent *dynamic concentric contractions,* the energy output for a given force is relatively high. According to Asmussen (1973), this type of exercise can probably be performed for long periods of time only if the developed strength does not exceed 10 to 20 percent of the maximal isometric strength.

Blood Supply

Everyone must have experienced the pain in the arms caused by carrying heavy luggage, which gradually increases in severity. After a few seconds of rest with the trunks idling on the ground, the pain disappears. One possible cause of the trouble is an impaired blood flow in the active arm muscles. In its contraction, the muscle consumes energy. Metabolites are formed, and oxygen, if available, is used with a concomitant production of carbon dioxide, water, and heat. A restoration of the internal equilibrium necessitates an adequate blood supply. During contraction, the active muscle swells and becomes hard. In maximal *static contractions* of the quadriceps femoris in humans, the pressure

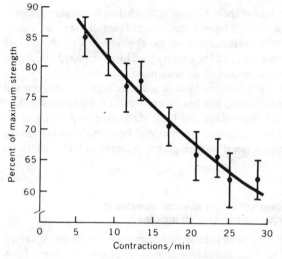

Figure 3-26
Percent of maximum isometric strength that can be maintained in a steady state during rhythmic contractions. Points are averages for finger muscles, hand muscles, arm muscles, and leg muscles, combined. Vertical lines denote ± standard error. (*Molbech, 1963.*)

within the muscle can be several hundred millimeters of mercury. Since the peak arterial blood pressure at rest is some 120 mm Hg (16 kPa) and during exercise below 200 mm Hg (26.6 kPa) in most cases, the blood flow through the active muscle will be partially or completely blocked. According to Edwards et al., (1972), the intramuscular pressure in the quadriceps muscles exceeds the systolic arterial pressure already at a force of 25 percent of MVC. Lind and coworkers have shown that at 5 and 10 percent of MVC, the blood flow (forearm) increased to a steady state and dropped immediately after exercise (Simonson and Lind, 1971). As mentioned, contractions at that level can be held for a very long time and the energy yield is most likely aerobic. At forces of 20 to 30 percent of MVC, the blood flow increased steadily during the activity and increased further immediately after the end of contraction. Apparently, there was a "blood flow debt" and the muscle fibers had to pay part of the energy cost by anaerobic processes. At forces exceeding 30 percent of MVC, there was a decrease in the blood flow and it was completely arrested at about 70 percent of MVC. As noted earlier, Edwards has reported that in the quadriceps, the circulation could be occluded at contraction forces greater than 20 percent of MVC.

With the blood flow occluded by a cuff just before and during a vigorous contraction, there was no difference in the initial forces developed as compared with the controls. This was the case with contraction forces from 60 to 70 percent and higher in Lind's studies; the blood flow was then occluded anyway. At lower forces, an occlusion of the local blood flow reduced the maximal isometric contraction time. It is not surprising that the reduction of endurance by external occlusion of blood flow is much more pronounced at lower forces than at forces closer to MVC. The normal occlusion of blood flow can be both a result of "nipping" of the arteries between moving and nonmoving tissues and an effect on the capillary flow due to the increased intramuscular pressure.

There are indications that stronger subjects are less able to maintain isometric contractions at the same percentage of their maximal strength than are weaker subjects, (Mundale, 1970; Thorstensson, 1976). From a mechanical point of view, an impediment of blood flow should be more dependent on the absolute than on the relative muscle forces; that is, the strong subject will be handicapped by an impaired circulation at a force that is relatively low in percent of the maximum.

With a blood flow below the level required for an adequate supply of oxygen and removal of carbon dioxide, metabolites, and heat, there must inevitably be a lack of oxygen and an accumulation of metabolites and heat. According to Ahlborg et al. (1972), the lactate accumulation at exhaustion is maximal between 30 to 60 percent of MVC. The accumulation rate is linear with respect to contraction forces above 20 percent of MVC.

What then causes fatigue when activating skeletal muscles in "all-out" performance, exercise up to some 10 minutes?

For many years, physiologists have been debating whether muscular fatigue is central or peripheral in origin. It has been attributed to a decrease in central excitatory drive, to conduction failure in the nerve fibers, to extinction of neuromuscular transmission and to decrease in contractile force. Figure 3-27 indicates sites where fatigue could interfere with performance. It has been pointed out and will be further stressed that mood—motivation can highly affect the performance. At present, there are no data indicating that the various synaptic connections should be subjected to fatigue. (The effects of inhibition are not a fatigue factor, for, by disinhibition, the effects can be removed.) From his review, Simonson (1971) concludes that transmission fatigue at the neuromuscular junction can be excluded. However, Stephens and Taylor (1972) have presented data on humans, which they interpret as evidence that in a maximal voluntary contraction, neuromuscular junction fatigue is most important at first (during the first minute), but later, contractile element fatigue increases. In view of the quantities of transmitter substance released at each transmission of an impulse (in contrast to the situation in synapses within the CNS), these authors' hypothesis seems reasonable. They also noted that "neuromuscular junction fatigue is believed to be most marked in high-threshold motor units, while contractile element fatigue more specifically affects low-threshold units." As stated earlier, the high-threshold motor units are activated at a high contraction rate, and consequently there must be a rapid turnover of acetylcholine. During the first minute of a maximal voluntary contraction, the force falls to about 50 percent. EMG recordings indicate that during this phase there is a loss of active motor units with a high-frequency rate of firing and recruitment of units with a lower rate (Lloyd, 1971).

There are characteristic changes of the EMG in muscle fatigue, indicating a change in both the impulse traffic, in the motor nerve; and the muscle reaction to the discharge. The amplitude increases and the rhythm slows down (see Komi, 1983). A grouping and synchronization of the discharges appear, which can at least partly be attributed to a decrease of the proprioceptive afferent impulses from muscle spindles, as shown by Kogi and Hakamada (1962). These authors found that the quotient of the electrically integrated amplitude of slower components divided by that of the faster components

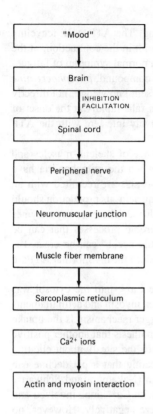

Figure 3-27
Control sequence for activation of voluntary muscle. Fatigue could operate at any step. (*Based on Edwards, 1981; 1983.*)

increased gradually and steadily in fatigue experiments of isometric-isotonic contractions of various strengths. The appearance of a high "slow wave" ratio was significantly related to the onset of a local fatigue sensation, to the feeling of pain, and to the subjects incapability of maintaining the intended tension.

In a maximal contraction, lasting for more than one minute, one must look for fatigue factors in the contractile element. In one way or another, an inefficient oxygen supply will "promote" fatigue, but is this the causal factor? At this stage, we will not discuss what limits the transport of oxygen from the air out to the active muscles. The links of interest in that transport will be analyzed in Chap. 4 and 5. Furthermore, we have not as yet presented details about the biochemical processes within the cell (this discussion follows in Chap. 12). However, we shall at this point attempt to discuss the question of fatigue on the basis of the information already presented.

As mentioned, ATP plays a key role as a primary energy producer for many of the processes behind a muscle contraction. Its concentration in the muscle cell is very low and in some compartments it could become critically low. So far the application of the muscle biopsy technique has not shown any dramatic decline even after exhausting exercise. The problem is that it takes a few seconds from the moment a piece of tissue is cut until all biochemical events are stopped by freezing. The ATP-ADP recycling is extremely fast. However, a lack of ATP at a critical point in the contraction, at the myosin head, would cause rigor (Chap. 2), which is not a normal symptom of fatigue. Let us take a look at the other energy yielding phosphate compound, *phosphocreatine* (PCr). It is a potent resynthesizer of ATP but it is, to present knowledge, not directly involved in the contractile mechanisms. Its concentration falls rapidly at the onset of vigorous exercise to very low values. This could negatively interfere with the ATP level at some crucial site(s) within the cell.

A classic candidate responsible for reduced performance of skeletal muscles and fatigue is an accumulation of lactic acid. This is true when the mitochondria have inadequate access to enough oxygen, the *anaerobic processes* are recruited with an inevitable accumulation of lactic acid. As a relatively strong acid, its production should increase the proton concentration, i.e., the pH becomes reduced. There are key enzymes of importance for both the anaerobic as well as the aerobic processes that can be inhibited by a reduced pH. Whether or not critical "bottle necks" can be created is presently not known. A reduced pH could, for instance, reduce the myofibrillar ATPase activity, a key factor for efficient muscular contraction.

In Chap. 2, it was pointed out that a release of free Ca^{2+} into the cytosol was necessary for establishing a cross-bridge formation between myosin and actin filaments and therefore a necessity for muscle contraction. The trigger mechanism is the uptake of calcium on specific sites of the troponin. There is a hypothesis that another positive ion, for example H^+, could compete with Ca^{2+} and block the sites without eliciting the cross-bridge formation. There are also studies indicating that a pH decline can reduce the Ca^{2+} release from the sarcoplasmic reticulum (Nakamura and Schwartz, 1972). Therefore, at many points in the chain reaction with Ca^{2+} ions involved and leading to cross-bridge formations, protons could interfere negatively. However, no factual data are available as yet proving that this is the case. The assumption that lactate formation interferes with the contractile and biochemical process is opposed by a recent proposal, which suggests that the hydrolysis of ATP, *not* lactate production is the dominant source of the intracellular acid accompanying an anaerobic energy yield (see Busa and Nuccitelli, 1984). If, in experiments, the pH in activated muscles is kept at a given level the muscles contract with a high power even if the lactate concentration is very high. It is also an interesting observation that the highest lactate concentration in muscle and blood is usually observed in well-trained athletes participating in important competitions, not when the same athletes perform a subjectively "all-out" test in the laboratory. Apparently, muscles can function despite a higher lactate level if the athlete is particularly well motivated.

Summary The ability of the muscle fibers to maintain a high force, and the individual's subjective feeling of fatigue, depend on the blood flow through the muscle. At the beginning of exercise, there is a time lag between blood demand and blood supply. In very short spells of isometric contraction, ATP and phosphocreatine can yield energy and the oxygen present in the muscle (bound to myoglobin) also makes possible an energy delivery from aerobic processes. A maximal contraction can, however, be sustained for only a few seconds. In isometric contractions with less than 15 percent of maximal force or with appropriately spaced pauses, blood flow can secure the supply of oxygen and energy-rich compounds, and remove the formed metabolites, thus exercise can proceed aerobically for long periods of time. At heavier loads, there is an impaired blood flow, because of blood vessel occlusion. Therefore, the oxygen need will exceed the oxygen supply, and the anaerobic processes must contribute markedly to the energy yield. The impaired blood flow limits not only the oxygen supply but also the removal of metabolites and heat. The exact factors that limit the performance are not known. One factor might be fatigue induced changes at the neuromuscular junction in maximal efforts sustained for less than a minute. At a force of roughly 50 percent of MVC (which can be sustained for about 1 min), accumulation of H^+ may negatively interfere with the contractile elements. In isometric contractions, demanding more than 15 percent but less than 50 percent MVC, there is no information available as to which factors might limit the performance.

It should be emphasized that different muscle groups behave differently because of mechanical design in structure and blood vessel arrangement, as well as composition with respect to fiber types.

The relation of developed strength to maximal voluntary contraction of the whole muscle may introduce erroneous conclusions about the relative force of the few muscle fibers studied by EMG or thermistor probe, or sampled by a biopsy needle. They may be submitted to a different strain; the local blood flow may vary in different parts of the engaged muscle group.

Finally, it should be kept in mind that a prolonged and strong, purely isometric contraction is primarily an artificial laboratory or training exercise that does not commonly occur in everyday life. However, for the understanding of the neuromuscular function such artificial experiments are essential.

Effect of Prolonged Exercise

In heavy, prolonged exercise maintained for hours, the energy output during maximal effort gradually decreases. After 1 hour's rest, a rate of exercise that normally could be tolerated for 6 min had to be terminated after about 4 min because of exhaustion. The peak lactate level in the blood was correspondingly decreased. It is believed that the limiting factor in this case must be sought at the cellular level in the exercising skeletal muscles, and that it may be anything from a change in the properties of the membrane of muscle fibers, a disturbed ATP-ADP "machine," etc., to a depletion of the glycogen stores or a reduced capacity to neutralize the metabolites produced.

In prolonged exercise requiring up to 90 percent of the individual's maximal oxygen uptake, it has been noted that the slow-twitch fibers in engaged muscle groups are the first to be glycogen-depleted. As exercise continues, fast-twitch fibers start to become recruited, and at last they too will be glycogen-depleted (see Burke and Edgerton, 1975; Piehl, 1974). A drop in tension as the slow fibers drop out may temporarily be prevented by an increased drive on high-threshold motoneurons induced by an afferent muscle spindle discharge. Their recruitment could compensate for the failing slow-twitch fibers. (In Chap. 12 the substrates utilized in the exercising muscles will be discussed as well as limiting factors in prolonged exercise).

"Fatigue" in Joints

Basmajian (1979) subjected a sitting person's upper limbs to a strong downward pull. The muscles crossing the shoulder joints and the elbow joints were electrically quiescent, but the subject felt local fatigue. Basmajian thinks that such fatigue originates from the painful feeling of tension in the articular capsule and ligaments.

Motivation

With the tools now available, the conductor of physiological exercise experiments need not rely exclusively on the subject's subjective statements for an evaluation of the relative load placed on the muscles, the circulation, or other functions. The production of lactic acid, the pH of the blood. EMG, heart rate, blood sugar level, tissue temperature, and other findings may be used as an indication of how the subject "should" feel. In healthy subjects, young as well as old, untrained and well-trained, the problem during maximal effort is, as a rule, that the muscles eventually stiffen and refuse to obey the subject's will. However, it has been repeatedly stressed that people usually perform better if motivation is maximized.

Cramp

The localized, sustained, and painful cramp that can occasionally throw a muscle into vigorous involuntary contraction, e.g., after prolonged monotonous exercise or even during sleep, cannot be explained at present. A stretch of the muscle usually relieves it from the cramp, probably due to inhibitory impulses from the stimulated Golgi tendon organs (Sherrington's lengthening reaction).

REFERENCES

Ahlborg, B. J., J. Bergström, L.-G. Ekelund, G. Guarnieri, R. C. Harris, E. Hultman, and L.-G. Nordesjö: Muscle Metabolism during Isometric Exercise Performed at Constant Force, *J. Appl. Physiol.,* **33:**224, 1972.

Allen, G. I., and N. Tsukahara: Cerebrocerebellar Communication Systems, *Physiol. Rev.,* **54:**957, 1974.

Allen, M.: Activity-Generated Endorphins. A Review of their Role in Sports Science, *Can. J. Appl. Sport. Sci.,* **8:**115, 1983.

Asmussen, E.: Growth in Muscular Strength and Power, in G. L. Rarick (ed.), "Physical Activity Human Growth and Development," p. 60, Academic Press, Inc., New York, 1973.

Asmussen, E., O. Hansen, and O. Lammert: The Relation Between Isometric and Dynamic Muscle Strength in Man, *Communications from the Testing and Observations Institute of the Danish National Association for Infantile Paralysis,* no. 20, 1965.

Basmajian, J. V.: "Muscles Alive," The Williams & Wilkins Company, Baltimore, 1979.

Bennett, M. R.: Development of Neuromuscular Synapses, *Physiol. Rev.,* **63:**915, 1983.

Björksten, M., and B. Jonsson: Endurance Limit of Force in Long-Term Intermittent Static Contractions, *Scand. J. Work Environ. & Health,* **3:**23, 1977.

Boyd, J. A.: The Isolated Mammalian Muscle Spindle, *Trends Neuro Sci.,* **3:**258, 1980.

Brooks, V. B.: Motor Programs Revised, in R. E. Talbott and D. R. Humphrey (eds.), "Posture and Movement," p. 13, Raven Press, New York, 1979.

Brown, M. C.: Sprouting of Motor Nerves in Adult Muscles: a Recapitulation of Ontogeny, *Trends Neuro Sci.*, **7**:10, 1984.

Buchthal, F., and H. Schmalbruch: Motor Unit of Mammalian Muscle, *Physiol. Rev.*, **60**:91, 1980.

Burke, D.: Muscle Spindle Function During Movement, *Trends Neuro Sci.*, **3**:251, 1980.

Burke, R. E., and V. R. Edgerton: Motor Unit Properties and Selective Involvement in Movement, *Exercise and Sport Sci. Rev.*, **3**:31, 1975.

Busa, W. B., and R. Nuccitelli: Metabolic Regulation via Intracellular pH, *Am. J. Physiol.*, **246**:R409, 1984.

Christensen, E. H.: Muscular Work and Fatigue, in K. Rodahl and S. M. Horvath (eds.), "Muscle as a Tissue," Chap. 9, McGraw-Hill Book Company, New York, 1960.

Clough, J. F. M., C. G. Phillips, and J. D. Sheridan: The Short-Latency Projection from the Baboon's Motor Cortex to Fusimotor Neurons of the Forearm and Hand, *J. Physiol.*, **216**:257, 1971.

Cooper, S., and J. C. Eccles: The Isometric Responses of Mammalian Muscles, *J. Physiol.*, **69**:377, 1930.

Couteaux, R.: Motor End Plate Structure, in G. H. Bourne (ed.), "The Structure and Function of Muscle," vol. 1, Academic Press, Inc., New York, 1960.

Cowan, W. M.: The Development of the Brain, *Sci. Am.*, **241**:112, 1979.

Coyle, E. F., D. C. Feiring, T. C. Rotkis, R. W. Cote III, F. B. Roby, W. Lee, and J. H. Wilmore: Specificity of Power Improvements Through Slow and Fast Ioskinetic Training, *J. Appl. Physiol.* **51**:1437, 1981.

Eccles, J. C.: "The Physiology of Nerve Cells," The John Hopkins Press, Baltimore, 1957.

Eccles, J. C.: The Synaps, *Sci. Am.*, **212**:56, 1965.

Eccles, J. C.: The Cerebellum as a Computer: Patterns in Space and Time, *J. Physiol.*, **229**:1, 1973a.

Eccles, J. C.: "The Understanding of the Brain," McGraw-Hill Book Company, New York, 1973b.

Eccles, J. C.: The Synapse: From Electrical to Chemical Transmission, *Ann. Rev. Neurosci.* **5**:325, 1982.

Edwards, R. H. T.: Human Muscle Function and Fatigue, *Ciba Found. Symp.*, **82**:1, 1981.

Edwards, R. H. T.: Biochemical Bases of Fatigue in Exercise Performance: Catastrophe Theory of Muscular Fatigue, in H. G. Knuttgen, J. A. Vogel, and J. Poortmans (eds.), International Series of Sport Sciences, vol. 13, "Biochemistry of Exercise," p. 3, Human Kinetics Publishing, Inc., Champaign, Ill., 1983.

Edwards, R. H. T., D. K. Hill, and M. J. McDonnell: Myothermal and Intramuscular Pressure Measurements in Man, *J. Physiol.*, **224**:58P, 1972.

Elias, M. H., and J. E. Pauly: "Human Microanatomy," 2d ed., Da Vinci Publishing Co., Chicago, 1961.

Evarts, E. V.: Motor Cortex Reflexes Associated with Learned Movement, *Science,* **179**:501, 1973.

Evarts, E. V.: Brain Mechanisms of Movement, *Sci. Am.* **241**:164, 1979.

Floyd, W. F., and P. H. S. Silver: The Function of the Erectores Spinae Muscles in Certain Movements and Postures in Man, *J. Physiol.*, **129**:184, 1955.

Forssberg, H., and L. Nashner: Ontogenetic Development of Postural Control in Man: Adaptation to Altered Support and Visual Conditions during Stance, *J. Neuroscience*, **2**:545, 1982.

Forssberg, H., and G. Svartengren: Hardwired Locomotor Network in Cat Revealed by a Retained Motor Pattern to Gastrocnemius after Muscle Transposition, *Neurosci. letters,* **41**:283, 1983.

Fox, C. A.: "Correlative Anatomy of the Nervous System," The Macmillan Company, New York, 1962.

Freund, H. J.: Motor Unit and Muscle Activity in Voluntary Motor Control, *Physiol. Rev.* **63:**387, 1983.

Garnett, R., and J. A. Stephens: Changes in the Recruitment Threshold of Motor Units Produced by Cutaneous Stimulation in Man, *J. Physiol.,* **311:**463, 1981.

Grafstein, B., and D. S. Forman: Intracellular Transport in Neurons, *Physiol. Rev.,* **60:**1167, 1980.

Granit, R.: Muscle Tone and Postural Regulation in K. Rodahl and S. M. Horvath (eds.), "Muscle as a Tissue," p. 190, McGraw-Hill Book Company, New York, 1962.

Granit, R.: Constant Errors in the Execution and Appreciation of Movement, *Brain,* **95:**649, 1972.

Granit, R.: The Functional Role of the Spindles—Facts and Hypotheses, *Brain,* **98** (part IV):531, 1975.

Granit, R., and R. E. Burke: The Control of Movement and Posture, *Brain Research,* **53:**1, 1973.

Grillner, S.: Locomotion in Vertebrates: Central Mechanisms and Reflex Interaction, *Physiol. Rev.,* **55:**247, 1975.

Grillner, S.: Control of Locomotion in Bipeds, Tetrapods, and Fish, in V. Brooks (ed.): "Handbook of Physiology: The Nervous system," *vol. 22, Chap. 26,* p. 1179, *American Physiological Society,* Distributed by Williams & Wilkins, Baltimore, 1981.

Grimby, L.: Firing Properties of Single Human Motor Units during Locomotion, *J. Physiol.,* **346:**195, 1984.

Grimby, L., and J. Hannerz: Disturbances in Voluntary Recruitment Order of Low and High Frequency Motor Units on Blockades of Proprioceptive Afferent Activity, *Acta Physiol. Scand.,* **96:**207, 1976.

Grimby, L., and B. Saltin: The Ageing Muscle, *Clin. Physiol.,* **3:**209, 1983.

Hannerz, J.: Discharge Properties of Motor Units in Relation to Recruitment Order in Voluntary Contraction, *Acta Physiol. Scand.,* **91:**374, 1974.

Harber, V. J., and J. R. Sutton: Endorphins and Exercise, *Sports. Med.,* **1:**154, 1984.

Harvey, R. J.: Cerebellar Regulation in Movement Control, *Trends Neuro Sci.,* **3:**281, 1980.

Hebb, C.: Biosynthesis of Acetylcholine in Nervous Tissue, *Physiol. Rev.,* **52:**918, 1972.

Henneman, E.: Skeletal Muscle, The Servant of the Nervous System, in V. B. Mountcastle (ed.): "Medical Physiology", 14th ed., vol. 1, p. 674, C. V. Mosby, St. Louis, 1980.

Hökfelt, T., O. Johansson, Å. Ljungdahl, J. M. Lundberg, and M. Schultzberg: Peptidergic Neurons, *Nature* (London), **284:**515, 1980.

Hubbard, A. W.: Homokinetics: Muscular Function in Human Movement, in W. R. Johnson (ed.), "Science and Medicine of Exercise and Sports," Harper & Row Publishers, Inc., New York, 1960.

Ikai, M., and A. H. Steinhaus: Some Factors Modifying the Expression of Human Strength, *J. Appl. Physiol.,* **16:**157, 1961.

Kandel, E. R., and J. H. Schwartz (eds.): "Principles of Neural Science," Elsevier North Holland Inc., New York, 1981.

Keele, C. A., E. Neil, and N. Joels: "Samson Wright's Applied Physiology," 13th ed., Oxford University Press, Oxford, 1982.

Kholmsgorova, N. V.: Correlation of Postural Components with Voluntary Movement, translation from *Fiziologiya Cheloveka,* **8:**642, 1982.

Kogi, K., and T. Hakamada: Slowing of Surface Electromyogram and Muscle Strength in Muscle Fatigue, *Rep. Inst. Sci. Labour* (*Tokyo*) **60:**27, 1962.

Komi, P. V.: Electromyographic, Mechanical, and Metabolic Changes during Static and Dynamic Fatigue, in "Biochemistry of Exercise," International Series of Sport Sciences, Human Kinetics Publishers Inc., Champaign, Ill., vol. 13, p. 197, 1983.

Krujevic, K.: Chemical Nature of Synaptic Transmission in Vertebrates, *Physiol. Rev.*, **54:**418, 1974.

Kuffler, S. W., C. C. Hunt, and J. P. Quilliam: Function of Medullated Small-nerve Fibers in Mammalian Ventral Roots: Efferent Muscle Spindle Innervation, *J. Neurophysiol.*, **14:**29, 1951.

Lambert, O.: The Relationship between Maximum Isometric Strength and Maximum Concentric Strength at Different Speeds, *Intern. Fed. Phys. Educ. Bull.*, **35:**13, 1965

Lewi-Montalcini, R.: Developmental Neurobiology and the Natural History of Nerve Growth Factor, *Ann. Rev. Neurosci.*, **5:**341, 1982.

Lloyd, A. J.: Surface Electromyography during Sustained Isometric Contractions, *J. Appl. Physiol.: Resp. Env. Exer. Physiol.*, **30:**713, 1971.

Matthews, P. B. C.: Muscle Spindles and Their Motor Control, *Physiol. Rev.*, **44:**219, 1964.

Matthews, P. B. C.: "Mammalian Muscle Receptors and Their Central Actions," The Williams & Wilkins Company, Baltimore, 1972.

McDonagh, M. J. N., C. M. Hayward, and C. T. M. Davies: Isometric Training in Human Elbow Flexor Muscles, *J. Bone Joint Surgery*, **65-B:**355, 1983.

Miledi, R., P. C. Molenaar, and R. L. Polak: Electrophysiological and Chemical Determination of Acetylcholine Release of the Frog Neuromuscular Function, *J. Physiol.*, **334:**245, 1983.

Milner-Brown, H. S., R. B. Stein, and R. Yemm: Changes in Firing Rate of Human Motor Units During Voluntary Isometric Contractions, *J. Physiol.*, **230:**371, 1974.

Milner-Brown, H. S., and R. B. Stein: The Relation between the Surface Electromyogram and Muscular Force, *J. Physiol.*, **246:**549, 1975.

Molbech, S.: Average Percentage Force at Repeated Maximal Isometric Muscle Contractions at Different Frequencies, *Communications from the Testing and Observations Institute of the Danish National Association for Infantile Paralysis*, no. 16, 1963.

Mundale, M. O.: The Relationship of Intermittent Isometric Exercise to Fatigue of Hand Grip, *Arch. Phys. Med. Rehabil.*, **51:**532, 1970.

Nakamura, Y., and A. Schwartz: The Influence of Hydrogen Ion Concentration on Calcium Binding and Release by Skeletal Muscle Sarcoplasmic Reticulum, *J. Gen. Physiol.*, **59:**22, 1972.

Nashner, L. M.: Adapting Reflexes Controlling the Human Posture, *Exp. Brain Res.*, **26:**59, 1976.

Nauta, W. J. H., and M. Feirtag: The Organization of the Brain, *Sci. Am.*, **241:**88, 1979.

Ohtsuki, T.: Decrease in Grip Strength Induced by Simultaneous Bilateral Exertion with Reference to Finger Strength, *Ergonomics*, **24:**37, 1981.

Peper, K., R. J. Bradley, and F. Dreyer: The Acetylcholine Receptor at the Neuromuscular Junction, *Physiol. Rev.*, **62:**1271, 1982.

Piehl, K.: Glycogen Storage and Depletion in Human Skeletal Muscle Fibers, *Acta Physiol. Scand.*, (Suppl., 402), 1974.

Rohmert, W.: Die Beziehung Zwischen Kraft und Ausdauer bei Statischer Muskelarbeit, Schriftenreihe Arbeitsmedizin, Sozialmedizin, Arbeitshygiene, band 22, p. 118, A. W. Gentner Verlag, Stuttgart, 1968.

Rudfjord, T.: Model Study of Muscle Spindles Subjected to Static Fusimotor Activation, *Kybernetik*, **10:**189, 1972.

Saltin, B., and P. D. Gollnick: Skeletal Muscle Adaptability: Significance for Metabolism and Performance, in Peachey, L. D., R. H. Adrian, and S. R. Geiger (eds.): "Handbook of

Physiology Section 10: Skeletal Muscle," *American Physiological Society* Distributed by The Williams & Wilkins Company, Baltimore, 1983.

Schreiner, K. E., and A. Schreiner: "Menneskeorganismen," 6th ed., J. Jansen (ed.), Universitets Anatomiske Inst., Oslo, 1964.

Simonson, E. (ed.): "Physiology of Work Capacity and Fatigue," Charles C. Thomas, Springfield, Ill., 1971.

Simonson, E., and A. R.. Lind: Fatigue in Static Work, in E. Simonson (ed.), "Physiology of Work Capacity and Fatigue," p. 241, Charles C. Thomas, Springfield, Ill., 1971.

Sperry, R. W.: Changing Priorities, *Ann. Rev. Neurosci.,* **4:**1, 1981.

Squire, L. R.: The Neuropsychology of Human Memory, *Ann. Rev. Neurosci.,* **5:**241, 1982.

Stein, R. B.: Peripheral Control of Movement, *Physiol. Rev.,* **54:**215, 1974.

Stent, G. S.: Strength and Weakness of the Genetic Approach to the Development of the Nervous System, *Ann. Rev. Neurosci.,* **4:**163, 1981.

Stephens, J. A., and A. Taylor: Fatigue of Maintained Voluntary Muscle Contraction in Man, *J. Physiol.,* **220:**1, 1972.

Stevens, C. F.: The Neuron, *Sci. Am.,* **241:**55, 1979.

Taylor, A., and J. A. Stephens: Study of Human Motor Unit Contractions by Controlled Intramuscular Microstimulation, *Brain Res.,* **117:**331, 1976.

Thompson, R. F., T. W. Berger, and J. Madden IV: Cellular Processes of Learning and Memory in the Mammalian CNS, *Ann. Rev. Neurosci.,* **6:**447, 1983.

Thorstensson, A.: Muscle Strength, Fibre Types and Enzyme Activities in Man, *Acta Physiol. Scand.,* (Suppl. 443), 1976.

Tornvall, G.: Assessment of Physical Capabilities, *Acta Physiol. Scand.,* **58** (Suppl. 201), 1963.

Tsukahara, N.: Synaptic Plasticity in the Mammalian Central Nervous System, *Ann. Rev. Neurosci.,* **4:**351, 1981.

Vallbo, Å. B.: Muscle Spindle Afferent Discharge from Resting and Contracting Muscles in Normal Human Subjects, in J. E. Desmedt (ed.), "New Development in Electromyography and Clinical Neurophysiology," vol. 3, p. 251, Karger, Basel, 1973.

Vilensky, J. A., and G. Gehlsen: Temporal Gait Parameters in Humans and Quadrupeds: How do They Change with Speed, *J. Human Movement Studies,* **10:**175, 1984.

Wawrzinoszek, M., and H. Kramer: Experimentelle Untersuchungen zur Maximalkraftentwicklung: Maximale Hardschlusskräfte bei Ein-und Beidhandbetätigung in Zwei Unterschiedlichen Arbeitsebenen, *Medizin und Sport,* **24:**105, 1984.

Winter, D. A.: "Biomechanics of Human Movement," John Wiley & Sons, New York, 1979.

CHAPTER 4

BODY FLUIDS, BLOOD AND CIRCULATION

CONTENTS

BODY FLUIDS

BLOOD
Volume
Cells
Plasma
Specific Heat
Viscosity
Buffer Action—Blood pH—CO_2 Transport

HEART
Cardiac Muscle
Blood Flow
Pressures During a Cardiac Cycle
Innervation of the Heart

HEMODYNAMICS
Mechanical Work and Pressure
Hydrostatic Pressure
Tension
Flow and Resistance

BLOOD VESSELS
Arteries
Arterioles
Capillaries

Capillary Structure and Transport Mechanisms
Filtration and Osmosis
Diffusion—Filtration
Veins
Vascularization of Skeletal Muscles
REGULATION OF CIRCULATION AT REST
Arterial Blood Pressure and Vasomotor Tone
The Heart and the Effect of Nerve Impulses
Control and Effects Exerted by the Central Nervous System
Mechanoreceptors in Systemic Arteries
Posture
Other Receptors
REGULATION OF CIRCULATION DURING EXERCISE
CARDIAC OUTPUT AND THE TRANSPORTATION OF OXYGEN
Efficiency of the Heart
Venous Blood Return
Cardiac Output and Oxygen Uptake
Oxygen Content of Arterial and Mixed Venous Blood
Stroke Volume
Heart Rate
Blood Pressure
Type of Exercise
Heart Volume
Age
Training and Cardiac Output
SUMMARY

The function of the individual cells within the body is dependent on the constancy of their internal and surrounding environment. Claude Bernard recognized that an evolution of higher forms of organisms could not have taken place without the establishment of a stable *milieu interne,* its composition being guarded by regulatory mechanisms.

The muscle cell is unique in regard to its ability to increase its metabolic rate. The maintenance of a constant *milieu interne* of the cell during the transition from rest to vigorous exercise necessarily represents, at times, a tremendous challenge to the circulation. The result may be that the muscle cell must cease its contractions or that it is forced to slow down the rate of activity as the changed composition and property of the fluid within the cell or surrounding it interfere with the various processes which are necessary for the cell to function and to perform under optimal conditions.*

---

*For a more comprehensive and detailed treatment of the subject, the following sources are recommended: Shepherd and Vanhoutte, 1979; Keele et al., 1982; Shepherd and Abboud, 1983.

For definitions, symbols and abbreviations, see Appendix.

BODY FLUIDS

Generally speaking, we are dealing with a large water pool of about 60 percent of the body weight in men, 50 percent in women. As adipose tissue contains very little water, the percentage of total water in the obese individual is lower than in the nonobese. The relationship between total water and fat-free body weight (lean body mass) is fairly constant; in an adult the total water is about 72 percent of the lean body mass (Hernandez-Peon, 1961).

We can divide the water space into three compartments: (1) The intracellular fluid is separated by the cell membrane from the extrastitial fluid (2), while the third water compartment, the intravascular fluid (3), is circulating within the blood vessels (Fig. 4-1). In the vascular system, only the walls of the capillaries and, to some degree, of the postcapillary venules permit an exchange of materials.

The fluid outside the cells, i.e., the extracellular fluid, constitutes about 30 percent of the total body water. Extracellular fluid has an ionic composition similar to that of the

BLOOD AND BODY FLUIDS

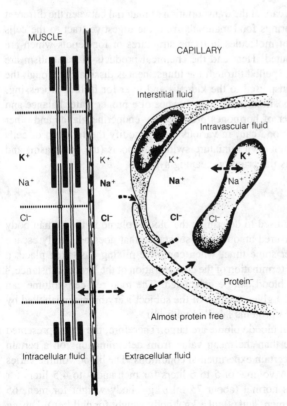

Figure 4-1
Schematic drawing of capillary and muscle cell walls. There is a high concentration of Na$^+$ and Cl$^-$ outside the cells (heavy letters) and a high concentration of K$^+$ inside the cells (including the red cell).

MUSCLE

CAPILLARY

Interstitial fluid

Intravascular fluid

K$^+$ K$^+$ K$^+$ K$^+$

Na$^+$ Na$^+$ Na$^+$ Na$^+$

Cl$^-$ Cl$^-$ Cl$^-$ Cl$^-$

Protein$^-$

Almost protein free

Intracellular fluid   Extracellular fluid

ancient sea water. The ions are found in approximately the same relative proportions, although the total ionic concentration of sea water of today is several times that of extracellular fluid. It has been postulated that this similarity suggests that extracellular fluid was orginally derived from the ancient oceans, which were more dilute than those of today (Hernandez-Peon, 1961).

The structure of the capillary wall allows free passage of all "normal" substances in the blood except the blood corpuscles and a markedly reduced passage for the smaller plasma proteins. On the other hand, there is a marked difference in concentration of various electrolytes between the extra- and intracellular fluid. As mentioned in the previous chapter, the membrane of the resting cell acts as a "barrier," especially for positively charged ions (cations), but processes within the cells must continuously work to maintain this barrier property by throwing out some intruding ions (mainly Na^+) and by holding back others (mainly K^+). It is somewhat of a paradox that the prerequisite for homeostasis in the body is based on a "heterostasis" at the cellular level.

BLOOD

The blood and the lymph take care of the transportation of material between the different cells or tissues. The blood brings food materials from the digestive tract to the cells for catabolism or synthesis of molecules in tissue structures or for depots which are later mobilized and redistributed. Heat and the chemical products of catabolism are removed, carbon dioxide is expelled through the lungs, heat is dissipated through the skin, and metabolites are transported to the kidneys and liver for further processing. Blood circulation has a key position in the maintenance of a proper water balance and fluid distribution. As a carrier of hormones produced by endocrine glands and other active chemical agents, the blood can, in various ways, modify the function of cells and tissues. All cells have their own miniature swimming pools (the cytoplasm) and the circulation of the blood is the primary cleaning system.

Volume

The most common procedure used to determine the blood volume in the human body is to inject into a vein a measured amount of a substance that does not easily escape from the blood vessels. After some time, when complete mixing has taken place, a blood sample is secured for determination of the concentration of the "tracer substance," and the subject's volume of blood can be calculated. The total red cell volume can be estimated by reinfusion of a given volume of the subject's erythrocytes labelled by the radioactive ^{51}Cr or plasma proteins labelled with ^{131}I.

The individual variations in blood volume are large. Therefore, the figure presented in textbooks is nothing more than the mean value from determinations on a certain number of individuals under certain experimental conditions. The blood volume varies with the degree of training. A volume of 5 to 6 liters for men and 4 to 4.5 liters for women can be considered as normal (about 75 ml · kg^{-1} body weight for men, 65 ml · kg^{-1} body weight for women, and 60 ml · kg^{-1} body weight for children). During heavy exercise there is a slight reduction in blood volume (see below). The blood volume may also be reduced when the individual is dehydrated, e.g., due to profuse sweating and limited fluid intake (see Chap. 13).

Cells

The solid elements of the blood which are visible under the ordinary microscope are red corpuscles, or *erythrocytes;* white cells, or *leukocytes;* and platelets, or *thrombocytes.* When a blood sample in a tube containing an anticoagulant is centrifuged, these elements are separated from the liquid portion of the blood, known as the *plasma.* The relative amount of plasma and corpuscles in blood is known as the *hematocrit.* For men the hematocrit is 47 (42–54) percent, and for women and children it is 42 (38–46) percent.

The erythrocyte is a biconcave circular disc without a nucleus. It has an average diameter of 7.3 μm, and a thickness of 1 μm in the center and 2.4 μm near the edge (Fig. 4-1). The cell count in an adult male is 5.4 (4.6–6.2) million per μl blood; in females and children it is 4.8 (4.2–5.4) million · μl^{-1}. The life span of a red cell is about 4 months. Red cells are formed in the bone marrow at a speed that normally matches the rate of destruction. It can be estimated that the formation of erythrocytes proceeds at a rate of 2 to 3 million per second. The production of the red cells (erythropoieses) is regulated by a substance, erythropoietin. If the kidney is exposed to low oxygen pressure, an enzyme, erythrogenin, is delivered to the blood. This enzyme transforms a plasma globulin into erythropoietin which will enhance the red cell production. The increase in the hemoglobin concentration in the blood of people acclimatized to high altitude is a consequence of an increased production of erythropoietin. The color of the red cell (and the blood) is due to its content of hemoglobin, which is a protein (globin) united with a pigment (hematin). This pigment contains iron; each Fe^{2+} atom can combine with one molecule of O_2. This combination is not a chemical oxidation but loose and reversible. In an oxygen-free medium the *hemoglobin* (Hb) is reduced, and with O_2 available, it forms *oxyhemoglobin* (HbO$_2$). Another property of Hb is its affinity for CO. The Fe^{2+} atom reacts with CO, and when given a choice it prefers CO (its affinity is about 250 times greater for CO than for O_2), resulting in a proportionally decreased capacity to take up O_2. This phenomenon explains the high toxicity of CO. (For details, see pp. 179–181.)

In adult men, the average Hb content is 158 g · liter^{-1} of blood; in women it is 139 g · liter^{-1}. From a statistical point of view, a normal value could be within 140 to 180 and 115 to 160 g · liter^{-1} for men and women respectively, with a range of 95 percent. Each gram of Hb can maximally combine with 1.34 · 10^{-3} liter of oxygen. With 150 g Hb per liter of blood, the fully saturated blood can carry 0.201 liter oxygen per liter plus some 0.003 liter oxygen dissolved in the plasma (oxygen pressure 100 mm Hg; 13.3 kPa.).

The concentration of Hb in blood can be determined according to two principles: (1) The blood sample is saturated with O_2, and the content of O_2 in a measured volume of blood is determined. The volume of dissolved O_2 can be calculated, and as the O_2-combining power of 1 g Hb is fixed to 1.34 · 10^{-3} liter, the concentration of Hb in the sample can easily be calculated. (2) The second method is based on the typical and specific light absorption spectra of Hb and its derivatives. A rough estimation of the Hb content of blood can be obtained from the hematocrit value, or from blood counts, with the presumption that each red corpuscle has a normal Hb content.

Plasma

The plasma occupies about 55 percent of the total blood volume. It contains about 9 percent solids and 91 percent water. The concentration of protein in the plasma is 60 to 80 g · liter^{-1}. Three types of proteins are present: albumin (4.8 percent), globulin (various fractions, 2.3 percent), and fibrinogen (0.3 percent). If blood is permitted to clot, the fibrinogen, in the presence of thrombin, forms fibrin. The fluid "squeezed" from the clot is called *serum*.

The proteins of the blood have several important functions: fibrinogen is necessary for blood coagulation, and the globulin (especially gamma globulin) is necessary for the formation of antibodies. All fractions, but especially the albumin, have the important transport functions of carrying ionic and nonionic substances to the sites of need or elimination. The blood proteins are active in the buffer action and constitute in a way a mobile reserve store of amino acids. Finally, they play an important role in the plasma volume and tissue fluid balance.

All plasma proteins can pass through the capillary wall in small amounts, the albumin most readily. However, as most of the protein molecules are kept within the capillaries, they will exert an osmotic pressure of about 25 mm Hg (3.3 kPa). This protein or colloidosmotic pressure is an important factor in the exchange of fluid between the intravascular and interstitial spaces.

The total amount of electrolytes in the plasma is 9 g · liter^{-1}. The main ions are Na$^+$ and Cl$^-$. The plasma normally contains about 5 mmol · liter^{-1} (100 mg · 100 ml^{-1}) of glucose. It also contains free fatty acids (FFA), amino acids, hormones, various enzymes, and about 25 different electrolytes in varying amounts.

Specific Heat

The specific heat of whole blood is 3.85 J (0.92 cal) per gram; that is, the change in temperature of 1 g blood by 1°C will require or deliver 3.85 J (0.92 cal), or about 3.8 kJ (0.9 kcal) per liter of blood.

Viscosity

Viscosity is the property of a fluid that gives rise to internal forces which will affect its flow. In tubes, like vessels, laminar layers of the fluid slide onto one another and move at different speeds, causing a velocity gradient in a direction perpendicular to the wall of the tube. This velocity gradient is called the rate of shear. The faster layer tends to drag along, and to be held back by the slower layer. The viscosity of blood is not a consequence of friction between the blood and the vessel wall but is due to the friction between adjacent laminae in fluid. In a homogeneous, so-called Newtonian fluid, the viscosity is constant at different rates of laminar flow. However, blood is definitely of a heterogeneous composition and its viscosity does vary depending on flow rates. It depends mainly on the plasma proteins and particularly on the corpuscular components. At moderate or high shear rates, blood with a normal hematocrit of 45 percent is 4 to 5 times as viscous as water, measured in vitro. An increase in hematocrit will increase the blood viscosity. The plasma viscosity in healthy subjects is about 2.2 in relation to water (the viscosity of water being 1). However, blood has an "anomalous viscosity." The red cells tend to accumulate along the axis of the blood vessels, which leaves a zone near the wall relatively free of cells. This axial accumulation of the erythrocytes may result in small side branches of a blood vessel containing a volume of red cells that is considerably less than that for mixed blood (effect of "plasma skimming"). Another effect is a lower viscosity of the blood than expected. The axial accumulation of cells is more pronounced with an increase in velocity of the blood. However, in the physiological range of flow, the axial accumulation is complete, and therefore, the effective viscosity is relatively constant. In vessels of the capillary size (diameter about 6 μm), the highly deformable red cells (diameter about 7 μm) can be squeezed through single file, with a bullet-like configuration. Therefore, varying quantities of red cells can pass through the capillary per unit time with little effect on blood viscosity.

There is a third factor that influences the effective viscosity of blood. In the very narrow vessels (arterioles and capillaries), the blood behaves as if the viscosity were reduced (Fåhraeus-Lindquist effect), which also contributes to reducing the demand on the heart to produce a driving force. This effect is of great importance during heavy muscular activity.

For reasons, mentioned above, when measured in vivo, the effective viscosity of the blood passing through capillaries is approximately 50 percent of that in large vessels and is reduced to about the same viscosity as that of plasma, or to 2.2 relative units.

The viscosity of the blood varies with its temperature. At 0°C, the viscosity is 2.5 to 3 times as great as at 37°C. This contributes to reducing the circulation in tissues exposed to cold, as in the case of frostbite.

In Summary Various factors reduce the effective viscosity of the blood passing through vessels with small diameters, so that the viscosity of blood traversing the capillaries differs little from that of plasma. The readily deformable nature of the red cells is one such factor. Furthermore, the red cells float along the axis of the vessel leaving a cell-free surrounding zone that acts as a lamina over which the cells can slide smoothly. In reality, the blood, when passing through narrow vessels will behave as if it were a Newtonian fluid. This means that its viscosity will not increase when (e.g., during heavy exercise) the cardiac output and the rate of blood flow through the muscle capillary bed increases dramatically. (See Burton, 1965; Keele et al., 1982.)

Buffer Action—Blood pH—CO_2 Transport

Apart from its many other functions, the blood also acts as a buffer (for a detailed discussion, see Davenport, 1969; Siggaard-Andersen, 1974). A buffer solution contains an acid or base that is only slightly ionized (weak) and a highly ionized salt of the same acid or base. If we take a weak acid (HA), we have an equilibrium

$$HA \rightleftharpoons H^+ + A^-$$

If a strong acid is added to the solution, there is an increase in the hydrogen concentration, pushing the reaction to the left. As long as the buffer salt can provide A^- ions, thereby forming undissociated HA, a change in the pH of the solution is prevented, or "buffered."

The pH of arterial blood is, at rest, 7.40, and of the mixed venous blood approximately 7.37. (At a pH of 7.40 the hydrogen concentration is 0.00004 mM.) In the catabolism of the cells, CO_2 is formed, and in the case of anaerobic metabolism there is a net production of H^+ with the hydrolysis of ATP as the dominant source (see Busa and Nuccitelli, 1984). The oxidation of P and S in protein leads to the formation of phosphoric and sulphuric acids. The predominantly acid nature of the metabolites could easily explain the shift in pH as blood passes the tissue's capillary beds.

The isoelectric point of the proteins in the blood is on the acid side of the blood pH. Thus, suspended in an alkaline solution, the proteins ionize as acids and form negatively charged anions ($NH_2-R-COOH \rightleftharpoons NH_2-R-COO^- + H^+$ where R symbolizes the protein radical; for simplicity we may write protein$^-$ for the anion). The proteins, therefore, act as hydrogen acceptors. With proteins ionized as weak acids, we have a buffer system: H protein $\rightleftharpoons H^+ +$ protein$^-$.

The number of ionizable groups in the protein is large, and as whole blood contains about 190 g protein per liter (with about 70 g in the plasma), its capacity to accept hydrogen without pH changes is considerable. In this respect, the potency of the plasma proteins is only one-sixth that of Hb.

There is another reaction of considerable physiological importance. $H-HbO_2$ is a stronger acid than reduced $H-Hb$; i.e., it dissociates more completely than does $H-Hb$. Hence, when blood is giving off O_2, the hydrogen-ion concentration of the blood falls, and pH rises.

Before we discuss the hydrogen exchange in the capillaries, we must draw attention to the second important buffer system of the blood. The CO_2 formed in the cells is dissolved and diffuses freely into the erythrocytes. Catalyzed by carbonic anhydrase present in the red cells, CO_2 forms, with water, H_2CO_3. This weak acid is dissociated as follows

$$CO_2 + H_2O \rightleftharpoons H_2CO_3 \rightleftharpoons H^+ + HCO_3^- \tag{1}$$

The equilibrium in the reactions in Eq. (1) is determined by the concentration of the various molecules and ions. If free H^+ can be removed from the system, the reaction will be pushed to the right. Potentially, this reaction gives place for more CO_2 in the solution without a change in the pH. Though more H_2CO_3 will be formed, this is only an intermediate step in the formation of H^+ and HCO_3^-. In this sense, the supply of a hydrogen acceptor will actually determine the final equilibrium. Protein anions may serve as such a hydrogen acceptor. The CO_2 is produced in the tissue and diffuses into the red cell. At the same time, O_2 is diffusing out to the tissue, where the O_2 concentration is lower than in the capillaries ($HbO_2 \rightarrow Hb + O_2$). Reduced Hb is a weaker acid than HbO_2, and when reduced, it "binds" some of the dissociated H^+ ions ($CO_2 + H_2O + Hb^- \rightarrow HCO_3^- + H$—Hb).

We may now summarize the main functions involved in *the CO_2 transport from the muscle tissue to the lungs* (Fig. 4-2):

1 The most important reactions in the O_2 and CO_2 exchange between blood and tissue are (*a*) the formation of a weak acid, reduced Hb, from the stronger one, HbO_2; (*b*) the formation of $H^+ + HCO_3^-$ from $CO_2 + H_2O$ with the carbonic anhydrase serving as an enzyme in the intermediate formation of H_2CO_3. The interplay between (*a*) and (*b*) can in a quantitative way be illustrated by the following example: For each millimole of oxyhemoglobin reduced, about 0.7 mmole of H^+ can be taken up, and consequently 0.7 mmole CO_2 can enter the blood without causing any change in

Figure 4-2
Schematic presentation of the processes occuring when carbon dioxide passes from the tissues into the erythrocytes. At the bottom the effect of oxygenation and reduction upon buffering action of the imidazole group of hemoglobin is illustrated. An increase in the acidity of the blood drives the reaction to the right, and oxygen is given off. (A decrease in the acidity of the solution would drive the reaction to the left, and oxygen would be taken up by reduced hemoglobin.) In other words, the reduction of oxyhemoglobin causes the hemoglobin to become a weaker acid and to take up hydrogen ions from the solution. (*From Davenport, 1969.*)

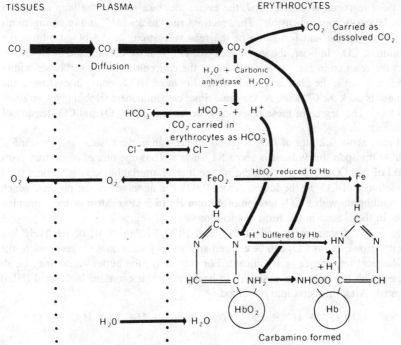

pH. If the CO_2 were derived only from fat combustion, 0.7 mole of CO_2 would be produced per mole of O_2 used [respiratory quotient (RQ) = 0.7]. This means that the formation of reduced Hb from the oxygenated HbO_2 would completely buffer the CO_2 uptake. However, at rest, 0.82 to 0.85 mole of CO_2 is formed per mole O_2 used. During heavy exercise, this figure is close to 1.00.

2 Some of the remaining CO_2 can combine directly with Hb forming carbaminohemoglobin (CO_2 + $HbNH_2$ $\rightleftharpoons$ $HbNHCOOH$), and the simultaneous reduction of HbO_2 greatly favors this formation.

3 There is an increase in the volume of dissolved CO_2, so that the venous blood at rest contains about 0.003 l CO_2 · l^{-1} more than the arterial blood.

Together, these three factors explain how the CO_2, produced at rest, can be transported with the very small change in pH of 0.03 to the acid side.

The net result of the reaction under item 1 is an increase in hydrogencarbonate ions (HCO_3^-) in the red cells. HCO_3^- can freely diffuse across the cell membrane and enter the plasma. Actually, there is a simultaneous decrease in the concentration of the anions as undissociated HHb is formed. The consequent excess of cations within the cell and lack of positive ions in the plasma cannot be compensated by a simple diffusion of a cation (in this case K^+) out to the plasma. The cellular activity "prohibits" an exchange of metallic cations. To restore the electrochemical equilibrium (Donnan equilibrium), the Cl^- ions diffuse into the cell, thereby replacing its loss of anions. This chloride shift allows about 70 percent of the formed HCO_3^- to be transported in the plasma.

If these reactions are understood, the events that take place in the lung capillaries should be easily comprehensible. The reactions run "backward," and in an ingenious way the O_2 uptake and formation of the relatively strong acid HbO_2 facilitate the exclusion of CO_2. In brief, the increase in free H^+ with the dissociation of H—HbO_2 forces the reaction in Eq. (1) to the left. As the concentration of bicarbonate within the cell decreases, the plasma is ready to send in more HCO_3^- ions. In exchange, the plasma gets back the Cl^- ions. At the same time, carbaminohemoglobin gives up some of its CO_2. The degree of these reactions is dependent on the O_2 and CO_2 tension of the alveolar air.

At rest, about 0.2 liter of CO_2 is produced per min in the tissues, and the cardiac output to transport this volume is about 5 l · min^{-1}. Thus one liter of blood transports 0.040 l of this CO_2. The CO_2 content of one liter of arterial blood is still about 0.5 l after delivery of CO_2 to the lungs. About 0.025 l is dissolved in the plasma, which is in equilibrium with a CO_2 tension of 40 mm Hg (5.3 kPa). Most of the remaining CO_2 is in the plasma in the form of bicarbonate.

The buffer system composed of CO_2 and HCO_3^- is, in itself, of relatively low capacity. However, as CO_2 can be expired and actually stimulates the respiration, the physiological importance is significant. For simplicity, the buffer action may be illustrated with the following example. In heavy muscular exercise, lactic acid (HLa) is formed. After diffusion into the blood,

$$Na^+ + HCO_3^- + H^+ + La^- \rightleftharpoons Na^+ + La^- + H_2CO_3 (\rightleftharpoons H_2O + CO_2)$$

the reaction goes to the right, as H_2CO_3 is a weaker acid (a stronger acceptor of H^+) than lactic acid. The increase in free CO_2 and the decrease in pH of the blood stimulate to a hyperventilation, thereby decreasing the CO_2 content of the blood and body. This causes the buffer base to decrease due to the release of "volatile" carbonic acid anhydride. As the lactic acid is later oxidized or transformed into glycogen, the blood becomes more alkaline, but this tendency is opposed by a withholding of CO_2 (depression of breathing). (It should be mentioned that the hydrolyses of ATP in the anaerobic metabolism evidently is the dominant source in the H^+ production, not the formation of lactic acid, see Busa and Nuccitelli, 1984.)

It was mentioned that the pH of the arterial blood is close to 7.4 at rest. In the muscle cell, however, it is much lower, or approximately 7.00. After exhaustive exercise, the arterial blood pH may fall below 7.00 and inside the exercising muscle it may decrease to 6.4 (Sahlin et al., 1978). It is mainly the anaerobic breakdown of glycogen and glucose that causes this pH change. The intracellular bicarbonate concentration has been reported to decline from 10.2 mmol $\cdot$ l^{-1} to about 3 mmol $\cdot$ l^{-1}. The intracellular pH was back to normal after 20 min of recovery, whereas the bicarbonate concentration was still well below normal (Sahlin et al., 1978). Exactly what factors are causing exhaustion in exercise of only a few minutes duration are not known. A reduced pH may be an important contributor, as discussed in Chap. 3 ("Muscle Fatigue"). It should be emphasized that it is technically difficult to measure the intracellular pH. The experimentally determined cellular pH is usually an average of the pH of many compartments within the cell, the interior of which is not homogeneous. The pH may be critically low at one site without this being revealed as a figure representing the average.

The powerful buffer capacity of the blood has been emphasized, and the various proteins play an important role in this function. As the protein content of many tissues is high (averaging about 15 percent), we should expect that such tissues have a good potential for contributing to the acid-base equilibrium. In fact, it appears that about 5 times as much acid is neutralized by other tissues than by the blood. Actually, the intracellular proteins can take a 50 percent share in buffering acids. The body of an average man might neutralize one equivalent (in other words, 1 liter of a 1.0 M solution) of a strong acid, such as HCl, before the blood pH would fall below 7.0, close to the lowest pH value compatible with life (Hitchcock, 1960). Without buffers the pH would be about 1.6!

The buffer capacity of blood (and tissues) must be considered as a "first aid" service not capable of maintaining a constant acid-base equilibrium for long periods. It is no problem to eliminate the potential acid produced in the cell respiration; in one day 12 moles of CO_2 is produced, with the potential to yield 12 moles of protons, but the CO_2 can easily be removed by expiration.

There is, however, the metabolic production of "non-volatile" acids, and the formation of phosphoric and sulphuric acids in the degradation of proteins. Depending on the diet, the production of protons can be 40 to 80 mmoles daily. With these protons remaining in the body, the blood pH would fall to about 2.0 in one day. (A very protein-rich diet can contribute to the formation of 150 mmoles per day.) Here the *kidney function* stands as a final guard over the body pH. This is accomplished by several mechanisms:

1 In the tubuli and the collecting ducts, a certain amount of H^+ is secreted into the lumen, causing an increase in the acidity of the urine. This increase may cause the pH of the urine to drop to as low as 4.5. If the CO_2 content of the body fluid increases, more H_2CO_3 is formed in the kidneys, and it is in turn dissociated into $H^+ + HCO_3^-$. The hydrogen ions diffuse through the tubuli wall into the tubular fluid; HCO_3^- diffuses more or less back into the blood.

2 In the tubuli there is also a secretion of ammonia, NH_3, which, together with H^+ may form NH_4^+, thereby trapping hydrogen ions. (Ammonia is actually generated from the amide glutamine, produced in the muscle tissue.) If this proton were to be eliminated as, for instance, $H^+ + Cl^-$, the acidity would be intolerable. But as $NH_3 + H^+ + Cl^-$, the elimination takes place in the form of a neutral salt NH_4Cl.

3 In the blood there are phosphates, like $Na^+ + Na^+ + HPO_4^{2-}$, which, during the passage through the tubuli, pick up hydrogen ions, with the formation of $Na^+ + H_2PO_4^-$ as the result. Na_2HPO_4 is more alkaline than NaH_2PO_4, and in the tubuli, the ratio Na_2HPO_4/NaH_2PO_4 may be varied according to the need for the elimination of H^+ (see Fig. 4-3).

Figure 4-3
To the left is illustrated how one H^+ ion is "trapped" from H_2CO_3 by the formation of $H_2PO_4^-$ from HPO_4^{2-}. In exchange another cation, Na^+, is leaving the Na_2HPO_4 complex, crossing the tubular cells into the extracellular fluid. The net effect is an ongoing passage of $Na^+ + H_2PO_4^-$ in the tubule. To the right is a simplified illustration of how NH_3 is secreted by the tubular epithelial cells with the formation of a NH_4^+ ion by the inclusion of H^+. In this case a Na^+ ion can also be transferred to the extracellular fluid and eventually back to the blood.

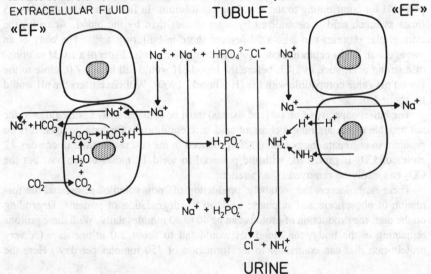

The formation and excretion of acidic phosphate, as well as the other two processes, assist the conservation of sodium by the body. On the other hand, excess alkali may be eliminated with the urine as bicarbonate or basic phosphate. If the amount of "nonvolatile" acids should rise, due to diet or metabolic disturbances, normal kidneys have the capacity to increase their rate of H^+ elimination. When the arterial blood pH is reduced it is called *acidosis;* when it is abnormally high it is called *alkalosis* (below or above 7.35 and 7.45 respectively).

Summary   The blood concentration of H^+ is low, with a pH around 7.40 at rest. The metabolic rate of H^+ production is high, however, but buffer systems can keep the pH fluctuations at a very moderate level. In the healthy individual, vigorous muscular activity may cause marked disturbances of the "normal" pH and the intracellular pH may drop to 6.4 and the arterial blood pH may decline to levels below 7.0.

An important aspect of the buffer systems is the dissociation of the weak carbonic acid H_2CO_3 into CO_2 and H_2O. An increase in the arterial CO_2 and H^+ concentrations stimulate an increased pulmonary ventilation. Certainly one cannot expire hydrogen ions, but the expiration of CO_2 has a similar effect, i.e., it keeps the blood H^+ concentration low.

The proteins in the blood and tissues form powerful buffer systems. The difference in acidity between Hb and HbO_2 minimizes the variation in the pH of the blood as it passes the tissue and takes up CO_2 and subsequently releases it through the lungs.

In the kidneys, excess H^+ produced in the metabolism of "nonvolatile" acids can be excreted as free ions, and bound to NH_4^+, and NaH_2PO_4.

In other words, there are three strategies of defence against increasing acidity: buffers, respiration, and renal excretion of acids.

HEART

It is assumed that the reader is familiar with the anatomy and basic physiology of the heart. (See Fig. 4-4a.)

Cardiac Muscle

Under a microscope the heart muscle shows transverse striation essentially similar to that of the skeletal muscle, but the nuclei are placed centrally within the cell; the fibers give off branches which anastomose with adjacent fibers, giving an impression of a syncytial continuity of the muscle fibers. However, electron microscope studies give a different and clearer picture of the structure. Apparently the heart muscle is subdivided into numerous cell territories by cell boundaries fanning transversely across the branches of the heart muscle tissue (Sjöstrand and Andersson-Cedergren, 1960). These boundaries are associated with the intercalated discs of the heart, which form wavy "membranes" about 10 nm apart. Because of the "cutting" of the muscle fibers, the heart muscle does not represent a true syncytium. The cell mass of the left ventricle is approximately three times that of the right ventricle. Basically, the mechanism behind

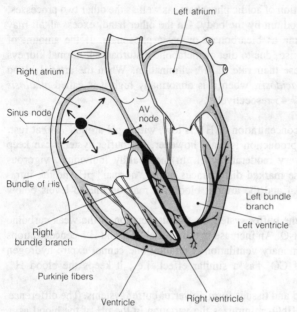

Figure 4-4a
Electrical impulses that cause regular, rhythmic contractions of the heart muscle originate in the sinus node. The impulses spread along the atria and are conducted to the ventricles by way of the atrioventricular node (AV node), the bundle of His, the left and right bundle branches, and the Purkinje fibers. (*From Shepherd & Vanhoutte, 1979.*)

the activation and contraction of the heart muscle fiber is similar to the events described for the skeletal muscle (Chap. 2).

The conductive system of the heart and the effect of the refractory period after an excitation give the heart its rhythmical activity, unceasing from early embryonic life until death.

Blood Flow

Normally the blood vessels in the heart do not form anastomoses if the diameter is larger than 40 μm (the size of small arterioles). Therefore, in experiments on animals, a sudden occlusion of an artery will be followed by an infarction of the tissue supplied by that artery. On the other hand, a gradual narrowing and final occlusion of a coronary artery will promote the development of a very rich network of anastomotic vessels. There are studies indicating that effective collateral vessels may develop more profusely if the animals are exercised, whereas training without concomitant coronary artery constriction apparently does not improve collateral blood flow (see Feigl, 1983). Clinically, the experience is that a development of collateral circulation may occur in cardiac patients (see Gregg, 1974). On the other hand, there are no convincing studies indicating that habitual physical activity will improve the collateral circulation in healthy individuals. The capillary network in the heart is extraordinarily rich, with 2500 to 3000 capillaries per mm^2. The capillary density increases by training (see Chap. 10). There is a crucial vascular compartment in between the larger arteries and the capillary network which is certainly functionally important, but difficult to evaluate as to whether or not it is affected by use and disuse, in general, and with regard to degree of regular physical activity, in particular.

The contraction of the heart muscle interferes mechanically with the blood flow through the heart. In the left ventricle of the dog at rest, there is a complete stop during the isometric contraction, and a backflow is actually established, but a peak flow is reached when the systolic pressure is at maximum, during the early ejection phase. A second peak flow is reached during early diastole, after which it drops gradually to about 70 percent of maximum just at the end of diastole (Gregg and Green, 1940) (Fig. 4-4b). In the right heart, where pressure is lower during systole, the fluctuations are not so pronounced as in the left ventricle.

At rest, the coronary artery flow during systole is less than 20 percent of the total blood flow per cardiac cycle. During exercise, however, up to 40 percent of the total coronary blood flow takes place during systole. It is not completely understood why the diastolic blood flow stops completely when the systolic blood pressure drops below about 40 mm Hg (5.3 kPa).

The coronary blood flow increases a few heart beats after the onset of exercise. Oxygen deficiency has a strong dilating effect on the arterioles of the heart. One causative factor for the vasodilation may be the nucleoside adenosine (Knabb et al., 1983). During the contraction of the heart muscle there is a breakdown of ATP to ADP as well as AMP. Some AMP is enzymatically catalyzed to adenosine which can leave the cardiac cell, diffuse through the interstitial space, and reach the arterioles to cause vasodilation. Other factors may contribute, e.g., CO_2 produced in the aerobic metabolism. K^+ diffusing out of the depolarized muscle cell may have a transient vasodilating effect at the onset of a tachycardia (increase in heart rate). There is, for some unknown reason, an increase in cardiac contractility and oxygen uptake secondary to an increased coronary blood flow. Normally, there is an excellent match between myocardial metabolism and the coronary perfusion. Even at rest, a relatively large part of the oxygen offered to the tissue, or 60–70 percent of the arterial content is utilized in the mitochondria; during maximal exercise the arteriovenous oxygen difference is further increased by 10 to 20 percent. There is no evidence for an exhaustion of the coronary vasodilation reserve in healthy people even when exercising at maximal rate. The heart's potential for an anaerobic energy yield is much more limited than that of the skeletal muscles. Apparently, it takes only about 8 contractions or 8 sec for the heart to deteriorate in its performance, if deprived of oxygen (see Gibbs, 1978). No evidence is reported of a significant reflex effect on coronary blood flow.

Studies of trained male subjects showed that the myocardial oxygen demand and coronary blood flow was relatively low at rest as well as during a given work rate. A reduced heart rate may be one explanation of these findings (Heiss et al., 1976).

Summary In healthy individuals, there is a very good matching of the coronary blood flow and the cardiac metabolic demands. There is much evidence that adenosine

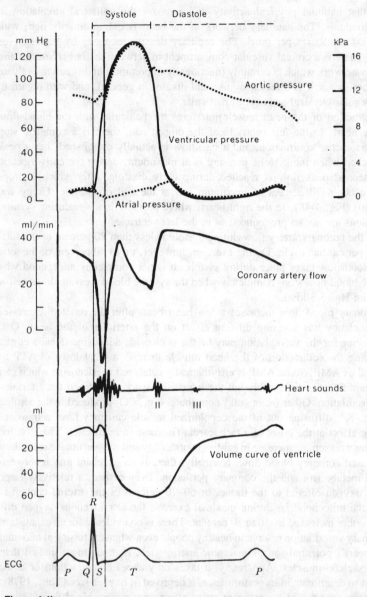

Figure 4-4b

Pressure variations in the left heart chambers and aorta during the cardiac cycle. See text for details. [The heart sounds can be objectively analyzed under various pathological conditions by phonocardiography. Closure of the atrioventricular valves contributes largely to the occurrence of the first sound (I), but so do vibrations in the tissues and turbulence of the blood flow; closure of the aortic valves is of prime importance for the occurrence of the second heart sound (II); vibrations of the chamber walls caused by movement of blood into the relaxed ventricle produce the third sound (III).] (*Modified in part from Wiggers, 1952.*)

acts as the local metabolic vasodilating transmitter between the cardiac cells and the coronary vascular smooth muscle cells. There is also an interesting observation that the total blood flow per cardiac cycle does not change much in the transition from rest to exercise. In other words, the heart rate is well correlated with the coronary blood flow. (For review, see Feigl, 1983.)

Pressures During a Cardiac Cycle

Figure 4-4*b* describes the pressure variations in the left heart chambers as well as in the aorta. The systole of the left ventricle starts with an isometric contraction, for the mitral valves close rapidly as the pressure in the ventricle exceeds that of the atrium (phase within the vertical lines). Within about 0.05 s, the ventricular pressure is brought up to and above the level in the aorta, and the aortic valves open. The isotonic contraction increases the pressure further. The peripheral resistance does not permit the same volume of blood to escape from the aorta as is ejected into it. Part of this volume is "stored" in the distended aorta and its large branches ("windkessel vessel"). Then, as the pressure falls in the ventricle during the relaxation of the muscle, the aortic valves close, and the elastic property of the aortic wall can propel the stored blood out into the arterial tree. The intermittent energy outbursts of the heart would give an intermittent flow if the vessels were rigid tubes. Part of the potential energy is, however, taken up by the elastic arterial wall and then released during diastole of the heart, keeping the hydraulic energy level close to the heart continuously high. During the systole, blood has returned to the large veins close to the heart and the atrium. There is, perhaps, a passive lengthening of the atrium as the ventricle contracts, which may facilitate the filling of the atrium. During diastole there is a period of rapid filling of the ventricle after the opening of the mitral valves. The period of diastasis follows, during which the filling is much less rapid. The next cycle begins with the atrial contraction, which more or less empties the atrium. At rest the atrial contraction is responsible for approximately 20 percent of the total ventricular filling.

With a heart rate of 75, the time for diastole is about 0.48 s and for systole 0.32 s (40 percent of the heart cycle); with a heart rate of 150, the periods are 0.19 and 0.21 s (52 percent), respectively. It should be noted that an increase in heart rate occurs mainly at the expense of the length of diastole and its period of diastasis. As the ventricle may eventually offer resistance to filling at the end of diastole, the pressure gradient between atrium and ventricle drops. An increase in heart rate can, in such a case, improve the filling of the heart and increase the cardiac output.

The physical events in the right heart are essentially similar to those just described, but the ventricular and pulmonary artery pressures during systole are about one-fifth of those in the left heart.

In the *aorta,* the pressure at rest varies between 120 mm Hg (16 kPa) during systole and 80 mm Hg (10.5 kPa) at the end of diastole. In the *pulmonary artery* the values are 25 and 7 mm Hg (3.3 and 0.9 kPa), respectively. It should be emphasized that the sphygmomanometer cuff technique does not always give the same value as direct measurement or measurement via a catheter in the artery. This holds true particularly during exercise.

Innervation of the Heart

Many *parasympathetic* nerve fibers from the vagus terminate in the region of the pacemaker, or sinoatrial node. When stimulated, they deliver acetylcholine, causing a slowing of the heart rate (inhibition). *Sympathetic* nerves are the efferent fibers from the upper part of the paravertebral ganglia which join the cardiac sympathetic nerves. They end in a dense network within the heart muscle. The transmitter substance is norepinephrine. A sympathetic activation of the adrenal medulla will release the chemically related epinephrine and also some norepinephrine. (A common name for epinephrine and norepinephrine is catecholamines). As mentioned in previous chapters the effects of transmitter substances are determined by the membrane receptors of the target cells. On the basis of purely pharmacological criteria, we can distinguish two groups of receptors, alpha (α), and beta (β) adrenergic receptors (with subgroups α_1 and α_2, β_1 and β_2). The α receptors are abundant on the cell membrane of all vascular smooth muscle cells, but are relatively sparse in the heart. Throughout the body α adrenergic activity causes a vasoconstriction. The effect on the heart vessels is, however, masked for the following reason: Norepinephrine and epinephrine activate β_1 receptors in the heart. This will (1) increase the heart rate by increasing the firing rate of the sinus node and increase the conduction velocity in the atria, the atrioventricular node, and the Purkinje system. The effect is therefore the opposite of a stimulation of the parasympathetic nerve. (The effect on heart rate is called *chronotropic action*.) Then (2) β_1 receptors activation will increase the myocardial contractility. The consequence of this will be an increased stroke volume by a reduced end-systolic volume of the heart. This effect, exerted by the catecholamines, is known as a *positive inotropic action*. Certainly the venous filling of the heart is decisive for the size of the consecutive stroke volumes. This will be discussed below.

Let us now turn back to the question of α activity and vasoconstriction. An increased sympathetic drive will, as just mentioned, elevate the heart rate and the heart beat becomes more forceful. This will increase the myocardial oxygen uptake *and* coronary blood flow, most likely with adenosine as a mediator. Therefore, the net effect will be a dilation of the coronary vessels. The vagus does not seem to influence the contractility of the heart muscle, but it may slightly depress the atrial contractility. Via this automatic innervation of the heart, the inherent rate of beating can be highly modified; the healthy heart in a young, fully grown individual can cover a range from about 40 at rest to 200 beats/min during heavy exercise.

β_2 adrenergic receptors are present in vascular smooth muscles, most abundant in resistance vessels (see below) in skeletal muscles, and also in the myocardium. An activation will dilate the vessels. The β_2 receptors are sensitive to epinephrine but insensitive to norepinephrine.

HEMODYNAMICS

Mechanical Work and Pressure

Chemical energy is transformed into mechanical (external work) energy plus heat by the contraction of the heart muscle. The mechanical efficiency of the heart, i.e., the external work divided by the total energy exchange, is rather low, or only some 10 percent at rest.

The external work is calculated from data on force times distance moved, or for a fluid, pressure times volume moved. The pressure can create kinetic energy and a flow of the fluid. The higher the velocity of the fluid, the higher the kinetic energy (related to velocity squared). On the other hand, it may be concluded that in the part of a vessel where the velocity is highest, the fluid pressure against the wall is smallest, since total hydraulic energy = pressure energy + kinetic energy (Bernoulli's principle). If the pressure is measured in a vessel with a catheter or tube connected to a manometer with the opening against the flow, the kinetic energy of flow is reconverted to pressure (*end pressure*) and the total energy is measured. With only a side hole in the catheter the *side* or *lateral pressure* is measured, and it will be less than the end pressure by an amount equivalent to the kinetic energy of flow, according to the formula given above. It can be calculated that, at rest, the kinetic factor in the aorta flow is only about 3 percent of the total work of the heart, but during exercise with a cardiac output 5 times the resting level, it may be an important part of the total work of the heart, possibly as much as 30 percent. This means that a simultaneous measurement of end pressure and side pressure in the aorta may show a difference of some 75 mm Hg (10 kPa), the end pressure being highest (Burton, 1965) (see discussion below). In the other arteries and in the smaller vessels, the kinetic energy factor is normally negligible, at least at rest. It should be pointed out that the sphygmomanometer cuff method for the measurement of blood pressure only measures the side pressure.

Hydrostatic Pressure

The pressure in a vessel is created by the continuous bombardment by the molecules of the fluid against the inner surface of the vessel. The pressure is equal at all points lying in the same horizontal plane in a static liquid. The hydrostatic pressure in a fluid at rest, under the influence of gravity, increases uniformly with depth under the free surface (Pascal's law). It is evident that in the supine position the hydrostatic pressure is of similar magnitude in all parts of the body, but in the standing person it is higher (perhaps 100 mm Hg or 13.3 kPa) in the arteries of the feet than in the head. The hydrostatic factor in a standing individual is modified by the action of the valves in the veins of the extremities. Arterial pressure (e.g., for diagnostic purposes) should be measured at the horizontal level of the heart, or the value should be corrected to correspond to heart level. For practical purposes, measurement of the blood pressure with a sphygmomanometer cuff around the arm of a seated subject gives satisfactory values.

When a person is tilted from a supine position to an erect position with the feet down, the veins, and to some extent the other vessels below heart level, will dilate passively as a result of the hydrostatic force. Temporarily the blood is pooled, but when the vessels are filled, the flow continues unhindered by the uphill flow.

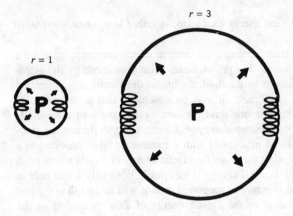

Figure 4-5
The vascular tension at a given transmural pressure P is roughly proportional to the radius of the vessel, r. (This is a simplification neglecting the thickness and properties of the wall.) Thus, if the radius is tripled, 3 times more tension will be required to maintain the same pressure.

Tension

The pressure within the heart and vessels will more or less distend the walls (Fig. 4-5). The resistance to this force, producing *tension*, depends on the thickness of the wall and its content of elastic and collagen fibers and of active smooth-muscle fibers. The elastic tissue can balance the pressure without any energy output, but the contraction of the muscles requires a continuous expenditure of energy. The tension T is roughly proportional to the pressure difference P on the inside and outside of the wall (transmural pressure) and the radius r of the vessel, or $T = aPr$, where a is a constant (Laplace's law). Therefore, a given pressure applied in a small vessel does not produce the same tension it does in a vessel with a larger radius (Fig. 4-5). The very thin membrane of the capillary, essential for the exchange of materials, can withstand the capillary blood pressure because its radius is so small.

The same law may, with some reservations, be applied to the heart. With a given tension developed by the heart muscle, a lower pressure is produced if the heart is dilated (increased in size, i.e., radii of curvature). If the diameter of the heart is doubled, the tension per unit length of ventricular wall must be about twice as great to produce the same pressure. The energy cost, i.e., the oxygen uptake, of the heart is related to the tension that must be developed and the time this tension is maintained. Therefore, an increase in the size of the heart increases the load on the heart.

It should be borne in mind that the heart muscle follows the same law as the skeletal muscle: Its ability to produce tension increases with length. By applying Laplace's law, one finds that the heart has to pay more to maintain a given pressure if a reduction of muscular strength is compensated for by a dilation (increase in length of the fibers).

The maximal active force developed by the cardiac muscle is attained at a sarcomere length of about 2.2 µm but it decreases rapidly both at sarcomere lengths below and when stretched beyond this optimum length. In skeletal muscle fibers there is more of a force plateau with changes in sarcomere length of approximately between 2.0 and 2.2 µm. The cardiac muscle is also relatively stiff and tension rises extremely rapidly as the muscle is stretched over a sarcomere length range for which resting tension is minimal in skeletal muscle (Sonnenblick and Skelton, 1974).

Flow and Resistance

Since the difference in the hydraulic energy of a fluid in two parts of a system cannot be resisted by the fluid, it flows with a rate proportional to the energy difference or pressure head. The resistance to flow of blood results from the inner friction or viscosity of the blood. There is a cohesive force between the blood and the wall of the vessel which "retards" the flow of the molecule layers close to the wall. The nearer the center of the vessel, the higher the speed of each lamina of fluid, and this "friction" phenomenon results in a maximal speed in the very center of the vessel. The hydraulic energy provided the blood by the contracting heart muscle is gradually spent and transformed to heat. Fluid flow can be *laminar* (streamlined); that is, each lamina of liquid slips over adjacent laminae without mixture or interchange of fluid. At a critical velocity (expressed by Reynolds number) the laminae of fluid move irregularly and start to mix with one another; the flow becomes *turbulent*. During turbulence the energy loss is larger, and for a given pressure gradient the flow is lower (Fig. 4-6). The velocity of flow in the ventricles of the heart and aorta is normally turbulent during the early phase of contraction. This turbulence produces vibrations, causing high-pitched sounds which can easily be heard by listening over the heart (Fig. 4-4*b*). If the velocity is abnormal, as in mitral stenosis, the heart sounds are abnormal. Similarly, a shunt between aorta and the pulmonary artery (open ductus Botalli) can give rise to a turbulent flow in that shunt vessel. In measurement of the blood pressure by applying a measured pressure over the brachial artery (with the sphygomomanometer cuff), the peak blood pressure suddenly overcomes the resistance of the gradually reduced

Figure 4-6
In turbulent flow, the energy loss is greater and the rate of flow slower than in laminar flow at a given pressure gradient. Length of arrows indicates rate of flow. (*From Haynes and Rodbard, 1962.*)

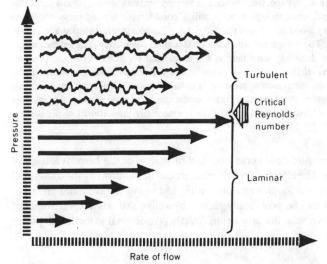

Pressure

Turbulent

Critical
Reynolds
number

Laminar

Rate of flow

compression. The blood flows with a high velocity through the narrowed artery, and a turbulence gives rise to sounds detectable by the stethoscope.

The radius of a tube is a deciding factor in the flow through it; the resistance to flow is a reverse function of its radius to the fourth power. A decrease to half the radius, other things being equal, will actually decrease the flow to one-sixteenth of the orginal value. A dilation of 10 percent increases the blood flow about 50 percent compared with the flow through the same vessel constricted. The variation in activity of the smooth muscles of the vessels gives a very sensitive instrument for the control of blood flow.

The factors of importance for flow can be described in more detail by presenting the Poiseuille-Hagen formula, although the equation is only applicable for nonpulsatile flow of Newtonian fluids. Actually, flow in the blood vessels is normally laminar except in and close to the heart. As already mentioned, blood passing through narrow vessels behaves like a Newtonian fluid. The flow (F) through a tube is actually proportional to the driving force (ΔP), inverse to the viscosity (η), proportional to the radius of the vessel raised to the 4th power (r^4), and inverse to the length of the vessel (l):

$$F = \Delta P \times \frac{\pi}{8} \times \frac{1}{\eta} \times \frac{r^4}{l} \text{ (Poiseuille-Hagen formula)}$$

The resistance to laminar flow (R) is proportional to pressure gradient per rate of flow. This simple formula can help to evaluate the resistance offered to the blood flow in the vascular bed. If the blood pressure in the right atrium is zero, an increase in mean aortic blood pressure (BP_{mean}) at constant cardiac output ($\dot{Q}$) gives evidence of an increased peripheral resistance to flow (as $R \times \dot{Q} = BP_{mean}$).

The complex composition of blood and the distensible blood vessels complicate the situation in some ways. If a given pressure gradient ("driving pressure") is maintained between artery and vein, the flow is not, as expected, constant at varying pressure levels. At low pressures the flow becomes zero, even though there is still a considerable driving pressure. When discussing the coronary blood flow it was pointed out that the flow stopped even at a relatively high arterial pressure. The pressure at which the vessels close completely is called the critical closing pressure. The decisive factor here is the transmural pressure. When the pressure outside the inner layer of the wall (tissue pressure and active muscular contraction in the vessel wall) exceeds the intravascular pressure, a collapsible vessel collapses (this happens "artificially" in the regular blood pressure measurement commented on earlier). At pressures well above the critical closing pressure, the flow is essentially proportional to the driving pressure.

Summary   The pressure-flow curve obtained in studies of the hemodynamics of circulation is not linear. The effect of driving pressure can be modified by the transmural pressure, especially in the arterioles and capillaries, and at low intravascular pressure. The resistance to flow can be profoundly altered by active and passive variations of the radius of the blood vessels, the flow being directly proportional to the fourth power of the radius (r^4).

BLOOD VESSELS

Anatomically, and to some extent functionally, the blood vessels may be classified into arteries, arterioles, capillaries, and veins. The exchange of water, electrolytes, gases, etc., can occur only across the semipermeable membrane of the capillaries and, to some degree, in the postcapillary venules; the other vessels are only transport channels. The amount of elastic fibers, collagen fibers, and smooth muscles varies in the different vessels; in the big arteries the walls are thickest, but in veins of the same size the walls are thin, with few elastic and muscle fibers. Endothelial cells cover the inner surface of the vessels; they are formed as flattened plates where pressure is high, and as cuboidal or rounded plates where pressure is low. The vascular system adapts to the forces acting on the walls, modifying their mechanical property.

It has been emphasized that the local blood flow is mainly determined by the pressure head and the diameter of the actual vessels. The capacity of the different vessels to influence vascular resistance, blood flow, and blood distribution is, in a functional way, described by the following subdivision: windkessel vessels (main arteries), resistance vessels (arterioles are the most important, but capillary and postcapillary sections are also of significance), precapillary sphincters (determining the functioning capillary surface area), shunt vessels (e.g., arteriovenous anastomoses), capacitance vessels (veins), and exchange vessels (capillaries) (Folkow and Neil, 1971). Most of this terminology is illustrated in Fig. 4-7.

Arteries

The arteries serve as windkessels, or pressure tanks, during the ejection of blood from the heart, and the elastic tissue tends to recoil when stretched. This property enables it to store and release energy generated by the heart and convert an intermittent flow to a continuous flow. The resistance to flow in the large arteries (and veins) is very small. The diameter of the vessels is large and the velocity of flow is high.

The expansion of the arterial wall during the ejection of blood causes a pressure wave that travels along the peripheral blood vessels at a speed of 5 to 9 m $\cdot$ s^{-1} (the velocity of the blood in the aorta at rest is 0.5 m $\cdot$ s^{-1}). The more elastic the arterial wall, the lower the speed of the pulse wave. In practical application, the frequency of this wave is counted as the pulse rate, since it can easily be felt over the radial or carotid arteries.

Arterioles

In arterioles the peripheral resistance is high, causing a marked pressure drop. The velocity is still high because the total cross section of the arterioles is not so large. The individual vessels are narrow, from 0.1 mm to 60 μm. In the wall are transversely oriented smooth-muscle fibers that can be activated by stimuli of their nerves' transmitter substances, or by other chemical factors. Figure 4-8, *A, B,* and *C,* shows schematically various degrees of constriction of an arteriole caused by contraction of

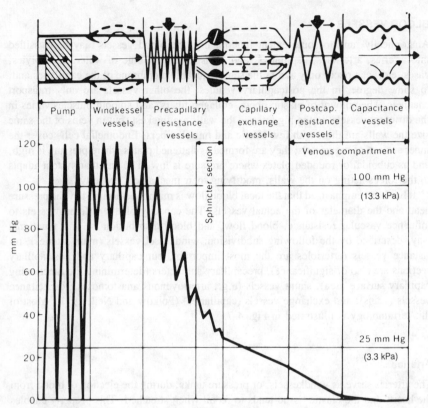

Figure 4-7
A schematic illustration of the functionally different, consecutive sections of the vascular bed, related to the blood pressure fall along the circuit (ordinate blood pressure in mm Hg). Note the marked pressure drop and absorption of the pulse amplitudes in the precapillary resistance vessels. The line at 25 mm Hg (3.3 kPa) represents the plasma protein osmotic pressure (colloid-osmotic pressure). A similar illustration can be made of the sections of the vascular bed in the "low pressure system" (pulmonary circulation), but the pressure in the right pump at rest is alternating between about 25 and 0 mm Hg (3.3 and 0 kPa), and in the pulmonary artery (windkessel vessels) between 25 and 10 mm Hg (3.3 and 1.3 kPa). (*Modified from Folkow and Neil, 1971.*)

the smooth-muscle fibers. It should be stressed that the vessels do not have muscles that can actively dilate them.

Figure 4-9 illustrates how the arterioles (and capillaries) are arranged in parallel-coupled circuits between the arteries and the veins. They can effectively alter the total resistance against the outflow of blood from the arteries and thereby the arterial blood pressure and work of the heart; they have a decisive influence on the distribution of blood flow to the various organs. An increase in caliber of the arterioles in a muscle will decrease the resistance in that area and hence increase the flow. If this local vasodilation is not compensated for by a vasoconstriction in another area or an increase

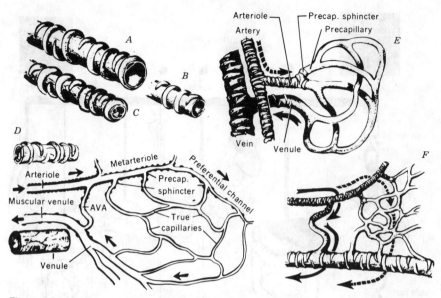

Figure 4-8
A schematic representation of the structural pattern of the capillary bed. The distribution of smooth muscle is indicated in the vessel wall; see text for discussion. AVA = arteriovenous anastomose. (*Modified from Elias and Pauly, 1960.*)

in cardiac output, the arterial blood pressure will inevitably fall. The arterioles are arranged in a series of parallel channels joining the capillary bed (and veins) with the arterial side, effectively "regulating" the blood flow through the organs and tissues. The pressure is still high when the blood enters the arterioles, but then drops to 30 to 40 mm Hg (4 to 5.3 kPa) (the pressure is somewhat higher in the kidneys). The systolic and diastolic pressure variations will usually disappear before the blood reaches the capillaries (Fig. 4-7).

Capillaries

The microarchitecture of the capillary network is different in different organs (Fig. 4-8). The precise pattern of the capillary arrangement is not completely known. It may be assumed to be as follows (see Zweifach, 1973):

1 The arterioles branch off into terminal or final arterioles. There the smooth-muscle fibers are arranged in circular or spiral alignment and constitute a single layer. The terminal arteriole ends in a capillary-like channel or "thoroughfare channel" (preferential channel), from which the capillaries branch (Fig. 4-8, *D*). In the proximal wall portions of the thoroughfare channel is a discontinuous coat of thin muscle cells. There is frequently a group of muscle fibers forming precapillary sphincters, especially at the origin of the capillary. The preferential channel finally loses its muscle coat

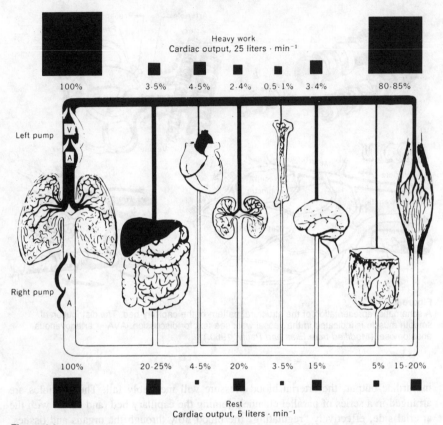

Figure 4-9
Schematic drawing showing how the arterioles and capillaries are arranged in parallel-coupled circuits between the arteries (top) and the veins. The cardiac output may be increased fivefold when changing from real to strenuous exercise. The figures indicate the relative distribution of the blood to the various organs at rest (lower scale) and during exercise (upper scale). During exercise the circulating blood is primarily diverted to the muscles. The area of the black squares is roughly proportional to the minute volume of blood flow. Not included is an estimated blood flow of 5 to 10 percent to fatty tissues at rest, about 1 percent during heavy work.

completely, but can be differentiated from adjoining capillaries by the presence of a somewhat thicker supporting connective-tissue coat. The true capillaries are thin-walled endothelial tubes with no muscle or fibrous fibers. They are kept in place by fine connective-tissue fibers. In skeletal muscles the true capillaries are at least 8 to 10 times as numerous as the preferential channels. In other tissues there are relatively fewer capillaries. The capillaries and the preferential channels drain into the venules, which in turn drain into the larger veins.

2 Another pattern of the microcirculation may be a net of arterioles which divide into anastomosing true capillaries (thin endothelial tubes without muscle cells) (Fig.

4-8, *E*). Precapillary sphincters can be seen. The capillaries converge into venules. No vessels resembling a thoroughfare channel are found.

3 Arteriovenous anastomoses or shunts are typical for the vascular bed in the skin but are also present in other tissues. Anatomically, this may be a vessel with a relatively thick muscle coat running from an artery or arteriole to a vein or venule (Fig. 4-8, *F*). The diameter can, in the latter case, be as small as 15 to 20 μm. Similar shunts may be located proximally to the terminal arterioles or form a direct continuation from one of the branches of an arteriole to the venous circulation. In both cases, this shunt provides a good possibility of maintaining blood flow and low resistance, but the capillary vascular bed is bypassed (dark arrows in Fig. 4-8, *F*).

Summary   The circulation of blood from arteries to veins can take various routes via thoroughfare channels, true capillaries, and arteriovenous or arteriolovenous shunts. The pressure gradient and the activity of the smooth-muscle cells in the walls or the precapillary sphincters decide which route the blood flow will take.

Capillary Structure and Transport Mechanisms

Let us now take a closer microscopic and ultramicroscopic look at the architecture of the capillary. Its length is less than 1 mm (down to 0.4 mm) and its inner diameter ranges from 3 to 20 μm; therefore, there is sometimes hardly room for an erythrocyte (diameter 7.2 μm), which may temporarily stop the flow and be squeezed out of shape as it is forced through. The capillary wall is made of very thin simple squamous cell plates of protein and lipid. The cells are linked to one another by so-called intercellular junctions. This space between adjacent cells is sealed to a greater or lesser extent. On the outer side of the endothelium is a "rigid jacket" consisting of amorphous matrix (basement membrane or basal lamina). The plasma-endothelial interface is also of functional importance, a composite of plasma proteins and other extracellular materials which covers the surface of the endothelial cells and fills the cracks between them. Such a "fuzzy" coating can contain heparin (an anticoagulant), lipoprotein lipase, and fibrinogen. (It should be pointed out that the inner surface of all vessels has a similar architecture.) The capillary wall has no smooth muscle cells or other contractile elements, but has elastic properties. The thin-walled venules following the capillaries have walls which are similar to those of the capillaries.

The most important processes for the exchange of substances across the capillary membrane are filtration and diffusion. Isotope studies have shown that gases, water, and molecules in water and lipid can rapidly diffuse back and forth in the direction of the concentration gradients through the capillary wall.

Lipid-soluble materials can pass through the cells by diffusion. Therefore O_2 and CO_2 can utilize the entire surface area of the capillary wall, and easily use the trans-endothelial route. Lipid-insoluble materials must utilize different mechanisms for passage, although there is some disagreement about which pathways are used (see Lewis, 1979; Crone, 1980; Simionescu, 1983). The interendothelial junction is one natural possibility for the passage of solutes through systems of "pores" or "slits," which provide aqueous channels for filtration and diffusion. There are small "pores" with a diameter less than 10 nm which water and small to moderate size molecules may use as pathways. For transport of larger molecules, like albumin and other proteins, large "pores" or "leak" pathways, which range in diameter size from about 60 nm and upwards, are possible routes (see Renkin, in: Lewis, 1979). The problem is that pores, slits, clefts, or "leaks" have not been morphologically identified well enough. Another complication is that the "fuzzy" coating on the inner surface of the vessel wall can alter the permeability of the pathways. The electrical charge of these structures (polianionic) and the charge of the potential "travellers" may also enhance or complicate a passage (see Haraldsson et al., 1983).

Another transport mechanism under discussion is a shuttle traffic with spherical membranous vesicles (diameter about 60 nm) passing back and forth between the inner and the outer endothelial surfaces. In some areas, e.g. in the renal glomeruli and some of the tubuli, in glands, and in the intestinal mucosa, there are thinned-out portions of the endothelium, the so-called fenestrae, with or without a diaphragm. These structures are most likely involved in the blood-tissue exchange of various substances. Non-fenestrated capillaries, however, are the most widely distributed capillaries in vertebrate animals, e.g., in skin, all types of muscles, and lung alveoli. In the central nervous system (CNS) the capillaries have intercellular tight-junctions which may, to a greater extent, restrict or even prevent passage of substances, in comparison to the more "communicating" junctions in the capillaries of the skeletal muscles. In the CNS there are probably carrier-mediated and active transports between blood and tissues, to force the blood-brain barrier.

According to some estimates, only 0.02 percent of the capillary surface area is devoted to "pores," and the ratio between small and large "pores" may be approximately 100:1.

Summary   The very thin walls of the capillaries (and venules) have a simple endothelium cell layer resting on a basal lamina. Lipid-soluble substances, which include gases, can diffuse freely through the endothelial cells. For the filtration and diffusion of water-soluble particles and larger molecules, the pathways are thought to be "pores" or other types of interruptions of various size in the lines of fusion between the surfaces of the cell membrane of adjacent endothelial cells, i.e. the intercellular junctions.

Filtration and Osmosis

Figure 4-10 shows the driving forces in a process so well outlined by Starling in 1896. Within the capillary the hydrostatic pressure (h.p.) is normally well above the pressure in the interstitial fluid. As the plasma proteins cannot pass through the capillary wall (which is a statement more didactic than true; within 1 to 2 days the entire quantity of plasma protein may be exchanged), they will exert a colloid-osmotic pressure much higher than that existing outside the capillary. For simplicity, the pressures outside the capillary are disregarded (or corrected for) in Fig. 4-10. The gradual change in colloid-osmotic pressure as water escapes or returns to the vessel is also omitted.

When entering the capillary, the net hydrostatic pressure may be 40 mm Hg (5.3 kPa) and this pressure drops gradually to 10 mm Hg (1.3 kPa) on the venous side of

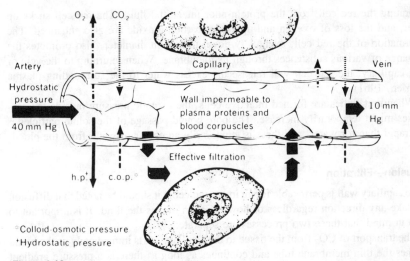

Figure 4-10
Driving force in capillaries. The hydrostatic pressure depends on the degree of vasoconstriction in precapillary vessels. In a resting skeletal muscle it may be only 15 mm Hg (2.0 kPa) at the beginning of the capillary. For details, see text.

the capillary. The net colloid-osmotic pressure is about 25 mm Hg (3.3 kPa, see Fig. 4-7). The net filtration pressure is then $40 - 25$ mm Hg ($5.3 - 3.3$ kPa) "proximally," giving an outflow of fluid, but it is $10 - 25$ ($1.3 - 3.3$ kPa) "distally," giving a pressure differential of 15 mm Hg (2.0 kPa). This means that fluid is sucked back with a force of 15 mm Hg (2.0 kPa). It is easy to see how small variations in pressures can profoundly change the exchanges of fluid. Three examples are given: (1) If the arterial pressure falls, for instance during blood loss, there is also a reduced capillary pressure. Thereby the balance between the hydrostatic pressure and protein osmotic pressure becomes disturbed, with an increased absorption of fluid from the interstitial fluid as a consequence. (2) A decrease in concentration of plasma protein (prolonged starvation or loss due to kidney disease causing albuminuria) lowers the plasma protein osmotic pressure and the return of fluid from the tissue. An increase in the extravascular fluid follows (edema). (3) A rise of the venous pressure increases the mean capillary pressure and hence the outward passage of fluid. Furthermore, the reabsorption of fluid at the venous side of the capillary is decreased (it may be $15 - 25 = -10$ mm Hg, i.e., $2.0 - 3.3 = -1.3$ kPa). It is easy to understand how a failure of the left ventricle of the heart can cause pulmonary edema. When the right ventricle eventually becomes uncompensated by the increased load, the edema "moves" to the systemic circulation area. The liver and dependent parts of the body, which are most affected by the hydrostatic pressure, are sensitive to an elevation of pressure in the systemic veins.

It was mentioned that many capillaries are "too small" for the red cells. Consequently the red cell becomes temporarily squeezed in the bulging capillary, with no plasma separating the red cell from the protein-poor interstitial fluid. The red cell sucks up water, and the loss of oxygen and uptake of carbon dioxide are also enhanced. The deformation of the red cell, eventually down to a 3μm diameter, also promotes the exchange of various substances through its membrane. When returning to the vein or lung capillary, the red cell will release the excess water to the surrounding plasma (Hansen, 1961).

When metabolies are formed in active cells, the osmotic pressure outside the capillaries might temporarily increase. The pore-bound passage of the solute particles is less rapid than for water, and there is temporarily an egress of water from the blood.

Diffusion—Filtration

If the capillary wall is permeable for an ion or molecule it should be noted that diffusion can take any direction regardless of how filtration moves the fluid. It is important to keep in mind that these two processes are separate.

The transport of CO_2 from the tissue to the capillary starts immediately as the blood reaches the thin membrane tube and continues as long as there is a pressure gradient for CO_2. In the same way, oxygen diffuses from the capillary throughout its length, irrespective of the direction of the net water flow across the membrane. As pointed out, the whole surface area of the capillaries is open for a free diffusion of these gases.

Veins

The collecting venules have a supporting coat of connective tissue and irregularly spaced smooth-muscle cells (Fig. 4-8, E and F). In the distal part of the venules a well-defined muscle layer is developed. Also, the larger veins have muscles in their walls which can constrict the lumen of the vessel. Valves are frequent in the veins of the extremities of the body.

The veins within a skeletal muscle will be compressed mechanically during the muscular contraction. Blood is squeezed from the veins, and because of the higher resistance on the capillary side and the design of the valves, the blood return to the heart is facilitated by this "muscle pump." During muscular relaxation, the venous pressure drops, and therefore blood may flow from the superficial veins, anastomosing the deep veins. The blood flow to an active muscle is higher during the period of relaxation than during contraction. The venous pressure has its minimum at the time of relaxation, which increases the arteriovenous pressure difference and thus the perfusion.

It has been calculated (since direct measurements are presently impossible) that the venous systems normally contain some 60 to 70 percent of the total blood volume. The veins are therefore often referred to as capacitance vessels (Fig. 4-7). The pulmonary veins may account for about 10 to 15 percent of the total blood volume. (At rest, the blood in the capillaries is estimated to be only 5 percent of the total; the spleen plays only a minor role in the human as a blood or red-cell depot. In the resting human being 8 to 11 percent of the blood volume is in the heart and 10 to 12 percent in the arteries.)

The resistance to flow in veins larger than 0.5 mm is very modest compared to the resistance to blood flow in the preceding vessels, perhaps only 10 percent of the total under normal conditions.

In an inactive individual, approximately 8,000 liters of blood flows through the capillaries per day. Only about 24 liters of this fluid is filtered from the capillaries (except for about 170 liters filtered across the glomeruli of the kidneys). At the distal end of the capillaries some 20 to 22 liters of fluid are reabsorbed. The remaining 2 to 4 liters return to the blood via the lymphatic system. (The lymphatic capillaries begin as blind, saccular ends.)

Vascularization of Skeletal Muscles

Nerves and vessels enter the muscle at a neuromuscular hilus often located at half-length of the muscle. The artery enters the muscle substance and branches freely in its course along the perimysium (Fig. 4-11). By anastomoses, a primary arterial network is established. Finer arteries arise and create a secondary network infiltrating the muscle tissue. The smallest arteries and terminal arterioles branch off, usually transversely to the long axis of the muscle fibers and at fairly regular intervals of 1 mm. The arterioles then supply the capillary network oriented parallel to the individual muscle fibers, but also form frequent transverse linkages over or under the intervening fibers, thereby forming a delicate oblong mesh. The veins have valves (from a caliber of 40 μm or larger) directing the blood flow toward the heart, and they follow the course of the arterioles and arteries. The capillary network at the motor endplate is especially well developed.

The exact capillary density in skeletal muscles in the human body is not known. Brodal et al., 1977, counted, with the aid of the electron microscope, the number of

Figure 4-11
Vascularization of skeletal muscle. Capillaries run parallel to the fibrils which make up the fiber.

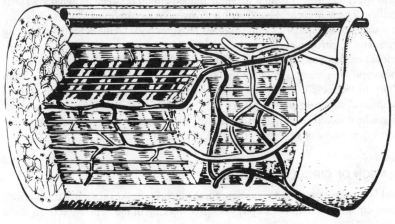

capillaries per mm^2 in cross sections of needle biopsies taken from the vastus lateralis of the quadriceps muscle from 12 untrained and 11 endurance-trained individuals. The mean number of capillaries per mm^2 was 585 ± 40 in the untrained and 820 ± 28 in the well-trained individuals. Correcting for shrinkage and artificial spaces between the fibers, they arrived at a mean number of capillaries per mm^2 of 325 for the untrained and 460 for the trained group (Andersen, 1975). Thus up to 500 capillaries/mm^2 may be open during heavy muscular activity. The total surface area of the capillary bed would then be about 200 m^2 in an individual with 30 kg of muscles. With an average capillary diameter of 6 μm, the area of the capillaries would be close to 2 percent of the total tissue cross-section area; the distance from a capillary to the most remote mitochondria would be about 13 μm if the capillaries are evenly distributed. The length of the capillaries varies markedly; as a mean length, 1 mm is a reasonable figure. The arterioles can close the inlet to the capillary, and at rest the number of open capillaries is markedly reduced. The opening or closing is supposed to be operated by local chemical factors of a hypoxic or metabolic nature, and only to a smaller degree by nerve activity (see below). In a given degree of metabolic activity in the tissue, the number of working capillaries may be fairly constant, but the individual capillary is intermittently open or closed.

The human skeletal muscle is composed of both slow twitch fibers (type I), and fast twitch fibers (type II) and the fibers of the different motor units are mixed (see Chap. 2). Therefore, two different fiber types often share common capillaries. On the average, the capillary density around the slow twitch fibers is higher (3–4 capillaries per fiber) than around the fast twitch fiber (2–3 capillaries per fiber). In the untrained muscle, the average tissue area which a capillary may supply is 20 to 30 percent larger for type IIb and 10 to 20 percent more for type IIa fibers as compared to the type I fibers. As mentioned, training will increase the capillary density; disuse of a muscle will reduce the density (see Chap. 10). (For a more detailed discussion of the vascularization of the skeletal muscle, see Saltin and Gollnick, 1983.)

It is still uncertain which type of capillary system is predominant in the muscle (Fig. 4-8): whether it is the thoroughfare channel (preferential channel) system or a system in which capillaries form a parallel-coupled circuit between terminal arterioles and venules. There are probably no real arteriovenous or arteriolovenous shunts.

The abundance of capillaries in the muscles provides good facilities for the supplying of oxygen and nutritive materials to the cells "bathing" in the interstitial fluid, and for the removal of products from metabolic activity. Krogh (1929) calculated that with 100 open capillaries/mm^2, a gradient in oxygen tension of 12 mm Hg (1.6 kPa) from a capillary to the most distant point would be sufficient during resting conditions to provide those distant parts with oxygen by diffusion. With all capillaries open, a lower oxygen gradient would be enough to drive a diffusion; the mitochondria are functioning with an oxygen pressure of 1 mm Hg (0.13 kPa) or even less.

REGULATION OF CIRCULATION AT REST

This could be a very confusing discussion, for the theories on "regulation of circulation" are almost equal to the number of physiologists working in the field. Moreover, there is no unanimous agreement on the terminology of *regulation* and *control*. From a demonstration of an effect on heart rate or vasomotor tone of a stimulation of some part of an animal, it may be tempting to construct a hypothesis concerning regulatory mechanisms. On the contrary, the noticed effect may represent an interference with, or a disturbance of, the normal function and by no means a reflection of a regulatory mechanism.

It should be remembered that by following nerve fibers, one can travel from one cell to almost any other cell within an animal's body. The anatomic basis for a regulation, or interference, in the function of cells and organs is therefore definitely present. Hormones and other chemical substances (including anesthetic agents) can exert a purposeful effect on some cells but, by accident, disturb the function of others.

The evolution of Homo sapiens has passed through many stages. Nerve tracts and organs have probably changed their functions, but remnants of once very essential functions and mechanisms may still remain, normally suppressed or transiently noticeable only in specific situations. Similar obsolete mechanisms may eventually be revived under special conditions. If so, this is most likely to occur when points in the central nervous system or nerves are artificially stimulated, which may rouse centra or tracts from "sleep" or inhibition. The effect thus displayed cannot, however, be considered as a mechanism of regulation. In research, artificial situations and tools are essential in studying the cells, organs, and even the intact organism. The most critical and crucial parts of a study are the interpretation and application of results obtained. With these difficulties and pitfalls in mind, we shall now discuss the regulation of circulation at rest and during exercise.

Any change in cellular activity should be met by a corresponding variation in local blood flow through the capillary bed. If an individual cell, in one way or another, could control its environment by varying the blood supply in balance with the actual nutritional demand, that cell would benefit. But other cells or tissues might suffer if some cells selfishly take more than their share, leaving others with little or nothing. Hence, coordinative mechanisms are essential if the distribution of blood is to be balanced properly. An active regulatory mechanism ensures that more active and less active cells, as well as more susceptible and less susceptible organs, are supplied according to their need and to the capacity of the whole circulation.

Arterial Blood Pressure and Vasomotor Tone

The blood pressure in the aorta is maintained by an integration of the following factors: (1) cardiac output, (2) peripheral resistance, (3) elasticity of the main arteries, (4) viscosity of the blood, and (5) blood volume. Evidently a regulation of the arterial blood pressure can use factors 1 and 2, since tools for 3 through 5 are normally not at its disposal for rapid modifications.

The local blood flow is mainly determined by the pressure head and the diameter of the actual vessels. The smooth muscles of the arterioles and veins in many regions continuously receive nerve impulses that keep the lumen of the vessels more or less constricted. This vasomotor tone is provided by the sympathetic vasoconstrictor fibers driven from the vasomotor area in the medulla oblongata. The transmitter substance is norepinephrine. As mentioned, the membrane receptors of alpha type have a fairly general distribution in the vascular tree, and the effect on the smooth muscles by the transmitter substance norepinephrine is a constriction of the vessels. There are β-receptors in some precapillary resistance sections, such as in skeletal muscle and myocardium where epinephrine can relax the smooth muscles. The heart and brain receive few vasomotor fibers; the supply to the abdominal organs (by splanchnic nerves) and skin is very rich. The muscles have an intermediate position. The vasomotor tone at rest can be demonstrated by section or blocking of the sympathetic nerve fibers in an animal. The arterioles dilate, and the arterial blood pressure falls. The effect of such an inhibition of vasomotor tone is very marked in the skin, but less pronounced in skeletal muscles. From a basal blood flow of 3 to 5 ml $\cdot$ 100 ml^{-1} $\cdot$ min^{-1} of muscle tissue, a doubling of the flow may occur. During exercise it can, however, amount to some 100 ml $\cdot$ min^{-1} or more. The smooth muscles in precapillary vessels may exhibit spontaneous rhythmic contractions creating a basal vascular tone. The intravascular pressure may be a stimulating factor. Some of the smooth muscle fibers, predominantly localized in the most narrow vessels, may actually serve as stretch receptors and pacemakers and thereby as triggers for the neighboring cells. Thus a wave of depolarization and contraction is propagated in the proximal direction (see Folkow and Neil, 1971; Grände et al., 1979). The vasomotor tone is important in keeping arterial blood pressure and cardiac output on an economical level. The splanchnic area could contain the whole blood volume after maximal dilation of the vessels. Also, the vascular bed of skin and muscles has a similarly large capacity. Fainting (vasovagal syncope) may be the result of a central inhibition of the vasomotor efferent impulses.

The cardiac output can certainly not exceed the venous return. A constriction of the postcapillary vessels (the capacitance vessels) with their large content of blood will increase the blood flow toward the heart, making an increase in cardiac output possible. A decrease (inhibition) in the vasomotor tone, which actually causes a relaxation of the smooth muscles in the vessel wall and therefore a vasodilation, can be obtained principally in three ways: (1) The smooth muscles can be affected by chemical substances liberated locally from neighboring cells or delivered from the blood. More or less effective as dilating agents are hypoxia, lowered pH, an excess of CO_2 and lactic acid, adenosine compounds, an increase in extracellular potassium, P_i, hyperosmolarity, prostaglandins (heat can exert a similar effect on the skin vessel, but only to a small degree in the skeletal muscles). (For discussion see Olsson, 1981.) It should be noted that smooth muscles in the precapillary vessels in the skeletal muscles are effectively relaxed by metabolites. A constriction of veins induced by sympathetic nerve activity is, on the other hand, well maintained even at extensive metabolite accumulation (Folkow and Neil, 1971). Vasodilation may also be caused (2) by a decreased discharge in the sympathetic vasomotor nerves and (3) by liberation of acetylcholine from the nerve endings of active sympathetic vasodilator fibers (cholinergic effect which is blocked by atropine). The final common path of those nerve fibers starts in the lateral spinal horns, i.e., they have from here on a similar anatomy to the sympathetic nerves. However, these sympathetic cholinergic vasodilator fibers are distributed only to the vessels of the skeletal muscles.

Summary   The peripheral resistance to blood flow is determined by the vasomotor tone. The degree of contraction of the smooth muscles in the arterioles is of special importance for the local blood flow as well as the total resistance. In some tissues (e.g., muscles) this tone is probably partly spontaneous, the smooth muscle contractions (myogenic activity) being triggered by mechanical stretch induced by the intravascular pressure, but a sympathetic vasomotor tone is superimposed. In other regions (such as the skin) this sympathetic vasomotor tone is definitely dominating. A vasodilation can occur after a local inhibition of the sympathetic effect or the myogenic activity (by metabolites, heat, activity in the sympathetic vasodilator fibers) or by central inhibition of the sympathetic impulse traffic. The spontaneously active mechanoreceptors in the precapillary vessels can induce vasoconstriction. With the capillary flow reduced, tissue metabolite concentrations will increase and exert a vasodilator influence on the sphincter mechanism. Soon the pacemakers will become activated again, etc. (For further discussion, see Nilsson, 1985.)

The Heart and the Effect of Nerve Impulses

The heart has its own pacemaker, the sinus node, initiating about 110 impulses/min if left alone. Normally, however, it is under the influence of nervous activity and at rest inhibitory impulses will dominate. Both sympathetic and parasympathetic nerve impulses can modify the heart rate. The parasympathetic activity from a cardio-inhibitory center via the vagus nerve (and acetylcholine) causes a slowing of the heart rate (bradycardia), and the sympathetic cardiac nerves (norepinephrine) as well as blood-borne epinephrine can produce an increased heart rate (tachycardia). In a resting subject, a blocking of the vagus nerve will cause the heart to beat faster, indicating a predominating parasympathetic tone. There are indications that the slowing of the heart rate during a standard rate of exercise after a period of physical training is induced by an increase in vagus tone and a reduction of the sympathetic drive (see Ekblom et al., 1973). The sympathetic nerves can increase the contractile force of the heart muscle fibers, but sympathetic nervous control of the vasomotor tone in the heart vessels is probably insignificant. As mentioned, the blood vessels of the heart dilate willingly when affected by hypoxia and metabolites.

(The adrenergic β-receptors dominate in the heart muscle, and it is possible pharmacologically to reduce or block those receptors to respond to sympathetic stimuli, e.g., by propranolol, see Tesch, 1985.)

Control and Effects Exerted by the Central Nervous System

As we have now analyzed the tools available for a variation and redistribution of blood flow within the body, the actual control and regulating mechanisms may be discussed.

Important and essential nuclei are located in the brain stem, particularly in the medulla oblongata. Traditionally, textbooks have exposed the students to a *medullary vasomotor area*, divided into a *vasoconstrictor center* and a *vasodepressor area*, the latter operating through inhibition of the sympathetic vasoconstrictor outflow. These terms may be convenient but they give a false impression of well-defined collections of neurons with precise effects. This is not so. There is no clear anatomical separation between the pressor and depressor areas in the brain stem, but an overlap in the distribution of nuclei which can excite sympathetic nerve fibers to the heart and blood vessels, and neurons which can inhibit these effects. Keele et al. (1982, p. 124) suggest the terminology "medullary cardiovascular centers" for the brain stem "centers" for vasomotor control.

The activators presumably release norepinephrine as a transmitter substance in their communication with preganglionic neurons in the spinal cord (adrenergic neurons). Another class of neurons in the brain stem belonging to the sympathetic system also sends their axons down towards preganglionic neurons but presumably they release serotonin (5-hydroxytryptamine; which are serotoninergic neurons). According to one theory, they can presynaptically inhibit the adrenergic neurons. The balance in activity between these two systems provides the major control of vasomotor tone by increasing or decreasing the impulse traffic in the sympathetic nerves leaving the spinal cord, the preganglionic neuron (see Shepherd and Vanhoutte, 1979). As mentioned, the adrenergic nerve activity constricts all vascular beds via α receptors (if not modified by other factors). In addition, it increases the heart rate and contractility (β-receptors). Antagonism to the sympathetic activity also stems from the "vagal nuclei," located in the brain stem. Their parasympathetic fibers innervate nerves that subserve smooth muscle cells, but the direct dilating effect on blood vessels seems to be weak. As mentioned, the transmitter substance, acetylcholine, reduces the heart rate by an inhibition of the sinus node and the atrioventricular node (by a hyperpolarization). There is an interesting interaction between the sympathetic and parasympathetic systems in organs that they both innervate. On the presynaptic endings of sympathetic fibers there may be specific receptors. When exposed to acetylcholine the release of norepinephrine is reduced. Conversely, catecholamines can reduce the release of acetylcholine. This effect is mediated by α adrenergic receptors in the presynaptic region of the parasympathetic nerve fibers. This is an example of a reciprocal innervation at the presynaptic level. It should be added that a high concentration of norepinephrine released into the synaptic cleft by an adrenergic fiber, may limit the subsequent release mediated by α receptors on the presynaptic nerve terminal (a negative feedback mechanism).

The neurogenic vasomotor tone of the blood vessels originates essentially in the brain stem, and a continuous, somewhat rhythmical discharge can be recorded in some areas. This discharge is probably caused by the influence of the chemical composition of the interstitial fluid that bathes the cells. The nuclei eliciting vasodilation are essentially relay stations without spontaneous activity, but they can be activated by incoming impulses from peripheral receptors and other areas in the CNS. Certainly, the situation is the same for the vasoconstrictor activators.

On higher levels of the CNS, there are areas, especially in the cerebral cortex and diencephalon, from which cardiovascular reactions can be elicited. Although these higher centers do not contribute to the continuous vasomotor tone, many adjustments are initiated primarily from the brain above the level of the medullary centers (see Korner, 1971). Of special interest are the *sympathetic cholinergic vasodilator fibers* (see Folkow and Neil, 1971). They can be traced from the motor cortex and followed through the anterior hypothalamus and mesencephalon (relay stations). The nerve fibers bypass the medullary cardiovascular centers and run to the lateral spinal horns and the sympathetic final common path. When stimulated, they can be activated in synergism with vasoconstrictor fibers. The combined effect is then a vasodilation of precapillary resistance vessels in the skeletal muscles and vasoconstriction of the vessels of the abdominal organs and skin. With very little shift in arterial pressure this automatic activation pattern leads to a remarkable and instantaneous redistribution of cardiac output to favor skeletal muscles. Simultaneously, accelerator fibers to the heart may also be stimulated, and the medulla of the suprarenals may give rise to a liberation of epinephrine. This hormone dilates the resistance vessels of the skeletal muscles and excites the smooth muscles of the capacitance vessels; norepinephrine strongly contracts both resistance and capacitance vessels. (These various responses are consequences of the different receptors, α and β.)

Similar reaction patterns are characteristic for emergency conditions, such as fear, and can be elicited by electrical stimulation of the hypothalamus and even by cutaneous stimulation.

Figure 4-12 describes the central vasomotor integration in a schematic way.

Mechanoreceptors in Systemic Arteries

The afferent input is far from being completely mapped out. Important afferent fibers come from mechanoreceptors in blood vessels and in the heart. The systemic arterial receptors are located in the tissue of the carotid sinus, aortic arch, right subclavian artery, and common carotid artery (Heymans and Neil, 1958; Kirchheim, 1976). Mechanical deformation of the walls of the vessels is the normal stimulus of the receptors, and they respond to the rate of rise of blood pressure as well as amplitude of pulse pressure.

The receptors are actually stretch receptors, and pressure as such is not the adequate stimulus. (The commonly used name "baroreceptors" is therefore somewhat misleading.) However, an increase in the intravascular pressure does expand the vessel wall and stretch the receptors. They respond with a discharge transmitted to the CNS. If the wall of the vessel, where the stretch receptors are located, becomes less distensible, owing to increased activity of smooth muscles in the wall or a progressive structural change (in a hypertensive patient or aging person), a given pressure would induce less deformation of the receptors and a reduced impulse output. In fact, pathways from suprabulbar areas are probably involved in a central resetting of the mechanoreceptor reflex during increased activity of other receptor groups, e.g., during exercise (see Korner, 1971). This can augment or depress the reflex responses mediated through mechanoreceptors.

Pulsatile pressure about a given mean blood pressure is more effective than a steady mean pressure to set up an impulse traffic in the afferent nerves (sinus branch of the glossopharyngeal nerve and afferent fibers of the vagus). The threshold at which a stimulus is effective varies for the different receptors. A recording of activity in the sinus nerve normally reveals a continuous discharge and a variation in the impulse traffic with each pulse beat.

The baroreceptors can report a fall as well as a rise in blood pressure to the

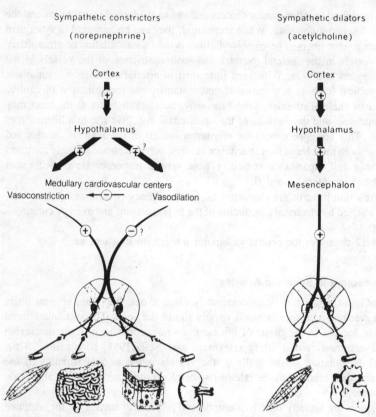

Sympathetic constrictors
(norepinephrine)

Sympathetic dilators
(acetylcholine)

Cortex

Cortex

Hypothalamus

Hypothalamus

Medullary cardiovascular centers

Mesencephalon

Vasoconstriction ◄——(−)—— Vasodilation

Figure 4-12
Central vasomotor integration. The inhibition on the vasoconstrictor activity may operate directly
on the center or eventually at the spinal level. The vessels in skeletal muscles and the
gastrointestinal tract are, at rest, easier to involve in a constrictor activity ("low-threshold areas")
than are the cutaneous and kidney vessels. For detailed description, see text.

cardiovascular centers, primarily the medullary vasomotor area. At rest, the baroreceptors exert a restraining influence on the cardiovascular system, causing a reflex
bradycardia and reflex inhibition of the medullary vasomotor center (Fig.4-13).

Posture

The physiological interplay of the various factors involved in the maintenance of an
adequate arterial blood pressure can be elucidated by the following experiment:

On a tilting table, a subject is tilted from supine to a head-up position (about a 60°
angle to the horizontal). Owing to the force of gravity, blood is pooled in the parts
of the body below the heart level. Thus, the venous return to the heart is temporarily

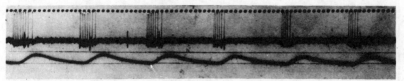

Figure 4-13
Action potentials from mechanoreceptors in the carotid sinus recorded from the sinus nerve in the cat. Dots at the top give the time, 50 Hz; the bottom line shows the blood pressure recorded in the carotid artery with calibration lines at 100 and 150 mm Hg respectively. Note that on the left side, when pressure is high, there are 9 to 10 large "spikes" per heartbeat. When the pressure is lower, there are only 5 spikes. (The spikes occur at the pressure rise, but there was some delay in the pressure recording.)

reduced. Consequently, the cardiac output decreases and so does the arterial blood pressure. The strain exerted on the baroreceptors is reduced, and fewer impulses are transmitted from them to the CNS. The impulse output from the parasympathetic neurons is diminished (which results in an increase in heart rate) and from the adrenergic sympathetic neurons in the brain stem there is an increased impulse traffic (the effect is a vasoconstriction in resistance vessels and capacitance vessels, especially in the splanchnic area, and an increase in heart rate). The precapillary vessels in skeletal muscles are also important targets for this baroreceptor reflex (see Rowell, 1974). Thus, the peripheral resistance becomes higher, the cardiac output can be restored to an adequate level, and the arterial blood pressure can increase. The variation in heart rate in a subject passively tilted to different body positions is illustrated by Fig 4-14 (Asmussen et al., 1939). If blood pressure cuffs are placed around the upper parts of the thighs and a pressure of about 200 mm Hg (26.6 kPa) is applied when the subject is in a horizontal or head-down position, the heart rate response to the head-up position is less pronounced (Fig. 4-14). The hydrostatic forces are acting on a shorter "column," as the blood is prevented from circulating to the legs. If the pressure within the cuffs is suddenly released, the fall in arterial blood pressure may be very pronounced, and eventually the subject faints as the blood, and thereby the oxygen supply to the brain, becomes inadequate. The capacity of the legs to retain blood has increased, for its arterioles dilate as anaerobic metabolites accumulate during the period of circulatory occlusion. Tilting of the subject to a head-down position will quickly restore the circulation and consciousness as the legs are drained. Figure 4-15 shows how the heart rate can be lowered some 10 beats/min if the subject in a head-up position (on a tilting table) contracts the leg muscles. The massaging effect of the repeatedly contracting muscles on the capillaries and veins enhances the venous return to the heart, and the heart rate is lowered. Most likely the nerves from the cardiovascular mechanoreceptors form an important link in the reflex chain.

The beneficial effect of the muscle pump on the venous return should certainly be stressed for people who work in a fixed sitting or standing position. Bandaged legs can partially reduce the hydrostatic shift of fluid to the legs in the upright position, and thereby the circulation is facilitated (Lundgren, 1946; Arenander, 1960). A sudden standstill after prolonged exercise, particularly in a hot environment, can cause fainting,

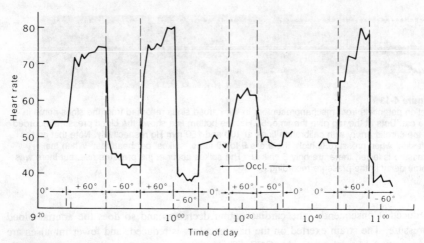

Figure 4-14
Variation in heart rate (ordinate) in a subject passively tilted to different body positions. +60° = tilting to a head-up position; −60° = head-down position; Occl. = experiments in which inflated blood pressure cuffs were placed around the upper parts of the thighs. For further details, see text. (*From Asmussen et al., 1939.*)

as the blood pools in the dilated vessels in exercised legs and in the skin. The unexpected fall of the tall soldier during a military parade can also be explained by an inappropriate distribution of blood.

There are two important factors counteracting an edema formation in the legs in erect position: (1) a pressure-induced facilitation of the rate of myogenic contractions of the smooth muscles of the arterioles and precapillary sphincters, decreasing the surface area of the capillary bed available for filtration; (2) an effective reabsorption of fluid via the red cells squeezed in the narrow capillaries. The close contact between red cells and tissue fluid will promote an uptake of water in the capillaries as well as the exchange of gases, as discussed earlier in this chapter.

Summary   A change in body position will inevitably affect the circulation as long as the individual stays under the influence of the pull of gravity. A head-up position will primarily increase the blood volume in the legs and decrease the central blood volume and cardiac output. Secondary variations in arterial blood pressure are reported from the mechanoreceptors in some arteries to inform the cardiovascular centers in the brain. The activity in sympathetic and parasympathetic nerves varies by reciprocal innervation in such a way that the arterial blood pressure and cardiac output return to a level fairly close to the one typical for the individual in a supine position. The nerves from the mechanoreceptors were appropriately called "buffer nerves" by Samson Wright.

Other Receptors

In the walls of the pulmonary artery there are *mechanoreceptors* with reflex effects on the systemic circulation and heart similar to those caused by the systemic arterial

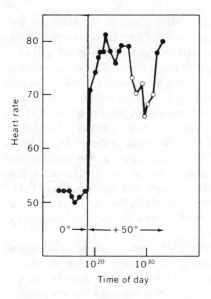

Figure 4-15
Effect of voluntary contraction of leg muscles on heart rate during passive standing. ● = passive standing; ○ = voluntary contraction of leg muscles. (*From Asmussen et al., 1939.*)

baroreceptors. A third group of mechanoreceptors are located in the walls of the atria and ventricles of the heart. When stimulated, they cause by reflex a vasodilation, bradycardia, and systemic hypotension, so that the load on the heart diminishes. Exceptions are other "low-pressure" receptors in the atria which, when stimulated, will reflexly accelerate the heart, and which, perhaps, may be responsible for the so-called Bainbridge reflex. The other cardiovascular reflexes are of a buffer, depressor, or negative feedback nature. Stretch receptors also serve as one sensory mechanism in a reflex regulation of blood volume by control of urine output (see Folkow and Neil, 1971). The filling volume of the cardiac atria, related to the circulating or thoracic blood volume, is likely to be the appropriate stimulus; a variation in the production of antidiuretic hormone from the hypophysis is likely to be the tool.

The *chemoreceptors* in the carotid and aortic bodies, stimulated by low oxygen tension in the circulating blood, not only influence the pulmonary ventilation, but also, indirectly, the circulation. In artificially ventilated dogs, stimulus of the chemoreceptors by hypoxia causes an increase in sympathetic vasoconstrictor discharge, a bradycardia, and decrease in cardiac output (Daly, 1964). In a spontaneously breathing animal, however, a hypoxia causes a tachycardia, probably secondary to the hyperventilation. Consequently, it can be concluded that the tachycardia of systemic hypoxia does not have a chemoreceptor reflex origin.

In this context it is of interest that one of the primary *adjustments which take place during a dive* is circulatory in nature (Scholander et al., 1962). Most animals, including man, display a diving bradycardia, which usually develops gradually. Thus during the dive, the frequency may drop to one-half of normal in some species and to one-tenth in others. As a rule, the bradycardia stays with the animal whether it exercises or not during the dive. Blood pressure is maintained at a normal, or even an elevated, level as a result of peripheral vasoconstriction. In the diving animal, the reduction and selective redistribution of the circulation can save oxygen for vital organs such as the brain and heart. In the human, the heart rate increases immediately at the start of a dive and slows down very markedly as the dive progresses, even if the diver exercises vigorously (Scholander et al., 1962; Olsen et al., 1962; Kooyman et al., 1981; Schmidt-Nielsen, 1983). Prompt return to normal sinus rhythm occurs with the first breath. This bradycardia during diving is not a result of apnea only, since it is reinforced by water submersion. Skin receptors, wetting of the nose, and asphyxial reflexes may contribute to the development of the bradycardia. The onset of the decline in heart rate is very rapid. However, it may be that the chemoreceptors gradually contribute to the bradycardia by a progressive hypoxic drive, since the drop in alveolar oxygen tension is very marked, at least during exercise combined with breath-holding (P.-O. Åstrand, 1960). In any case, there is a striking similarity in the circulatory response of isolated chemoreceptors in dogs to a stimulus of hypoxia and the effect induced by diving on the circulation in the intact animal.

Phylogenetically, the chemoreceptors may be a very ancient receptor mechanism of importance for diving animals, but they assumed a new function when the animal left the water for good. *In summary* stimulation of arterial stretch receptors inhibits the sympathetic discharge, but chemoreceptor activation enhances the sympathetic discharge. Both of them, however, excite the vagal cardio-inhibitory "center."

Afferent impulses to the cardiovascular centers from receptors in the skin and joints have been suggested but have not as yet been shown to exist. *Receptors in the skeletal muscles* can be stimulated by stretch and they respond to muscular contraction ("ergoreceptors"). Group III and some group IV afferent fibers transmit the impulse traffic. Free nerve endings in the muscles serve as receptors which respond to pain (nociceptors). There is evidence that they can be activated by chemical factors such as serotonin, bradykinin, potassium, hyperosmotic lactate, and phosphate. As is discussed below, they may be stimulated by metabolites and may be of some importance in the regulation of the cardiovascular function during exercise.

Any damage to superficial tissues causes a regional vasodilation response. The afferent fibers of pain receptors branch, and impulses pass along those branches to arterioles, causing a dilation (antidromic activation or *axon reflex*).

REGULATION OF CIRCULATION DURING EXERCISE

The conclusions of the preceding discussions are as follows: The blood flow in the precapillary vessels of the muscles is "regulated" locally, for the most part, by the level of metabolism in the muscle cells. The actual nutritional demand is the determining factor and will be superimposed on any nervous influence; the vessels then become functionally sympathectomized. The blood flow in the splanchnic area is, on the other hand, under control of the CNS. The flow is varied so that the systemic arterial blood pressure is maintained at an adequate level to supply brain, heart, and some other vital organs with blood. In this case, neural vasoconstrictor activity can be superimposed upon the local dilator control. A stimulation of the vasoconstrictor fibers can increase the resistance in the vessels of the gastrointestinal tract about 20 times. The vessels of the skin also subserve centrally controlled mechanisms. Impulses from the medullary cardiovascular centers are important, but the final control is probably exerted by the temperature-regulating centers, at least while the subject is at rest or during submaximal exercise.

At rest, the kidneys receive about 25 percent of the cardiac output. There are no tonic impulses from the CNS to renal blood vessels, but electrical stimulation of the renal nerves causes intense renal vessel constriction with associated changes in blood flow (down to 250 ml/min) and in excretion of water and electrolytes (Pappenheimer, 1960). Exercise, postural changes, and circulatory stress in general may cause profound alterations in renal function, mediated through the hemodynamic effects of renal nerves.

This summary gives the main tools available for the regulation of the circulatory system during exercise. The approximate blood distribution to the various organs is illustrated in Fig. 4-9.

The changes in heart function and circulation, from the moment muscular exercise begins (or even before), are initiated from the brain levels above the medullary centers (probably the cerebral cortex and diencephalon). This is a *central command* and it is speculated that the central activity that recruits the motor units also stimulates the medullary and spinal neuronal circuits that elicit the cardiovascular responses to exercise. In addition, the mechanical and chemical events in the exercising muscle are thought to contribute by reflex mechanisms to the cardiovascular adjustment by some sort of *peripheral command* (see Mitchell et al., 1983). The complicated responses to various cardiopulmonary reflexes must also be considered when discussing the regulation of circulation during exercise, and certainly there are many unanswered questions. At any rate, by a reciprocal innervation there is simultaneous increase in the sympathetic activity and decrease in the parasympathetic impulse traffic. The skeletal muscles may receive an increased share of the cardiac output because of their innervation with sympathetic cholinergic vasodilator fibers. The activity in the sympathetic adrenergic vasoconstrictor fibers reduces the blood flow to skin, kidneys, and splanchnic area (Fig. 4-16). Rowell (1974) calculated that at maximal vasoconstriction of the splanchnic and renal blood vessels, about 2.2 liters $\cdot$ min^{-1} can be redistributed to the activated muscles. This could increase their oxygen uptake by about 0.5 liters $\cdot$ min^{-1} without any additional increase in the cardiac output. It should be emphasized that a vasoconstriciton of precapillary vessels will cause secondarily a reduction in the transmural (distending) pressure in the postcapillary vessels. The previously, more or less distended veins collapse easily due to the passive recoil of the venous wall (see Rothe, 1983). The reduction in blood flow to the splanchnic tissues and kidneys (and the increase in heart rate) during exercise is more related to the severity of the exercise in relation to the individual's maximal oxygen uptake than to the absolute rate of oxygen uptake (see Rowell, 1974: Clausen, 1976). Rowell noted a linear decline in the splanchnic blood flow down to about 30 percent of the flow at rest, during exercise and/or heat stress (see also Fig. 13-8). Actually, the splanchnic blood flow in humans is inversely related to the heart rate under a wide variety of stresses (about 0.6 percent reduced flow per heart beat increase). One may speculate that the sympathetic branches that control the heart rate and splanchnic vascular resistance are parallel. The balance in the redistribution of blood can be such that there is no (or only a slight) decrease

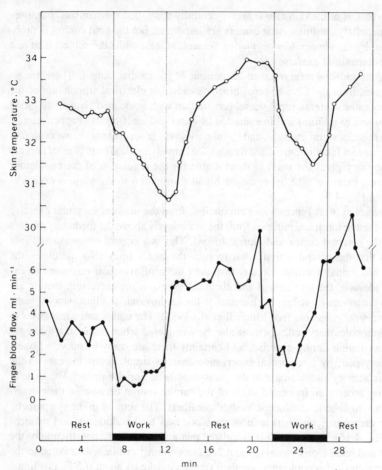

Figure 4-16
Blood flow in index finger (measured by means of a plethysmographic method) and skin temperature on the same finger at rest and during two periods of work on a cycle ergometer (1080 kpm · min⁻¹, or 180 watts). (*From Christensen and Nielsen, 1942.*)

in the systolic blood pressure during the initial period of exercise, despite a marked dilation of the resistance vessels in the muscles (Holmgren, 1956). The splanchnic bed and the kidneys "are particularly suited for short-term adjustments of the systemic vascular resistance because their blood supply is greatly in excess of the metabolic requirements of their tissues, and their normal function can be temporarily held in abeyance", (Shepherd and Vanhoutte, 1979). Mainly passive constriction of the veins (capacitance vessels), pumping action of the exercising muscles, and forced respiratory movements assist the venous return to the heart; causing what may be termed as an increased *preload*. Since the heart escapes from the vagal inhibition, and sympathetic impulses can increase the force and frequency of the muscle contractions, the heart gains capacity to take care of the increased inflow of blood, and if necessary, pumps it out against an elevated resistance, the *afterload*. The volume of blood in the heart, the end-dialostic volume, increases more during exercise in the supine position than when upright (Mitchell et al., 1958; see Poliner et al., 1980).

Figure 4-17 illustrates the changes in the composition of the interstitial fluid in the

Figure 4-17
Major changes in the composition of the interstitial fluid during contraction of muscle cells. When the muscle is inactive (left) the arterioles are constricted, the concentration of metabolites and carbon dioxide in the interstitial fluid is low, and little oxygen is used. When the muscles become active (right): (1) the depolarization of the cell membrane (CM) increases the K^+ concentration in the extracellular space; (2) the regeneration of adenosine triphosphate (ATP) by the mitochondria (Mi) augments the production of carbon dioxide, which diffuses to the extracellular space; (3) the anaerobic production of ATP in the cytoplasm results in the formation of lactic acid, which slowly diffuses out of the cell; (4) the increased amounts of lactic acid and carbon dioxide causes an increase in H^+ concentration of the extracellular fluid and thus a decrease in pH; (5) the breakdown of ATP to diphosphate (ADP) and monophosphate (AMP) and to adenosine, with liberation of inorganic phosphate (P_i), augments the concentration of adenosine and adenine nucleotides in the extracellular space; and (6) the osmolarity of the extracellular fluid increases. Each of these changes can cause relaxation of contracted smooth muscle cells, and it is likely that their combination is responsible for the adjustment of the blood flow to the metabolic needs of the tissues. (Increased concentrations or osmolarity are symbolized by larger letters.) (*From Shepherd and Vanhoutte, 1979.*)

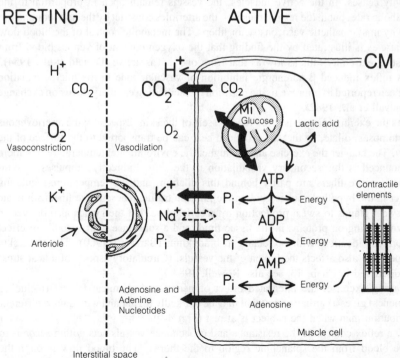

skeletal muscle from a state of rest to that of exercise. Substances released by the activated skeletal muscle cells interact with specific vascular receptors on the adrenergic nerve endings in the resistance blood vessel walls to inhibit the release of norepinephrine. The smooth muscle cells determining the diameter of the resistance vessels can be hyperpolarized, which means relaxed, by such substances, including the mentioned pacemaker muscle cells in the arterioles. With increasing muscular activity the resistance vessels will progressively become freed from sympathetic vasoconstrictor control. This local control of blood flow is a very important factor in securing an efficient blood supply to the exercising muscles. Some metabolites play more of a key role in this control than others but so far the mechanism behind the exercise-induced vasodilation is not clear. To these changes in the interstitial fluid, caused by exercise, as illustrated in Fig. 4-17, can be added release of prostaglandins with a potential to produce vasodilation (see Faber et al., 1982). There is evidence that an increase in potassium concentration and hyperosmolarity only contribute to the early phase of vasodilation (see Olsson, 1981; Mohrman, 1982). An activity in sympathethic cholinergic vasodilator fibers seems to be of short duration. However, it makes physiological sense to interpret the initial vasodilation in muscles as a state of general preparedness; arterioles and thoroughfare channels are flooded by a rapid bloodstream just at the doors of the capillaries. In the metabolically active areas, the capillary sphincters open and the nutritive vessels can immediately be perfused by blood (part of an emergency reaction). After perhaps some 10 s, the sympathetic vasodilator activity ceases. In the active muscles, the vessels remain open in proportion to the metabolic rate, but in the resting muscles, the arterioles constrict by the now-dominating activity in sympathetic vasoconstrictor fibers. The metabolic control of the blood flow in muscles is illustrated by the finding that the oxygen content of venous blood from a resting extremity falls as low as that from one that is active (Donald et al., 1954).

A reflex induced β-adrenergic inhibition of vascular tone in the microcirculation has been reported to improve tissue nutrition by facilitating capillary diffusion exchange (Lundvall et al., 1982).

As the exercise proceeds, the blood vessels of the skin, especially the arteriovenous anastomoses, dilate, so that the produced heat can be transported to the surface of the body. The heavier the exercise and the higher the environmental temperature, the more pronounced is this secondary vasodilation in the skin. Indirectly, impulses in sympathetic nerve fibers are partly behind this dilation, and the temperature-regulating center in the hypothalamus is guiding the impulse traffic. These nerve fibers stimulate the sweat glands to sweat production by acetylcholine, but the glands also deliver an enzyme acting on proteins in the tissue fluid, and a substance with vasodilator effect is formed, identical with or related to bradykinin (Barcroft, 1960). The local skin temperature also affects the lumen of the vessels. (Circulatory aspects of a heat stress are discussed in Chap. 13; see also Rowell, 1983.)

During exercise, the integrated effect of neural and chemical factors (including hormones) gives a cardiac output that may be markedly higher and with quite a different distribution than when the subject is at rest (Fig. 4-9).

The reflex control of the resistance and capacitance vessels acts within seconds to move blood from the splanchnic region to the thorax. The blood flow through the exercising muscles increases by a capillary recruitment and an increase in flow velocity (see Honig and Odoroff, 1981).

Summary We can schematically describe the circulatory response to exercise by considering four stages:

1 At rest the skeletal muscles receive only some 15 percent of the minute blood flow, and their arterioles are constricted by a continuous vasoconstrictor activity and some sort of a spontaneous vascular tone. Few capillaries are open, but the individual capillaries open and close alternatively. The heart rate is kept down by a parasympathetic outflow via the vagal nerve.

2 When, or even before, exercise begins, there are an inhibition of the parasympathetic activity and an increased sympathetic impulse traffic. According to one hypothesis, signals from higher levels of the CNS formulate a central command which initiate these autonomic nervous responses. Receptors in the activated skeletal muscles send afferent impulses to the medullary cardiovascular centers, acting in concert with the central command. The heart is freed from its inhibition and supported by the adrenergic activity it beats faster and with increased force. Sympathetic cholinergic vasodilator fibers may dilate arterioles in the muscles, thereby increasing their blood flow. On the other hand, sympathetic adrenergic vasoconstrictor fibers act on the vessels of abdominal organs and skin so that a decreasing share of the cardiac output flows through those tissues. The veins become constricted passively, and by activity in the constrictor fibers. This constriction of veins, together with the pumping action of the working muscles and the forced respiratory movements, facilitates the blood return to the heart, making an increased cardiac output possible.

3 The appropriate adjustment of the circulation occurs. In the activated muscles, the increased metabolism causes changes in the environment which locally dilate arterioles and open capillaries. The sympathetic vasodilator fibers are probably inactive or without effect, and in the resting muscles, the arterioles constrict. Hormones, including circulating catecholamines contribute in the constriction of vessels in nonactive areas.

4 For temperature balance within the body, the produced heat is transported to the skin, since skin vessels become dilated.

The raised capillary pressure leads to a net outward filtration of fluid, the flow of which is facilitated by a simultaneous increase of the capillary area. The mean capillary pressure may increase from 15–20 to 25–35 mm Hg (2.0–2.7 to 3.3–4.7 kPa) in activated skeletal muscles. In addition, the osmolarity increases in the exercising muscles due to the breakdown of larger molecules into smaller units. This will contribute to an increased fluid volume in those muscles. Conversely, in tissues where the precapillary vessels are constricted the mean capillary pressure becomes lower. This factor, plus an increased arterial osmolarity, favor a mobilization of extravascular fluid, and the plasma volume can be relatively well maintained. Lundvall et al. (1972) calculated that the total fluid loss into the active muscle mass was about 1,100 ml during heavy exercise (bicycling) but that loss was partly compensated for by a fluid absorption of some 500 ml from inactive tissues. Sjögaard and Saltin (1982) report that the extracellular fluid increased approximately 100 percent in muscles that had done short-term intensive exercise. The intracellular water increased about 10 percent (but its total water pool is much larger than that of the interstitial fluid). There is a linear relationship between the bulk of plasma volume lost from the vascular system and the exercise intensity. At a given oxygen uptake, a relatively larger percentage decrease in plasma volume is noted for arm compared with leg exercise (Miles et al., 1983). During arm exercise, the mean arterial blood pressure is higher than in leg exercise and as a consequence the net outward filtration in the dilated capillaries is enhanced. (On the other hand the total capillary surface area is smaller in the exercising arm muscles than in leg cycling.) It may take as long as 50–60 min before the plasma volume is restored (Harrison et al., 1975). Bjurstedt et al., (1983) report a marked impairment of the orthostatic tolerance during the first half-hour of recovery after exhaustive exercise. A reduced plasma volume may be one of several factors behind a change in the reaction to a 70° head-up tilt test.

CARDIAC OUTPUT AND THE TRANSPORTATION OF OXYGEN

Efficiency of the Heart

Studies on cardiac output and the energy output and work efficiency of the heart have shown that (1) a given stroke volume can be ejected with a minimum of myocardial shortening if the contraction starts at a larger volume; (2) energy losses in the form of friction and tension developed within the heart wall are also at a minimum in a dilated heart; (3) the stretched muscle fiber can, within limits, provide a higher tension than the unstretched one; (4) loss of energy is larger when the contraction occurs rapidly, i.e., with heart rate high as compared with a slower contraction rate; (5) on the other hand, the greater the volume of the heart, the higher is the tension of the myocardial fibers necessary to sustain a particular intraventricular pressure (as a consequence of Laplace's law, as discussed earlier in this chapter). The energy need for a contraction is closely related to the tension that has to be developed.

There are a few reports on the energy output and mechanical efficiency of the heart muscle when heart rate and stroke volume are altered. In the isolated heart, it has been found that an increase in cardiac output caused by an increased stroke volume markedly improved the efficiency of the heart work (blood pressure and heart rate were kept constant) (Evans, 1918). This means that the increase in oxygen uptake was not proportional to the increased performance.

If the cardiac output was maintained unaltered, variations in heart rate were followed by an almost linear variation in oxygen uptake of the heart; the high heart rate was associated with a low mechanical efficiency.

The *individual with a high capacity for oxygen transport* because of natural endowment and/or training is characterized by a large stroke volume and a slow heart rate. Cardiac output and systolic/diastolic blood pressures at given rates of exercise or at rest are not noticeably different from normal. Of the factors listed above, the first four tend to act in the individual's favor as far as each single heart beat is concerned, while the fifth factor acts against the person. However, as relatively few contractions are performed per minute, the total energy cost to maintain a given work level (flow times pressure) may be relatively low and the efficiency high. At maximal work rates, the person with the large diastolic heart volume and stroke volume may have as high a heart rate as the individual with a small diastolic filling. In that situation, factors 1 through 4 still favor the fit person, but factor 5 tends to decrease the efficiency when performance per unit time is considered.

It should be borne in mind that at a given maximal heart rate, the heart which has the capacity to provide the largest stroke volume can attain the highest cardiac output. One condition is a good diastolic filling, which inevitably leads to the drawback of increased tension required to produce the pressure.

A rich capillary network in the myocardium would provide the capacity to meet the demand of an increased metabolism. For the exercising skeletal muscles, however, it is essential that the increased demand for blood flow through the myocardium does not consume too many of the extra liters of blood that the heart eventually can manage to pump out. Apparently the energy requirement of the heart muscle is normally highly correlated to the cardiac output. The coronary blood flow is 4 to 5 percent of the cardiac output at rest as well as during vigorous exercise (see Fig. 4-9). There is only a slight increase in the a-$\bar{v}$ oxygen difference in the heart with increasing cardiac output. As mentioned, this a-$\bar{v}$ difference is high already at rest. The arteriovenous difference is as high as 16 to 17 volume percent at rest. During exercise, it may increase to 180 to 190 ml $\cdot$ 1^{-1} blood flow (oxygen content of the arterial blood 20 vol percent or higher).

From a general viewpoint, it is considered an advantage if a given level of cardiac output can be maintained with a low heart rate, i.e., a large stroke volume. The reason why a training with "overload" apparently is necessary to improve the efficiency of the heart is presently unknown.

For the *cardiac patient*, the situation is different if a large diastolic filling is combined with a small stroke volume and a high heart rate. The dilation of the heart muscle gives improved ability for the muscle fibers to produce tension as long as they are stretched, as in factor 3. Factors 1 and 2 help in keeping the efficiency high. Factors 4 and especially 5 tend to reduce the efficiency. Undoubtedly the heart has to pay for its compensating of the basically reduced myocardial strength. Sooner or later there will be a critical equation for energy requirement and energy supply of the heart, and the vascular bed in the myocardium is the key in that formula. At rest, the capillary blood flow may cover the need, but during even mild exercise, a discrepancy arises, and the patient is forced to stop because of symptoms such as angina pectoris, which is a pain probably caused by hypoxia in the myocardium. It should be emphasized that with a high heart rate, the tension time, i.e., the time during which the heart muscle is contracted per minute, is prolonged, and the blood flow through the heart muscle is therefore reduced (Fig. 4-4b).

Apparently there are two types of cardiac hypertrophy: one physiological type that is accompanied by a normal or augmented contractile state, in which the maximal rate at which myosin hydrolyzes ATP, and the peak velocity of muscle shortening, are either normal or elevated. Then there is the pathological hypertrophy in which case the rate of myosin ATPase activity and the velocity of muscle shortening are reduced. In both situations it is a question of biochemical and physiological adaptations to cellular alterations in response to external demands (see Wikman-Coffelt et al., 1979).

According to Kitamura et al. (1972), there is in young, healthy male subjects a high correlation (r = 0.88 to 0.90) between coronary blood flow and myocardial oxygen consumption versus the product of the heart rate and systolic blood pressure measured in the aorta. This product is frequently used as an estimate of the load on the heart during exercise "stress" testing. (It should be emphasized, however, that the systolic blood pressure measured indirectly over a peripheral artery, is not identical with the aortic pressure, see "Blood Pressure".)

Summary A large dilated heart may provide tension when the muscle fibers are stretched, but if tension decreases markedly as the muscle shortens, the frequency of contractions must be high. A high heart rate lowers the mechanical efficiency of the heart and increases the oxygen uptake for a given cardiac output. An impaired circulation in the myocardium can further complicate the situation. The difference in fitness of the athlete's large heart and that of the cardiac patient is evident.

Venous Blood Return

Certainly, the cardiac output cannot exceed the flow of blood returning to the heart. In the supine position, the hydrostatic pressure on the venous side of most of the capillaries is about 10 mm Hg (1.3 kPa) but the pressure gradient to the right atrium is increased by the negative intrathoracic pressure. At rest, this pressure is about 5 mm Hg (0.7 kPa) less than the ambient barometric pressure because the elastic tissue of the lungs is expanded to the size of the thoracic cavity and the recoil of the tissue exerts a tension on the thin-walled vessels within the thorax. The inspiration increases this pulling force, and blood is sucked into the thorax. At the same time, the abdominal veins are compressed as the diaphragm contracts. The variations in intrathoracic and intra-abdominal pressures with breathing will significantly enhance the venous return because a backflow in the veins is hindered by the capillary resistance and venous valves. During an expiratory effort with the glottis closed (Valsalva's maneuver), the intrathoracic pressure increases, impairing both the venous return and the cardiac output; this is usually the case in weight lifting. Hyperventilation followed by Valsalva's maneuver may cause fainting (for the explanation, see Chap. 5). There is evidence that the ventricular action does contribute to the subsequent filling of the ventricle by exerting a suction force on the atrium (see Folkow and Neil, 1971).

A third important factor improving the venous return is dynamic activity of the skeletal muscles (Figs. 4-15, 4-18). As pointed out earlier in this chapter, the muscle pump is very effective in propelling the blood toward the heart. When the leg muscles contract rhythmically, there is a decrease in the blood volume of the legs, indicating the emptying of blood; and blood can thus be driven from a segment against a resistance of 90 mm Hg (12.0 kPa) (Barcroft and Swan, 1953). The action of the muscle pump is especially important in the erect position. If exercise is started (walking), the pressure in the veins of a foot can decrease from 100 mm Hg (13.3 kPa) in passive standing

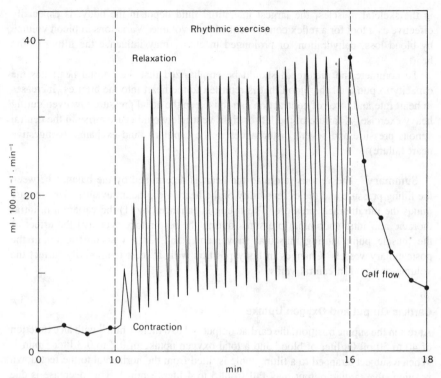

Figure 4-18
Schematic representation of changes in blood flow through the calf muscle during strong rhythmic contractions. During contraction the blood within the muscles is emptied toward the heart but at the same time the inflow is greatly reduced. (*From Barcroft and Swan, 1953.*)

to about 20 mm Hg (2.7 kPa). The increase in venous return to the heart as exercise starts will promote the possibility of an immediate increase in cardiac output. An increased filling pressure in the heart (increased preload) may stimulate the stretch receptors in the atria and in a reflex manner accelerate the heart and improve its stroke. The bigger the muscle mass involved, the more pronounced is this effect. As pointed out above, when the exercise stops, the blood stays temporarily in the dilated vascular bed, and the decrease in venous return to the heart may cause such a drop in cardiac output and arterial blood pressure that fainting occurs. (Potentially about one liter of blood may be present in the muscles during maximal vasodilation, Folkow and Neil, 1971.) This transient pressure drop is more likely to occur if the skin vessels are dilated to secure heat elimination and the person remains motionless in a standing position.

At rest, the venous system (capacitance vessels) contains about 65 to 70 percent of the total blood volume. By constriction of the venules and veins, close to half of that blood volume may be mobilized and emptied toward the heart. Vasomotor activity in the skeletal muscles, the largest interstitial fluid depot in the body, is especially effective as a tool for a reflex control of the blood volume. Variations in blood volume by blood loss, dehydration, or prolonged inactivity may influence the filling of the heart.

To complete the picture, it should be emphasized that the normal heart has the capacity to pump all the blood that is returned to the heart into the arteries. Increases in heart rate and force of contraction can thus keep the atrial pressure low even during heavy exercise. A failure of the left or right ventricle would cause a rise in the central venous pressure and severe disturbances in the capillary fluid exchange (congestive heart failure).

Summary   The venous return to the heart is determined by the balance between the filling pressure and the distensibility of the heart, i.e., the intraventricular pressure minus the intrathoracic pressure. The filling is enhanced by (1) the variation in intrathoracic and intra-abdominal pressures during the respiratory cycle,(2) the effect of the muscle pump during muscular movements, and (3) a vasoconstriction in the postcapillary vessels. Changes in body position will, at least temporarily, affect the volume of blood in central veins.

Cardiac Output and Oxygen Uptake

At rest in the supine position, the cardiac output is 4 to 6 liters $\cdot$ min^{-1} with an extraction of 40 to 50 ml O_2/liter of blood and a total oxygen uptake of 0.2 to 0.3 liter $\cdot$ min^{-1}. When a subject strapped to a tilting table is tilted from the horizontal to the feet-down position, the cardiac output may fall from 5 to 4 liters $\cdot$ min^{-1}. This decrease is due to the previously discussed venous pooling. Stroke volume is reduced and heart rate is usually increased. Activation of the muscle pump propels the blood toward the heart, and the heart rate may even decrease as stroke volume increases (Fig. 4-15). In the passive feet-down position, the oxygen uptake is unchanged, and hence the arteriovenous O_2 difference is increased.

During exercise, the cardiac output increases with the increase in oxygen uptake, but not linearly, if a range from rest value up to maximal is considered (Fig. 4-19). On 11 women and 12 men, well-trained, twenty to thirty years of age, the cardiac output, oxygen uptake, heart rate, and oxygen content of arterial blood were determined at rest sitting on a cycle ergometer and on four to five different work rates up to the maximum that could be maintained for 4 to 6 min. Dye dilution technique with indocyanine green was used for determination of cardiac output. Two to three measurements were done at each load, the first after about 5 min of exercise (except for maximal load). The mean values were used to calculate cardiac output, stroke volume, and $a\text{-}\bar{v}O_2$ difference (P.-O. Åstrand et al., 1964). The physically very fit man can increase his oxygen uptake from 0.25 to 5.00 liters $\cdot$ min^{-1} or more when exercising on a cycle ergometer or treadmill. This increase is, let us assume, met by an increase in heart rate from 50 to 200, and in stroke volume from 100 to 150 ml, which means that during maximal exercise the cardiac output has increased from 5.0 to 30.0 liters $\cdot$ min^{-1}, or 6 times the resting level. Since the oxygen uptake increased 20 times,

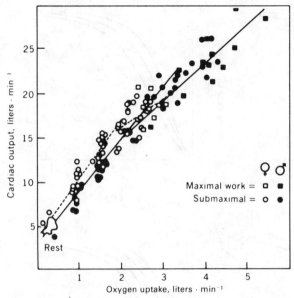

Figure 4-19
Individual values on cardiac output in relation to oxygen at rest, during submaximal, and during maximal exercise on 23 subjects sitting on a cycle ergometer. Regression lines (broken lines for women) were calculated for experiments where the oxygen uptake was (1) below 70 percent and (2) above 70 percent of the individual's maximum. (*From P.-O. Åstrand et al., 1964.*)

the a-$\bar{v}O_2$ difference must have changed from 50 to 165 ml · liter^{-1} of blood, or 3.3 times the resting level (and 3.3 times 6 is close to 20). This better utilization of the oxygen transported by the blood is reached principally in two ways: (1) The blood flow is redistributed during exercise so that skeletal muscles, with their pronounced ability to extract oxygen, may receive 80 to 85 percent of the cardiac output, as compared with some 15 percent at rest. (2) The oxygen dissociation curve is shifted so that more oxyhemoglobin is reduced than normally at a given pressure for oxygen, i.e., the percentage of saturation is less. This "Bohr effect" is easier to understand if one studies Fig. 4-20. In the active muscles, the temperature may exceed 40°C and the pH may be lower than 7.0, so there is really not a fixed relation between oxygen tension and oxygen saturation of the hemoglobin.

At normal pH (7.40), CO_2 tension (40 mm Hg), and blood temperature (37°C), the blood keeps about 33 percent HbO_2 at an oxygen tension of 20 mm Hg (5.3 kPa). At a pH of 7.2 and temperature of 39°C, the percentage of HbO_2 at the same oxygen tension is reduced to about 17, which means that 1 liter of blood with an O_2 capacity of 200 ml can deliver 26 ml more oxygen without changes in the pressure gradient for oxygen between capillary and muscle cell. With 24 liters of blood circulating the working muscles per minute, this extra oxygen delivery amounts to about 0.6 liter · min^{-1}, or 12 percent of the total uptake, in the example given above. With pH 7.0 and muscle temperature 40°C, the extra oxygen uptake

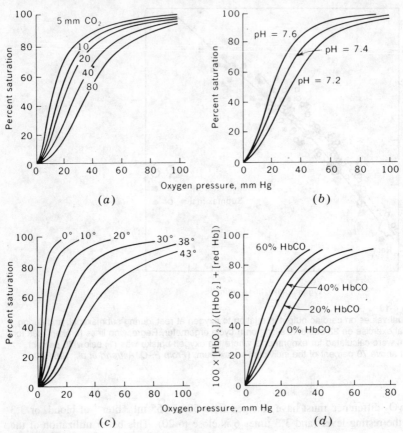

Figure 4-20
Effects of CO_2, pH, temperature (°C), and CO on the oxygen dissociation curve of the blood. Percent of saturation = percent of HbO_2. (*From various sources; see Ruch and Patton, p. 331, 1974.*)

would, however, not be further increased because the oxygen content of the arterial blood would be affected negatively by the low pH and high temperature. (The effect of 2,3-DPG on the O_2 dissociation curve will be discussed in the next chapter.)

With carbon monoxide present, carboxyhemoglobin (HbCO) is formed. Such a conversion affects the oxyhemoglobin dissociation curve with a shift to the left. The effect of carbon monoxide on oxygen transport is therefore twofold: it reduces the amount of hemoglobin available for oxygen transport, and it interferes with the unloading of oxygen in the tissues. For smokers, the effect of CO on the hemoglobin becomes a real handicap during exercise.

The shift in the dissociation curve is a result of the heat production by the exercising muscle cells and the formation of CO_2 and free protons during the heavy exercise. The effect of CO_2 in releasing O_2 from the blood is actually twofold: CO_2 lowers the pH of the blood, and by combining with hemoglobin, reduces its affinity for O_2.

By the two mechanisms, based on (1) a regulation of the circulation and (2) an inherent characteristic of hemoglobin, the oxygen uptake can be elevated 20 times, but the cardiac output has to increase to only 30 liters $\cdot$ min^{-1}, not $20 \times 5 = 100$ liters $\cdot$ min^{-1}.

Oxygen Content of Arterial and Mixed Venous Blood

During exercise, there is a *hemoconcentration* of the blood, which is partly explained by the mentioned withdrawal of fluid to the active muscle cells and by the interstitial fluid (receiving the metabolites produced in the cells). Hence the osmotic pressure is highest within and close to the metabolically active cells. The raised capillary pressure and surface area also lead to an increased outward filtration. In the experiments on 23 subjects (P.-O. Åstrand et al., 1964), the oxygen capacity of the arterial blood was about 10 percent higher during maximal exercise than at rest. The actual oxygen content of the blood drawn from the brachial artery was 3 percent higher during heavy exercise than at rest. This hemoconcentration makes the blood more viscous, but it also increases the transportation capacity per liter of blood for both oxygen and carbon dioxide.

There is evidently a discrepancy between the increase in oxygen-binding capacity of the blood and the extra oxygen actually taken up during strenuous exercise. In other words, there is a slight reduction in saturation of the arterial blood during maximal exercise, despite a normal or even elevated oxygen tension in the lung alveoli. The arterial pH may, however, be below 7.2 and the blood temperature markedly elevated, and therefore the shift in the oxygen dissociation curve to the right gives a noticeable effect on the oxygen saturation even at a high oxygen tension in the blood. There is always an admixture of venous blood from the heart muscle and lung tissue to the arterial blood. These factors taken together can reduce the oxygen saturation of arterial blood from, say, 97 percent at rest to about 90 percent during maximal exercise. This shift is a disadvantage in the lungs, but the overall effect is, as discussed above, an improved oxygen delivery due to the advantage at the tissue level, both in active and nonactive areas.

The increased extraction of oxygen from the arterial blood as exercise becomes heavier is illustrated in Fig. 4-21. During maximal exercise, the venous blood leaving the muscles has a very low oxygen content. In this study, the calculated oxygen content in *mixed venous blood* averaged about 20 ml $\cdot$ l^{-1} of blood for both women and men. At rest, the blood flow through most tissues is luxurious as far as the oxygen need is concerned, since other functions determine the flow distribution (e.g., through the vascular bed in kidneys, intestines, and skin; for the oxygen supply to the kidneys, about 50 ml of blood per min would be enough, but at rest the actual flow exceeds 1 liter $\cdot$ min^{-1}). As emphasized in previous discussions, the blood flow, during exercise is redistributed with the primary object of supplying metabolically active tissues with oxygen and removing the produced carbon dioxide. This blood flow has the potential to handle the transport of substrates, hormones, waste products, and the heat produced by metabolism.

A summary of some studies in which factors in the Ficks formula were systematically

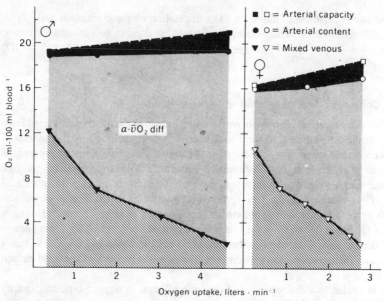

Figure 4-21
Oxygen-binding capacity and measured oxygen content of arterial blood; calculated oxygen content of mixed venous blood at rest and during exercise up to maximum on cycle ergometer. Mean values for five female (right part) and five male subjects (left), twenty to thirty years of age, and with high maximal aerobic power. (*From data presented by P.-O. Åstrand et al., 1964.*) During maximal exercise, the arterial saturation is about 92 percent as compared with 97 to 98 percent at rest, and the venous oxygen content is very low and similar for women and men.

varied will illustrate how the oxygen transport reacted to such variations when the subject was performing maximal exercise of about 5 min duration.

Saltin et al. (1968) report the oxygen content in the femoral vein during maximal running on the treadmill to be 1.4 vol percent (average of four subjects). The oxygen tension was 12 mm Hg and the pH 7.09. They noted that the difference in oxygen content between femoral venous and mixed venous blood (containing about 2 vol percent of oxygen) was small, particularly after training (see Saltin and Rowell, 1980).

It is reasonable to assume that some correlation should exist between the oxygen content of the arterial blood and the cardiac output at a given oxygen uptake. For some unknown reason, women have about 10 percent lower concentration of Hb in the blood than men. In the mentioned study (P.-O. Åstrand et al., 1964), the cardiac output required to transport 1.0 liter of oxygen was 9.0 liters for women (O_2 content in arterial blood: 167 ml $\cdot$ l^{-1}), and 8.0 liters for men (O_2 content: 192 ml $\cdot$ l^{-1}) during sub-maximal exercise with an oxygen uptake of 1.5 liters $\cdot$ min^{-1}. The cardiac output in males is more effective in its oxygen-transporting function than in women, and this difference can be explained by the Hb content of the blood.

In other experiments, the relationship between oxygen uptake during maximal exercise and the oxygen content of arterial blood has been further analyzed. The subjects were exposed to acute hypoxia by reducing the ambient pressure of the inspired air to simulate an altitude of 4,000 meters (P_{bar} = 460 mm Hg; 61.2 kPa). The cardiac output during submaximal exercise was higher at high altitude than at sea level, but during maximal effort, no difference in cardiac output was observed. The oxygen uptake during maximal exercise was, however, reduced in proportion to the decrease in oxygen content of the arterial blood, or to about 70 percent of what it was at sea level (Stenberg et al., 1966; Hartley et al., 1973).

With part of the hemoglobin blocked by carbon monoxide (up to 20 percent), the oxygen transport at a given submaximal rate of work can be maintained. The heart rate is increased and the cardiac output is at control level or somewhat higher. During maximal exercise the oxygen uptake is reduced more or less in proportion to the varied oxygen content of the arterial blood. However, with 15 percent HbCO the cardiac output averaged 5 percent lower than in the control experiment (see Ekblom et al., 1975).

An increased oxygen tension in the inspired air will increase the maximal oxygen uptake and improve the performance (see Ekblom et al., 1975; Fagraeus, 1974; Nielsen and Hansen, 1937). In the studies by Ekblom et al. (1975) on eight subjects breathing 50 percent oxygen in nitrogen at sea level showed an average 12 percent increase in maximal aerobic power (in uphill running) (Fig. 4-22a). The cardiac output was in maximal running similar in both hyperoxia and in the control.

With controlled blood loss and reinfusion of red cells, the effect of acute variations in hematocrit can be studied. The effect of blood loss is a deterioration of physical performance, which is related to the reduced maximal oxygen uptake. A reinfusion of red cells (equivalent to 800 ml of blood) in subjects who had recovered after blood loss could dramatically (overnight) improve the maximal oxygen uptake and the performance to supernormal values (in average an increase in maximal oxygen uptake of 9 percent) (Fig. 4-22b). In five subjects running at maximal speed, which could be maintained for about five minutes, the oxygen content of the arterial blood was in average 16 percent higher after reinfusion of red cells compared with the situation after blood loss. The difference in maximal oxygen uptake was actually about 14 percent (but the individual variations were large). The maximal heart rate and stroke volume respectively were more or less identical in the different experiments (Ekblom et al., 1976). (See Fig. 15-9.) The positive effect on maximal oxygen uptake and performance of such a "blood doping" has been repeatedly confirmed, if the reinfused blood volume is 800 ml or higher (or the equivalent in packed red blood cells). Thomson et al. (1982) report an increase in maximal oxygen uptake from 4.0 to 4.5 $1 \cdot min^{-1}$ after an autologous blood reinfusion elevating the hematocrit from 42.4 to 46.2 percent (average on 4 subjects). During submaximal exercises the cardiac output was no different from control level, but during maximal exercise they found a small increase in cardiac output which together with the increased content of O_2 in the arterial blood (about 20 ml $\cdot$ 1^{-1}) explained the improved maximal aerobic power.

An induced erythrocythemia can also increase the hypoxia tolerance during physical exercise (Robertson et al., 1982).

A plasma expansion can increase the stroke volume and cardiac output (see Spriet

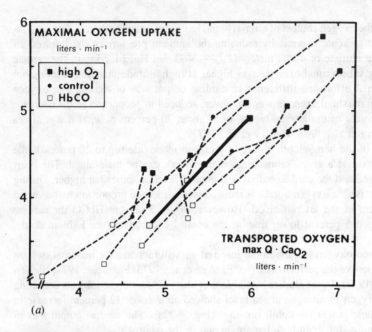

(a)

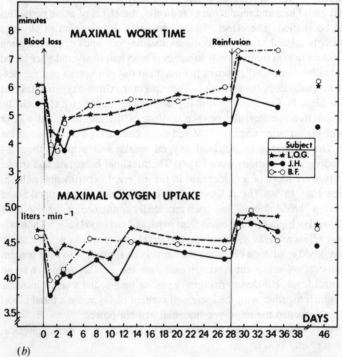

(b)

et al., 1980; Fortney et al., 1981). Kanstrup and Ekblom (1982) found that a plasma expansion by infusion of dextran (a 700 ml increase in plasma volume, on the average) increased the maximal stroke volume and cardiac output just compensating for the reduced hemoglobin concentration so that the normal maximal aerobic power could be attained. An increase in blood volume is the only factor which so far has been shown to make a "supranormal" cardiac output possible, probably due to an enhanced filling of the heart. In one way or another the total amount of hemoglobin seems to be decisive for the maximal oxygen uptake (see Fig. 9-5; Kanstrup and Ekblom, 1984). (It should be noted that "blood doping" when applied on competing athletes is a violation against the rules of the International Olympic Committee and most International Federations for different sport events. The problem is that so far it is not feasible to prove whether or not "blood doping" has been practiced.)

The purpose of this brief summary is to illustrate that the maximal oxygen uptake (maximal aerobic power) in exercise engaging large muscle groups is apparently not limited by the capacity of the muscle mitochondria to consume oxygen. Slight variations in the volume of oxygen offered to the tissue $\dot{Q}$ · Ca_{O_2} will produce almost proportional changes in the volume of oxygen consumed. Exercise with the arms (in swimming) as well as with one leg (bicycling) does include muscle groups which are also engaged in normal swimming and two-leg work respectively. It is remarkable, however, that the combined exercise does not dramatically increase the maximal oxygen uptake (see Clausen, 1976; Davies and Sargent, 1974; Holmér, 1974; Blomqvist and Saltin, 1983).

It should be emphasized that a period of physical conditioning will increase the volume of mitochondria in trained muscles increasing their aerobic energy potential (see Chap. 10). The crucial question is whether or not there are enzymes in the skeletal muscles which may serve as a bottleneck for maximal aerobic energy yield. So far, the central circulation and the capillary bed available for perfusion have been considered to be the limitation for an individual's maximal oxygen uptake. Saltin and Gollnick (1983) have estimated that the potential of enzyme systems of the skeletal muscles to consume oxygen, by far exceeds the maximum actually attained.

Stroke Volume

Factors affecting the stroke volume are (1) the venous return to the heart and (2) the distensibility of the ventricles. The degree of diastolic filling has an anatomic limitation (children-adults, women-men), but within a range various factors, some of which were

---

Figure 4-22
(a) Relation between oxygen uptake and transported oxygen. $\dot{Q} \times Ca_{O_2}$, during maximal running. Individual values on eight subjects (broken lines) and means (solid lines), at hypoxia induced by about 15 percent HbCO, control experiments, and hyperoxia with the subjects inhaling 50 percent oxygen in nitrogen, respectively. (*From Ekblom et al., 1975.*) (b) Exercise time at a standard maximal run (maximal work time) and maximal oxygen uptake during control (day "O"), after 800 ml blood loss, and after reinfusion of the packed red cells (day "28") in three subjects. (*From Ekblom et al., 1972a.*)

discussed above, affect the stretching of the muscle fibers. The final factors determining the stroke volume are (3) the force of contraction in relation to (4) the pressure in the artery (aorta or pulmonary artery).

The heart adjusts itself to changing conditions by an inherent self-regulatory mechanism. Starling (1896) using his famous lung-heart preparations, found that the normal heart tended to empty itself almost completely. It was distended to a greater diastolic volume in response to either a greater venous return or an increase of the arterial pressure. In the latter case, there was a transient decrease in stroke volume, but as the force of contraction increased with a greater initial length of the muscle fibers, the stroke volume and cardiac output became normal. By stimulation of the sympathetic cardiac nerves, the contraction force increased from the same initial length, and the arterial resistance could be overcome despite a greater extent of myocardial shortening.

It has repeatedly been emphasized that the increase in length of a muscle fiber (within limits) will improve its force generating potential. In addition, catecholamines will elicit a similar effect plus an increase in the heart rate. What, then, may be the mechanism behind these effects? In Chap. 2, the fact that in the resting muscle, troponin exerts an inhibitory effect on actin via tropomyosin, was described. In this situation, the actin and myosin filaments cannot interact and no cross-bridges are formed. A depolarization of the sarcolemma will cause a release of calcium from the sarcoplasmic reticulum and as a result the inhibitory effect on actin of the troponin-tropomyosin complex will be removed and the cross-bridges are formed. Ca^{2+} will also activate ATPase which will break down ATP, thereby providing the energy yield for the muscle contraction. Then follows the re-uptake of Ca^{2+} by the sarcoplasmic reticulum, the muscle will relax and the troponin-tropomyosin complex is again a hindrance for contraction. The same events that take place when the skeletal muscle contracts and relaxes occur in the heart muscle. There are findings suggesting that catecholamines can increase the rate of calcium release and also the rate at which calcium is removed from troponin. In addition, there may be an increase in the amount of calcium stored in the sarcoplasmic reticulum. If so, more calcium is available for delivery to the contractile proteins in subsequent contractions. Cyclic AMP seems to be involved as a mediator. Such mechanisms can explain the effects known to be produced by catecholamines, namely enhanced rate of tension rise, augmentation of contractility, and elevated heart rate (see Tada et al., 1978). Regarding the muscle length-tension relationship, there are data indicating that the binding constant of troponin for calcium is modified when the fiber length is changed (Allen and Kurihara, 1979), and this will also influence the amount of calcium released from the sarcoplasmic reticulum.

The results from Starling's studies of the isolated heart were also considered applicable in the intact animal. Most earlier textbooks concluded that the diastolic volume of the heart was smaller at rest but increased during exercise, when the venous return increased and the arterial pressure was elevated. The same end-systolic volume could be maintained or was dependent on the strength of the heart muscle. The well-trained person was characterized by a small residual volume of blood in the heart after systole at rest as well as during exercise. The net effect was a substantial increase in stroke volume during exercise, according to these standard texts.

The present concept on the stroke volume in the human during exercise accepted by most cardiophysiologists can be summarized as follows: When the position is changed from supine to standing or sitting, there is a diminuition in end-dialostic size of the heart and a decrease in stroke volume. If muscular exercise is then performed, the stroke volume increases to approximately the same size or to a higher level as obtained in the recumbent position. (See Poliner et al., 1980). The discrepency between earlier and recent studies is in most cases not due to inconsistent results, but must be blamed on erroneous conclusions and bold extrapolations of the data to conditions not examined. One complicating factor is that the species easily available for experimentation may give data which are not applicable to other species.

The importance of the *central blood volume* for the stroke volume was demonstrated in 1939 by Asmussen and Christensen. Subjects in the sitting position were exercising with their arms. In some experiments, the subjects lay down with their legs elevated for about 10 min before the exercise started. The circulation to the legs was then arrested by pressure cuffs around the thighs. When assuming a sitting position following this procedure, there was approximately 600 ml of blood less in the legs compared with sitting without occlusion of the blood flow to the legs. It was noted that the cardiac output was about 30 percent higher when the legs were "blood free," i.e., when the central blood volume was high, as compared with experiments with blood pooling in the legs. The high cardiac output was due to a high stroke volume, for the heart rate was actually lower than in exercise with reduced central blood volume (and low cardiac output).

In Fig. 4-23, individual data on stroke volume are plotted, expressed in percentage of the maximum reached. The maximum oxygen uptake of the subject is similarly set to 100 percent, and the submaximal loads are defined in percent of this maximum. At rest, the stroke volume is, for most subjects, between 50 and 70 percent (mean 63 percent) of the maximum measured during exercise. Hence, during exercise in the sitting position, there is a definite increase in stroke volume with an "optimum" reached when oxygen uptake exceeds 40 percent of maximal aerobic power. The heart rate at that load is 110 to 120. Methodological errors and biological variations result in a variation in calculated stroke volume of ± 4 percent as oxygen uptake further increases. There is no tendency toward a decrease in stroke volume at the peak load. No significant correlation was found between maximal heart rate and eventual decrease in stroke volume during the maximal exercise. In our opinion, this rules out the hypothesis that a high heart rate (about 200) during exercise should interfere with the filling of the heart.

In this discussion of the variation in stroke volume, it should be emphasized that in experiments on animals, reflexes that evoke tachycardia also evoke simultaneously, either directly or indirectly, an increase in contractility of the heart muscle, so that stroke power and stroke work increase (Sarnoff and Mitchell, 1962).

Braunwald et al. (1967) observed in experiments on human subjects that an increase in heart rate increases the velocity of the shortening of the myocardial fibers. They emphasize that the exercise tachycardia as such contributes to an improvement of the contractile state of the myocardium. They conclude that "the normal cardiac response to exercise involves the integrated effects on the myocardium of simple tachycardia, sympathetic stimulation, and the operation of the Frank-Starling mechanism. During

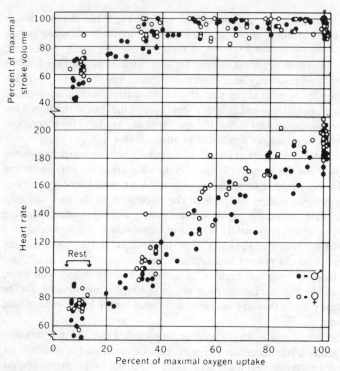

Figure 4-23
Stroke volume in percentage of the individual's maximum, and heart rate at rest and during exercise. The oxygen uptake on the abscissa is expressed in percentage of the subject's maximum. Circled dot at "100 percent" represents 11 of the 23 subjects. Measurements were made with the subjects in the sitting position (same subjects as in Fig. 4-19).

submaximal levels of exertion, cardiac output can rise even when one or two of these influences are blocked. However, during maximal levels of exercise, the ventricular myocardium requires all three influences in order to sustain a level of activity sufficient to satisfy the greatly augmented oxygen requirements of the exercising skeletal muscles." (See below.)

Heart Rate

In many types of exercise, the increase in heart rate is linear with the increase in rate of exercise. There are exceptions and those exceptions are perhaps more frequent among untrained subjects. When the subject performs very heavy exercise, the $a\text{-}vO_2$ difference may increase so that the oxygen uptake increases relatively more than the cardiac output. The evaluation, from submaximal rates of exercise, of an individual's maximal oxygen uptake or capacity to perform work is, in most test procedures, based on a registration of the heart rate during steady state and then an extrapolation to a fixed heart rate or to an assumed maximal heart rate (see Chap. 8). There are many pitfalls in this method. Here, the following should be noted: (1) The standard deviation for maximal heart rate during exercise is ± 10 beats $\cdot$ min^{-1}. Hence, for twenty-five-year-old individuals, women or men, the maximal heart rate for 5 out of 95 subjects may be below 175 or above 215, since the maximum is about 195 on an average. (2) There is a gradual decline in maximal heart rate with age, so that the ten-year-old attains 210, the sixty-five-year-old only about 165 beats $\cdot$ min^{-1} (Fig. 4-24) (Robinson, 1938; P.-O. Åstrand, 1952; I. Åstrand, 1960; Pollock et al., 1978; Hollmann and Hettinger, 1980, p. 621). Futhermore, longitudinal studies have shown a wide individual scatter in the decline in maximal heart rate with age (I. Åstrand et al., 1973).

When 50 percent of the maximal aerobic power is used, the heart rate in the twenty-five-year-old man is about 130, but the same relative work rate and feeling of strain are experienced at a heart rate of 110 for the sixty-five-year-old man (Fig. 4-24) (I. Åstrand, 1960). For women, the 50 percent oxygen uptake is attained at a heart rate of about 140 beats $\cdot$ min^{-1} at the age of twenty-five.

Prolonged exercise in a hot environment causes a higher heart rate than exercise at a low room temperature. Emotional factors, nervousness, and apprehension may also affect the heart rate at rest and during exercise of light and moderate intensity. During repeated maximal exercise, the heart rate is, however, remarkably similar under various conditions, with a standard deviation of ± 3 beats per min (P.-O Åstrand and Saltin, 1961).

The heart rate at a given oxygen uptake is higher when the exercise is performed with the arms than with the legs (Christensen, 1931; Stenberg et al., 1967; Vokac et al., 1975). Static (isometric) exercise also increases the heart rate above the value

Figure 4-24
The decline in maximal heart rate with age, and heart rate during a submaximal work rate. Mean values from studies on 350 subjects. The standard deviation in maximal heart rate is about ± 10 beats/min in all age groups. (*From Åstrand and Christensen,, 1964.*)

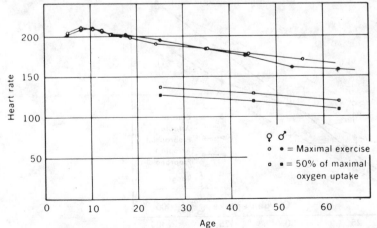

expected from the metabolic rate (see discussion under "Blood Pressure"). The mechanism for these differences in heart-rate response to exercise is not understood. However, the elevated heart rate is usually accompanied by a decreased stroke volume. It is known that a variation in heart rate at a given oxygen uptake at rest and during submaximal exercise often produces a change in stroke volume, so that the cardiac output is maintained at an appropriate level. (This information is based on studies in patients with artificial pacemakers or irregular heart rate, and in subjects submitted to various drugs influencing the heart rate.) (Bevegård and Shepherd, 1967; Braunwald et al., 1967.)

Fig. 4-25 presents data on subjects submitted to submaximal (cycle ergometer) and maximal (treadmill) exercise four times: (1) control; (2) after infusion of 10 mg propranolol blocking the adrenergic β-receptors in the heart; (3) after infusion of 2 mg atropine blocking the parasympathetic impulse traffic; (4) after double blockade (propranolol and atropine). At a given oxygen uptake the heart rate varied about 40 beats · min⁻¹, taking the extremes, but the cardiac output was almost similar in the four situations, since the stroke volume compensated for the changes in heart rate. (In the propranolol experiments there was in average 1.5 to 2 liters · min⁻¹ reduction in cardiac output.) It should be noted that the subjects reached their normal maximal oxygen uptake despite a reduction in maximal heart rate from 195 to about 160 beats · min⁻¹. The performance time was significantly shorter after β-blockade and the intra-arterial blood pressure was reduced (Ekblom et al., 1972b). Sable et al.

Figure 4-25
Relationship between heart rate (means and ranges) and relative oxygen uptake
($\dot{V}_{O_2} \cdot \dot{V}_{O_2}max^{-1} \cdot 100$) in five sets of experiments (four subjects) during normal conditions
(control) and after blockade of receptors. Resting heart rate was recorded in the sitting position.
(*From Ekblom et al., 1972b.*)

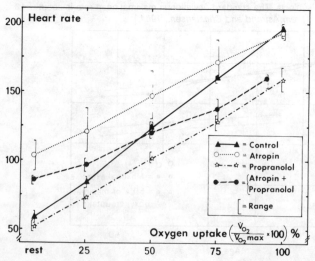

(1982), also report a "normal" maximal aerobic power after β-blockade. In other studies, however, a reduction in maximal oxygen uptake as a consequence of β-blockade, has been reported (see Hughson et al., 1984).

The regulation of the circulation in exercise is probably guided primarily by factors sensitive to an adequate cardiac output to secure the oxygen supply to the exercising skeletal muscles. There is a remarkable constancy in the relationship between oxygen uptake and cardiac output. If, for some reason, the stroke volume of the heart is reduced there is a compensatory increase in the heart rate. One exception is hypoxic conditions when the cardiac output at a given oxygen uptake is elevated. Heart rate and stroke volume are the variables, and the stroke volume is more likely to be directly influenced by such factors as venous return or peripheral vascular resistance in order to secure an adequate oxygen supply to the exercising skeletal muscles.

Pendergast et al. (1980), studied the kinetics of the increase in oxygen uptake and cardiac output at the *onset* of exercise. They conclude that the cardiac output and muscular blood flow do not play any role in determining the $\dot{V}_{O_2}$ kinetics. They maintain that the limitation must be at the mitochondrial level. On the other hand, Hughson and Morrissey (1983) emphasize the role of oxygen transport, as opposed to oxygen utilization, as the major rate-limiting step in the adaptation of the oxygen uptake at the onset of constant load exercise. Obviously, further evidence is needed in order to settle this question.

Blood Pressure

There are some methodological aspects to be considered when defining the blood pressure, especially the arterial blood pressure. Measurement of the pressure in the aorta with a catheter supplied with side openings gives a significantly lower pressure during exercise than does measurement with an open-ended catheter directed upstream, (see discussion earlier in this chapter). The total energy of the blood is the sum of the kinetic energy and the pressure energy. The side-hole catheter is measuring only the pressure energy (side pressure), but the catheter with the opening directed upstream also includes the kinetic energy. Since the kinetic energy factor of the blood in the aorta is high during exercise with a pronounced increase in cardiac output, the two catheters should give quite different pressure readings. Marx et al. (1967), have convincingly shown that this is the case. The mechanoreceptors in the walls of arteries cannot sense the kinetic energy factor of the passing blood. The lateral distending pressure gives some information of the stretch to which these receptors are exposed but a sympathetic vasoconstrictor activity involving the vessel wall will make it stiffer and less deformed by a given intravascular pulsative pressure.

A simultaneous measurement of intra-arterial blood pressure in a peripheral artery and in the aorta during exercise gives a significantly higher systolic end pressure in the peripheral artery, but the mean and diastolic pressures are about the same as in the aorta (see Marx et al., 1967). The systolic pressure in a peripheral artery is higher in a resting than in an exercising limb (P.-O. Åstrand et al., 1965). The progressive increase in systolic pressure (and pulse pressure) along an artery is at least in part due to a distortion in the transmission because of summation of the centrifuge wave and the reflected waves from the periphery. The importance of the wave reflection increases when the peripheral resistance is high, as is the case in a resting limb.

As a result of the vasodilation in the vascular bed in the active muscles, the peripheral resistance to blood flow is reduced during exercise, but the elevation in cardiac output causes the blood pressure to rise. If the *cardiac output* ($\dot{Q}$) during the exercise is 4 times the resting value, giving a 25 percent increase in arterial mean pressure (P_{mean}), it follows that the resistance to flow (R) is reduced to more than one-third of what it was at rest, since $\dot{Q} \times R = P_{mean}$. The blood pressure obtained in a peripheral artery at rest, 120 mm Hg (16 kPa) during systole, and 80mm Hg (10.6 kPa) during diastole, may exceed 175 and 100 mm Hg (23.3 and 13.3 kPa), respectively, during heavy exercise. The diastolic blood pressure, when measured with a blood pressure cuff, is constant or falls slightly with increasing rate of exercise (Hollmann and Hettinger, 1980). In this case the cuff method apparently gives a slightly different picture than the intra-vascular recording of the diastolic pressure.

It should be noted that the arterial blood pressure is significantly higher in arm exercise than in leg exercise (Fig. 4-26). The high blood pressure at a given cardiac output, when the work is performed by the arms, induces an increased stroke work of the heart. Therefore, for untrained individuals or for cardiac patients, it may be hazardous to exercise hard with the arms, e.g., to shovel snow, dig in the garden, or carry heavy trunks. The relatively high blood pressure in exercise with small muscle groups is probably due to a vasoconstriction in the inactive muscles. The larger the activated muscle groups, the more pronounced is the dilation of the resistance vessels. The lower peripheral resistance is reflected in a lower blood pressure (since $\dot{Q} \times R = P_{mean}$).

In isometric exercise we have a similar situation with a considerable ventricular afterload. The arterial blood pressure, both systolic and diastolic, and heart rate increase abruptly with sustained isometric effort at 15 percent of the maximal voluntary contraction (MVC) or higher. Usually there is no steady state in these functions, but a gradual increase until the end of contraction. The blood pressures increase more or less linearly with the developed force in a given muscle group. The larger the muscle mass involved, the more pronounced the pressure and heart rate response (see Kilbom and Persson, 1981; Mitchell et al., 1981; Seals et al. 1983). With large muscle groups contracting at great force, the systolic pressure may well exceed 300 mm Hg (40 kPa), and the diastolic pressure rise beyond 150 mm Hg (26.6 kPa). The stroke volume appears to remain constant, despite the increase in afterload. The increase in heart rate causes an elevated cardiac output (Kilbom and Persson, 1981). The reason why the arterial blood pressure at a given cardiac output is higher during isometric contraction than during dynamic exercise is not known. From a teleological viewpoint, one might say that a higher pressure will facilitate the blood flow through a muscle with a high intramuscular pressure. The greater mental effort involved in maintaining the isometric force, and the accumulation of trapped metabolites in the muscle, may contribute to an exaggerated sympathetic drive.

If during recovery from isometric contractions, the blood flow through the engaged muscles is stopped by an inflated cuff, the heart rate returns to control level. Blood pressure also drops, but remains above the precontraction level until the cuff is released

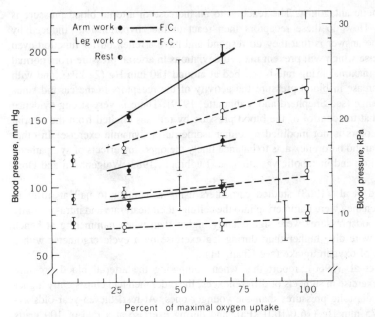

Figure 4-26
Effect of exercise on blood pressure (end pressure). Regression lines of arterial systolic, mean, and diastolic blood pressures, respectively, in relation to oxygen uptake (in percentage of the maximum) during arm and leg exercise in the sitting position for 13 subjects. F.C. = femoral artery catheter. The figure summarizes data from 23 submaximal and 13 maximal work rates with arm work (cranking) and 44 and 13 experiments, respectively, with leg work. The vertical heavy lines represent ± 1 SD (standard deviation) around the regression line. The dots and thin lines represent the mean ± 1 SEM (standard error of the mean) for three groups of values at different levels of oxygen consumption. (The systolic pressure measured in a peripheral artery is higher than in the aorta, but the mean and diastolic pressures are similar in the two vessels.)
(*From P.-O. Åstrand et al., 1965.*)

(Mitchell et al., 1981). Bonde-Petersen and Suzuki (1982), report a similar pattern during recovery after dynamic cycle exercise, i.e., the mean arterial pressure remains elevated during occluded recovery, whereas the heart rate tends to recover at the same rate as in the control without occlusion.

These findings support the previously discussed hypothesis that the reflex control of the cardiovascular function, including the arterial blood pressure, during exercise, is triggered from at least two more-or-less independent mechanisms. Mention was made of a "central command" related to the activation of the motor units, and a peripheral control mechanism mediated by muscle afferents reporting the metabolic changes in the contracting muscles.

In the control of arterial blood pressure, the mechanoreceptors in some artery walls play, as mentioned, a key role in that they perform some sort of a "buffer" function. If at rest, there is a pressure drop, the reduced activity, in e. g., the carotid sinus receptors, will activate the sympathetic system with a reciprocal inhibition of the parasympathetic antagonist. The reaction to an increase in arterial blood pressure is the reverse. How do these receptors then react to the "hypertension" induced by exercise? The answer is that they do respond and have a reflex buffer function even during exercise which will prevent marked deviations in arterial pressure from normal values. The maximal firing rate is reached at around 180 mm Hg (24 kPa), and with a further increase in blood pressure the activity of the receptors in the carotid sinus does not change (see Shepherd and Vanhoutte, 1979). There is very strong evidence suggesting that the control of the blood pressure by reflexes elicited from the arterial mechanoreceptors is not modified by either isometric or dynamic exercise. In other words, the carotid baroreflex acts to balance finely the opposing effects of sympathetic vasoconstriciton and metabolic vasodilation (Ludbrook, 1983; Walgenbach and Donald, 1983).

I. Åstrand et al. (1968) studied carpenters using a hammer to nail at different heights. When they were hammering into the ceiling, their heart rate and intra-arterially measured blood pressures were significantly higher than when hammering at bench level. They were also higher than during leg exercise on a cycle ergometer with a similar level of oxygen uptake (see Chap. 11).

Reindell et al. (1960), report that when comparing the arterial blood pressure response to exercise in subjects of different ages, the older men had consistently higher systolic and diastolic pressures than the younger ones. At rest, the 25-year-olds averaged 125/75 mm Hg (16.0/10.0 kPa), and during exercise at a rate of 100 watts (oxygen uptake about 1.5 liters · min^{-1}), the pressures were 160 and 80 mm Hg (21.3 and 10.6 kPa) in systole and diastole, respectively. For the 55-year-old group, the increase was from 140/85 mm Hg (18.6/11.3 kPa) at rest up to 180/90 mm Hg (23.9/12.0 kPa), the work rate being the same. Similar results are reported by Gerstenblith et al. (1976), in their review.

Type of Exercise

The cardiac output at a given submaximal oxygen uptake is in many types of exercise similar, e.g., in exercise, with arms, cycling with one or two legs, combined arm and leg exercise, walking, running, and swimming (Clausen, 1976; Davies and Sargent, 1974; Hermansen et al., 1970; Holmér, 1974; Stenberg et al., 1967). The cardiac output at a given oxygen uptake during submaximal exercise is consistently 1 to 2 liters less in the erect position than when the subject is recumbent; the heart rate is about the same (Reeves et al., 1961; Poliner et al., 1980). The compensation for the lower cardiac output must, by definition, be an increased $a\text{-}\bar{v}O_2$ difference when erect. It should be emphasized that the maximal oxygen uptake during cycling in the supine position is lower than during exercise in the sitting position on a cycle ergometer (Fig. 4-27). Similarly, the cardiac output is somewhat lower during maximal exercise with the legs in the supine position. This difference in response may be explained in the following way: Rhythmic muscular contractions will squeeze out blood from the veins lowering the average venous pressure considerably and hence raising the effective perfusion pressure of flow. This is evident in exercise in the upright position as the arterial pressure at the calf level is then raised by some 70 to 80 mm Hg (9.3 to 10.6

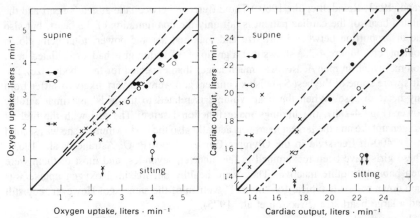

Figure 4-27
Oxygen uptake (to the left) and cardiac output; comparison between the highest individual values attained in the sitting (abscissa) and supine (ordinate) position for arm exercise (x), leg exercise (○), and combined arm and leg exercise (●). Line of identity and lines corresponding to 10 percent deviation are drawn. Symbols with arrows give the mean of the different groups. Note that leg exercise in the supine position did not bring oxygen uptake and cardiac output to a maximum, but when the arm muscles were also exercised, the oxygen uptake and cardiac output did increase to the same level as in leg exercise, or in combined arm and leg exercise in the sitting position. (*From Stenberg et al., 1967.*)

kPa) as compared with the supine position, i.e., in proportion to the distance from the heart because of the increase in hydrostatic pressure. However, the pressure in the calf veins is maintained at a low level by the "milking" action of the muscle pump. Folkow et al. (1971) noted that the calf blood flow in man could be 50 to 60 percent larger when a standard heavy rhythmic exercise was performed in the upright position as compared with the reclining position. Combined arm and leg exercise in the sitting or supine position reveals almost the same values for maximal oxygen uptake, heart rate, and cardiac output as exercise in the sitting position with only the leg muscles (Fig. 4-27). It has been emphasized that in one way or another, the central circulation seems to limit the maximal cardiac output, and thereby also the maximal oxygen uptake.

In prolonged exercise in a neutral thermal environment the cardiac output is normally well maintained, but there is a progressive rise in heart rate and fall in stroke volume. Furthermore, there is a gradual reduction in blood pressures (systemic arteries, pulmonary artery, right ventricular end-diastolic pressures). The mechanism of these cardiovascular shifts are, so far, not established (see Rowell, 1983).

Heart Volume

The size of the heart can be visualized by means of roentgenograms, and its volume can be computed by application of empirical formulas. A high correlation has been established between heart volume and various parameters, such as blood volume, total amount of hemoglobin, and stroke volume in healthy younger individuals (see Sjöstrand, 1953, Reindell et al., 1960.) The difference between the athlete's heart and the dilated heart of the cardiac patient is not only the configuration of the heart, but also the disproportion between heart size and maximal aerobic power, total hemoglobin content, etc. Figure 4-28a shows that a group of 30 young girls had a calculated heart volume which in many cases was much larger than expected for their body size. The girls were some of the best Swedish swimmers and were not very likely to suffer from any heart disease. When the heart volume is related to the girls' maximal aerobic power (Fig. 4-28b), the findings make functional sense. The girl with the highest oxygen uptake and the largest heart was actually also the best swimmer; she was second in the 400-m free-style in the Olympic Games, 1960 (P.-O. Åstrand et al., 1963). These girls have been reexamined in the nineteen-seventies, and most of them were then habitually quite inactive. They were healthy but maximal oxygen uptake was then down to levels typical for "normal" women of the same age. However, they still had a large heart size (Eriksson et al., 1975).

In active athletes there is a correlation between the demand placed on the oxygen transport system by the event in question, and the estimated heart volume. On the average, the heart volume calculated from roentgenograms has been found to be largest (above 900 ml) for well-trained athletes engaged in events calling for endurance (bicyclists, canoeists, cross-country skiers, long-distance runners). The average for middle-distance runners, swimmers, and soccer, and tennis players was between 800 and 900 ml, and for boxers, fencers, gymnasts, jumpers, sprinters, throwers, and

Figure 4-28
Relationship between heart volume and (a) calculated body surface area and (b) maximal O_2 uptake in 30 young well-trained girl swimmers. Shadowed area gives the 95 percent range for "normal" girls. (*Modified from P.-O. Åstrand et al., 1963.*)

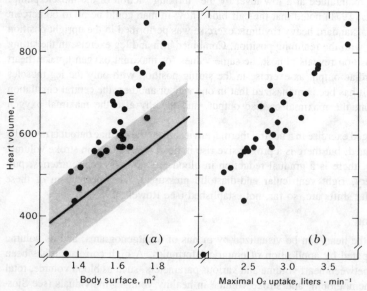

untrained controls, below 800 ml. The individual variations are large, however (see Reindell et al., 1960; Rost and Hollmann, 1983.)

There are data from longitudinal studies of former endurance athletes who have later on become relatively inactive which confirm the results from the study of the girl swimmers. Many of these previous athletes still have a large heart (see Blomqvist and Saltin, 1983). There are exceptions, however, thus Rost and Hollmann (1983), report a heart volume of 1,700 ml in a professional bicyclist; his maximal aerobic power was above 6 l · min^{-1}. Four years after cessation of training his heart volume was reduced to 980 ml. In former endurance trained athletes the correlation between maximal oxygen uptake and heart volume is low. This may, at least in part, be explained by their different levels of habitual physical activity. With the development of echocardiographic and radionuclear techniques, the tools are now available for more detailed dimensional analyses, particularly on the left ventricle. From a review of the literature, Blomqvist and Saltin (1983) conclude that the results support the older radiographic observation.

Summary   Endurance training which demands a high oxygen uptake (and therefore inevitably a large stroke volume) is generally accompanied by an increase in left ventricular end-diastolic volume. The ventricular wall thickness may not change or it may increase slightly. Apparently, there is an increase in the weight of the heart. In athletes engaged in strength training, involving isometric effort, an increase in wall thickness is noted without any change in ventricular volume. As discussed under "Blood Pressure," isometric exercise engaging large muscle groups exposes the heart to high afterloads without change in stroke volume. Thus, there is definitely a difference in the demands on the heart when comparing strength training with endurance training. (See also Peronnet et al., 1981; Dickhuth et al., 1983.)

The weight of the athlete's heart does not generally exceed 500 g, which is regarded as the limit of physiological hypertrophy (see Rost, 1982). In patients with myocardial disease, the heart weight may exceed 1000 g.

Age

At a given rate of exercise/oxygen uptake, the older individual attains, on the average, the same heart rate as the younger one (I. Åstrand, 1960; Strandell, 1964), or it may be slightly lower (Hollmann and Hettinger, 1980; Jonsson and I. Åstrand, 1979). The latter finding is not easy to explain. On the other hand Strandell (1964) found that the cardiac output at a given work rate was about 2 liters · min^{-1} lower in the sixty- to eighty-year-old men at any level oxygen uptake compared with the young ones. Stroke volume was also significantly lower for the older men (about 20 percent). (Age changes in myocardial function were summarized by Gerstenblith et al., 1976.) Eriksson (1972) studied 11 to 13 year-old boys and noted that the cardiac output during submaximal exercise was 1 to 2 liters · min^{-1} less than for adult young men.

As already mentioned, the heart rate reached during maximal exercise decreases with age. The value typical for the ten-year-old girl or boy is 210, for the twenty-five-year-old 195, and for the fifty-year-old 175 beats · min^{-1} (Fig. 4-24). Therefore the decrease in circulatory capacity in the old individual is more marked than predicted from heart rate, stroke volume, and cardiac output observed during submaximal exercise if the norms are the same as when evaluating young individuals. The old man has a larger heart volume, calculated from roentgenograms taken in supine position, than does the young man; blood volume and total amount of hemoglobin are not different in young and old men (Strandell, 1964; Grimby and Saltin, 1966). These findings should be related to the decrease in maximal stroke volume, cardiac output, and maximal aerobic power in old men (see Fig. 9-7a).

Whether the decrease in maximal heart rate with age is a consequence of an arteriosclerosis in the vessels of the heart is not known. The oxygen cost of a cardiac performance involving a high heart rate is great, and therefore the load on the heart is reduced by a lowered ceiling for heart rate. The lower heart rate during maximal exercise in the old individual is probably not a direct response to hypoxia, since breathing pure oxygen instead of room air during the exercise does not further elevate the heart rate (I. Åstrand et al., 1959).

Age changes in the sinus node and conductive system in the heart are possible explanations for the gradual decline in heart rate with age. The individual variation in maximal heart rate is of the same order of magnitude in old as in young people, i.e., S.D. = $\pm$ 10 beats · min^{-1}.

It is well documented that with age there is, in most individuals, an increase in arterial blood pressure. This is also so during exercise at a given exercise rate, as reported by Reindell et al. (1960).

Unfortunately, our knowledge of the cardiovascular response to exercise in the aged is rather scanty. One of the reasons for this is the problem of selecting a random sample. It is relatively simple to define a "normal healthy subject" when dealing with a young population. But, who may be termed a "normal healthy subject" when approaching the age of 70? Such a group of older individuals is already a very selected group of subjects in comparison with younger counterparts.

Training and Cardiac Output

It is an old observation that the heart rate at rest and during standard exercise, as well as during recovery is lowered with a training of the oxygen-transporting system. Published reports on the cardiac output during standard exercise repeated during a course of training indicate that cardiac output is maintained at the same level or is slightly reduced, i.e., the a-$\bar{v}$ O_2 difference is widened (see Rowell, 1974). (The training effects are further discussed in Chap. 10.)

Top athletes in endurance events are characterized by a very high maximal oxygen uptake. Their maximal values for circulatory parameters must therefore be high compared with those of less athletic individuals. Ekblom and Hermansen (1968) have collected data obtained during maximal exercise on the treadmill in athletes, using the dye dilution technique. Some data are presented in Table 4-1. Both an intensive training and superb natural endowments contribute to the remarkable circulatory capacity for oxygen transportation in these subjects. The highest reported figure on maximal oxygen uptake, as far as we know, is 7.4 1 · min^{-1}. This subject's cardiac ouput was not

TABLE 4-1

Subject	Maximal values				
	$\dot{V}O_2$, liters · min⁻¹	Cardiac output, liters · min⁻¹	Heart rate	Stroke volume, ml	a-$\bar{v}O_2$ difference, ml · l⁻¹
G.P.	6.00	39.8	188	212	151
C.R.	5.77	37.8	188	201	153
A.H.	5.60	34.4	189	182	163
C.S.	5.50	36.2	198	183	152
B.T.	5.64	38.0	193	197	148
L.R.	6.24	42.3	206	205	148
Mean	5.79	38.1	194	197	153

Figure 4-29
Cardiac output, heart rate, stroke volume, and arteriovenous oxygen difference during maximal exercise in relation to maximal oxygen uptake in top athletes who were very successful in endurance events (stars), well-trained but less successful athletes (filled squares), and twenty-five-year-old habitually sedentary subjects (unfilled circles). (*From Ekblom, 1969.*) Also included are maximal values on the male subjects presented in Fig. 4-19 (unfilled squares).

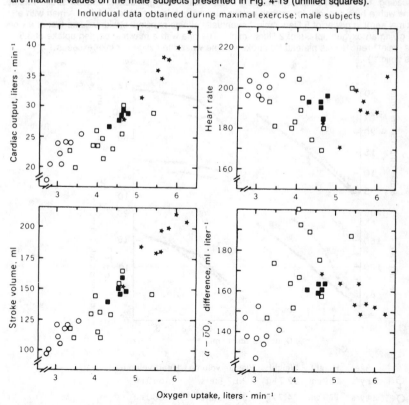

Individual data obtained during maximal exercise: male subjects

Oxygen uptake, liters · min⁻¹

measured, but if his a-$\bar{v}O_2$ difference was 153 ml $\cdot$ l^{-1} of blood (Table 4-1), the cardiac output would be 48.4 l $\cdot$ min^{-1}, i.e. his heart managed to pump some 97 liters of blood per minute.

Figure 4-29 summarizes data on 32 subjects with maximal aerobic power ranging from 2.8 to 6.2 liters $\cdot$ min^{-1} of O_2 uptake. This figure shows a clear relationship between maximal cardiac output and oxygen uptake. It also shows that it is the stroke volume which to a large extent determines the maximal cardiac output. The most pronounced difference between the sexes is the smaller stroke volume and higher heart rate during exercise of a given severity for women compared with men. Actually, a similar difference in stroke volume and heart rate is usually observed when comparing individuals of the same age but with a low and high performance capacity, respectively (Fig. 4-30).

Figure 4-30
The figure is based on average values from measurements on 11 women and 12 men, all of them relatively well trained and working on a cycle ergometer in the sitting position (*P.-O. Åstrand et al., 1964*). The individual data are presented in Figs. 4-19 and 4-23. (Since the abcissa gives the oxygen uptake in absolute values, the calculated mean curves can be misleading. The less fit subjects have both a low maximal oxygen uptake and low stroke volume. Those with a high capacity for oxygen uptake also have a larger stroke volume. A man with a maximal aerobic power of 5 liters $\cdot$ min^{-1} eventually attains maximal stroke volume first at a work rate giving an oxygen uptake of 2 liters $\cdot$ min^{-1}. The one with a maximal oxygen uptake of 3.5 liters $\cdot$ min^{-1} reaches his plateau for stroke volume when the oxygen uptake exceeds 1.3 liters $\cdot$ min.$^{-1}$)

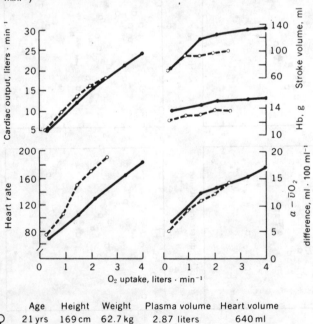

	Age	Height	Weight	Plasma volume	Heart volume
○ ♀	21 yrs	169 cm	62.7 kg	2.87 liters	640 ml
● ♂	24 yrs	179 cm	74.7 kg	3.70 liters	880 ml

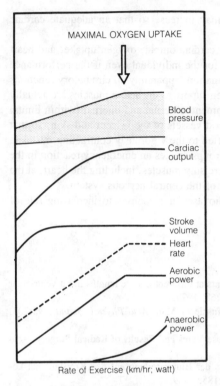

MAXIMAL OXYGEN UPTAKE

Blood pressure

Cardiac output

Stroke volume

Heart rate

Aerobic power

Anaerobic power

Rate of Exercise (km/hr; watt)

Figure 4-31
A schematic diagram showing some of the major cardiovascular responses to exercise of increasing intensity.

A reduced maximal stroke volume is a mechanism limiting the maximal cardiac output, and therefore the maximal aerobic power in many patients with coronary heart disease. During submaximal exercise the cardiac output at a given oxygen uptake is, however, often the same as in healthy subjects of the same age (Bruce et al., 1974; Clausen, 1976; McDonough et al., 1974; Faulkner, 1979). The effects of training and habitual inactivity respectively on various parameters will be discussed, as mentioned, in Chap. 10.

SUMMARY

This discussion of the oxygen-transporting system during muscular activity reveals that a regulation of the circulation at rest and during submaximal exercise is not guided only by the metabolic rate; various other factors may influence the circulatory response to exercise.

Of probable primary importance for the regulatory mechanisms is the relation between volume of oxygen supplied to the metabolically active tissue (cardiac output times the oxygen content of arterial blood) and the oxygen demand of the tissue. Within limits, other demands can be met by compensatory mechanisms; for example, if the stroke volume is reduced, the heart rate may increase so that an adequate cardiac output is still maintained.

During maximal exercise, however, the cardiac output, oxygen uptake, and heart rate are remarkably fixed to values typical for the individual even if the performance is made under adverse conditions. In this situation, apparently all circulatory functions of decisive importance for a maximal oxygen supply to the active muscles are actually devoted to this task. Irrespective of environment, external and internal (within limits) maximal vasoconstriction occurs in the blood vessels of the viscera and skin, so that practically the entire cardiac output is diverted to the vigorously contracting muscles. Maximal exercise involving large muscle groups creates an emergency reaction in the circulatory adjustment which favors the exercising muscles, including the heart, at the expense of all other tissues with exception of the central nervous system.

A summary of some of the major cardiovascular responses to increasing rate of exercise is presented in Fig. 4-31.

REFERENCES

Allen, D. G., and S. Kurihara: Calcium transients at different muscle lengths in rat ventricular muscle proceedings, *J. Physiol* (Lond), **292:**68, 1979.

Andersen, P.: Capillary Density in Skeletal Muscle of Man, *Acta Physiol. Scand.*, **95:**203, 1975.

Arenander, E.: Hemodynamic Effects of Varicose Veins and Results of Radical Surgery, *Acta Chir. Scand.*, (Suppl. 260):1, 1960.

Asmussen, E., and E. H. Christensen: Einfluss der Blutverteilung auf den Kreislauf bei köperlicher Arbeit, *Scand. Arch. Physiol.*, **82:**185, 1939.

Asmussen, E., E. H. Christensen, and M. Nielsen: Pulsfrequenz und Körperstellung, *Scand. Arch. Physiol.*, **81:**190, 1939.

Åstrand, I.: Aerobic Work Capacity in Men and Women with Special Reference to Age, *Acta Physiol. Scand.*, **49**(Suppl. 169):1960.

Åstrand, I., P.-O. Åstrand, and K. Rodahl: Maximal Heart Rate during Work in Older Men, *J. Appl. Physiol.*, **14:**562, 1959.

Åstrand, I., A. Guharay, and J. Wahren: Circulatory Response to Arm Exercise with Different Arm Positions, *J. Appl. Physiol.*, **25:**258, 1968.

Åstrand, I., P.-O. Åstrand, I. Hallbäck, and Å. Kilbom: Reduction in Maximal Oxygen Uptake with Age. *J. Appl. Physiol.* **35:**649, 1973.

Åstrand, P.-O.: "Experimental Studies of Physical Working Capacity in Relation to Sex and Age," Munksgaard, Copenhagen, 1952.

Åstrand, P.-O.: Breath Holding during and after Muscular Exercise, *J. Appl. Physiol.*, **15:**220, 1960.

Åstrand, P.-O., and B. Saltin: Oxygen Uptake during the First Minutes of Heavy Muscular Exercise, *J. Appl. Physiol.* **16:**971, 1961.

Åstrand, P.-O., L. Engström, B. O. Eriksson, P. Karlberg, I. Nylander, B. Saltin, and C. Thorén: Girl Swimmers, *Acta Paediat.* (Suppl. 147), 1963.

Åstrand, P.-O., T. E. Cuddy, B. Saltin, and J. Stenberg: Cardiac Output during Submaximal and Maximal Work, *J. Appl. Physiol.*, **19:**268, 1964.

Åstrand, P.-O., B. Ekblom, R. Messin, B. Saltin, and J. Stenberg: Intra-arterial Blood Pressure during Exercise with Different Muscle Groups, *J. Appl. Physiol.*, **20:**253, 1965.

Barcroft, H.: Sympathetic Control of Vessels in the Hand and Forearm Skin, *Physiol. Rev.*, **40**(Suppl. 4): 1960.

Barcroft, H., and H. J. C. Swan: Sympathetic Control of Human Blood Vessels, Edward Arnold (Publishers) Ltd., London, 1953.

Bevegård, B. S., and J. T. Shepherd: Regulation of the Circulation during Exercise in Man, *Physiol. Rev.*, **47**:178, 1967.

Bjurstedt, H., G. Rosenhamer, U. Balldin, and V. Katkov: Orthostatic Reactions during Recovery from Exhaustive Exercise of Short Duration, *Acta Physiol. Scand.*, **119**:25, 1983.

Blomqvist, G., and B. Saltin: Cardiovascular Adaptations to Physical Training, *Ann. Rev. Physiol.*, **45**:169, 1983.

Bonde-Petersen, F., and Y. Suzuki: Heart Contractility at Pressure Loads Induced by Ischemia of Exercised Muscle in Humans, *J. Appl. Physiol.: Resp. Environ. Exer. Physiol.*, **52**:340, 1982.

Braunwald, E., E. H. Sonnenblick, J. Ross, Jr., G. Glick, and S. E. Epstein: An Analysis of the Cardiac Response to Exercise, *Circulation Res.*, 20 and 21:44, 1967.

Brodahl, P., F. Ingjer, and L. Hermansen: Capillary Supply of Skeletal Muscle Fibers in Untrained and Endurance-Trained Men, *J. Am. Physiol.*, **232**:H705, 1977.

Bruce, R. A., F. Kusumi, M. Niederberger, and J. L. Petersen: Cardiovascular Mechanisms of Functional Impairment in Patients with Coronary Heart Disease, *Circulation*, **49**:696, 1974.

Burton, A. C.: Hemodynamics and the Physics of the Circulation, in T. C. Ruch and H. D. Patton (eds.), "Physiology and Biophysics," pp. 523–542, W. B. Saunders Company, Philadelphia, 1965.

Busa, W. B., and R. Nuccitelli: Metabolic Regulation via Intracellular pH, *Am. J. Physiol.*, **246**(15): R409, 1984.

Christensen, E. H.: Beiträge zur Physiologie schwerer körperlichter Arbeit. Minutenvolumen und Schlagvolumen des Herzens während schwerer körperlicher Arbeit, *Arbeitsphysiol.*, **4**:453, 470, 1931.

Christensen, E. H., and M. Nielsen: Investigation of the Circulation in the Skin at Beginning of Muscular Work, *Acta Physiol. Scand.*, **4**:162, 1942.

Clausen, J. P.: Circulatory Adjustments to Dynamic Exercise and Effect of Physical Training in Normal Subjects and Patients with Coronary Artery Disease, *Progr. Cardiovas. Diseases*, **18**:459, 1976.

Crone, C.: The Zweifach International Award 1979 Ariadne's Thread. An Autobiographical Essay on Capillary Permeability, *Microvasc. Res.*, **20**:133, 1980.

Daly, M. de B.: Reflex Circulatory and Respiratory Responses to Hypoxia, in F. Dickens and E. Neil (eds.), "Oxygen in the Animal Organism," p. 267, Pergamon Press, New York, 1964.

Davenport, H. W.: "The ABC of Acid-Base Chemistry," 5th ed., The University of Chicago Press, Chicago, 1969.

Davies, C. T. M., and A. J. Sargent: Physiological Responses to One- and Two-leg Exercise Breathing Air and 45% Oxygen, *J. Appl. Physiol.*, **36**:142, 1974.

Dickhuth, H.-H., A. Nause, J. Staiger, T. Bonzel, and J. Keul: Two-dimensional Echocardiographic Measurements of Left Ventricular Volume and Stroke Volume of Endurance-trained Athletes and Untrained Subjects, *Int. J. Sports Med.*, **4**:21, 1983.

Donald, K. W., J. K. Bishop, and O. L. Wade: A Study of Minute to Minute Changes of Arteriovenous Oxygen Content Difference, Oxygen Uptake and Cardiac Output and Rate of Achievement of a Steady State during Exercise in Rheumatic Heart Disease, *J. Clin. Invest.*, **33**:1946, 1954.

Ekblom, B.: Effect of Physical Training on Oxygen Transport System in Man, *Acta Physiol. Scand.* (Suppl. 328), 1969.

Ekblom, B., and L. Hermansen: Cardiac Output in Athletes, *J. Appl. Physiol.*, **25**:619, 1968.

Ekblom, B., A. N. Goldbarg, and B. Gullbring: Response to Exercise after Blood Loss and Reinfusion, *J. Appl. Physiol.*, **33**:175, 1972a.

Ekblom, B., A. N. Goldbarg, Å. Kilbom, and P.-O. Åstrand: Effects of Atropine and Propranolol on the Oxygen Transport System during Exercise in Man, *Scand. J. Clin. Lab. Invest.*, **30**:35, 1972b.

Ekblom, B., Å. Kilblom, and J. Soltysiak: Physical Training, Bradycardia and Autonomic Nervous System, *Scand. J. Clin. Lab. Invest.* **32**:251, 1973.

Ekblom, B., R. Huot, E. M. Stein, and A. T. Torstensson: Effect of Changes in Arterial Oxygen Content on Circulation and Physical Performance, *J. Appl. Physiol.*, **39**:71, 1975.

Ekblom, B., G. Wilson, and P-O. Åstrand: Central Circulation During Exercise After Venesection and Reinfusion of Red Blood Cells, *J. Appl. Physiol.*, **40**:379, 1976.

Elias, H. M., and J. E. Pauly: "Human Microanatomy," Da Vinci, Chicago, 1960.

Eriksson, B. O.: Physical Training, Oxygen Supply and Muscle Metabolism in 11-13-year Old Boys, *Acta Physiol. Scand.*, (Suppl. 384), 1972.

Eriksson, B. O., A Lundin, and B. Saltin: Cardiopulmonary Function in Former Girl Swimmers and the Effects of Physical Training, *Scand. J. Clin. Lab. Invest.*, **35**:135, 1975.

Evans, C. L.: The Velocity Factor in Cardiac Work, *J. Physiol.*, **52**:6, 1918.

Faber, J. E., P. D. Harris, and F. N. Miller: Microvascular Sensitivity to PGE_2 and PGI_2 in Skeletal Muscle of Decerebrated Rat, *Am. J. Physiol.*, **243(6)**:H844, 1982.

Fagraeus, L.: Cardiorespiratory and Metabolic Functions during Exercise in Hyperbaric Environment, *Acta Physiol. Scand.*, (Suppl. 414). 1974.

Faulkner, J. A.: Cardiac Rehabilitation: Major Concerns in Basic Physiology, in M. L. Pollock and D. H. Schmitt (eds.), "Heart Disease and Rehabilitation," p. 663, Houghton Mifflin Professional Publishers, Boston, 1979.

Feigl, E. O.: Coronary Physiology, *Physiol. Rev.*, **63**:1, 1983.

Folkow, B., U. Haglund, M. Jodahl, and O. Lundgren: Blood Flow in the Calf Muscle of Man during Heavy Rhythmic Exercise, *Acta Physiol. Scand.*, **81**:157, 1971.

Folkow, B., and E. Neil: "Circulation," Oxford University Press, London, 1971.

Fortney, S. M., E. R. Nadel, C. B. Wenger, and J. R. Bove: Effect of Acute Alterations of Blood Volume on Circulatory Performance in Humans, *J. Appl. Physiol.: Resp. Environ. Exer. Physiol.*, **50**:292, 1981.

Gerstenblith, G., E. G. Lakatta, and M. L. Weisfeldt: Age Changes in Myocardial Function and Exercise Response, *Progr. Cardiovas. Diseases*, **19**:1, 1976.

Gibbs, C. L.: Cardiac Energetics, *Physiol. Rev.*, **58**:174, 1978.

Grände, P-O., P. Borgström, and S. Mellander: On the Nature of Basal Vascular Tone in Cat Skeletal Muscle and its Dependence on Transmural Pressure Stimuli, *Acta Physiol. Scand.*, **107**:365, 1979.

Gregg, D. E.: The Natural History of Coronary Collateral Development, *Circ. Res.*, **35**:335, 1974.

Gregg, D. E., and H. D. Green: Registration and Interpretation of Normal Phasic Inflow into Left Coronary Artery by Improved Differential Manometric Method, *Am. J. Physiol.*, **130**:114, 1940.

Grimby, G., and B. Saltin: Physiological Analysis of Physically Well-trained Middle-aged and Old Athletes, *Acta Physiol. Scand.*, **179**:513, 1966.

Hansen, T.: Osmotic Pressure Effect of the Red Cells: Possible Physiological Significance, *Nature*, **190**:504, 1961.

Haraldsson, B., C. Ekholm, and B. Rippe: Importance of Molecular Charge for the Passage of Endogenous Macromolecules Across Continuous Capillary Walls, Studied by Serum Clearance of Lactate Dehydrogenase (LHD) Isoenzymes, *Acta Physiol. Scand.*, **117:**123, 1983.

Harrison, M. H., R. J. Edwards, and D. R. Leitch: Effect of Exercise and Thermal Heat Stress on Plasma Volume, *J. Appl. Physiol. Resp. Environ. Exer. Physiol.*, **39:**925, 1975.

Hartley, L. H., J. A. Vogel and M. Landowne: Central, Femoral, and Brachial Circulation during Exercise in Hypoxia, *J. Appl. Physiol.*, **34:**87, 1973.

Haynes, R. H., and S. Rodbard: Arterial and Arteriolar Systems, Biophysical Principles and Physiology, chap. 2, p. 26, in D. I. Abramson (ed.), "Blood Vessels and Lymphatics," Academic Press, Inc., New York, 1962.

Heiss, H. W., J. Barmeyer, K. Wink, G. Hell, F. J. Cerny, J. Keul, and H. Reindell: Studies on Regulation of Myocardial Blood Flow in Man. I. Training Effects on Blood Flow and Metabolism of the Healthy Heart at Rest and during Standardized Heavy Exercise, *Basic Res. Cardiol.*, **71:**658, 1976.

Hermansen, L., B. Ekblom, and B. Saltin: Cardiac Output during Submaximal and Maximal Treadmill and Bicycle Exercise, *J. Appl. Physiol.*, **29:**82, 1970.

Hernandez-Peon, R.: Physiology in Body Fluids, in D. S. Dittmer (ed.), "Blood and Other Body Fluids," Federation of American Societies for Experimental Biology, Washington, D.C., 1961.

Heymans, C., and E. Neil: "Reflexogenic Areas of the Cardiovascular System," Churchill, London, 1958.

Hollmann, W., and Th. Hettinger: "Sportsmedizin-Arbeits-und Trainingsgrundlagen," 2nd ed., F. K. Schattauer Verlag, Stuttgart; 1980.

Holmér, I.: Physiology of Swimming Man, *Acta Physiol. Scand.* (Suppl. 407), 1974.

Holmgren, A.: Circulatory Changes during Muscular Work in Man, *Scand J. Clin. Lab. Invest.* (Suppl. 24), 1956.

Honig, C. R., and C. L. Odoroff: Calculated Dispersion of Capillary Transit Times: Significance for Oxygen Exchange; *Am. J. Physiol.*, **240(2):**H199, 1981.

Hughson, R. L., and M. A. Morrissey: Delayed Kinetics of $\dot{V}_{O_2}$ in the Transition from Prior Exercise. Evidence for O_2 Transport Limitation of $\dot{V}_{O_2}$ Kinetics: a Review, *Int. J. Sports Med.*, **4:**31, 1983.

Hughson, R. L., C. A. Russel, and M. R. Marshall: Effect of Metropol on Cycle and Treadmill Maximal Exercise Performance, *J. Cardiac Rehab.*, **4:**27, 1984.

Jonsson, B. G., and I. Åstrand: Physical Work Capacity in Men and Women Aged 18 to 65, *Scand. J. Soc. Med.*, **7:**131, 1979.

Kanstrup, I.-L., and B. Ekblom: Acute Hypervolemia, Cardiac Performance and Aerobic Power during Exercise, *J. Appl. Physiol.: Resp. Environ. Exer. Physiol.*, **52:**1186, 1982.

Kanstrup, I.-L., and B. Ekblom: Blood Volume and Hemoglobin Concentration as Determinants of Maximal Aerobic Power, *Med. Sci. Sports Exer.*, **16:**256, 1984.

Keele, C. A., E. Neil, and N. Joels: "Samson Wright's Applied Physiology," Oxford University Press, Oxford, 1982.

Kilbom, Å., and J. Persson: Circulatory Response to Static Muscle Contractions in Three Muscle Groups, *Clin. Physiol.*, **1:**215, 1981.

Kirchheim, H. R.: Systemic Arterial Baroreceptor Reflexes, *Physiol. Rev.*, **56:**100, 1976.

Kitamura, K., C. R. Jorgensen, F. L. Gobel, H. L. Taylor, and Y. Wang: Hemodynamic Correlates of Myocardial Oxygen Consumption during Upright Exercise, *J. Appl. Physiol.*, **32:**516, 1972.

Knabb, R. M., S. W. Ely, A. N. Bacchus, R. Rubio, and R. M. Berne: Consistent Parallel Relationships among Myocardial Oxygen Consumption, Coronary Blood Flow, and Pericardial Infusate Adenosine Concentration with Various Interventions and β-Blockade in the Dog, *Circ. Res.*, **53**:33, 1983.

Kooyman, G. L., M. A., Castellini, and R. W. Davis: Physiology of Diving in Marine Mammals, *Ann. Rev. Physiol.*, **43**:343, 1981.

Korner, P. I.: Intergrative Neural Cardiovascular Control, *Physiol. Rev.*, **51**:312, 1971.

Krogh, A.: "The Anatomy and Physiology of Capillaries," rev. ed., Yale University Press, New Haven, Conn., 1929.

Lewis, D. H. (ed.): Lymph Circulation, *Acta Physiol. Scand.*, (Suppl. 463), 1979.

Ludbrook, J.: Reflex Control of Blood Pressure During Exercise, *Ann. Rev. Physiol.*, **45**:213, 1983.

Lundgren, N.: The Physiological Effects of Time Schedule Work on Lumber Workers, *Acta Physiol. Scand.*, **13** (Suppl. 41):1946.

Lundvall, J., S. Mellander, H. Westling, and T. White: Fluid Transfer between Blood and Tissues during Exercise, *Acta Physiol. Scand.*, **85**:258, 1972.

Lundvall, J., J. Hillman, and D. Gustafsson: β-adrenergic Dilation Effects in Consecutive Vascular Sections of Skeletal Muscle, *Am. J. Physiol.*, **243***(5):* H819, 1982.

Marx, H. J., L. B. Rowell, R. D. Conn, R. A. Bruce, and F. Kusumi: Maintenance of Aortic Pressure and Total Peripheral Resistance during Exercise in Heat, *J. Appl. Physiol.*, **22**:519, 1967.

McDonough, J. R., R. A. Danielson, R. E. Wills, and D. L. Vine: Maximal Cardiac Output during Exercise in Patients with Coronary Artery Disease, *Am. J. Cardiol.*, **33**:23, 1974.

Miles, D. S., M. N. Sawka, R. M. Glaser, and J. S. Petrofsky: Plasma Volume Shifts during Progressive Arm and Leg Exercise, *J. Appl. Physiol.: Resp. Environ. Exer. Physiol.*, **54**:491, 1983.

Mitchell, J. H., B. J. Sproule, and C. B. Chapman: Factors Influencing Respiration during Heavy Exercise, *J. Clin. Invest.*, **37**:1693, 1958.

Mitchell, J. H., B. Schibye, F. C. Payne, III, and B. Saltin: Response of Arterial Blood Pressure to Static Exercise in Relation to Muscle Mass, Force Development, and Electromyographic Activity, *Circ. Res.*, **48** (Suppl.I):70, 1981.

Mitchell, J. H., M. P. Kaufman, and G. A. Iwamoto: The Exercise Pressor Reflex: Its Cardiovascular Effects, Afferent Mechanisms, and Central Pathways, *Ann. Rev. Physiol.*, **45**:229, 1983.

Mohrman, D. E.: Lack of Influence of Potassium or Osmolarity on Steady-state Exercise Hyperemia, *Am. J. Physiol.*, **242***(6):*H 949, 1982.

Nielsen, M., and O. Hansen: Maximale Körperliche Arbeit bei Atmung O_2-reicher Luft, *Scand. Arch. Physiol.*, **76**:37, 1937.

Nilsson, H.: Adrenergic Nervous Control of Resistance and Capacitance Vessels, *Acta Physiol. Scand.*, **124** (Suppl. 541), 1985.

Olsen, C. R., D. D. Fanestil, and P. F. Scholander: Some Effects of Breath Holding and Apneic Underwater Diving on Cardiac Rhythm in Man, *J. Appl. Physiol.*, **17**:461, 1962.

Olsson, R. A.: Local Factors Regulating Cardiac and Skeletal Muscle Blood Flow, *Ann. Rev. Physiol.*, **43**:385, 1981.

Pappenheimer, J. R.: Central Control of Renal Circulation, *Physiol. Rev.*, **40** (Suppl. 4):35, 1960.

Pendergast, D. R., D. Shindell, P. Cerretelli, and D. W. Rennie: Role of Central and Peripheral Circulatory Adjustments in Oxygen Transport at the Onset of Exercise, *Int. J. Sports Med.*, **1**:160, 1980.

Peronnet, F., R. J. Ferguson, H. Perrault, G. Ricci, and D. Lajoie: Echocardiography and the Athletes's Heart, *Phys. Sports Med.*, **9**:102, 1981.

Poliner, Z. R., G. J. Dehmer, S. E. Lewis, R. W. Parkey, C. G. Blomqvist, and J. T. Willerson: Left Ventricular Performance in Normal Subjects: a Comparison of the Response to Exercise in the Upright and Supine Position, *Circulation*, **62**:528, 1980.

Pollock, M. W., J. H. Wilmore, and S. M. Fox: "Health and Fitness Through Physical Activity," John Wiley & Sons, New York, 1978.

Reeves, J. T., R. F. Grover, S. G. Blount, Jr., and G. F. Filley: Cardiac Output Responses to Standing and Treadmill Walking. *J. Appl. Physiol.*, **16:**283, 1961.

Reindell, H., H. Klepzig, H. Steim, K. Musshoff, H. Roskamm, and E. Schildge: "Herz Kreislaufkrankheiten und Sport." Johann Ambrosius Barth, Munich, 1960.

Robertson, R. J., R. Gilcher, K. F. Metz, G. S. Skrinar, T. G. Allison, H. T. Bahnson, R. A. Abbott, R. Becker, and J. E. Falkel: Effect of Induced Erythrocythemia on Hypoxia Tolerance during Physical Exercise, *J. Appl. Physiol.: Resp. Environ. Exerc. Physiol.*, **53:**490, 1982.

Robinson, S.: Experimental Studies of Physical Fitness in Relation to Age, *Arbeitsphysiol.*, **10:**251, 1938.

Rost, R.: "The Athlete's Heart", *Eur. Heart J.*, **3**(suppl. A); 193, 1982.

Rost, R., and W. Hollmann: Athlete's Heart—a Review of its Historical Assessment and New Aspects, *Int. J. Sports Med.*, **4:**147, 1983.

Rothe, C. F.: Reflex Control of Veins and Vascular Capacitance, *Physiol. Rev.*, **63:**1281, 1983.

Rowell, L. B.: Human Cardiovascular Adjustments to Exercise and Thermal Stress, *Physiol. Rev.* **54:**75, 1974.

Rowell, L. B.: Cardiovascular Adjustments to Thermal Stress, in "Handbook of Physiology The Cardiovascular System III," Chapter 27, p. 967, 1983.

Ruch, T. C., and H. D. Patton (eds.): "Physiology and Biophysics, Circulation, Respiration and Fluid Balance," vol. II, W. B. Saunders Company, Philadelphia, 1974.

Sable, D. L., H. L. Brammel, M. W. Sheehan, A. S. Nies, J. Gerber, and D. L. Horwitz: Attenuation of Exercise Conditioning by β-Adrenergic Blockade, *Circulation,* **65:**679, 1982.

Sahlin, K., A. Alvestrand, R. Brandt, and E. Hultman: Intracellular pH and Bicarbonate Concentration in Human Muscle during Recovery from Exercise, *J. Appl. Physiol.*, **45:**474, 1978.

Saltin, B., G. Blomqvist, J. H. Mitchell, R. L. Johnson, Jr., K. Wildenthal, and C. B. Chapman: Response to Submaximal and Maximal Exercise after Bedrest and Training, *Circulation* **38**(Suppl. 7):1968.

Saltin B., and L. B. Rowell: Functional Adaptation to Physical Activity and Inactivity, *Federation Proc.*, **39:**1506, 1980.

Saltin, B. and P. D. Gollnick: Skeletal Muscle Adaptability: Significance for Metabolism and Performance, In "Handbook of Physiology. Skeletal Muscle," sect. 10, chap. 19, p. 555, American Physiological Society, Bethesda, Md., 1983.

Sarnoff, S. J., and J. H. Mitchell: The Control of the Function of the Heart, in W. F. Hamilton and P. Dow (eds.), "Handbook of Physiology, Section 2: Circulation," Vol. 1, p. 489, American Physiological Society, Washington, D. C., distributed by the Williams & Wilkins Company, Baltimore, Maryland, 1962.

Schmidt-Nielsen, K.: "Animal Physiology: Adaptation and Environment," 3rd ed., Cambridge University Press, Cambridge, 1983.

Scholander, P. F., H. T. Hammel, H. LeMessurier, E. Hemmingsen, and W. Garey: Circulatory Adjustment in Pearl Divers, *J. Appl. Physiol.*, **17:**184, 1962.

Seals, D. R., R. A. Washburn, P. G. Hanson, P. L. Painter, and F. J. Nagle: Increased Cardiovascular Response to Static Contraction of Larger Muscle Groups, *J. Appl. Physiol.: Resp. Environ. Exer. Physiol.*, **4:**434, 1983.

Shepherd, J. T., and P. M. Vanhoutte: "The Human Cardiovascular System," Raven Press, New York, 1979.

Shepherd, J. T., and F. M. Abboud (eds.): Handbook of Physiology, Sect. 2: Cardiovascular system, Vol. 3 "Peripheral circulation and organ blood flow", Part 1–2, American Physiological Society, 2nd ed., Bethesda, Md., 1983.

Siggaard-Andersen, O.: "The Acid-Base Status of the Blood," Munksgaard, Copenhagen, 1974.

Simionescu, N.: Cellular Aspects of Transcapillary Exchange, Physiol. Rev., 63:1536, 1983.

Sjögaard, G., and B. Saltin: Extra-and Intracellular Water Spaces in Muscles of Man at Rest and With Dynamic Exercise, Am. J. Physiol., 243:(3):R271, 1982.

Sjöstrand, T.: Volume and Distribution of Blood and Their Significance in Regulating Circulation, Physiol. Rev., 33:202, 1953.

Sjöstrand, F. S., and E. Andersson-Cedergren: Intercolated Discs of Heart Muscle, in G. H. Bourne (ed.), "Structure and Function of Muscle," vol. 1, chap. 12, p. 421, Academic Press, Inc., New York, 1960.

Sonnenblick, E. H., and C. L. Skelton: Reconsideration of the Ultrastructural Basis of Cardiac Length-tension Relations, Circ. Res., 35:519, 1974.

Spriet, L. L., N. Gledhill, A. B. Froese, D. R. Wilkes and E. C. Meyers: The Effect of Induced Erythrocythemia on Central Circulation and Oxygen Transport during Maximal Exercise, Med. Sci. Sports Exer., 12:122, 1980.

Starling, E. H.: J. Physiol., 19:312, 1896.

Stenberg, J., B. Ekblom, and R. Messin: Hemodynamic Response to Work at Simulated Altitude, 4,000 m, J. Appl. Physiol., 21:1589, 1966.

Stenberg, J., P.-O. Åstrand, B. Ekblom, J. Royce, and B. Saltin: Hemodynamic Response to Work with Different Muscle Groups in Sitting and Supine, J. Appl. Physiol., 22:61, 1967.

Strandell, T.: Circulatory Studies on Healthy Old Men, Acta Med. Scand., 175(Suppl. 414):1964.

Tada, M., T. Yamamoto, and G. Tonomura: Molecular Mechanism of Active Calcium Transport by Sarcoplasmic Reticulum, Physiol. Rev., 58:1, 1978.

Tesch, P. A.: B-blockade and Exercise Performance, Sports Med., to be published, 1985.

Thomson, J. M., J. A. Stone, A. D. Ginsburg, and P. Hamilton: O_2 Transport during Exercise Following Blood Reinfusion, J. Appl. Physiol.: Resp. Environ. Exer. Physiol., 53:1213, 1982.

Vokac, Z., H. J. Bell, E. Bautz-Holter, and K. Rodahl: Oxygen Uptake/Heart Rate Relationship in Leg and Arm Exercise, Sitting and Standing, J. Appl. Physiol., 39:(1):54, 1975.

Walgenbach, S. C., and D. E. Donald: Inhibition by carotid baroreflex of exercise-induced increases in arterial pressure, Circ. Res., 52(3):253, 1983.

Wiggers, C. J.: "Circulatory Dynamics," Grune & Stratton, Inc., New York, 1952.

Wikman-Coffelt, J., W. W. Parmley, and D. T. Mason: The Cardiac Hypertrophy Process: Analyses of Factors Determining Pathological Vs. Physiological Development, Circ. Res., 45:687, 1979.

Zweifach, B. W.: Microcirculation, Ann. Rev. Physiol., 35:117, 1973.

CHAPTER **5**

RESPIRATION

CONTENTS

MAIN FUNCTION
ANATOMY AND HISTOLOGY
  Airways
  Blood Vessels
  Nerves
"AIR CONDITION"
FILTRATION AND CLEANSING MECHANISMS
MECHANICS OF BREATHING
  Pleurae
  Respiratory Muscles
  Total Resistance to Breathing
VOLUME CHANGES
  Terminology and Methods for the Determination of "Static" Volumes
  Age and Sex
  "Dynamic" Volumes
COMPLIANCE
AIRWAY RESISTANCE
PULMONARY VENTILATION AT REST AND DURING EXERCISE
  Methods
  Pulmonary Ventilation During Exercise
  Dead Space
  Tidal Volume-Respiratory Frequency
  Respiratory Work (Respiration as a Limiting Factor in Exercise)

209

Do the Respiratory Muscles "Steal" Oxygen from Other Tissues?
Studies Indicating Fatigue of the Respiratory Muscles After Prolonged
  Respiratory Exertion
DIFFUSION IN LUNG TISSUES, GAS PRESSURES
VENTILATION AND PERFUSION
OXYGEN PRESSURE AND OXYGEN-BINDING CAPACITY OF THE
  BLOOD
REGULATION OF BREATHING
  Rest
  Central Chemoreceptors
  Peripheral Chemoreceptors
  Exercise
  A Cerebral Drive?
  The Role of Muscle Spindles
  Combined Effects of Neurogenic and Chemical Stimuli
  Hypoxic Drive During Exercise
BREATHLESSNESS (DYSPNEA)
SECOND WIND
HIGH AIR PRESSURES, BREATH HOLDING, DIVING
  High Air Pressures
  Breath Holding—Diving

MAIN FUNCTION

The living cell uses oxygen for its metabolism, and in the process, carbon dioxide is produced. Thus, the concentration of oxygen within the cell is lowered, and oxygen will tend to diffuse toward the place of combustion. Similarly, the carbon dioxide produced will tend to diffuse away. The exchange of O_2 and CO_2 is dependent on the distance the molecules have to travel and the pressure gradient. In the single-cell organism the surface can effectively be utilized for the respiratory exchange. By diffusion, oxygen can easily reach every point within the cell, and CO_2 can be eliminated. The calculations of Krogh (1941) indicated that when metabolism is fairly high, diffusion can provide sufficient oxygen only to organisms with a diameter of 1 mm or less. A spherical cell with a 1-cm radius and an oxygen uptake of 100 ml $\cdot$ kg^{-1} $\cdot$ hr^{-1} would need an external oxygen pressure of 25 atm to secure the oxygen supply to its center by diffusion. In multicelled animals, the problem of gas transport is solved in various ways; e.g., numerous airways, like tracheae in insects, or specialized organs, like lungs or gills, are developed, exposing an enlarged respiratory surface to the external medium to effect exchange of oxygen and carbon dioxide. With the gas exchanger located distant from the sites of metabolism, two transport systems carry O_2 and CO_2 between the respiratory surface and the cells in which the metabolism proceeds: a circulatory system supplies the blood, and a respiratory system supplies the air to the lungs. The spongelike structure of the gas-exchange area provides an enormous contact surface between air and blood. This air-tissue-blood interface of the average adult human lungs is estimated to be some 70 to 90 m², with a thickness of the tissues varying from 0.2 μm to several μm (average, some 0.7 μm). Fresh air and "new" blood must continuously be supplied to the many million gas-exchange units, since the store of oxygen in the body is very limited and the central nervous system and heart muscle do not tolerate any lag in the supply of oxygen.

In this chapter, we shall discuss the exchange of oxygen and carbon dioxide between ambient air and blood. For a comprehensive discussion of the many aspects of respiration, the reader is referred to the *Handbook of Physiology* (Fenn and Rahn, eds., 1964–1965) and other sources (Widdicombe, 1974; Jacques, 1979; Loeppky and Riedesel, 1981).

The ideal gas exchanger involves four factors. (1) It should provide *a large contact area* between the air and blood with a very thin membrane separating the two media, since diffusion is directly proportional to the area but inversely related to the thickness of the membrane. The membrane should cause a minimal resistance to gas flow. (2) The inspired air must become *saturated with water vapor and heated* to tissue temperature to *protect the delicate membranes from injury;* any particles and agents in the air which may be harmful should be removed during the passage through the airways, and if introduced, should be expelled. (3) Variations in oxygen and carbon dioxide concentrations in the blood leaving the lungs should vary only within small limits, and therefore, *the distribution of gas and blood in the many exchange units should be closely matched.* (4) *The gas exchange, and therefore the perfusion of the units, must be proportional to the uptake of oxygen and production of carbon dioxide by the cells.* These demand some sort of regulative mechanism linking the needs of the distant cells with the external respiration. The anatomic location of airways and lungs inevitably causes respiration to be influenced by movements of the trunk. Speaking and singing modify the respiration.

Point (4) of the factors listed is of special interest in this connection, since muscular exercise may increase the oxygen uptake of the body some 20 times the resting level, with a similar rise in CO_2 production. The respiration also plays an important role in the *maintenance of the pH of the blood at normal levels.* Hydrogen ions cannot be exchanged between air and blood, but the acid-base equilibrium of the blood (and tissue) is closely associated with the CO_2 content and pressure in the blood (see Chap. 4). During heavy exercise, there is, in the anaerobic metabolism, a high rate of proton (H^+) production and more CO_2 is driven out from the hydrogencarbonate ions (HCO_3^-), which together with the excess proton concentration in the blood, will stimulate respiration. Elimination of CO_2 during hyperventilation will limit the decline in the blood pH.

Before we discuss respiration during exercise, a summary of the basic respiratory function will be presented.

ANATOMY AND HISTOLOGY

Airways

Figure 5-1 illustrates the respiratory tree and its subdivision into finer airways until they finally terminate in numerous blind pouches, the alveoli. Figure 5-2 presents

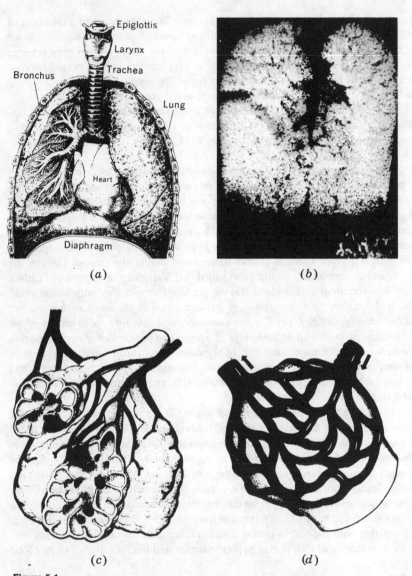

Figure 5-1
(a) Principal organs of breathing. In this drawing, the ribs, the large arteries from the heart, and part of one lung have been cut away.
(b) A cast of the complete air spaces of the lung, showing the millions of air sacs at the end of the bronchioles. Inset: the terminal portion of a bronchiole magnified. (*From C. M. Fletcher, BBC Publication 3 s., 1963.*)
(c) Schematic drawing of a bronchiole and its air sacs and alveoli.
(d) An alveolus embraced by the capillary network.

schematically the general architecture of the airways. Via nose or mouth, the inspired gases pass through the pharynx, larynx, and trachea into the bronchial tree. The airways branch by asymetric dichotomy and the two daughter branches in turn become parent branches, etc. They terminate in the *alveoli,* polyhedral or cuplike outpouchings of the finer airways. Their diameter is up to 300 μm. The surface of the alveolus is not smooth but is corrugated by the capillaries and by various subcellular structures, like nuclei, bulging into the alveolus (Fig. 5-3). The alveoli share their walls with their neighbors and form the spongelike texture of the lungs (Fig. 5-1).

Figure 5-2
General architecture of conductive and transitory airways. The z column designates the order of generations of branching; *T,* the terminal generation. A more detailed discussion appears in the text. (*Weibel, 1963.*)

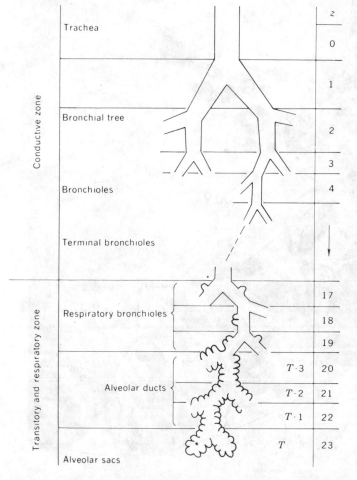

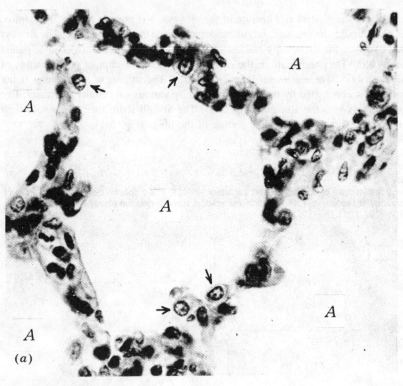

(a)

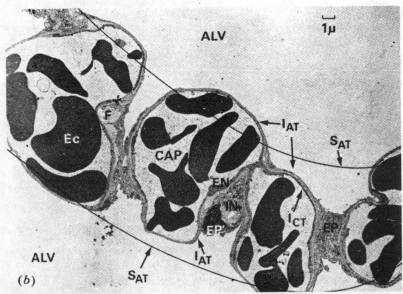

(b)

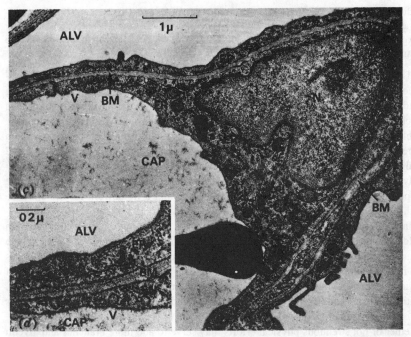

Figure 5-3
(a) Histologic section through one complete alveolar outline (center) and portions of four adjacent alveoli (A); magnification ×650. (*Modified from Krahl, 1964.*) (b,c,d) Electron micrographs of interalveolar septum of rat lung in cross section showing the barrier composed of alveolar epithelial (*EP*), capillary endothelial cells (*EN*), and some interstitial elements. (*From Weibel, 1964.*)
(b) Magnification ×5,500. Note the difference between the estimated alveolar surface (S_{AT} and the real "corrugated" air-tissue interface (I_{AT}), I_{CT} = tissue-blood interface; CAP = capillaries; IN = interstitium; F = fibrous elements; Ec = erythrocytes.
(c) Magnification ×23,500. BM = basal membranes; N = nucleus; EP = endoplasmic reticulum; V = pinocytotic vesicle.
(d) Magnification ×59,000 of the thin portion of air-blood barrier with four membranes.

In the adult man, about 23 such generations of branches can be traced as the airways subdivide into the periphery of the lung and distribute them among the numerous respiratory units. The first 16 generations roughly constitute the "conductive zone" where practically no gas exchange occurs between blood and air. Then follow the "transitory and respiratory zones": generations 17 to 19 of branching form the respiratory bronchioles with a diameter of about 1 mm, which subdivide to produce about 1 million alveolar ducts. The last of a short series of alveolar ducts terminates in rotundate enclosures: the alveolar ducts with a diameter of about 400 μm. The cylindrical surface of the respiratory bronchioles bears a smaller number of variously spaced alveoli, but the alveolar ducts and sacs are fully alveolated. Therefore, they lack proper walls but open out on all sides into alveoli, some 300 million altogether in the adult (Fig. 5-1). (There are considerable normal variations in those numbers—from 200 million up to 600 million—related to the individual's body height, but the alveoli have similar dimensions in large and small adult human lungs. Angus and Thurlbeck, 1972). The respiratory and transitory zones, including the alveoli, amount to about 90 percent of the lung volume; about 65 percent of the air in the lungs is in the actual alveoli at three-fourths the maximal inflation of the lungs (Weibel, 1972).

The bronchi and bronchioles that are more than about 1 mm wide have a discontinuous cartilaginous support in the wall. Muscle fibers in circular or crisscrossing bundles are incorporated into a complex connective tissue framework of collagenous reticular and elastic fibers. The inner surface is covered with a ciliated epithelium. Goblet cells occur singly or in groups between the epithelial cells and produce a secretion. In the finest bronchioles, the mucus-secreting elements become sparse and finally absent. They are lacking cartilages, and the ciliated cells also disappear gradually. Muscle fibers as well as elastic, collagenous, and reticular fibers provide the supporting latticework of the interalveolar septum and a framework for the entrance of the alveoli and alveolar sacs and ducts.

In the fetus, the alveoli contain no air, but the first influx of air in the newborn child provides a force that stretches the original cuboidal epithelium lining of the alveoli into an extremely thin layer of squamous cells. The lung of the newborn is not fully developed. The airways have subdivided into only some 17 generations of branchings and the number of alveoli is less than one-tenth of that found in the adult. As time goes on, additional branches are added and many more alveoli are formed as new ramifications grow out, so that before ten years of age the adult number is reached. Whether or not strenuous physical efforts may serve to add more branches and alveoli to the mature lung is not known (see below). The relative alveolar volume decreases with age.

Despite the latticework fibers supporting the alveolar walls, the air-tissue-blood interface would provide special problems due to the surface tension which is created. This tension tends to decrease the surface wall to a minimum so that the alveoli may collapse. The alveoli are, however, lined with an insoluble surface film of lipoprotein, about 5 nm thick. It is produced in the alveoli and keeps the alveoli open and free from transudate from the blood by lowering the surface tension. This function is especially important as the volume decreases, e.g., during forced expiration, which would otherwise empty the small alveoli. During quiet breathing, there may actually be a reduced effect of the surface film and an occasional collapse of alveolar units. Forced inflation of the lungs may cause more material to be provided for the lining film and open the alveoli. A yawn or deep breath may exert such a beneficial function. The surface film thus stabilizes the small alveolar spaces and enables the lung to retain air at low inflation pressure (Pattle, 1965).

Blood Vessels

The pulmonary arteries enter the lungs with the stem bronchi and provide arterial partners to the airways as they subdivide toward the respiratory zones of the lungs. The arterioles follow the bronchioles, alveolar ducts, and sacs, and provide short twigs to capillary networks enveloping the alveoli surrounding the particular airway terminal and to any other alveoli in the immediate vicinity. Each alveolus may be covered by a capillary network consisting of almost 2,000 segments; the capillary networks in the lungs are the richest in the body and are more or less continuous throughout large parts of the lungs (Fig. 5-4). The air-blood "barrier" is formed by the continuous alveolar epithelial and capillary endothelial cells with a tenuous interstitium with fibrous elements in between the two cell layers. The thickness of the "barrier" can vary from about 0.2 μm to several μm, and the variations are caused by various structures scattered throughout the continuous cell layers (Fig. 5-3). It is evident that the capillary network should be regarded as a sheet of blood floating along the alveolar surface, the sheet merely connecting the two walls. The capillary surface area is, in fact, of the same order of size as the alveolar surface area. Thus, most of the space in the interalveolar septa is occupied by capillaries, and their blood flow is intercalated between two air chambers. The blood is separated from the air on either side only by a very thin tissue barrier.

Nerves

Vagal efferent fibers go to the smooth muscles of the bronchial tree as far as the terminations of the alveolar ducts and sacs, and to the bronchial mucous glands. Nerve impulses stimulate the smooth muscles to contract and activate the glands. *Vagal afferent fibers* carry impulses from special stretch receptors scattered in lungs and

Figure 5-4
(a) Network of capillaries in alveolar walls. Magnification ×375. (*From Miller, 1947.*)
(b) Blood-filled capillary network in interalveolar septum of human lung. Larger, dark vessel to the right (arteriole) gives off short precapillaries which open at once into pulmonary capillaries. Magnification ×650. (*From Krahl, 1964.*)

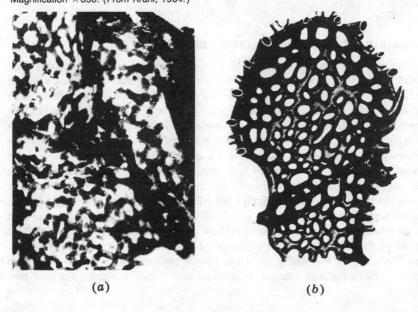

(a)   (b)

pleura. Sympathetic fibers act as bronchodilators. There are many free nerve endings in the bronchial walls and pleura, but their function is not known.

"AIR CONDITION"

The inspired air may be cold or hot, dry or moist, but owing to the rich blood supply of the mucous membranes of the nose, the mouth, and the pharynx, the air temperature becomes adjusted to body temperature and moistened before reaching the alveoli. Recent evidence indicates that when inhaling large volumes of cold air, the point at which the air reaches body temperature moves progressively deeper into the lungs. Under extreme conditions, i.e., when inspiring air with a temperature of $-40°C$, thermal transfer can be measured in airways less than 2 mm in diameter (see McFadden, 1983). Air saturated with water vapor at 37°C has a $P_{H_2O} = 47$ mm Hg (6.3 kPa); the content of water is then 44 g $\cdot$ m^{-3}. At low temperatures the water content in the air is low. Even if saturated, the air at 0°C contains only 5 g $H_2O \cdot$ m^{-3}. In a normal climate, about 10 percent of the total heat loss of the body at rest or during exercise takes place through the respiratory tracts by the air conditioning of the inspired air. At -15 to $-20°C$ the percentage would be about 25 (see Chap. 13). The respiratory tract serves as a regenerative system: the heating and humidifying of inspired air cool the mucosa. But during expiration, some of the heat and water are recovered by the mucosa from the passing alveolar air. Body heat and water are conserved. On a very cold day, this condensation of water vapor may result in excessive accumulation of water in the nostrils, leading to a runny nose! A cross-country skier breathing 100 liters/min of air at $-20°C$ must, in 1 hr, add about 250 ml of water to this air. Not all of this water volume is expired, however, thanks to the regenerative system. (See Ferrus et al., 1984).

FILTRATION AND CLEANSING MECHANISMS

If living in a city, we may inhale billions of particles of foreign matter every day. Particles larger than about 10 μm are effectively removed from the inspired air in the nose, where they are trapped by the hair or the moist mucous membranes. Those particles which escape these obstacles usually settle on the walls of the trachea, the bronchi, and the bronchioles. Therefore only a few very small particles are likely to reach the alveoli, and this part of the lung is practically sterile. Alveolar macrophages perform a vital function of maintaining the alveoli clean and sterile. They are of hematogenous origin but thanks to their extraordinary amoeboid mobility they can pass into the alveoli and move freely over the airspace surfaces. They are able to phagocytose large quantities of foreign material, such as dust particles and bacteria. The macrophages and the phagocytosed remains can be cleared into the digestive tract via the airways or removed via the blood flow or lymphatics. It should be pointed out that the phagocytic capacity can be diminished by certain influences such as smoking. (See Weibel, 1972.) Actually, the lungs have a high metabolic activity. They can remove substances like serotonin, bradykinin, and certain prostaglandins, and they can synthesize and release a variety of substances, such as prostaglandins, peptides, and enzymes (see Said, 1982).

As mentioned, the epithelium of the airways within the lungs consists of *ciliated cells*. In the conductive zone each cell carries up to 300 cilia about 6 to 7 µm in length, at the free cell surface. The cilia of many thousands of cells beat in an organized, whiplike fashion in strokes, like oars of a boat, with a rapid upward propulsive stroke followed by a slower recovery downward stroke. This goes on continuously day and night. The cilia are covered by a continuous surface of watery mucus. By the ciliar activity, this fluid carpet with all the entrapped particles moves toward the larynx at a speed of well over 1 cm · min^{-1}. This mucus is either expectorated or swallowed. The ciliary escalator is remarkably resistant to noxious influences. However, cigarette smoke has a deleterious effect on the ciliar function. The cilia slow down or stop their beating when exposed to the smoke.

From time to time we may sneeze or cough, and with our explosive blast (the air moves with a speed approaching the speed of sound), foreign particles may be expelled.

MECHANICS OF BREATHING

Pleurae

The lungs increase and decrease their volume with the reciprocating movement of the bellowslike pump, the thorax. The thoracic cavity is covered by the very thin parietal pleura, and the lungs by the pulmonary (visceral) pleura. These very thin membranes of single layers of flat epithelial cells on fibrous connective sheets continue uninterrupted from one pleural surface to the other across the pulmonary hilus. The two pleurae surfaces are held close together with a thin fluid film in between, providing smooth lubricated surfaces. If the thorax is opened so that the atmospheric pressure prevails in the intrapleural space, the elastic recoil of the lungs causes them to collapse, and the chest expands a little, since a retractive force of the lungs is normally counterbalanced by an outward spring of the chest cage (pneumothorax). Such an injury, of course, makes the lung involved incapable of any respiratory function. Normally, however, the pleurae are in close, but friction-free, contact with each other. Any volume changes in the thorax are completely transmitted to the lungs. The two pleural surfaces may be compared with two flat sheets of glass placed face to face with a thin layer of water between the two opposing surfaces. While the two sheets of glass can easily be slid back and forth, a great force is required to move the two sheets away from one another by forces acting perpendicular to the glass surfaces.

> It would appear plausible that gas and fluid from the blood might collect in the intrapleural space because of the discrepancy in the size of the thoracic cavity and the lungs: the opposing forces of the lung and chest wall tend to separate the two pleurae with a pressure of a few centimeters of water lower than atmospheric pressure. Gas is, however, absorbed because the sum of the gas tensions in venous blood (and pleural liquid) is less than arterial (and atmospheric) pressure (see Fig. 5-14). Water from the space is effectively absorbed, since the colloid osmotic pressure of the plasma proteins in the pulmonary capillaries easily matches the slightly lower hydrostatic pressure in the pulmonary vessels.

Respiratory Muscles

Figure 5-5 shows the contours of the thorax and lungs at the end of an expiration and an inspiration, respectively. During quiet breathing at rest, the diaphragm is the principal muscle driving the inspiratory pump: the abdominal muscles relax, the abdomen protrudes, the thoracic volume increases, and the lungs expand. The contraction of the diaphragm causes its dome to descend some 1.5 cm, and the intra-abdominal pressure increases. This movement is actually of the same order of magnitude both in the so-called costal and in the diaphragmatic type of breathing. During deep breathing, the vertical movements of the diaphragm may exceed 10 cm. The external intercostal muscles assist in the inspiration, especially during exercise. The fibers slope obliquely downward and forward from the caudal margin of one rib to the cranial margin of the rib below. The lower insertion is located more distant from the center of rotation than the upper one. When the fibers contract, the force exerted by the muscle is equal at both insertions, but the longer leverage of the lower rib gives a torque that raises rather than lowers the upper rib. The net effect is a lifting of the ribs when the external intercostal muscles contract. The elevation of the ribs, rotating around the axis of their necks, increases the dimensions of the rib cage in both transverse and dorsoventral directions, similar to that of the handle of a bucket when lifted. When the inspiratory muscles relax during quiet breathing, the elastic recoil forces in the lung tissue, thoracic wall, and abdomen restore the chest to the resting position without any help of the expiratory muscles. During exercise or forced breathing at rest, with a ventilation exceeding 2 or 3 times the resting value, these recoil pressures are supplemented by activity of the expiratory muscles. The internal intercostal fibers have a direction

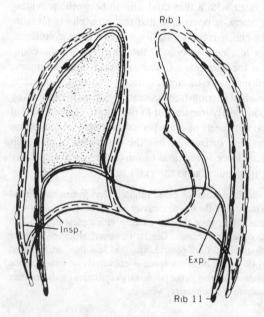

Figure 5-5
Diagram of frontal section of thorax, based on roentgenograms, showing changes in size of thorax and position of heart and diaphragm with respiration. (*From Braus, 1956.*)

opposite to those of the external intercostal muscles. Therefore the function of the inner layer of the intercostal muscles is to facilitate expiration. The muscles of the abdominal wall are essentially expiratory muscles, but do not become engaged forcefully until the pulmonary ventilation reaches high levels.

At a ventilation exceeding 50 liters $\cdot$ min^{-1} and especially at very high ventilation, accessory muscles may assist. The sternocleidomastoids and scaleni are the most important ones during forced inspiration. When the athlete grasps for support after an exhausting spurt, this posture may facilitate the action of the respiratory muscles.

The activity of the inspiratory muscles increases progressively throughout inspiration and they actually continue to contract while being stretched during the early part of expiration. Part of the work done during inspiration is "stored" in the elastic structures of the system and is then available to supply part of the power for the expiration. If an expiration decreases the lung volume below the resting level of the system, the chest wall recoils outwardly, causing a passive inspiration back to the resting volume.

During exercise, the inspiratory and expiratory muscles are activated reciprocally, especially the expiratory ones in the last part of expiration.

Total Resistance to Breathing

The respiratory muscles work mainly against an airway resistance and a pulmonary tissue and chest wall resistance.

The work done against inert forces to accelerate tissues and gases is in this connection negligible. Most of the tissue resistance is offered by elastic forces. But the collagen fibers, providing the supporting framework for the delicate structures of the lungs, also contribute to the resistance when a volume change occurs. Thanks to the soft, yielding tissues of the lungs, the resistance is low. Of the total pulmonary resistance, only about 20 percent is a tissue resistance and 80 percent is airway resistance. The resistance of the upper airways is about half that of the entire respiratory system (see Bartlett, 1979).

At high flow velocities, as during heavy exercise, the air flow is turbulent in the trachea and the main bronchi, giving a high flow resistance. Owing to the large total cross area of the finest air tubes, the air flow in this region is low and therefore laminar. In fact the greatest part of resistance to airflow within the lungs lies in airways greater than 2 mm internal diameter and particularly in the medium-sized bronchi (the four to ten airway generations, Fig. 5-2). An air flow of 1 liter $\cdot$ s^{-1} requires a pressure drop along the airways of less than 2 cm H_2O.

At rest, the oxygen cost of the breathing is only a small fraction of the total resting energy turnover. It has been estimated to be about 0.5 to 1.0 ml $\cdot$ liter^{-1} of moved air. With pulmonary ventilation of 6 liters $\cdot$ min^{-1}, the oxygen uptake of the respiratory muscles would be up to 6 ml, compared with a total resting oxygen uptake of the body as a whole of about 250 to 300 ml. With the high pulmonary ventilation during heavy exercise, the energy cost per liter ventilation becomes progressively greater, and the oxygen cost of breathing may be up to 8 ml $\cdot$ liter^{-1} when the pulmonary ventilation exceeds 100 liters $\cdot$ min^{-1}. The air resistance when breathing through the nose is 2 to 3 times greater than that obtained by breathing through the mouth. It is therefore natural for almost everyone also to breath through the mouth (or oronasally) when performing heavy exercise. At a high pulmonary ventilation, approaching 100 liter $\cdot$ min^{-1}, the oral minute volume accounts for more than 50 percent of the total ventilation (see Niinimaa, 1983).

Summary   When the respiratory muscles are relaxed (resting volume), the chest wall is retracted by the elastic recoil of the lungs. The beginning of an inspiration is assisted by the recoil of the chest wall but the lung tissue is further stretched. During deeper inspiration, there is a retractive force from both the chest wall and the lungs. Part of the energy provided by the inspiratory muscle, mainly the diaphragm and the external intercostal muscles, is "stored" in the elastic structures and is utilized during the expiration. An expiration below the resting volume increases the outward recoil of the chest wall. In any volume position, the lungs and the chest cage behave like opposing springs, and at the resting volume, the forces exerted exactly counterbalance each other. The work done by the respiratory muscles is mainly devoted to doing elastic work and to overcoming the airway resistance.

VOLUME CHANGES

Terminology and Methods for the Determination of "Static" Volumes

Figure 5-6 should be consulted for the terminology. When the respiratory muscles are relaxed, there is air still left in the lungs. This air volume is the *functional residual*

Figure 5-6
Diagram of lung volumes and capacities. (*From Pappenheimer, 1950.*)

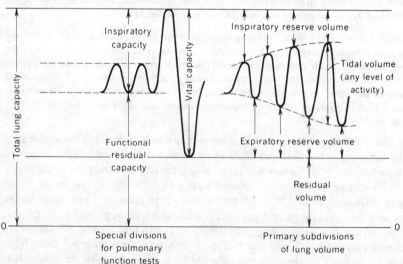

capacity (FRC). A forced maximal expiration brings the volume down to the *residual volume* (RV) by expiration of the *expiratory reserve volume*. (Actually, the limit of a maximal expiration is not only the capacity of the expiratory muscles to compress the thoracic cage. Many small airways become occluded during the forced expiration, and the lungs, with trapped gas, are also compressed.) A maximal inspiration from the functional residual capacity adds the *inspiratory capacity,* and the gas volume contained in the lungs is then the *total lung capacity* (TLC). The maximal volume of gas that can be expelled from the lungs following a maximal inspiration is called the *vital capacity* (VC). It follows that vital capacity plus the residual volume constitute the total lung capacity. The volume of gas moved during each respiratory cycle is the *tidal volume* (V_T).

The vital capacity and its subdivisions are commonly measured with the help of a spirometer. With the subject connected to the spirometer via a wide-bore tube, any change in lung volume is reflected in a volume displacement in the spirometer. A calibration factor translates this displacement recorded on a kymograph into liters.

The functional residual capacity can be measured with the closed-circuit methods (*gas-dilution method*). A closed spirometer contains a small, known amount of helium (or hydrogen). After a normal expiration, the subject is connected to the spirometer and rebreathes from the system. The expired carbon dioxide is absorbed by soda lime. Oxygen is added to the circuit at a rate to keep the volume at the end of expiration at a constant level. This refilling can be adjusted automatically. The concentration of the indicator gas falls in the spirometer and rises in the lungs. The final concentration is a simple function of the added gas volume, i.e., the functional residual capacity. The principle for this method is clarified by Fig. 5-7. The concentration of the indicator is analyzed continuously, for example with a katharometer, and a constant reading for

Figure 5-7
A spirometer with a measured volume of gas (V_S) contains helium in a small, analyzed concentration (He_1). After a normal expiration (lung volume = functional residual capacity = V_{FRC}), the subject rebreathes from the spirometer until a homogeneous gas mixture is attained with a new and lower helium concentration (He_2) due to its dilution with the air in the lungs. V_{FRC} can now be calculated (see text).

$$V_{FRC} = V_S (He_1 - He_2^{-1}) He_2^{-1}$$

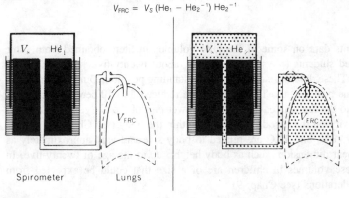

Spirometer   Lungs

about 2 min indicates a complete mixing. If the subject then performs a maximal expiration, followed by a maximal inspiration to total lung capacity, the recordings permit calculation of residual volume and the subdivisions discussed above (Fig. 5-6). Normally, about 5 min of rebreathing is enough for complete mixing of the indicator gas within spirometer-lungs, but in patients with an impaired lung function, up to 20 min of rebreathing may be necessary. The reason for using helium or hydrogen as indicator gas is that these gases are absorbed by the lung tissues and blood to only a negligible degree.

If He_1 and He_2 are the initial and final concentrations, respectively, of helium, and V_s is the volume of gas in the spirometer to the point of the subject's mouth, the functional residual capacity, V_{FRC}, can be calculated from the formula

$$V_s \cdot He_1 = (V_{FRC} + V_s)He_2 \text{ or } V_{FRC} = \frac{V_s(He_1 - He_2)}{He_2}$$

The lung volumes are expressed at BTPS, i.e., gas volume at body temperature and ambient pressure (P_B), saturated with water vapor ($P_{H_2O} = 47$ mm Hg; 6.3 kPa) and therefore the gas volumes recorded by the kymograph must be recalculated. For a spirometer temperature of t°C with a water pressure of P_{H_2O}, we have

$$FRC = V_{FRC} \cdot \frac{310}{273 + t} \cdot \frac{P_B - P_{H_2O}}{P_B - 47} \text{ liters (BTPS)}$$

If the subject is connected with the spirometer after a maximal expiration and then rebreathes deeply three times before being disconnected, the residual volume after a maximal expiration can be directly determined by the application of the same formula.

Besides the gas-dilution method, there are other methods of obtaining fairly accurate measurements of the absolute gas volumes and air spaces in the airways: the gas washout and the body plethysmography methods. The results obtained by these methods are closely comparable and reproducible with a coefficient of variation of roughly ± 5 percent. (For methods see West, 1979.)

Age and Sex

Table 5-1 presents data on some of the lung volumes in liters obtained from some fairly well-trained students (physical education), about twenty-five years old (P.-O. Åstrand, 1952). These data were obtained in the standing position. During tilting from standing to supine position, the TLC and VC are reduced 5 to 10 percent because of a shift of blood to the thoracic cavity from the lower part of the body. This illustrates the effect of gravity on the blood distribution within the body (see Chap. 4).

Vital capacity, RV, and TLC are related to body size and vary approximately as the cube of a linear dimension, such as body height, up to the age of twenty-five. In other words, these volumes in children are of a size that could be expected from theoretical considerations (see Chap. 9).

TABLE 5-1
DATA ON LUNG VOLUME (LITERS) IN 27 FEMALE AND 26 MALE FORMER PHYSICAL EDUCATION STUDENTS. IN 1949, THE AVERAGE AGE WAS 22 AND 26 YEARS, RESPECTIVELY. MAXIMAL TIDAL VOLUME AND PULMONARY VENTILATION ($\dot{V}_E$) WERE MEASURED DURING EXERCISE

Function	Females			Males		
	1949	1970	1982	1949	1970	1982
Vital capacity	4.26	4.25	4.05	5.55	5.39	4.92
Residual volume	1.10	1.71	1.64	1.45	2.04	2.12
Total lung capacity	5.36	5.96	5.69	7.00	7.43	7.04
Maximal tidal volume	2.24	2.26	2.23	3.38	3.31	3.23
Maximal $\dot{V}_E$ $l \cdot min^{-1}$	92.5	90	88.5	121	121	114

The individual dimensions are, however, not exclusively decisive for the size of the lung volumes. The lung volumes are about 10 percent smaller in women than in men of the same age and size. For the average person, the vital capacity is up to 20 percent smaller than the values listed in Table 5-1. Training during adolescence will eventually increase the VC and TLC. After the age of about thirty, the residual volume and functional residual capacity increase and the vital capacity usually decreases. There are observations from a longitudinal study that well-trained individuals attained the same vital capacity at the age of forty to forty-five as twenty years earlier (I. Åstrand et al., 1973). Twelve years later a slight reduction in VC was observed, see Table 5-1 (to be published). The ratio of RV/TLC · 100 in the young individual is about 20 percent, but for the fifty- to sixty-year-old individual, this ratio increases to about 40 percent, an increase which can be accounted for almost entirely by changes in lung elasticity with age (Turner et al., 1968; Viljanen, 1982).

Athletes have similar or slightly higher values for VC and TLC compared with the data in Table 5-1. The highest recorded value for VC is 9.0 liters for a Danish rower, recorded by Secker and Jackson (personal communication).

The vital capacity has previously been proposed as one method to assess physical fitness. In a group of about 190 individuals seven to thirty years of age, a significant correlation was found between vital capacity and maximal oxygen uptake (Fig. 5-8). A closer examination of the individual figures reveals, however, that individuals with a vital capacity of approximately 4 liters may have a maximal oxygen uptake from about 2.0 to 3.5 liters · min^{-1}. From this and similar studies, it is evident that vital capacities of 6.0 liters may be associated with oxygen uptake capacities varying from about 3.5 to 5.5 liters · min^{-1}. This example shows that one function may appear closely related to another if the data are derived from persons of greatly different size.

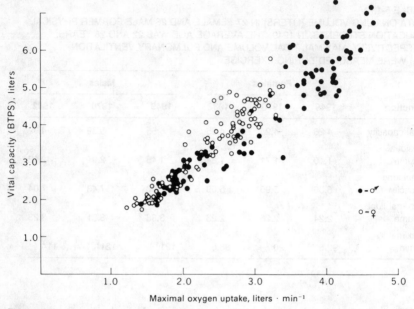

Figure 5-8
Individual data on vital capacity measured in standing position in relation to maximal oxygen uptake during running or cycling in 190 subjects from seven to thirty years of age. (*From P.-O. Åstrand, 1952.*)

However, the scattering of the data may still be considerable and sufficiently large to make any prediction of an individual's maximal oxygen uptake from such parameters as vital capacity rather unreliable. The conclusion may be drawn, however, that an oxygen uptake of 4.0 liters · min^{-1} or more does require a vital capacity of at least 4.5 liters.

The measurement of the vital capacity as part of a larger test battery may yield valuable information, especially concerning the distensibility of the respiratory system. Certain pathological conditions are associated with a reduced vital capacity.

"Dynamic" Volumes

Dynamic Spirometry, that is, the determination of ventilatory capacity per unit time, is also used to assess an individual's respiratory function. The subject breathes into a low resistance spirometer, and its displacements are recorded with the aid of a kymograph.

For the determination of *forced expiratory volume* (FEV), the subject first takes a deep breath and inspires maximally. The subject then exhales as forcefully and completely as possible. In this way it is determined how much of the person's vital capacity can be exhaled in the course of 1 s($FEV_{1.0}$), and this volume is expressed as a percentage of the individual's entire vital capacity. A normal figure for a twenty-five-year-old individual is about 80 percent. The maximal flow is limited by the rate by which the muscles are able to transform chemical energy into mechanical energy and also by a rising flow resistance. Thus, $FEV_{1.0}$ is reduced in persons who have any airway obstructions.

An evaluation of the mechanical properties of the lungs and the chest wall can also be made by determining the *maximal voluntary ventilation* (MVV) (also referred to as maximal breathing capacity). The subject is asked to breathe as rapidly and as deeply as possible during a given time interval, usually 15 s. The individual differences in MVV are large. In the case of healthy twenty-five-year-old men, the mean value is about 140 liters $\cdot$ min^{-1}, with a range from 100 to 180 liters $\cdot$ min^{-1}. For women, the normal values range from about 70 to 120 liters. The pulmonary ventilation during maximal work is somewhat lower than that obtained during the determination of MVV.

COMPLIANCE

The lungs and the thorax are partly made of elastic tissue. During inspiration these tissues are stretched. Because of their elastic nature, they return to their resting position as soon as the inspiratory muscles are relaxed. The more rigid these tissues are, the greater muscular force must be applied in order to achieve a given change in volume. The relation between force and stretch or between pressure and volume can be measured. Thus a measure is obtained of the tissue's elastic resistance to distension, or its so-called compliance. With the aid of a balloon placed in the intrathoracic esophagus, the pressure may be measured at the end of a normal expiration and again after the subject has inhaled a known volume of gas. These measurements may be repeated at different volume changes. The volume changes in liters produced by a unit of pressure change in centimeters H_2O gives the lung compliance. If a pressure change of 5 cm H_2O produces a change in lung volume of 1 liter, the lung compliance is 1.0 liters/5 cm H_2O, or 0.2 liter/cm H_2O, which is the normal value at quiet breathing. With a respiratory depth of about 0.5 liter, the pressure variations in overcoming the resistance are, in consequence, a few cm of water. At lung volumes closer to maximal inspiration, or maximal expiration, a greater pressure is required for a given volume change, that is, the compliance is reduced. If, owing to pathological changes, such as interstitial or pleural fibrosis, the lungs are more rigid and less distensible, the compliance is also reduced, and the respiratory work is increased.

In the foregoing, the principle for the measurement of the compliance of the *lungs* has been discussed. It is also of interest to assess the compliance of the thoracic cage. This can be estimated by measuring the compliance of the respiratory system as a whole and then subtracting the compliance of the lungs alone. The chest wall compliance decreases markedly with age.

AIRWAY RESISTANCE

In addition to overcoming the elastic resistance of the respiratory system, part of the energy of the respiratory muscles has to be applied to overcome two types of nonelastic resistance: a tissue viscous resistance due to friction, and a resistance to the movement of air in the air passages. This airway resistance may be doubled by bronchial smooth-muscle contraction or reduced to half the normal resistance by bronchodilation. The airway resistance may also be increased by mucous edema or by intraluminal secretion. The factors causing this bronchoconstriction may be local or they may be a reflex response to inhaled fine, inert particles, smoke, dust, noxious gases, or to the action of the parasympathetic system. The effect of the sympathetic system and epinephrine on bronchial tone is to dilate the airways. The increased sympathicus-tonus during muscular effort thus tends to lower the airway resistance. Individuals with reactive airways may develop an "exercise-induced airway constriction." Drying and cooling the inspired air will increase the response, while humidification and warming of the air can prevent the constriction. Exercise per se is not essential for the development of obstruction, and a heat loss and concomitant airway cooling relate directly and probably causally to both the occurrence and severity of the airway constriction (McFadden and Ingram, 1983).

In this connection, it should be pointed out that inhalation of the smoke from a cigarette within seconds causes a two- to threefold rise in airway resistance which may last 10 to 30 min (Comroe, 1966b). At rest, this increased airway resistance is not noticeable. In order to give rise to subjective symptoms of distress, the airway resistance has to be increased 4 to 5 times the normal value. However, during muscular effort, with its increased demand on pulmonary ventilation, the effect of tobacco smoking becomes apparent. The causative factor is not nicotine but particles that have a smaller diameter than 1 μm and that affect the sensory receptors in the airway path. The chronic effect of tobacco smoking is an increased secretion in the respiratory tract and a narrowing of the air passages. FEV as well as MVV may be reduced. It is a common observation that athletes involved in events requiring endurance never smoke. This abstention may be explained by the fact that cigarette smoking reduces the respiratory function and increases the amount of carboxyhemoglobin. The latter reduces the oxygen-transporting capacity of the blood. A sprinter, shot-putter, or diver may be unaffected by cigarette smoking, however, since the requirement for aerobic power in these athletic events may be insignificant. It was mentioned above that smoking also reduced the bactericidal effect of the alveolar macrophages.

PULMONARY VENTILATION AT REST AND DURING EXERCISE

The pulmonary ventilation is the mass movement of gas in and out of the lungs. The pulmonary ventilation is mainly regulated so as to provide the gaseous exchange required for the aerobic energy metabolism. Some gaseous exchange does take place through the skin, and some gas is lost in the urine and other secretions, but the volume of gas thus exchanged is negligible.

Methods

The gas volumes are usually measured very accurately with a water-filled spirometer. Room air (or a mixture of gases) is inhaled through a respiratory valve. The expired air is either collected in a bag (Douglas bag) or collected directly in the spirometer.

The volume may also be measured by other means, with the aid of a gas meter or flow meter, for example. The latter is constructed on the principle that the pressure gradient along a rigid tube of uniform cross section is linearly related to the flow of a gas or fluid, as long as the flow is linear. A flow resistance, which may be a fine mesh screen of a dimension which will not affect respiration, is inserted in the tube. With a differential pressure manometer, the pressure difference across the resistance can be measured when gas is flowing. The volume of gas flow is obtained by graphic or electric integration.

The amounts of inhaled and exhaled air are usually not exactly equal, since the volume of inspired oxygen in most situations is larger than the volume of carbon dioxide expired. Pulmonary ventilation usually means the volume of air which is *exhaled* per minute. It is exceedingly important that the mouthpiece, respiratory valve, tubes, and stopcocks are so constructed that they cause a minimum of increased airway resistance during heavy physical exertion. Thus, corrugated external breathing tubes should not be used, since they may give rise to turbulence. The diameter of the tubes and all openings should be about 30 mm or *wider*.

Pulmonary Ventilation during Exercise

Figure 5-9 shows how ventilation ($\dot{V}_E$) increases during increasing rates of exercise up to the maximal level. From a resting value of about 6.0 liters $\cdot$ min^{-1}, the ventilation increases to 100, 150, and in extreme cases, to 200 liters $\cdot$ min^{-1} (Saltin and P.-O. Åstrand, 1967) (BTPS = gas volume at normal body temperature and ambient barometric pressure, saturated with water vapor.) The increase is semilinear, with a relatively greater increase at the heavier exercise intensities. (The explanation for this is discussed in the section "Regulation of Respiration.")

Figure 5-10 presents data on maximal pulmonary ventilation for about 225 subjects, from four to about thirty years of age, collected during maximal running for about 5 min. A positive correlation exists between maximal $\dot{V}_E$ and $\dot{V}_{O_2}$, but it is evident that maximal ventilation cannot be used for prediction of maximal oxygen uptake. Maximal pulmonary ventilation is actually not a well-defined parameter. Figure 5-9 illustrates a marked increase in ventilation during heavy exercise without any further increase in oxygen uptake. A well-motivated subject may continue to exercise at very high rates of exercise despite strain (as judged from blood lactic acid concentration), and the person attains a high ventilation; the less motivated one just quits exercise when still submaximal in a physiological sense.

If pulmonary ventilation is expressed in relation to the magnitude of oxygen uptake, it is 20 to 25 liters/liter O_2 at rest and during moderately heavy exercise, but it increases to 30 to 40 liters/liter O_2 during maximal exercise. In children under ten years of age, the values are about 30 liters during light exercise and up to 40 liters/liter O_2 uptake during maximal exercise (Fig. 5-10).

Figure 5-11 presents mean values for maximal pulmonary ventilation during exercise (running or cycling) in different age groups. The lower ventilation in the older individuals is associated with a reduced maximal oxygen uptake (Fig. 7-13). In the subjects followed-up over a 33-year period, the maximal pulmonary ventilation declined only about 5 percent, on average (Table 5-1).

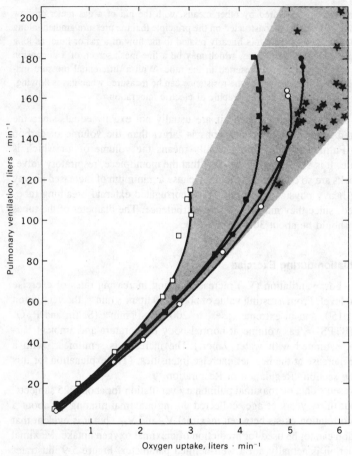

Figure 5-9
Pulmonary ventilation at rest and during exercise (running or cycling). Four individual curves are presented. Several rates of exercise gave the same maximal oxygen uptake. Exercise time from 2 to 6 min. Stars denote individual values for top athletes measured when maximal oxygen uptake was attained. (*Data from Saltin and P.-O. Åstrand, 1967.*) Individuals with maximal oxygen uptake of 3 liters · min⁻¹ or higher usually fall within the shadowed area. Note the wide scattering at high oxygen uptakes.

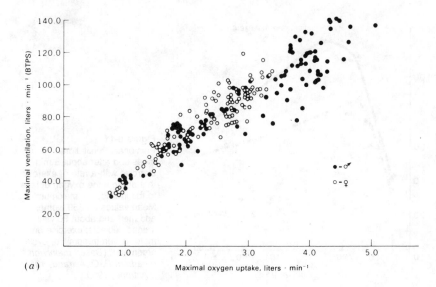

(a)

Maximal oxygen uptake, liters · min⁻¹

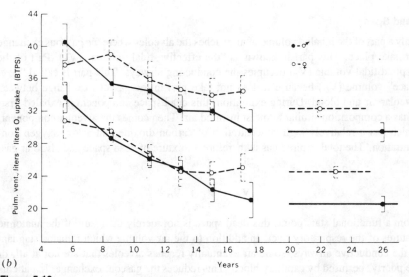

(b)

Years

Figure 5-10
Data on 225 subjects from four to about thirty years of age. (a) Maximal pulmonary ventilation in relation to maximal oxygen uptake measured during running on a motor-driven treadmill for about 5 min.
(b) Average values of ventilation per liter oxygen uptake in relation to age. The upper curves show maximal values (attained during running), and the lower ones, submaximal values during running or cycling, with an oxygen uptake which was 60 to 70 percent of the subject's maximal aerobic power [same subjects as in (a)]. Vertical lines denote ±2 SE (standard error of the mean). (*From P.-O. Åstrand, 1952.*)

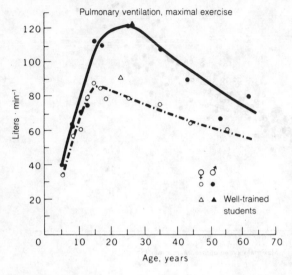

Figure 5-11
Pulmonary ventilation measured after about 5 min exercise with a rate of exercise that brought the oxygen uptake to the individual's maximum. Mean values on 350 women and men and about 80 well-trained subjects; exercise on motor-driven treadmill or cycle ergometer. (*Based mainly on data from P.-O. Åstrand, 1952; I. Åstrand, 1960.*)

Dead Space

Only a part of the inhaled volume of air reaches the alveoles where the gaseous exchange can take place. This part is known as "the effective tidal volume" (V_A). Part of the inspired tidal volume (V_T) occupies the conducting airways. This part is called "dead space" volume (V_D) because it does not take part in the gaseous exchange between alveolar air and blood. During expiration, this dead space component is exhaled first. It has a composition similar to moist inspired air. Then comes the alveolar component, which has a relatively high concentration of carbon dioxide and a low oxygen concentration. The total expired gas is therefore a mixture of dead space and alveolar gas, or

$$V_T = V_A + V_D$$

From a functional standpoint, this dead space is not merely the result of the anatomic features of the respiratory tract. In addition to the air volume which remains stagnant in the conductive airways, some air eventually reaches alveoles that are not at all, or are poorly, perfused by capillary blood. This reduces the gaseous exchange. In patients suffering from pulmonary disease, an unfavorable relationship between ventilation and perfusion may increase the physiological dead space.

The volume of the dead space may be estimated with the aid of Bohr's formula, which is based on the fact that the expired volume of oxygen at each respiration ($V_T \cdot FE_{O_2}$*) is equal to the sum of the volume of oxygen contained in the dead space compartment ($V_D \cdot FI_{O_2}$)

*F = Fraction of oxygen in the expired, inspired, and alveolar air respectively.

and the volume of oxygen coming from the alveolar air ($V_A \cdot FA_{O_2}$). We therefore arrive at the following formula:

$$V_T \cdot FE_{O_2} = V_D \cdot FI_{O_2} + V_A \cdot FA_{O_2}$$

Since $V_A = V_T - V_D$, the formula may be simplified as follows:

$$V_D = V_T \frac{FE_{O_2} - FA_{O_2}}{FI_{O_2} - FA_{O_2}}$$

If the oxygen content of the inspired air is 21 percent, the oxygen content of the expired air 16 percent, the oxygen content of the alveolar air is 14 percent, and the depth of respiration, V_T, is 500 ml:

$$V_D = 500 \frac{16 - 14}{21 - 14} = 143 \text{ ml}$$

The same calculation can be made on the basis of CO_2.

With a depth of respiration (tidal volume) of 500 ml at rest, the dead space constitutes approximately 150 ml. The rest of the tidal volume reaches the alveoles. It should be noted, however, that the first portion of the inhaled air is, in reality, the respiratory air which remained in the dead space compartment from the previous respiration. The "fresh" air which is pulled down into the alveoles is diluted into a relatively large volume, i.e., the functional residual capacity. The variations in the gas concentration are therefore relatively small in the alveoles during rest and normal breathing.

Owing to methodological difficulties, it is not easy to measure the exact behavior of the dead space during exercise. Asmussen and Nielsen (1956) estimated that with a tidal volume of 3 liters, the dead space was 300 to 350 ml, whereas Bargeton (1967) concluded from his data that the increase in dead space with increasing tidal volume is very slight and can be taken as a constant for moderate changes in tidal volume. Relatively speaking, the dead space is reduced with increasing tidal volume. If, at a ventilation of 6.0 liters $\times$ min^{-1}, the respiratory frequency is 10, and the dead space 0.15 liter, the alveolar ventilation is

$$6.0 - 0.15 \cdot 10 = 4.5 \text{ liters} \cdot \text{min}$$

If the respiratory rate, on the other hand, is 20, and the gross ventilation and dead space are assumed to be unchanged, the alveolar ventilation is only

$$6.0 - 0.15 \cdot 20 = 3.0 \text{ liters} \cdot \text{min}^{-1}$$

Animals which depend on evaporative heat loss from the respiratory tract for their temperature avoid hyperventilation of the alveoles, thanks to a high respiratory rate and a low alveolar ventilation (panting).

From the pulmonary ventilation data that are given in Fig. 5-9, part of the volume does not participate in the gas exchange. The method of concealment, often described in adventure stories, by hiding submerged in water and breathing through a snorkel represents a considerable complication of the gas exchange. The tube (snorkel) represents an extension of the respiratory dead space, and the tidal volume has to be increased by an amount equal to the volume of the tube if the alveolar ventilation is to be maintained unchanged. The breathing may therefore become very laborious.

A second complication of this diving is the increased load on the inspiratory muscles. Within the lungs, there is the same pressure as at the water surface, i.e., atmospheric pressure. The outside of the thorax, however, is subjected to atmospheric pressure *plus* the pressure of the column of water above the diver. At a depth of 1.0 m, this extra pressure will be 0.1 atm, or about 76 mm Hg (10.1 kPa). The highest pressure the inspiratory muscles can overcome is just above 70 mm Hg (9.3 kPa), and, therefore, a depth of 1.0 meter would be the maximum that can be tolerated even if the problem of extra dead space could be solved by a system of valves.

Tidal Volume-Respiratory Frequency

By definition, the pulmonary ventilation equals the frequency of breathing multiplied by the mean expired tidal volume, or

$$\dot{V}_E = f \cdot \overline{V}_T$$

At rest, the respiratory frequency is between 10 and 20. Inspiration occupies less than half the total cycle, the rise in flow being more abrupt than the fall. During exercise of low intensity, it is primarily the tidal volume that is increased. In many types of exercise, this may amount to approximately 50 percent of the vital capacity when the rate of exercise is moderately heavy or heavy (Fig. 5-12). The respiratory frequency is also increased, especially in the case of heavy exercise. Children about five years of age may have a respiratory frequency of about 70 at maximal exercise, twelve-year-old children about 55, and twenty-five-year-old individuals 40 to 45 (Fig. 5-12) (P.-O. Åstrand, 1952). In well-trained athletes with high aerobic power, respiratory frequencies up to about 60 · min⁻¹ are not unusual (see Clark et al., 1983).

The increase in tidal volume is brought about through the utilization of both the inspiratory reserve volume and the expiratory reserve volume (see Fig. 5-6). Inspiration and expiration become more equal in both time and pattern. Naturally, the vital capacity limits the tidal volume, but rarely more than 60 percent of the vital capacity is utilized.

When the body was submerged in water, the VC was, in one study, reduced by 10 percent and the expiratory reserve volume was less than 1 liter as compared with 2.5 liters in air. The increase in V_T was affected in water exclusively by utilization of the inspiratory reserve volume (Holmér, 1974). Similarly, the upper limit for the respiratory frequency is determined by the rate at which the neuromuscular system can generate alternating movements. Studies have indicated that an individual spontaneously balances the depth of respiration and respiratory frequency in such a way that a certain ventilation takes place at optimal efficiency, that is, with the utilization of a minimum of energy by the respiratory muscles (Milic-Emili et al., 1960). The

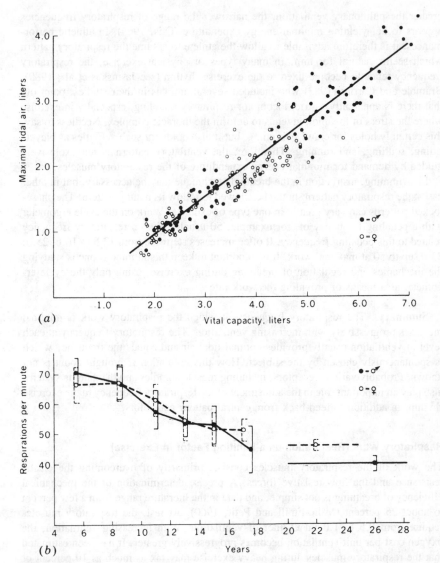

Figure 5-12
(a) Highest tidal volume measured during running at submaximal and maximal speed (exercise time about 5 min) related to the individual's vital capacity measured in standing position. Altogether, 190 subjects from seven to thirty years of age. On an average, 50 to 55 percent of the vital capacity is used as maximal tidal air.
(b) Respiratory frequency during running at a speed that brings the oxygen uptake up to maximum; average values (±2 SE) for 225 subjects from four to thirty years of age. (*From P.-O. Åstrand, 1952.*)

greater the pulmonary ventilation, the narrower the range of respiratory frequencies appears to be, yielding minimal energy expenditure (Otis, 1964). In athletic performances, it is therefore advisable to allow the athlete to assume the respiratory pattern which seems natural for him. In many types of physical exercise, the respiratory frequency tends to become fixed to the exercise rhythm (see Jasinskas et al., 1980). Bramble and Carrier (1983) who included several animals in their studies, point out that there is apparently a strict locomotor-respiratory coupling, especially in exercise where the stress of locomotion tends to deform the thoracic complex. Needless to say, this certainly holds for crawl swimming, but it also holds for such activities as bicycle riding, sculling, and running. Therefore, the ventilatory pattern is not exclusively guided by demand for minimal energy expenditure of the respiratory muscles.

In swimming, instruction in the breathing technique may be necessary, but in other cases, the respiratory pattern should be allowed to follow its natural pattern. The phase-locked patterns can vary greatly in one type of exercise. Work on the cycle ergometer with a pedaling frequency of, for example, 50 usually gives a respiratory frequency related to this pedaling frequency. It often increases stepwise from 12.5 to 16.6, 25.0, 33.0, up to 50 at maximal work. It is important to keep this in mind if one is studying the mechanics and regulation of breathing during exercise, using only the cycle ergometer as a means of providing the work rate.

Summary   The respiratory frequency at which the respiratory work is minimal increases progressively with increasing ventilation. The respiratory frequency at each level of ventilation usually provides optimal tidal air and equals the frequency which is spontaneously chosen by the subject. How this regulation is brought about is not known. Probably various receptors, including muscle spindles and the gamma system, also play an important role in the adjustment of the respiratory frequency to the exercise rhythm, in addition to feed-back from central pattern generators.

Respiratory Work (Respiration as a Limiting Factor in Exercise)

The work of the respiratory muscles consists primarily of overcoming the elastic resistance and the flow-resistive forces. A precise determination of the mechanical efficiency of breathing is not simple, and data in the literature range from a few percent to about 25 percent (Milic-Emili and Petit, 1960). At rest, the respiratory muscles require from 0.5 to 1.0 ml O_2/liter of ventilation. With increasing ventilation, the oxygen cost per unit ventilation becomes progressively greater. It has been estimated that the respiratory muscles during heavy exercise may tax as much as 10 percent or more of the total oxygen uptake (Liljestrand, 1918; Nielsen, 1936; Otis, 1964). Bye et al. (1983), have calculated that oxygen uptakes as high as about 1 liter $\cdot$ min^{-1} by the respiratory muscles may not be unrealistic in exercise forcing the pulmonary ventilation to very high levels.

A question of considerable importance is whether or not hyperventilation may limit the body's total oxygen uptake. The answer is probably negative for the following reasons: (1) After the maximal oxygen uptake is reached, it is still possible for the subject to continue to exercise at higher rates because of the anaerobic processes. At the same time, the pulmonary ventilation is markedly increased, without any distinct ceiling being reached except in some extreme cases. (The "normal" pattern is illustrated in Fig. 5-9). (2) During extremely heavy exercise which can be tolerated for only a few minutes at the most, the pulmonary ventilation is greater than at a somewhat lower but still maximal load, which may be tolerated for about 6 min. The oxygen uptake is nevertheless the same in both cases (P.-O. Åstrand and Saltin, 1961). (3) At maximal exercise, it is possible voluntarily to increase the ventilation further, showing that the ability of the respiratory muscles to ventilate the lungs evidently is not exhausted during spontaneous respiration. One example: Five well-trained young subjects exercised on a cycle ergometer at, or very close to, maximal oxygen uptake for 10 min (max $\dot{V}_{O_2}$ 4.46 liters · min^{-1}, 63 ml · kg^{-1} · min^{-1}). Their MVV (30 sec) was 211 liters · min^{-1} and their pulmonary ventilation during maximal cycling was, when breathing spontaneously, on the average 159 liters · min^{-1}, or 76 percent of the MVV. Following another protocol, they hyperventilated voluntarily during the 5th minute of the 10 min standard exercise. This increased the pulmonary ventilation to a mean value of 179 liters · min^{-1}, or 85 percent of MVV. In other words, there was a reserve capacity available. It should be noted that the hyperventilation increased the oxygen uptake ($+ 0.10$ liters · min^{-1}), and the blood lactate concentration increased from 11.1 to 12.7 mmoles · liter^{-1}, both changes being significant ($P < 0.01$). (Unpublished data from Hartley, Lacour, and P.-O. Åstrand). Well-trained and very fit athletes can apparently utilize some 95 percent of this MVV during exercise, but less fit subjects normally, only attain 60 to 70 percent of their MVV (see Folinsbee et al., 1982). (4) At heavy rates of exercise, the pulmonary ventilation depends on the volume of CO_2 produced. The alveolar oxygen tension increases and the carbon dioxide tension decreases, a fact that indicates an effective gas exchange in the lungs (Fig. 5-13). The oxygen tension of the arterial blood is maintained or only slightly reduced except in highly trained individuals with high maximal aerobic power (see Dempsey et al., 1982). These authors actually suggest that the failure to increase pulmonary ventilation adequately, and the very rapid transit time of red cells through the pulmonary capillary bed, could together result in the failure of efficient diffusion of oxygen from air to blood. The size of the contact area may also be critical when large volumes of oxygen are "needed" (see also Weibel and Taylor, 1981, p. 61).

These considerations refer to bicycling and running. The situation appears to be comparable for these two types of exercise.

Do the Respiratory Muscles "Steal" Oxygen from Other Tissues?

To some extent, the pulmonary ventilation may be a limiting factor even though the maximal capacity of the respiratory muscles is not fully taxed. Thus, if the energy demand of the respiratory muscles, in order to increase the pulmonary ventilation, necessitates such a marked increase in the oxygen consumption that all the achieved increase in oxygen content of the alveolar air (and increase in oxygen content of the arterial blood) is entirely utilized by the respiratory muscles themselves, then none of this extra oxygen will benefit the rest of the active muscles of the body. In other words, an increase in pulmonary ventilation beyond a certain point would not be

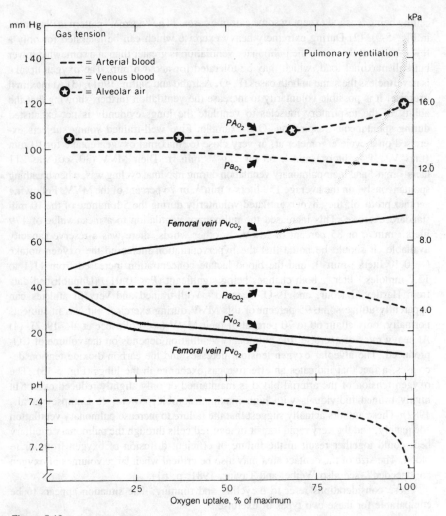

Figure 5-13
Oxygen and carbon dioxide tensions in blood and alveolar air at rest and during various levels of exercise up to and exceeding the load necessary to reach the individual's maximal oxygen uptake (= 100 percent). At bottom, arterial pH. Curves are based on data from different authors and unpublished studies. (The line denoting pulmonary ventilation refers to an individual with maximal aerobic power of 4.0 liters · min⁻¹ if the figures on the ordinate are valid.)

physiologically useful, since all the additional oxygen thus gained would be required for the work of breathing (Otis, 1964; Bye et al., 1983). It is even conceivable that the oxygen utilization by the respiratory muscles may become so great that the oxygen supply to other tissues is reduced. However, such a critical limit is probably not reached in normal individuals. It is likely that the blood flow in the vessels of the respiratory muscles is maximal even at a ventilation below the maximal ceiling and that the oxygen content of the blood is more or less completely extracted. A further increase in ventilation beyond this point is probably met by anaerobic processes. A point in favor of this view is the fact that the oxygen uptake reaches a distinct plateau during extremely heavy exercise, even if the rate of exercise, and the pulmonary ventilation is further increased. It has already been mentioned that voluntary hyperventilation supplemented the spontaneous ventilation during maximal or near maximal exercise. It increased the peak lactate value and slightly elevated the oxygen uptake. Even though the full capacity of the respiratory muscles may not be utilized during heavy exercise, the maximal force which these muscles may develop is limited by the rate at which chemical energy can be transformed into mechanical energy.

Studies Indicating Fatigue of the Respiratory Muscles After Prolonged Respiratory Exertion.

In their review, Bye et al. (1983), point out that respiratory muscle fatigue invariably occurs with extreme ventilatory demands, e.g., during resistance breathing. The question is whether this may occur in an exercise situation. Keeping the energy cost of the respiratory muscles at a level acceptable to the muscles directly involved in the external task to be performed has to be balanced against the inevitable consequences of a reduction in oxygen tension in the arterial blood.

It should be noted that a substitution of air with a 79 percent helium—21 percent oxygen mixture (with reduced density) increases the maximal pulmonary ventilation and oxygen uptake significantly (see Brice and Welch, 1983). At the present it is difficult to explain such findings. Loke et al. (1982) report that respiratory muscle fatigue is evidently a consequence of marathon running (both the strength of the respiratory muscles and the MVV declined following the race). Martin et al. (1982), noticed that the short-term maximal running performance decreased after 150 min of isocapnic hyperventilation at two-thirds of the subjects' MVV. They conclude that a reduced ventilatory muscle performance explained their finding. Bye et al. (1984), have also presented recent data indicating that diaphragmatic fatigue (evaluated from recorded EMG) was established in such subjects performing exhausting exercise at 80 percent of their maximal power output on a cycle ergometer.

Summary Under normal conditions, with the possible exception of individuals with very high aerobic power, and therefore exceptional demands on the respiratory function, the potential of the respiratory muscles to move air into and out of the lungs does not seem to limit the maximal oxygen uptake. One complicating factor is that the improved oxygenation of the blood passing the lungs by an exaggerated pulmonary ventilation also leads to a proportionally increased oxygen demand by the respiratory muscles. Therefore, the net effect of an extra respiratory effort may be questioned.

DIFFUSION IN LUNG TISSUES, GAS PRESSURES

The role of respiration is to provide the gaseous exchange between the blood and the ambient air. This is accomplished by the flowing of blood through capillaries of extremely small caliber; these are located only a few microns from the alveolar air, which is a derivate of the ambient air (Fig. 5-3). The gas exchange between the capillary blood and the alveolar air is achieved by the process of diffusion (for details, see Forster and Crandell, 1976; West, 1979; Cassidy et al., 1980).

The diffusion takes place as a movement of gas molecules from a region of higher to one of lower chemical activity. The partial pressure of the gas is a measure of this activity. The normal pressure of oxygen, carbon dioxide, and nitrogen in atmospheric air (P_{Bar} = 760 mm Hg; 101 kPa), in alveolar air, and in mixed venous blood and arterial blood at rest is given in Fig. 5-14. (If, for example, the oxygen concentration in the alveolar air is 15 percent of the dry gas, its partial pressure is

$$P_{O_2} = {}^{15}/_{100} (760 - 47) = 107 \text{ mm Hg or:}$$
$$P_{O_2} = 15/100 (101.1 - 6.3) = 14.2 \text{ kPa}$$

since the partial pressure of water vapor is 47mmHg; 6.3 kPa.)

Blood flow through a tissue is not always determined by the metabolic activity in the tissue in question. Thus, the oxygen uptake of tissues such as the kidneys and skin is small compared with the magnitude of the blood flow through these tissues. For this reason, the partial pressure of oxygen in the venous blood remains high and the CO_2 pressure relatively low. Comparatively more O_2 is utilized in the muscle, and here a greater amount of CO_2 is produced, so that the partial pressures of these gases in the venous blood are different from those of the above-mentioned organs.

It should be noted that the total gas pressure in venous blood is considerably lower than in arterial blood (706 mm Hg as against 760 mm Hg; 94 and 101 kPa respectively; see Fig. 5-14). In this way, accumulation of gas in the intrapleural space in the thorax is avoided, despite the opposed recoil of the lungs and chest wall. If gas is trapped behind an occlusion of an airway, it becomes absorbed into the pulmonary circulation because of this subatmospheric gas pressure of venous blood. Under normal conditions, the diffusion processes are so rapid that the gases in the blood leaving the pulmonary capillaries are approximately in equilibrium with the gases in the alveoli. The gas exchange between pulmonary air and blood is achieved entirely by the process of diffusion. No other processes, such as secretion, are involved. An analysis of the gas concentrations and pressures in the expired air at the end of the expiration gives an approximate idea of the gas pressure in the arterial blood. One may simply collect the last portion of the expiratory air volume, either at the end of a single forced expiration (Haldane-Priestley method) or from several repeated respiratory cycles (end-tidal sampling technique), for an analysis of the gaseous composition of expired air. With the

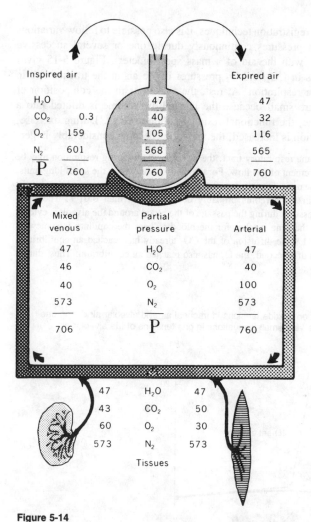

Figure 5-14
Typical values of gas tensions in inspired air, alveolar air (encircled), expired air, and blood, at rest. Barometric pressure, 760 mm Hg (100 mm Hg = 13.3 kPa); for simplicity, the inspired air is considered free from water (dry). Tension of oxygen and carbon dioxide varies markedly in venous blood from different organs. In this figure, gas tensions in venous blood from the kidney and muscle are presented.

aid of modern analytic and registration techniques, it is also possible to follow variations in gas concentrations and pressures continuously during one or several successive respirations (for example, with the aid of a mass spectrometer). Figure 5-15 gives examples of the variations in CO_2 and O_2 pressures in the air in the trachea and in the alveoli during a single respiration. At rest, the variations in the composition of the gases in the alveoli are small because the inhaled air volume is diluted into a relatively large gas volume, the functional residual volume (FRV). During exercise, when the depth of respiration is increased, the variations become considerably larger.

Because of the length of the respiratory tract, the gas movement during respiration may be considered as a mass movement of gas flow. For the distribution within the small lung units, a molecular diffusion is the main determinant. Because of the small dimensions of the alveoli, a complete mixing within the alveolus probably occurs in less than 0.01 s. The rapid equalization of the gas pressures during the passage of the blood around the alveoli is evident from Fig. 5-16. At rest, the time it takes for the blood to pass the capillary is somewhat less than 1 s, but after 0.1 s the diffusion of the CO_2 already has reached an equilibrium. After a further few tenths of a second, the O_2 has also reached an equilibrium. Thus, during

Figure 5-15
Variations in oxygen and carbon dioxide tensions in tracheal air and alveolar air during one single breath at rest. Note the very small fluctuations in gas tensions of the alveolar air.

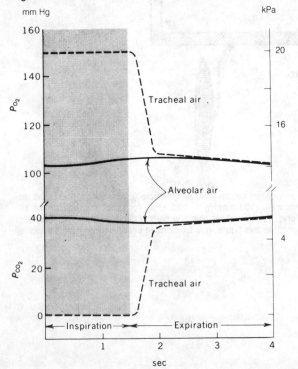

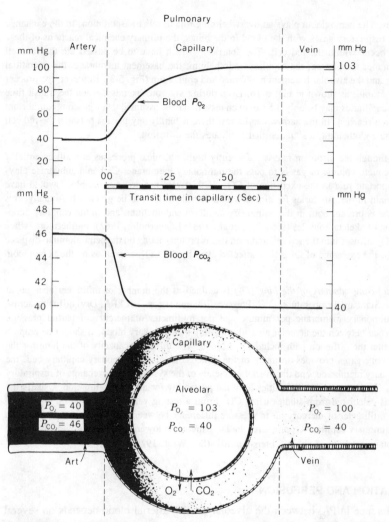

Figure 5-16
Change in the partial pressures of oxygen and carbon dioxide as blood passes along the pulmonary capillary. Note that in the first part of the capillary, the blood is already equilibrated with the alveolar gas. (*From Cherniack et al., 1972.*)

normal resting conditions, the blood in the pulmonary capillaries is almost completely equilibrated with the alveolar oxygen and carbon dioxide pressures. The size of the CO_2 molecule is larger than that of the O_2 molecule, which actually slows the rate of diffusion. On the other hand, the CO_2 is about 25 times more soluble in liquids than the O_2, so that the net effect is that the CO_2 diffuses about 20 times more rapidly in aqueous liquids than does oxygen. Both CO_2 and O_2 are carried by the blood mainly in reversible chemical combinations. The hemoglobin plays an overwhelming role in this transportation. In the exchange of the respiratory gases with the blood in the lungs, the primary chemical reactions of these gases occur within the red cell. The "barriers" which have to be passed are the red cell membrane, the plasma, the capillary endothelium, the basement membrane, the interstitial tissue, and the alveolar basement membrane and epithelium (Fig. 5-3). However, the process is very rapid, as shown in Fig. 5-16. Even during vigorous exercise, when the transit time in the capillaries may be only 0.5 s or even less, it may be assumed that a gaseous equilibrium has been reached. In the narrow capillaries, there is hardly any plasma between the red cell and the endothelium, a situation that facilitates the diffusion.

Even though the diffusion rate is sufficiently high, chemical processes are still essential if an adequate volume of gas is to pass the pulmonary membrane. Carbonic anhydrase plays an important role in the exchange of CO_2. If this were not present, the blood would have to remain in the capillaries for almost 4 min for all the CO_2 to be given off. Actually, this enzyme is present both in the pulmonary capillary endothelium, and in the capillary endo-thelium of skeletal muscles (see Effros et al., 1981; Lönnerholm, 1982). Furthermore, when the CO_2 diffuses into the red cell and then forms protons acting on the hemoglobin to displace the O_2, the exchange of O_2 is thus affected even by the CO_2, as well as by the hemoglobin itself.

The *diffusing capacity of the lung* (D_L) is defined as the number of milliliters of a gas at STPD (Standard Temperature, $0°C$, Pressure 760 mm Hg $= 101$ kPA, Dry) diffusing across the pulmonary membrane per minute and per millimeter of mercury of partial pressure difference between the alveolar air and the pulmonary capillary blood. It should be empha-sized that the diffusion path includes the blood. The diffusing capacity of the lung for the respiratory gases provides an index of the dimensions of the pulmonary capillary bed, the pulmonary membrane, and the overall efficiency of the system in the exchange of respiratory gases. Certain technical difficulties limit the possibility for an exact estimation. Usually the $D_{L_{O_2}}$ is calculated from studies using CO. From a resting value of 20 to 30 ml $O_2 \cdot min^{-1}$ and a millimeter of mercury mean pressure difference between the alveolar O_2 pressure and the pulmonary capillary O_2 pressure, the $D_{L_{O_2}}$ increases toward 75 ml in individuals with an O_2-uptake capacity of about 5 liters $\cdot min^{-1}$. (See West, 1979.)

VENTILATION AND PERFUSION

The difference in P_{O_2} between the alveolar air and arterial blood depends on several factors: the membrane component plays a role; a certain amount of admixture of bronchial and cardiac venous blood occurs; and finally, there is the effect of the passage of some blood through poorly ventilated alveoli.

Since CO_2 diffuses about 20 times more rapidly than does O_2, one cannot speak of any diffusion obstacle for CO_2. The inhaled air is not equally distributed to all the alveoli, and the composition of the gases is therefore not uniform throughout the lungs. The pulmonary capillary bed has a common blood supply, the mixed venous blood, but different areas of the lungs have an uneven perfusion. The composition of the gas in various parts of the alveolar space depends on the ventilation as well as on the blood flow, or the ratio $\dot{V}_A/\dot{Q}$.

Under extreme conditions, the $\dot{V}_A/\dot{Q}$ ratio may vary from zero (when there is perfusion but no ventilation) to infinity (when there is ventilation but no perfusion).

When the ratio is zero, the tensions of O_2 and CO_2 of the arterial blood are equal to those in mixed venous blood, since there is no net gas exchange in the capillaries. In the latter case, no modification of the inspired air takes place. Although these extreme situations rarely occur under normal conditions, the various parts of the lung have a wide range of ventilation-perfusion ratio. The "alveolar air" actually represents various contributions from several hundred million alveoli, each possibly having slightly different exchange ratios and gas composition.

In other words, the alveolar gas tensions vary from moment to moment and from place to place within the lungs because of regional inhomogeneity of ventilation and blood perfusion; the supply of air and blood is not perfectly matched in the lungs, even in a healthy individual in any posture (Rahn and Farhi, 1964; West, 1979).

There are mechanisms which to some extent compensate for an uneven ventilation in relation to the blood flow in the capillary bed of the alveoli: (1) inadequately ventilated alveoles have a low P_{O_2}, which in turn causes an alveolar vasoconstriction and reduced blood flow; (2) reduced blood flow produces a reduction in the alveolar P_{CO_2}, which causes a constriction of the bronchioles and, therefore, reduced gas flow. It appears, however, that there is no effective regional variation in the vasomotor tone, and it is apparent that the matching of ventilation against flow is not perfect. The small alveolar-arterial oxygen pressure difference, which actually exists even in normal individuals, is primarily a consequence of this unequal distribution. In the lungs, the blood flow is greatly affected by body position. In the discussion of the lung volumes, mention was made of the fact that the vital capacity increases in the standing position, compared with the lying position, owing to the reduction in blood volume in the thorax in the upright position. The effect of gravity on the distribution of the blood, as well as on the perfusion, is such that in the upright position, the perfusion per unit lung volume is about 5 times greater at the base than at the apex of the lung. It is true that the ventilation per unit lung volume changes in the same direction, but only slightly. As a result the $\dot{V}_A/\dot{Q}$ ratio becomes much higher in the upper lobes than in the lower lobes in the erect posture, or above 3 at the top and below 1 at the bottom of the lungs (West, 1962). This is illustrated by Fig. 5-17. As indicated, a high $\dot{V}_A/\dot{Q}$ ratio means either an overventilation or an underperfusion. The result is a highly variable composition of the alveolar·air in the different parts of the lung. It has been calculated that the P_{O_2} of the uppermost alveoli may be as high as above 130 mm Hg (17.3 kPa) and of the lowest alveoli below 90 mm Hg (12.0 kPa). Once again it is apparent that the lung cannot be considered as a homogeneous unit. It should be pointed out that overventilated alveoli cannot completely compensate for a desaturation of arterial blood from underventilated alveoli. The shape of the O_2 dissociation curve is such that an increase in arterial P_{O_2} above the "normal" 100 mm Hg (13.3 kPa) adds little to the oxygen content of the blood, however, a reduction in P_{O_2} will affect the oxygen content considerably more.

When the position of the body is changed from upright to supine, the perfusion of the upper lung zone increases markedly at the expense of the lower zone. In the recumbent position, the calculated $\dot{V}_A/\dot{Q}$ therefore becomes quite uniform in the different pulmonary lobes. (See Amis et al., 1984.)

Even during light exercise in the sitting or standing position, the $\dot{V}_A/\dot{Q}$ ratio also becomes more uniform throughout the lungs. Certainly both upper and lower zone blood flows increase, but the former increases relatively more. The slight increase in pulmonary artery pressure that accompanies exercise is one factor that changes the balance between arterial, capillary,

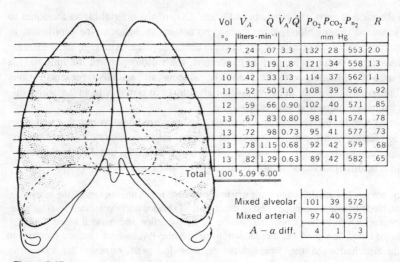

Vol	$\dot{V}_A$	$\dot{Q}$	$\dot{V}_A/\dot{Q}$	P_{O_2}	P_{CO_2}	P_{n_2}	R
%	liters·min⁻¹				mm Hg		
7	.24	.07	3.3	132	28	553	2.0
8	.33	.19	1.8	121	34	558	1.3
10	.42	.33	1.3	114	37	562	1.1
11	.52	.50	1.0	108	39	566	.92
12	.59	.66	0.90	102	40	571	.85
13	.67	.83	0.80	98	41	574	.78
13	.72	.98	0.73	95	41	577	.73
13	.78	1.15	0.68	92	42	579	.68
13	.82	1.29	0.63	89	42	582	.65
Total 100	5.09	6.00					

Mixed alveolar	101	39	572
Mixed arterial	97	40	575
$A - a$ diff.	4	1	3

Figure 5-17
Effects of observed distribution of ventilation and perfusion on regional gas tension within the lung of a normal man in sitting position. The lung is divided into nine horizontal slices, and the position of each slice is shown by its anterior rib marking. Table shows relative lung volume (Vol), ventilation ($\dot{V}_A$), perfusion ($\dot{Q}$), ventilation-perfusion ratio ($\dot{V}_A/\dot{Q}$), gas tensions (P_{O_2}, P_{CO_2}, P_{N_2}), and respiratory exchange ratio (R) of each slice. Lower table shows differences between mixed-alveolar and mixed-arterial gas tensions which would result from this degree of nonuniformity of $\dot{V}/\dot{Q}$ ratios. (*From West, 1962.*)

and venous pressures on one side and the pressure outside the vessels on the other, favoring the perfusion of the upper zones of the lungs. The pressure within the pulmonary artery may vary between 7 and 20 mm Hg (0.9 and 2.7 kPa) during a cardiac cycle at rest but between 15 and 35 mm Hg (2.0 and 4.7 kPa) during exercise with an oxygen uptake of 2.0 liters · min⁻¹. The systolic pressure may exceed 50 mm Hg (6.6 kPa) during maximal exercise. These pressure variations actually mean that the distribution of the blood flow to different parts of the lungs will vary not only with body position but also with the cardiac cycle. Even during heavy exercise, some parts of the lungs may be unperfused during part of the diastole.

What is the effect of this difference in the topographical distribution of air and blood flow? In the final analysis, this interference with the overall gas exchange is, surprisingly, rather insignificant. West (1962) states that the lung, with its uneven distribution (Fig. 5-17), wastes only some 3 percent of its ventilation and 1 percent of its blood flow, compared with an ideal lung. Wasted ventilation, including the effect of dead space, interferes particularly with CO_2 exchange, whereas wasted blood flow affects mainly the oxygen, and the effect may be calculated to be an impairment of the exchange of these gases by a few percentage points.

It is not clear to what extent the oxygen uptake may be affected by these factors during maximal exercise. It is evident, however, that an increased gravitational field as well as a drop in the blood pressure regardless of the cause, i.e., a vasovagal syncope, would more seriously affect the $\dot{V}_A/\dot{Q}$ ratio and therefore the gas exchange in the lungs.

Summary   Considerable regional inequality in ventilation of the alveoli and in blood perfusion of the capillaries exists in the lungs, causing differences in the gas exchange in different parts of the lungs. At rest in the supine or prone position, however, the ventilation distribution is rather well adjusted to follow this perfusion. In the erect posture, hydrostatic forces cause a progressive decrease in perfusion from the bottom to the top of the lung without corresponding variations in the ventilation. Therefore, the upper lobes are relatively underperfused. During exercise, the $\dot{V}_A/\dot{Q}$ ratio becomes more uniform, and the increase in pulmonary arterial pressure is at least one important contributor to this change.

OXYGEN PRESSURE AND OXYGEN-BINDING CAPACITY OF THE BLOOD

Figure 4-20 illustrates how the oxygen-binding capacity of the blood is affected by the partial pressure of oxygen in the blood. At $P_{O_2} = 100$ mm Hg (13.3 kPa), 98 percent of the hemoglobin is normally saturated with oxygen. The oxygen saturation curve is such that about half the hemoglobin is in the form of HbO_2 and half in the form of reduced Hb at P_{O_2} in the order of 26 mm Hg (3.5 kPa; 37.0°C, pH 7.40). The HbO_2 dissociation curve of the blood is affected by the CO_2 pressure, the pH, and blood temperature. Also, salts have their effects, a fact that at least partially explains the difference in curves for various species of animals. P_{CO_2} affects pH and thereby the oxygen saturation curve, but it also has a specific effect in that CO_2 combines with Hb to form carbamino compounds ($HbNH_2 + CO_2 \rightleftharpoons HbNHCOOH$), thereby reducing the capacity of Hb to bind oxygen (Fig. 4-2). During muscular exercise the CO_2 increases, as does the temperature locally in the muscle, while at the same time the pH is lowered. This causes the liberation of O_2 to increase at a given O_2 tension. Because of this feature of hemoglobin, an effective O_2 diffusion gradient may be maintained between the capillaries and the O_2-utilizing cell. With regard to the effect of an increase in temperature, the diffusion increases about 2 percent/°C. These factors aid in the unloading of oxygen to the active muscles, but are, however, of less quantitative importance than the opening of additional capillaries in the muscles. In this manner, the distance between the capillary and the muscle cell is reduced, which shortens the diffusion distance. Owing to the shape of the oxygen saturation curve, a change in pH, P_{CO_2}, and blood temperature plays a relatively small role at a P_{O_2} around 100 mm Hg (13.3 kPa). During very strenuous exercise, the oxygen saturation of arterial blood may, however, be reduced below 95 percent without a corresponding decrease in P_{O_2} (Fig. 4-21). At high altitude, where the alveolar P_{O_2} is lower, the arterial saturation is still more affected in a negative direction by a lowered pH and increased P_{CO_2} and temperature.

An organic phosphate compound, diphosphoglycerate or 2,3-DPG, is normally present in the red cells. Its effect is a shift of the O_2 dissociation curve to the right, thus assisting the unloading of O_2 to peripheral tissues. (For references, see Adamson and Finch, 1975.) Prolonged exercise training causes a slightly decreased Hb affinity for oxygen, which may be due to an increase in the erythrocyte 2,3-DPG concentration. Mairbäurl et al. (1983), noticed an increase in the oxygen tension with 1.3 mm Hg

(0.17 kPa) at 50 percent O_2 saturation of Hb in their fit subjects. However, they explain the reduced Hb-O_2 affinity after training as a shift in the average age of erythrocytes toward a younger range. There are reports that the 2,3-DPG concentration also increases at altitude, but in a recent report on natives living at a barometric pressure of about 430 mm Hg (57 kPa), the natives had concentrations that were not significantly different from the sea-level controls (Winslow et al., 1981). In patients with severe iron-deficient anemia, O_2 delivery is assisted by a shift of the O_2-dissociation curve to the right, due in part, to 2,3-DPG (see Ohira et al., 1983). In other words, there are many stress situations that, by various mechanisms, may affect the O_2 dissociation curve and thus enhance the oxygen yield at a given O_2 pressure. From an evolutionary standpoint, the beneficial effect of a shift of the oxygen dissociation curve to the right was useful, no doubt, under conditions prevailing at, or near sea level. Unfortunately, at high altitudes, this shift may create disadvantages, in that a diffusion limitation may be a determining factor in exercise tolerance (Bencowitz et al., 1982). In this situation, a rightward shift of the O_2 dissociation curve, inevitably becomes a handicap when the venous blood passes the alveoli, exposed to a relatively low oxygen tension.

Each Hb molecule has four iron atoms. In reality, the ratio between Hb and HbO_2 may vary as follows: Hb_4, Hb_4O_2, Hb_4O_4, Hb_4O_6, Hb_4O_8, in which Hb_4 is the completely reduced hemoglobin. Only in the case of Hb_4O_8 is it 100 percent saturated. Under normal conditions, these combinations are mixed in proportions which are determined by such factors as P_{O_2}. As far as the volume CO_2 bound to Hb in the carbaminohemoglobin form is concerned, it varies with the degree of O_2 saturation and is greater at low O_2 saturation and less at high O_2 saturation. This may seem like a fortunate concidence: when the O_2 uptake is great, the O_2 saturation becomes low. At the same time, CO_2 is formed and the capacity to transport CO_2 from the tissues to the lungs is increased. The transport of CO_2 and O_2 was also discussed in Chap. 4. In this connection, it should be recalled that reduced hemoglobin is a weaker acid than oxyhemoglobin, and when reduced, it mops up H^+ ions more readily and thereby helps to prevent too large a reduction in pH through the formation of acid metabolites. (See also Hlastala, in Loeppky and Riedesel, 1981, p. 91.)

The lower part of the oxygen dissociation curve is, so to speak, a reserve which is utilized during muscular work or pathological conditions. There are very few reported measurements of oxygen content and partial pressure of oxygen in a vein draining a muscle which is activated at maximum. Saltin et al. (1968) report from studies on four subjects running at maximum for about 5 min on a treadmill a P_{O_2} that averaged 12 mm Hg (1.6 kPa) in blood collected from the femoral vein (O_2 content, 1.4 vol percent; pH, 7.09; and P_{CO_2}, 70 mm Hg or 9.3 kPa).

Myoglobin in the muscle cell, is particularly rich in the slow-twitch fibers (Type 1 fibers). Besides its ability to store oxygen, even at low oxygen tension, it facilitates the O_2 diffusion within both heart and skeletal muscle to approximately twice the level observed when this protein is inactivated. Favorable for this diffusion is a low intracellular oxygen tension, certainly present during vigorous exercise, as well as a gradient of the myoglobin-oxygen complex to provide a driving force for a facilitated diffusion, and a mobility within the cell of the myoglobin-O_2 to permit diffusion of the oxygen carrier (see Scholander, 1960; de Konig et al., 1981; Wittenberg and Wittenberg,

1981; and Cole et al., 1982; Livingston et al., 1983). The diffusion of oxygen is accelerated by an elevation of the tissue temperature. This temperature effect of diffusion is greatly enhanced in the presence of myoglobin (Stevens and Carey, 1981).

From an O_2 content of 20 vol percent in arterial blood, it may eventually drop during heavy exercise to below 1 vol percent because of the abundance of capillaries in muscle tissue with short diffusion distances, a pH around 7.0, and a temperature exceeding 40°C. The mitochondria are apparently functioning efficiently even if the intracellular P_{O_2} is less than 1 mm Hg (0.13 kPa); at the end of a capillary in a vigorously contracting muscle, the O_2 tension may probably be less than 10 mm Hg; 1.3 kPa (Chance et al., 1964).

REGULATION OF BREATHING

The object of breathing is to ensure the exchange of oxygen and carbon dioxide between the blood and atmospheric air. It would therefore appear logical if these gases were to partake in the regulation of the breathing. Such is actually the case. A change in the P_{O_2}, P_{CO_2}, and H^+ concentration in the arterial blood results in a change in ventilation in such a manner as to moderate the primary change (negative feedback). Stretch receptors in the lungs and muscles, as well as a number of other factors affect respiration. As in the case of changes in the circulation under different conditions, the question is often: What is the regulating, and what is the disturbing effect? In respect to P_{O_2}, P_{CO_2} and H^+ concentration in the blood, a change in pulmonary ventilation may influence these factors. The temperature of the blood affects respiration, but a change in pulmonary ventilation does not cause a substantial change in the body temperature of the human. In some animals, such as the dog, respiration also plays a part in temperature regulation. The effect of an increased temperature of the blood in their case also has a much more pronounced effect on respiration than in the case of human beings. Catecholamines affect respiration, but a change in ventilation certainly does not affect the content of these substances in the blood. Emotion affects respiration (as in the sighing of a mourning person, the rapid breathing of an excited person). A number of different theories have been advanced concerning the regulation of breathing, but none of them has as yet fully explained how the respiratory volume is adjusted to meet the demand at rest and during physical work. The interested student will find a series of review articles in Fenn and Rahn, 1964–1965; Widdicombe, 1974; von Euler and Lagercrantz, 1979; Whipp and Ward, 1980; Dempsey and Forster, 1982; Loeppky and Riedesel, 1981; Loeschke, 1982; von Euler, 1983).

Rest

In the central nervous system, a large number of physical, chemical, and nervous variables are integrated. P_{O_2}, P_{CO_2}, and H^+ concentration, or chemical changes related to them, appear to be the most prominently controlled chemical variables. A lowering of the arterial P_{O_2} stimulates the breathing via *peripheral chemoreceptors* in the carotid and aortic bodies whence impulses go to the brain stem. An increase in P_{CO_2} and H^+

concentration also represents a stimulation leading to an increased ventilation, but this effect is primarily elicited from *medullary chemosensitive receptors,* located on the ventral surface of the medulla, separated from the blood by the blood-brain barrier.

In the medulla there are widely dispersed clusters of neurons. The existance of anatomically well-localized concentrations of expiratory and inspiratory neurons has not been demonstrated, however. A respiratory center as such cannot be defined, it is rather a question of a *central pattern generator* guiding the respiration. There are neurons with an inherent rhythmicity, but so far it is unknown where this rhythmicity originates. The inspiratory activity is characterized by a sudden onset, followed by a slowly augmenting "ramp-shaped" increase until it is "switched off". During the early phase of the expiration, there is again an activity in the inspiratory motor neurons, apparently controlling the events of the early steps of the expiratory phase. Only then is the expiratory motor activity recruited, and only if the pulmonary function is under some sort of stress. The stronger the demand on an active expiration, as during exercise, the earlier the expiratory motor activity is brought into play. Apparently, the activity in neurons causing an inspiration also activates some kind of an off-switch mechanism. When the inspiratory activity has proceeded for an adequate period of time to produce a sufficient tidal volume, a threshold is reached, and the inspiration is terminated. Various inputs can adjust the off-switch threshold up or down, thus the depth and rate of breathing can be regulated. As we will see, an increase in the CO_2 and H^+ concentrations in the blood and a reduction of its oxygen tension will raise the threshold that will delay the off-switch, thereby increasing the tidal volume. There are powerful inputs from nuclear areas in the rostral pons. Furthermore, when the lung tissues are stretched in the course of inspiration, stretch receptors in the inspiratory bronchioles are stimulated and, via afferent vagus fibers, inhibit the inspiratory activity, and the inspiration is stopped (the Hering-Breuer reflex). If the vagus is cut in an animal, the respiratory frequency becomes slower and the respiratory depth is increased. These events are schematically illustrated in Fig. 5-18.

This report on the volume situation in the lungs to the respiratory central pattern generator in the brain stem is apparently not needed in humans until the tidal air is 1–1.5 liters or higher. (For a detailed description of the present status regarding the function of the central pattern generator, see von Euler, 1983.) During the expiratory phase there is an inhibitory activity that suppresses the inspiration. The duration and strength of this inhibition will control the expiratory length. In other words, during the expiratory phase, afferent inputs from the peripheral and central chemoreceptors driving the inspiratory integrator are in some way "gated" during this phase of respiration and "opened" only during the inspiratory cycle.

Numerous neurons discharging with respiratory modulations can be found throughout the brain stem from caudal medulla to rostral pons.

The respiratory muscles, especially the intercostal muscles, are amply supplied with muscle spindles. Central respiratory drive descends by a common path to the spinal level, being then distributed to both α and γ motoneurons. Thereby we have the tool for a coactivation of the α-γ system. A second and separate input to the respiratory motoneurons originates from the pyramidal and extrapyramidal systems (see Chap.

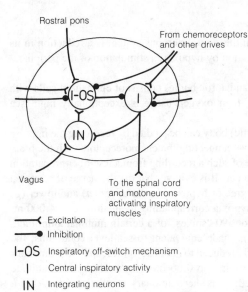

Rostral pons

From chemoreceptors
and other drives

Vagus

To the spinal cord
and motoneurons
activating inspiratory
muscles

——< Excitation

——• Inhibition

I-OS Inspiratory off-switch mechanism

I Central inspiratory activity

IN Integrating neurons

Figure 5-18
Schematic model diagram illustrating
one hypothesis for an inspiratory off-
switch mechanism of the central pattern
generator in the medulla. There is an
input to the off-switch neurons from
pulmonary stretch receptors (via
vagus), from the rostral pons, and from
a recurrent feedback from central
inspiratory activators relayed over
integrator neurons. The inspiratory off-
switch mechanism and central
inspiratory activators also receive other
impulses, both excitatory and inhibitory.
(*Modified from von Euler, 1983.*)

3). It should be emphasized that the respiratory muscles can be volitionally controlled
and they are, in almost all of our daily activities engaged in many "nonrespiratory"
activities. On the spinal level different descending and afferent inputs interact and the
interneuronal networks can exert reciprocal effects on the antagonistic muscles. This
is the place for a final integration; there are many situations that require an instantaneous
adjustment to trunk movements and postural changes to enable an immediate compensation for any changes in muscle length, direction of muscle forces, or other factors
of chest wall mechanics (see von Euler, 1983).

Central Chemoreceptors

The pulmonary ventilation at rest is chiefly regulated by the chemical state of the
blood, particularly its CO_2 tension which is highly related to the H^+ concentration.
An increase in the H^+ and CO_2 concentration (hypercapnia) in the arterial blood causes
increased ventilation. This response is closely correlated with changes in the local
extracellular fluid H^+ concentration, as measured on the ventrolateral surface of the
coroid plexus of the fourth ventricle and on, or 1 to 2 mm below, the medullary
surface (see Dempsey and Forster, 1982). The exchange of CO_2 and HCO_3^- between
the blood and this tissue is very rapid (but is slow into the cerebrospinal fluid). The
nature of the receptors and their contacts is poorly understood. Loeschke (1982)
concludes that the chemosensitivity mechanism is probably a cholinergic synapse on
which H^+ acts like acetylcholine and can be replaced by ACh. It is assumed that CO_2
does not have an independent effect per se, but acts via H^+ ions.

Peripheral Chemoreceptors

The direct effect of lack of oxygen on the central nervous system is a reduction in its function. For the stimulation of respiration by hypoxia, a stimulation of the peripheral chemoreceptors is required.

The chemoreceptors in the carotid and aortic bodies consist of epithelioid cells with a rich innervation and blood supply; their oxygen usage is exceptionally high (see Biscoe, 1971).

The function of the carotid (and aortic) body can be studied by recording the impulse activity electrically in the afferent nerve connecting the chemoreceptors with the brain stem. Figure 5-19 presents an example of such a recording from a few-fiber preparation of the afferent carotid sinus nerve in a cat. It is evident that the discharge of scattered impulses increases with increasing degree of hypoxia. In the upper (a) and lower (g) tracing, the animal is exposed to an hypoxia corresponding to an altitude of 4,000 m. The discharge is relatively strong, about 90 "spikes" of a certain minimal amplitude. When the oxygen tension of the air is made equivalent to sea-level conditions, the activity in the chemoreceptors is quickly reduced to about 25 spikes, and the pulmonary ventilation decreases from 0.85 liter $\cdot$ min^{-1} to 0.66 liter $\cdot$ min^{-1} (c). At an altitude of 6,000 m, the discharge is increased, as is the pulmonary ventilation (d). After a change to oxygen breathing, the chemoreceptors almost completely cease to fire (a single spike is seen in the tracing) (f).

It has been shown conclusively that the decisive factor in the hypoxic stimulation of the chemoreceptors is a lowering of P_{O_2}, not a reduced oxygen content as such. Thus, in laboratory animals, up to 80 percent of the hemoglobin may be bound to CO without the respiration being affected, providing the P_{O_2} is kept at a normal level (Heymans and Neil, 1958; Lahiri et al., 1981b).

The discharge of the chemoreceptors increases if the blood flow through them is reduced. Stimulation of sympathetic fibers to the carotid bodies as well as blood-borne epinephrine and norepinephrine may reduce the blood flow to the chemoreceptor tissue (Lee et al., 1964; Neil and Joels, 1963). Therefore, chemoreceptors may send an increasing number of nerve impulses centrally, despite an unaltered oxygen tension of the arterial blood, if there is an increased sympathetic activity, as after hemorrhage or during exercise (see below). On the other hand, there are also depressant sinus nerve efferents which may increase the blood flow and diminish the chemoreceptor activity (Biscoe, 1971).

There are considerable individual variations in the reaction to hypoxia. Some persons may be brought to unconsciousness without pulmonary ventilation being significantly

Figure 5-19
Action potentials recorded from a few fibers in the carotid sinus nerve in a cat subjected to different altitudes in a low-pressure chamber and oxygen breathing at 4,000-m simulated altitude. Tracheal oxygen tension to the left. Records on each film strip from top downward: time (50 Hz), electroneurogram and arterial blood pressure (calibration lines at 150 and 100 mm Hg or 20.0 and 13.3 kPa respectively). Right column gives pulmonary ventilation and number of action-potential spikes of an arbitrarily chosen minimal height. Note the relation between chemoreceptor activity and ventilation. (*From P.-O. Åstrand, 1954.*)

	number of spikes	vent. liters · min⁻¹
(a) 4,000 m P_{O_2} 87 mm 11.6 kPa	93	0.847
(b) Sea level P_{O_2} 149 mm 19.8 kPa	50	0.664
(c) Sea level P_{O_2} 149 mm 4 min later	24	0.974
(d) 6,000 m P_{O_2} 68 mm 9.0 kPa	139	
(e) Oxygen P_{O_2} 415 mm 55.2 kPa	11	0.705
(f) Oxygen P_{O_2} 415 mm 4 min later	1	
(g) 4,000 m P_{O_2} 87 mm 11.6 kPa	90	0.861

affected. In others, the pulmonary ventilation may be increased to double or more during hypoxia.

What it is that actually stimulates these peripheral chemoreceptors is not clear. If the oxygen tension in the arterial blood is high, as is the case when breathing 95 percent O_2, the arterial P_{CO_2} (by breathing CO_2) may increase and its pH drop markedly without the chemoreceptors being stimulated. So, if an hypoxic drive does not exist but the oxygen saturation is adequate, these chemoreceptors then seem rather insensitive to changes in P_{CO_2} and H^+ concentration. At low oxygen tension, the discharge from the chemoreceptors is increased further, however, if P_{CO_2} at the same time is high and the pH low (Hornbein and Roos, 1963; Neil and Joels, 1963). Thus, hypercapnia potentiates the effect of hypoxia, possibly by a change in the intracellular H^+ concentration or reduction in the blood flow to the area where the chemoreceptors are located (see Widdicombe, 1974, p. 247; Eyzaguirre and Fidone, 1980). Heeringa et al. (1979), report that in anesthetized cats, the peripheral chemoreceptors contributed from 20 to 50 percent of the slope of the total ventilatory response to CO_2, under steady state conditions. The aortic chemoreceptors are much less responsive to CO_2 stimulus than the carotid chemoreceptors (Lahiri et al., 1981a).

Normally, the P_{O_2} in persons breathing ordinary air at sea level is low enough to contribute a small but significant tonic ventilatory stimulus from the peripheral chemoreceptors (Dejours, 1981). Most subjects exhibit a sustained increase in pulmonary ventilation only when breathing 16 percent O_2 and a marked increase in ventilation only when the inspired O_2 has dropped to about 10 percent or less. Thus, the chemoreceptors have different thresholds, some of which are not stimulated markedly until the P_{O_2} has fallen to 40 or 50 mm Hg; 5.3 or 6.6 kPa (8 to 10 percent O_2 inhaled). The chemoreceptors are very resistant to anoxia and can therefore maintain breathing reflexively for long periods of time (see also Dempsey and Forster, 1982).

Since a hypoxic drive via the peripheral chemoreceptors increases ventilation, more CO_2 will be expired than is produced. The result of this hyperventilation is that the arterial P_{CO_2} sinks (hypocapnia) and the pH rises. This means a reduced respiratory drive via Pa_{CO_2} and H^+ concentration. If this hypoxic stimulus is now eliminated by the breathing of oxygen-rich air, the ventilation is immediately reduced. The CO_2 is thereby accumulated, which tends to offset the effect of interrupting the P_{O_2}-dependent stimulation.

Summary   The central and peripheral inputs are integrated in the respiratory central pattern generators so that P_{O_2}, P_{CO_2}, the acid-base characteristics of the blood, and the internal environment of the body are maintained at a constant level. The inspiratory and expiratory generators in the medulla oblongata are inherently rhythmic and are, in part, reciprocally coordinating the inspiratory and expiratory muscles that vary the rate and depth of breathing. Normal respiration is maintained by an interaction between this generator and neurons in the pons. Inspiration is interrupted by a dampening or cessation of the stimulating effect from chemical or other driving forces on the inspiratory generator and by an off-switch mechanism in the medulla. Stretch receptors in the lungs and muscle spindles in the respiratory muscles may participate in this switch from inspiration to expiration.

The H^+ ion concentration and, especially, P_{CO_2} in the blood continuously affect respiratory activity by sending drive inputs to cells in the ventrolateral surface of the medulla. In other words, changes in carbonic acid concentration in blood and other body fluids ameliorates the acid-base inbalance caused by the excess or deficit of the nonvolatile acids. Deviation from a normal Pa_{CO_2} of 40 mm Hg (5.3 kPa) results in a ventilatory response minimizing the deviation. During hypoxia an increase in ventilation occurs because of discharge from peripheral chemoreceptors in the carotic and, to a smaller extent, aortic bodies. A number of structures in the medulla oblongata may influence respiration. Activity of the reticular formation also increases ventilation. This effect forms a link between the proprioceptors in the muscles and joints, the cortex of the brain, and different centers of importance for the motor function, as well as the respiratory neurons. Emotion, voluntary actions, and reflexes such as swallowing can easily alter the pattern of respiration. We are all aware of how speech controlling mechanisms take over the control of the duration and rate of expiration in order to fit the requirements for phrasing, loudness, and articulation. (See also Doust and Patrick, 1981.)

Exercise

This topic has attracted respiratory physiologists for almost a century, and one has been searching for a "work factor" that could explain exercise hyperpnea. Efforts have been made to quantitate various factors of chemical and nervous nature with regard to their participation in the regulation of breathing during exercise. So far, these efforts have not been too promising. Here we shall present a somewhat different approach compared with the classical concepts, and we shall discuss a neurogenic factor as the primary activator of the respiratory muscles during exercise, with a secondary feedback mechanism of chemical nature which will regulate and adjust the respiratory volume mainly according to the composition of the arterial blood. (See also Whipp and Ward, 1980; Whipp, 1983.)

The pulmonary ventilation increases during muscular exercise almost rectilinearly with the increase in O_2 uptake up to a certain level, after which the increase in ventilation becomes steeper. Over the whole range, pulmonary ventilation is actually more related to the CO_2 volume exhaled than to the O_2 uptake (Figs. 5-9 and 5-13). As we have already stated, there is a great deal of controversy concerning the mechanism underlying this increase in ventilation during exercise. During submaximal physical activity the arterial P_{CO_2}, P_{O_2}, and H^+ concentrations are roughly at the same level as at rest (Fig. 5-13). During very heavy exercise, the anaerobic contribution to the energy yield is inevitably coupled with the production of H^+ ions. Thus pH decreases and may be as low as 7.0 in the arterial blood. The relative hyperventilation that follows elevates the alveolar P_{O_2}, but the arterial P_{O_2}, actually drops toward 90 to 85 mm Hg (12.0 to 11.3 kPa) as compared with the normal value of 95 to 100 mm Hg (12.6 to 13.3 kPa). Pa_{CO_2} drops toward 35 mm Hg (4.6 kPa) or even lower values (Fig. 5-13). A similar drop in Pa_{O_2} and pH at rest should cause a rather moderate increase in ventilation, and a lowering of Pa_{CO_2} would in itself cause a reduced respiratory activity. *The chemical changes in the composition of the arterial blood* *cannot, then, explain per se the ventilation of 100 to 250 liters · min⁻¹ observed during heavy muscular exercise.*

The threshold at which the ventilation increases proportionally more than the O_2 uptake does actually vary from person to person. The individual's potential to supply the exercising muscles with oxygen is of decisive importance. A person who has a low maximal aerobic power reaches this threshold value at a lower O_2 uptake than does a person who has a high maximal $\dot{V}_{O_2}$ (Fig. 5-9). During exercise with small muscle groups (the arms, for instance), the ventilation at a given O_2 uptake is greater than it is when larger muscle groups are engaged, for instance, with exercise involving the leg muscles (Stenberg et al., 1967). (An example: during cycle ergometer exercise with the arms, the ventilation was 89 liters · min⁻¹ at an oxygen uptake of 2.8 liter · min⁻¹, but only 67 liters · min⁻¹ when the work was performed with the legs. The lactate concentration of the blood was 11 and 4 mmol · liter⁻¹, respectively. The heart rate was 176 and 154 respectively.)

A Cerebral Drive?

The possibility of impulses from motor centers in the cerebral cortex or from the active muscles and joints involved in the exercise reaching the respiratory generators has been discussed (Krogh and Lindhard, 1913; Asmussen, 1967; Whipp, 1983). Such impulses may conceivably bring about an alteration in the threshold or "set point" for the sensitivity of the respiratory neuron pools for CO_2, H^+, and O_2 mediated by peripheral chemoreceptors.

It should be emphasized that the cerebral hemispheres normally contribute an important component to the volume of breathing during wakefulness, and this cerebral drive will maintain the rhythm of respiration even when metabolic stimuli are temporarily removed (e.g., after a period of hyperventilation). There is a voluntary and behavioral system connected with the act of breathing located in the somatomotor and limbic forebrain structures, which adapts man to vocalization and conditions him for the expected metabolic demands of exercise (see Plum, 1970). As mentioned, the efferent impulses can descend directly to the motoneurons of the respiratory muscles, but they can also make connections with the medullary reticulum. On this lower level of the brain, there are the neural systems subserving the metabolic needs of the body. As a third system we could consider the effects on the respiratory muscles by the postural demands governed by the cerebellum.

The Role of Muscle Spindles

Campbell (1964), von Euler (1974), and other investigators have emphasized the importance of the muscle spindles for the activity of the respiratory muscles. Campbell points out that the respiratory muscles are voluntary muscles subject to all the spinal and supraspinal mechanisms that affect tone, posture, and movement. In the anterior horn cells, there is an integration of respiratory and nonrespiratory drives. Muscle spindles are numerous in the intercostal muscles but much less numerous and less evenly distributed in the diaphragm. It should be recalled (from Chap. 3) that an increased activity in the fusimotor γ fibers produces a contraction of the muscle spindle's (intrafusal) muscle, and if the parent muscle (the extrafusal muscle) at the same time is not activated via its α fibers, the receptors in the muscle spindles become deformed and produce an afferent impulse which, via the dorsal root, reaches the spinal cord. This afferent impulse stimulates the α motoneuron, the entire muscle is activated, and the difference in length between the receptor muscle and the muscle as a whole is reduced sufficiently so that the afferent signals cease. This γ motor spindle system represents a "follow-up-length-servo" system. Thus, any difference in length change between intra- and extrafusal muscles is perceived by this receptor of the α motoneuron through excitatory synapses so that the misalignment will be reduced and the demanded length, and therefore the demanded volume, will be assumed. The importance of this mechanism for ventilation is shown by the fact that adding a resistance on the inspiratory side increases the force of contraction of the inspiratory muscles, thanks to an augmentation of their motoneuron discharge by the stretch reflex. In other words, the afferent nerves are a feedback design leading from the muscle spindles which provides information as to how successfully the respiratory muscles have accomplished their task.

Combined Effects of Neurogenic and Chemical Stimuli

The system of γ and α motoneurons, linked together via the afferent nerves of the muscle spindles and affected by supraspinal mechanisms, especially the reticular formation, may form the basis for the regulation of the ventilation of the lungs. In connection with exercise, the activity in the reticular formation is increased, partially through impulses from cortex cerebri and other higher centers, and partially through afferent impulses from the muscle spindles in the activated muscles, etc. Since the reticular formation activity is unaffected by the conditions in respiratory generators, it is conceivable that this may increase the activity also in the γ and α motoneurons to the respiratory muscles, with an increased volume change in the thorax as a consequence. It would be natural if the respiratory movements and respiratory frequency were adjusted according to the exercise rhythm, which usually is the case: as mentioned, the respiratory-frequency is adjusted in accordance with stride frequency, pedal frequency in cycling, rowing, etc. A change in posture and rhythm may result in a new pattern of breathing with reference to respiratory frequency and depth. The momentary increased ventilation noticed in connection with the onset of exercise must then be adjusted according to the requirement for the elimination of CO_2 and uptake of O_2. Here, chemical regulation enters into the picture. The α-γ system has a possible tendency to produce a hyperventilation, especially during the beginning of the exercise until the produced CO_2 has reached the lungs. Through a negative feedback elicited from the respiratory centers, if Pa_{CO_2} tends to drop, the γ and α activity driving the respiratory muscles may then be inhibited by the respiratory generators. When the CO_2 reaches the lungs without being eliminated in sufficient quantity, the P_{CO_2} of the arterial blood will rise and the inhibition becomes diminished.

This is to some extent a hypothetical discussion which postulates that in connection with muscular exercise, there is also an increased α-γ activity to the respiratory muscles. The system has, however, every possibility to produce a rhythmic, coordinated switching between inspiration and expiration, partly determined by the exercise rhythm. It should be recalled that the respiratory muscles are involved in most activities not only to provide adequate pulmonary ventilation, but also to stabilize the trunk in posture or movements. A synchronization between the respiratory movements and exercise rhythm is therefore important. The α-γ system provides the possibility for such a coordination. [Actually there are two types of intercostal fusimotor neurons driven from different central nervous areas: rhythmic ones, with a functional link with the respiratory drive, and in addition, tonically firing ones, much more readily influenced by various reflexes from postural influence and affected by the cerebellum and other structures (von Euler, 1966). This combination of systems makes it possible to integrate respiratory and postural movements at the spinal level and to correct "actual" length of the intercostal muscles to "wanted" length in accordance with the demands of breathing and also with a view to postural engagement.]

Figure 5-20 is a simplified diagram presenting components involved in exercise hyperpnea. *The neurogenic factors cannot be considered as a regulating stimulus, but act as an activator. The chemical stimuli are of decisive importance for the finer*

Figure 5-20
Schematic summary of the discussion on regulation of breathing during exercise. The respiratory muscles are activated via their gamma (γ) and alpha (α) motoneurons (filled-line arrows). In a similar way, the other exercising muscles are activated (dotted-line arrows). The central pattern generator is influenced directly or indirectly by the chemical composition of the arterial blood, mainly its P_{CO_2}, P_{O_2}, and pH. These centers can then facilitate or inhibit the motoneurons of the respiratory muscles, depending on the effectiveness of the gas exchange in the lungs. Particularly critical is the CO_2 refill from the muscles and CO_2 output from the lungs. (The afferent nerve impulses to motor centers include impulses from receptors located in tendons, muscle spindles and joints.)

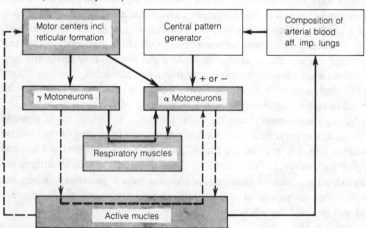

adjustment of the ventilation. If one exercises very hard for 5 s, rests for 5 s, and so forth, the ventilation may eventually rise to 100 liters · min^{-1} but no difference in ventilation is detected between the periods of rest and exercise (Christensen et al., 1960). A variation in afferent impulses from the exercising muscles is in this case of less importance in that the input of CO_2 to the lungs is continuously high. The greater the CO_2 production, the greater the "independence" of the respiratory muscles.

There are many intriguing experimental findings to consider when analyzing the neurogenic respiratory drive during exercise (Asmussen, 1967). In negative exercise, the force developed by the contracting muscles may be 5 to 7 times greater than in positive exercise, but the pulmonary ventilation per liter oxygen uptake, is similar in the two types of activities. Prolonged static effort with maintained high muscular force does not represent an especially strong respiratory drive. The very moderate ventilatory response to the first seconds of maximal effort, e.g., a sprint, or when running up a flight of stairs at maximal speed should be mentioned.

However, the afferent impulses from exercising limbs and the stimulation from various parts of the brain due to the increased motor activity must be considered as coordinated activators of the respiratory muscles, but their motoneurons are subjected to various degrees of inhibition from the central pattern generator, depending on the chemical composition of the blood.

It is noteworthy that any change in respiratory frequency that adapts to the rhythm of movements is, within limits, automatically followed by a change in tidal air to provide an adequate alveolar ventilation. (We have a similar situation in the regulation of cardiac output: at rest and during not too severe exercise, a change in heart rate is matched by a variation in stroke volume so that the cardiac output is maintained constant.)

Hypoxic Drive during Exercise

If during exercise the normal air is replaced by oxygen, the ventilation drops within seconds, and this effect is particularly marked during heavy exercise (Asmussen and Nielsen, 1946; Dejours, 1964). The reaction is so rapid that an effect via the chemoreceptors in the carotid and aortic bodies has to be assumed. This ventilation-reducing effect of O_2 breathing is marked, however, even if the arterial P_{O_2} is at a normal level, and this fact has confused respiratory physiologists.

It has already been mentioned that the chemoreceptors have a relatively large blood flow and high oxygen uptake. They are innervated by sympathetic nerve fibers which supply the arterioles of the carotid and aortic bodies with vasoconstrictor fibers. When these fibers are activated, there is a reduction in the blood flow through the chemoreceptor areas, possibly by the blood being diverted through adjacent arteriovenous anastomoses. It is possible that a change in the P_{CO_2} and H^+ concentration and hypoxia of the arterial blood may contribute to modify the blood flow to the epithelioid cells of the chemosensitive areas. Actually, the O_2 tension inside specific tissue zones in the carotid body is much lower than that predicted from the a-v O_2 difference in the carotid body (Acker and Lübbers, 1980).

During exercise, the sympaticus activity is increased and the concentration of epinephrine and norepinephrine in the blood increases. In fact, Biscoe and Purves (1965) report immediate increase in nerve activity of the cervical sympathetic and the postgaglionic nerve to the carotid body by exercise (in the cat), and these changes were abolished by femoral and sciatic nerve section. It is conceivable that the blood flow to the chemoreceptor cells is reduced to such an extent that the O_2 supply to the metabolically very active cells becomes unsatisfactory or that the removal of products of tissue metabolism becomes insufficient. This might then produce an excitation of the chemoreceptors and a stimulation of respiration despite a normal, or only slightly different from normal, systemic arterial P_{O_2} (Hornbein and Roos, 1962: Lee et al., 1964; Whalen and Nair, 1975).

The heavier the exercise in relation to the performance capacity of the individual, the greater the sympaticus activity. Thereby, the chemoreceptor drive may also increase, despite the elevated perfusion pressure. This mechanism may also contribute to the increase in ventilation during pronounced emotion as well as during exercise involving small muscle groups, when the ventilation is relatively high at a given oxygen uptake. It should be recalled that during hypoxic conditions, the effect of CO_2 and H^+ concentration is a marked reinforcement of the discharge from the peripheral chemoreceptors.

In the following, we shall present a short resumé of *the time course of the increase in pulmonary ventilation from the start of exercise,* and offer some comments on possible mechanisms in the regulation of the ventilation. As mentioned, there is an abrupt increase in ventilation taking place at the first respiratory cycle following the transition from rest to exercise, and the increase is relatively independent of the rate of exercise. It lasts for some 15 to 20. This is phase 1, and everything speaks in favor of a neural irradiation from higher levels of the brain (see Whipp, 1983). Eldridge et al. (1981), on the basis of experiments performed on walking, unanesthetized decorticated cats, suggest that there are hypothalamic command signals primarily responsible for the driving of both locomotion and respiration during exercise. DiMarco et al. (1983), using high-level decerebrated cats, and Favier et al. (1983), using dogs as experimental animals, also support the existance of a neurally mediated factor during the early stages of exercise hyperpnea. According to one hypothesis, the sudden increase in blood return to the thorax region could in one way or another, e.g., via chemoreceptors on the low pressure side, stimulate to increased pulmonary ventilation. This hypothesis has no experimental support (see Fordyce et al., 1982).

The subsequent exponential rise in pulmonary ventilation, is related to the severity of the exercise. This is phase 2 of the increase in ventilation. Its delay in onset can be explained if peripheral chemoreceptors act as important stimulators. The vascular transit time for metabolites to be transported from the exercising muscles to reach these receptors supports this theory. As mentioned, a local hypoxia may develop in critical areas of the chemoreceptors. During heavy exercise, the anaerobic metabolism will produce protons which will, via the blood, reach the carotid body and strengthen the hypoxic stimulus. Support for the importance of the carotid body during phase 2 is provided by the finding that patients with resected carotid bodies have substantially slow ventilatory kinetics as compared with controls (see Honda et al., 1979; Whipp,

1983). Secondly, induced hypoxic subjects have a faster response in phase 2, while subjects breathing 100 percent O_2, which functionally inactivates the carotid bodies, have a slower response (Whipp, 1983).

Phase 3, is the steady-state condition. During exercise with an intensity that will not continuously lower the blood pH (i.e. below the so-called anaerobic threshold), the CO_2 tension in the arterial blood seems to be more closely regulated than its oxygen tension. But, even in this steady state condition, there is often a gradual increase in pulmonary ventilation during prolonged exercise with no change in arterial pH, P_{CO_2}, lactate, or $\dot{V}_{CO_2}$ (Martin et al., 1981). They calculated that an increase in the dead space ventilation explained the demand for an elevation of the ventilation.

It should be recalled that stretch receptors in the lungs are supposed to activate the off-switch mechanism with a potential to shorten the inspiratory duration and therefore to increase the respiratory frequency. With an increase in tidal volume this mechanism may be more active.

During heavy exercise, there is a steeper increase in pulmonary ventilation, as illustrated in Fig. 5-9. The ventilatory compensation for the metabolic acidosis becomes progressively inadequate. Both central and peripheral chemoreceptors are probably driving the central pattern generator. There may also be additional effects of circulating catecholamines, blood temperature, blood osmotic pressure, etc. Simon et al. (1983) report a study which suggests that exercise hyperventilation is not necessarily proportional to the increase in plasma lactate. There are also interesting reports that patients who cannot produce lactic acid (McArdle's disease) exhibit a normal hyperventilatory response during a progressive exercise test (see Hagberg et al., 1982).

Summary During exercise, the ventilation increases relatively rectilinearly, but this increase is relatively steeper during very heavy exercise. How this increase in ventilation is elicited is unknown. The comparatively small changes which are observed in the P_{O_2}, P_{CO_2}, and H^+ concentration in the arterial blood cannot explain the increase in ventilation. It is suggested as a hypothesis that muscular activity as such, through afferent impulses from the central nervous system, increases the activity in the γ and α motoneurons of the respiratory muscles and through spinal and supraspinal reflex centers, and thus in a closely coordinated manner produces an increase in the frequency and depth of respiration, often in pace with the muscular movements. The actual regulation of the respiratory volume then takes place through a negative feedback mechanism, primarily determined by the CO_2 production in relation to CO_2 elimination during expiration. In this manner, the P_{CO_2} of the arterial blood will, via a respiratory pattern generator, determine the magnitude of the ventilation. During anaerobic exercise, the H^+ concentration of the blood will increase, which represents a further stimulation of respiration. The peripheral chemoreceptors in the carotid and aortic bodies will also stimulate respiration, possibly because an increased sympaticus activity will reduce the blood flow to the chemoreceptor areas so that the local P_{O_2} drops in spite of an almost normal value for P_{O_2} in the arterial blood. During maximal exercise, there is a reduction in Pa_{O_2} which should further increase ventilation. The hyperventilation which follows may even produce a drop in Pa_{CO_2}, but the arterial pH inevitably drops (Fig. 5-13).

BREATHLESSNESS (DYSPNEA)

Dyspnea is difficult, labored, uncomfortable breathing (Comroe, 1966a). The reason why respiration may become consciously troublesome is not clear. The magnitude of the ventilation is not the determining factor: a patient may experience a pulmonary ventilation of 10 liters $\cdot$ min^{-1} as extremely disturbing, whereas the athlete is not consciously troubled by a ventilation as high as 200 liters $\cdot$ min^{-1}. In certain cases, afferent vagus impulses may be the cause of the dyspnea (for example, through the collapse of some alveoli), but, more commonly, impulses from muscle spindles, tendon organs, and thoracic joint receptors appear to give rise to conscious awareness and distress. It may be a question of length/tension, that is, tension appropriateness. Altered afferent signals of the muscle spindles and receptors in the chest wall reaching subcortical and cortical levels may cause unusual sensations. From experience we learn what to expect and how it should be felt or experienced in a certain situation, and the respiration itself, which normally proceeds unconsciously, may give rise to distress when it requires a conscious modification. Light activity at high altitude requires a ventilation which at sea level is associated with heavy exercise, and this difference may be experienced as a distress. A person unaccustomed to exercise may experience a ventilation of 75 liters $\cdot$ min^{-1} as unpleasant, but after suitable training this may even appear as a pleasant sensation! (See DiMarco et al., 1982; Killian and Campbell, 1983.)

SECOND WIND

During the first minutes of exercise, the load may appear very strenuous. One may experience dyspnea, but this distress eventually subsides; one experiences a "second wind." The factors eliciting the distress may be an accumulation of metabolites in the activated muscles and in the blood because the O_2 transport is inadequate to satisfy the requirement.

By what mechanism this changed environment is brought to consciousness is not known. During heavy exercise, there is actually at the commencement of the activity a hypoventilation due to the fact that there is a time lag in the chemical regulation of the respiration. It is then actually a matter of a length/tension inappropriateness in the intercostal muscles. When the second wind occurs, the respiration is increased and adjusted according to requirement.

It appears that the respiratory muscles are forced to work anaerobically during the initial phases of the exercise if there is a time lag in the redistribution of blood. A stitch in the side may then develop. This is probably the result of hypoxia in the diaphragm. It is most common in untrained persons and is particularly apt to occur if heavy exercise is performed shortly after a large meal, when the circulatory adjustment at the commencement of exercise is slower. As the blood supply to the respiratory muscles is improved, the pain disappears. This theory is not entirely satisfactory. It is more common when running than in cycling and swimming. An alternative trigger of this stitch could be a stimulation of pain receptors in the abdominal region of a mechanical origin. A bouncing effect on the abdominal organs is certainly evident in jogging/running. However, it is puzzling that this type of problem does not follow a strict and reproducible course. It was previously believed that the pain was caused by an emptying of the blood depots in the spleen and the contractions taking place in the spleen. In humans, the spleen serves no such depot function, however. Furthermore, persons who have had their spleen removed may still experience such pain.

Well-trained athletes who have warmed up adequately prior to a muscular effort seldom experience such pain.

HIGH AIR PRESSURES, BREATH HOLDING, DIVING

High Air Pressures

Although human beings can become acclimatized to low air pressures, there is no way to become acclimatized to high air pressures such as are encountered in deep sea diving and during escape from a submarine when the survivor attempts to get from the inside of the craft where the pressure is normal to the surface through the sea where the air pressure is higher. For every 10 m (33 ft) of sea water the diver descends, an additional pressure of 1 atm is acting upon the body. As the pressure increases, more gases can be taken up by the body and dissolved in the various tissues. At a depth of about 10 m, twice as much gas will be dissolved in the diver's blood and tissues as at sea surface. This is apt to give the diver trouble, mainly because of the nitrogen. The trouble with nitrogen is that it diffuses into various tissues of the body very slowly, and once dissolved, it also leaves the body very slowly when the pressure once more is reduced to the normal atmospheric pressure. This is especially bad when the pressure is suddenly reduced from several atmospheres, as may be the case during submarine escape or deep sea diving. Then the nitrogen is released from the tissues in the form of insoluble gas bubbles. These bubbles congregate in the small blood vessels where they obstruct the flow of blood. This then gives rise to symptoms such as pains in the muscles and joints, and even paralysis may develop if the bubbles become trapped in the central nervous system. These symptoms are known as "the bends." Obviously, the severity of the symptoms depends on the magnitude of the pressure, which means the depth to which the person has descended under water, the length of time spent at that depth, and the speed of ascent to the surface.

The bends can be avoided to a large extent by a slow return to normal pressure so as to allow time for the tissues to get rid of their excess nitrogen without the formation of bubbles. Another way to avoid the bends is to prevent the formation of these bubbles by replacing atmospheric nitrogen with helium, which is less easily dissolved in the body. This is done by having the diver breathe a helium-oxygen gas mixture. Another advantage with this is that it is more apt to prevent the so-called nitrogen narcosis, which occurs when air is breathed at 3 atm or more and which results in an onset of euphoria and impaired mental activity with lack of ability to concentrate. With increasing pressures, the individual is progressively handicapped and may be rendered helpless at 10 atm. Pilots of high-flying aircraft may also suffer from bends if there is a sudden loss of pressure in the pressurized cabin, but the symptoms in these cases are usually not so severe as in the divers. In any case, they usually do not occur at altitudes lower than 10,000 m (30,000 ft).

Prolonged breathing of 100 percent oxygen may be quite harmful, for irritation of the respiratory tract may occur after 12 hr, and frank bronchopneumonia after 24 hr. Matalon and Egan (1981) have produced data on the solute permeability of the alveolar epithelium of rabbits, indicating that exposure to 100% O_2 causes an increase in the permeability of the blood-gas barrier, allowing molecules that are normally restricted in the alveolar space to appear in the blood. This increase occurs before the onset of arterial hypoxemia, or other cardiovascular changes. They conclude that increasing the length of O_2 exposure increases the solute permeability of the alveolar epithelium, which precedes the appearance of pulmonary edema. In most individuals, no harmful effects result from breathing mixtures with less than 60 percent oxygen, while newborn infants are particularly susceptible to oxygen poisoning and may suffer harmful effects with oxygen concentrations over 40 percent. The remarkable thing is that oxygen poisoning apparently is no problem when breathing 100 percent oxygen at altitudes over 5,500 m, irregardless of how long. Oxygen poisoning, therefore, is not much of a problem in aviation medicine, but it is indeed an important problem in deep sea diving where it may even affect the brain function when pure oxygen is used at depths greater than about 10 m (33 ft), but there are great individual variations in sensitivity to 100 percent oxygen. The onset of symptoms may be hastened by vigorous physical activity at great depths; it starts with muscular twitchings and a jerking type of breathing, and it ends in unconsciousness and convulsions. The exact cause is unknown, but it is assumed that it is a matter of interference with certain enzyme systems in the tissues.

Furthermore, when breathing pure oxygen at a pressure of 3 atm or more, the oxygen dissolved in the blood covers the oxygen need of the body at rest. Therefore, at rest, the hemoglobin of the venous blood is still saturated with oxygen. This interferes with the CO_2 transportation (Chap. 4). The result will be gradual CO_2 retention in the tissue and decrease in pH (For further details, see Lambertsen, 1971; West, 1979; Elliott, 1981; Halsey, 1982.)

Breath Holding-Diving

Well-nourished individuals may survive without difficulty for weeks without food and for days without water. But they can live only a few minutes without oxygen. The body's ability to store oxygen is extremely limited. The blood may contain up to about 1.0 liter. In the lungs after a normal inspiration, there may be about 0.5 liter oxygen; after a maximal inspiration, it may be about 1.0 liter. (A total capacity of 7.0 liters and a concentration of O_2 of 15 percent $= 15/100 \cdot 7.0 = 1.05$ liters.)

The O_2 bound to myoglobin may amount to as much as 0.5 liter, but it can be considered only as a local store and cannot be utilized by other tissues. The hyperbolic slope of the O_2-myoglobin dissociation curve binds the oxygen effectively until its partial pressure drops to extremely low values.

When one holds one's breath at rest, a total of about 600 ml O_2 is available and can be utilized. This is enough to last about 2 min. "Normal" maximal breath-holding time is 30 to 60 s. The arterial P_{O_2} then drops to about 75 to 50 mm Hg (10 to 6.6 kPa) and the P_{CO_2} rises to 45 to 50 mm Hg (6.0 to 6.6 kPa). This elevated CO_2

pressure plays a greater role in forcing the individual to discontinue the breath holding than does the reduced O_2 pressure. The hypoxic drive is also an important factor, as demonstrated by the fact that breath holding may be extended if it is preceded by O_2 breathing. Under these conditions, a further increase of Pa_{CO_2} of 5 to 10 mm Hg (0.65 to 1.30 kPa) may be tolerated. Other factors also affect the capacity for breath holding: (1) if one holds one's breath with a lung volume near the total lung capacity (maximally inflated), the respiratory standstill may be extended until the P_{CO_2} has reached a value about 10 mm Hg (1.3 kPa) higher than with a lung volume near the residual level (this is even the case with O_2 breathing; for this reason, a different hypoxic drive may be excluded). The reason is probably that afferent impulses from the receptors in the lungs and thoracic cage produce a greater stimulus to inspiration in the expiratory position. (2) Breath holding may be prolonged by swallowing; swallowing constitutes a momentary inhibition in inspiration. (3) Single or repeated breaths after the breaking point, without change in alveolar air composition, make a new breath holding possible, and higher Pa_{CO_2} and lower Pa_{O_2} are obtained compared with the first trial. (4) Total paralyses of the muscles by curate prolonged the breath-holding time, and it totally abolished any sensation of breathlessness whatever in the subjects despite a highly elevated arterial CO_2 pressures (Godfrey and Campbell, 1970). Apparently a disproportion between the force developed in the respiratory muscles and a motor effect is in some way transmitted to sensation by afferent impulses from the muscles and the chest wall. (See also Mithoefer, 1965; Whitelaw et al., 1981.)

If a subject voluntarily hyperventilates forcefully for about a minute, the pulmonary air is exchanged more often and the composition of its gases more closely approaches that of the inspired air. In this manner, the alveolar P_{O_2} may increase to about 135 mm Hg (18 kPa), the P_{CO_2} may drop below 20 mm Hg (2.7 kPa), and the arterial blood will assume similar gas pressure. Because of the shape of the O_2 saturation curve (Fig. 4-20), the O_2 content of the arterial blood is hardly affected, however, and the pulmonary air will receive an addition of only 40 ml · liter^{-1} pulmonary air (19 vol percent O_2 instead of 15).

A more important effect of hyperventilation in this connection is the reduction in the CO_2 content of the body. Thus the breath holding may be extended to several minutes by providing more room for CO_2. The breaking point occurs probably at a P_{CO_2} of about 40 mm Hg and a P_{O_2} of 45 mm Hg (5.3 and 6.0 kPa respectively). This is still within the safe range as far as the critical O_2 pressure for the function of the nervous system is concerned, which in the case of the O_2 pressure of the arterial blood is in the order of 25 to 30 mm Hg (3.3 to 4 kPa).

If a maximal breath holding is performed during muscular exercise (P.-O. Åstrand, 1960), the breath-holding time is shorter than at rest, but owing to the fact that O_2 uptake and CO_2 production occur at a higher rate, the composition of the arterial blood and the pulmonary air will change more rapidly. At the breaking point, the alveolar P_{CO_2} may be as high as 75 mm Hg (10 kPa) and the P_{O_2} may drop toward 40 mm Hg (5.3 kPa) (the arterial gas pressures are probably at the same level). If a person breathes pure oxygen prior to the breath holding, the $P_{A_{CO_2}}$ may exceed 90 mm Hg (12 kPa).

If a forceful hyperventilation precedes a breath holding during heavy exercise, breath holding may be prolonged until convulsions or even fainting occurs (P.-O. Åstrand, 1960). The alveolar P_{O_2} may then drop toward 20 mm Hg (2.7 kPa), which explains the symptoms. (In experiments of this type in the laboratory, the subjects were secured in a harness in order to prevent their fall off the cycle ergometer!)

The various possible explanations will not be discussed here. However, on the basis of these experiments, the risk involved in prolonged diving and deep diving without equipment will be emphasized (Lanphier and Rahn, 1963).

If such a dive is performed following a marked hyperventilation, one may apparently hold the breath until unconsciousness occurs. It should be emphasized that the total pressure of the alveolar air at sea level is about 760 mm Hg (101 kPa) but at a depth of 10 m it is doubled (to twice the atmospheric pressure) by the pressure of the water on the thorax. An O_2 percentage which at sea level corresponds to an alveolar O_2 tension of about 25 mm Hg or 3.3 kPa (3.5 vol percent) produces at a depth of 2 m (total pressure = 912 mm Hg; 121 kPa) a pressure of 30 mm Hg or 4 kPa $[3.5 \cdot 100^{-1} \cdot (912 - 47) = 30]$. As long as the diver remains at a depth of 2 m, the O_2 tension may thus be adequate to meet the requirement of the nerve cells. But when the diver approaches the surface, the oxygen tension may drop below the critical level. The danger is greater if the ascent is slow, in that the O_2 utilization then causes the O_2 pressure to drop further. Several cases have been described when divers who were attempting to beat a record, or who for various other reasons have taken their time during prolonged diving, have lost their lives or have been rescued in the nick of time (Craig, 1961; Davis, 1961). Often, hyperventilation had proceded the diving: the diver then exhibited a "strange" behavior or simply ceased to swim and sank. Cases have been described when the swimmer had had no major difficulty holding his breath but then suddenly had lost consciousness.

It is thus justifiable to warn against extremely deep diving and prolonged diving without effective supervision. The diver should avoid hyperventilation prior to the dive (taking, at the most, five deep respirations). During diving without equipment, oxygen breathing prior to the dive may definitely improve the performance. It represents an improvement if the diver's lungs contain 5 liters O_2 instead of barely 1 liter during the dive. In other types of work, the effect is negligible, however, unless the oxygen is inspired during the actual performance. In a few respirations, the extra volume of oxygen is washed out, since no extra stores of oxygen can be deposited in the body. In this connection, another effect of hyperventilation should be stressed. The washing out of CO_2 in connection with the hyperventilation increases the pH. The effect of this alkalosis and reduced Pa_{CO_2} is a vasoconstriction, including the vessels in the brain. Dizziness and cramps may be the result. If one holds one's breath after a hyperventilation against a closed glottis and at the same time contracts the abdominal muscles (Valsalva maneuver), the cardiac output is reduced. This in combination with the vasoconstriction in the cerebral blood vessels may produce an oxygen deprivation in the CNS sufficient to cause a transitory loss of consciousness.

Summary   During breath holding while exercising, a greater CO_2 content in the blood is tolerated. Since oxygen is rapidly utilized during work, there is a risk that the breath holding will end in unconsciousness. This is particularly true if the breath holding is preceded by marked hyperventilation. The possibility of unconsciousness represents a risk during prolonged and deep diving, especially if the ascent from considerable depths takes place slowly. The intrathoracic pressure increases with the depth of the water. A given oxygen pressure in the alveoli and in the blood may therefore be adequate for the normal function of the central nervous system at a depth of a few meters, but it may become critically low as the diver approaches the surface.

The respiratory adaptation to prolonged hypoxia, i.e., acclimatization to high altitude, will be discussed in Chap. 15.

REFERENCES

Acker, H., D. W. Lübbers: "The Physiology of Chemoreceptors in the Carotid Body." XXIII International Congress of Physiological Sciences, vol. XIV, p. 7, Budapest, 1980.

Adamson, J. W., and C. A. Finch: Hemoglobin Function, Oxygen Affinity, and Erythropoietin, *Ann. Rev. Physiol.*, **37**:351, 1975.

Amis, T. C., H. A. Jones, and J. M. B. Hughes: Effect of Posture on Inter-regional Distribution of Pulmonary Perfusion and $\dot{V}_A/\dot{Q}$ Ratios in Man. *Resp. Physiol.*,**56**:169, 1984.

Angus, G. E., and W. M. Thurlbeck: Number of Alveoli in the Human Lung, *J. Appl. Physiol.*, **32**:483, 1972.

Asmussen, E.: Exercise and Regulation of Ventilation, *Circulation Res.* **20**:1–132, 1967.

Asmussen, E., and M. Nielsen: Studies on the Regulation of Respiration in Heavy Work, *Acta Physiol. Scand.*, **12**:171, 1946.

Asmussen, E., and M. Nielsen: Physiological Dead Space and Alveolar Gas Pressures at Rest and during Muscular Exercise, *Acta Physiol. Scand.*, **38**:1, 1956.

Åstrand, I.: Aerobic Work Capacity in Men and Women with Special Reference to Age, *Acta Physiol. Scand.*, **49**:(Suppl. 169), 1960.

Åstrand, I., P.-O.: Åstrand I. Hallbäck, and Å. Kilbom: Reduction in Maximal Oxygen Uptake with Age, *J. Appl. Physiol.*, **35**:649, 1973.

Åstrand P.-O.: "Experimental Studies of Physical Working Capacity in Relation to Sex and Age," Munksgaard, Copenhagen, 1952.

Åstrand P.-O.: A Study of Chemoceptor Activity in Animals Exposed to Prolonged Hypoxia, *Acta Physiol. Scand.*, **30**:335, 1954.

Åstrand P.-O.: Breath Holding during and after Muscular Exercise, *J. Appl. Physiol.*, **15**:220, 1960.

Åstrand P.-O., and B. Saltin: Oxygen Uptake during the First Minutes of Heavy Muscular Exercise, *J. Appl. Physiol.*, **16**:971, 1961.

Bargeton, D.: Analysis of Capnigram and Oxygram in Man, *Bull. de Physio-Pathologie Respiratoire*, **3**:503, 1967.

Bartlett, D., Jr.: Effect of Hypercapnia and Hypoxia on Laryngeal Resistance to Airflow, *Respir. Physiol.*, **37**:293, 1979.

Bencowitz, H. Z., P. D. Wagner, and J. B. West: Effect of Change in P_{50} on Exercise Tolerance at High Altitude: A Theoretical Study, *J. Appl. Physiol.: Resp. Environ. Exer. Physiol.*, **53**:1487, 1982.

Biscoe, T. J.: Carotid Body: Structure and Function, *Physiol. Rev.*, **51**:437, 1971.

Biscoe, T. J., and M. J. Purves: Carotid Chemoreceptor and Cervical Sympathetic Activity during Passive Third Limb Exercise in the Anaesthetized Cat, *J. Physiol. (London)*, **178**:43P, 1965.

Bramble, D. M., and D. R. Carrier: Running and Breathing in Mammals, *Science.*, **219:**251, 1983.

Braus, H.: "Anatomie des Menschen," Bd. 2, Aufl. 3, p. 134, Springer-Verlag OHG, Berlin, 1956.

Brice, A. G., and H. G. Welch: Metabolic and Cardiorespiratory Responses to He-O_2 Breathing during Exercise, *J. Appl. Physiol.: Resp.: Environ. Exer. Physiol.*, **54:**387, 1983.

Bye, P. T. P., G. A. Farkas, and Ch. Roussos: Respiratory Factors Limiting Exercise, *Ann. Rev. Physiol.*, **45:**465, 1983.

Bye, P. T. P., S. A. Esau, K. R. Walley, P. T. Macklem, and R. L. Pardy: Ventilatory Muscles during Exercise in Air and Oxygen in Normal Men, *J. Appl. Physiol.: Resp. Environ. Exer. Physiol.*, **56:**464, 1984.

Campbell, E. J. M.: Motor Pathways, in W. O. Fenn and H. Rahn (eds.), "Handbook of Physiology," sec. 3, Respiration, vol. I, p. 535, American Physiological Society, Washington, D.C., 1964.

Cassidy, S. S., M. Ramanathan, G. L. Rose, and R. L. Johnson, Jr.: Hysteresis in the Relation between Diffusing Capacity of the Lung Volume, *J. Appl. Physiol.: Resp. Environ. Exer. Physiol.*, **49:**566, 1980.

Chance, B., B. Schoener, and F. Schindler: The Intracellular Oxidation-Reduction State, in F. Dickens and E. Neil (eds.), "Oxygen in the Animal Organism," p. 367, Pergamon Press, New York, 1964.

Cherniak, R. M., L. Cherniak, A. Haimark, and V. Cherniak; "Respiration in Health and Disease," W. B. Saunders Company, Philadelphia, 1972.

Christensen, E. H., R. Hedman, and B. Saltin: Intermittent and Continuous Running, *Acta Physiol. Scand.*, **50:**269, 1960.

Clark, J. M., F. C. Hagerman and R. Gelfand: Breathing Patterns during Submaximal and Maximal Exercise in Elite Oarsman, *J. Appl. Physiol.: Resp. Environ. Exer. Physiol.*, **55:**440, 1983.

Cole, R., P. C. Sukanek, J. B. Wittenberg, and B. A. Wittenberg: Mitochondrial Function in the Presence of Myoglobin, *J. Appl. Physiol.: Resp. Environ. Exer. Physiol.*, **53:**1116, 1982.

Comroe, J. H., Jr.: Some Theories of the Mechanism of Dyspnoea, in J. B. L. Howell and E. J. M. Campbell (eds.), "Breathlessness," p. 1, Blackwell Scientific Publications, Ltd., Oxford, 1966a.

Comroe, J. H., Jr.: The Lung, *Sci. Am.*, **214:**56, 1966b.

Craig, A. B., Jr.: Causes of Loss of Consciousness during Underwater Swimming, *J. Appl. Physiol.*, **16:**583, 1961.

Davis, J. H.: Fatal Underwater Breath Holding in Trained Swimmers, *J. Forensic Sci.*, **6:**301, 1961.

Dejours, P.: Control of Respiration in Muscular Exercise, in W. O. Fenn and H. Rahn (eds.), "Handbook of Physiology," sec. 3, Respiration, vol. I, p. 631, American Physiological Society, Washington, D.C., 1964.

Dejours, P.: "Principles of Comparative Respiratory Physiology," 2nd ed., Elsevier, North Holland, Inc., New York, 1981.

Dempsey, J. A., and H. V. Forster: Mediation of Ventilatory Adaptations, *Physiol. Rev.*, **62:**262, 1982.

Dempsey, J. A., P. Hanson, D. Pegelow, and A. Claremont: Limitations to Exercise Capacity and Endurance: Pulmonary System, *Can. J. Appl. Sports Sci.*, **7:**4, 1982.

DiMarco, A. F., D. A. Wolfson, S. B. Gottfried, and M. D. Altose: Sensation of Inspired Volume in Normal Subjects and Quadriplegic Patients, *J. Appl. Physiol.: Resp. Environ. Exer. Physiol.*, **53:**1481, 1982.

DiMarco, A. F., J. R. Romaniuk, C. von Euler, and Y. Yamamoto: Immediate Changes in Ventilation and Respiratory Pattern Associated with Onset Cessation of Locomotion in the Cat, *J. Physiol.*, **343**:1, 1983.

Doust, J. H., and J. M. Patrick: The Limitation of Exercise Ventilation during Speech, *Respir. Physiol.*, **46**:137, 1981.

Effros, R. M., G. Mason, and P. Silverman: Assymetric Distribution of Anhydrase in the Alveolar-Capillary Barrier, *J. Appl. Physiol.: Resp. Environ. Exer. Physiol.*, **51**:190, 1981.

Eldridge, L. F., D. E. Millhorn, and T. G. Waldrop: Exercise Hyperpnea and Locomotion: Parallel Activation from the Hypothalamus, *Science*, **211**(4484):844, 1981.

Elliot, D. H.: Underwater Physiology, in, O. G. Edholm and J. S. Weiner (eds.): "The Principles and Practice of Human Physiology," Chap. 6, Academic Press, London, 1981.

Euler, C. von: Proprioceptive Control on Respiration, in R. Granit (ed.), "Muscular Afferents and Motor Control," p. 197, Nobel Symposium I, John Wiley & Sons Inc., New York, 1966.

Euler, C. von: On the Role of Proprioceptors in Perception and Execution of Motor Acts with Special Reference to Breathing," in L. D. Pengelly, A. S. Rebuck, and E. J. M. Campbell (eds.):, "Loaded Breathing," p. 139, Longman Canada Limited, Ontario, 1974.

Euler, C. von: On the Central Pattern Generator for the Basic Breathing Rhythmicity, *J. Appl. Physiol.: Resp. Environ. Exer. Physiol.*, **55**:1647, 1983.

Euler, C. von., and H. Lagercrantz (eds.): "Central Nervous Control Mechanisms in Breathing," Pergamon Press, Oxford, 1979.

Eyzaguirre, C., and S. J. Fidone: Transduction Mechanisms in Carotid Body: Glomus Cells, Putative Neurotransmitters, and Nerve Endings, *Am. J. Physiol.*, **239**:C135, 1980.

Favier, R., D. Desplanches, J. Frutoso, M. Grandmontagne, and R. Flandrois: Ventilatory Transients during Exercise: Peripheral or Central Control? *Pflügers Arch.*, **396**:269, 1983.

Fenn, W. O., and H. Rahn (eds.): "Handbook of Physiology," sec. 3, Respiration, vols. I and II, American Physiological Society, Washington D.C., 1964–1965.

Ferrus, L., D. Commenges, J. Gire, and P. Varène: Respiratory Water Loss as a Function of Ventilation and Environmental Factors, *Resp. Physiol.*, **56**:11, 1984.

Folinsbee, L. J., E. S. Wallace, J. A. Bedi, J. A. Gliner and S. M. Horvath: Exercise Respiratory Pattern in Elite Athletes and Sedentary Subjects, *Am. Rev. Respir. Dis.*, **125**:240, 1982.

Fordyce, W. E., F. M. Bennett, S. K. Edelman, and F. S. Grodins: Evidence in Man for a Fast Neural Mechanism during the Early Phase of Exercise Hyperpnea, *Respir. Physiol.*, **48**:27, 1982.

Forster, R. E., and E. D. Crandell: Pulmonary Gas Exchange, *Ann. Rev. Physiol.*, **38**:69, 1976.

Godfrey, S., and E. J. M. Campbell: The Role of Afferent Impulses from the Lung and Chest Wall in Respiratory Control and Sensation, in R. Porter (ed.), "Breathing: Hering-Breuer Centenary Symposium," p. 219, Churchill, London, 1970.

Hagberg, J. M., E. F. Coyle, J. E. Caroll, J. M. Miller, W. H. Martin, and M. H. Brooke: Exercise Hyperventilation in Patients with McArdle's Disease, *J. Appl. Physiol.: Resp. Environ. Exer. Physiol.*, **52**:991, 1982.

Halsey, M. J.: Effects of High Pressure on the Central Nervous System, *Physiol. Rev.*, **62**(4):1341, 1982.

Heeringa, J., A. Berkenbosch, J. de Goede, and C. N. Olievier: Relative Contribution of Central and Peripheral Chemoreceptors to the Ventilatory Response to CO_2 during Hyperoxia, *Respir. Physiol.*, **37**:363, 1979.

Heymans, C., and E. Neil: "Reflexogenic Areas of the Cardiovascular System," Churchill, London, 1958.

Holmér, I.: Physiology of Swimming Man, *Acta Physiol. Scand.*, (Suppl. 407) 1974.

Honda, Y., S. Myojo, S. Hasegawa, T. Hasegawa, and J. Severinghaus: Decreased Exercise Hyperpnea in Patients with Bilateral Carotid Chemoreceptor Resection, *J. Appl. Physiol.: Resp. Environ. Exer. Physiol.,* **46:**908, 1979.

Hornbein, T. F., and A. Roos: Effect of Mild Hypoxia on Ventilation during Exercise, *J. Appl. Physiol.,* **17:**239, 1962.

Hornbein, T. F., and A. Roos: Specificity of H Ion Concentration as a Carotid Chemoreceptor Stimulus, *J. Appl. Physiol.,* **18:**580, 1963.

Jacques, J. A. (ed.): "Respiratory Physiology", McGraw Hill, Maidenhead, 1979.

Jasinskas, C. L., B. A. Wilson, and J. Hoare: Entrainment of Breathing Rate to Movement Frequency during Work at Two Intensities, *Respir. Physiol.,* **42:**199, 1980.

Killian, K. J., and E. J. M. Campbell: Dyspnea and Exercise, *Ann. Rev. Physiol.,* **45:**465, 1983.

deKonig, J., L. J. Hoofd, and F. Kreuzer: Oxygen Transport and the Function of Myoglobin, *Pflügers Arch.,* **393:**211, 1981.

Krahl, V. E.: Anatomy of the Mammalian Lung, in W. O. Fenn and H. Rahn (eds.), "Handbook of Physiology," sec. 3, Respiration, vol. I, p. 213, American Physiological Society, Washington, D.C., 1964.

Krogh,A.: "The Comparative Physiology of Respiratory Mechanisms," University of Pennsylvania Press, Philadelphia, 1941.

Krogh, A., and J. Lindhard: Regulation of Respiration and Circulation during the Initial Stages of Muscular Work, *J. Physiol. (London),* **47:**112, 1913.

Lahiri, S., A. Mokashi, E. Mulligan, and T. Nishino: Comparison of Aortic and Carotid Chemoreceptor Responses to Hypercapnia and Hypoxia, *J. Appl. Physiol.: Resp. Environ. Exer. Physiol.,* **51:**55, 1981a.

Lahiri, S., E. Mulligan, T. Nishino, A. Mokashi, and R. O. Davies: Relative Responses of Aortic Body and Carotid Body Chemoreceptors to Carboxyhemoglobinia, *J. Appl. Physiol.: Resp. Environ. Exer. Physiol.,* **50:**580, 1981b.

Lambertsen, C. J. (ed.): "Proceedings of the Fourth Symposium on Underwater Physiology," Academic Press, Inc., New York, 1971.

Lanphier, E. H., and H. Rahn: Alveolar Gas Exchange during Breath-hold Diving, *J. Appl. Physiol.,* **81:**471, 1963.

Lee, K. D., R. A. Mayou, and R. W. Torrance: The Effect of Blood Pressure upon Chemoreceptor Discharge to Hypoxia and the Modifications of This Effect by the Sympathetic-adrenal System, *J. Quart. Exp. Physiol.,* **49:**171, 1964.

Liljestrand, G.: Untersuchungen über die Atmungsarbeit, *Skand. Arch. Physiol* **35:**199, 1918.

Livingston, D. J., G. N. LaMar, and W. D. Brown: Myoglobin Diffusion in Bovine Heart Muscle, *Science* **220**(4592):71, 1983.

Loeppky, J. A., and M. L. Riedesel (eds.): "Oxygen Transport to Human Tissues," Elsevier Biomedical, North Holland, Inc., 1981.

Loeschke, H. H.: Central Chemosensitivity and the Reaction Theory, *J. Physiol.,* **332:**1, 1982.

Loke, J., D. A. Mahler, and J. A. Virgulto: Respiratory Muscle Fatigue after Marathon Running, *J. Appl. Physiol.: Resp. Environ. Exer. Physiol.,* **52:**821, 1982.

Lönnerholm, G.: Pulmonary Carbonic Anhydrase in the Human, Monkey, and Rat, *J. Appl. Physiol.: Resp. Environ. Exer. Physiol.,* **52:**352, 1982.

Mairbäurl, H., E. Humpeler, G. Schwaberger, and H. Pessenhofer: Training-dependent Changes of Red Cell Density and Erythrocytic Oxygen Transport, *J. Appl. Physiol.: Resp. Environ. Exer. Physiol.,* **55:**1403, 1983.

Martin, B., J. Edward, J. Morgan, C. W. Zwillich, and J. V. Weil: Control of Breathing During Prolonged Exercise, *J. Appl. Physiol.: Resp. Environ. Exer. Physiol.,* **50:**27, 1981.

Martin, B., M. Heintzelman, and Hsiun-ing Chen: Exercise Performance after Ventilatory Work, *J. Appl. Physiol.: Resp. Environ. Exer. Physiol.*, **52**:1581, 1982.

Matalon, S., and E. A. Egan: Effects of 100% O_2 breathing on permeability of alveolar epithelium to solute, *J. Appl. Physiol.: Resp. Environ. Exer. Physiol.*, **50**(4):859, 1981.

McFadden, E. R., Jr.: Respiratory Heat and Water Exchange: Physiological and Clinical Implications, *J. Appl. Physiol.: Resp. Environ. Exer. Physiol.*, **54**:331, 1983.

McFadden, E. R., Jr., and R. H. Ingram, Jr.: Exercise Induced Airway Obstruction, *Ann. Rev. Physiol.*, **45**:453, 1983.

Milic-Emili, G., and J. M. Petit: Mechanical Efficiency of Breathing, *J. Appl. Physiol.*, **15**:359, 1960.

Milic-Emili, G., J. M. Petit, and R. Deroanne: The Effects of Respiratory Rate on the Mechanical Work of Breathing during Muscular Exercise, *Intern. Z. Angew. Physiol.*, **18**:330, 1960.

Miller, W. S.: "The Lung," Charles C. Thomas, Springfield, Ill., 1947.

Mithoefer, J. C.: Breath Holding, in W. O. Fenn and H. Rahn (eds.), "Handbook of Physiology," sec. 3, Respiration, vol. II, p. 1011, American Physiological Society, Washington, D.C., 1965.

Neil, E., and N. Joels: The Carotid Glomus Sensory Mechanism, in D. J. C. Cunningham and B. B. Lloyd (eds.), "Regulation of Human Respiration," p. 163, Blackwell Scientific Publications, Ltd., Oxford, 1963.

Nielsen, M.: Die Respirationsarbeit bei Körperruhe und bei Muskelarbeit, *Skand. Arch. Physiol.*, **74**:299, 1936.

Niinimaa, W.: Oronasal Airway Choice during Running, *Respir. Physiol.*, **53**:129, 1983.

Ohira, Y., D. R. Simpson, V. R. Edgerton, G. W. Gardner, and B. Senewiratne: Characteristics of Blood Gas in Response to Iron Treatment and Exercise in Iron-Deficient and Anemic Subjects, *J. Nutr. Sci. Vitaminol.*, **29**:129, 1983.

Otis, A. B.: The Work of Breathing, in W. O. Fenn and H. Rahn (eds.), "Handbook of Physiology," sec. 3, Respiration, vol. I, p. 463, American Physiological Society, Washington, D.C., 1964.

Pappenheimer, J. R.: Standardization of Definitions and Symbols in Respiratory Physiology, *Fed. Proc.*, **9**:602, 1950.

Pattle, R. E.: Surface Lining of Lung Alveoli, *Physiol. Rev.*, **45**:48, 1965.

Plum, F.: Neurological Integration of Behavioural and Metabolic Control of Breathing, in R. Porter (ed.), "Breathing: Hering-Breuer Centenary Symposium," p. 159, Churchill, London, 1970.

Rahn, H., and L. E. Farhi: Ventilation, Perfusion and Gas Exchange: the $\dot{V}_A/Q$ Concept, in W. O. Fenn and H. Rahn (eds.), "Handbook of Physiology," sec. 3, Respiration, vol. I, p. 735, American Physiological Society, Washington D. C., 1964.

Said, S. I.: Metabolic Functions of the Pulmonary Circulation, *Circ. Res.*, **50**:325, 1982.

Saltin, B., and P.-O. Åstrand: Maximal Oxygen Uptake in Athletes. *J. Appl. Physiol.*, **23**:353, 1967.

Saltin, B., G. Blomqvist, J. H. Mitchell, R. L. Johnson, Jr., K. Wildenthal, and C. B. Chapman: Response to Submaximal and Maximal Exercise after Bedrest and Training, *Circulation*, **38**(Suppl. 7):1968.

Scholander, P. F.: Oxygen Transport through Hemoglobin Solution, *Science*, **131**:585, 1960.

Simon, J., J. L. Young, B. Gutin, D. K. Blood, and R. B. Case: Lactate Accumulation Relative to the Anaerobic and Respiratory Compensation Thresholds, *J. Appl. Physiol.: Resp. Environ. Exer. Physiol.*, **54**:13, 1983.

Stenberg, J., P.-O. Åstrand, B. Ekblom, J. Royce, and B. Saltin: Hemodynamic Response to Work with Different Muscle Groups, Sitting and Supine, *J. Appl. Physiol.*, **22**:61, 1967.

Stevens, E. D., and F. G. Carey: One Why of the Warmth of Warm-bodied Fish, *Am. J. Physiol.*, **240**:R151, 1981.

Turner, J. M., J. Mead, and M. E. Wohl: Elasticity of Human Lungs in Relation to Age, *J. Appl. Physiol.*, **25**:664, 1968.

Viljanen, A. (ed.): Reference Values for Spirometric, Pulmonary Diffusing Capacity and Body Plethysmographic Studies, *Scand. J. Clin. Lab. Inv.*, **42**(Suppl. 159), 1982.

Weibel, E. R.: "Morphometry of the Human Lung," Springer-Verlag OHG, Berlin, 1963.

Weibel, E. R.: Morphometrics of the Lung, in W. O. Fenn and H. Rahn (eds.), "Handbook of Physiology," sec. 3, Respiration, vol. I, p. 285, American Physiological Society, Washington, D.C., 1964.

Weibel, E. R.: Morphological Basis of Alveolar-Capillary Gas Exchange, *Physiol. Rev.*, **53**:419, 1972.

Weibel, E. R., and C. R. Taylor (eds.): Design of the Mammalian Respiratory System, *Respir. Physiol.*, **44**:1, 1981.

West, J.: Regional Differences in Gas Exchange in the Lung of Erect Man, *J. Appl. Physiol.*, **17**:893, 1962.

West, J. B.: "Respiratory Physiology—the Essentials," The Williams & Wilkins Company, Baltimore, 1979.

Whalen, W. J. and P. Nair: Some Factors Affecting Tissue P_{O2} in the Carotid Body, *J. Appl. Physiol.*, **39**:562, 1975.

Whipp, B. J.: Ventilatory Control During Exercise in Humans, *Ann. Rev. Physiol.*, **45**:415, 1983.

Whipp, B. J., and S. A. Ward: Ventilatory Control Dynamics during Muscular Exercise in Man, *Int. J. Sports Med.*, **1**:146, 1980.

Whitelaw, W. A., B. McBride, J. Amar, and K. Corbet: Respiratory Neuromuscular Output During Breath Holding, *J. Appl. Physiol.: Resp. Environ. Exer. Physiol.*, **50**:435, 1981.

Widdicombe, J. G. (ed.): "MTP International Reviews of Science Physiology, Series One, Respiratory Physiology," Butterworths, London, 1974.

Winslow, R. M., C. C. Monge, N. J. Stratham, C. G. Gibson, S. Carache, J. Whittembury, O. Moran, and R. L. Berger: Variability of Oxygen Affinity of Blood—Human Subjects Native to High Altitude. *J. Appl. Physiol.: Resp. Environ. Exer. Physiol.*, **51**:1411, 1981.

Wittenberg, J. B., and B. A. Wittenberg: Facilitated Oxygen Diffusion by Oxygen Carriers, in D. L. Gilbert (ed.), Oxygen and Living Processes, an Interdisciplinary Approach, p. 177, Springer-Verlag, New York, 1981.

6

SKELETAL SYSTEM

CONTENTS

THE MANY FUNCTIONS OF BONE
EVOLUTION OF THE SKELETAL SYSTEM
STRUCTURE
FUNCTION
GROWTH
REPAIR–ADAPTATION
JOINTS
LIGAMENTS AND TENDONS
PATHOPHYSIOLOGY OF THE BACK

THE MANY FUNCTIONS OF BONE

The skeletal system provides the mechanical levers for the muscles so that their contraction may cause the body to move. It is the supporting framework that prevents the entire body from collapsing into a heap of soft tissue; it is the protecting shell or casing for such vital, viable organs as the brain, lungs, heart, and pelvic organs; and it contains within its structures the factories for formed elements of the blood. In addition, it is the great calcium and phosphorus reserve of the body, constantly being drawn upon or added to. Bone is a vital living tissue, continuously undergoing changes of building up and tearing down.

For a comprehensive review of the osseous system, the reader is referred to such volumes as: *Bone as a Tissue,* edited by Rodahl, Nicholson, and Brown (1960); *The*

Biochemistry and Physiology of Bone, vols. 1–3, edited by G. H. Bourne (1971), and *Biology of Bone,* by N. M. Hancox (1972). The purpose of this chapter is to present a summary of the structure and functions of bone as they pertain to physical performance, work stress, and exercise.

EVOLUTION OF THE SKELETAL SYSTEM

In the history of vertebrates, bones and muscles are intimately related in the embryonic development and in physical activity. Primitive multicelled animals probably had only two body layers: one inside layer and one outside layer. With growth in size, a third layer developed in between; muscles and bones are products of this middle layer, from which the circulatory system also originates.

The need for a rigid supporting framework developed early in many of the multi-celled animals, especially as they moved from sea to land (Romer, 1957). In most animals, there is some kind of connective tissue filling the spaces between different organs, helping to bind them together. Generally such tissues contain numerous fibers felted together into a supporting structure. With the increase in size and mobility, characteristic for most chordates, came the need for more rigid support, such as the skeletal structures. Many of the nonchordates solved the problem with the development of an outside supporting structure, such as the shell of the mollusk or the hardened superficial armor of the lobster, whereas, in vertebrates, the major skeletal structures are internal, covered by muscles and skin. This arrangement not only facilitates mobility, but also enhances the accessibility of the stored mineral reserves in the bones for participation in metabolic processes and chemical reactions in the body.

The first type of skeletal material which developed in the vertebrates was the *notochord,* already typically developed in such primitive forms as the amphioxus; it consists of a tough but flexible rod running the length of the body, along the back, and below the nerve cord (Romer, 1957). This notochord not only helps to stiffen the body and support the various organs but also serves as a point of attachment for the muscles of the trunk. Although the notochord long persisted as a functional element in the vertebrates, there appeared new skeletal materials, in the form of cartilage and bone, which have assumed and expanded the role of support originally provided by this notochord. Both these substances are formed from the connective tissue, but they differ considerably in their nature and appearance.

STRUCTURE

The *cartilage* is a translucent material formed by round cells (cartilage cells) embedded in a substance which binds them together. It is firm yet elastic, flexible, and capable of rapid growth. It is therefore a useful supporting structure as long as the stress is moderate, and it is a common supporting structure in lower vetebrates. In the shark, for example, it is the only skeletal material in addition to the notochord. On the other hand, in land vertebrates, such as the adult human being, it plays only a limited part in the skeletal makeup and is found only where elasticity is required, such as at the ends of the ribs, as discs between the joints of the backbone, and at the joint surfaces of the limbs.

Bone, like cartilage, is a derivative of the original connective tissue, but it is a highly specialized form and is different from connective tissue proper in that it is very hard. It consists of *osteocytes,* which are cells with long, branching processes that occupy cavities called *lacunas* in a dense matrix made up of collagenous fibers laid down in an amorphous ground substance called *cement,* which is impregnated with calcium phosphate complexes (Fig. 6-1). The ratio between inorganic and organic substances in the bone varies during the life span. The ratio is about 1:1 in the child. In young adults, there is about 4 times more inorganic material than organic, and in the old individual the ratio is about 7:1. Therefore the bone is more fragile in the old than in the young person.

A series of in vitro studies have shown that the development of bone and cartilage cells may be affected by extrinsic factors in the environment. Thus, an experiment by C. A. I. Basset (1962) showed that the formation of normal bone from a tissue culture of primitive fibroblasts depended on adequate oxygenation. When high oxygen pressure was introduced in addition to compaction of the culture, bone formation resulted. With low oxygenation, cartilage was formed from the primitive fibroblasts.

FUNCTION

The cellular components of bone are associated with specific functions. The *osteoblasts* are involved in the formation of bone, the *osteocytes* with the maintenance of bone as a living tissue, and the *osteoclasts* with the destruction and resorption of bone. These cells are closely interrelated, and transformation may occur from one form to

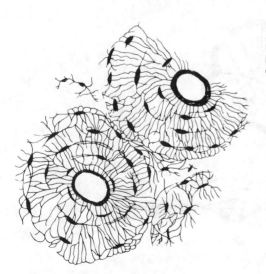

Figure 6-1
Schematic drawing of the structure of bone, showing osteocytes with long branching processes that occupy cavities in the dense bone matrix.

the other. Thus, an osteocyte may actually be an osteoblast that has been surrounded by calcified interstitial substance. It may undergo a further transformation and assume the form of an osteoclast.

Many bones, or parts of bones, are compact, solid structures, nourished through small canals carrying blood vessels and through tiny tubules connecting cell spaces with one another and with the canals. However, if all bones were solid and compact, they would be unnecessarily heavy in proportion to the strength requirements. The large bones are therefore hollow. They are solid only at the surface, and sufficient bone bars and braces extend from the solid bone exterior into the hollow interior to provide reinforcement in ways more or less similar to the manner in which many bridges are engineered (Fig. 6-2). So wisely is the order of nature arranged that this space in the hollow bone shafts is not wasted, but filled with bone marrow which is used as a factory for red blood cells and capable of producing about 2 to 3 million erythrocytes per second.

Under normal conditions, a state of equilibrium exists between the calcium of the blood and that of the bones, but in disease this balance may be seriously altered. The calcium is absorbed from the food in the small intestine and is excreted by the large intestine and, to a lesser extent, by the kidney. The regulators of calcium metabolism are the parathyroid glands and vitamin D; the former regulate the interchange of calcium between the blood and the bones.

Excess parathyroid hormone results in a rise in the blood calcium with a corresponding fall in the calcium of the bones and a loss of calcium from the body by

Figure 6-2
Schematic drawing showing reinforcing bars and braces, similar to those of a bridge, of the large bones.

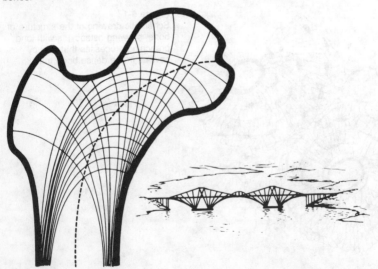

increased excretion. Vitamin D governs the absorption of calcium from the intestine, but excess intakes of this vitamin cause a mobilization of calcium from the bones and a deposition of calcium in soft tissues.

The maintenance of the normal mineral metabolism of the bones also depends upon the longitudinal pressure on the long bones, brought about by the stress of gravity on the upright, ambulatory human frame (Jansen, 1920; Rodahl et al., 1966; Greenleaf, 1984; Drinkwater, 1985). In fact, all changes in function of a bone are attended by alterations in its internal structure (Wolff's law). Pressure will stimulate the appositional bone growth. Increased weight bearing will result in increased thickness of the bone and density of the shaft (Lanyon and O'Connor, 1980). The examinations of Atkinson, Weatherell, and Weidmann (1962) and Eisenberg and Gordan (1961) confirm this fact. Atkinson and his coworkers examined the density of the femur in individuals with active and with sedentary habits after fifty years of age. The bone density was slightly higher in the active group. Eisenberg and Gordan found, by measuring the dynamics of growth in the human with nonradioactive strontium, that muscular exercise accelerates the rate of bone deposition. Conversely, elimination of the "normal effects of stress and strain" leads to the familiar picture of disuse osteoporosis.

In her review of the literature on osteoporosis, Drinkwater (1985) discusses the possible effects of sex, age, diet, physical activity, and differences in hormonal function. She points out that young amenorrheic athletes with estradiol levels similar to those of postmenopausal women have significantly lower vertebral bone density than comparable cyclic athletes, and questions the effectiveness of exercise preventing osteoporosis in the absence of adequate estrogen stimulation.

Montoye et al. (1980), examined 61 male veteran tennis players aged 55 and over (mean age, 64), who had played tennis regularly for 40 years on the average. The volume and circumferance of the hand and forearm were significantly greater in the dominant playing limb. Bone width and mineral content of the ulna, radius, and humerus were greater in the dominant side. The length, total area, and cortical cross-sectional area of the metacarpal bones were significantly greater on the dominant side.

The beneficial effect of physical exercise as a prophylactic measure against vertebral bone loss in middle-aged women, has been emphasized by Krølner et al., (1982). The results of a study by White and associates (1985), support the hypothesis that mechanical loading due to exercise may be effective in preventing postmenopausal osteoporosis. They compared the effect of walking and aerobic dancing on the bones in 73 recently postmenopausal women with a control group who did not exercise, based on photon absorptiometry of the distal radius. The period of observation was six months. The control group and the walking group lost statistically significant amounts of bone mineral content, but the dancing group did not. The control group did not show a significant increase in the bone width, but both the dancing and walking groups did. Plasma estrogen levels were not influenced by exercise. Evidently, too little attention has been paid to the concept that disuse is a very common cause of osteopenia (Radin, 1982). There is also evidence that blood plasma calcium concentrations may rise abruptly and significantly as a result of intense muscular activity in vertebrates which have osseous skeletons. The source of the excess calcium is bone (Ruben and Bennett, 1981).

The pumping action of the muscles on the circulation and the rate of blood flow through the extremities can be estimated from denervation and devascularization experiments (Troupp, 1961). A study on the longitudinal growth of the femur and tibia of rabbits following interruption of vascular and nervous supply demonstrated that retardation of growth is unequivocally more significant following vascular blocking than following nerve paralysis. It was suggested that vascular insufficiency causes hypoxia in the cartilage cells of the epiphysis, thus contributing to growth inhibition by inhibiting the transformation of bone from cartilage cells.

Studies of the effects of prolonged bed rest on calcium metabolism (Rodahl et al., 1966) have shown that the increased urinary calcium excretion that occurs during prolonged confinement to bed is not due to inactivity but to the absence of longitudinal pressure on the long bones. The increased urinary calcium excretion was unaffected by heavy cycle ergometer exercise performed in the supine position for 1 to 4 hours daily and by 8 hours of inactive sitting in a wheelchair. But 3 hours' standing per day in addition to recumbent bed rest caused the urinary calcium excretion to return toward normal values (Fig. 6-3). When a pressure equivalent to the body weight of the individual confined to horizontal bed rest was exerted along the longitudinal axis of his body by heavy springs fixed to a shoulder harness and the foot of the bed for 3

Figure 6-3
Urinary output of calcium during (a) prolonged inactive bed rest; (c) bed rest combined with 4 hr/day supine exercise; (b) bed rest combined with 8 hr/day quiet sitting; and (d) bed rest combined with 3 hr/day standing. (*From Rodahl et al., 1966.*)

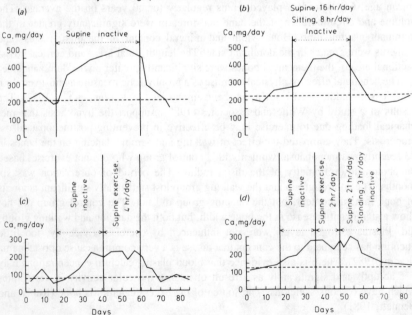

hours a day, the urinary calcium elimination also returned toward normal values in one of two subjects.

If the urinary calcium elimination may be taken as an indication of bone mineralization, it appears that gravitational stress on the long bones is essential for normal bone growth. This is supported by studies of calcium metabolism in astronauts during prolonged space flights. Thus Whedon et al. (1975), studying mineral and nitrogen metabolism during the United States Skylab flights lasting from 28 to 84 days, observed that urinary calcium excretion increased as in bed rest studies by 80 to 100 percent. The urinary calcium excretion showed a gradual rise during the first 3 weeks of the flight, after which it remained constant at the elevated level for the duration of the space flight, whether it lasted 28 or 84 days.

It is thus clear that the development of greater sturdiness and strength of bone is brought about by subjecting it to increased pressure. This process may be slower than the increase in muscle strength resulting from a program of weight lifting. In such training, the training intensity or the increase in training load should be sufficiently gradual to allow for the development of skeletal strength to keep pace with the increase in muscle strength.

GROWTH

Bone growth is a unique physiological process, consisting of a constant construction and reconstruction which continue beyond the point of actual growth and development, so that nothing is stable and final in bone except the external shape. A typical long bone, such as the femur, consists of a shaft (the *diaphysis*) and two enlarged ends (the *epiphyses*). The part of the shaft which is next to the epiphysis is called the *metaphysis*. The surface of the long bone is covered by a sheet of connective tissue (the *periosteum*). The shaft of the bone consists of a dense outer layer (the *cortex*), a middle layer of spongy bone, and the inner portion, which is occupied by the marrow cavity. The joint surface of the epiphysis is covered with smooth, glistening white hyaline cartilage. The articular cartilage has to absorb the shocks transmitted to the joint during motion and reflects the wear and tear of the joint.

In providing rigidity, the bone has to sacrifice the flexibility and the expandable traits enjoyed by the cartilage. This naturally presents a problem of growth. Thus, for example, the thigh bone of a two-year-old child has to double its length by adulthood. Although the bone may become stouter simply by adding more bone to its surface layers, it cannot stretch, nor can new bone be added to its ends, for such changes would interfere with its articulate surfaces that form part of the hip joint and the knee joint. This problem is solved by the insertion of a special growth section, the epiphyseal plate, close to both ends of the shaft. On the surface of this plate are layers of actively growing cells. It is here that the increase in length of the bone occurs without interference with the joint surfaces. When the individual is fully grown, the epiphyses fuse with the shaft, and growth terminates.

All the elements of the internal skeleton are first formed in cartilage in the human embryo. Later, the cartilage is destroyed and replaced by bone. It is in this replacement process that a thin film of cartilage is left, separating the ends from the shaft of the bone in the form of the epiphyses, described earlier. Lyle J. Micheli of the Childrens' Hospital Medical Center in Boston, Mass. (Micheli, 1983) has warned against the participation of young children in intensive and specialized training and competition. He is referring to reports of seven-year-old children training for running the marathon race, and reports on serious problems due to repeated microtrauma including irreversible injuries of the epiphysial plate of the bones in the legs. In this respect, it should be kept in mind that chronological age is not a very good yardstick for maturity in young athletes, in that there may be as much as a seven-year difference between the onset of puberty in the earliest girl and the latest boy (See Fig. 7-12).

REPAIR-ADAPTATION

Following a fracture of a bone which was initially preformed in cartilage, the first attempt at repair, starting within the first few weeks following the fracture, results in the formation of a mass of fibrocartilage known as *callus*. The bridging of a fracture by new bone occurs in a fashion resembling that of a fixed-arch bridge, according to the principle of cantilevering frequently seen in contemporary architecture (Fig. 6-4). The new bone grows out upon the surface of the cortex on each side of the fracture line and envelops the fibrocartilaginous callus to form an arch of new bone over the fracture gap. The new bone comes down like ribs let down from the arch of the bridge to suspend the deck. This bone gradually replaces the cartilage toward the fracture gap. Finally the "deck" is laid down between the fracture ends and provides permanent union. The superstructure disappears, leaving only the bone required for union of the fracture ends. In the case of a fracture of a long bone in a young individual, this process is usually completed within a period of about 3 to 6 months.

Owing to the readiness with which bone formation and bone absorption can occur, a remodeling of bone is continually taking place. Isotope studies using P^{32} or Ca^{45} indicate that over 20 percent of the calcium and phosphate ions of bone are involved in a fairly rapid exchange in adults (LeBlond and Greulich, 1956). As mentioned, bone tissue is surprisingly sensitive to demands made upon it and responds readily to these demands, so that every change in the function of bone is followed more or less by a definite change in the internal architecture. Bone bars not stressed will disappear and new ones will be created where altered mechanical forces increase the demand for sturdiness. Thus, as a result of training or a fracture, the lines of stress in a bone

Figure 6-4
Schematic drawing showing similarity between the process of bone repair and the construction of a fixed-arch bridge. (*Modified from McLean and Urist, 1955.*)

may change with the altered direction of mechanical forces, as if it were the most plastic of structures.

Absorption of bone may occur from many causes and leads to bone rarefaction, or *osteoporosis*. It occurs commonly in old age, from disuse, and as a local phenomenon caused by such processes as inflammation, tumors, and aneurysms.

JOINTS

Joints are formed where two or more bones of the skeleton meet one another (Fig. 6-5). The function of the joint determines its character and structure. In areas like the skull, it is important that no movement should be permitted between contiguous bones; in the vertebral column, on the other hand, a slight degree of mobility is desirable, provided that it can be obtained without loss of strength or sturdiness. In other situations, the provision of a more or less wide range of movement is essential. In the first case, *fibrous joints* are provided; these are articulations in which the surfaces of

Figure 6-5
Schematic drawings of a cartilaginous joint (*a*), simple synovial joint (*b*), synovial joint with articular disk (*c*), and a sagittal section at the lumbar region of the vertebral column (*d*).

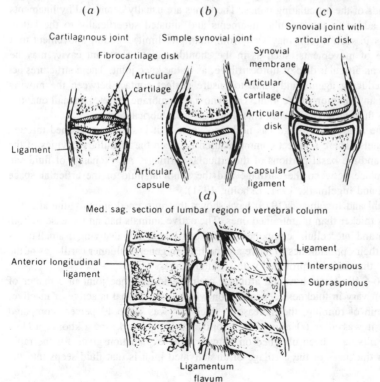

(*a*) Cartilaginous joint
Fibrocartilage disk
Articular cartilage
Ligament

(*b*) Simple synovial joint
Articular capsule
Articular cartilage

(*c*) Synovial joint with articular disk
Synovial membrane
Articular cartilage
Articular disk
Capsular ligament

(*d*)
Med. sag. section of lumbar region of vertebral column
Anterior longitudinal ligament
Ligament
Interspinous
Supraspinous
Ligamentum flavum

the bones are fastened together by intervening fibrous tissue and in which there is no appreciable motion. In the second case, the connection medium between the bones concerned is white fibrocartilage; such joints are capable of a limited range of movement and are termed *cartilaginous joints* (Fig. 6-5a). In the third case, the opposed bones are separated from one another by a space lined by a special membrane, which is termed *synovial membrane;* such a joint possesses a more or less wide range of movement and is known as a *synovial joint* (Fig. 6-5b,c).

In the cartilaginous joints of the spinal column, the opposed bony surfaces are covered with hyaline cartilage and are connected to one another by flattened discs of fibrocartilage of a more or less complex structure. The bones are also connected by bands of white fibrous tissue known as *ligaments,* which do not form a complete capsule around the joint. A limited degree of movement is made possible by the compressibility of the cartilaginous disc and the degree of leverage which is available (Fig. 6-5d).

Most of the joints of the body, including all joints of the limbs, belong to the synovial group. In these joints (Fig. 6-5), the contiguous bony surfaces are covered with articular cartilage separated by a joint cavity, which in a healthy individual is merely a tiny space. The joint is completely surrounded by an articular capsule consisting of a capsular ligament lined with a synovial membrane. The synovial membrane lines the whole of the interior of the joint, with the exception of the cartilage-covered ends of the articulating bones. The bones are usually connected by ligaments which are added to the capsular ligaments and situated superficially to the latter. Movements in such a joint may vary from a simple, limited gliding movement to a wide range of movements, such as in the shoulder joint. The joint cavity may be divided by an articular disc of fibrocartilage, as in the knee joint. These structures act as shock-reducing agents and serve to ensure perfect contact between the moving surfaces in any position of the joint. The synovial membrane secretes a small quantity of a viscid fluid termed *synovia.* This fluid acts as a lubricant.

Since the articular cartilage has no blood vessels, it has to be nourished through other channels. There is direct communication between the medullary cavities of the epiphyses and the basal portions of the articular cartilage. An exchange of fluid can also take place between the cartilage and the synovial fluid of the articular space (Holmdahl and Ingelmark, 1951; Ekholm, 1951).

Holmdahl and Ingelmark (1948) have shown that in trained animals, the articular cartilage is thicker than in untrained ones. The active animals had an increase in both the cellular and intercellular components of the cartilage. Animals living in small cages restricting their opportunities to move around had, in general, thinner cartilages of the knee joints than animals provided with spacious cages.

Figure 6-6 illustrates how the articular cartilage in the knee joint in a matter of minutes can vary in thickness depending on whether the animal is active or inactive. After 10 min of running, the increase in thickness was 12 to 13 percent compared with what it was after 60 min of immobilization (Ingelmark and Ekholm, 1948). Similar results have been obtained on humans. The explanation given for the rapid increase in thickness of the cartilage of an activated joint is that fluid seeps into the

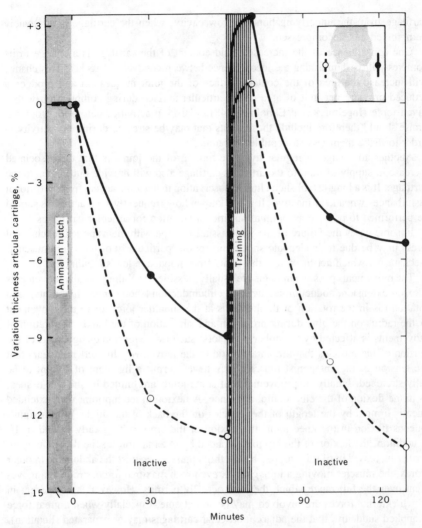

Figure 6-6
Rabbits were taken from their spacious hutches where they could run freely. At "0" minutes, x-ray photographs were taken for measurements of the thickness of the articular cartilage of the fibular and tibial ends of the knee joints (for symbols, see inserted figure). Measurements were repeated after 30 and 60 min of rest in a position which did not stress the rabbits' knees with their body weight. Then the rabbits were exercised on a motor-driven treadmill for 10 min, speed 40 m · min⁻¹. Measurements were repeated after exercise and further periods of rest. Note a significant decrease in cartilage thickness after inactivity and a rapid increase during exercise. (Figures represent the mean of about 50 animals.) (*Modified from Ingelmark and Ekholm, 1948.*)

cartilage from the underlying bone marrow cavity, when the cartilage is alternately compressed and decompressed.

One consequence of the increased fluid content of the cartilage is a change in its compressibility, diminishing the incongruence between condyle and socket. This change will increase the area of the contact surface of the joint in question and produce a reduced pressure per unit of area of the articular surface during compression with a given force (Ingelmark and Ekholm, 1948). Warm-up activities before vigorous exercise should therefore include those joints that may be stressed during the activity in order to make them less susceptible to trauma.

Another advantage in regular dynamic motion of the joints is that the associated increase in supply of fluid to the articular cartilage also will provide nutriments to the cartilage. It is a long-established clinical observation that a joint suffers from nutritional disturbances when kept inactive. It is not known how frequently such activities should be performed to provide an optimal nutritional situation for articular cartilages.

In some cases the failure of the joint tissues to cope with their applied mechanical stress may be due to inadequate shock absorption, particularly of the neuromuscular mechanisms which act to protect the joints from impulsive loads (Radin, 1982).

The movements possible in joints are usually classified as gliding angular movements (flexion-extension, abduction-adduction), circumduction (shoulder and hip joints), and rotation (as in the rotation of the humerus at the shoulder joint, or in the movement of the radius on the ulna during pronation and supination of the hand). Limitation of movements is affected by a number of factors, such as the tension of ligaments or the tension of the muscles that are antagonistic to the movement. In fact, it appears that the tension of the antagonist muscles may never permit a ligament of a joint to be fully stretched. Finally, the movements of some joints are limited by the soft tissues, as in the flexion of the elbow, hip, and knee. A flexion in the hip joint with extended knee is limited by the length of the muscles on the back of the thigh. With a simultaneous flexion in the knee joint, the flexion of the hip can be greatly extended. If, in addition, the flexion of the hip joint is aided by external forces, the flexion may be further increased until it is stopped by the thigh pressing against the abdomen. In other words, the muscles moving a joint cannot, even with maximal force, produce a movement over the full range which the joint actually permits. However, a movement in which external forces are involved may be so extreme, especially when a great force is applied suddenly, that the adjacent articular cartilages may be separated (luxation). At the same time, bone, ligaments, joint capsules, soft tissues, and blood vessels may be damaged.

Since the limiting factor for flexibility is often the length of the muscles, training that produces a lengthening of these muscles will increase the joint flexibility.

In synovial joints where the bones are connected only by ligaments and muscles, the articular surfaces are in constant apposition in all positions of the joint. The maintenance of this apposition is facilitated by atmospheric pressure and cohesion, but the muscles play a far more important role. The balance in tone between the different muscle groups which act on the joint is responsible for maintaining the articular surfaces in constant apposition, so that the stability of any joint depends on the tonus of the muscles that act on it.

Movable joints are innervated by the nerves which supply the muscles that act on them, and it is reasonable to assume that this arrangement establishes local reflex arcs which ensure stability. The part of the articular capsule which is rendered taut in the contraction of a given muscle or group of muscles is innervated by the nerve or nerves supplying their antagonists. This condition may serve to prevent overstretching or tearing of the ligaments.

Studies of different joints both in humans and animals have shown that nerve fibers to a joint arrive both via special branches from motor nerves, periostal nerves, and often cutaneous nerves (Brodal, 1972). The contribution of these various fibers to the overall nerve supply of a joint varies greatly from one joint to another. Thus, the anterior aspects of the knee joints are mainly supplied by nerve fibers coming from nerve branches supplying the adjacent muscles, whereas the ankle joint receives relatively few such fibers. Similarly, the total number of nerve fibers supplying a joint may vary greatly. Thus, the nerve supply to the knee joint is much greater than to the shoulder joint. It has been suggested that the reason for this fact may be that the muscle mass surrounding the shoulder joint may play a role in the perception of the joint position, thus augmenting or partially replacing the function of the afferent nerve fibers of the joint (Brodal, 1972).

In general, there are four types of nerve endings in the joints (Fig. 6-7). Three of them are nerves ending in specialized end organs, as in skin receptors, often surrounded by a capsule of connective tissue, while one of them ends as free-branching nerve endings. Of the former, one type is primarily specialized to give information about changes in joint position (type 1), and a second type to give information about the speed of movement in the joint (type 2). Both types are located in the parts of the

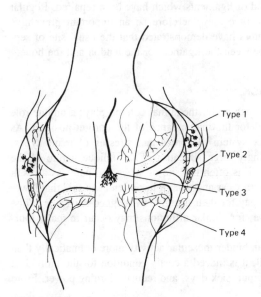

Type 1

Type 2

Type 3

Type 4

Figure 6-7
Schematic drawing showing the four different types of nerve endings in the joints. (*By courtesy of P. Brodal.*)

joint capsule where bending and stretching take place. A third type of receptor (type 3) appears to be particularly adjusted to register the actual position of the joint, and it is located in the ligaments of the joints. The fourth type of receptor consists of free-branching nerve endings which are pain-sensitive fibers similar to those found elsewhere in the body. The synovial membrane has no nerve fibers of its own, as is also true of the joint cartilage.

Afferent impulses from the joints reach the somato-sensoric region of the cortex. For this reason, it has been postulated that the joint innervation is of major importance for the conscious perception of the position and motions of the joint, whereas afferent impulses from the muscles, which are not consciously perceived, are significant for the subconscious, reflectoric control of motor activity (Brodal, 1972). Skoglund (1956) has shown that afferent impulses from the joints affect the activity of the alpha-motoneurons. This effect is reciprocal on the muscles which act as antagonists on a joint; that is, during passive extension, the extensors are inhibited while the flexors are facilitated. Skoglund (1956) has also provided evidence suggesting that the joint receptors are able to provide information pertaining to the resistance against a movement.

LIGAMENTS AND TENDONS

In wild primates, Noyes et al. (1974) have shown that 8 weeks of immobilization caused significant impairment of the functional capacity of ligaments. After 20 weeks of resumed activity following immobilization, there was only partial recovery of ligament strength.

Tipton et al. (1975) have shown that the mechanical stress produced by regular, prolonged exercise or training causes increased strength of the junctions between ligaments or tendons and bones, and of ligaments which have been repaired. Regular physical exercise of the endurance type may therefore be an important preventive measure, since numerous investigators have demonstrated that the usual site of separation is in the transitional zone between the ligament or the tendon and the bone.

PATHOPHYSIOLOGY OF THE BACK

Of all the bony structures of the human body, the spinal column plays a unique role in that it serves as a sustaining rod for the maintenance of the upright position. As such, it is subjected to a complex system of forces and stresses of many types. Frequently these forces are amazingly large. Because of the unique nature of the structure and function of the spine, it is often the site of aches and pains as a result of numerous processes associated with wear and tear. It has often been said that lower backache is the price humans must pay for their upright, two-legged existence. This, however, is not necessarily the case, for similar afflictions may occur in many four-legged animals.

There are few ailments which can hinder muscular activity more dramatically than backache. Apart from being painful, it is indeed a costly condition for the individual and society in terms of lost work output, sick days, and reduced earning power. From a health standpoint, lower backache ranks as one of the most common medical complaints. As a matter of fact, back pain, in one form or another, is experienced by almost all individuals at one time or another. Low backache is also quite common in athletes, especially those whose training includes heavy weight lifting.

Hult (1954) has presented results of examinations of about 1,200 individuals representing different professions. They were divided into two main groups, those engaged in physically light work (such as white-collar workers, workers in light industry, and retail trade employees) and in physically heavy work (including longshoremen, construction workers, and employees in heavy industry). The subjects ranged from twenty-five to sixty years of age. Subjective symptoms revealed by careful notations of case histories could, on the whole, be classified as belonging to one of three syndromes: two common ones, the stiff neck-brachialgia syndrome (in 51 percent of the individuals) and the lumbar insufficiency-lumbago-sciatica syndrome (in 60 percent), and a less common one, the dorsal spine syndrome (in 5 percent of the subjects).

Figure 6-8 (upper part) shows (1) the marked increase in symptoms with age; and (2) a slightly higher percentage of individuals with symptoms of lower back troubles among heavy workers (64 percent) compared with those involved in light work (53 percent). The incidence of symptoms in the upper back was the same in those doing light and heavy work (51 percent).

The results of clinical and roentgenologic examinations are also summarized in Fig. 6-8 (lower part). The trend is the same as for the subjective symptoms, with the age factor being of decisive importance for the occurrence of roentgen signs of disc degeneration, which was as high as about 90 percent in the fifty-five- to fifty-nine-year age group. Such signs appear with consistently higher frequency in those subjects with heavy occupations, even if the differences between the two groups is moderate. Disc degeneration should be interpreted as a more or less physiological process which begins early in some individuals but eventually develops in all persons regardless of occupation.

In patients with a history of lumbago or sciatica, the symptoms were provoked by an accident in only 20 percent of the cases. In an additional 15 to 20 percent, they appeared in connection with heavy lifting or a similar strain. This means that in 60 to 65 percent of those who had had attacks of lumbago or sciatica, the symptoms appeared without such external causative factors.

Static deformities, such as differences in leg length, kyphosis, lordosis, and scoliosis, had no demonstrable significance in the origin of low back trouble in these groups. Nor was there any correlation between low back trouble and body type (evaluated according to Rohrer's index), flat feet, or foot fatigue.

There was some, but not a striking, relation between subjective symptoms and clinical-roentgenologic findings. Thus, disc degeneration could develop and even attain an advanced stage without ever having given rise to pain.

Weight lifters did not in any respect differ significantly from the other subjects.

It is estimated that in Sweden, approximately 2 million working days annually are lost because of back trouble. The occurrence of such a high frequency of symptoms from the back is not at all surprising in view of the complexity of the spinal column, its complicated joint systems, and the very heavy stress to which it may be exposed.

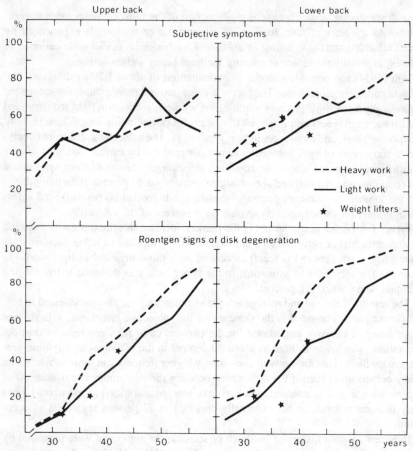

Figure 6-8
Data on approximately 1,200 individuals divided into two main groups according to occupation. Included are data from 56 weight lifters. The upper part of the figure shows results of case histories dealing wih symptoms emanating from the upper or lower extremities and the spine. The lower part of the figure shows frequencies of objective incidence of disk degeneration and x-ray-anatomic anomalies. The graphs to the left present cases of stiff neck-brachialgia syndrome ("upper back"); the graphs to the right, lumbar insufficiency-lumbago-sciatica syndrome ("lower back"). (*Data from Hult, 1954.*)

An interesting analysis of the tolerance of the spinal column has been published by Morris et al. (1961). They point out that the spinal column may be considered to have both an intrinsic and extrinsic stability, the former being provided by the alternating rigid and elastic components of the spine bound together by a system of ligaments, while extrinsic stability is provided by the paraspinal and other trunk muscles. The stability of the ligamentous spine, which may be considered as a modified elastic rod, depends largely on the action of the extrinsic support provided by the trunk muscles.

It has been shown that the critical load value at which buckling of the isolated ligamentous spine occurs, fixed at the base, is much less than the weight of the body above the pelvis (Lucas and Bresler, 1960).

In order to explain the discrepancy between the force which the spine as such can tolerate and the much larger forces to which the back is subjected in actual life situations, Morris and his coworkers investigated the role of the compartments of the trunk, i.e., thorax and abdomen, in helping to provide stability of the spine. They then showed that the discrepancy apparently can be explained by the role played by the trunk.

In providing extrinsic stability of the spine, Morris et al. (1961) considered that if the nucleus pulposus of the fifth lumbar disc is considered as the fulcrum of movement and if a heavy weight is lifted with the hands, the arms and trunk form a long anterior lever (Fig. 6-9). The weight being lifted and the weight of the head, arms, and upper part of the trunk are counterbalanced by the contraction of the deep muscles of the back acting through a much shorter lever arm, that is, the distance from the center of the disc to the center of the spinous process. With these factors in mind, they computed the force that results when a 77 kg man lifts a 90 kg weight, and they concluded that the force on the lumbosacral disc is about 8,800 N, if the role of the trunk is omitted.

Figure 6-9
Schematic drawing of a 77-kg man lifting 90 kg. (*From Morris et al., 1961.*) The nucleus pulposus of the fifth lumbar disk is considered as the fulcrum of movement. The arms and trunk form a long anterior lever. The weight being lifted is counterbalanced by the contraction of the deep muscles of the back acting on a much shorter lever (the distance from the center of the disk to the center of the spinous process). If the role of the trunk is omitted, the force on the lumbosacral disk would be about 9000 N, which is considerably more than segments of the isolated ligamentous spine can withstand without structural failure. When this does not happen, it is because the contracted trunk muscles convert the abdominal and thoracic cavities into semirigid cylinders which relieve the load on the spine itself.

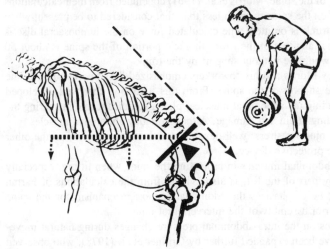

This force is considerably more than segments of the isolated ligamentous spine can withstand without structural failure; compression tests of two vertebral bodies with their intervening disc from subjects under forty years of age resulted in failure of the segments of the spine under compressive forces ranging from 4,400 to 7,600 N. In older subjects, this figure was sometimes as low as 1,300 N.

In a series of experiments, Morris et al. were able to show that since the spinal column is attached to the sides of, and within, the two chambers of the abdominal and thoracic cavities, the action of the trunk muscles converts these chambers into nearly rigid cylinders containing air, liquid, and semisolid material. Thus, these cylinders are capable of transmitting part of the forces generated in loading the trunk, thereby relieving the load on the spine itself. When large forces are applied to the spine, such as when lifting a weight of 90 kg, there is generalized contraction of the trunk muscles, including the intercostals, the muscles of the abdominal wall, and the diaphragm. The action of the intercostals and the muscles of the shoulder girdle renders the thoracic cage a rigid structure firmly bound to the thoracic part of the spine. When inspiration increases intrathoracic pressure, the thoracic cage and spine become a solid, sturdy unit capable of transmitting large forces. By the contraction of the diaphragm, attached at the lower margin of the thorax and overlying the abdominal wall, especially the transversus abdominis, the abdominal contents are also compressed into a semirigid cylinder. The force of weights lifted by the arms is thus transmitted to the spinal column by the shoulder girdle muscles, principally the trapezius, and then to the abdominal cylinder and to the pelvis, partly through the spinal column but also through the rib cage. When larger forces are involved, increased rigidity of the rib cage and increased compression of the abdominal contents are achieved by increased activity of the trunk muscles, resulting in increased intracavitary pressures.

This is all brought about by a reflex mechanism: When a load is placed on the spine, the trunk muscles are involuntarily called into action to fix the rib cage and to compress the abdominal contents. The intracavitary pressures are thereby increased, aiding in the support of the spine. Morris et al. (1961) concluded from their calculations that the actual force on the spine is much less than that considered to be present when this support by the trunk is omitted. The calculated force on the lumbosacral disc is about 30 percent less, and that of the lower thoracic portion of the spine is about 50 percent less than it would be without support by the trunk.

Thus, the study by Morris and his coworkers emphasized the important role of the trunk muscles in the support of the spine. From this it follows that well-developed trunk muscles, including the abdominal muscles, play a significant role in sparing the spine and thus avoiding strain and damage. Although flabby abdominal muscles may expose the spine to injurious stress, well-developed abdominal muscles, on the other hand, are a valuable protective device.

The role of the abdominal muscles in protecting the spine when lifting, especially during the very early part of the lift, is also evident from the calculations of Farfan (1973). In addition, he emphasizes the additional support mechanism for the spine offered by the posterior assembly of the intervertebral joints.

These observations on the intra-abdominal pressure changes during natural movements in humans have been expanded further by Grillner et al. (1978), who observed increases in intra-abdominal pressures in excess of 100 mg Hg (13.3 kPa) in subjects after a jump down from a moderate height of 0.4 m.

Human beings have had to carry loads from the beginning of their existence and have had time to learn how to perform this task in keeping with the most efficient bioengineering principles. The methods used by primitive native tribes in carrying loads are well known (Fig. 6-10 shows one method).

It has been generally advocated that in lifting and carrying loads, one should bring the center of gravity of the load as close as possible to the axis of the body to minimize the force movement. Although this principle should be followed more often than is commonly done, it should be borne in mind that, in many real-life situations, it is not always practicable to do so. Ekholm et al. (1982), have calculated the compressive forces on the L5—S1 disc to be about 4,400 N when lifting 12.8 kg with straight knees, but only about 3,400 N when lifting with bent knees, straight back, and the burden moved between the knees and close to the pelvis.

Often a series of motions are involved in the execution of a given task, including bending, lifting, turning, and walking with the load. However, experience has shown that symptoms of lower backache are most apt to occur when lifting is combined with a twisting or turning motion. It is therefore wise to arrange the situation so that the individual is facing the direction of movement before picking up the load. It is equally important to have a secure foundation for the feet during such motions, since lower backache very commonly occurs during sudden bodily movements associated with slipping of the foothold. In view of the findings of Morris and his coworkers (1961), it appears reasonable to assume that under such sudden motions, there may not be sufficient time for a reflex contraction of the trunk muscles to occur, thus leaving the spinal column more vulnerable.

These considerations underline the significance of well-trained trunk muscles, including the abdominal muscles, in the protection of the spine. It cannot be overemphasized that in the case of backaches or damage to the back, prevention is the most important aspect, since the back is an inherent weakness in the human body. This fact brings out the need for intelligent indoctrination, starting with children of school age,

Figure 6-10
A physiological method of carrying a load, practiced by many peoples.

that will induce a complete understanding of the mechanical aspects of the spinal column and methodical instruction in the proper techniques involving the use of the back in all kinds of daily activities as well as in special industrial tasks. By such early indoctrination, it may well be possible to avoid some of the lower back pain common in adults. For a review of the effect of exercise on degenerative back disease see Wickström (1978).

Nachemson (1976) and his group have been measuring the intradiscal pressure (in the third lumbar disc) together with EMG and intra-abdominal pressure in various exercises and occupational activities. As expected, good lumbar and also arm support are important factors in decreasing pressures as is avoidance of lifting heavy objects with great distance between the load and the body. Compared with the disc pressure or load in the upright standing position, reclining reduces the pressure by 50 to 80 percent, while unsupported sitting increases the load by 40%, forward leaning and weight lifting by more than 100%, and the position of forward flexion and rotation by 400% (Nachemson, 1981).

Schultz et al. (1982) have developed a calculation technique by which the load on the spine can be predicted for different physical activities, without actually measuring the loads imposed on the spine.

Similar principles are also involved in other joints of the body, such as the neck and the knee. In the case of the knee joint, for example, the importance of the quadriceps muscle in stabilizing the joint when nearly extended is well recognized. In the treatment of hydrops in the knee joint, it is often overlooked that although bed rest and immobilization cause a resorption of the fluid, the concomitant atrophy of the quadriceps muscle caused by immobilization may result in recurrence of the hydrops, at times even more severe than before as the result of damage caused by inadequate stabilization by the atrophied quadriceps muscle. The correct treatment would be to put the patient to bed in order to relieve the joint of the burden of the body weight, but at the same time to train the leg muscles. In this connection, it should again be noted that the nutrition of the cartilage may be enhanced by allowing some activity of the joint in question.

Not infrequently, lumbar scoliosis is seen as a compensatory measure when the pelvis is tilted to one side, the lower limbs being of unequal length. To compensate for this, it is only by curving the lumbar spine through an angle equal to the pelvic tilt that the trunk can be held vertical. Usually, this compensatory scoliosis disappears automatically when the pelvic tilt is corrected by the use of an insole of proper thickness.

Observations among young Scandinavian girls engaged in elite calisthenics have indicated that extreme bending of the spine may cause damage, and that such exercises, therefore, should be omitted.

Kumar (1980) measured the intra-abdominal pressure in normal volunteers while lifting a weight of 10 kg from ground to knee, hip, and shoulder levels in the sagittal, lateral, and oblique planes. At the same time he recorded the EMG activity of the erector opinae and external oblique muscles. He found a significant difference between the responses in the three planes: the sagittal plane activities evoked least response. Intra-abdominal pressures, and EMG activity in the erector spinal and external oblique muscles were highly significantly correlated in each of the three planes. In subsequent studies, Kumar and Davis (1983) have examined spinal loading in static and dynamic postures recording both erector spinal EMG and intra-abdominal pressure.

REFERENCES

Atkinson, P. J., J. A. Weatherell, and S. M. Weidmann: Changes in Density of Human Femoral Cortex with Age, *J. Bone Joint Surg.* **44B**:496, 1962.

Basset, C. A. I.: Current Concepts of Bone Formation, *J. Bone Joint Surg.*, **44A**:1217, 1962.

Bourne, G. H. (ed.): "The Biochemistry and Physiology of Bone," vols. 1–3, Academic Press, Inc., New York, 1971.

Brodal, P.: Leddinnervasjon—et forsömt kapittel? *Fysioterapeuten*, **39**:65, 1972.

Drinkwater, B.L.: Osteoporosis and the Female Athlete, unpublished, 1985.

Eisenberg, E., and G. S. Gordan: Skeletal Dynamics in Man Measured by Nonradioactive Strontium. *J. Clin. Invest.*, **40**:1809, 1961.

Ekholm, R.: Articular Cartilage Nutrition, *Acta Anat.*, **11**(Suppl. 15–2): 1951.

Ekholm, J., U. P. Arborelius, and G. Németh: The Load on the Lumbo-sacral Joint and Trunk Muscle Activity during Lifting, *Ergonomics*, **25**(2):145, 1982.

Farfan, H. F.: "Mechanical Disorders of the Low Back," Lea & Febiger, Philadelphia, 1973.

Greenleaf, J.E.: Physiological Responses to Prolonged Bed Rest and Fluid Immersion in Humans. *J. Appl. Physiol.: Resp. Environ. Exerc. Physiol.*, **57**:619, 1984.

Grillner, S., J. Nilsson and A. Thorstenson: Intra-abdominal Pressure Changes during Natural Movements in Man, *Acta Physiol. Scand.*, **103**:275, 1978.

Hancox, N.J.: "Biology of Bone," Cambridge University Press, London, 1972.

Holmdahl, D. E., and B. E. Ingelmark: Der Bau des Gelenkknorpels unter Verschiedenen Funktionellen Verhältnissen, *Acta Anat.*, **6**:309, 1948.

Holmdahl, D. E., and B. E. Ingelmark: The Contact between the Articular Cartilage and the Medullary Cavities of the Bones, *Acta Anat.*, **12**:341, 1951.

Hult, L.: Cervical, Dorsal and Lumbar Spinal Syndromes, *Acta Orthop. Scand.* (Suppl. 17), 1954.

Ingelmark, B. E., and R. Ekholm: A Study on Variations in the Thickness of Articular Cartilage in Association with Rest and Periodical Load, *Uppsala Läkareförenings Förhandlingar*, **53**:61, 1948.

Jansen, M.: "On Bone Formation: Its Relation to Tension and Pressure," Longmans, Green & Co. Inc., New York, 1920.

Krølner, B., E. Tøndevold, B. Toft, B. Berthelsen, and S. Ports Nielsen: Bone Mass of the Axial and the Appendicular Skeleton in Women with Colles' Fracture: its Relation to Physical Activity, *Clin. Physiol.*, **2**:147, 1982.

Kumar, S.: Physiological Responses to Weight Lifting in Different Planes, *Ergonomics*, **23**(10):987, 1980.

Kumar, S., and P. R. Davis: Spinal Loading in Static and Dynamic Postures: EMG and Intra-Abdominal Pressure Study, *Ergonomics*, **26**(9):913, 1983.

Lanyon, L. E., and J. A. O'Connor: Adaptation of Bone Artificially Loaded at High and Low Physiological Strain Rates, *J. Physiol.*, **303**:36P, 1980.

LeBlond, C. P., and R. C. Greulich: Autoradiographic Studies of Bone Formation and Growth, in G. H. Bourne (ed.), "The Biochemistry and Physiology of Bone," Academic Press, Inc., New York, 1956.

Lucas, D. B., and B. Bresler: "Stability of the Ligamentous Spine," *Technical Report Series II*, no. 40, University of California, Biomechanics Laboratory, Berkeley, December 1960.

McLean, F. C. and M. R. Urist: "Bone: An Introduction to the Physiology of Skeletal Tissue," The University of Chicago Press, Chicago, 1955.

Micheli, L. J.: Overuse Injuries in Children's Sports: The Growth Factor, *Orthopedic Clinics of North America*, **14**(2):337, 1983.

Montoye, H. J., E. L. Smith, D. F. Fardon, and E. T. Howley: Bone Mineral in Senior Tennis Players, *Scand. J. Sports Sci.*, **2**(1):26, 1980.

Morris, J. M., D. R. Lucas and B. Bresler: Role of the Trunk in Stability of the Spine, *J. Bone Joint Surg.*, **43A**:327, 1961.

Nachemson, A. L.: the Lumbar Spine an Orthopedic Challenge, *Spine*, **1**:59, 1976.

Nachemson, A. L.: Disc Pressure Measurements, *Spine*, **6**:93, 1981.

Noyes, F. R., P. J. Torvik, W. B. Hyde, and J. L. De Lucas: Biomechanics of Ligament Failure. *J. Bone Joint Surg.*, **56a**(7):1406, 1974.

Radin, E. L.: The Effects of Ageing on the Skeleton, in A. Viidik (ed.), "Lectures on Gerontology, vol. 1: On Biology of Ageing," p. 379, Academic Press, London, 1982.

Rodahl, K., J. T. Nicholson, and E. M. Brown (eds.): "Bone as a Tissue," McGraw-Hill Book Company, New York, 1960.

Rodahl, K., N. C. Birkhead, J. J. Blizzard, B. Issekutz, Jr., and E. D. R. Pruett: Fysiologiske Forandringer under Langvarig Sengeleie, *Nord. Med.*, **75**:182, 1966.

Romer, A. S.: "Man and the Vertebrates," The University of Chicago Press, Chicago, 1957.

Ruben, J. A., and A. F. Bennett: Intense Exercise, Bone Structure and Blood Calcium Levels in Vertebrates, *Nature*, **291**:411, 1981.

Schultz, A., G. Andersson, R. Ørtengren, K. Haderspeck, and A. Nachemson: Loads on the Lumbar Spine. Validation of a Biomechanical Analysis by Measurements of Intradiscal Pressures and Myoelectric Signals, *J. Bone Joint Surg.*, **64a**:713, 1982.

Skoglund, S.: Anatomical and Physiological Studies of Knee-joint Innervation in the Cat, *Acta Physiol. Scand.*, **36**(Suppl. 124):1956.

Tipton, C. M., R. D. Matthes, J. A. Maynard, and R. A. Carey: The Influence of Physical Activity on Ligaments and Tendons, *Med. and Sci. in Sports*, **7**:165, 1975.

Troupp, H.: Nervous and Vascular Influence on Longitudinal Growth of Bone; an Experimental Study on Rabbits, *Acta Ortho. Scand.*, **51**(Suppl.):1, 1961.

Whedon, G. D., L. Lutwak, P. Rambaut, M. Whittle, C. Leach, J. Reid, and M. Smith: Mineral and Nitrogen Metabolic Studies on Skylab Flights and Comparison with Effects of Earth Long-term Recumbency, *Proc. COSPAR Symposium on Gravitational Physiology*, Varna, Bulgaria, May 30–31, Bulgarian Academy of Sciences Press, Sofia, 1975.

White, M. K., R. B. Martin, R. A. Yeater, R. L. Butcher, and E. L. Radin: The Effects of Exercise on the Bones of Postmenopausal Women, *International Orthopaedics*, 1985 (in press).

Wickström, G.: Effect of Work on Degenerative Back Disease, *Scand. J. Work Environ. Health*, **4**(Suppl. 1):1, 1978.

CHAPTER 7

PHYSICAL PERFORMANCE

CONTENTS

DEMAND—CAPABILITY
AEROBIC PROCESSES
  Intensity and Duration of Exercise
  Recovery
  Intermittent Exercise
  Prolonged Exercise
  Muscular Mass Involved in Exercise

ANAEROBIC PROCESSES
  Power and Capacity for High-Energy Phosphate Breakdown
  Power and Capacity of Anaerobic Glycogen Breakdown
LACTATE PRODUCTION, DISTRIBUTION, AND DISAPPEARANCE
  Time Course of the Blood Lactate Concentration
  What is the Fate of the Lactate Produced?
  How Does Continued Light Exercise Influence the Lactate Removal.
  Effects of Metabolism on Tissue and Blood pH

INTERACTION BETWEEN AEROBIC AND ANAEROBIC ENERGY
    YIELD
  Anaerobic Threshold

MAXIMAL AEROBIC POWER—AGE AND SEX
  Age

Maximal Aerobic Power—Absolute Values
Maximal Aerobic Power—Related to Body Size
A High Maximal Oxygen Uptake Does Not Guarantee Top Performance
ANAEROBIC POWER, SEX, AND AGE
MUSCLE STRENGTH, SEX, AND AGE
Age
Sex
PERFORMANCE IN SPORTS, WOMEN VERSUS MEN
CONCLUDING REMARKS

DEMAND—CAPABILITY

Athletic competition represents the classical test of physical fitness or performance capacity. Under such conditions the performance may be measured objectively in centimeters or seconds, or it may be judged subjectively, as in gymnastics, figure skating, or diving. The individual's performance is the combined result of the coordinated exertion and integration of a variety of functions. The demands of the actual event must be perfectly matched by the individual's capabilities in order to achieve top performance and championship. It is impossible to present one formula that takes into account all aspects of a person's maximal performance, since the demands set by different types of activities vary greatly. However, the following factors may serve as a frame of reference for our discussion.

Natural endowment (genetic factors) probably plays a major role in a person's performance capacity, at least for those persons aspiring to the levels required for the attainment of Olympic medals. The individual's reponse to training is also associated with an endowed genotype. Thus, it appears that up to 70 percent of an individual's maximal force, power, or capacity is a matter of genetical factors (see Bouchard and Malina, 1983). Since the possible genetic combinations are astronomical in number, it is an interesting question whether a country must have a population of 100,000, 1 million, 10 million, or more, to "breed" an individual with proper endowment for top results. The more popular an event, the greater is the chance that an individual with the suitable constitution will participate and thus discover his or her ability. Obviously, the environment and geographic location are also important. If an individual with the perfect endowment for skiing grows up in a place where skiing is impossible, this endowment may be wasted from an athletic standpoint. The fact that an increasingly large number of naturally endowed persons enter the ranks of competitive athletes may in part explain the gradual improvement of athletic records.

Granted the endowment, however, definite improvement in performance may be achieved by *training*, and all factors listed in Fig. 7-1 as contributing to physical performance capacity may be modified. The very intense training programs currently employed in many fields of athletic performance contribute greatly to the improved results. Another factor explaining the gradual improvement in athletic achievement

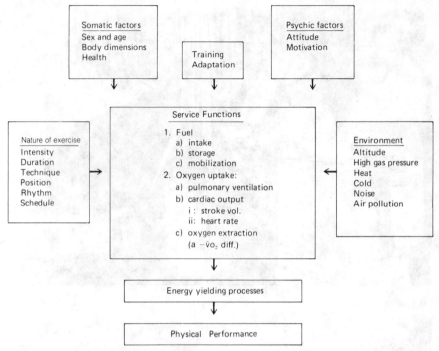

Figure 7-1
Factors influencing the power and capacity for aerobic muscular activity.

over the years is the better techniques applied and the superior equipment which is becoming available through technical progress.

The athletes themselves are mainly concerned with improving their ability to cut off seconds or add centimeters to their records. The scientist is interested in analyzing why the results improve or vary from time to time. Therefore, the scientific objective is (1) to evaluate quantitatively the influence of the various factors upon the performance capacity in different tasks (performance requirements); (2) to examine how these factors vary with sex, age, and body size (capacity profile); and (3) to study the effect of such factors as training and environment. It is realistic to conclude that scientists have merely begun a systematic research on the performance capacity and the many factors involved. The most advanced information concerns the energy output by aerobic processes. This may be explained by the fact that methods for quantitative measurements of energy output by the human combustion engine have long been available, actually ever since Lavoisier's discovery of the oxygen utilization by living animals. We shall therefore begin the more detailed analysis of physical performance with a discussion of the oxygen uptake during submaximal and maximal exercise and the maximal aerobic power (the individual's maximal oxygen uptake).

(It should be emphasized that *capacity* denotes total energy available, and that *power* means energy per unit of time.)

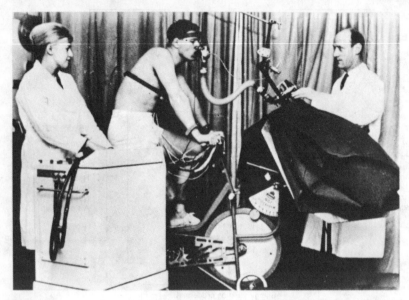

(a)

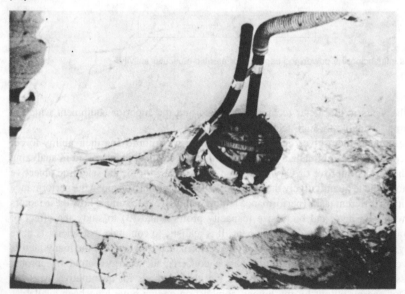

(b)

Figure 7-2
Application of the Douglas bag method for measuring aerobic energy output during different types of exercise. The skier shown in (c) carries a three-way stopcock and a stopwatch on his chest for the recording of time during which the expired air is collected in the Douglas bag. The stopwatch automatically starts and stops when the stopcock is turned.

(c)

Figure 7-2 cont.

AEROBIC PROCESSES

The complexity of the capacity for aerobic muscular exercise is illustrated by Fig. 7-1. For each liter of O_2 consumed, about 20kJ, range 19.7 to 21.2 (5 kcal, range 4.7 to 5.05) will be delivered; hence, the higher the oxygen uptake, the higher the energy aerobic output. The oxygen uptake during exercise may be measured with an accuracy of ± 0.04 liter $\cdot$ min^{-1} ($\dot{V}_{O_2} > 1$ liter $\cdot$ min^{-1}). Figure 7-2 gives examples of how the classical Douglas bag method can be applied when studying the aerobic energy output during exercise.

Intensity and Duration of Exercise

Figure 7-3*a* shows how oxygen uptake increases during the first minutes of exercise to a "steady state" where the oxygen uptake corresponds to the demands of the tissues. When the exercise stops, the oxygen uptake gradually decreases to the resting level; and the so-called "oxygen debt" is paid off (see below).

The slow increase in oxygen uptake at the beginning of exercise is explained by the sluggish adjustment of respiration and circulation, i.e., the sluggish adjustment of the oxygen-transporting systems to exercise; during the first 2 to 3 min there is an *oxygen deficit*. The attainment of this state coincides roughly with the adaptation of cardiac output, heart rate, and pulmonary ventilation. A steady state condition denotes a work situation where oxygen uptake equals the oxygen requirement of the tissues; consequently, there is no accumulation of lactic acid in the body. Heart rate, cardiac output, and pulmonary ventilation have attained fairly constant levels.

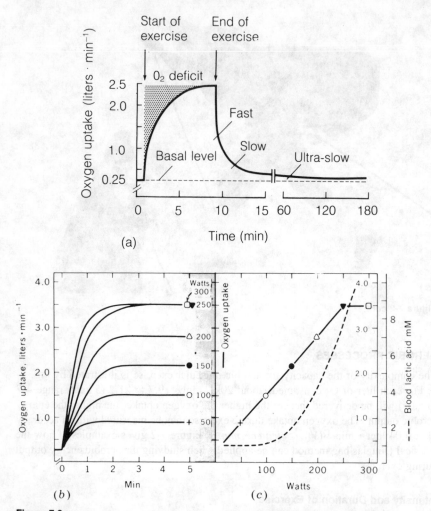

Figure 7-3
(a) During the first minutes of exercise there is an oxygen deficiency while the oxygen uptake increases to a level adequate to meet the oxygen demand of the tissue. At the cessation of exercise, there is a gradual decrease in the oxygen uptake and several components can be identified. Note the change in time scale. (*Modified from Newsholme and Leech, 1983.*)
(b) Schematic demonstration of the increase in oxygen uptake during exercise on cycle ergometer with different intensities (noted within shadowed area) performed during 5 to 6 min.
(c) Oxygen uptake in the above-mentioned experiments, measured after 5 min and plotted in relation to rate of exercise. Note that 250 watts (1500 kpm · min⁻¹) brought the oxygen uptake up to this subject's maximum and that 300 watts did not further increase the oxygen uptake; the increased rate of exercise was possible thanks to anaerobic processes. Maximal aerobic power = 3.5 liters · min⁻¹. (For simplicity, the work rate which is sufficient to bring the oxygen uptake to the subject's maximum, in this case 250 watts, may be written $WL_{max} \dot{V}_{O_2}$.) Peak lactate concentrations in the blood at each experiment have been included (blood samples secured immediately after exercise and every second min up to 10 min of recovery).

However, in the steady state situation there may be a slight gradual increase in the aforementioned parameters, which may be a consequence of an increasing body temperature (see Chap. 13) and an increase in the utilization of free fatty acids as a substrate (which costs more oxygen per unit of energy yield, see Chap. 12). It should be noted that during prolonged exercise there is also a change in the body's store of energy and water if not replaced. However, everything considered, the steady state concept as defined above is very useful when discussing the energy-yielding processes during exercise.

In light exercise, the energy output during the first minutes of exercise can be delivered aerobically, since oxygen is stored in the muscles bound to myoglobin and in the blood perfusing the muscles. During more severe exercise, anaerobic processes must supply part of the energy during the early phase of exercise.

The breakdown of energy-rich phosphate compounds, ATP and phosphocreatine, PCr, is certainly anaerobic and essential but, their quantitative role is limited. Therefore the anaerobic breakdown of glycogen (glycogenolysis) and glucose (glycolysis) to lactic acid has an important potential to support the aerobic processes when they cannot provide enough energy for ATP production (see below and Chap. 12). Actually, in exercise that engages large muscle groups, e.g., running, which demands an oxygen uptake higher than 50 percent of the individual's maximum and, which is performed for some minutes, lactate produced in the activated muscles escapes and appears in the blood (see Fig. 7-3c). (The well-trained person can exercise at a higher percentage without an increase in the blood-lactate concentration than an untrained individual.) The heavier the exercise, the more important the anaerobic energy yield and the muscle and blood-lactate concentrations increase. As the exercise becomes subjectively more strenuous, a decrease in the body's pH affects muscular tissues, respiration, and other functions.

Figure 7-3c illustrates the linear increase in oxygen uptake, measured after about 5 min of exercise with different work rates up to a point where the maximum for oxygen transport and uptake in the tissues appears to be reached. In this case, the maximal oxygen uptake is 3.5 liters $\cdot$ min^{-1}, which is this subject's *maximal aerobic power*. There are two main criteria showing that it actually represents the subject's maximal aerobic power. (1) There is no further increase in oxygen uptake despite further increase in the rate of exercise; and (2) the blood lactate concentration exceeds 8 to 9 mM. (In children and old subjects it can be difficult to attain such high values, however.) The assumption is that large muscle groups are involved in the exercise and that the exercise time is above about 3 min. As will be discussed in Chap. 10, it is not possible to establish a "levelling off" of the oxygen uptake in all subjects (point 1). Furthermore, most individuals attain a slightly higher oxygen uptake (on the average 6 percent) in a maximal running test compared with cycling. Therefore, one must be careful when evaluating "maximal oxygen uptake" data.

From a methodological viewpoint, it is important to emphasize that maximal oxygen uptake is attained at rates of exercise that are not necessarily maximal. From Fig. 7-3c, it is obvious that 250 watts were enough to reveal the subject's maximum, 3.5 liters $\cdot$ min^{-1}, but the subject was not exhausted by the rate of exercise. This individual could actually endure 300 watts for the same period of time. *An all-out test is not necessary for the assessment of an individual's maximal aerobic power.*

The heavier the rate of exercise, the steeper is the increase in the oxygen uptake (and heart rate). This is illustrated by Fig. 7-4. After a 10-min period of exercise at 50 percent of maximal oxygen uptake, intensities of 300 to 450 watts were applied until exhaustion. The tolerated exercise time varied from 6 min (300 watts) to less than 2 min (heaviest load). The oxygen uptake at the end was the same in all experiments, or about 4.1 liters $\cdot$ min^{-1}. However, after 1 min of extremely heavy exercise, the oxygen uptake was 4.0 liters $\cdot$ min^{-1} at the "supermaximal" intensity but only 3.0 liters $\cdot$ min^{-1} during the less extreme but still heavy work rate of 300 watts, which could not be tolerated for more than 6 min (P-O. Åstrand and Saltin, 1961a).

There are several implications from these experiments: (1) In studies where maximal oxygen uptake is to be measured, the collection of expired air or other measurements may start after about 1 min of exercise, provided that the rate of exercise is extremely heavy (supermaximal) and is preceded by a warming-up period. For many reasons, however, it is wise to aim at an exercise period of about 5 min. (2) An exercise time of 1 min or even less may maximally load the oxygen-transporting system. (3) It is an interesting question why the ability to increase the oxygen uptake maximally within 1 min is not utilized in exercise that can be prolonged to 5 to 10 min and which has a marked oxygen deficit during the first minutes. Such an increase would be an advantage, for the sooner the aerobic processes could come into full swing, the less

Figure 7-4
Curves showing increase in oxygen uptake during heavy exercise following a 10-min warm-up period. Arrows indicate time when the subject had to stop due to exhaustion. Figures indicate work load on the cycle ergometer. The subject could continue the power of 275 watts (1650 kpm $\cdot$ min^{-1}) for more than 8 min. (From P.-O. Åstrand and Saltin, 1961a.)

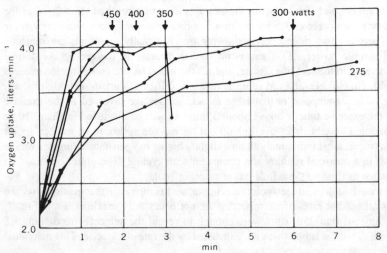

demand there would be on the anaerobic processes, and less lactic acid would therefore accumulate.

The kinetics of the increase in oxygen uptake during the first minutes of an exercise, which lead to a steady state situation have a time constant of about 30 s, about 20 s when a given moderate exercise is preceded by a warming-up period. Apparently, the $\dot{V}_{O_2}$ kinetics are different during maximal and "supermaximal" exercises.

In repeated determination of maximal oxygen uptake on the same subject, the standard deviation is 3 percent, which includes biological and methodological variables (see Chap. 8).

Recovery

It may take 60 min or more before the oxygen uptake, and rate of aerobic metabolism, return to the level attained in the subject resting before the exercise. After relatively heavy exercise one may identify three phases:

1 First a fast exponential component in the decline in oxygen uptake with a half-time of about 30 s. Most likely this rapid phase is associated with an aerobic replenishment of the ATP and PCr stores and a refilling of the oxygen stores of the body (myoglobin and hemoglobin). During the first minutes of recovery there is a rapid resynthesis of the energy-rich phosphates. Actually, there is a close relationship between the reduction in ATP and PCr on the one hand, and both oxygen deficit and the fast component of the "oxygen debt" on the other (see Knuttgen and Saltin, 1973). A realistic figure for refilling depleted oxygen stores is about 0.5 liter and the oxygen cost for the ATP and PCr production is up to 1 to 1.5 liter (body weight about 70 kg, nonobese person).

2 Then comes a more complex slow component which, after a supermaximal exercise, has a half-time of about 15 min. In the classical literature, it has been attributed to the energy cost of the elimination of the lactate produced, i.e., a payment of a *lactic oxygen debt*. This concept has been heavily criticized by Brooks and his group. They suggest that the term should be deleted and replaced by "excess postexercise oxygen consumption," EPOC (Brooks and Fahey, 1984, pp. 191–215).

Inevitably most of the energy yielded in the metabolism is converted to heat and there is an increase in tissue temperature, which elevates the metabolic rate (about 13 percent elevation per degree centigrade). The increase in sympathetic nerve activity will *via* epinephrine and norepinephrine also stimulate the metabolism. An elevated oxygen demand of the activated respiratory muscles and heart will also contribute. It is not known whether or not a bout of exercise gives rise to other factors that may elevate the metabolism during recovery (hormones like thyroxine, metabolites, disturbed ion balance).

As mentioned, when the oxygen demand during exercise exceeds the oxygen supply, the breakdown of glycogen to lactate will support the metabolism. In very heavy exercise lasting for minutes or more, substantial amounts of glycogen are used up. During recovery some of that lactate is reconverted to glycogen, a process that demands energy (see "What is the Fate of the Lactate Produced").

Altogether, the oxygen uptake during recovery from maximal exercise of some 5 min duration may, in extreme cases, amount to almost 40 liter during the following 60 min. At rest, the individual would, during the same period of time, consume about 18 liters of oxygen.

3 After sustained exercise there is a slightly elevated metabolism lasting for several hours, eventually for at least 24 hours. It is suggested that the oxygen consumed during this ultra-slow phase is due to a stimulation of substrate cycles (see Newsholme and Leech, 1983, pp. 699–700).

Summary In many types of exercise, the oxygen uptake increases roughly linearly with an increase in the rate of exercise. The maximal oxygen uptake or maximal aerobic power is defined as the highest oxygen uptake the individual can attain during exercise engaging large muscle groups while breathing air at sea level (exercise time 2 to 6 min, depending on the type of exercise).

During heavy exercise, anaerobic processes contribute to the energy yield not only at the beginning of exercise but continuously throughout the exercise period. An accumulation of metabolites will eventually necessitate the termination of the exercise.

In very heavy exercise, the maximal oxygen uptake and maximal heart rate may be attained within 1 min, providing that a sufficient warm-up period precedes the maximal effort.

During recovery, there is a gradual decline in the oxygen uptake with a fast component that is evidently associated with the energy cost to replenish the ATP and PCr pool and to refill the body's oxygen stores (no doubt this is a payment of an oxygen debt). Then follows a slow component, and eventually an ultra-slow phase, the duration of which depends on the severity and duration of the exercise. The slow component is at least partly caused by an elevation in tissue temperature, circulating catecholamines, and if the exercise has been highly anaerobic, aerobic processes are involved in a resynthesis of glycogen from lactate.

(The kinetics of the oxygen uptake at the beginning of exercise and during recovery are discussed by di Prampero, 1981; di Prampero et al., 1983.)

Intermittent Exercise

Muscular work in industrial or recreational activities is very seldom maintained for very long at a steady rate. For this reason, a steady state, as discussed earlier, is rarely attained. The classical laboratory studies, with subjects exercising continuously for 5 min or longer on the treadmill or cycle ergometer, in many ways, represent artificial situations. Nevertheless, such procedures have distinct advantages when one is studying the physiology of exercise or studying patients, for they provide standardized conditions and permit comparisons to be made on repeated occasions. They may also simulate the demands placed on the body in many sport events. However, from both a practical and a theoretical point of view, it is equally important to study the effect of intermittent exercise, which better mirrors the type of muscular activities encountered in industry or at home and in most types of ordinary exercise or recreational activity.

Some of the most important principles will be discussed by presenting some experiments concerning intermittent exercise (I. Åstrand et al., 1960a,b; Christensen et al., 1960; Essén, 1978).

1 A subject whose maximal oxygen uptake was 4.6 liters · min^{-1} could exercise at 350 watts for about 8 min. Since the oxygen need was approximately 5.2 liters · min^{-1}, the anaerobic processes had to provide part of the energy. When the rate of exercise was reduced to 175 watts the exercise could easily be prolonged to 60 min, the final heart rate was 135 and oxygen uptake 2.45 liters · min^{-1}, and the blood lactate concentration did not increase above resting level. The total oxygen uptake during the hour was 145 liters.

2 In another experiment with the same subject, the rate of exercise was again 350 watts, but now exercise periods of 3 min were alternated with 3-min rest periods. The subject could proceed with great difficulty for 1 hr, and the same total amount of work was performed as in Experiment 1. The oxygen uptake and heart rate were now maximal, as was the peak blood lactate concentration (13.2 mM). The total energy output during the second experiment was about 10 percent higher than in the first one.

3 When the heavy exercise periods were shortened by introducing the rest periods more frequently, the *total* oxygen uptake over the hour was not markedly reduced. The subjective feeling of strain was less severe, however, and peak oxygen uptake, heart rate, and blood lactate concentration were now lower. Hence, with intermittent exercise and rest for 30 s respectively, the heart rate did not exceed 150, the blood lactate was only 2.2 mM, and the total oxygen uptake was 154 liters during the hour. (The subject's maximal heart rate was 190.) (Some of these data are presented in Table 10-2.)

Figure 7-5a illustrates another set of experiments with the same subject. He exercised on the cycle ergometer with an extremely heavy intensity of 412 watts. When exercising continuously at this rate, he became exhausted within about 3 min. When exercising intermittently for 1 min and resting for 2 min, he could continue for 24 min before being totally exhausted, and the blood lactate concentration rose to 15.7 mM. In another experiment, the periods of exercise were reduced to 10 s and the rest periods to 20 s. Now he could complete the intended production of 247 kJ within 30 min with no severe feeling of strain, and his blood lactate concentration did not exceed 2 mM, indicating an almost balanced oxygen supply to his heavily stressed muscles. With periods of exercise and rest of 30 and 60 s respectively, intermediate results were obtained.

A prolongation of the rest periods, with the ratio between exercise and rest changed to 1:4, gave, of course, a decreased total work output but had scarcely any beneficial effect on the subject's fatigue. The critical factor was the length of the exercise periods, and the duration of the rest pauses and the total time spent resting during the 30-min period were only of secondary importance.

Figure 7-5b attempts to explain the findings above. When a person exercises intermittently for short periods, in this case of 10 s duration, at an extremely high energy output, the aerobic metabolism is apparently adequate despite an insufficient transport of oxygen during the burst of activity. There is, at least, no continuous increase in blood lactate concentration. Gradually there is a vasodilation in the active muscles,

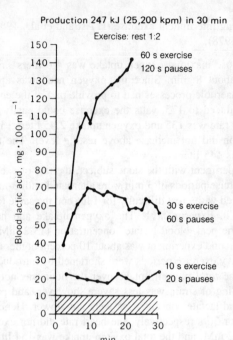

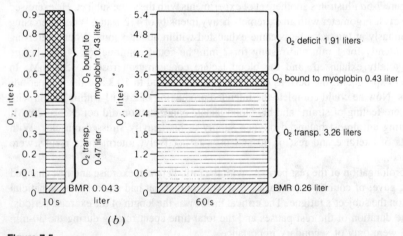

Figure 7-5
(a) The blood lactate concentration in a total work production of 247 kJ (25,200 kpm) in 30 min. The exercise was accomplished with a work rate of 412 watts (2520 kpm · min⁻¹), the exercise periods being 10, 30, and 60 s, and the corresponding rest periods 20, 60, and 120 s respectively. (*From I. Åstrand et al., 1960b.*)
(b) The oxygen requirement for 10- and 60 s power of 412 watts. The schematic drawing indicates the basal metabolic rate (BMR), the calculated fraction of O_2 bound to myoglobin, transported by the blood, and the O_2 deficit. (*From I. Åstrand et al., 1960b.*)

which will secure an improved blood supply. In addition there is an oxygen store in the myoglobin that can be consumed during the bout of exercise. During the following period of rest, this depot is refilled with oxygen. Saltin et al. (1976), and Essén (1978) have repeated this type of intermittent exercise protocol, supplemented with studies of the metabolic events in the activated muscles (samplings were secured with needle biopsy technique). During the exercise (5 to 20 s), there was a reduction in the ATP and PCr concentration which, however, was restored during the period of rest, evidently by aerobic processes. In intermittent exercise at the same work rate as in continuous exercise, less glycogen is utilized and the lactate concentration in the muscle is much lower. It should be emphasized that approximately 13 times more ATP can be replenished when glycogen is metabolized aerobically as compared with the efficiency in anaerobic breakdown of glycogen to lactate (see Chap. 12). In the intermittent exercise, lipids also contribute more in the energy yield than in continuous exercise at the same intensity, which is also glycogen saving.

Apparently, in intermittent exercise with short exercise periods, one can endure very high rates of exercise aerobically and therefore with little lactate production. It is, however, essential that the exercise periods are kept sufficiently brief (around 15 s) to prevent the oxygen supply from being exhausted and the anaerobic lactate production from being too great. By spacing the exercise so that running periods lasted for 10 s and resting for 5 s, a subject could prolong the total exercise + rest period to 30 min without undue fatigue at a speed that normally exhausted him after about 4 min continuous running (Christensen et al., 1960). This type of periodic activity is important in athletic training (see Chap. 10) but has also its application in manual labor and many leisure activities.

Dynamic exercise is certainly an intermittent type of activity, and its superiority over static exercise as an endurance exercise may be explained partly on the basis of the muscle pump and the alternating emptying and filling of the oxygen stores during alternating muscle contraction and relaxation.

Summary   The buffering effect of an oxygen store may mean, practically speaking, that a great amount of exercise can be performed at an extremely heavy rate, with a relatively low peak demand on the circulation and respiration, by the introduction of properly spaced, short exercise and rest periods ("micropauses"). The heavier the work rate, the shorter should be the exercise periods. This physiological concept has at least two important applications:

1 It may explain why older or physically disabled individuals, in spite of a reduced maximal aerobic power, can remain in jobs involving heavy manual labor such as forestry, farming, and construction, or enjoy physically demanding hobbies. As long as they are free to choose the optimal length of the exercise and rest periods, the acute loads on the respiration and circulation may not exceed the limits of their reduced capacity. It should be emphasized, however, that if the pace is determined by a machine, even a less heavy peak load, but with relatively long activity periods, may overtax the capacity of the worker whose physical performance capacity is limited.

2 If the aim of a training program is to increase muscle strength, the highest load on the muscle fibers may be obtained within a given period of time, if periods of rest are frequently interspaced between activity periods of 5 to 10 s. On the other hand, training of the oxygen-transporting system will be easier to accomplish if the exercise periods are prolonged to at least 2 to 3 min. This type of exercise would also adapt the tissues to high lactate concentrations, provided the exercise is severe (see Chap. 10).

Prolonged Exercise

Moderately well-trained individuals may walk or run for about 1 hr with an oxygen uptake up to about 50 percent of the $\dot{V}_{O_2}$ max, maintaining the oxygen uptake, heart rate, and cardiac output at approximately the same level as attained after about 5 min of exercise. The lactate concentration in active muscles and in the arterial blood is not elevated, indicating a steady state (Fig. 7-6a). When the exercise time is further prolonged, there is a progressive increase in oxygen uptake and heart rate, and the subject becomes more or less fatigued. Figure 7-6b illustrates an experiment in which exercise was performed continuously for seven 50-min periods at an oxygen uptake of 50 percent of the subject's maximal oxygen uptake. The subjects rested for 10 min in between, and after 4 hr of exercise (cycling, alternating with running) they had a break of 1 hr for lunch. The most fit subject, with a maximal oxygen uptake of 5.60 liters · min^{-1}, exercised with an average oxygen uptake of 2.75 liters · min^{-1}; one subject with a maximal aerobic power of 2.25 liters · min^{-1} exercised with a work rate requiring an oxygen uptake of 1.15 liters · min^{-1}. The four subjects participating in the experiments could fulfill the task, but they were fatigued. It appears that a 50 percent demand on the aerobic power is too high for a steady state if the physical activity is continuous for an entire working day. Actually, in manual labor a 40 percent ceiling is advisable (Chap. 11).

The well-trained individual can maintain steady state (as defined above) at a higher relative work rate than the 50 percent, indicating a more efficient oxygen transport and oxygen and substrate utilization in their active muscles. Elite cross-country skiers can exercise at 85 percent of their maximal aerobic power at least for 1 hr, oxygen uptake being 4.5 liters · min^{-1} or even higher (P.-O.Åstrand et al., 1963).

Well-trained athletes, including marathon runners can exercise for hours with an oxygen uptake around 70 to 80 percent of their maximum with little or no increase in blood lactate concentration (see Costill, 1970; Hedman, 1957; Maron et al., 1976). Davies and Thompson (1979) studied "ultradistance" runners who were running for 24 hr. They noticed a gradual decline in the engagement of the aerobic potential. It dropped from about 90 percent at the beginning to slightly below 50 percent of the maximal aerobic power at the end. The energy output was estimated to be close to 78 MJ (18,600 kcal).

In Chap. 3, we discussed muscular fatigue, particularly in activities demanding maximal or close to maximal forces. In prolonged activities, a relatively small fraction of maximal strength is usually applied. Still a gradually declining power output, despite efforts to continue at a high intensity, may be caused by disturbances in the neuromuscular function, but other factors may also be involved. During prolonged heavy exercise, the water balance may be disturbed and the stores of available energy,

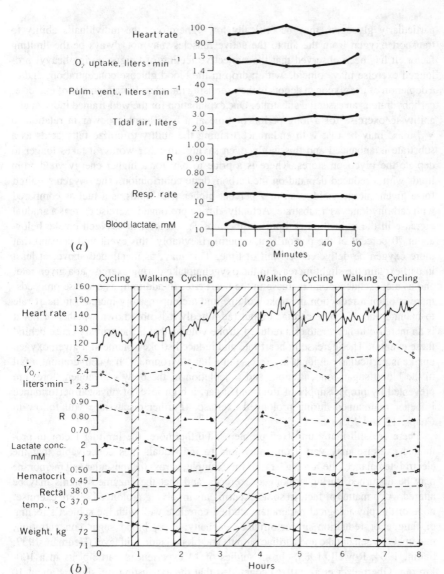

Figure 7-6
(a) Metabolic parameters during 1-hr exercise in a subject exercising at an intensity demanding close to 50 percent of his maximal aerobic power (1.5 versus 2.94 liters $O_2 \cdot min^{-1}$). (*Data from I. Åstrand et al., 1959.*)
(b) Metabolic parameters in one subject during an experiment consisting of seven exercise periods of 50 min each. The shaded columns represent rest periods. The subject's maximal aerobic power was 4.6 liters $O_2 \cdot min^{-1}$. (*From I. Åstrand, 1960.*)

particularly glycogen, may be critically low. Therefore, the individual's ability to transport oxygen from the air to the active muscles may not always be the limiting factor. It has been observed that the subjective feeling of fatigue during heavy, prolonged exercise may coincide with a drop in the blood glucose concentration, and/or a depletion of the glycogen depots in the exercising muscles. For a high rate of exercise, carbohydrate is an essential substrate. One explanation for the well-trained individual's ability to exercise for a long period of time at a high aerobic power in relation to $\dot{V}_{0_2}$ max, may be that with endurance training the ability to utilize fatty acids as a substrate is enhanced and this is glycogen saving. In other words, it takes longer to deplete the glycogen stores. There is a price to pay for a higher energy yield from lipids with a reduced demand on the carbohydrate contribution. The oxygen required for a given energy yield is up to 7 percent higher with lipids as a fuel as compared with carbohydrates as a substrate. Actually, during prolonged exercise there is a gradual increase in the utilization of lipids in the muscle metabolism (oxygen uptake below about 70 percent of the person's maximum). Inevitably, this event will demand that more oxygen be delivered per unit of time. This may, as mentioned above, at least in part explain the slight increase in the oxygen uptake during exercise at a given rate, and as a consequence, an increase in the heart rate. In addition, as exercise proceeds, there is often a reduction in stroke volume and a compensatory increase in heart rate to maintain an adequate cardiac output (admittedly it is not known whether or not it is an increase in heart rate or a reduced stroke volume that is the primary cause behind these events). The increase in heart rate and reduced stroke volume at a given oxygen uptake is particularly evident in exercise in a hot environment. If a dehydration, a fall in the blood sugar concentration, and a depletion of the body's glycogen stores are prevented by proper supply of fluid and sugar, a high level of physical performance is better maintained during prolonged exercise. (Further discussions will follow in Chaps. 12 and 13.)

There are still many unsolved problems. Furthermore, the limiting factor in prolonged exercise may vary from individual to individual. It is conceivable that the electrolyte balance, the K^+/Na^+ ratios for example, across the muscular cell membrane may be disturbed during prolonged exercise, and that the enzyme systems may be altered. As a matter of fact, in some experiments involving prolonged severe exercise, none of the physiological parameters studied correlate well with the subject's feeling of fatigue or reduction in performance capacity, e.g., blood sugar concentration, maximal oxygen uptake and cardiac output, and blood lactic acid level (Hedman, 1957; Saltin, 1964; Rowell, 1983). From studies of 14 experienced participants in a 100-km run, Oberholzer et al. (1976) concluded that the exhausting run, lasting about 10 hours, caused no distinguishable destructions in mitochondrial or cellular ultrastructure. A decrease in the activities of some extra- and intramitochondrial enzymes could be attributed to a reduced synthesis and/or enhanced loss of enzyme proteins (analysed on biopsies taken out of the vastus lateralis muscle; however, in running this muscle is less active than, e.g., the gastrocnemius).

Motivation is undoubtedly an important factor determining the endurance during heavy exercise. Well-trained, highly motivated subjects may maintain the oxygen uptake at a maximal level for at least 15 min, although most individuals feel forced to stop after 4 to 5 min at a work rate which taxes the oxygen-transporting systems to a maximum. Furthermore, physical performance may be subject to diurnal (A. Rodahl et al., 1976), as well as seasonal variations (Erikssen and Rodahl, 1979; see Shephard, 1984a).

Summary It is obvious that the individual's maximal aerobic power plays a decisive role in his or her physical performance. If a given task demands an oxygen uptake of 2.0 liters $\cdot$ min^{-1}, the man with a maximal O_2 uptake of 4.0 liters $\cdot$ min^{-1} has a satisfactory safety margin, but the 2.5-liter individual must exercise close to his maximum, and consequently his internal equilibrium becomes much more disturbed. In prolonged exercise, motivation, state of training, water balance, and depots of available energy are important for the performance capacity. It should be pointed out that a technique and efficiency factor is of decisive importance for the energy cost of a given task, more so in activities that demand skill like swimming and cross-country skiing.

Muscular Mass Involved in Exercise

The demand on the oxygen-transporting functions varies with the size of the active muscles. Since isometric contractions hinder the local blood flow and dynamic exercise facilitates the circulation, it follows that a greater oxygen uptake can be obtained during dynamic exercise. Usually, exercise involves both static and dynamic muscle contractions. Static exercise produces a relatively high heart rate and arterial blood pressure; this result may complicate a task evaluation based on the measurements of heart rate and blood pressure (Chap. 4).

For all practical purposes, the maximal oxygen uptake is approximately the same whether it is measured while running on a treadmill or during cross-country skiing or bicycling (see Chap. 8). In maximal work on a cycle ergometer in the supine position, the oxygen uptake is, however, only about 85 percent of the value obtained in the sitting position. But, if the subject exercises with both legs and arms simultaneously in the supine position, the oxygen uptake, cardiac output, and heart rate go up to the values typical for maximal exercise in the upright position (Stenberg et al., 1967). One plausible explanation for the lower aerobic power for maximal cycling in the supine position, despite an optimal venous return to the heart, is the less favorable position, since the body weight cannot be utilized during the critical stages of the pedaling. Secondly, the blood perfusion of the activated leg muscles is enhanced when exercising in the upright position (Chap. 4).

In a group of about 70 fairly well-trained women and men, no difference was noted in maximal oxygen uptake in the two types of exercise: exercise on the cycle ergometer and uphill running on a motor-driven treadmill at an inclination of 1.75 percent (P.-O. Åstrand, 1952). Kasch et al. (1976) established maximal oxygen uptake on 12 well-trained students using two treadmill protocols: horizontal and inclined (>4 percent). No significant differences were found in maximal oxygen uptake, heart rate, and pulmonary ventilation. Rowell (1974) reports a significantly higher maximal oxygen uptake during running (grade 5 to 15 percent), and P.-O. Åstrand and Saltin

(1961b) noticed a 5 percent difference, maximal $\dot{V}_{O_2}$ being higher during running than during cycling. Hermansen (1973) found, on the average, 7 percent higher maximal oxygen uptake in 3° uphill treadmill running than in cycling. The difference was greatest in well-trained individuals. (See also Fig. 8-1.) It is possible that running at high speed on the treadmill with a low grade is technically so difficult that some subjects fail before reaching maximal oxygen uptake. Another explanation for a higher oxygen uptake when running at high grades and at low to moderate speeds may be that more muscles are active in the uphill running. However, the combined arm-plus-leg work should necessarily include more muscles than leg exercise only; yet oxygen uptake is not further increased, compared with leg exercise. On the other hand, slightly higher maximal oxygen uptakes are obtained by "ski-walking," that is, using ski poles as in skiing, while walking uphill at an inclination of 12° (21 percent) at a speed of 60 to 160 m · min^{-1} (Hermansen, 1973).

In our experience, the maximal oxygen uptake is not consistently higher in running than in cycling, and the small difference sometimes noticed has scarcely any practical consequence. To obtain maximal values when using the cycle ergometer, motivation and "stimulation" may be particularly important, owing to more pronounced local fatigue in the legs (knee region) when cycling. Furthermore, the exercise position is critical when using the cycle ergometer as a tool. The exercise seat should be high enough and the subject should be positioned almost vertically above the pedals. Otherwise the exercise position will be more or less similar to exercise in the supine position. On the whole, there are many reports that treadmill running may give slightly higher maximal oxygen uptakes than cycling when the inclination is 3° or higher, in which case the oxygen-transporting capacity may be fully taxed without increasing the speed too much.

Senay et al. (1980), observed decidedly different fluid dynamics in the two forms of leg exercise: progressive treadmill walking and progressive ergometer cycling. They observed no relationship between maximal oxygen uptake and vascular volume fluid shift during treadmill exercise, and training did not alter the response. A significant relationship was found, however, for exercise level and hemoconcentration during cycle ergometer exercise. Protein left the vascular volume during cycle ergometer exercise, but not during treadmill walking. The reason for this difference is not clear, however.

In arm exercise, the maximal oxygen uptake is about 70 percent of what is attained in leg exercise. The intra-arterial blood pressure during arm exercise is higher than in leg exercise at a given oxygen uptake or cardiac output (Fig. 4-26) and the heart rate is also higher. The consequence is a heavier load on the heart. For patients with heart disease or for completely untrained older individuals, heavy exercise with the arms (such as digging, shoveling snow) may therefore be hazardous. This may in part be due to the Valsalva effect during such maneuvers. The subject is apt to hold his or her breath while lifting the load, increasing the intrathoracic pressure which, in turn, may hinder the normal venous return to the heart.

When combining arm and leg exercise (cranking and cycling) the highest oxygen uptake that can be attained depends upon the relative load on the arms. In a study by

Bergh et al. (1976) it was noticed that the oxygen uptake was the same in maximal running as in arm plus leg exercise when the arm work rate was 20 to 30 percent of the total rate of work (the total oxygen requirement exceeded the subject's maximal oxygen uptake). Subjects with strong arm and shoulder muscles could be submitted to relatively heavier arm exercise and still reach the maximum attained during uphill running. Otherwise, a typical finding was that the maximal oxygen uptake became reduced to 90 percent or below the maximum during running, when the arm work rate was 40 percent of the total rate of work. However, the difference in oxygen uptake in leg exercise and arm plus leg exercise is much smaller than expected from the difference in mass of active muscles in the two procedures (see discussion in Chap. 4). The central circulation may in one way or the other impose a limiting factor for the aerobic power. There are advantages when a large muscle mass is activated. Let us analyse Fig. 7-7. A work rate of 350 watts could be tolerated for about 3 min if only the leg muscles were involved. However, with 100 watts for the arms and 250 watts for the legs (= 350 watts), the exercise time could be prolonged to 6 min, even if the oxygen uptake (and cardiac output) did not increase further (P.-O. Åstrand and Saltin, 1961b; Stenberg et al., 1967). Evidently the organism (inclusive heart) could

Figure 7-7
Increase in oxygen uptake at start of exhausting exercise following a 10-min warm-up period. Left: illustrates leg exercise only; right: exercise with the same external power but with both arms and legs involved. This given exercise could be tolerated twice as long with arms and legs activated. Calculations of energy demand and yield are explained in text. (From data presented by P.-O. Åstrand and Saltin, 1961b; subject 1-P.-O.Å.)

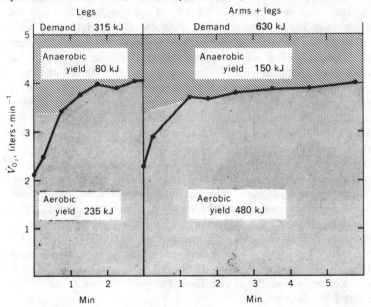

tolerate a prolongation of the exercise period when a larger mass of skeletal muscles were activated. The subjective feeling of strain is related more to the metabolic rate per square area of muscle than to the total metabolism. Therefore, a training of the oxygen-transporting system is more efficient and is psychologically less strenuous, the larger the muscular mass involved in dynamic activities.

It may be concluded that the assessment of the individual's maximal oxygen uptake, i.e., maximal aerobic power, should be made with the subject exercising in the upright position (running or cycling) with or without arm exercise added. If the subject is tested when exercising in the supine position, both arms and legs should be involved in the exercise.

ANAEROBIC PROCESSES

During light exercise, the required energy may be produced almost exclusively by aerobic processes, as mentioned, but during more severe exercise, anaerobic processes are brought into play as well. Anaerobic, energy-yielding metabolic processes play an increasingly greater role as the severity of the exercise increases. As discussed above, the aerobic power during exercise can be followed quite accurately by measuring the oxygen uptake. Since we lack a similar tool for direct measurement of the anaerobic power, some sort of indirect methods have to be applied when studying the kinetics, power, and capacity of these processes.

It should be recalled that the energy yield from the breakdown of ATP and PCr is indispensible, but that, quantitatively, the available stores of these high-energy phosphates alone can only cover the energy requirement for less than 10 s during maximal effort. The kinetics of these processes of breakdown and resynthesis are very fast. Actually, processes of key importance in the anaerobic energy yield occur within fractions of a second, but with the available methods applicable in human experiments (using muscle biopsies) seconds may elapse between sampling and bringing the biochemical events in the sample to a halt. It can be calculated that the rate of turnover of ATP in a sprinting man is approximately 2.7 mmol $\cdot$ s^{-1} $\cdot$ kg^{-1} muscle; in a high jump it may be as high as 7 mmol $\cdot$ s^{-1} $\cdot$ kg^{-1}. With only 5 mmol $\cdot$ kg^{-1} available, it obviously has to be quickly resynthesized, as in the case of a 100 m race (see Newsholme and Leech, 1983). PCr is the initial source of energy in this process. For an analysis of the anaerobic biochemistry, it may be convenient to discuss the breakdown of high-energy phosphate ($\sim$P) separately from the anaerobic glycogen breakdown to lactate (often named alactate or alactic, and lactate or lactic period, respectively). There is an overlap in these processes as well as in the kinetics of the aerobic energy yield, but with an exercise period of about 5 s the alactic power will dominate. (There are reports indicating that there is a delay of up to 6 s before the glycogenolysis sets in, see di Prampero (1981). However, Hultman and Sjöholm (1983) claim that 20 percent of the energy yield during 1.26 s, near maximal muscle contraction comes from a degradation of glycogen to lactate. In the following, the calculations of the alactic power are based on a nonlactate contribution in maximal exercise lasting some 6 s, not preceeded by warm-up).

Power and Capacity for High-Energy Phosphate Breakdown

In some activities the developed (external) power may be measured. Davies and Rennie (1968) had their subjects perform a high jump from a force platform. From the time course of the vertical velocity of the center of gravity of the body, they could calculate the average peak power output to be 3900 watts for their male, and 2350 watts for their female subjects. This peak power, developed during 0.2 s in the jump, is actually 15 times larger than the subject's maximal aerobic power, developed for 5-8 min during cycling. However, the mechanical efficiency in a jump is not known, and therefore its metabolic cost cannot be estimated.

Margaria and his group have developed a method by which one can estimate the maximal anaerobic power output (see di Prampero, 1981). The subjects climb a normal flight of stairs at maximal speed taking two to three steps at a time. The peak speed is attained within 2 to 3 s, and can be maintained up to the 6th s; from then on it declines. From speed, vertical distance climbed, and body weight, the power output can be calculated. By this method they have obtained values which are about 25 percent of the peak "external" power reported by Davies and Rennie (1968). The variation can be explained by the difference in the two types of exercise: in the high jump, it is a matter of one muscular contraction of the two legs simultaneously; in the stair climbing, the values obtained represent an average involving a series of contractions using one leg at a time.

Assuming that the mechanical efficiency is 25 percent when climbing stairs, an "external" power 1000 W would require a power of 4000 watts from the energy-yielding processes. With an exercise period lasting only a few seconds, the anaerobic breakdown of the high-energy phosphate compounds will dominate, and the stair climbing test measures mainly the alactic anaerobic power. One crucial problem is that, in maximal exercise it is impossible to measure the mechanical efficiency accurately. Therefore, a calculation of the power of the metabolic processes for a given external power output is rather hazardous. As discussed in Chap. 3, a utilization of the elastic properties of a muscle can greatly improve the efficiency, e.g., when running at high speed, and therefore an extrapolation of energy demands, at maximal exercise from data on submaximal steady state conditions, has its limitations (see the example in Fig. 2-16). However, it has been estimated that the energy cost of a 100 m sprint is about 33 kJ, and if the time is 10 s the power is 3300 watts. It is estimated that approximately 85 percent of the energy comes from anaerobic processes, the remaining 15 percent from aerobic sources.

Another approach to a calculation of the alactic power is to calculate the total energy bound in the ATP and PCr compounds. With the concentrations known, one can arrive at the maximal anaerobic alactic energy yield (capacity) per kg muscle if these phosphate compounds were completely exhausted. With an ATP concentration of 5 mmol $\cdot$ kg^{-1} fresh muscle and a PCr concentration of 17 mmol $\cdot$ kg^{-1}, the potential total energy yield is approximately 0.9 kJ (0.2 kcal) (See Chap. 12). In 20 kg of muscle, the total energy content will be 18 kJ (4 kcal). The equivalent oxygen requirement for an aerobic energy yield of that order is just 0.8 liters. In other words, it is only a small oxygen deficit that can be covered by the breakdown of ATP and

PCr. Actually, it is less than the theoretical value. During maximal exercise, there is an exponential decline in PCr in proportion to the severity of the exercise down to values as low as about 2 mmol $\cdot$ kg^{-1}. The change in ATP concentration is, as mentioned, difficult to study in the *in vivo* situation. It may be reduced to 40 percent of the level in a resting muscle.

di Prampero (1981) estimates the maximal alactic capacity of the resting muscle to be the equivalent of an oxygen utilization of about 37 ml $\cdot$ kg^{-1} body weight. For an individual weighing 70 kg this would give a total of 2.6 liters of oxygen, which seems too high.

Power and Capacity of Anaerobic Glycogen Breakdown

Figure 7-7 summarizes experiments performed on a cycle ergometer and illustrates one model used to estimate the size of the anaerobic energy yield during "supramaximal" exercise. With a correction for the alactic contribution, we can calculate the contribution from the anaerobic breakdown of carbohydrates.

By an extrapolation from the steady state oxygen uptake during different rates of submaximal exercise, the oxygen and thereby the energy demand of the supramaximal rate of exercise can be estimated. Actually, it was the same subject as presented in Fig. 7-4. The work rate of 350 watts required 5.0 liters $\cdot$ min^{-1} and was maintained for 3 min. Therefore the total energy demand during this period of time was $5 \times 21 \times 3 = 315$ kJ (75 kcal), the energy yield per liter of oxygen taken up in the mitochondria being about 21 kJ or 5 kcal. (A similar procedure, applied to the subject whose performance data are presented in Fig. 7-3, will arrive at an oxygen demand of about 4.2 liter $\cdot$ min^{-1} when exercising at the external power of 300 watts.)

Let us now try to analyze how the energy demand of 315 kJ was covered. The oxygen uptake was continuously measured during the 3 min and found to be 10.7 liters. It is calculated that an additional 0.5 liter was utilized from stores, bound to myoglobin and hemoglobin, refilled after the exercise. Thus, the aerobic energy yield can be estimated to be 235 kJ (11.2×21). The deficit was therefore $315 - 235 = 80$ kJ, and this energy must have been derived anaerobically. A breakdown of ATP and phosphocreatine may yield 20 kJ at the most, i.e., it may substitute for about 1 liter of oxygen. The remaining deficit of about 60 kJ must have been provided by glycogenolysis and glycolysis with a formation of lactic acid.

Since about 220 kJ are released for each 6-carbon unit of glycogen, which is converted into lactate (Chap. 12), a production of 2 moles or 180 g of lactate should yield 220 kJ. For a release of 60 kJ, the lactate production must then be about 0.55 mol (50 g).

As mentioned before, the subject also performed the same work rate with both arms and legs. Under these conditions, the exercise could be prolonged to 6 min before exhaustion (Fig. 7-7). At submaximal exercise, the mechanical efficiency is not significantly different from ordinary cycling. (If anything, the oxygen uptake tends to be higher in exercise performed with both arms and legs than in leg exercise alone; it is also conceivable that the mechanical efficiency becomes lower at very heavy exercise since muscles that are at a mechanical disadvantage have to contribute. Therefore, the calculated energy demand is probably a minimal figure.) The energy requirement is therefore assumed to be 105 kJ · min⁻¹, or 630 kJ altogether, during the 6 min. The measured oxygen uptake of 22.3 liters, supplemented by 0.5 liter from oxygen stores within the body, covers 480 kJ, leaving 150 kJ for the anaerobic processes. A subtraction of 30 kJ as a contribution from high-energy phosphate compounds leaves 120 kJ from glycogenolysis, i.e., a formation of 1.1 mol, or 100 g lactic acid. (Since, in this case, more muscles were exercising, the oxygen from myoglobin and the energy yield from ATP and PCr is somewhat larger than in the leg exercise. Quantitatively, however, the figures chosen for the oxygen stores and energy yield from high-energy phosphates are not critical for the overall picture.)

The glycogen concentration in the human muscle in individuals on a mixed diet is 80 mmol glucosyl units · kg⁻¹ fresh weight. Assuming that 25 kg of muscles are involved in arms and legs exercise, the available glycogen content in those muscles will be 2 mol (360 g). If so, approximately 25 percent of the available glycogen store was utilized anaerobically in the "Fig. 7-7 experiment" (since a breakdown of 0.55 mol glycogen ends up in the formation of 1.1 mol lactate).

A calculation of the average breakdown of glycogen gives about 90 mmol · min⁻¹ or 3.6 mmol · kg⁻¹ fresh muscle · min⁻¹. A degradation rate of approximately 2.5 mmol glucosyl units · kg⁻¹ in exercise at 90 to 100 percent of the fit individual's maximal aerobic power has been noted (see Hultman and Sjöholm, 1983). The total metabolic power was 1750 watt with a share of about 330 watt to the glycogenolysis. (In Fig. 7-7 the anaerobic energy yield during the 5 to 6 min interval is about 20 percent of the total.)

Figure 7-8 presents data from the study of a subject who exercised for 2.63 min with an external power close to 400 watts with a calculated energy demand of 295 kJ (70 kcal). The total oxygen uptake during cycling was 7 liters. Adding 0.5 liters from oxygen stores, the aerobic energy yield will be 160 kJ (38 kcal). Therefore, the anaerobic contribution was 135 kJ (32 kcal). Assuming that a degradation of glycogen yielded 105 kJ (25 kcal), the production of lactate would have been about 0.92 mol (83 g). In this case the average anaerobic power was about 650 watts. The maximal degradation rate of glycogen can be calculated to be 175 mmol · min⁻¹.

In maximal exercise of short duration the rate of glycogen breakdown is higher. In his review, di Prampero (1981) presents the maximal power for the "lactic mechanism" for an average individual. The oxygen equivalent is 75 ml · kg⁻¹ · min⁻¹. With a body weight of 75 kg the total will be 5.6 liters · min⁻¹ and the power as high as 1950 watts.

A degradation of one mol glucosyl of the glycogen unit yields energy which can resynthesize 3 ATP. The equivalent amount of glucose only covers the formation of 2 ATP and therefore it is not likely that glucose is an important substrate during very intense anaerobic exercise. A second advantage with glycogen is that it is stored in the muscle fibers. During short-term exercise, there would not be time to transport substrates from the liver and the fatty tissues; the muscles must be able to function on their own resources.

The *blood lactate concentration* has been used in the evaluation of the anaerobic power and capacity. An increase in concentration means that the uptake of lactate by

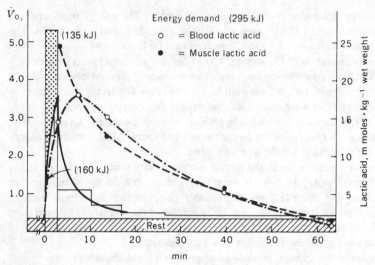

Figure 7-8
Calculated energy requirement for a 2.63-min exercise on a cycle ergometer (column represents
295 kJ, 70 kcal) and measured oxygen uptake during exercise and during 60 min recovery.
Horizontal lines denote the level of oxygen uptake measured at rest before exercise. Calculated
aerobic energy yield during exercise: 160 kJ (38 kcal); anaerobic energy yield: 135 kJ (32 kcal)
(dotted area). Lactate concentration was analyzed in blood samples and pieces of skeletal
muscle obtained by needle biopsy. (*Data by courtesy of B. Diamant, K. Karlsson, and B. Saltin.*)

the blood exceeds the lactate removal. The problem is that we lack information about
the total water pool in the body available for an uptake of lactate. There is also an
uneven distribution of lactate between the extra- and intracellular water (due to a
difference in pH). Therefore, the weighted mean concentration at equilibrium of this
water soluble molecule is not the same throughout the different compartments. di
Prampero (1981) proposes a formula for the calculation of the lactate production from
the peak lactate concentration and the subject's body weight, realizing the limitations
of such a formula, however. Another piece of information which is lacking, is the
rate by which lactate is chemically removed during exercise and recovery (see below).
In the experiments presented in Fig. 7-7, the peak blood lactate concentration after 3
min of maximal exercise was almost identical with the peak value attained during
recovery after the 6 min ride (17.1 and 17.4 mM respectively). The estimated lactate
production was, however, very different or 0.55 and 1.1 mol, respectively. This
illustrates the poor correlation existing between peak blood lactate concentration and
quantity of lactate produced.

With data on the *muscle lactate* kinetics one comes closer to the actual metabolic
events. However, we are also facing here the problem that one cannot estimate how
large a muscle mass is exercising and, secondly, is the muscle sample representative
for all of the muscles that are activated? At high rates of exercise, the type II fibers

(fast-twitch fibers) are recruited and their metabolic profile is different when compared with the type I fibers (slow-twitch fibers).

There are protocols proposed with the aim of measuring the anaerobic capacity. The subject is made to exercise with the highest possible power during 30 to 60 s on a cycle ergometer and the total work done as well as power developed during defined time periods are calculated. Certainly the anaerobic processes dominate the energy yield, and it can be assumed that a 30 to 60 s work output correlates with the absolute anaerobic capacity. These protocols will be discussed in Chap. 8.

Summary Under highly standardized conditions, as in the experiments just discussed, one can estimate the rate of energy output. With the aerobic energy yield measured (calculated from the oxygen uptake), the anaerobic contribution can be computed. It can be divided into an alactic part (energy from the breakdown of ATP and PCr) and a lactic part (energy from the breakdown of glycogen). An increase in the lactate concentration in an active muscle shows that the lactate formation rate exceeds the lactate removal rate. However, the change in concentration may differ within the muscle itself. A calculation of the lactate production is also impossible, as long as one does not know the muscle mass involved, and the amount of lactate removed to the blood or by chemical processes within the muscle itself prior to the collection of the muscle tissue sample. Similarly, determinations of the blood lactate concentration may give some information as to whether or not the glycolysis contributes to the ATP resynthesis. However, an estimation of the rate of lactate production and its magnitude is not possible in our opinion.

In maximal exercises of a few seconds duration, the developed metabolic power can exceed 4000 watts with the energy-rich phosphate compounds being the source of the energy yield. This is the alactic phase. With some seconds delay, the breakdown of carbohydrates to lactate, dominated by glycogen as the substrate, can produce a power that may come close to 2000 watts; but with a prolongation of the exercise time, the power of this anaerobic process gradually declines. At the end of a 6 min arm and leg exercise, the anaerobic power was calculated to be about 350 watts, supporting the 1400 watts produced by the aerobic metabolism (Fig. 7-7). The capacity of the energy-rich phosphates is very limited, being in the order of 0.9 kJ · kg^{-1} muscle. With 25 kg of muscles in maximal exercise, the potential is then 22 kJ equivalent to 1 liter of oxygen, providing all ATP and PCr is utilized, which is not realistic. The capacity of the anaerobic carbohydrate metabolism is difficult to estimate; in the experiment presented in Fig. 7-7 it was calculated to be 120 kJ (29 kcal) equivalent to about 6 liters of oxygen consumed in an aerobic metabolism. In athletes specialized in events demanding maximal effort of a few minutes duration, it may be up to 200 kJ (45 kcal).

The weak point in all these calculations is that we do not have exact figures for mechanical efficiency during maximal cycling. In most other exercises it is even more difficult to predict the total energy demands, which is a necessity for an exact measurement of the anaerobic energy yield.

LACTATE PRODUCTION, DISTRIBUTION, AND DISAPPEARANCE

Time Course of the Blood Lactate Concentration

The blood lactate concentration is relatively simple to determine. It is an index of anaerobic metabolism but it does not inform us about the anaerobic power. Increased lactate level, in muscle and blood indicates an anaerobic supplement to the aerobic production of ATP. Not surprisingly, under hypoxic conditions, such as at high altitude, the oxygen deficit and the blood lactate concentration are higher at a given work rate compared with normoxic conditions. When exercising under hyperoxic conditions the picture is reversed (see P.-O. Åstrand, 1954; Knuttgen and Saltin, 1973; di Prampero, 1981; Pedersen, 1983). Some glycolysis seems to be lactate producing also in muscles which are active in fully aerobic conditions (see Connett et al., 1984). It should be pointed out that even at rest, the blood lactate concentration is about 1 mM. Figure 7-3 illustrates how the peak blood lactate concentration starts to increase when the exercise rate becomes more severe. The events can be summarized as follows:

1 During light exercise, the oxygen store in the muscle plus the oxygen supplied as the respiration and the circulation adapt to the exercise will completely cover the oxygen need. Most of the ordinary daily occupations belong to this category.

2 During exercise of moderate intensity, anaerobic processes contribute to the energy output at the beginning of the exercise until the aerobic oxidation can take over and completely cover the energy demand. Produced lactate diffuses into the blood and can be traced in the venous blood draining the muscle, and eventually in the arterial blood if the quantity of lactate produced is high enough. As the exercise proceeds, the blood lactate concentration falls again to the resting level and the exercise can be continued for hours.

3 During heavier exercise, the lactic acid production, and therefore, the rise in blood lactate concentration are higher and remain high throughout the exercise period. The length of time that the work rate may be endured will, to some extent, depend on the subject's motivation.

4 During very severe exercise, there is a continuously growing oxygen deficit and an increase in the lactate content of the blood because of the predominantly anaerobic metabolism. The exercise cannot be continued for more than a few minutes as a rule, because the subject's muscles can no longer function. The work rate has exceeded some sort of a threshold and the condition is no longer one of a "steady state." These stages were well analyzed and described by Bang (1936), but there is at present a renewed interest in the transition from stage 3 to 4.

In the first place there is the discussion on terminology: is it an anaerobic or aerobic threshold? One suggested term is "onset of blood lactate accumulation." A more detailed discussion will follow later on in this chapter.

Figure 7-9 illustrates how the arterial lactic acid concentration increases during and after severe exercise, followed by a slow decline back to the resting level. The lactate is produced in the muscles during the actual exercise, but there is a time lag for the

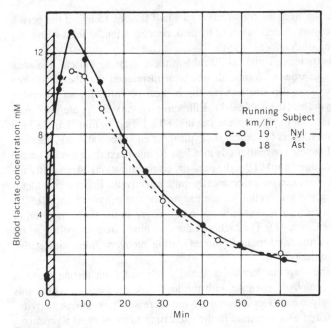

	Running km/hr	Subject
o--o	19	Nyl
●—●	18	Åst

Figure 7-9
Blood lactate concentration after severe exercise of 2 min duration (shaded column) in two subjects. Peak values occur several minutes following the cessation of exercise (*From I. Åstrand, 1960.*)

diffusion from the active muscles and redistribution within the body. For a determination of peak lactate concentration in the blood, samples must be taken at intervals during the first 5 to 10 min of the recovery period. It should also be noted that it takes up to 60 min or even longer before the resting level is again reached, so that if the effect of a stepwise increasing work rate is studied, the samples secured at the end of the last exercise period not only reflect the anaerobic component of the last exercise intensity; they are also affected by the preceding work rates. Furthermore, in competitive events where the lactic acid production is high, the time between heats should be at least 1 hr to allow time for the blood lactate to return to resting values. Obviously, this decision should be based on well-established physiological principles, and it should not be left to the organizing committee's arbitrary judgment as to whether or not the competitor should have to start the next competition with a high tissue lactate concentration.

Figure 7-8 illustrates how the lactate concentration at the end of exercise is much higher in the active muscles than in the blood, but after some 5 min of recovery the concentrations in the two compartments run parallel. After exhausting exercise involving glycolysis, the peak lactate concentration is attained after 5 to 8 min. The lactate is assumed to have reached equilibrium in 85 percent of the total body water.

The elimination of lactate from the blood has then a half time of 15 min if the person was resting during recovery, independent of its peak concentration, at least within the range 4 to 16 mM (see di Prampero, 1981).

With lactate concentrations as high as 30 mM in the exercising muscle, the peak blood concentration was around 20 mM. It has been mentioned that the fast-twitch fibers, particularly the Type IIb fibers, have the highest potential for an anaerobic breakdown of carbohydrates (see Table 2-1). Elite cross-country skiers are also characterized by a high maximal oxygen uptake (around 80 ml $\cdot$ kg^{-1} $\cdot$ min^{-1}) and a high percentage of slow-twitch fibers (75 percent). Typical for endurance athletes, they can attain very high blood lactate concentrations in a 5 to 10 min maximal exercise on the treadmill, or on the average 15 mM (Bergh, personal communication). Apparently the slow-twitch fibers can also exercise anaerobically quite effectively. Endurance athletes have a relatively low activity potential for the enzyme lactate dehydrogenase (LHD) in their fibers, but this enzyme is normally not rate limiting in the transformation of pyruvate to lactate (see Chap. 12). The highest values on blood lactate reported so far are in samples drawn from well-trained athletes during recovery from competitive events of 1- to 2-min duration.

There are reports showing that lactate is actually released from resting muscles (arms) during heavy, prolonged exercise with the legs. The mechanism behind this release is probably a glycogenolysis induced by the increased sympathetic activity. This lactate can be utilized as a substrate in the exercising muscles or as a precursor for a glucose formation in the liver (see Ahlborg et al., 1981).

What is the Fate of the Lactate Produced?

In an aerobic metabolism of glycogen down to CO_2 and H_2O, the energy yield for ATP production is 2813 kJ (672 kcal) per six-carbon unit. Of this potential, only about 8 percent is utilized in the anaerobic breakdown to lactate. However, the produced lactate is not wasted. Without any loss of energy, the process of pyruvate-transformation to lactate can be operated in reverse. From that point, there are two alternative routes: (1) the pyruvate can be oxidized, or (2) be a substrate for a synthesis of glucose and glycogen. When oxidized it yields the remaining 92 percent as energy, and the heart muscle, kidney cortex (Newman et al., 1937; Keul et al., 1967, see Newsholme and Leech, 1983), and skeletal muscles (both resting and exercising) can use lactate as a substrate (see Ahlborg et al., 1976; Brooks and Fahey, 1984, pp. 191–219).

It has been well established that lactate produced during exercise can be resynthesized back to glycogen in the liver. It is an open question as to what extent such a synthesis (glyconeogenesis) can take place directly in mammalian muscles (Krebs, 1964). Previously, several of the essential enzymes for these pathways were believed not to exist in the muscle. However, recent investigations have shown that a set of key enzymes for one of several pathways does exist in the required amounts in the muscle. There are two opposite opinions about the fate of lactates:

1 Hermansen and Vaage (1977) report from studies on humans that most of the lactate produced during repeated 1-min maximal running periods interspersed with 4-

min rest periods was resynthesized to glycogen during recovery at rest, evidently directly in the muscles. The uptake of glucose in the muscle studied could not explain the increase in its glycogen content.

According to the classical concept about 20 percent of the lactate produced during exercise is reoxidized to pyruvate and then dissimilated to CO_2 and H_2O, and the remaining lactate is taken up by the liver and forms glucose, which can be reconverted into glycogen or be delivered to the blood. The muscles can then utilize this glucose in its glycogenesis to restore their glycogen depots (see Krebs, 1964). From the calculations by Hermansen and Vaage, approximately 75 percent of the lactate was reconverted into glycogen but the pathway via the liver (the so-called Cori cycle) was not utilized.

2 Brooks and his group (1973; see Brooks and Fahey, 1984, pp. 191–215) have concluded from their studies on rats that most of the lactate is oxidized to CO_2 and H_2O and that only a smaller part of the lactate is converted to glycogen.

There are data indicating that the truth regarding the fate of lactate is somewhere in between the two opposing views, at least in humans. With the oxygen uptake measured, not only during the period of exercise, as illustrated in Fig. 7-7, but also during 60 min of recovery, one can calculate the total energy yield. With lactate, as the *only* substrate for all tissues in their aerobic metabolism during recovery, at the most about 50 percent of the lactate was removed by the Krebs citric cycle and respiratory chain. There is, during recovery, a significant increase in the muscle's glycogen content and this process is energy demanding, i.e., in a way there is a payment of a debt.

How Does Continued Light Exercise Influence the Lactate Removal?

Newman et al. (1937), noticed that the removal of lactate, accumulated in the body after exhausting exercise, was enhanced if, during recovery, the subject continued to exercise, but at a lower intensity, which normally did not produce any lactate. This observation has been confirmed, and Belcastro and Bonen (1975) report that the optimal removal rate occurred when the "recovery" exercise gave an oxygen uptake of less than 50 percent of the maximum. McLellan and Skinner (1982) recommend an active recovery intensity of approximately 10 percent of maximal oxygen uptake. It is logical that lactate is utilized as a substrate in the active muscles, partly replacing glycogen, glucose, and free fatty acids.

There are observations showing that if a standardized exhaustive exercise is performed with blood and muscle lactate concentrations elevated in advance (because other muscle groups have been previously exercised to exhaustion), then the performance time is reduced (see Karlsson et al., 1975). This should be kept in mind when one is warming-up before an athletic event. At least from a theoretical point of view, it appears that if interval training is conducted with the aim of elevating the lactate concentration to very high values, one should rest in between each bout of exercise. On the other hand, if the main purpose of the training is to attain a high oxygen uptake and cardiac output, one can tolerate more repetition if the lactate level is relatively low. Evidently, light activity during the "resting" period provides the optimal effect. The logical recovery for an ice-hockey player when off the ice during the game might be to exercise on a cycle ergometer.

Effects of Metabolism on Tissue and Blood pH

The hydrolysis of one ATP produces about one proton at the muscle cell pH. However, during the oxidative phosphorylation all the products of ATP hydrolysis (ADP, P_i and H^+) are reutilized. This means that there is no net accumulation of H^+ in the aerobic metabolism. Hochachka and Mommsen (1983) emphasize: "This closely balanced system dissipates during anaerobic metabolism, but how and why the breakdown occurs is controversial. In part, the controversy arises from a common belief that glycolysis yields lactic acid, a fairly strong acid (pK_a, 3.9) which then dissociates into lactic anions and H^+. This oversimplified statement, which we found in four or five recent and successful biochemistry textbooks is in fact rather misleading." In contrast to the situation in aerobic metabolism, in anaerobic glycolysis, hydrogen ions are only partially reutilized when glycogen is converted to lactate. In humans, exercising at maximal rates, muscle ATP concentrations can fall from 5 mmol to about 2.5 mmol $\cdot$ kg^{-1} or even lower. This drop in ATP levels represents a potential increase in the proton concentration from 0.1 μmol to 2.5 mmol $\cdot$ kg^{-1} (Hochachka and Mommsen, 1983). From their review Busa and Nuccitelli (1984) also conclude that lactic acid accumulation is not the source of the intracellular pH drop during anaerobic glycolysis. From a theoretical point of view, this is interesting and important. However, the result of the sequence ATP hydrolysis plus rephosphorylation by energy from the breakdown of glycogen is a constant production of 2 mol H^+ for each mol of glucosyl unit (because of the opposite pH dependencies of H^+ production by glycolysis and by ATP hydrolysis). In a way, the net effect is what the traditional formula reveals:

$$\text{glycogen or glucose} \rightarrow 2 \text{ lactate} + 2 \text{ H}^+$$

It is not surprising that the muscle pH becomes reduced during anaerobic exercise, from about 7.0 to 6.5 or even lower. Secondarily, the arterial blood pH can fall from 7.4 to below 7.0. As discussed above, the rate of the glycolysis is determined by the need to resynthesize ATP anaerobically and the rate of ATP hydrolysis and lactate formation are coupled. It is not surprising that there is a high correlation between lactate concentration and pH values in blood samples taken at rest, during, and after exercise (steady state as well as maximal) (Keul et al., 1967). These authors point out that because of the buffer systems of the blood, a 10-fold increase in lactate concentration causes only a 1.42-fold increase in the H^+ concentration.

As discussed in Chap. 5, one effect of the change in blood pH is a hyperventilation, i.e., the pulmonary ventilation per liter oxygen uptake increases (see Fig. 5-9). Beyond a certain point it does not increase linearly with the oxygen uptake, but exponentially. More CO_2 is exhaled than produced in the aerobic metabolism. During recovery, the situation is the reverse.

It is suggested that an accumulation of protons can be the factor that causes fatigue and failure to continue exercises involving the anaerobic metabolism (see discussion under "Muscular Fatigue" in Chap. 3). It is then tempting to investigate whether a manipulation with the tissue and blood pH prior to an exhausting exercise of one or a few minutes duration would modify the exercise at the given work rate. A metabolism alkalosis can be induced by ingestion of sodium bicarbonate and an acidosis by ammonium chloride. In the 1930s, positive effects of an intake of sodium bicarbonate were reported (see Gledhill's review, 1984). The following studies on the effect of alkalosis prior to a standard exercise did not significantly improve the performance (McCartney et al., 1983, exercise time 30 s; Katz et al., 1984, exercise time 100 s). On the other hand Gledhill (1984) reports a significantly faster 800 m race after "soda loading" (control 2:09.9; sodium bicarbonate 2:05.8). In his review, Gledhill quotes studies that indicate that bicarbonate ingestion can improve performance in anaerobic events such as 400 to 1500 m races. There is an ethical problem, however; Does the use of sodium bicarbonate or chemicals with a similar effect on the body's pH in various compartments, fall under the general definition of doping agents? So far no such agents are on the doping control list. An analysis of a urine sample will reveal an abnormally high concentration of, e.g., sodium bicarbonate.

INTERACTION BETWEEN AEROBIC AND ANAEROBIC ENERGY YIELD

This is an attempt to summarize parts of the preceding discussion. Table 7-1 presents the contribution to energy output from aerobic and anaerobic processes, respectively, in maximal efforts in exercise involving large muscle groups. The individual's maximal aerobic power is set to 5 liters $\cdot$ min^{-1} = about 100 kJ $\cdot$ min^{-1} and maximal anaerobic capacity to 200 kJ $\cdot$ min^{-1}, equivalent to 9 liters of oxygen uptake in aerobic exercise. It is assumed that 100 percent of the maximal oxygen uptake can be maintained for

TABLE 7-1

Process	Exercise time, maximal effort							
	10 sec	1 min	2 min	4 min	10 min	30 min	60 min	120 min
Anaerobic								
kJ	100	170	200	200	150	125	80	65
kcal	25	40	45	45	35	30	20	15
percent	85	65–70	50	30	10–15	5	2	1
Aerobic								
kJ	20	80	200	420	1000	3000	5500	10,000
kcal	5	20	45	100	250	700	1300	2400
percent	15	30–35	50	70	85–90	95	98	99
Total								
kJ	120	250	400	620	1150	3125	5580	10,065
kcal	30	60	90	145	285	730	1320	2415

10 min, 95 percent for 30 min, 85 percent for 60, and 80 percent for 120 min. For nonathletic men, the figures will be roughly 50 percent of the values listed.

For exercise periods up to 2 min, the anaerobic power is more important than the aerobic contribution; at about 2 min there is a 50:50 ratio, and with prolonged exercise the aerobic power becomes gradually more dominating. This is graphically illustrated in Figure 7-10.

It is very rare that an individual possesses top power for both aerobic and anaerobic processes. Therefore, the analysis in Table 7-1 should not be interpreted and applied too literally. A maximal aerobic power of 100 kJ · min^{-1} may be coupled with a maximal anaerobic yield of 100 kJ. For this individual, the proportional participation of anaerobic and aerobic processes will be different compared with the tabulated data. (For example, a man should be advised to compete over longer distances since his relatively low maximal anaerobic power is then less of a handicap, and he should try to get rid of his rivals before starting the finish!)

An analysis of the energetic demands of different sport events and the athlete's capabilities to fulfill these requirements may help him or her in both training and the selection of suitable events. One factor to consider is that endurance athletes have skeletal muscles with a high proportion of slow twitch fibers and sprinters are characterized by a dominance of fast twitch fibers. However, in many events there is not such a strict pattern in fiber composition (see Costill et al., 1976; Saltin and Gollnick, 1983). So far a fiber typing is not a good selective instrument for picking potential athletes.

A trained individual can exercise at a relatively high oxygen uptake in relation to the $\dot{V}_{O_2}$ max without any continuous elevation in blood lactate concentration. For an untrained person, the critical point may be at 50 percent of the V_{O_2} max. For endurance-trained, top athletes this relative work rate may be as high as 85 percent of the maximal oxygen uptake (see below under "Anaerobic Threshold").

There are still many unanswered questions concerning which factors do govern the selection of pathways for the energy yielding processes, i.e., whether the anaerobic

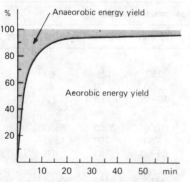

Time from start until complete exhaustion

Figure 7-10
Relative contribution in percentage of total energy yield from aerobic and anaerobic processes respectively during maximal efforts of up to 60 min duration for an individual with high maximal power for both processes. Note that a 2-min maximal effort hits the 50 percent mark, meaning that both processes are equally important for success.

or aerobic processes will prevail. This will be discussed further in Chap. 12. At any rate, in one way or another, the availability of oxygen plays a decisive role in the interrelation between aerobic and anaerobic energy yield.

It has been pointed out (Chap. 4) that the maximal volume of oxygen that can be offered to the exercising skeletal muscles may be decisive for their aerobic power. From a teleological point of view, it is very efficient that low threshold, slow-twitch fibers are recruited during exercise of moderate intensity. They have the enzyme potential for aerobic exercise, and the myoglobin enhances the oxygen diffusion within the cells. During very heavy exercise, the high threshold, fast-twitch fibers also become activated, but now the oxygen supply is critical. The myoglobin in the slow-twitch fibers diverts the oxygen to these fibers, preventing the fast-twitch fibers from taking it. On the other hand, the fast-twitch fibers are well equipped for anaerobic intensive exercise. From a mechanical point of view, they are specialized for very intensive exercise. When these fast-twitch fibers are required, as in the case of very vigorous exercise, or at the very beginning of exercise, only a limited amount of oxygen is available for them. It would be something of a waste of resources to have their metabolic repertoire developed as replicas of the slow-twitch fibers, merely duplicating their function. With training, however, the potential of the central circulation to transport oxygen to the tissues is developed. With more oxygen available, it makes sense that fast-twitch fibers can improve their aerobic potential; they will not be competing with the slow-twitch fibers. (For a more detailed discussion, see Chap. 10.)

The density of mitochondria is less in the fast-twitch fibers than in the slow-twitch fibers. Theoretically, this will provide more space for the contractile elements, the actin, and myosin filaments. Available data does not, however, indicate a greater force per cross-sectional area in individuals who have a higher percentage of fast-twitch fibers when compared with persons with a higher percentage of slow-twitch fibers.

The problem of liberating NADH from hydrogen by the production of lactate is indeed a practical mechanism. Lactate can diffuse out of the lactate-producing muscle cells, and it is an excellent substrate in the aerobic energy yield as well as for the glucose and glycogen resynthesis. With the anaerobic processes running at a high rate, an inevitable consequence is the accumulation of protons that may affect the physical performance in a negative way. So far, it is an open question as to how far one can modify the consequences of a given anaerobic power output by training and by chemical manipulation.

Anaerobic Threshold

An anaerobic threshold concept was introduced in order to define the point when metabolic acidosis and associated changes in gas exchange in the lungs, occur during graded exercise (Wasserman et al., 1973). Actually, Bang (1936) pointed out that there was a metabolic phase when the exercising person would encounter an accumulation of lactate in the blood as the exercise proceeded. Recently, considerable efforts have been made to establish the oxygen uptake in relation to the person's maximal aerobic power when the blood lactate concentration gradually starts to increase during continuous exercise. The work rate at this "breaking point" is named *anaerobic threshold,* or *onset of blood lactate accumulation* (OBLA). (See Jones and Ehrsam, 1982; Karlsson and Jacobs, 1982.)

For the student of the physiology of exercise, this concept poses several problems. In the first place there is a problem of terminology: are we talking about a threshold for a single muscle fiber, a whole muscle group, a threshold for a regulatory system involving chemoreceptors etc., or, are we talking about the onset of blood lactate accumulation, or some other lactate effects, or a pH-dependent function?

Definitely some muscle fibers are utilizing an anaerobic metabolism for the ATP resynthesis even before the muscle, as a whole, has reached its maximal oxygen uptake. The capillary density varies within the muscle and so does the myoglobin concentration. In particular, the Type IIb fibers are put at a disadvantage. As previously mentioned, a lactate formation can take place in resting muscles and in muscles exercising under fully aerobic conditions. An increased sympathetic activity, accompanying intense exercise, can enhance this production. There is a work rate at which the lactate uptake in the blood exceeds the removal so that the lactate concentration increases gradually. Only a relatively small increase in the intensity of exercise may reduce the time to exhaustion from say 1 hr to 15 min. A threshold value often used is an arterial lactate concentration of 4mM. However, there are marked individual variations in this threshold value. Stegmann and Kindermann (1982) report one example. Out of 19 rowing athletes, 15 failed to endure the work rate on the cycle ergometer that in the first test trial gave a lactate concentration of 4mM. They had to stop after 14.4 min, work rate 242 watts, lactate 9.6 mM. The individual anaerobic threshold was also established and the mean value was 193 watts. The blood lactate value in this "threshold test" was 2.3 mM. The work rate 190 watts could now be continued for the intended time, 50 min, with a lactate concentration of 3.75 mM at the end. The perceived exertion was rated by the Borg scale (1982). The values were 18 (close to "very, very heavy") at the end of the "4.0 mM" experiment, and 14.0 (between "somewhat heavy" and "heavy") at the "2.3 mM" ride. The established individual threshold was 4.0 mM lactate for 3 subjects and 6.1 mM for the remaining subject.

There are individual patterns in the kinetics of lactate production and disappearance. It should also be pointed out that the increase in blood lactate concentration with a stepwise or ramp-wise increasing work rate is rather smooth, making it difficult to define a threshold. It is therefore not surprising that the variability of the threshold for a given subject is large (average range 16 percent in the study reported by Yeh et al., 1983). The threshold is also modified by such factors as the nutritional state and speed of movement (Hughes et al., 1982). Individual variations in the lactate kinetics and methodological problems can explain the finding in the study by Tesch et al., (1982). Here their 10 healthy subjects cycled at the 4 mM OBLA mark, the lactate concentration in m. vastus lateralis had a range from 2.1 to 12.6 mmol · kg^{-1} muscle wet weight with a mean value of 6.9 mmol. In the study by Jacobs and Kaiser (1982), the muscle lactate concentration had an individual range from 4.5 to 14.4 mmol · kg^{-1} (mean value 8.3 mmol) at the power output that gave a blood concentration of 4 mM. They found no difference in the lactate concentration in pools of Type I and Type II fibers.

In some laboratories, 2.5 mM is used as a threshold value, others use something in between this value and 4 mM. Different protocols are used in the determination of the threshold. Some use a stepwise increase in work rate, others use a continuous increase and with different rates of increase in metabolic demands. Such variations make it difficult to interpret and compare data from different laboratories.

As already mentioned, an alternative noninvasive method to establish the anaerobic or lactate threshold has been used, i.e., by determining at what rate of exercise/oxygen uptake a nonlinear increase in pulmonary ventilation occurs. The theoretical basis for this is that the lactate accumulation in the blood should reduce its pH and thereby increase the chemoreceptor drive to the respiratory central generators. There are, however, similar methodological problems in defining a reproducible respiratory threshold, as with the anaerobic threshold. Secondly, other factors than pH can contribute to the nonlinear increase in pulmonary ventilation with increasing oxygen uptake. As mentioned in Chap. 5, local hypoxia in the peripheral chemoreceptor area may be more pronounced in severe exercise. This will stimulate the peripheral chemoreceptors and cause an increase in pulmonary ventilation, yet it has nothing to do with the blood lactate level. An impulse traffic from higher brain centers can also modify the ventilation, more so if the exercise requires "mental" effort. There is an interesting observation that patients with McArdle's syndrome, who, due to a lack of a key enzyme, cannot utilize glycolysis and form lactate, nonetheless react with an "abrupt" increase in pulmonary ventilation, similar to an anaerobic threshold, (see Hagberg et al., 1982).

In the study by Stegmann and Kindermann (1982), the perceived exertion was, as mentioned, around 14 at the anaerobic threshold, determined individually for each person. Purvis and Cureton (1981) give similar mean values, of 13 to 14 ("somewhat hard") for their female and male subjects at anaerobic threshold, ranging from 11 to 16. For both groups, the percentage of the maximal oxygen uptake was around 60 ± about 8. This relatively wide scatter is not surprising. The endurance-trained person can exercise closer to her/his maximal oxygen uptake than someone who is untrained. One example: Hurley et al. (1984), report a 26 percent increase in maximal aerobic power in their 8 subjects after a 12-week training program. A blood lactate of 2.5 mM was attained at 68 ± 4 percent of the maximal oxygen uptake before, and at 75 ± 3 percent after training. In 8 competitive runners the relative oxygen uptake was even higher, or 83 ± 2 percent at the blood lactate concentration 2.5 mM. (See also Denis et al., 1982.) A change in the capillary density, myoglobin function, and enzyme pattern in the muscles may contribute to the training-induced improvement in endurance.

Summary The concept *anaerobic threshold* or *onset of blood lactate accumulation* is based on an exponential increase in blood lactate concentration when exceeding a certain rate of exercise/oxygen uptake. It is usually determined during incremental exercise for the establishment of the point mentioned. Indirectly, a similar breaking point of pulmonary ventilation versus oxygen uptake has been applied. The assumption is that these two points are highly correlated. In many laboratories the testing has been standardized with the goal of finding the rate of exercise or oxygen uptake at which the blood lactate concentration reaches a value somewhere between 2.5 and 4 mM.

It is evident that the theoretical considerations behind this concept need further clarification. It is difficult to establish a well-defined "point." The work rate at which a nonlinear increase in ventilation occurs need not be the same exercise rate at which the lactate concentration increases. There are individual variations in the highest lactate level which can be tolerated during prolonged exercise (see Yeh et al., 1983; Green et al., 1983).

These critical notes do not mean that it is of no interest to study, in normals as well as in patients, how intensively they can exercise without an accumulation of lactate, or to follow the lactate response in subjects submitted to a given rate of exercise or oxygen uptake. On the contrary, it provides important information especially about the individual's aerobic potential, and about the effect of training. The threshold concept, as such, however, rests on an unstable foundation.

The importance of the threshold concept as a coaching guide awaits scientific evaluation. At any rate, the experienced endurance athletes know quite well themselves what rate of speed can be tolerated without fatigue due to lactate accumulations. Thus, there is hardly any need for the coach to tell them on the basis of blood lactate analyses.

MAXIMAL AEROBIC POWER—AGE AND SEX

Age

Chronological age is not a very good reference point when analyzing biological data, particularly in the case of children and teenagers. It is an inevitable evolutionary consequence that individuals within a species are different in many ways. In this aspect "men are not born equal." Tanner, who is one of the pioneers in this field, has established the general framework of the biological age (see Tanner, 1980). By regular measurements of physical characteristics such as height and weight at least twice a year, and observations on the development of secondary sex characteristics (pubic hair, development of scrotum, testis, and penis in boys; breasts and time of first menstruation in girls), the maturity of the young individual may be followed (Fig. 7-11). In addition, radiogram analysis provides information about the maturation of the bones in the hands (Tanner et al., 1975). It is evident that a longitudinal recording of the child's height, which is certainly easy to do, gives a good picture of the onset of puberty. During the first years in life the child grows rapidly, followed by a somewhat slower rate of growth for about a ten-year period. Then comes a second spurt, when the increase in height for girls may be, on the average, about 7cm in one year, and 10 cm for the boys (Northern Europe, North America) (Fig. 7-12). The girls first menstruation normally occurs closely after this accelerated growth period, with a time lag of about one year. As mentioned, the development of the sex characteristics both in girls and in boys is also related to this adolescent growth spurt, which is often referred to as the *peak height velocity*. A *peak weight velocity* is usually observed some 6 months later.

As an illustration of the fact that neither men nor women are born equal, Fig. 7-12 shows how the peak height velocity may, in girls, occur as early as at the age of nine and a half years for one girl, but not until the age of fifteen years for another.

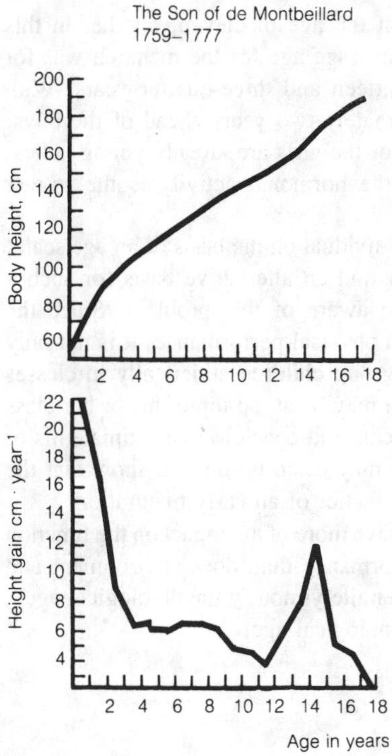

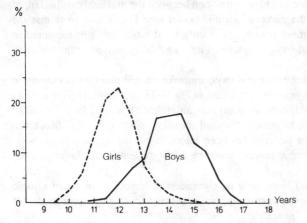

The Son of de Montbeillard
1759–1777

Body height, cm

Height gain cm · year⁻¹

Age in years

Figure 7-11
Illustrates the growth in height of a French boy recorded by his father (De Montbeillard). The upper curve shows the gradual gain during an 18-year period, the lower curve the height gain each year. Note the accelerated growth around the age of 14-15 years ("peak height velocity"). (*From Tanner, 1962.*)

Figure 7-12
Distribution of age at Peak Height Velocity age (PHVage) for girls (n = 358) and boys (n = 373). (*Modified from Lindgren, 1978.*)

%

Girls   Boys

Years

In boys, one had his adolescent growth spurt at the age of eleven, another in this group of subjects at the age of seventeen. The average age for the menarch was for 50 percent of the girls twelve and a half to thirteen and three-quarter years. With regard to sexual maturity, the girls are approximately two years ahead of the boys. At the age of about thirteen years, the majority of the girls are already young ladies, while the boys are still kids. The pattern of the hormonal activity is the trigger mechanism for this development.

While it is quite practical to characterize an individual on the basis of an age scale, it is biologically unsound. Yet it is not easy to find an alternative basis for such a classification. At any rate it is important to be aware of this problem. Since the adolescent growth spurt has a profound effect on physical performance, it is not only unfair, but it may also possibly be harmful to group children athletically in classes according to age. One early developed individual may be at the top of his or her class athletically for a period of time, causing the parents and coach to overestimate his or her athletic talent. In time his or her classmates may catch up only to show that the early success was not due to talent but simply a matter of an early maturation.

In the aging individual, the genetic code may have more of an impact on the function of systems of key importance for physical performance than does environment and life style. However, a change in life style can definitely modify the "biological age", both upward and downward, at almost any chronological age.

Maximal Aerobic Power—Absolute Values

It should be recalled that the maximal aerobic power is defined as the highest oxygen uptake the individual can attain during exercise while breathing air at sea level. To evaluate whether or not the subject's maximal oxygen uptake has been attained, objective criteria should be used, such as measured oxygen uptake lower than expected from the work rate, and/or blood lactic acid concentration higher than about 8 mM.

The information provided by the assessment of maximal oxygen uptake is a measure of (1) the maximal energy output by aerobic processes, and (2) the functional capacity of the circulation, since there is a high correlation between the maximal cardiac output (and stroke volume) and the maximal aerobic power (see Fig. 4-29). Note that the high correlation between these parameters is only valid when they are measured in liters $\cdot$ min^{-1}. The maximal oxygen uptake in ml $\cdot$ kg^{-1} body weight $\cdot$ min^{-1} is quite useless in this evaluation.

Direct measurements of the maximal oxygen uptake on 350 individuals ranging in age from four to sixty-five years are presented in Fig. 7-13. All subjects were healthy and moderately well-trained; none of them was an athlete. It should be emphasized that it is almost impossible to present "normal material" since it is very difficult to define what is normal. This material has been selected, but the age and sex factors, which modify maximal aerobic power should be fairly evident in this homogeneous group of subjects.

Before puberty, girls and boys show no significant difference in maximal aerobic power. Thereafter, the woman's power is, on an average 65 to 75 percent of that of

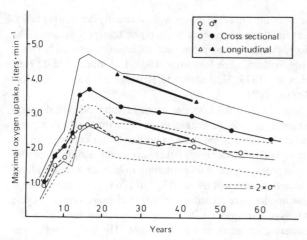

Figure 7-13
Mean values for maximal oxygen uptake (maximal aerobic power) measured during exercise on treadmill or cycle ergometer in 350 female and male subjects four to sixty-five years of age. Included are values from a group of 86 students trained in physical education (*From P.-O. Åstrand and Christensen, 1964*), and data from a follow-up study of 35 female and 31 male students belonging to the same group. (*From I. Åstrand et al., 1973.*)

the men. In both sexes, there is a peak of eighteen to twenty years of age, followed by a gradual decline in the maximal oxygen uptake. At the age of sixty-five, the mean value is about 70 percent of what it is for a twenty-five-year old individual. The maximal oxygen uptake for the sixty-five-year old man (average) is the same as that typical for a twenty-five-year old woman.

The individual variation should be noticed. Many older subjects have a maximal power that is higher than that found in many much younger individuals. In Fig. 7-13, the " ± 2 standard deviation line" for male subjects coincides closely with the average values for women, and the 95 percent range is actually ± 20 to 30 percent of the mean value at a given age.

Similar data have been reported from other countries. Sex differences and changes in maximal oxygen uptake with age are very similar to those just described, but the absolute values vary. This can be explained by the different methods used in the selection of subjects and variations in the average body size, e.g., in Sweden as compared with Japan. A random sample is difficult to study successfully, particularly in the older age groups. For subjects with different occupations, the mean values tend to vary with the nature of the occupation. It is not surprising that forestry workers have higher maximal oxygen uptakes than white-collar employees, for instance (see I. Åstrand, 1967b). Similarly kibbutz dwellers, mostly engaged in agricultural activities have a higher aerobic power than city dwellers (Epstein et al., 1981).

These differences in maximal aerobic power, as well as in many other parameters, are perhaps partly due to selection, since those who are endowed with a strong constitution are probably over-represented in occupations with physically demanding tasks. Furthermore, such jobs may in themselves train the oxygen transport system. The more mechanized the society, the less pronounced such differences may be between individuals from different occupations. (See Robinson, 1938; I. Åstrand, 1967b; Hermansen, 1973; Drinkwater et al., 1975; Hollmann and Hettinger, 1980, p. 365; Kobayashi, 1982; Pate and Kriska, 1984).

Training consisting of three training sessions per week of some 30 min duration commonly results in an average of 10 to 20 percent increase in maximal aerobic power. It is evident that the natural endowment is the most important factor determining the individual's maximum. In the 1940's, Gunder Hägg held many world records in middle- and long distance running. No physiological data pertaining to him are available from this period. His body weight at that time was 69 to 70 kg. In 1963, at the age of forty-five his heart rate and oxygen uptake were recorded while he was exercising on a cycle ergometer. His maximal O_2 was 4.0 liters · min^{-1}, and his maximal heart rate 181 beats · min^{-1}. His blood lactate concentration was 13.8 mM. His body weight was 94.5 kg. Although he had not trained since 1946, he had a very high maximal aerobic power (see Fig. 7-13). This is an example of how an untrained individual, given a favorable endowment, can maintain a very high aerobic power. In the introduction of this chapter, it was mentioned that up to some 70 percent of the individual differences in maximal oxygen uptake are probably due to constitutional differences. Thus, the maximal oxygen uptake does not in itself reveal whether or not an individual has been physically active in the preceding years. A simple activity history is actually more informative about habitual training, than the data from a so-called stress test.

Figure 7-13 includes data originally obtained from a group of 86 students in physical education. The mean values are definitely higher than those for the average physically active woman or man, but the difference between the female and male students in maximal power is of the same magnitude as in other groups (females, 70 percent of the male maximum). Twenty-one years later, 66 of these students were restudied and without exception the maximal oxygen uptake was now reduced, on the average, by about 20 percent. Thirteen years later they were again tested at maximal exercise, running and cycling. Of the entire group, 53 were able to participate in this 3rd test. Their mean age was now 55 and 59 years for women and men, respectively. Surprisingly, over the intervening 13 years, the women had maintained their maximal aerobic power (mean value); for the men there was an average decline of 4 percent. There were marked individual variations, however, in both groups. Actually, for some of the subjects, the same maximum was now recorded as had been recorded 33 years before when they were physical education students. They were still very physically active (data to be published).

The highest values for maximal oxygen uptake so far reported are 7.4 liters · min^{-1} for a male, and 4.5 liters · min^{-1} for a female cross-country skier (see Fig. 10-1).

It is not surprising that the maximal oxygen uptake increases during childhood and adolescence. During this period there is a growth in all tissues of importance for strength and power. The sex differences will be discussed below. But what about the gradual decline in maximal aerobic power beyond the age of twenty?

There are many age-induced changes in tissue and organ functions which may explain this decline. A decline in maximal heart rate is quite evident (see Fig. 4-24). The lower maximal heart rate with higher age certainly must reduce the maximal cardiac output and hence the oxygen-transporting potential. In one of the few longitudinal studies just mentioned, including subjects with a relatively high level of habitual physical activity, the maximal heart rate declined from an average of 195, when they were in their mid-twenties, to 176, 33 years later, a trend that fits Fig. 4-24. However, in this group there is not a good correlation between the decline in maximal oxygen uptake and the change in the individual's maximal heart rate. We have to face detrimental effect of age in almost all functions of importance for the maximal aerobic power, as discussed in other chapters. In addition, on the whole, the aging person becomes less physically active, which will inevitably reduce her or his maximal aerobic power. Saltin and Grimby (1968) have studied middle-aged and old athletes and compared the data obtained with those from former athletes of the same ages who now live a sedentary life. The two groups were about equal as far as their performance in orienteering (including cross-country running) was concerned. It was therefore concluded that there was, at the time, no significant difference in maximal oxygen uptake. As can be seen from Table 7-2, the nonactive former athletes have now fallen behind as far as their maximal aerobic power is concerned, particularly in relation to body weight. In all groups, the mean values are higher than normally found for the

TABLE 7-2
DATA ON PERFORMANCE OF PRESENT AND FORMER ATHLETES IN ORIENTATION RACING.*

	Age						
	20–39	40–49		50–59		60–69	
Function	active N = 9	active N = 15	non-active N = 10	active N = 14	non-active N = 14	active N = 4	non-active N = 5
Max O₂ uptake							
liters · min⁻¹	5.4	4.0	3.3	3.4	2.9	2.7	2.6
ml · kg⁻¹ · min⁻¹	77	57	44	38	38	43	37
Heart volume, ml		1,050	835	940	915	830	865
Max heart rate		175	182	176	175	165	170
Cholesterol, mg · 100 ml⁻¹		222	231	251	277	286	266
Neutral fat serum, mM		0.85	1.56	0.95	1.44	1.10	1.85
Blood pressure, mm Hg	135/83	128/82	137/81	133/82	138/83	123/86	
ECG IV: 1–3**	2	1	4	2	1	0	

*The youngest group is currently competing. The active older athletes are still training and competing regularly. The inactive subjects had, when young, competed successfully with those of the same age who are still active; they discontinued their training more than 10 years ago because of lack of time. N denotes the number of subjects in each group.
**Classified according to the Minnesota Code (I. Åstrand et al., 1967a).
Source: Data from Saltin and Grimby. 1968.

same age groups. In the sedentary group, this is due to natural endowment (a highly selected group), whereas in those who are still active, this endowment is further developed by regular physical training. In both groups, there is a decline in maximal heart rate with age.

There is another consequence of a lower maximal heart rate. Studies on 33 building workers (bricklayers, carpenters, laborers) from thirty to seventy years of age showed that the mean heart rate during occupational activity was correlated with the individual's maximal heart rate (Fig. 7-14). For subjects with a maximal heart rate of 185 beats/min, mostly the younger workers, the mean heart rate during occupational activity was 110, and those with a maximum of 150 had a mean of 90 beats $\cdot$ min^{-1}. The maximal oxygen uptake ranged from 2.2 to 3.6 liters $\cdot$ min^{-1}. In general, the worker utilized the same percentage of his maximum in the work operations irrespective of his maximal oxygen uptake. In other words, the older worker with a lower maximal aerobic power keeps a slower tempo than the younger one, but the relative load is the same for the two workers, or about 40 percent. The person with a high maximal heart rate can do a day's work at a higher mean heart rate than a person with a low maximum., but the relative strain may be the same on the two persons (I. Åstrand, 1967a).

Maximal Aerobic Power—Related to Body Size

Wilmore (1979) describes the difference between the North American male and female as follows: "At full maturity, the average female is approximately 13 cm shorter than the average male, 15 to 18 kg lighter in total weight, 18 to 23 lighter in lean body weight, and considerably fatter, i.e., 25% vs. 15% relative body fat."

Obviously, differences in body size are important. And how much of the sex differences in maximal aerobic power can be explained by differences in body size? Furthermore, in work and exercise where the body is lifted (as in walking, running,

Figure 7-14
Individual values for the relationship between mean heart rate during occupational activity (building) and maximal heart rate attained during exercise on a cycle ergometer. The heart rate was recorded by telemetry. The estimated oxygen uptakes during occupational work are presented on the right, together with their symbols. (*From I. Åstrand, 1967b.*)

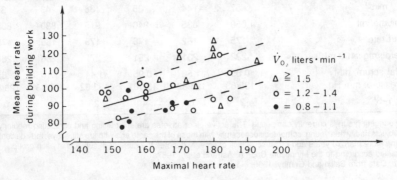

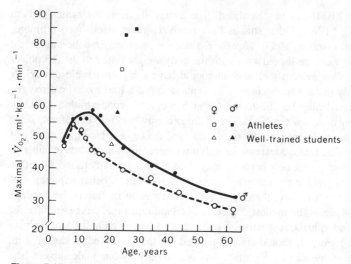

Figure 7-15
Mean values for maximal oxygen uptake expressed in ml · kg⁻¹ · min⁻¹. Same subjects as presented in Fig. 7–13. The standard deviation is between 2.5 and 5 ml O_2 · kg⁻¹ · min⁻¹.

climbing), the oxygen uptake potential related to body weight is certainly more relevant than the liter · min⁻¹ figure. Figure 7-15 presents the data from the same study as in Fig. 7-13, but the maximal aerobic power is expressed in milliliters of O_2 per kilogram gross body weight (ml · kg⁻¹ · min⁻¹; can also be written as ml/(kg · min); too frequently one finds in the literature; "ml/kg/min" or "ml/kg · min," both expressions being mathematically incorrect. Also there is a general agreement in data reported from different countries regarding the slopes and relative positions of the curves, but the levels may differ. For young girls and boys there is no significant difference in the maximal aerobic power until the age of approximately ten years, but from then on the females reach, on the average, 75 to 85 percent of the male's maximum. Again, the variation in this sex difference can be explained by the sampling procedure.

The highest maximal aerobic powers reported so far are 94 ml · kg⁻¹ · min⁻¹ for a male, and 77 ml · kg⁻¹ · min⁻¹ for a female cross-country skier.

Another way to express the individual's maximal oxygen uptake is to relate it to the dimensions of the body or to various organs. It may be of both theoretical and clinical value to examine whether the maximal oxygen uptake is proportional to heart size, muscular mass, lung volume, etc. Since fatty tissue is metabolically fairly inert but can constitute a large proportion of the body weight, it may be important to exclude it when evaluating the oxygen-transporting capacity.

When the weight of adipose tissue, estimated on the basis of hydrostatic weighing, is subtracted from the gross body weight of the well-trained students in Fig. 7-13 (12 kg or 20.3 percent of the body weight for the women and 8 kg or 10.6 percent for the men, Döbeln, 1956), the maximal oxygen uptake per kilogram fat-free body weight (lean body mass, LBM) can be calculated. The averge figure is the same for both groups: 71 ml · kg LBM^{-1}. Other studies have reached similar results (see Wilmore, 1979; Sparling and Cureton, 1983). However, since the aerobic metabolic rate at rest or during maximal exercise should vary with the body weight raised to the 2/3 power (Chap. 9), it becomes evident that women should have a higher aerobic power per kilogram lean body mass than the men. The explanation of a lower maximal oxygen uptake than expected may be due to the lower hemoglobin concentration found in women. The lower maximal aerobic power of women may therefore be natural, since their dimensions are different from those of men, and the oxygen-binding capacity of the blood is lower. The relative increase in body-fat content in women starts at puberty. In reality she has to carry not only her lean body mass but also her fatty tissue; therefore the maximal oxygen uptake per kilogram gross body weight is a better expression of her potential to move her body, than the maximum related to fat free body weight.

Figure 7-16 illustrates the modest degree of overlap in maximal oxygen uptake for female and male individuals, respectively, particularly when given in liters per minute (heavier shadowed area). It should also be pointed out that the Swedish student with an aerobic power of 69 ml · kg^{-1} · min^{-1} was beaten only by one male subject (she was a Swedish champion in cross-country running). At higher ages, however, there is a much larger scatter in the data and more women can successfully compete with the men (unpublished data collected by Åstrand and co-workers).

A High Maximal Oxygen Uptake Does Not Guarantee Top Performance

An interesting question is why the best performance, in events demanding a high aerobic power, is usually obtained by athletes 20 to 30 years of age, when the highest maximal oxygen uptake is usually reached before the age of twenty, see Figs. 7-13 and 7-15. There are, however, other factors to be considered. In general, physical activity is more regular and vigorous for those below than for those above twenty years of age, at least as long as physical education is compulsory in schools. This may explain the results presented in the figures mentioned. On the other hand, if training is continued, the maximal aerobic power can certainly be maintained or even further increased for another 10-year period. Finally, the performance also depends on technique, tactics, motivation, and other factors, and intensive training and experience over the years make a gradual improvement possible.

Figure 7-17 illustrates how physical performance is related to the maximal oxygen uptake in exercises with large muscle groups vigorously involved for some minutes or longer. No one can attain top results in such exercises without a high aerobic power. On the other hand, a high power does not guarantee a good performance, since technique, state of training, and psychological factors may have a modifying influence in a positive or negative direction. It is tempting to take a look at how accurately one may predict a person's performance in an all-out run taxing the oxygen transport system maximally, on the basis of the measured maximal oxygen uptake (ml · kg^{-1} · min^{-1}). Many such studies have been made. Shephard (1984b) has reviewed 37 such reports and noted coefficients of correlation from 0.04 to 0.90. (By choosing subjects with a wide range in $\dot{V}_{O_2}$ max, it is easier to obtain a high correlation;

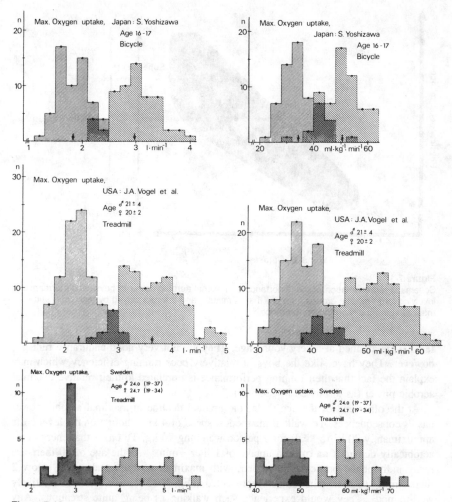

Figure 7-16
Variation in maximal oxygen uptake in liters · min⁻¹ (left panels) and related to body weight (right). The 120 Japanese subjects were randomly selected from school classes; the 184 subjects in the study by Vogel et al. (1977) were examined when entering the U.S. army; the Swedish subjects were 44 students in physical education, randomly selected. The arrows denote the mean value for women and men respectively. The deeper shadowed area indicates an overlap between females and males. n denotes the number of subjects in each column.

however, a reliable prediction *in the individual case* demands a correlation well above 0.90. A "significant correlation" is in itself useless in this context.) An important conclusion is that data on maximal oxygen uptake do not reveal the person's potential to perform well in aerobic-power demanding events. More examples of this truism are presented in other sections of this book.

It should be noted that the peak maximal oxygen uptake, ml · kg⁻¹ · min⁻¹ for

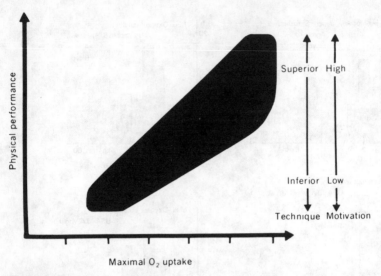

Figure 7-17
Schematic representation of the importance of maximal aerobic power in physical performance involving large muscle groups for more than a minute, and the role played by technique and motivation in modifying top performance.

females, is attained at a very young age. They are not very good endurance runners, however. They have, like the boys, a relatively poor running efficiency which may explain the fact that their running performance is unrelated to their relative maximal aerobic power (as discussed in Chap. 3).

At the other end of the age scale, the gradual decline in maximal aerobic power has its consequences. To walk at the modest speed of 4 km $\cdot$ hour^{-1} on the level costs approximately 20 kJ (5 kcal) for a person weighing 70 kg. To cover that energy cost aerobically demands an oxygen uptake of 1 liter $\cdot$ min^{-1}. At the age of 60 there are many individuals, particularly women, with maximal oxygen uptake well below 2 liters $\cdot$ min^{-1} (note that the subjects represented in Fig. 7-13 were selected and relatively well-trained). They would experience such walking as being quite strenuous. Any overweight will increase the energy cost in proportion to the additional number of kilograms. The oxygen requirement for a 100 kg person walking will be about 1.5 liters $\cdot$ min^{-1}; in this case the walk may end up being an all-out effort. It is easy to understand that aging obese persons will gradually reduce their habitual physical activity with a decreasing fitness as a consequence; there is a typical *circulus vitiosus*. There are two solutions to this problem: (1) the obese person may devote more effort to losing weight while still middle-aged, and (2) increase the maximal oxygen uptake by training, thus expanding the safety margin between the oxygen requirement of daily routine activities, and the maximal capacity.

Summary Particularly when discussing physical performance in relation to age, it is essential to realize that chronological age is a poor reference scale. For instance, in a class of twelve- to thirteen-year old children there may be one child who is biologically sixteen years old (a girl) and one who is just nine (a boy). The adolescent growth spurt may start as early as at the age of nine, or as late as at the age of seventeen with the average girl maturing about two years earlier than the average boy.

The maximal oxygen uptake expressed in liters · min^{-1} informs us about a person's potential for activity requiring a high aerobic power. It is highly correlated with the individual's maximal stroke volume and cardiac output (which limits the maximal aerobic power). The maximal oxygen uptake in ml · kg^{-1} · min^{-1} is indicative of a person's potential to move the body during such activities as running or climbing stairs, etc. for several minutes or longer. The actual oxygen uptake that can be maintained is at a certain percentage of the $\dot{V}_{O_2}$ max, this percentage becoming reduced, the longer the exercise is carried on. As a predictor of aerobic performance there is, in many events, a certain minimal requirement, but other factors such as technique, motivation, and in prolonged activity, muscle enzyme functions, muscle capillary density, and the supply of substrates will play an important role determining the performance level.

The maximal oxygen uptake (maximal aerobic power) increases with age up to between sixteen to seventeen years in females, and eighteen to twenty years in males. Beyond these ages there is a gradual decline so that the sixty-year-old individual attains about 70 percent of the $\dot{V}_{O_2}$ max from what it was at the age of twenty-five. Before the age of ten, there is no significant difference between girls and boys; thereafter, the average difference in maximal oxygen uptake between women and men amounts to 25 to 35 percent. Related to the body weight, the sex difference in aerobic power after puberty is 15 to 25 percent. A calculation of the maximal oxygen uptake per kilogram fat-free body weight, or related to muscle mass, blood volume, or other such parameters, makes it possible to analyze dimensions versus function. When expressed per kg fat-free body weight, the maximal aerobic power for women and men is very similar, particularly when they are well-trained.

Training and lack of regular physical activity can significantly modify the maximal oxygen uptake. Top athletes in endurance events have a $\dot{V}_{O_2}$ max that is about twice as high as that of an average person. The basis for such high levels is a combination of constitution and effective training. The gradual decline in $\dot{V}_{O_2}$ max with age beyond twenty is, at least, partly due to a reduction in maximal heart rate. Inactivity is another factor that decreases the functional range of the oxygen transporting system. Inactivity reduces the stroke volume and perhaps the efficiency of the regulation of the circulation during exercise. The margin between the aerobic demands of various daily activities and the maximal aerobic power becomes narrower, and sooner or later this will create problems. Particularly vulnerable is the obese person because any overweight increases the energy cost of moving the body.

ANAEROBIC POWER, SEX, AND AGE

Earlier in this chapter it was emphasized that we lack reliable methods to measure anaerobic power. According to di Prampero the average man is 15 to 30 percent superior to the average women in maximal alactic anaerobic power (calculated per kg body weight). The sixty-year-old person has an alactic anaerobic power which is 60 percent of the value for one who is twenty years old. Applying the Margaria staircase sprint, di Prampero and Cerritelli (1969) calculated the anaerobic energy output in African natives, and reported it as equivalent to oxygen delivery per kg body weight. Their figure was 88 ml $\cdot$ kg^{-1} $\cdot$ min^{-1} for eight-year-old girls and boys. It increased to 143 and 165 ml $\cdot$ kg^{-1} $\cdot$ min^{-1} respectively at the age of 20. This power reflects mainly the potential of the high-energy phosphate compounds. Since children have a similar concentration of ATP and PCr in their muscles as do adults, this improvement in a sprint lasting some 6 s is surprising. It may perhaps essentially reflect an improved technique with age, including greater ability to make use of the elastic properties of the muscles. In a 30 s all-out test, teenage girls produce 20 to 25 percent less work than do the boys. Part of this difference is due to the small muscle mass of the girls.

As mentioned above, the peak blood lactate concentrations in the muscles and the blood are *not* reflecting anaerobic power or capacity. They will, however, provide information about the internal environment after exhausting exercise. In most studies, children and older individuals do not attain as high blood lactate values as young adults. For 68 boys and girls, the peak recovery lactate value after about 5 min maximal treadmill exercise was, on the average, 9 mM which is lower than a "normal" maximum (Åstrand, 1952). Robinson (1938), in his classical study of male subjects, made a similar observation. Eriksson (1972) reports a lower muscle lactate concentration in eleven-year-old boys (about 10 mM $\cdot$ kg^{-1} wet muscle) than usually noted in older subjects after an all-out effort (20 mM or higher). He observed that the concentration of a key enzyme in the glycolysis, phosphofructokinase (PFK) was lower in children than in adults. This may be one factor explaining the children's lower peak lactate concentration. Actually, after a period of training the PFK concentration increased and so did the lactate concentration after maximal exercise. In all such tests, motivation always plays a decisive role, and children may, in general, be more reluctant to push themselves to their "anaerobic limit". However, high lactate values in children have been observed (Cumming et al., 1980; see Shephard, 1982, p. 97).

There is no significant sex difference in the peak lactate concentrations.

MUSCLE STRENGTH, SEX, AND AGE

In Chaps 2 and 3 many aspects of muscle strength were discussed. In this context it is important to remember that in any test of muscular strength, the day-to-day variation is usually on the order of some ± 10 percent. The correlation between the strength of different muscle groups in the same individual is low, moderate, or fairly high, depending on which muscle groups are compared.

Age

Figure 7-18 presents curves for maximal muscle strength in relation to age for males and females. A peak is usually reached at the age of twenty for men and a few years earlier for women. The strength of the 65-year-old person is, on the average, 75 to 80 percent of that attained between the ages of twenty to thirty, with a further decline to about 60 percent in leg and back muscles, and to 70 percent in arm muscles from thirty to eighty years of age. In other words, the rate of the decline with age in the

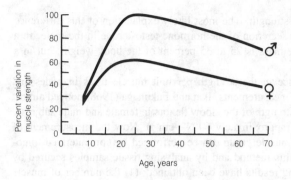

Figure 7-18
Changes in maximal isometric strength with age in women and men. (*By courtesy of E. Asmussen.*)

strength of the leg and trunk muscles is in both sexes greater than in the strength of arm muscles (see Hollmann and Hettinger, 1980, p. 202; Grimby and Saltin, 1983). Some studies have provided strength-curves that are flatter than the ones presented in Fig. 7-18 up to the age of fifty (see Larsson et al., 1979). As in the case of maximal aerobic power, training may improve muscular performance by increasing muscle strength, and by increasing the engagement of muscle synergists in the activities of everyday life. This may greatly influence the test results.

The decline in maximal muscle strength with age seems to parallel the reduction in muscle mass. The individual muscle fibers have a relatively "normal" cross-sectional area during adult life, up to the age of eighty years, with some diminution in the area of the fast-twitch fibers (type II) (see Grimby and Saltin, 1983). The reason for the reduced muscle mass, and hence the reduced strength in older persons is because of a loss of muscle fibers, perhaps down to some 60 percent of the initial number. As discussed in Chap. 3, a loss of motoneurons is most likely the cause of this type of degeneration. It is not surprising that older individuals have a relatively poor power potential for the performance of a vertical jump (Davies et al., 1983).

In the case of children, Asmussen (1973) concludes that age affects muscle strength by (1) increased size of anatomical dimensions; (2) the impact of aging in itself (one extra year before puberty increases the strength by 5 to 10 percent of the average strength of the same group, a gain that at least partly is attributed to the maturation of the central nervous system); and (3) development of the child's sexual maturity; for boys the male sex hormones are of special importance for the development of their muscle strength (see below). As a matter of fact, about one third of the increase in body height occurs between the ages of six and twenty, but during the same period, four-fifths of the development of strength takes place.

Sex

An increase in muscular strength is definitely a consequence of growth. Of young children, up to the age of ten to twelve years, there is no significant difference in strength between girls and boys, although there is a trend for the boys to be stronger. After this age boys become continuously stronger for some years, while the girls do not improve much in muscle strength. The most likely explanation of this difference in development is the greater secretion of the hormone testosterone in the male. In a nonobese woman, the muscle mass is 25 to 35 percent of the body weight, but in a man it is 40 to 45 percent.

Is there then any difference in the strength per unit muscle mass in males and females? Applying ultrasonic measurements, Ikai and Fukunaga (1968) noticed almost the same maximal isometric strength of the elbow flexors in female and male subjects, aged twelve to twenty years (approximately 60 N · cm^{-2}). Now the measurement of the cross-sectional area of a muscle group can be performed by computerized tomographic scanning. Applying this method and by analysing tissue samples secured by muscle biopsies, the following results have been obtained: (1) the number of muscle fibers in a given muscle group is the same in body-builders and in female and male physical education students; and (2) the maximal tension developed per unit of cross-sectional area does not differ between the female and male students or the body-builders (see Fig. 7-19). Therefore, the difference in maximal strength is apparently mainly explained by a difference in size of the individual muscle fiber. Costill et al. (1976) present data on fiber size in untrained women and men and, on the average, the cross-sectional area for slow-twitch fibers in women is approximately

Figure 7-19
Relationship between muscle cross-sectional area times body height (proportional to the lever arm) in body builders (BB) and physical education students, females (PEF) and males (PEM). Muscle cross-section image of the left thigh, obtained through computed tomography scanning. The medial extensor (ME), lateral extensor (LE), and non-extensor (NE) muscle groups as well as bone have been outlined. The contraction was isokinetic at an angular velocity of 30B · s^{-1}. The total area of ME plus LE was for BB 125.0 cm^2, for PEM 88.3, and for PEF 66.5 cm^2 (i.e., 75 percent of the figure for their male colleagues). The mean muscle fiber area was 8400 mm^2, 6200, and 4400 mm^2 respectively. (*From Schantz et al., 1983.*)

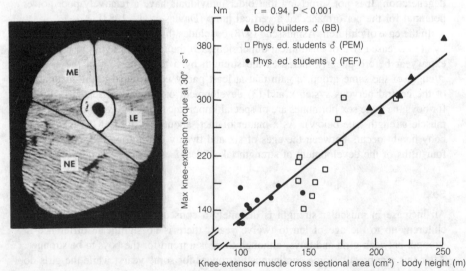

70 percent of the men's size; for fast-twitch fibers it is about 85 percent. The scatter is very large, however. There is no sex difference in the proportion of the fiber types (see Chap. 2).

In absolute figures, the women's maximal strength in isometric or dynamic contractions of the leg muscles (e.g. the knee flexors or extensors) is, on the average, 65 to 75 percent of the men's maximum. For trunk muscles, the percentages are 60 to 70. In elbow flexion and extension, the females are at a greater disadvantage, since they can attain only about 50 percent of the male maximum. The area of arm muscles is not in proportion to the women's body size. One may speculate that the women's leg muscles are relatively well trained because women have less muscle mass in relation to their body weight than do men. (See Nygaard, 1981; Schantz et al., 1983; Maughan, 1984.)

Figure 7-20 illustrates the large difference in muscle strength between females and males. The subjects studied in 1977 by Vogel et al., (Fig.7-16) were also submitted to isometric strength tests. The maximal strength for elbow flexors was 46 percent

Figure 7-20
Data on 230 female and male subjects 17 to 18 years of age on maximal strength in arm, trunk, and leg engagements respectively. The data are presented in a similar manner as in Fig. 7-16. The strength is given in arbitrary units. (*Data by courtesy of M. Miyashita.*)

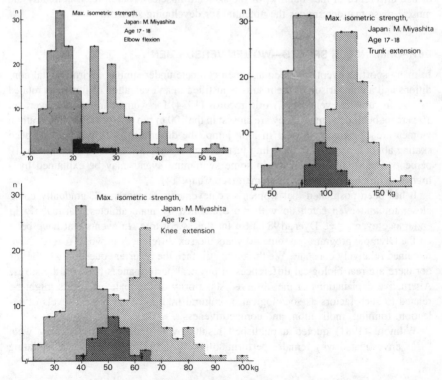

lower in the females; for knee flexors it was 34 percent, and for knee extensors, 23 percent lower than the figures recorded for the male group.

Summary The maximal strength per unit of cross-sectional area of the skeletal muscle is about the same in women and men, irrespective of age. In Chap. 3 , it was pointed out that at the beginning of a strength training program an increase in strength may be observed without any change in the cross-sectional area of the muscles involved. A recruitment of more motor units in the trained person may explain this finding. Therefore the correlation between maximal strength and the cross-sectional area of the engaged muscles is not 1 : 0. Differences between individuals in the internal muscle architecture, in limb lengths, and in joint structures may prove to be important factors capable of influencing strength (Maughan, 1984).

The aging person is facing a reduction in the number of motoneurons and therefore a decline in muscle mass. An inevitable consequence of this is a decrease in muscular strength so that, on the average, the sixty-five-year-old person can develop 75 to 80 percent of the strength attained when he or she was young. As women have a smaller muscular mass than men, it is not surprising that the maximal muscular strength is lower for them. In women, the maximal leg muscle strength is about 70 percent of that of the males, but in the case of elbow flexors, the maximal strength of the women is, on the average, only about 50 percent of that of the men. Evidently there is no sex or age difference in the "quality" of the skeletal muscles. Therefore, it is mainly the muscle mass that determines the potential for developing strength.

PERFORMANCE IN SPORTS—WOMEN VERSUS MEN

In many sporting events, women and men compete under similar environmental conditions and a comparison of the results is justified in an evaluation of fitness as related to sex. In running events, the world records (1984) for women are, on the average, 10 percent behind the men. They are closest in the 100 m run (8.5 percent); in marathon women are 12 percent slower. In long jump, the difference is 25 percent, in speed skating about 8 percent, in bicycling just below 12 percent, and in swimming 6 to 10 percent. The smaller difference in some swimming events may be explained by a higher swimming efficiency in women (see Chap. 14).

It has been postulated that women's achievements in sports will gradually come closer to, and even catch up with those attained by male athletes, particularly in endurance events (see Dyer, 1984.) So far this is not true. In events that have been on the Olympic program for some 40 years the sex differences in world records have remained relatively constant. We therefore still face the recurrent question whether or not there are real biological differences in physical performances between the sexes. Alternative explanations of the observed superiority of the males in sports might be related to such factors as sociological or cultural traditions and biases, levels of selection, training, motivation, and competitiveness.

Wilmore (1981) quoted unpublished results by Grimditch and Sockolov who: " . . . investigated why females perform so poorly in the softball throw. Postulating this difference to be the result of insufficient practice and experience, they recruited over 200 males and females, three to twenty years of age, to throw the softball for distance with both the dominant and nondominant arm. As they had predicted, there was absolutely no difference between males and females throwing with the nondominant arm up to the age of ten to twelve years, a pattern identical to that on the other motor tasks, whereas, with the dominant arm, the males were able to throw over twice the distance of the females at all ages. Thus, the softball throw for distance using the dominant arm appears to be biased by the previous experience and practice of the male."

Boys apparently practice ball throwing much more than do the girls. The cause of this difference in behavior is still an open question. Silva et al. (1984), have followed 954 girls and boys from birth to the age of seven years and submitted them to a "basic motor ability test". At this age the boys gained significantly higher scores than did the girls on the gross motor measurements, particularly in tasks requiring strength where shoulder girdle and arm muscles were involved, while the girls gained significantly higher scores in tasks requiring finer motor involvement (Gross motor performance factor was defined by long jump, agility run, target throwing, push ups, and face down to standing subtest; the fine motor performance factor was measured by the tapping board, balance, bead stringing, and hamstring stretch subtests). The authors suggest that different interest and therefore practice may explain the observed sex differences.

Apparently women's performances in sports events are closer to the men's results than mirrored in the laboratory tests. The female top athlete in running has a relative body fat content similar to the male runner, i.e., below 10 percent, in some cases down to 6 percent (see Wilmore, 1979). A smaller amount of fat to carry when running is certainly an advantage.

CONCLUDING REMARKS

On the whole, it appears that when selecting population samples at random older age groups are less physically active than younger age groups. Furthermore, females of all ages are underrepresented in the sporting field, both in terms of participation and in terms of the number of events that they participate in (see Loy et al., 1978; Engström, 1980; Claeys, 1982). Since training can improve muscle strength and metabolic power, various testing procedures do not reveal the real potential of the subject studied.

Figures 7-16 and 7-20 are quite representative and illustrate that few females can compete with the males in aerobic power and muscle strength, at least between the ages of fifteen to some fifty years, the level of training in terms of volume and relative intensity being the same. In many countries there is a policy to mix girls and boys in classes of physical education. In events demanding a high aerobic energy output and muscle strength this will complicate the educational process as the range of these functions within each sex is large to start with.

Women may apply for jobs in occupations such as police (traffic officers), fire departments, or military service. If the job requires heavy manual labor and lifting of heavy items (e.g., a person from a burning car or apartment), it is reasonable that the selection of personnel, to a large extent, be based on the applicant's physical fitness, regardless of sex (see Ergonomics, 1984).

We are back to the evolutionary fact: we were not born equal. For the ultimate survival of the offspring it is certainly an advantage that female and male counterparts are at least partly different. However, in the animal kingdom, it is not necessarily the male who is the more powerful protector, or the most successful hunter. For some birds of prey, for instance, the female is superior to the male in body size and power.

REFERENCES

Ahlborg, G., L. Hagenfeldt, and J. Wahren: Influence of Lactate Infusion on Glucose and FFA Metabolism in Man, *Scand. J. Clin. Lab. Invest.*, **36**:193, 1976.

Ahlborg, G., R. Hendler, and P. Felig: Lactate Production by Resting Muscle during and after Prolonged Exercise: Mechanism of Redistribution of Muscle Glycogen, *Clin. Physiol.*, **1**:608, 1981.

Asmussen, E.: Growth in Muscular Strength and Power, in G. L. Rarick (ed.), "Physical Activity Human Growth and Development," p. 60, Academic Press, Inc., New York, 1973.

Åstrand, I.: Aerobic Work Capacity in Men and Women with Special Reference to Age, *Acta Physiol. Scand.*, **49**(Suppl. 169):1960.

Åstrand, I.: Degree of Strain during Building Work as Related to Individual Aerobic Work Capacity, *Ergonomics*, **10**:293, 1967a.

Åstrand, I.: Aerobic Working Capacity in Men and Women in Some Professions, *Försvarsmedicin*, **3**:163, 1967b.

Åstrand, I., P.-O. Åstrand, and K. Rodahl: Maximal Heart Rate during Work in Older Men, *J. Appl. Physiol.*, **14**:562, 1959.

Åstrand, I., P.-O. Åstrand, E. H. Christensen, and R. Hedman: Intermittent Muscular Work, *Acta Physiol. Scand.*, **48**:443, 1960a.

Åstrand, I., P.-O. Åstrand, E. H. Christensen, and R. Hedman: Myohemoglobin as an Oxygen-store in Man, *Acta Physiol Scand.*, **48**:454, 1960b.

Åstrand, I., P.-O. Åstrand, I. Hallbäck, and A. Kilbom: Reduction in Maximal Oxygen Uptake with Age, *J. Appl. Physiol.*, **35**:649, 1973.

Åstrand, P.-O.: "Experimental Studies of Physical Working Capacity in Relation to Sex and Age," Ejnar Munksgaard, Copenhagen, 1952.

Åstrand, P.-O.: The Respiratory Activity in Man Exposed to Prolonged Hypoxia, *Acta Physiol. Scand.*, **30**:343, 1954.

Åstrand, P.-O., and B. Saltin: Oxygen Uptake during the First Minutes of Heavy Muscular Exercise, *J. Appl. Physiol.*, **16**:971, 1961a.

Åstrand, P.-O., and B. Saltin: Maximal Oxygen Uptake and Heart Rate in Various Types of Muscular Activity, *J. Appl. Physiol.*, **16**:977, 1961b.

Åstrand, P.-O., J. Hallbäck, R. Hedman, and B. Saltin: Blood Lactates after Prolonged Severe Exercise, *J. Appl. Physiol.*, **18**:619, 1963.

Åstrand, P.-O., and E. H. Christensen: Aerobic Work Capacity, in F. Dickens, E. Neil, and W. F. Widdas (eds.), "Oxygen in the Animal Organism," p. 295, Pergamon Press, New York, 1964.

Bang, O.: The Lactate Content of the Blood during and after Muscular Exercise in Man, *Skand. Arch. Physiol.*, **74**(Suppl. 10):51, 1936.

Belcastro, A. N., and A. Bonen: Lactic Acid Removal Rates during Controlled and Uncontrolled Recovery Exercise, *J. Appl. Physiol.*, **39**:932, 1975.

Bergh, U., I.-L. Kanstrup, and B. Ekblom: Maximal Oxygen Uptake during Exercise with Various Combinations of Arm and Leg Work, *J. Appl. Physiol.*, **41**:191, 1976.

Borg, G. A. V.: Psychophysical Bases of Perceived Exertion, *Med. Sci. Sports Exc.*, **14**:377, 1982.

Bouchard, C., and R. M. Malina: Genetics of Physical Fitness and Motor Performance, *Exercise and Sport Sciences Reviews*, **11**:306, 1983.

Brooks, G. A., K. E. Brauner, and R. G. Cassens: Glycogen Synthesis and Metabolism of Lactic Acid after Exercise, *Am. J. Physiol.*, **224**:1162, 1973.

Brooks, G. A., and T. D. Fahey: Exercise Physiology: "Human Bioenergetics and its Applications," John Wiley & Sons, New York, 1984.

Busa, W. B., and R. Nuccitelli: Metabolic Regulation via Intracellular pH, *Am. J. Physiol.*, **246**:R409, 1984.

Christensen, E. H., R. Hedman, and B. Saltin: Intermittent and Continuous Running, *Acta ·Physiol. Scand.*, **50**:269, 1960.

Claeys, U.: "Sport in European Society," Council of Europe, Committee for the Development of Sport, Strasbourg, CDDS(82) 25-E Part I, 1982.

Connett, R. J., T. E. J., Gayeski, and C. R. Ḥonig: Lactate Accumulation in Fully Aerobic Working Dog Gracilis Muscle, *Am. J. Physiol.*, **246**:H120, 1984.

Costill, D. L.: Metabolic Response during Distance Running, *J. Appl. Physiol.*, **28**:251, 1970.

Costill, D. L., J. Daniels, W. Evans, W. Fink, G. Krahenbuhl, and B. Saltin: Skeletal Muscle Enzymes and Fiber Composition in Male and Female Track Athletes, *J. Appl. Physiol.*, **40**:149, 1976.

Cumming, G. R., L. Hastman, J. McCort, and S. McCullough: High Serum Lactates Do Occur in Young Children after Maximal Work, *Int. J. Sports Med.*, **1**:66, 1980.

Davies, C. T. M., and R. Rennie: Human Power Output, *Nature*, **217**:770, 1968.

Davies, C. T. M., and M. W. Thompson: Aerobic Performance of Female Marathon and Male Ultramarathon Athletes, *Europ. J. Appl. Physiol.*, **41**:233, 1979.

Davies, C. T. M., J. White, and K. Young: Electrically Evoked and Voluntary Maximal Isometric Tension in Relation to Dynamic Muscle Performance in Elderly Male Subjects, Aged 69 years, *Eur. J. Appl. Physiol.*, **51**:37, 1983.

Denis, C., R. Fouquet., P. Poty, A. Geyssant, and J. R. Lacour: Effect of 40 weeks of Endurance Training on the Anaerobic Threshold, *Int. J. Sports Med.*, **3**:208, 1982.

di Prampero, P. E.: Energetics of Muscular Exercise, *Rev. Physiol. Biochem. Pharmacol.*, **89**:143, 1981.

di Prampero, P. E., and P. Cerritelli: Maximal Muscular Power (Aerobic and Anaerobic) in African Natives, *Ergonomics*, **12**:51, 1969.

di Prampero, P. E., Boutellier, U., and P. Pretsch: Oxygen Deficit and Stores at Onset of Muscular Exercise in Humans, *J. Appl. Physiol.: Resp. Env. Exer. Physiol.*, **55**:146, 1983.

Döbeln, W. von: Human Standard and Maximal Metabolic Rate in Relation to Fat-free Body Mass, *Acta Physiol. Scand.*, **37**(Suppl. 126):1956.

Drinkwater, B, L., S. M. Horvath, and C. L. Welles: Aerobic Power of Females, Ages 10 to 68, *J. Gerontol.*, **30**:385, 1975.

Dyer, K.: Catching up the Men, Olympics' 84, *New Scientist*, **103**:25, 1984.

Engström, L.-M.: Physical Activity of Children and Youth, *Acta Paediatr. Scand.*, (suppl. 283):101, 1980.

Epstein, Y., G. Keren, R. Udassin and Y. Shapiro: Way of Life as a Determinant of Physical Fitness, *Eur. J. Appl. Physiol.*, **47**:1, 1981.

Ergonomics: Special Issue: Women at Work, **27**(5):463, 1984.

Erikssen, J., and K. Rodahl: Seasonal Variation in Work Performance and Heart Rate Response to Exercise, *Eur. J. Appl. Physiol.,* **42:**133, 1979.

Eriksson, B. O.: Physical Training, Oxygen Supply and Muscle Metabolism in 11-13 year old Boys, *Acta Physiol. Scand.* (Suppl. 384), 1972.

Essén, B.: Studies on the Regulation of Metabolism in Human Skeletal Muscle using Intermittent Exercise as an Experimental Model, *Acta Physiol. Scand. (Suppl. 454),* 1978.

Gledhill, N.: Bicarbonate Ingestion and Anaerobic Performance, *Sports Med.,* **1:**177, 1984.

Green, H. J., R. L. Hughson, G. W. Orr, and D. A. Ranney: Anaerobic Threshold, Blood Lactate, and Muscle Metabolites in Progressive Exercise, *J. Appl. Physiol.: Resp. Env. Exer. Physiol.,* **54:**1032, 1983.

Grimby, G., and B. Saltin: The Ageing Muscle, *Clin. Physiol.,* **3:**209, 1983.

Hagberg, J. M., E. F. Coyle, J. E. Carroll, J. M. Miller, W. H. Martin, and M. H. Brooke: Exercise Hyperventilation in Patients with McArdle's Disease, *J. Appl. Physiol.: Resp. Env. Exer. Physiol.,* **52:**991, 1982.

Hedman, R.: The Available Glycogen in Man and the Connection between Rate of Oxygen Intake and Carbohydrate Usage, *Acta Physiol. Scand.,* **40:**305, 1957.

Hermansen, L.: Oxygen Transport during Exercise in Human Subjects, *Acta Physiol. Scand.* (Suppl. 399): 1973.

Hermansen, L., and O. Vaage: Lactate Disappearance and Glycogen Synthesis in Human Muscle after Maximal Exercise, *Am. J. Physiol.,* **2:**E422, 1977.

Hochachka, P. W., and T. P. Mommsen: Protons and Anaerobiosis, *Science,* **219**(4588):1391, 1983.

Hollmann, W., and Th. Hettinger: "Sports medizin-Arbeits-und Trainings-grundlagen," F. K. Schattauer Verlag, Stuttgart, 1980.

Hughes, E. F., S. C. Turner, and G. A. Brooks: Effects of Glycogen Depletion and Pedaling Speed on "Anaerobic Threshold," *J. Appl. Physiol.: Resp. Exer. Env. Physiol.,* **52,**1598, 1982.

Hultman, E., and H. Sjöholm: Substrate Availability, in Knuttgen, H. G., J. A. Vogel, and J. Poortmans (eds.): "Biochemistry of Exercise," International Series on Sport Sciences, vol. 13, p 63, Human Kinetics Publishers, Inc., Champaign, Il, 1983.

Hurley, B. F., J. M. Hagberg, W. K. Allen, D. R. Seals, J. C. Young, R. W. Cuddihee, and J. O. Holloszy: Effect of Training on Blood Lactate Levels during Submaximal Exercise, *J. Appl. Physiol.: Resp. Env. Exer. Physiol.,* **56:**1260, 1984.

Ikai, M., and T. Fukunaga: Calculation of Muscle Strength per Unit Cross-sectional Area of Human Muscle by Means of Ultrasonic Measurement, *Int. Z. Angew. Physiol. Einschl. Arbeitsphysiol.,* **26:**26, 1968.

Jacobs, J. and P. Kaiser: Lactate in Blood, Mixed Skeletal Muscle, and FT or ST Fibres during Cycle Exercise in Man, *Acta Physiol. Scand.,* **114:**461, 1982.

Jones, N., and R. Ehrsam: The Anaerobic Threshold, in R. Terjung (ed.), *Exercise and Sports Sciences Reviews,* **10:**49, 1982.

Karlsson, J., F. Bonde-Petersen, J. Henriksson, and H. G. Knuttgen: Effects of Previous Exercise with Arms or Legs on Metabolism and Performance in Exhaustive Exercise, *J. Appl. Physiol.,* **38:**763, 1975.

Karlsson, J., and I. Jacobs: Onset of Blood Lactate Accumulation during Muscular Exercise as a Threshold Concept, *Int. J. Sports Med.,* **3:**190, 1982.

Kasch, F., J. P. Wallace, R. R. Huhn, L. A. Krogh, and P. M. Hurl: $\dot{V}_{O_2}$ max during Horizontal and Inclined Treadmill Running, *J. Appl. Physiol.* **40:**982, 1976.

Katz, A., D. L. Costill, D. S. King, M. Hargreaves, and W. J. Fink: Maximal Exercise Tolerance After Induced Alkalosis, *Int. J. Sports Med.*, **5:**107, 1984.

Keul, I., D. Keppler, and E. Doll: Standard Bicarbonate, pH, Lactate and Pyruvate Concentrations during and after Muscular Exercise, *German Medical Monthly* **12:**156, 1967.

Knuttgen, H. G., and B. Saltin: Oxygen Uptake, Muscle High-Energy Phosphates, and Lactate in Exercise under Acute Hypoxic Conditions in Man, *Acta Physiol. Scand.*, **87:**368, 1973.

Kobayashi, K.: "Aerobic Power of the Japanese," Kyorin Shoin Publisher, Tokyo, 1982.

Krebs, H.: Gluconeogenesis, The Croonian Lecture, 1963, *Proc. Roy. Soc. London,* **159:**545, 1964.

Larsson, L., G. Grimby, and J. Karlsson: Muscle Strength and Speed of Movement in Relation to Age and Muscle Morphology, *J. Appl. Physiol.: Resp. Env. Exer. Physiol.*, **46:**451, 1979.

Lindgren, G.: Growth of Schoolchildren with Early, Average and Late Ages of Peak Height Velocity, *Ann. Human. Biol.*, **5:**253, 1978.

Loy, J. W., B. D. McPherson, and G. Kenyon: "Sport and Social Systems," Addison-Wesley Publishing Company, Reading, MA., 1978.

Maron, M., S. M. Horvath, J. E. Wilkerson, and J. A. Gliner: Oxygen Uptake Measurements during Competitive Marathon Running, *J. Appl. Physiol.*, **40:**836, 1976.

Maughan, R. J.: Relationship between Muscle Strength and Muscle Cross-sectional Area, *Sports Med.*, **1:**263, 1984.

McCartney, N., G. J. F. Heigenhauser, and N. L. Jones: Effects of pH on Maximal Power Output and Fatigue during Short-term Dynamic Exercises, *J. Appl. Physiol.: Resp. Exer. Env. Physiol.*, **55:**225, 1983.

McLellan, T. M., and J. S. Skinner: Blood Lactate Removal during Active Recovery Related to the Aerobic Threshold, *Int. J. Sports Med.*, **3:**224, 1982.

Newman, E. V., D. B. Dill, H. T. Edwards, and F. A. Webster: The Rate of Lactic Acid Removal in Exercise, *Amer. J. Physiol.*, **118:**457, 1937.

Newsholme, E. A., and A. R. Leech: "Biochemistry for the Medical Sciences," John Wiley & Sons, New York, 1983.

Oberholzer, F., H. Claassen, H. Moesch, and H. Howald: Ultrastrukturelle, biochemische und energetische Analyse einer extremen Dauerleistung (100 km-Lauf), *Schw. Z. Sportmedizin,* **24:**71, 1976.

Nygaard, E.: Skeletal Muscle Fiber Characteristics in Young Women, *Acta Physiol. Scand.*, **112:**299, 1981.

Pate, R. R., and A. Kriska: Physiological Basis of the Sex Difference in Cardiorespiratory Endurance, *Sports Med.*, **1:**87, 1984.

Pedersen, P. K.: Oxygen Uptake Kinetics and Lactate Accumulation in Heavy Submaximal Exercise with Normal and High Inspired Oxygen Fractions, in H. G. Knuttgen, J. A. Vogel, and J. Poortmans (eds.): "Biochemistry of Exercise," International Series on Sport Sciences, vol. 13, p. 415, Human Kinetics Publishers, Inc., Champaign, Il, 1983.

Purvis, J. W., and K. J. Cureton: Ratings of Perceived Exertion at the Anaerobic Threshold, *Ergonomics,* **24:**295, 1981.

Robinson, S.: Experimental Studies of Physical Fitness in Relation to Age, *Arbeitsphysiol.*, **10:**251, 1938.

Rodahl, A., M. O'Brien, and R. G. R. Firth: Diurnal Variation in Performance of Competitive Swimmers, *J. Sports Med. Phys. Fitness,* **16:**72, 1976.

Rowell, L. B.: Human Cardiovascular Adjustments to Exercise and Thermal Stress, *Physiol. Rev.,* **54:**75, 1974.

Rowell, L. B.: Cardiovascular Adjustments to Thermal Stress, Handbook of Physiology, sect. 2: The Cardiovascular System, vol. III: Peripheral Circulation and Organ Blood Flow, part 2, chap. 25, p. 967, American Physiological Society, Bethesda, MD., 1983.

Saltin, B.: Aerobic Work Capacity and Circulation at Exercise in Man, *Acta Physiol. Scand.*, **62**(Suppl. 230):1964.

Saltin, B., and G. Grimby: Physiological Analysis of Middle-aged and Old Former Athletes: Comparison with Still Active Athletes of the Same Ages, *Circulation*, **38**:1104, 1968.

Saltin, B., B. Essén, and P. K. Pedersen: Intermittant Exercise: its Physiology and Some Practical Application, in E. Jokl, R. L. Anand, and H. Stoboy (eds.), "Advances in Exercise Physiology," p. 23, S. Karger, Basel, 1976.

Saltin, B., and P. D. Gollnick: Skeletal Muscle Adaptability: Significance for Metabolism and Performance, in L. D. Peachey, R. H. Adrian, and S. R. Geiger (eds.), "Handbook of Physiology, section 10: Skeletal Muscle," p. 555, American Physiological Society, Williams & Wilkins, Baltimore, 1983.

Schantz, P., E. Randall-Fox, W. Hutchinson, A. Tyden, and P.-O. Åstrand: Muscle Fiber Type Distribution Muscle Cross-sectional Area and Maximal Voluntary Strength in Humans, *Acta Physiol. Scand.*, **117**:219, 1983.

Senay, L. C. Jr., G. Rogers, and P. Jooste: Changes in Blood Plasma during Progressive Treadmill and Cycle Exercise, *J. Appl. Physiol.: Resp. Exer. Env. Physiol.*, **49**(1):59, 1980.

Shephard, R. J.: "Physical Activity and Growth," Year Book Medical Publishers, Inc., Chicago, 1982.

Shephard, R. J.: Sleep Biorhythms and Human Performance, *Sports Med.*, **1**:11, 1984a.

Shepard, R. J.: Tests of Maximum Oxygen Intake: A Critical Review, *Sports Med.*, **1**:99, 1984b.

Silva, P. A., J. Birkbeck, D. G. Russel, and J. Wilson: Some Biological, Development, and Social Correlates of Gross and Fine Motor Performance in Dunedin Seven Year Olds, *J. Human Movement Studies*, **10**:35, 1984.

Sparling, P. B., and K. J. Cureton: Biological Determinants of the Sex Difference in 12-min Run Performance, *Med. Sci. Sports Exerc. (US)*, **15**(3):218, 1983.

Stegmann, H., and W. Kindermann: Comparison of Prolonged Exercise Tests at the Individual Anaerobic Threshold and the Fixed Anaerobic Threshold of 4 mmol · 1^{-1} Lactate, *Int. J. Sports Med.*, **3**:105, 1982.

Stenberg, J., P.-O. Åstrand, B. Ekblom, J. Royce, and B. Saltin: Hemodynamic Response to Work with Different Muscle Groups, Sitting and Supine, *J. Appl. Physiol.*, **22**:61, 1967.

Tanner, J. M.: Growth of Adolescence," 2nd ed., Blackwell Scientific Publications, Oxford, 1962.

Tanner, J. M.: Some Methodological Problems in the Analysis of Human Growth, Quetelet to the Present, in M. Ritzén et al., (eds.), "The Biology of Normal Human Growth," p. 309, Raven Press, N.Y., 1980.

Tanner, J. M., R. H. Whitehouse, N. Cameron, W. A. Marshall, M. J. R. Healy, H. Goldstein: "Assessment of Skeletal Maturity and Prediction of Adult Height (TW2 Method)," 2nd ed., Academic Press, London, 1983.

Tesch, P., W. L. Daniels, and D. S. Sharp: Lactate Accumulation in Muscle and Blood during Submaximal Exercise, *Acta Physiol. Scand.*, **114**:441, 1982.

Vogel, J. A., M. U. Ramos, and J. F. Patton: Comparison of Aerobic Power and Muscle Strength between Men and Women Entering the U.S. Army, *Med. Sci. Sports*, **9**:58, 1977.

Wasserman, K., B. J. Whipp, S. N. Koyal, and W. L. Beaver: Anaerobic Threshold and Respiratory Gas Exchange during Exercise, *J. Appl. Physiol.*, **35**:236, 1973.

Wilmore, J. H.: The Application of Science to Sport: Physiological Profiles of Male and Female Athletes, *Can. J. Appl. Spt. Sci.*, **4:**103, 1979.

Wilmore, J. H.: Women in Sport: an Introduction to the Physiological Aspects, in J. Borms, M. Hebbelinck, and M. Venerando (eds.), "Women and Sport," p. 109, Karger, Basel, 1981.

Yeh, M. P., R. M. Gardner, T. D. Adams, F. G. Yanowitz, and P. O. Crapo: "Anaerobic Threshold": Problems of Determination and Validation, *J. Appl. Physiol.: Resp. Exer. Env. Physiol.*, **55:**1178, 1983.

8

CHAPTER

EVALUATION OF PHYSICAL PERFORMANCE ON THE BASIS OF TESTS

CONTENTS

PHYSICAL FITNESS TESTS
TESTS OF MAXIMAL AEROBIC POWER
   Type of Exercise
   Measurement of Oxygen Uptake
   Treadmill Experiments
   Cycle Ergometer Experiments
   Step Test
   Maximal Oxygen Uptake in Various Sports
   Prediction From Data Obtained at Rest or Submaximal Test
   A Simple Submaximal Cycle Ergometer Test
   How is the "Normal" Test Conducted?
   Choice of Work Rate
   Can the Maximal Oxygen Uptake Be Predicted Accurately From Data
     Recorded During a Submaximal Exercise Test?
   Submaximal or Maximal Test?
   Endpoint for the Termination of a Test
   Evaluation of Test Results
EVALUATION OF THE ANAEROBIC POWER
MEASUREMENTS OF MUSCULAR STRENGTH
EXERCISE ELECTROCARDIOGRAM

Generally speaking, there have been two main approaches to the assessment of physical performance: (1) physical fitness tests with scoring of actual performance in situations that represent basic performance demands, and (2) studies of cardiopulmonary function at rest and/or during exercise.

PHYSICAL FITNESS TESTS

Since most of the so-called fitness tests, including evaluation of flexibility, skill, strength, etc., are related to special gymnastic or athletic performance, they are not really suitable for an analysis of basic physiological functions. Practice and training in the performance of the actual test may greatly influence the results.

The fact that there may be a significant correlation between the results from complicated test batteries when applied to a group of individuals, or that the scores are related to certain parameters characteristic of the subjects, does not necessarily mean a direct relationship. Such data may cause confusion rather than solve problems. From a physiological and medical viewpoint, any test battery for the evaluation of physical fitness is rather meaningless unless it is based on sound physiological considerations. The widespread use of such test batteries in physical education can be justified from an educational and psychological viewpoint. It may help the teacher or coach to stimulate the athlete's interest in training. Furthermore, any progress can be evaluated objectively. The selection of such activities and tests should therefore be based on pedagogic and psychological considerations adapted to local facilities. If they cannot be justified from these viewpoints, it is better to exclude them from the curriculum altogether. Too often the tests are incorrectly reputed to serve a physiological purpose. Actually, from a physiological viewpoint, the application of a test battery may sometimes be unsuitable, since the performance of the tests usually demands maximal exertion of a subject who may be completely untrained.

For a review of the commonly used physical fitness tests, the reader is referred to the fitness test manual published by the American Association for Health, Physical Education, and Recreation (1976), and to Mathews (1974).

TESTS OF MAXIMAL AEROBIC POWER

Type of Exercise

In laboratory experiments, three methods of producing standard work rates have been mainly applied: running on a treadmill, exercising on a cycle ergometer, and using a so-called step test. In Chap. 7, the general methodological criteria were discussed in some detail: the exercise should involve large muscle groups, and the measurement of the oxygen uptake should be initiated when the exercise has lasted a few minutes to permit the oxygen uptake to reach its maximum. Ideally, a definite plateau should be reached despite an increase in the rate of exercise. Such a plateau is, however, often difficult to establish in practice, as discussed in Chap. 7.

A critical question to be answered is which type of exercise will yield the individual's highest oxygen uptake. It appears that by running on the treadmill uphill ($\geqq 3°$ inclination) the O_2 uptake may be brought to a maximum whereas running horizontally may result in a somewhat lower maximal O_2 uptake (Taylor et al., 1955). "Ski-walking" on a motor-driven treadmill with an inclination of 12° at speeds between 60 and 160 · min^{-1} (the subject walking with slightly bent knees and using ski-poles as in cross-country skiing) produced a significantly higher maximal oxygen uptake than in maximal uphill treadmill running (Hermansen, 1973).

On the average, cycling produces a lower O_2 uptake, at least compared with running uphill. In studies in which objective criteria have been used to determine whether the maximal O_2 had been reached for the type of exercise in question, the reported values for running are on the average 4 to 8 percent higher than for cycling. In extreme cases, one has noted a maximal aerobic power in cycling averaging some 20 percent less than treadmill values (see Harrison et al., 1980a).

Table 8-1 summarizes the mean values for maximal oxygen uptake in various types of exercise in "normal" subjects as well as in the individuals trained for special activities.

At present, it is not possible to explain why most individuals attain a somewhat higher oxygen uptake when running uphill compared with exercise on the cycle ergometer. This phenomenon can scarcely be caused by the activation of a larger muscle mass during running uphill since simultaneous exercise with both arms and legs does not increase the maximal aerobic power compared with exercise using the legs only (see Bergh et al., 1976; for exceptions, see Table 8-1). The higher exercise tempo during running may enhance the blood perfusion and venous return, but if this is true, running uphill should not be more advantageous than running horizontally.

During cycling, the subject may often experience a feeling of local fatigue or a sensation of pain in the thighs or knees, which may be disturbing. This discomfort may cause the effort to be interrupted before the oxygen-transporting organs have been fully taxed. The exercise position is of critical importance. The subject should be sitting almost vertically over the pedals. The seat should be high enough so that the leg is almost completely stretched when the pedal is in its lowest position. In Swedish studies (Hermansen and Saltin, 1969), there was no difference between trained and

TABLE 8-1
MEAN VALUES FOR MAXIMAL OXYGEN UPTAKE ATTAINED IN VARIOUS
TYPES OF EXERCISE IN ORDINARY AND SPECIALLY TRAINED SUBJECTS.

Type of exercise	Ordinary subjects	Specially trained subjects
Running uphill	100	100
Arms and legs	100	100–115
Running horizontally	95–98	
Cycling, upright	92–96	100–108
Cycling, supine	82–85	
One leg, upright	65–70	75–80
Arms	65–70	105–115
Step test	97	
Rowing	100	100–115
Skiing	100	100–112
Swimming	85	100

All values are expressed as percentage of the maximal oxygen uptake attained at uphill treadmill running, which represents the 100% reference value. For literature references, see Table 1 in Åstrand, 1976, and Chap. 14.

untrained subjects when comparing cycling and running. However, in this material even the untrained subjects were familiar with bicycling. In persons who have never ridden a bicycle before, a *maximal test* on the cycle ergometer may be undesirable as a method to assess maximal oxygen uptake. Motivation and stimulation of the subject are especially important in the case of cycling. When a person runs on the treadmill, it is, so to speak, a matter of all or nothing: the subject is forced to follow the speed of the belt or jump off. On the cycle ergometer it is usually possible to continue to exercise at a reduced rate in most types of cycle ergometers. Figure 8-1 summarizes results of studies by Hermansen and Saltin (1969). Despite the 7 percent difference

Figure 8-1
Individual values for some functions studied in 55 subjects during maximal running uphill on a treadmill ($\geqq 3°$) and work on a cycle ergometer (50 rpm) in a sitting position for about 5 min. Line of identity is drawn. A trend is noted for a somewhat higher maximal oxygen uptake during running compared with cycling, but pulmonary ventilation, heart rate, and blood lactic acid concentration were not different. It should be emphasized that a pedaling rate of 60 rpm may give a higher maximal oxygen uptake than may occur with a pedaling rate of 50 rpm. *(From Hermansen and Saltin, 1969.)*

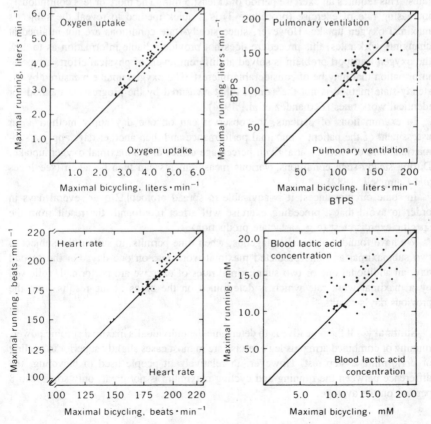

in maximal oxygen uptake between running and cycling, the maximal pulmonary ventilation, heart rate, and blood lactate concentration were not significantly different in the two procedures.

Normally, a test of maximal aerobic power starts with a submaximal rate of exercise which also serves as a warming-up activity. After this, the load may be increased in one of several ways: (1) The load may be immediately increased to a level which in preliminary experiments has been found to represent the predicted maximal load for the subject; this level is maintained for 3 to 6 min. (2) The load may be increased stepwise with several submaximal, maximal, or "supermaximal" loads, the subject exercising for 5 to 6 min at each work rate, with or without resting periods between each load. (3) The load may be increased stepwise every or every other minute until exhaustion.

When any one of these procedures is carefully conducted, they result in the same maximal oxygen uptake (authors' observations; Binkhorst and Leeuwen, 1963; McArdle et al., 1973; Stamford, 1976). From a *physiological* viewpoint, the second method (2) is preferable (see Fig. 7-3*b*). It is often of interest to obtain steady-state conditions when measuring oxygen uptake, pulse rate, ventilation, etc., at submaximal work rates. This requires an exercise period of at least 5 min. The more or less continuously increasing rate of exercise (procedure 3) is a quick method to reveal the subject's maximal oxygen uptake. However, since steady-state conditions are not attained at changing work rates, this procedure does not provide reliable information as to how the oxygen transport problem is solved at different levels of physical effort, a type of information that may be of considerable interest. The oxygen uptake measured by the steady-state method is not the same as that measured by the progressive method for identical work rates (Fernandez et al., 1974).

In examinations of patients, the observer can on one day apply method 3 for assessment of the patient's maximal performance and then another day, apply a submaximal rate of exercise at a given percentage of her or his maximal oxygen uptake. During steady-state conditions, various measurements can then be taken (see Jones and Campbell, 1982).

In some circumstances, it is advisable to spread protocol 2 over several days in order to avoid that a preceding exercise will affect (confound) the result from the specific exercise test (e.g., a lactate production).

We have found it quite advantageous, when time permits, to submit the subject to two submaximal and one predicted maximal work rate on one day. On the second day, an additional one or two submaximal rates of exercise are performed, followed by a maximal work rate which is determined on the bases of the results from the previous day's results.

Summary If the objective is to determine the individual's maximal aerobic power, running or combined arm plus leg exercise are in most cases slightly superior to cycling or performing a step test. However, in relatively fit people used to bicycling, the difference between the running and cycling maximum is not great, only some 4 to 8 percent on the average.

Measurement of Oxygen Uptake

The classical method for the determination of oxygen uptake, the Douglas bag method, rests on a very secure foundation. It is theoretically sound, and it is well tested under a wide variety of circumstances. In all its relative simplicity, it is unsurpassable in accuracy.

A disadvantage with the method is that the subject is somewhat hampered by the equipment required for the collection of the expired air. This limits the subject's freedom of movement. Furthermore, it merely provides an average figure for the oxygen uptake of, say, 60 s depending on the length of the time in which the expired air is collected.

Figure 8-2 shows the adaptation of this method for experiments using the cycle ergometer (right) and the treadmill (left). The inner area of the mouthpiece is 400 mm^2, the inner diameters of the tubes in the valve, stopcock, and bag are 28 mm, and of the connecting tube (smooth, not corrugated), 35mm. The tubes should be as short as possible. The resistance in this system with a connecting tube of 0.5-m length, with a flow rate of 100 liters $\cdot$ min^{-1} is 1 cm H_2O; of 200 liters $\cdot$ min^{-1}, 3 cm H_2O; of 300 liters $\cdot$ min^{-1}, 6 cm H_2O; and 400 liters $\cdot$ min^{-1}, 10 cm H_2O. The room temperature should be between 18 and 20°C, and the relative humidity between 40 and 60 percent. The oxygen content of the room air should not be below 20.90 percent.

Respiratory rate is registered via a Marey capsule. Two Douglas bags are connected to the four-way stopcock to enable continuous collection of air. The subject is connected to the bag during an inspiration. The turning of the stopcock starts a stopwatch that can be read down to .01 min (photograph at the center of Fig. 8-2). When no less

Figure 8-2
Arrangements of respiratory valves, stopcocks, and Douglas bags for the collection of expired air during experiments using a treadmill (left) and a cycle ergometer (right). With the aid of the arrangement shown in the center of the figure, the handle opens the stopcock for one of the two Douglas bags. At the same time, the corresponding stopwatch is automatically started. When the handle is moved to close the opening to the first bag, the stopwatch is automatically stopped. With the simultaneous opening of the stopcock aperture, leading the expired air into the second bag, the corresponding stopwatch is automatically started. In this manner, the time for the collection of the expired air is taken automatically.

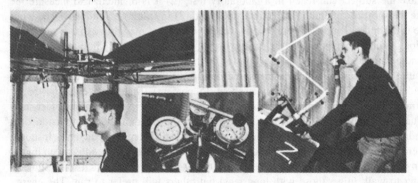

than about 50 liters of expired air has been collected in the bag, the stopcock is turned to the second bag or back to room air. The stopcock is always turned during inspiration; the second turning of the stopcock automatically stops the stopwatch. The volume of expired air is measured in a balanced spirometer and the composition of the air is analyzed by the Haldane or micro-Scholander or by electronic gas analyzers checked against the aforementioned manometric methods. When using electronic gas analyzers, care must be taken to correct for water pressure in the expired gas in relation to the gas used for calibration. Otherwise, the error may be as large as 25 percent in the calculation of oxygen uptake (Beaver, 1973). Figure 7-2 provides examples of the adaptation of the same method in experiments carried out in the field.

With a mass-spectrometer available, the analyses of concentrations of gases in normal air can be expanded to foreign gases like helium, argon, and acetylene which makes it possible to estimate the cardiac output during exercise with a noninvasive technique, that has some advantages as compared with the CO_2 rebreathing method (see Kanstrup and Hallbäck, 1981).

There are more or less sophisticated methods available to measure the pulmonary ventilation, e.g., using a flow meter and letting the expired air pass a mixing chamber from which a gas sample is sucked through an analyzer for measurements of the concentrations of the different gases. In this case, there is always the problem of correcting for the different response times of the two systems, i.e., that the gas concentrations obtained at a given time are related to the relevant gas flow rate. With proper technique one can measure the oxygen uptake, breath by breath.

With the classical method of measuring the oxygen uptake, as illustrated in Fig. 8-2, for the purpose of obtaining a steady-state value during submaximal work rates, the noseclip, mouthpiece, and respiratory value are given to the subject approximately 4 min after the beginning of the exercise and the collection of the expired air is initiated about 1 min later. (If the *metabolic* respiratory quotient is to be determined, a much longer period of exercise is required before the collection of expired air is started.)

Treadmill Experiments

The treadmill has an advantage when (1) the aim is to determine the highest oxygen uptake the subject can reach in a laboratory test, (2) it is of interest to measure the time the subject can endure a maximal trial, and (3) in studies on children. The work rate is dependent on the subject's body weight, a fact that may prove disadvantageous in longitudinal studies. Since the energy output per kilogram and kilometers per hour is more variable than it is at a given power on the cycle ergometer, the oxygen uptake should, whenever possible, be measured during walking and during running on the treadmill (coefficient of variation is about 15 percent) (Mahadeva et al., 1953; I. Åstrand, 1960). This requirement may limit application of the treadmill. Older individuals may have some difficulty in walking on the treadmill. The provision of a handrail for support will make the metabolic demand still more unpredictable. It should be emphasized that the energy cost of running (jogging) at a low speed is much higher than when walking (Fig. 14-5). Therefore, it may be a speed at which some subjects prefer to walk (e.g., those with long legs) but others will prefer to run. The energy demand will be quite different and it cannot be predicted from the speed alone. One example is presented in Fig. 8-3. Note that at a given speed and slope the oxygen uptake for this particular subject could vary from 2.2 to 3.2 liters · min^{-1} depending on his technique (with or without support; walking or jogging).

The treadmill is expensive and relatively immobile; electric power supply is a must. Recordings of ECG and other measurements may be more complicated during walking and running than during cycle ergometer exercise; however the treadmill is very useful and a "must" in a laboratory for basic research in exercise physiology.

It is evident from the discussion above that there is an advantage in keeping the speed of the belt constant and that the increase in work rate is established by an elevation of the slope. The initial energy demand of an optimal inclination is dependent on the subject's aerobic fitness. Unfortunately, this fitness is impossible to assess from examinations of the subject when resting. It is more predictable during a submaximal exercise test, but still subject to misjudgement. It is in a way a trial and error situation. An optimal exercise time to establish a subject's maximal aerobic power is 5 to 10 min (the period of warming up excluded).

Table 8-2 presents one protocol that has been successfully applied. It is a running protocol developed for relatively well-trained people and athletes. In order to select a suitable starting speed and slope for a maximal run on the treadmill, the subject's maximal oxygen uptake (ml · kg^{-1} · min^{-1}) is first predicted from the heart rate at a submaximal rate of exercise on treadmill or cycle ergometer (see below), then the data presented in Table 8-2 are applied.

1 Starting at this initial speed and incline, the incline is increased by 1.5 degrees (2.67 percent) every third min, keeping the speed constant. Under these conditions, not even top athletes are able to run for more than 7 min. Experience has shown that it is advisable to give women a somewhat lower relative starting work rate than is given to men (Table 8-2). Immediately preceding the maximal run, the subject is given a 10-min warm-up on the treadmill at a speed/slope corresponding to 50 percent of

Figure 8-3
Factors which influence the oxygen uptake when testing a subject on the treadmill at a given slope and speed. *(From P-O, Åstrand, 1982.)*

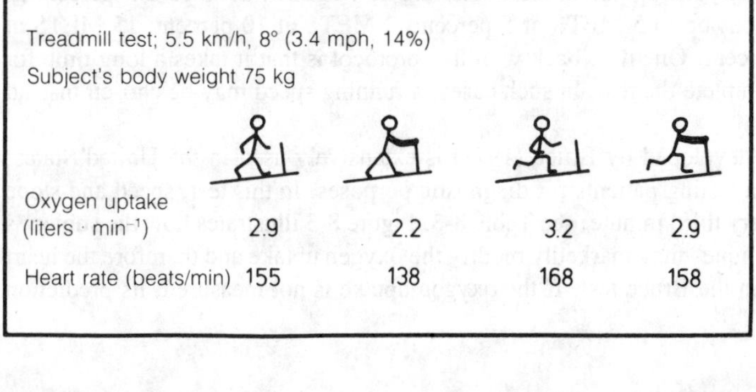

Treadmill test; 5.5 km/h, 8° (3.4 mph, 14%)

Subject's body weight 75 kg

Oxygen uptake (liters · min^{-1})	2.9	2.2	3.2	2.9
Heart rate (beats/min)	155	138	168	158

TABLE 8-2
STARTING WORK RATE USED FOR THE MAXIMAL RUN ON THE TREADMILL

Predicted max $\dot{V}_{O_2}$ ml · kg^{-1} · min^{-1}	Starting speed and inclination for max run on treadmill*							
	♂ Speed uphill				♀ Speed uphill			
	km · hr^{-1}	Degree	mph	%	km · hr^{-1}	Degree	mph	%
<40	10.0	3.0	6.2	5.25	10.0	1.5	6.2	2.62
40–50	12.5	3.0	7.8	5.25	10.0	3.0	6.2	5.25
55–75	15.0	3.0	9.3	5.25	12.5	3.0	7.8	5.25
>75								
A	15.0	4.5	9.3	7.9				
B	17.5	3.0	10.9	5.25				
C	20.0	1.5	12.5	2.62				

*A = cross-country skiers, skaters, etc.; B = cross-country runners, orientation runners, etc.; C = track runners.
*Source: Saltin and Åstrand: J. Appl. Physiol., **23**:353, 1967.*

the selected starting work rate. The subjects are not allowed to hold onto the treadmill railing during the run.

2 Another procedure, which in our experience has proved satisfactory, is to let the subject run in bouts of 3 min, with 4- to 5-min rest periods between each running bout, keeping the treadmill inclination constant at 3°, while increasing the speed by 15 m · min^{-1} each time. It is advisable at first to allow the subject to become accustomed to treadmill running, and in particular, to learn how to jump on and off the running treadmill with ease. The actual test should be started at a running speed which is not far from the subject's maximal level, yet no greater than a rate that the subject will feel confident of keeping up with for several minutes.

A widely used method of establishing an individual's maximal aerobic power was described by Balke (1954). The principle of his protocol has survived the test of time. The treadmill speed is kept constant in a standard test procedure at around 5 km · hour^{-1} (3.0–3.5 mph). The slope of the treadmill is increased every, or every other minute in steps of 2.5 percent (1.4°). With a speed of 4.8 km · hour^{-1} (3 mph) the increase in energy demand is about 3.5 ml · kg^{-1} min^{-1} per step, which is one MET unit (= the approximate oxygen uptake at rest). Starting at 4 METS, at slope 2.5 percent as an example; it goes up to 5 METS at 5 percent; 7 METS at 10 percent; 15 METS at a slope of 30 percent. One drawback with this protocol is that it takes a long time for fit subjects to complete the test. In such cases, a running speed may be chosen instead of walking.

The protocol developed by Bruce (1971) is extensively used in the United States, particularly when testing patients for diagnostic purposes. In this test, speed and slope are increased every third minute, see Table 8-3. Figure 8-3 illustrates how the subject's individual "technique" may markedly modify the oxygen uptake and therefore the heart rate at stage III in the Bruce test. If the oxygen uptake is not measured, its prediction

TABLE 8-3
ROBERT BRUCE'S TREADMILL PROTOCOL (1971)

Stage	I	II	III	IV	V
Minutes	3	3	3	3	3
Speed					
mph	1.7	2.5	3.4	4.2	5.0
km · hour^{-1}	2.7	4.0	5.5	6.8	8.0
Grade					
percent	10	12	14	16	18
degree	5.7	6.9	8.0	9.1	10.3
METS*	4	6–7	8–9	15–16	21

*1 MET, a metabolic unit, is 3.5 ml O_2 uptake · kg^{-1} · min^{-1}.

from the subject's weight and the speed and inclination of the treadmill evidently has its limitation at stages where walking or jogging is the alternative.

There are quite a few different treadmill protocols available; the end result of these tests provides data on maximal aerobic power of similar magnitude (see Pollock et al., 1978, pp. 289–298).

It should be pointed out that running on a treadmill at a given speed and slope has the same efficiency as running on a track with a similar surface and similar conditions, neglecting air resistance; there is no sex difference in the energy demand.

Formulas are available for calculating the energy cost of walking and running, respectively (Passmore and Durnin, 1955; Bobbert, 1960; Margaria et al., 1963; Van der Walt and Wyndham, 1973; Van Baak, 1979).

Cycle Ergometer Experiments

The term "bicycle ergometer", has traditionally been used, but most ergometers have just one wheel, so that "cycle ergometer" is a more appropriate term.

In many cases, the cycle ergometer is the preferred instrument for use in routine studies of physical power and adaption to exercise. Energy output can be predicted with greater accuracy in cycling than for any other type of exercise. Within limits, the mechanical efficiency is independent of body weight. This is a definite advantage in studies which require repeated examinations over the years. The work rate can, however, be selected simply according to the subject's gross body weight, calculated lean body mass, etc. (for example, 1 or 2 watts · kg^{-1} initially). A cycle ergometer operated with a mechanical brake is inexpensive (e.g., the Monark cycle ergometer). It is easy to move from place to place, and is not dependent on the availability of electrical power. Since the subject on the cycle ergometer exercises in a sitting or lying position with arms and chest relatively immobile, it is simple to obtain good ECG tracings and to perform studies with indwelling catheters. During *submaximal* exercise, a pedal frequency of from 40 to 50 revolutions/min produces the lowest O_2

uptake, i.e., the greatest mechanical efficiency, and therefore also a relatively low pulse rate (Grosse-Lordemann and Müler, 1937; Eckermann and Millahn, 1967).

The variation in O_2 uptake with different pedal frequencies at a standard work rate should be kept in mind if the O_2 uptake is not measured, but merely calculated from the rate of exercise used. Cycle ergometers producing a constant load, even with relatively large variations in pedal frequency, have certain advantages. Nevertheless, it should be clearly realized that the O_2 uptake is not strictly determined by the external power, but varies with the pedal frequency. Respiration is also affected by the pedal frequency (Chap. 5). Usually the subject is asked to try to maintain a certain pedal frequency, such as 50 or 60 rpm. (Acoustical signals, such as those produced by a metronome, are easier to follow than visual signals.) It should be emphasized that it is by no means essential that the chosen pedal frequency is optimal, as long as the mechanical efficiency is known.

In *heavy or maximal* exercise, the optimal pedal frequency is higher, e.g., the use of lower gear when travelling uphill during bicycle racing (P.-O. Åstrand, 1953). Sixty revolutions per minute may produce a somewhat higher maximal O_2 uptake than 50 rpm (Hermansen and Saltin, 1969; McKay and Banister, 1976). Thus, one should change from 50 rpm to 60 rpm in the case of heavy work rate, if the objective is to measure maximal oxygen uptake. The type of cycle ergometer that produces a constant power regardless of pedal frequency has the following drawback; during maximal exercise when the subject is tired and is unable to maintain the tempo, the force increases for each pedal revolution, chances are that the subject may be forced to discontinue exercise before it is possible to complete critical measurements. With the use of a cycle ergometer, in which the work rate varies with the pedal frequency (as in the case of the Monark), the power drops when the tempo no longer can be maintained, but the subject can in any case continue. (It is important to be able to record the pedal frequency continuously if the work rate is critical for the experiment.)

Inbar et al., (1983) have drawn attention to the importance of the crank-length of the cycle ergometer for obtaining maximal leg power output during tests of short duration, in the case of subject populations which are heterogenous in body size. They found, however, that for a homogenous population, the conventional 17.5 cm crank is close to the calculated optimum for power production. (See also P.-O. Åstrand, 1953.)

A quick and quite effective procedure to measure a subject's maximal aerobic power is the following: the initial power selected should be relatively low and gradually increased until a heart rate around 140 beats · min^{-1} for subjects less than fifty years of age, and approximately 120 beats · min^{-1} for older subjects. The perceived exertion will most likely be at the 13–14 level when evaluated from Borg's scale (see below). This is the warming-up period. The steady-state heart rate is recorded. The maximal oxygen uptake is then predicted from heart rate and the power setting on the ergometer, e.g., using the Åstrand & Åstrand nomogram (Fig. 8-4; explained in more detail below). A power is then chosen that will require an oxygen uptake approximately 10 to 20 percent higher than the predicted maximum, using Table 8-4. If the subject at the end of the first minute on this selected power has difficulty in keeping up the pedaling rate and starts to hyperventilate markedly, the power is lowered slightly to

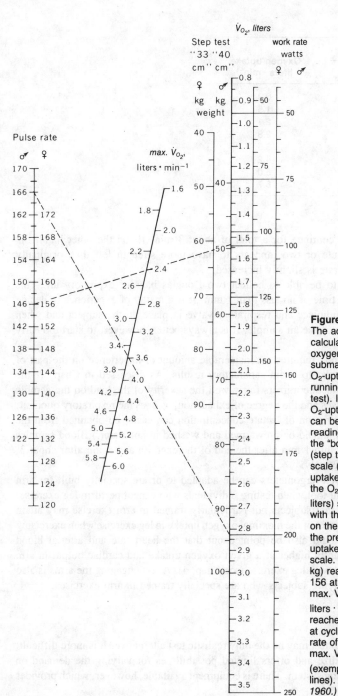

Figure 8-4
The adjusted nomogram for calculation of maximal oxygen uptake from submaximal pulse rate and O_2-uptake values (cycling, running or walking, and step test). In tests without direct O_2-uptake measurement, it can be estimated by reading horizontally from the "body weight" scale (step test) or "work rate" scale (cycle test) to the "O_2 uptake" scale. The point on the O_2-uptake scale (V_{O_2}, liters) shall be connected with the corresponding point on the pulse rate scale, and the predicted maximal O_2 uptake read on the middle scale. A female subject (61 kg) reaches a heart rate of 156 at step test; predicted max. $\dot{V}_{O_2}$ = 2.4 liters · min^{-1}. A male subject reaches a heart rate of 166 at cycling test on a work rate of 200 watts; predicted max. $\dot{V}_{O_2}$ = 3.6 liters · min^{-1} (exemplified by dotted lines). *(From I. Åstrand, 1960.)*

TABLE 8-4

| Work rate | | Oxygen uptake, |
watts	kpm · min⁻¹	liters · min⁻¹
50	300	0.9
100	600	1.5
150	900	2.1
200	1200	2.8
250	1500	3.5
300	1800	4.2
350	2100	5.0
400	2400	5.7

allow the subject to continue for a total of about 3 min. If, on the other hand, the subject, after a minute or two, appears to have more strength left than originally predicted, the work rate is slightly increased.

The objective is to be able to collect two Douglas bags of expiratory air. This requires a collection time of no less than 1 min. It is a matter of experience to decide when to put the mouthpiece and respiratory valve in place on the subject and when to start the collection of the air sample. It is always better, however, to start too early than too late.

This "quick method" requires a considerable amount of experience on the part of the investigator in order to attain satisfactory results. As discussed in Chap. 7, the peak value of blood lactate mirrors fairly well the severity of the load on the aerobic and anaerobic processes and the degree of exhaustion, as does the respiratory quotient. Blood for the determination of lactate concentration can easily be obtained from the finger tip. (The hand should be prewarmed and washed in hot water.) Blood samples should be secured within 1 min after the end of the exercise and again after about 3, 6, and 10 min respectively.

Many types of cycle ergometers can be adapted to or are specially built for arm exercise. It is a useful tool when testing individuals who cannot perform leg exercise. Table 8-1 illustrates that subjects not particularly trained in arm exercise may attain approximately 70 percent of the maximal oxygen uptake in leg exercise when exercising with the arms. It should also be pointed out that the heart rate and arterial blood pressure are significantly higher at a given oxygen uptake and cardiac output in arm exercise as compared with leg exercise (see Chap. 4). A test engaging the arms is also of interest when studying subjects who are specially trained in arm exercises.

Step Test

In field studies, a step test may be the only realistic test alternative. It is more difficult, however, to standardize and offers limited possibilitites for varying the demand on the oxygen transporting system. There is equipment available, however, which provides automatic adjustment of the stepping height from 2 up to 50 cm. At maximal effort, the stepping pace must be high, and there is always the risk that the subject may stumble when approaching a maximal rate of work. Besides, in the unaccustomed subject, the aftermath is sore muscles. It is even more difficult to perform recordings such as ECG on exercising subjects during a step test than is the case during the treadmill test. (Gradational step tests have been described by Hettinger & Rodahl, 1960; Maritz et al., 1961; Nagle et al., 1965; Kasch et al., 1966; Shephard, 1966.)

Maximal Oxygen Uptake in Various Sports

In many sporting events, the oxygen uptake during maximal effort can be measured by applying the Douglas bag method, e.g., in bicycling, canoeing, cross-country skiing, rowing, speed skating, swimming, and walking. Special ergometers for simulated rowing, skiing, and "real" swimming ("swimming flume") have been developed. In Chap. 14, some events will be further discussed with regard to measurements of maximal oxygen uptake.

Summary Ideally, any test of maximal oxygen uptake should meet at least the following general requirements: (1) The exercise must involve large muscle groups, (2) the rate must be measurable and reproducible, (3) the test conditions must be such that the results are comparable and reproducible, (4) the test must be tolerated by all healthy individuals, and (5) the mechanical efficiency (skill) required to perform the task should be as uniform as possible in the poplulation to be tested.

The magnitude of the external power can be expressed exactly, and it may be reproduced with a high degree of accuracy in tests using the cycle ergometer, with less accuracy in exercise on the treadmill. However, the use of both cycle ergometer and treadmill is advisable. As a research tool, we have good experience with the application of the cycle ergometer tests for submaximal exercise but prefer the treadmill test for maximal exercise. The running should be uphill with an inclination of 3° or more. On a cycle ergometer, the subject should sit in a position above the pedals. The seat should be sufficiently high, and the pedal frequency should be 50 to 60 rpm; the higher frequency is particularly advisable for subjects with limited muscular strength. By preference, the work rate should be selected so that the subject can proceed for at least 3 min, to be on the "safe" side. Repeatedly, it has been emphasied that an objective criterion for an attained maximal oxygen uptake is a final "leveling off" of the oxygen uptake despite an increasing rate of exercise, i.e., a failure of a higher work rate to increase the aerobic metabolic rate significantly. But this criterion may, in many cases, be difficult to attain. The heart rate attained at the end of an exercise performed for the purpose of measuring the subject's maximal oxygen uptake is a poor basis for an evaluation of whether or not the exercise actually represented a maximal load on the oxygen transporting system. It is quite common to use an average maximal heart rate for a given age group as a reference when evaluating whether or not a subject is exercising at a level demanding maximal aerobic power. It was mentioned in Chap. 4 that within a group of 100 subjects having an average maximal heart rate of 180 beats · min^{-1}, 2 to 3 persons will stop at a maximal heart rate of 160 or below, and 2 to 3 subjects may exceed 200 in heart rate (SD = ±10 beats · min^{-1}).

Prediction from Data Obtained at Rest or Submaximal Test

Although the aerobic power in terms of maximal oxygen uptake can be determined with a reasonable degree of accuracy, the method is rather time-consuming. It requires fairly complicated laboratory procedures and demands a high degree of cooperation from the subject. Although this is the method of choice for any scientific investigation, it is by no means a routine method that may be conveniently applied in the office of a physician.

The practicing physician (especially the industrial physician), the coach, the physical therapist, or anyone else interested in physical performance is nevertheless often faced with the need to assess a person's circulatory fitness. This requirement has created the need for a simple test for such an evaluation of an individual, based upon a submaximal "stress" test. In treating older individuals, as well as certain other patients, the physician may be reluctant to expose the patient to the risk of an exhausting maximal work rate. This is true whether one is considering job placement, fitness for continued employment, or retirement.

Rest

No objective measurements made on resting individuals will reveal their capacity for physical exercise or their maximal aerobic power. Even a simple questionnaire may provide more useful information that can be obtained from measurements made at rest. A low heart rate at rest, a large heart size, or similar parameters may indicate a high aerobic power, but may, on the other hand, be a symptom of disease.

A significant correlation between any parameter and maximal oxygen uptake indicates a direct or indirect dependence. Figure 9-5 (lower part) illustrates such a relationship. The range of the observed values is of decisive importance for the numerical value of the correlation coefficient, but for an evaluation of the individual case, the *standard deviation* from the regression line is the critical factor. The deviation may be large, i.e., a prediction of one parameter from the other is very uncertain, despite a high correlation between the parameters in question. From the data presented in Fig. 9-5 (lower part), we find that the correlation coefficient between maximal oxygen uptake and total amount of hemoglobin (Hb$_T$) is as high as 0.970, but the standard deviation of hemoglobin weight is still as high as 10.5 percent at an oxygen uptake of 2.6 liters · min^{-1}. The figure shows that one girl with 350 g Hb$_T$ can transport up to 2.7 liters · min^{-1}, but another girl with the same Hb$_T$ reaches an oxygen uptake of only 1.9 liters · min^{-1}. This is to be expected statistically, since 350 g Hb represents an oxygen uptake of 2.25 liters · min^{-1} with such a standard deviation that 95 out of 100 subjects of a similar group are expected to fall within the range of 1.8 to 2.7 liters · min^{-1} and only 5 subjects will lie outside these values.

In conclusion, it may be stated that the correlation between parameters is of interest when analyzing biological interactions, but for an evaluation of one *individual* on the

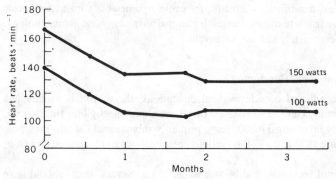

Figure 8-5
Decrease in the heart rate tested at two fixed work rates in the course of $3^1/_2$ months of training.

basis of indirect methods, the standard deviation from the regression line indicates the accuracy of the prediction.

A Simple Submaximal Cycle Ergometer Test

It has been repeatedly pointed out that there is a high correlation between the cardiac output during exercise and the oxygen uptake (in liters $\cdot$ min^{-1}). Since the cardiac output is the product of heart rate and stroke volume it is evident that a low heart rate at a given oxygen uptake is most likely associated with a large stroke volume. In Chap. 4, it was also pointed out that such a heart functions with a higher efficiency than a heart which produces a given cardiac output with a high heart rate and small stroke volume. This physiological fact forms an important basis for submitting people to submaximal exercise tests. When exercise is performed on a cycle ergometer there are relatively small individual variations in mechanical efficiency. Therefore, the oxygen uptake can be predicted from the external power to which the subject is subjected (see Table 8-4). Figure 8-5, illustrates the effect of training on the heart rate at two work rates. The oxygen uptake at a given power was constant throughout the study (100 watts $-$ 1.5 liters $\cdot$ min^{-1}; 150 watts—2.1 liters $\cdot$ min^{-1}). Most likely, the cardiac output did not change and the decline in heart rates is associated with an increase in stroke volume.

In our experience, the submaximal exercise test has proven to be a very useful tool for evaluating whether or not a training program has been effective in improving the individual's circulatory capacity. It has been widely applied in top athletes, in trained and untrained adults, and in children. In this test, the individual is his or her own control; it is a matter of comparing the individual's own performances in repeated tests over months or years. In the simple test on the cycle ergometer, a counting of the heart rate is all that is needed for the evaluation. A gradually decreasing heart rate at a standard work rate as the training progresses stimulates the individual's efforts to continue to improve his or her circulatory capacity further or to maintain a given heart rate level. In fact, in Sweden several hundred thousand tests have been conducted in sport clubs, factories, and offices. Actually, the cycle ergometer is a mandatory piece of equipment in Swedish schools used to teach exercise physiology and human biology. But it must be an ergometer, not an "exercycle."

How is the "Normal" Test Conducted?

When the test is conducted outside medical institutions, the subject is usually just asked about her or his health. The risks involved are almost negligible. [In Sweden, a retrospective study involving 370,000 tests, primarily submaximal but also maximal, was conducted during 1973 and 1974; no serious complications were reported (Jonsson, 1981).]

The subject should feel well, and be free of infection. Several hours should have elapsed between the last meal and the test. The subject should not have been engaged in any physical activity heavier than the test load the last few hours prior to the test. Smoking should not be allowed during the 2 hours immediately before the test. If possible, the room temperature should be 18 to 20°C and the room should be adequately ventilated. If the temperature is higher, an electric fan may be used.

Choice of Work Rate

For trained, active individuals, the risk of overstraining in connection with a test is very slight. For women, a suitable power is 75 to 100 watts, and for men, 100 to 150 watts. If the heart rate exceeds about 130 beats per min, the load can be considered adequate and the test can be discontinued after 6 min. If the heart rate is slower than about 130 beats per min after some minutes of exercise, the power should be increased by 50 watts. If the purpose is a multistage test, the work rate may be increased by 50 watts every 6 min for as long as the heart rate remains below about 150 beats per min. The final exercise period may be continued for 6 min even if the heart rate should exceed 150 beats per min.

For persons who may be expected to be physically unfit, persons who are completely untrained, and older individuals, lower work rates should be chosen and an initial load of 50 watts might be suitable.

A convenient pedalling frequency is 50 rpm paced by a metronome set at "100."

It is recommended that if a physician is not present, tests on persons over forty years of age should be discontinued if the heart rate exceeds 150 beats per min. The load for such persons should not exceed 100 watts for female subjects and 150 watts for male subjects. In the event of pressure or pain in the chest, or marked shortness of breath or distress, the test must be discontinued immediately.

Provided that the exercise is not too heavy, respiration and circulation will increase during the first 2 to 3 min, after which a steady state is attained. The increase in pulse rate can easily be established by counting the pulse once every minute. In any case, after 4 to 5 minutes of exercise the pulse rate has generally reached a steady state. As a rule, a period of about 5 to 6 min is therefore sufficient for the pulse rate to adapt to the task being performed. The pulse rate should then be taken every minute and the mean value of the pulse rate at the fifth and sixth minutes is taken as the heart rate response to the exercise demand. If the difference between these last two pulse rates exceeds 5 beats $\cdot$ min^{-1}, the exercise time should be prolonged one or more minutes until a constant level is reached. The pulse rate is most easily felt over the carotid artery just below the mandibular angle. For the inexperienced, it is rather difficult to count the pulse rate; the metronome confuses, the subject is in motion, and the pulse rate may vary. Practice under experienced supervision is important. The pulse rate may be measured during the last 15 to 20 s of every minute. The most exact value is obtained by taking the time for 30 pulse beats. A person trained in taking pulse rates, whether it be by palpation or by auscultation, obtains values that are in close agreement with those obtained by ECG recordings, provided the pulse rate is not irregular.

It has proven to be useful to ask the subject about the *subjective rating of perceived exertion* at the end of each exercise state. Borg has suggested a scale consisting of 15 grades from 6 to 20, see Table 8-5. There are exceptions, but in many situations the heart rate mirrors the physical strain experienced subjectively although the heart rate level at a given grade varies individually (see Ekblom and Goldbarg, 1971; Borg, 1982). The questionnaire may be modified to serve as a rating of general fatigue, leg fatigue, dyspnea, etc. The decline in heart rate as a consequence of more physical activity as illustrated in Fig. 8-5 is also accompanied by a reduced perceived exertion when the subject is exposed to a standard rate of exercise.

TABLE 8-5
A 15-GRADE SCALE FOR
RATINGS OF PERCEIVED
EXERTION, THE RPE SCALE,
(SEE BORG, 1982).

Score	Subjective Rating
6	
7	Very very light
8	
9	Very light
10	
11	Fairly light
12	
13	Somewhat hard
14	
15	Hard
16	
17	Very hard
18	
19	Very very hard
20	

Can the Maximal Oxygen Uptake Be Predicted Accurately from Data Recorded during a Submaximal Exercise Test?

The answer is definitely no, but the question warrants some further comments. For a long time there has been a need for a test by which one could select people for physically demanding tasks. One example is the classical Harvard fitness step test (see Johnson et al., 1942). In this, and in some other tests, the heart rate was counted during early recovery from a standardized exercise test. In the same individual, the correlation coefficient between the heart rate *during* submaximal exercise and the heart rate 1 to 1 1/2 min *after* the exercise is high. In one study, the correlation coefficient was 0.96 with a deviation from the regression line of 5 percent. For a group of subjects, the *r*-value for the steady-state heart rate and the recovery rate was reduced to 0.77 and the deviation increased to 10 percent (I. Ryhming, 1953). From these data, it is evident that the recovery heart rate gives only a rough idea of the heart rate attained *during* exercise if the results are compiled from different subjects.

Most modern circulatory exercise tests are based on a linear increase in heart rate with increasing oxygen uptake (or work rate), as illustrated by Fig. 8-6. If the maximal heart rate were fixed in *Homo sapiens,* from a few measurements of the heart rate during a multistage submaximal exercise protocol one could extrapolate and predict the subject's oxygen uptake at the maximal heart rate. As discussed in Chap: 4 there are, however, marked individual variations in maximal heart rate within a given age group, and it decreases with age. Therefore, an extrapolation to an "age predicted maximal heart rate may give quite erroneous results. Let us examine in more detail the reliability of predicting the maximal aerobic power and potential of the oxygen transporting system from the submaximal heart rate response to a given rate of exercise.

Figure 8-6
The relationship between heart rate and O_2 uptake on the cycle ergometer.

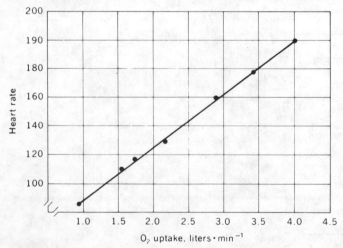

O, uptake, liters · min $^{-1}$

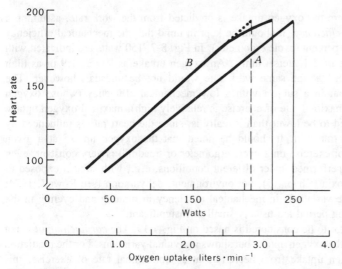

Figure 8-7
The increase in heart rate with increasing work rate (and oxygen uptake) is linear within a wide range. In some subjects *(B)*, the oxygen uptake may increase relatively more than the heart rate as the work rate becomes very heavy. The prediction of maximal oxygen uptake by an extrapolation to the subject's presumed maximal heart rate (195 in this case) suggests a maximum of 2.9 liters · min⁻¹ (dotted line), but the actual maximal aerobic power is 3.2 liters · min⁻¹. The individual's maximal heart rate is therefore a critical factor in an extrapolation.

1 *The linear increase in heart rate* with increase in oxygen uptake is, as mentioned, a typical feature. There are many exceptions, however. In some cases, the oxygen uptake increases relatively more than the heart rate as the work rate becomes very heavy (subject *B* in Fig. 8-7). One possible explanation of this phenomenon may be that an efficient redistribution of blood, giving the exercising muscles an appropriate share of the cardiac output, is not brought about until the very heavy work rates are reached. The consequence is that in this subject, the maximal oxygen uptake will be underestimated by an extrapolation from the heart rate response to submaximal loads.

2 *The maximal heart rate* declines with age (Fig. 4-24). Therefore, if old and young subjects are included in the same study, the circulatory capacity of the older subjects will be consistently overestimated compared with that of the younger subjects. By introducing an age factor, a correction can be made (see below). However, the standard deviation for maximal heart rate within an age group is about ± 10 beats · min⁻¹; thus, 50 percent of the tested subjects will be more or less overestimated and the remainder underestimated. In Fig. 8-7, subject *A* was assumed to have a maximal heart rate of 195, maximal oxygen uptake = 3.5 liters · min⁻¹. If the maximal heart rate was 170, the maximal oxygen uptake would be only 2.9 liters · min⁻¹. An extrapolation of the heart rate to 215 for a subject with the same slope for the relationship between heart rate and oxygen uptake as subject *A* would reach an oxygen uptake of 4.0 liters · min⁻¹.

3 In cases where the oxygen uptake is predicted from the work rate, assuming a fixed *mechanical efficiency*, it should be kept in mind that the mechanical efficiency may vary by $\pm$ 6 percent (cycle ergometer). In Fig. 8-7, 150 watts are indicated with a mean O_2 uptake of 2.1 liters $\cdot$ min^{-1}. An oxygen uptake as low as 1.9 or as high as 2.3 liters $\cdot$ min^{-1} at the same work rate would not be unusual, however. The consequence is that in a subject with a low mechanical efficiency (whose oxygen uptake at the submaximal rate of exercise is relatively high), maximal oxygen uptake would be predicted to be lower than it really is, since the heart rate is influenced by the extra oxygen transport. It should be noted that the oxygen uptake at a given *submaximal* rate of exercise on a cycle ergometer or treadmill is very constant, even if the exercise is performed under different conditions, e.g., with subject exposed to hypoxia or hyperoxia (Chap. 15), hot environment, dehydration (see Rowell, 1974; Saltin, 1964). The variations in mechanical efficiency in running and cycling in the course of a training period are usually small or insignificant.

4 The last factor to be considered is based on Fig. 4-19. The *cardiac output* is not strictly related to the oxygen uptake but shows individual variations. For the prediction of maximal oxygen uptake from heart rate at a submaximal rate of exercise, this variation does not matter. The oxygen uptake (measured or predicted) per heart beat is actually evaluated during the test. When consideration is given to the maximal heart rate, the maximal oxygen uptake is "calculated". However, if the test is conducted for an evaluation of cardiac performance, e.g., stroke volume, the individual variation in cardiac output and arteriovenous oxygen difference must be taken into consideration.

Under standardized conditions, the variation from day to day in heart rate at a given oxygen uptake is less than 5 beats $\cdot$ min^{-1}, provided the state of training is the same. The mean value for a group of subjects undergoing repeated tests remained exactly at the same level under these conditions.

However, a number of situations may cause a marked increase in the heart rate at a given submaximal rate of exercise, without being accompanied by a reduction in the individual's maximal oxygen uptake. The following examples may be mentioned:

1 Dehydration during heavy exercise or during exposure to heat (Chap. 13)

2 Prolonged heavy exercise (Chaps. 12 and 13)

3 Exercise with a muscle mass involved which is less than in, for instance, running and cycling (Chap. 4)

4 Exercise after ingestion of alcohol (Chap. 15)

In some of these situations the potential for a top physical performance is actually reduced. The excessive heart rate response to a given submaximal work rate is often a better criterion for a reduced physical fitness than the person's maximal oxygen uptake.

It should also be stressed that fear, excitement, and related emotional stress may cause a marked elevation of heart rate at a submaximal work rate without either maximal O_2 uptake or performance capacity being affected. The heavier the work rate, however, the less pronounced is this nervous effect on the heart rate. It is usually recommended that the test load should be sufficiently high so as to bring the heart rate up to, or above, 150 beats per min in the case of younger subjects.

In some instances, the heart rate at a standard work rate may be unchanged while the maximal O_2 uptake and the performance capacity are reduced; for example: (1) following acclimatization for a certain period at high altitude (Chap. 15); (2) during semistarvation (Keys et al., 1950).

These examples illustrate further the danger of drawing conclusions concerning maximal oxygen uptake from heart rate data obtained during submaximal tests.

One method of predicting an individual's maximal oxygen uptake has been critically examined over the years in a number of studies, i.e., the method based on the nomogram presented by Åstrand & Åstrand (P-O. Åstrand and Ryhming, 1954; I. Åstrand, 1960). Therefore, it may be justifiable to devote some attention to this particular method. The original data were collected on 86 female and male students in physical education. Not unexpectedly, the scatter of the heart rate at a given oxygen uptake was considerable (see Fig. 8-8). Note that in this group of relatively well-trained students, heart rates from 140 to 220 were recorded at an oxygen uptake of 3.0 liters · min^{-1}. A heart rate of 180 beats · min^{-1} represented an oxygen uptake of only 2.0 liters · min^{-1} for some female subjects and as much as 5.0 liters · min^{-1} for some male subjects. One major explanation of this scatter was the differences in the subjects' maximal aerobic power. It was noted that when young men were exercising at 50 percent of their maximal oxygen uptake, their heart rates were on the average 128 beats · min^{-1}; for women, the heart rates were 138. When the subjects were exercising against a heavier power

Figure 8-8
Heart rates in relation to oxygen uptake for 86 adult female and male subjects. Maximal as well as submaximal values are represented *(From P.-O. Åstrand, 1952.)*

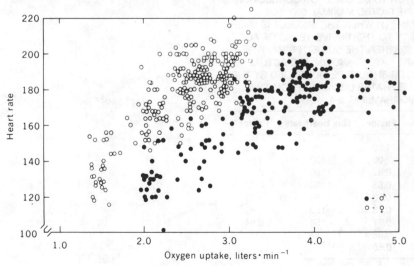

demanding an oxygen uptake of 70 percent of their maximal aerobic power, the average heart rates were 154 for males and 164 for females, with a standard deviation of 8 to 9 beats · min^{-1}. Based on these data, a nomogram was constructed (Fig. 8-4). From the steady-state value of the heart rate and oxygen uptake, the maximal oxygen uptake can be predicted. If the oxygen uptake is not measured during the exercise it can be predicted from the work rate during cycling and from body weight in the case of a particular step test. It should be emphasized that this nomogram is not based on any sophisticated theory, and that the individual's maximal heart rate originally was not considered in the construction of the nomogram. It was noted that the predicted maximal oxygen uptake of persons over approximately 30 years of age was in most cases overestimated. This could be explained on the basis of the reduction in the maximal heart rate with age (Fig. 4-24). A correction factor for age and also for the individual's maximal heart rate, when known, was introduced, see Table 8-6 (I. Åstrand, 1960).

The standard error of the method for such a prediction of maximal oxygen uptake from submaximal exercise tests was, in the studies behind the nomogram, 10 percent in relatively well-trained individuals of the same age, but up to 15 percent in moderately trained individuals of different ages when the age correction factor of maximal oxygen uptake was applied. Untrained persons are often underestimated; the extremely well-trained athletes are often overestimated. The consequence of a standard error of 15 percent is as follows: with a maximal aerobic power predicted to 3.0 liters · min^{-1}, the actual O_2 uptake for 5 out of 100 subjects is then less than 2.1 or higher than 3.9 liters · min^{-1}. It is important to keep in mind this limitation in accuracy; *this drawback holds true for any submaximal cardiopulmonary test described so far.* The validity of

TABLE 8-6
FACTOR TO BE USED FOR CORRECTION
OF PREDICTED MAXIMAL OXYGEN
UPTAKE (1) WHEN THE SUBJECT IS OVER
THIRTY TO THIRTY-FIVE YEARS OF AGE
OR (2) WHEN THE SUBJECT'S MAXIMAL
HEART RATE IS KNOWN. THE ACTUAL
FACTOR SHOULD BE MULTIPLIED BY THE
VALUE THAT IS OBTAINED FROM THE
NOMOGRAM, FIG. 8-4

Age	Factor	Max heart rate	Factor
15	1.10	210	1.12
25	1.00	200	1.00
35	0.87	190	0.93
40	0.83	180	0.83
45	0.78	170	0.75
50	0.75	160	0.69
55	0.71	150	0.64
60	0.68		
65	0.65		

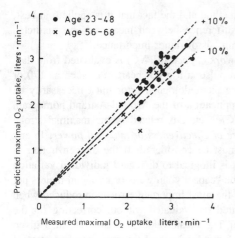

Figure 8-9
Estimated maximal O_2 uptake calculated from the Åstrand nomogram in relation to measured maximal O_2 uptake in 22 male subjects, ages twenty-three to forty-eight (●), and 9 male subjects, ages fifty-six to sixty-eight (×). Broken lines denote a deviation of ±10 percent from the "ideal line." *(Rodahl and Issekutz, 1962.)*

the nomogram has been tested in other laboratories. In some cases, there has been good agreement between the actually measured and predicted maximal O_2 uptake from the nomogram (see Kavanagh and Shephard, 1976; Fig. 8-9). In other studies, the subject's maximal oxygen uptake has been found to be underestimated when the nomogram was applied (see Chase et al., 1966; Harrison et al., 1980a). Siconolfi et al. (1982) have suggested the following equations as a modification of the nomogram:

For males: $y = 0.348(X_1) - 0.035(X_2) + 3.011$ (1)

$\qquad$ ($r = 0.86$; standard error of the estimate $= 0.359$ liter $\cdot$ min^{-1})

For females: $y = 0.302(X_1) - 0.019(X_2) + 1.593$ (2)

$\qquad$ ($r = 0.97$; standard error of the estimate $= 0.199$ liter $\cdot$ min^{-1})

where y is the $\dot{V}_{O_2}$ (liter $\cdot$ min^{-1}), X_1, is the $\dot{V}_{O_2}$ (liter $\cdot$ min^{-1}) from the nomogram, Fig. 8-4 (not corrected for age), and X_2 is the age (years).

Maritz et al. (1961) and Wyndham et al. (1966) have applied a similar principle in their predictions. Their subjects perform four submaximal exercise loads (usually step tests). From measurements of heart rate and oxygen uptake the researchers fit a straight line to the four pairs of plots and extrapolate to a mean maximal heart rate for the population in question. Margaria et al. (1965) have also introduced a nomogram based upon similar concepts, a test which has been modified by Harrison et al. (1980b) by adapting the step test, using a step, the height of which was adjusted to fit the length of the subject's leg (in the Margaria test it is a constant 40 cm step height, 15 steps $\cdot$ min^{-1} for 5 min and 25 $\cdot$ steps $\cdot$ min^{-1} for another 5 min).

It would be reasonable to assume that some variation in the accuracy may be encountered from one group of subjects to another, with regard to both the difference between predicted and measured maximal oxygen uptake and the scatter of the data.

Regarding the considerable error of the method and the fact that it is applied, at best, only as a screening test, a consistent difference between measured and predicted maximal O_2 uptake of a few 100 ml $\cdot$ min^{-1} is of minor importance.

In some test procedures the *physical work capacity* (PWC) is evaluated from data on work rate, oxygen uptake, or oxygen pulse at a given heart rate, such as 170 or 150 beats $\cdot$ min^{-1} (Sjöstrand, 1947). The methodological error must necessarily be about the same in these tests as in the application of the Åstrand-Åstrand nomogram (I. Åstrand, 1960). The calculated PWC$_{170}$, etc., is related to the maximal stroke volume of the heart, but it is *no measure of effectiveness or maximal power*. In any such estimate, the maximal heart rate must be considered. If the oxygen uptake is measured during the submaximal test, it is illogical to disregard individual variation in mechanical efficiency by expressing the "capacity" in watts or kilopond meters per minute at a heart rate of 170. However, the most important error is introduced when individuals of different ages are compared or evaluated without correcting for the decline in maximal heart rate with age. The mean value for heart rate at a given submaximal oxygen uptake is the same for individuals of the same sex and state of training regardless of age (from twenty-five up to at least seventy years of age) (I. Åstrand, 1960). By definition, therefore, the calculated PWC$_{170}$ or oxygen pulse at a given rate of exercise is the same. The real performance capacity, however, declines with age. Furthermore, the subjective feeling of strain is higher at a given heart rate, the older the individual.

Studies have also confirmed that there is a low correlation between the oxygen uptake or work rate achieved at a heart rate of 170 or 150 and the measured maximal oxygen uptake, cardiac output, heart size, or blood volume in individuals from twenty to seventy years of age (Strandell, 1964).

The conclusion is that an evaluation of the maximal effect of the oxygen-transport system, based on studies at submaximal work rates or oxygen uptake, should be performed with the utmost caution, especially when persons of different age groups are considered. Figure 9-7 presents mean values for various functional parameters in relation to age. It is evident that the decline in physical performance is not related to a similar change in heart size, blood volume, or heart rate during a standard work rate, etc.

Submaximal or Maximal Test?

Particularly in the clinical examination of patients or during a health checkup, it is often important to include an exercise test in order to examine the cardiovascular system under functional stress. Usually ECG and blood pressure monitoring, specific objective changes, signs, and symptoms will guide the conductor of the test as to when the safety of the subject being tested may be threatened. There are those who are reluctant to push the subject to a maximal effort and the test is terminated at a predetermined end-point, mostly based on the function of the heart. One such criterion for the termination of the test is a heart rate of 200 minus the subject's age in years. The rationale behind these criteria is a simple formula, not far from the truth, assuming the maximal heart rate to be 220 minus the subject's age. By this procedure the risk of submitting a subject to a maximal performance is relatively small (it has repeatedly been emphasized that the standard deviation in maximal heart rate is approximately $\pm$ 10 beats $\cdot$ min^{-1}). Another cut-off point for an automatic termination of a test is 85 percent of the predicted maximal heart rate based on the subject's age (see Pollock et al., 1978, p. 299). This level of exercise demands approximately 80 percent of the subject's maximal oxygen uptake, see Fig. 8-10.

The inevitable problem when using a submaximal exercise test with a fixed heart rate in predicting the cardiovascular fitness of the subject is the aforementioned fact that a heart rate of 160 may be maximal for one person, but 40 beats $\cdot$ min^{-1} or more below the maximum for another one. Considering the wide scatter in maximal heart rates reported in the literature (see Åstrand et al., 1973; Cooper et al., 1977), it is quite meaningless to be too sophisticated in the choice of the target heart rate.

There is a definite trend, supported by Bruce's experience (1971), to continue a test until the subject/patient has alarming symptoms or has to stop because of fatigue. This end-point is termed the functional maximum but the measured oxygen uptake is not necessarily the subject's maximal oxygen uptake.

Blomqvist and others (Blackburn, 1969, p. 281) have reported that the degree of ST-depression, if present in the ECG, will increase with higher O_2 uptake (and cardiac output) up to maximum. Pollock et al., (1978, p. 287) report from studies conducted at the Aerobics Institute in Dallas, Texas, that of the 7059 males who were subject to a maximal exercise test, approximately 15 percent were considered abnormal or questionable. Among the 552 abnormal tests, 34 percent did not become abnormal until after the heart rate exceeded 85 percent of the predicted maximal heart rate. However, more "false-positive cases" may appear in a maximal test while some "false-negative" cases probably may result from a submaximal test.

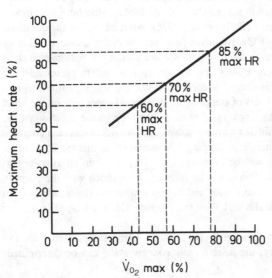

Figure 8-10
As illustrated in Fig. 8-6 the heart rate (HR) during exercise, e.g. when walking, running, or cycling, increases for the average person linearly with the oxygen uptake. In this figure it is illustrated that the relationship is not strictly on a percentage basis. When the heart rate is 85 percent of the maximum, the oxygen uptake is slightly below 80 percent of the maximal aerobic power. *(From Pollock et al., 1978, p. 123.)*

If the purpose of a test is to predict the maximal oxygen uptake because the methods to measure oxygen uptake are not available, a multistage submaximal test followed by a maximal spurt can be useful. The steady-state heart rates are plotted against the predicted submaximal oxygen uptakes (when using the cycle ergometer the data in Table 8-4 can be applied). An interpolation to the subject's measured maximal heart rate, determined at the end of about a 3-min spurt, gives a more reliable maximal aerobic power than an extrapolation to the maximal heart rate typical for the person's age.

If equipment is not available for a recording or a monitoring of the heart rate, it is recommended that the length of time for 30 heart beats is measured with a stopwatch, counting the pulse at the carotid artery by palpation or counting the heart rate with the use of a stethoscope at the apex of the heart. (At any rate it is advisable to learn how to count the heart rate manually because any recording system may fail unexpectedly.) If the heart rate is needed in connection with sport activities, e.g., during training, taking the length of time for ten heart beats, starting the counting immediately after the exercise, gives a good picture of the heart rate during the last part of the exercise.

In Chap. 7, it was mentioned that many studies have been devoted to the question of how accurately can the maximal oxygen uptake be predicted from running tests, e.g., the distance that can be covered in a given time, or the time it takes to run a given distance. The reported correlation coefficients ranged from 0.04 to 0.90 (for reference see Shephard, 1982, in Chap. 7). Cooper (1982, pp. 141–143) presents tables with "Fitness Category" from "very poor" to "superior" based on the performance in a 12-min running test, 1.5-mile run test, 3-mile walking test, 12-min swimming or cycling test. The main purpose of these tests is to follow one's own score over the years.

In Canada, a "safe, simple, self-administered fitness test, the purpose of which would be motivational rather than to accurately evaluate fitness," has been developed, the *"Canadian Home Fitness Test"* (Bailey et al., 1976). After answering the questions in a Physical Activity Readiness Questionnaire, the person can evaluate whether or not she or he can safely go through the test. The subject climbs the bottom two steps of a standard staircase, step height 20.3 cm (8 inches) at a rhythm presented by a long-playing record. The pace is chosen according to sex and age and after a 3-min warming-up, one or two 3-min bouts of stepping are performed depending on the heart rate taken from 5 to 15 s during recovery after each stage. From the duration of the stepping and the heart rate one obtains a fitness score. The individual can then compare the fitness level from time to time. The stepping rate is chosen so that the demand on the oxygen transporting system will be approximately 70 percent in an average Canadian of the same age and sex. This test is an interesting approach with the purpose of reaching a large population in an health and fitness education. (Information from Recreation Canada, National Health and Welfare, Ottawa, Canada KIA OX6.)

Blood Lactate, Respiratory Quotient By securing blood samples after the exercise and during early recovery, the *peak lactate concentration* can be determined.

As mentioned, concentrations of 8 mM or higher support the assumption that the subject's aerobic power was taxed to its maximum.

In Chap. 7 the rate of exercise in relation to the person's maximal oxygen uptake, when the blood lactate begins to accumulate (the popular "anaerobic threshold" concept) was discussed. An individual who is not very well-trained can exercise at about 50 percent of her or his maximal aerobic power without a noticeable increase in blood lactate concentration. Therefore, if this person can exercise with a work rate demanding an oxygen uptake of 1.5 liters $\cdot$ min^{-1} with no signficant increase in the blood lactate, the maximal oxygen uptake is probably some 3 liters $\cdot$ min^{-1} or higher.

It has been mentioned that a person involved in manual labor who is more or less free to set the pace, normally accepts exercising with an energy output that is approximately within 40 percent of the maximal aerobic power. If such a person can exercise for at least five minutes at a rate requiring an oxygen uptake of 2.5 liters $\cdot$ min^{-1}, that person can most likely exercise under steady-state conditions for hours with an oxygen uptake of at least 1 liter $\cdot$ min^{-1}. Another approach would be to let the person exercise on a cycle ergometer or a treadmill with the mentioned oxygen demand. If she or he does not develop clinical symptoms or signs, does not feel fatigued, there is no increase in blood lactate concentration, and the heart rate is low ($<$120 beats $\cdot$ min^{-1}), then that person can probably safely take part in an activity that demands an oxygen uptake of 1 liter $\cdot$ min^{-1}.

During heavy exercise of short duration (up to 5 min), the respiratory quotient (R) exceeds 1.0. Issekutz et al. (1962) have shown that the R-value obtained during standardized exercise conditions may be used as a predictor of maximal aerobic power. It was noted that ΔR (exercise R minus 0.75) increased logarithmically with the exercise rate, and maximal oxygen uptake was reached at a ΔR value of 0.40. This observation offers the possibility of predicting a person's maximal O_2 uptake from the measurement of R during a single 5-min cycle ergometer test at a submaximal work rate. In other sections of this book, it has been emphasized that an accumulation of blood lactate is correlated with an increase in the pulmonary ventilation beyond the demand of a sufficient oxygen supply to the exercising muscles. More or less by definition such a situation induces a hyperventilation with a decrease in the body's CO_2 content as a consequence. The $Co_2 : O_2$ ratio in the expired air will sooner or later exceed 1.0 as the exercise approaches maximum. This fact as well as a determination of the blood lactate concentration may be utilized as a rough method for assessing the effectiveness of an individual's oxygen-transporting system during standardized exercise. Such measurements are particularly useful when testing the aerobic exercise potential of patients who for medical reasons are kept on drugs (e.g., β-blockers) or have irregular heart rates. In such cases, the heart rate attained during submaximal exercise is a poor guide in the evaluation of the function of the oxygen-transporting system.

End-Point for the Termination of a Test

Especially when examining patients with established or suspected cardiovascular disease, it is important to have some predetermined rules as a guide for deciding when to terminate an exercise test. Pollock et al. (1978) have listed the following indications for discontinuing the test:

1 Failure of the monitoring system.

2 Progressive angina (chest pain).

3 Two-millimeter horizontal or downsloping ST-depression on the ECG. (As long as a patient has no symptoms, some laboratories will not use this criterion as a reason for terminating the test.)

4 Sustained supraventricular tachycardia.

5 Ventricular tachycardia.

6 Exercise induced left or right bundle branch block.

7 Any significant drop (10 mm Hg; 1.3 kPa) in systolic blood pressure.

8 Light headedness, confusion, ataxia, pallor, cyanosis, nausea, or any sign of peripheral circulatory insufficiency.

9 Inappropriate bradycardia (slow heart rate).

10 Excessively high blood pressure: systolic greater than 260 mm Hg (36 kPa), diastolic greater than 120 mm Hg (16 kPa).

11 Presence of dangerous dysrhythmias (irregular heart beats) such as frequent premature ventricular contractions (PVC's) and multifocal PVC's (premature beats that are triggered from more than one area of the heart).

Heart rate, blood pressure, and ECG monitoring should continue on a regular basis during exercise and recovery (at least up to 5 min after the end of exercise). Emergency equipment and qualified personnel should be available.

Evaluation of Test Results

An evaluation of the cardiac performance should be based on the total oxygen uptake (in liters $\cdot$ min^{-1}), since this is correlated with cardiac output, myocardial oxygen consumption, and blood flow. In a group of individuals, with different body weights the oxygen uptake corrected for weight (ml $\cdot$ kg^{-1} $\cdot$ min^{-1}) is unrelated to the actual load on their hearts. In longitudinal studies, the subject's body weight may vary considerably. The ml $\cdot$ kg^{-1} figure for maximal oxygen uptake is a good predictor of the subject's potential to move and lift the body, but again, it may not mirror the cardiac performance. As an example, a subject with a body weight of 80 kg, has a maximal heart rate 180; $\dot{V}_{O_2}$ max 2.0 liters $\cdot$ min^{-1}; cardiac output, 14 liters min^{-1}; stroke volume 78 ml. His maximal oxygen uptake expressed as ml $\cdot$ kg^{-1} $\cdot$ min^{-1} is 25 equivalent to 7 METS. He is retested after having lost 10 kg in body weight. His weight is now 70 kg, maximal heart rate 180, as before, $\dot{V}_{O_2}$ 2.0 liters $\cdot$ min^{-1}; cardiac output 14 liters $\cdot$ min^{-1}; stroke volume 78 ml, but his measured maximal MET-value is now 8, and his maximal oxygen uptake 29 ml $\cdot$ kg^{-1} $\cdot$ min^{-1}, a difference of 14 percent from the previous test, merely because of the weight loss. Yet the cardiac performance is unchanged.

In Chap. 7, it was pointed out that the maximal oxygen uptake and for that matter, the subject's response to a submaximal or maximal exercise test actually is of limited help in judging the subject's potential for running, skiing, swimming, etc. A superb technique, high motivation, and a good state of training, giving the potential to exercise close to maximal oxygen uptake for long periods of time, can significantly compensate for an inadequate maximal aerobic power. Actually, some marathon runners of world elite class have quite a modest maximal oxygen uptake, i.e., below 70 ml $\cdot$ kg^{-1} $\cdot$ min^{-1}. On the other hand, they have a very good running technique and can exercise for hours at a high percentage of their maximal oxygen uptake. In the marathon runners, the speed that was attained at the "anaerobic threshold" or the onset of blood lactate accumulation (the OBLA-speed) speed was highly correlated with their marathon time, $r > 0.90$ (see Sjödin and Svedenhag, 1985). This is logical since the threshold is dependent on the running economy, the maximal oxygen uptake, as well as the ability to exercise close to this maximum for hours.

Summary

1 Every type of exercise is, in a sense, a unique situation. However, all forms of muscular exercise do increase the metabolic rate and therefore it is of particular interest to be able to analyze the involvement of the oxygen transporting system. The oxygen uptake provides an accurate measure of the aerobic power, and it is highly related to the cardiac output. The $\dot{V}_{O_2}$ max is, under standardized conditions, a highly reproducible measure of the individual's aerobic fitness. It is, however, subject to variations under certain conditions (i.e., after prolonged inactivity, after training, as a consequence of cardiac disease). The main factor behind such variations in V_{O_2} max is a proportional change in the stroke volume. Therefore, a recording of the heart rate during exercise at a given oxygen uptake will reflect these variations in longitudinal studies. Generally speaking, a high heart rate is usually associated with a low stroke volume. However, from this information it is not possible to tell whether this exercise response is caused by genetic factors, lack of training, impaired heart function, or other factors.

2 A multistage exercise test on a treadmill or cycle ergometer will provide a measurement of the rate of work an individual is able to tolerate without symptoms or ECG abnormalities.

3 For a prediction of the subject's ability to move the body, the maximal oxygen uptake per kilogram body weight should be calculated. However, an evaluation of the cardiac performance should be based on the total oxygen transport (liter $\dot{V}_{O_2}$ $\cdot$ min^{-1}), for this is correlated to the cardiac output, the myocardial oxygen consumption, and the blood flow. Variations in the body-fat content are not followed by similar changes in the dimensions of muscles, heart, blood volume etc., and in the demands for local blood flow. In other words, in a heterogenous group of individuals the $\dot{V}_{O_2}$ $\cdot$ kg^{-1} $\cdot$ min^{-1} value is *unrelated* to the actual load on their hearts. One good measure of the cardiac performance is the ratio of oxygen uptake/heart rate, i.e. the oxygen pulse.

4 There is no one test protocol ideal for all situations. It is recommended that one adapts the initial rate of work and the increment in work rate to the assumed maximal aerobic power of the person to be tested. The exercise time at each stage should if possible be at least 3 min; the larger the increments in rate of work, the longer the time at each stage. If the main purpose of the exercise test is to establish the V_{O_2} max or symptom-limited exercise tolerance one can apply a nonsteady-state protocol with 1–2 min at each stage.

5 For the investigator who is willing to accept the small but definite risk involved, a multistage test carried to symptom-limited work rate or to maximal power, will provide the clearest results, particularly for a differentiation between normals and coronary heart disease patients.

6 An alternative is to terminate the multistage exercise test at a heart rate close to 200 minus the age (in years) of the subject. For most individuals this means a submaximal test. The third alternative is to simulate on a cycle ergometer or a treadmill the metabolic rate of the subject's job or recreational activities. Notes about perceived exertion, objective measurements of blood pressure, blood lactate concentration, and recording of the ECG will give essential information about the subject's potential to meet those demands.

Repeated single-stage or multistage tests give excellent indication of variations in physical fitness (changes in body weight must be considered in a treadmill test).

EVALUATION OF THE ANAEROBIC POWER

Theoretical as well as practical aspects of the anaerobic energy yield during exercise at various work rates are discussed in Chap. 7. Unfortunately, at the present, we do not have any satisfactory methods to measure the anaerobic power with a desirable degree of accuracy. It is true that providing the mechanical efficiency in cycle ergometer exercise is constant, the energy requirement during exhausting exercise may be predicted by extrapolation. By measuring the oxygen uptake continuously during the entire exercise period, the oxygen deficit may be calculated and used as a measure of the anaerobic energy yield. The problem is, however, that in most events or types of exercise, one cannot estimate the total energy demand with sufficient accuracy when dealing with maximal or near-maximal effort. Consequently, the anaerobic power cannot be determined. An increase in the lactate concentration in the muscle or in the blood will not give a quantitative measurement of the total lactate production, for we do not know the total volume of body fluid into which the lactate is dissolved. This is a handicap when considering how to train the anaerobic power most effectively, because we are unable to measure the anaerobic power objectively.

However, there are tests which to some extent reflect the maximal anaerobic power of a subject. In Chap. 7, the application of a force plate was mentioned to measure the maximal force and power that a subject can develop. In a maximal standing vertical jump, the height to which the mass of the subject's body is lifted, gives an idea of the mechanical power output (Sargeant's jump).

The Margaria staircase test, which makes it possible to estimate the power developed during a 5 to 6 s maximal effort, was also briefly discussed.

A laboratory test with 30 to 60 s sprint bouts on a cycle ergometer has been the subject of increasing interest. The Wingate test consists of an exhaustive cycling exercise against a resistance that is related to the subject's body weight. The subject is asked to make as many revolutions as possible for a total of 30 s, and the number of half-revolutions is recorded at 5-s intervals. The peak mechanical power, the average power, and the decline in this power from the peak recording down to the last 5-s period are calculated (for references and discussion see Jakobs, 1980; Jakobs et al., 1982). The physiological bases for these measurements are: (1) the maximal 5-s power is assumed to be correlated with the power generated by the intramuscular high-energy phosphate compounds, ATP and PCr (compare the Margaria staircase test), and (2) the 30-s average power output should reflect the anaerobic capacity. There may be a high correlation between those two factors, but a 30-s bout of exercise will not exhaust the anaerobic capacity.

When discussing the force-velocity curve for a maximally contracting muscle (in connection with Fig. 2-13) it was emphasized that the selected force/speed of contraction in relation to the maximal potential was critical. Actually, for the achievement of optimal human power output, cycling force and velocity should be equal and close to one third of their maximal values (see Davies et al., 1982). Sargeant et al. (1981) noticed that the velocity for the greatest mechanical power output was 110 revolutions · min^{-1} (mean value over a complete revolution of 840 watts in their male subjects). Bergh (1985) has suggested a braking force of 10 percent of the subject's body weight, which is very close to the optimum reported by Nadeau and Brassard (1983), 50 N for their female and 70 N for their male subjects.

Usually, the maximal power when tested on a cycle ergometer is attained during the first 2–3 s of the exercise. Without sophisticated apparatus, it is difficult to measure the total force, i.e., the force exerted to accelerate the mass of the wheel and to overcome the braking force. However, Bergh found a close relation between this total force · s^{-1} and the peak velocity of the wheel or pedals. The latter is much easier to measure.

The simplest apparatus is a mechanically braked, calibrated cycle ergometer with a speedometer (e.g., the Monark pendulum-braked or "weight" ergometer). The peak velocity is quite simple to read on the speedometer and the number of revolutions can be counted and timed. However, it is certainly an improvement to modify a regular cycle ergometer by adding toe clips, and by installing dual microswitches on the cranks, and using racing handlebars and a reinforced and lengthened seat stem.

In this and similar all-out tests on a cycle ergometer the mean coefficient of variance in data, such parameters as peak torque, mechanical power, average power, ride time, fatigability, and peak blood lactate concentration, is reported to be down to 5 to 6 percent under optimal conditions (see Coggan and Costill, 1984; Evans and Quinney, 1981).

Significant correlations have been noted between the 30-s results and the performances in running shorter distances, e.g., 40 m up to 300 m ($r = 0.7 - 0.8$; see Shephard, 1982, p. 66). As already emphasized, such a correlation coefficient is not high enough for a good prediction of *one* individual's running ability from a cycle ergometer test.

Bosco et al. (1983) have proposed a test for the measurement of the mechanical power during a series of vertical rebounce jumps, e.g., in 4 successive 15-s periods. The sum of the flight times is measured with a digital timer. On the basis of this information and the number of jumps measured, their formula will provide the mechanical power per unit mass. In the authors' hands, the reproducibility was high

($r = 0.95$). Due to the simplicity of the test it may be applied both in the laboratory and under field conditions.

Summary The method of measuring the maximal power output during a 30-s cycle ergometer test is assumed to reflect the maximal muscular strength of the engaged muscle groups and the power of the high energy phosphate compounds (the peak power is usually attained sometime during the first 5 s). The average power reflects the anaerobic capacity but it is not a measure of this capacity. The decline in power during the test, e.g., measured at 5-s intervals, provides a fatigue curve under these standardized conditions.

MEASUREMENTS OF MUSCULAR STRENGTH

The following comments are meant to supplement the discussions in Chaps. 2 and 3.

The isometric strength applied in a particular muscular effort may be measured with a calibrated dynometer, a method that may be particularly useful in field studies. In the laboratory much more sophisticated apparatuses are available. When an evaluation of a maximal dynamic strength development is essential, the mentioned dependence of the force-velocity relationship represents a problem. For this reason, instruments such as the Cybex apparatus have been developed by which the speed of contraction can be accurately controlled. In a way it tests an "artificial" contraction because movements are not normally isokinetic. Whether this has any practical implications for the interpretation of the data is difficult to assess.

In the isokinetic strength test, one measures the muscular force times the lever arm, i.e., the torque (in Nm). If it is of interest to compare data obtained on subjects of different body size a measure of the force may be desirable. Since the length of the lever is very difficult to measure, one may divide the Nm value by the subject's body height, assuming that it is proportional to the muscle's lever (see Chap. 9).

When discussing tests of anaerobic power, it was mentioned that a vertical jump and a "sprint" on the cycle ergometer may be applied as a test of knee extensor muscle strength. A "fatigue test" is also developed for the isokinetic muscle contraction. The subject is instructed to perform 50 repeated contractions with maximal effort without rest in between. The average peak torque for the last three contractions is given in percent of the highest peak torque recorded (or the mean of the first three contractions). The decline in torque is taken as a fatigue index (see Thorstensson, 1976).

The maximal muscular force can be measured in various types of exercise with resistance weight training machines, (e.g., bench press, "curl", "squat") but for obvious reasons such activities are very difficult to standardize. The same is true for the pull-up and push-up.

EXERCISE ELECTROCARDIOGRAM

It is beyond the scope of this text to provide a manual covering this topic. In the literature published thus far there is no general agreement as to the use of exercise ECG-recording to provide information with regard to (1) the severity of ischemic heart disease, (2) recommendations about the person's habitual physical activity, and (3) termination of an exercise test. As far as the dependability of the exercise ECG test is concerned Erikssen et al. (1976), performed coronary angiography in 105 men, in whom coronary heart disease was suspected on the basis of a positive exercise ECG and other criteria such as: (1) a WHO questionnaire on angina pectoris positive on interview, (2) typical angina during a near maximal cycle exercise test, (3) a Minnesota Code 1.1 on a resting ECG. Thirty-three of these 105 men had normal coronary angiograms, three had minimal disease of one or more vessels, eighteen had one-vessel disease, twenty-five had two-vessel disease, and twenty-six had three-vessel disease, i.e., 69 had pathological angiograms. The results showed that all coronary heart disease-suggestive criteria, including the exercise ECG, had approximately the same diagnostic performance: approximately one false-positive for every two true-positives.

For further discussion of exercise testing and ECG interpretations see Ellestad (1980); Jones and Campbell (1982); Froelicher (1983); Zeppilli, et al. (1983); Hanson (1984).

REFERENCES

American Association for Health, Physical Education and Recreation: The AAHPER Youth Fitness Test Manual, rev. ed., Washington, D.C., 1976.

Åstrand, I: Aerobic Work Capacity in Men and Women with Special Reference to Age, *Acta Physiol. Scand.*, **49** (Suppl. 169), 1960.

Åstrand, I., P.-O. Åstrand, I. Hallbäck, and Å. Kilbom: Reduction in Maximal Oxygen Uptake with Age, *J. Appl. Physiol.*, **35**(5):649, 1973.

Åstrand, P.-O.: "Experimental Studies of Physical Working Capacity in Relation to Sex and Age," Munksgaard, Copenhagen, 1952.

Åstrand, P.-O.: Study of Bicycle Modifications Using a Motor Driven Treadmill-bicycle Ergometer, *Arbeitsphysiol.*, **15**:23, 1953.

Åstrand, P.-O.: Quantification of Exercise Capability and Evaluation of Physical Capacity in Man, *Progress in Cardiovascular Diseases,* **19**:51, 1976.

Åstrand, P.-O.: How to Conduct and Evaluate an Exercise Stress Test, in I. Kass (ed.), "International Approaches to Issues in Pulmonary Disease," monograph 18, p. 20, International Exchange of Information Rehabilitation, World Rehabilitation Fund, Inc., New York, 1982.

Åstrand, P.-O., and Irma Ryhming: A Nomogram for Calculation of Aerobic Capacity (Physical Fitness) from Pulse Rate during Submaximal Work, *J. Appl. Physiol.,* **7**:218, 1954.

Bailey, D. A., R. J. Shephard, and R. L. Mirwald: Validation of a Self-administered Home Test of Fitness, *Can. J. Appl. Sports. Sci.,* **1**:67, 1976.

Balke, B.: Optimale Körperliche Leistungsfähigkeit, ihre Messung und Veränderung infolge Arbeitsermüdung, *Arbeitsphysiol.,* **15**:311, 1954.

Beaver, W. L.: Water Vapor Corrections in Oxygen Consumption Calculations, *J. Appl. Physiol,* **35**:928, 1973.

Bergh, U.: The Cycle Ergometer as a Tool for Muscle Strength Measurements, in U. Bergh and H. Robertsson (Eds.) 10th Int. Congress of Biomechanics, Umeå, Sweden, p. 27, 1985. 1985.

Bergh, U., I.-L. Kanstrup, and B. Ekblom: Maximal Oxygen Uptake during Exercise with Various Combinations of Arm and Leg Work, *J. Appl. Physiol.,* **41**:91, 1976.

Binkhorst, R. A., and P. van Leeuwen: A Rapid Method for the Determination of Aerobic Capacity, *Intern. Z. Angew. Physiol.*, **19**:459, 1963.

Blackburn, H. (ed.): "Measurements in Exercise Electrocardiography," Charles C Thomas, Springfield, Ill., 1969.

Bobbert, A. C.: Energy Expenditure in Level and Grade Walking, *J. Appl. Physiol.*, **15**:1015, 1960.

Borg, G. A. V.: Psychophysical Bases of Perceived Exertion, *Med. Sci. Sports Exc.*, **14**(5):377, 1982.

Bosco, C., P. Luhtanen, and P. V. Komi: A Simple Method for Measurement of Mechanical Power in Jumping, *Europ. J. Appl. Physiol.*, **50**:273, 1983.

Bruce, R. A.: Exercise Testing of Patients with Coronary Heart Disease, *Ann. Clin. Res.*, **3**:323, 1971.

Chase, G. A., C. Grave, and L. B. Rowell: Independence of Changes in Functional and Performance Capacities Attending Prolonged Bed Rest, *Aerospace Med.*, **37**:1232, 1966.

Coggan, A. R., and D. L. Costill: Biological and Technological Variability of Three Anaerobic Ergometer Tests, *Int. J. Sports Med.*, **5**(3):142, 1984.

Cooper, K. H.: "The Aerobics Program for Total Well-being," Bantam Books, Toronto, 1982.

Cooper, K. H., J. G. Purdy, S. R. White, M. L. Pollack, A. C. Linnerud: Age-Fitness Adjusted Maximal Heart Rates, in D. Brunner and E. Jokl (eds.), "The Role of Exercise in Internal Medicine," p. 78, S. Karger, Basel, 1977.

Davies, C. T. M., J. Wemyss-Holden, and K. Young: Maximal Power Output during Cycling and Jumping, *J. Physiol.*, **322**:43P, 1982.

Eckermann, P., and H. P. Millahn: Der Einfluss der Drehzahl auf die Herzfrequenz und die Sauerstoffaufnahme bei konstanter Leistung am Fahrradergometer, *Int. Z. angew. Physiol. einschl. Arbeitsphysiol.*, **23**:340, 1967.

Ekblom, B., and A. N. Goldbarg: The Influence of Physical Training and Other Factors on the Subjective Rating of Perceived Exertion, *Acta Physiol. Scand.*, **83**:399, 1971.

Ellestad, M. H.: "Stress Testing: Principles and Practices," 2nd ed., F. A. Davies, Philadelphia, 1980.

Erikssen, J., I. Enge, K. Forfang, and O. Storstein: False Positive Diagnostic Tests and Coronary Angiography Findings in 105 Presumably Healthy Males, *Circulation*, **54**(3):371, 1976.

Evans, J. A., and H. A. Quinney: Determination of Resistance Settings for Anaerobic Power Testing, *Can. J. Appl. Sports Sci.*, **6**(2):53, 1981.

Fernandez, E. A., J. G. Mohler, and J. P. Butler: Comparison of Oxygen Consumption Measured at Steady State and Progressive Rates of Work, *J. Appl. Physiol.*, **37**:982, 1974.

Froelicher, V. F.: "Exercise Testing and Training," Le Jacq Publishing Co., New York, 1983.

Grosse-Lordemann, H., and E. A. Müller: Der Einfluss der Tretkurbellänge auf das Arbeitsmaximum und den Wirkungsgrad beim Radfahren, *Arbeitsphysiologie*, **9**:619, 1937.

Hanson, P.: Clinical Exercise Testing, in R. H. Strauss (ed.), "Sports Medicine," p. 13, W. B. Saunders, Philadelphia, 1984.

Harrison, M.-H., G. A. Brown, and L. A. Cochrane: Maximal Oxygen Uptake: Its Measurement, Application, and Limitations, *Aviat. Space Environ. Med.*, **51**(10):1123, 1980a.

Harrison, M. H., D. L. Bruce, G. A. Brown, and L. A. Cochrane: A Comparison of Some Indirect Methods for Predicting Maximal Oxygen Uptake, *Aviat. Space Environ. Med.*, **51**(10):1128, 1980b.

Hermansen, L.: "Oxygen Transport during Exercise in Human Subjects," *Acta Physiol. Scand.* (Suppl. 399): 1973.

Hermansen, L., and B. Saltin: Oxygen Uptake during Maximal Treadmill and Bicycle Exercise, *J. Appl. Physiol.*, **26**:31, 1969.

Hettinger, Th., and K. Rodahl: Ein Modifizierter Stufentest Zur Messung Der Belastungsfähigkeit Des Kreislautes, *Deutsche Medizinische Wochenschrift,* **14:**553, 1960.

Inbar, O., R. Dotan, T. Trousil, and Z. Dvir: The Effect of Bicycle Crank-length Variation upon Power Performance, *Ergonomics,* **26**(12):1139, 1983.

Issekutz, B., Jr., N. C. Birkhead, and K. Rodahl: Use of Respiratory Quotients In Assessment of Aerobic Work Capacity, *J. Appl. Physiol.,* **17:**47, 1962.

Jacobs, I.: The Effects of Thermal Dehydration on Performance of the Wingate Anaerobic Test, *Int. J. Sports Med.,* **1:**21, 1980.

Jakobs, I., O. Bar-Or, J. Karlsson, R. Dotan, Per Tesch, P. Kaiser, and O. Inbar: Changes in Muscle Metabolites in Females with 30-s Exhaustive Exercise, *Med. Sci. Sports Exc.,* **14**(6):457, 1982.

Johnson, R. E., L. Brouha, and R. C. Darling: A Test of Physical Fitness for Strenuous Exertion, *Rev. Canad. Biol.,* **1:**491, 1942.

Jones, N. L., and E. J. M. Campbell: "Clinical Exercise Testing," 2nd ed., W. B. Saunders, Philadelphia, 1982.

Jonsson, B. G.: A Statistical Evaluation of Data Gathered in Connection with Exercise Tests, (thesis), Stockholm, 1981.

Kanstrup, I.-L., and I. Hallbäck: Cardiac Output Measured by Acethylene Rebreathing Technique at Rest and during Exercise. Comparison of Results Obtained by Various Calculation Procedures, *Pflügers Arch.,* **390:**179, 1981.

Kasch, F. W., W. H. Phillips, W. D. Ross, J. E. L. Carter, and J. L. Boyer: A Comparison of Maximal Oxygen Uptake by Treadmill and Step-Test Procedures, *J. Appl. Physiol.,* **21:**1387, 1966.

Kavanagh, T., and R. J. Shephard: Maximum Exercise Tests on "Postcoronary" Patients, *J. Appl. Physiol.,* **40:**611, 1976.

Keys, A., J. Brozek, A. Henschel, O. Michelsen, and H. L. Taylor: "The Biology of Human Starvation," pp. 675, 735, University of Minnesota Press, Minneapolis, 1950.

Mahadeva, K., R. Passmore, and B. Woolf: Individual Variations in the Metabolic Cost of Standardized Exercises: The Effect of Food, Age, Sex, and Race, *J. Physiol.,* **121:**225, 1953.

Margaria, R., P. Cerretelli, P. Aghemo, and G. Sassi: Energy Cost of Running, *J. Appl. Physiol.* **18:**367, 1963.

Margaria, R., P. Aghemo, and E. Rovelli: Indirect Determination of Maximal O_2 Consumption in Man, *J. Appl. Physiol.,* **20:**1070, 1965.

Maritz, J. S., J. F. Morrison, J. Peter, N. B. Strydom, and C. H. Wyndham: A Practical Method of Estimating an Individual's Maximal Oxygen Intake, *Ergonomics,* **4:**97, 1961.

Mathews, D. K.: "Measurement in Physical Education," W. B. Saunders, Philadelphia, 1974.

McArdle, W. D., F. I. Katch, and G. S. Pechar: Comparison of Continuous and Discontinuous Treadmill and Bicycle Tests for Max $\dot{V}_{O_2}$, *Med. Sci. Sports,* **5:**156, 1973.

McKay, G. A., and E. W. Banister: A Comparison of Maximum Oxygen Uptake Determination by Bicycle Ergometry at Various Pedaling Frequencies and by Treadmill Running at Various Speeds, *Europ. J. Appl. Physiol.,* **35:**191, 1976.

Nadeau, M., and A. Brassard: The Bicycle Ergometer for Muscle Power Testing, *Can. J. Appl. Sports Sci.,* **8**(1):41, 1983.

Nagle, F. J., B. Balke, and J. P. Naughton: Gradational Step Tests for Assessing Work Capacity, *J. Appl. Physiol.,* **20:**745, 1965.

Passmore, R., and J. V. G. A. Durnin: Human Energy Expenditure, *Physiol. Rev.,* **35:**801, 1955.

Pollock, M. L., J. H. Wilmore, and S. M. Fox III: "Health and Fitness through Physical Activity," American College of Sports Medicine Series, John Wiley & Sons, New York, 1978.

Rodahl, K., and B. Issekutz, Jr.: Physical Performance Capacity of the Older Individual, in K. Rodahl and S. M. Horvath (eds.), "Muscle as a Tissue," McGraw-Hill Book Company, New York, 1962.

Rowell, L. B.: Human Cardiovascular Adjustments to Exercise and Thermal Stress, *Physiol. Rev.*, **54**:75, 1974.

Ryhming, I.: A Modified Harvard Step Test for the Evaluation of Physical Fitness, *Arbeitsphysiol.*, **15**:235, 1953.

Saltin, B.: Aerobic Work Capacity and Circulation at Exercise in Man, *Acta Physiol. Scand.*, **62**(Suppl. 230): 1964.

Sargeant, A. J., E. Hoinville, and A. Young: Maximum Leg Force and Power Output during Short-term Dynamic Exercise, *J. Appl. Physiol.: Resp. Env. Exer. Physiol.*, **51**(5):1175, 1981.

Shephard, R. J.: The Relative Merits of the Step Test, Bicycle Ergometer, and Treadmill in the Assessment of Cardio-Respiratory Fitness, *Intern. Z. Angew. Physiol.*, **23**:219, 1966.

Shephard, R. J.: "Physical Activity and Growth," Year Book Medical Publishers, Inc., Chicago, 1982.

Siconolfi, S. F., E. M. Cullinane, R. A. Carleton, and P. D. Thompson: Assessing $\dot{V}_{O_2}$ in Epidemiologic Studies: Modification of the Åstrand-Ryhming Test, *Med. Sci. Sports Exc.*, **14**(5):335, 1982.

Sjödin, B., and J. Svedenhag: Applied Physiology of Marathon Running, *Sports Med.*, **2**:83, 1985.

Sjöstrand, T.: Changes in the Respiratory Organs of Workmen at an Ore Melting Works, *Acta Med. Scand.* (Suppl. 196):687, 1947.

Stamford, B. A.: Step Increment versus Constant Load Tests for Determination of Maximal Oxygen Uptake, *Europ. J. Appl. Physiol.*, **35**:89, 1976.

Strandell, T.: Circulatory Studies on Healthy Old Men with Special Reference to the Limitation of the Maximal Physical Working Capacity, *Acta Med. Scand.*, **175**(Suppl. 414): 1964.

Taylor, H. L., E. Buskirk, and A. Henschel: Maximal Oxygen Intake as an Objective Measure of Cardiorespiratory Performance, *J. Appl. Physiol.*, **8**:73, 1955.

Thorstensson, A.: Muscle Strength, Fibre Types and Enzyme Activities in Man, *Acta Physiol. Scand.*, (Suppl. 443), 1976.

Van Baak, M. A.: "The Physiological Load during Walking, Cycling, Running and Swimming, and the Cooper Exercise Programs," Krips Repro, Meppel, 1979.

Van der Walt, W. H., and C. H. Wyndham: An Equation for Prediction of Energy Expenditure of Walking and Running, *J. Appl. Physiol.*, **34**:559, 1973.

Wyndham, C. H., N. B. Strydom, W. P. Leary, and C. G. Williams: Studies of the Maximum Capacity of Men for Physical Effort, *Intern. Z. Angew. Physiol.*, **22**:285, 1966.

Zeppilli, P., M. Sassara, M. M. Pirrami, M. L. Caputi, and R. Fenici: Physiological and Pharmacological Tests in the Electrocardiographic Investigation of Athletes: a Modern Approach to an Old Problem, *J. Sports Med.*, **23**:240, 1983.

CHAPTER 9

BODY DIMENSIONS AND MUSCULAR EXERCISE

CONTENTS

STATICS
DYNAMICS
    Force
    "Chin-Ups"
    Time
    Acceleration
    Frequencies
    Locomotion
    Running Speed
    Jumping
    Maximal Running Speed for Children
    Kinetic Energy
    External Exercise
    Energy Supply
    Maximal Aerobic Power in Children
    Maximal Aerobic Power in Women
    Maximal Cardiac Output
    Heart Weight, Oxygen Pulse
    Heart Rate
    Secular Increase in Dimension
    Old Age

The thrill of watching athletic competitions is partially caused by the fact that it is impossible to predict who is going to win.

It is impossible from appearance alone to tell who is an athletic champion. On the other hand, it is often possible to exclude those who obviously *cannot* reach top results in certain sport events, such as shot-putting, rowing, and American football. A tiny individual hardly has a chance in these events.

It is of interest to consider the human resources in relation to muscular exercise from a biological viewpoint. If we compare animals of different size, it is evident that certain dimensions and functional capacities are determined by fundamental mechanical necessities. In addition, it may be a matter of biological adaptation (see Weibel and Taylor, 1981).

STATICS

If we take two geometrically similar cubes of different size, the relationship between the surface and the volume of the two cubes can easily be calculated when only the scale factor between the sides of the cubes is known. If this length scale is $L:1$, the surface ratio is $L^2:1$, and the volume ratio $L^3:1$.

If we consider two geometrically similar and qualitatively identical individuals, we may expect all linear dimensions (L) to be proportional. The length of the arms, the legs, the trachea, and the individual muscles will have a ratio $L:1$. If we compare two boys, one 120 cm high, the other 180 cm high, the scale factor will be such that all lengths, levers, ranges of joint motions, and muscular contractions during a specific motion will be related as 120:180 or as 1:1.5 (Fig. 9-1).

Cross sections of, for instance, a muscle, the aorta, a bone, the trachea, the alveolar surface, or the surface of the body are then related as $120^2:180^2$, or $1^2:1.5^2$, i.e., 1:2.25. Volumes, such as lung volumes, blood volumes, or heart volumes, should similarly be related as $120^3:180^3$, or $1^3:1.5^3$, i.e., 1:3.375. The sample applies to mass measured in units of weight, since the density of biological materials, generally speaking, is independent of size.

The force of gravity acts on the mass (M) of the body. If the limbs supporting this mass are proportional to the area of their cross section, there is a disproportion between the mass (which is proportional to L^3) and the support (which is proportional to L^2). In larger animals, an adjustment has taken place: elephants and rhinoceroses have relatively thick, short, and straight legs. In body geometry, these larger animals differ from small animals, which have slender limbs compared with their body mass. In relation to total weight, the human has over twice as much bone as the mouse (Thompson, 1943). Those of the larger animals that have relatively slender limbs have "compressed" bodies compared with their length (giraffe, antelope). In large animals, the design of their bodies is dominated by the requirement of supporting their weight, which was pointed out by Galilei in 1638.

DYNAMICS

If we pursue these theoretical considerations further, it is evident that the maximal force a muscle can develop (F), generally speaking, is proportional to $(\propto)$ its surface.

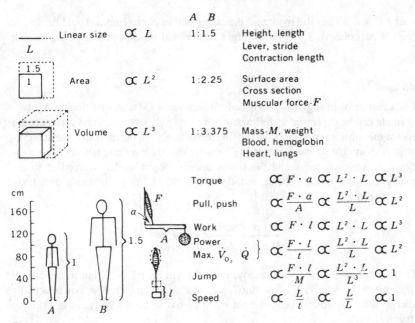

Figure 9-1
Schematic illustration of the influence of dimensions on some static and dynamic functions in geometrically similar individuals. A and B represents two persons with body height 120 and 180 cm respectively (α = proportional too.) See text for explanation. *(Partly modified from Asmussen and Christensen, 1967.)*

It would therefore be expected that the 1.5-times taller body should be able to produce a 2.25-times larger muscular strength, i.e., he should be able to lift a 2.25-times heavier weight. The advantage of the 1.5-times longer levers (a) for the taller boy's muscles to work on is offset by the fact that the weight to be lifted also has a 1.5-times longer lever (A). (According to Fig. 9-1:

$$F \cdot a = M \cdot A$$
$$M = \frac{F \cdot a}{A}$$

In accordance with the discussion above, $F \propto L^2$, $a \propto L$; $A \propto L$. The above equation can thus be expressed as follows:

$$M \propto \frac{L^2 \cdot L}{L} \propto L^2.)$$

Force

The magnitude of the work to be performed is determined by the developed force ($\propto L^2$) and the distance over which the force is applied ($\propto L$). Consequently,

$W \propto L^2 \cdot L \propto L^3$. Thus the work that the larger boy in our example should be able to perform is accordingly 3.375 times larger than would be expected from the smaller boy.

"Chin-ups"

If it is a matter of lifting one's own body with a mass (M), as in chinning the bar, the formula can be expressed as follows: $F \cdot a = M \cdot A$, where a and A represent the levers for the muscles and the body weight, respectively. In other words, the achievement is proportional to the force of the muscles and their levers, but inversely proportional to the body mass and the levers upon which it works. According to the reasoning we have applied above, it follows that the ability to lift one's own body (i.e., do a chin-up) is proportional to

$$\frac{F \cdot a}{M \cdot A} \text{ or } \frac{L^2 L}{L^3 L} \propto \frac{1}{L} \propto L^{-1}$$

Thus, in the case of the smaller boy, the ratio will be 1:1 = 1, but in the case of the larger boy the ratio is 1:1.5 = 0.67, i.e., the larger and stronger boy is actually handicapped by his greater body weight when he has to lift his body, as in chinning the bar.

Time

In many dynamic events an expression of the time scale (t) is necessary. It may be a matter of energy transfer per unit time, the time between two steps or between two heartbeats. It might be of interest to examine this time scale for similar animals of different sizes. If the ratio is expressed as $t{:}1$, the relationship between time and acceleration (a) according to the definition of the acceleration is $a \propto L \cdot t^{-2}$. The connection between the time scale and the length scale may then be calculated by the following formula (Döbeln, 1966):

$$\text{Force} = \text{acceleration} \cdot \text{mass}$$
$$F = a \cdot M \qquad a = \frac{F}{M} \quad a \propto \frac{L^2}{L^3}$$
$$\propto \frac{1}{L} \quad (\propto L^{-1})$$

Returning to the formula $a \propto L \cdot t^{-2}$ or $t^2 \propto L \cdot a^{-1}$, we may now replace a with L^{-1} as shown above, and we arrive at the following formula:

$$t^2 \propto \frac{L}{L^{-1}} \text{ or } t^2 \propto L^2 \quad t \propto L$$

In other words: *The time scale is proportional to the length scale.*

Acceleration

Since a is proportional to L^{-1}, taller (and heavier) persons are handicapped when accelerating their body mass.

Frequencies

The shorter the time between two steps, heartbeats, etc., the higher the frequency (f), which is another expression for the passing of time. This may be expressed as follows: $f \propto 1 \cdot t^{-1}$, and accordingly, $f \propto 1 \cdot L^{-1} (\propto L^{-1})$.

According to this reasoning, we might expect that the frequency of limb motion should vary as an inverse function of limb length. As pointed out by Hill (1950), this is generally the case. A hummingbird moves its wings about 75 to 100 times $\cdot$ s^{-1} while flying forward, a sparrow some 15 times $\cdot$ s^{-1}, and a stork only 2 or 3 times. These frequencies are roughly in inverse proportion to the linear size of the birds. "If the sparrow's muscles were as slow as the stork's, it would be unable to fly. If the stork's muscles were as fast as the hummingbird's, it would be exhausted very quickly" (Hill, 1950). The maximal force of a contracting voluntary muscle is roughly constant in different animals, being on the order of 60 N per square centimeter cross-sectional area in an isometric contraction.

However, the speed of contraction varies enormously between different muscles and different animals. The balance between muscle strength, length of levers, and speed of contraction is very delicate. If humans had muscles contracting as fast as those moving wings of a hummingbird, they would soon break the bones and tear the muscles and tendons. It is possible to swing a rod made of fragile material back and forth if it is short and relatively thick. But if it is long and with a thickness just proportional to L^2, the inertia will cause it to break if moved at the same speed. For the same reason, a small motor may run at higher revolutions per minute than a large one, and the strength of the material is an important factor determining the maximal speed. For similar reasons, a smaller creature can tolerate a greater quickness of movement than a larger one. In fact, the margin of safety is quite small, so that it does happen occasionally that muscles tear, tendons break, and bones splinter during unusual strain, such as during strenuous athletic events. *Without altering the general design,* it would be extremely hazardous for the athlete's locomotor organs if the muscles could be altered to allow him or her to run suddenly, say, 25 percent faster.

Locomotion

Diamond (1983), who questions why evolution never evolved the wheel, has compared the energy cost of transporting the body by locomotion in terms of energy cost per gram per kilometer for a variety of creatures, from a cockroach to a human being. He finds that transport costs are lower for swimming than flying, and lower for flying than running, independent of the number of legs. The most efficient known vehicle is a human on a bicycle, requiring only one-fifth of the energy of the same human walking. According to Denny (1980) the greatest cost of locomotion ever measured is that of a slug crawling.

Running Speed

The speed that may be attained in moving the body is determined by the length of stride and the number of movements per unit of time ($\propto 1 \cdot L^{-1}$), among other things. Thus, for similar animals of different size, the maximal speed is proportional to $L \cdot L^{-1} = 1$, which means that the speed is the same. Short limbs with short strides move more rapidly and can therefore cover as much ground as do longer limbs moving more slowly.

It is well known that "athletic animals" of different size may achieve approximately the same maximal speed. A blue whale of 100 tons and a dolphin of 80 kg attain the same "steady-state" speed of about 15 knots, and a maximal speed of about 20 knots (37 km $\cdot$ hr^{-1}). The speeds of a whippet, a greyhound, and a racehorse, very similar in general design, are nearly the same, or about 65 km $\cdot$ hr^{-1} (40 mph) (Hill, 1950). Gazelles and antelopes with wide variation in size are all able to reach a maximal speed of about 80 km $\cdot$ hr^{-1} (50 mph).

Since Hill (1950) presented his prediction of how performance can be expected to change as a function of body size on the basis of his mathematical model, experimental evidence has shown that it is not quite true that all quadrupeds have the same maximal speed, or that stride frequency is inversely proportional to limb length,, or that the rate of oxygen uptake strictly increases with the cube of running speed (Taylor et al., 1970; Schmidt-Nielsen, 1972; Heglund et al., 1974). McMahon (1975), comparing animals ranging in size from mice to horses running at the transition point between trotting and galloping, found that "elastic similarity" (that is, similarity in the structures of animals so that they are similarly threatened by elastic failure under their own weight) provides a better correlation with published data on body and bone proportions, body surface area, resting metabolic rate, and basal heart rate than does geometric similarity.

Jumping

In the broad jump and high jump it is also a question of the maximal muscular force that may be developed and the distance which the muscle can shorten before the body leaves the ground. We should, therefore, expect the performance to be proportional to $L^2 \cdot L \cdot L^{-3} = 1$ (Fig. 9-1), i.e., a small and a large animal should be equally able to lift its own center of gravity. In broad jumping, kangaroos, jackrabbits, horses, mule deer, and impala antelopes actually seem to be equals to the record-holding man (Hill, 1950). Borelli drew similar conclusions some 300 years ago (1685). In high jumping, in which the aim is to lift the body as high as possible, the larger animal has an advantage, however, since its center of gravity before the jump is already at a higher level (Hill, 1950).

The outstanding high jumper is without exception a tall individual. This person is usually not geometrically similar to the average individual, but is long-legged, with a body weight that is not proportional to L^3 but lower.

Maximal Running Speed for Children

We might expect that the 180-cm body would perform better in high jump than the 120-cm-tall body, which it actually does. Their ability to move their centers of gravity

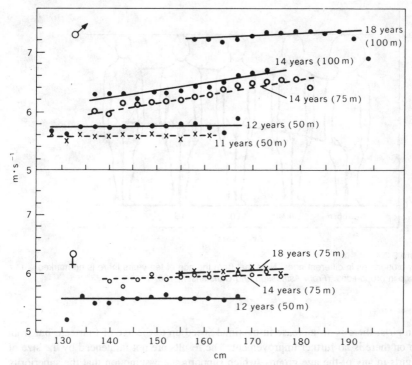

Figure 9-2
Maximal speed in relation to body height for girls and boys of different age; almost 100,000 subjects are included in the statistics. See text for explanation. *(Modified from Asmussen and Christensen, 1967.)*

vertically, on the other hand, should not be different, nor should their maximal speeds. In an analysis of speed in children of different size, Asmussen has divided them into age groups and plotted maximal speed, calculated from the best time on 50 to 100 m, in relation to body height (Fig. 9-2). From the age of about ten, the body proportions remain approximately the same. We may therefore consider the children represented in Fig. 9-2 as geometrically similar (Fig. 9-3).

For eleven- to twelve-year-old boys there is no significant variation in speed with body size, which would not be expected from the discussion above. The somewhat better performance of the twelve-year-olds over the eleven-year-old boys may be due to maturity of the neuromuscular function, improving the coordination. Even better are fourteen-year-old boys, but one also finds that the taller boys can run faster than the shorter boys. There is a further improvement in coordination with age, but this is also probably due to sexual maturity. Their male sex hormones may have influenced their muscular strength in a positive direction. The smaller fourteen-year-old boys may not have reached puberty, in contrast to the taller boys. In the eighteen-year-old group, there is again hardly any variation in results despite a large difference in body height. At this age, all the boys have passed puberty and are sexually mature.

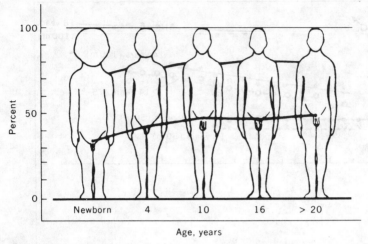

Figure 9-3
Body proportions in different ages. Note that from the age of ten years there is no marked change in proportions. *(From Asmussen and Christensen, 1967.)*

In girls, there is an increase in maximal speed up to the age of fourteen, but from then on there is no further improvement. The results are not influenced by the size of the girls in any of the age groups, which supports the assumption that the superiority of the taller fourteen-year-old boys is due to the effect of male sex hormones.

This independence of maximal speed with body height is in contrast to the greater muscle strength in taller children. From an anatomic viewpoint, this is just what would be expected: the muscle force should increase in proportion to L^2, but the speed should be independent. As already discussed, the results obtained for boys of different size do not strictly follow the results predicted from body dimensions. Apparently, biological factors can modify muscular dynamics. We have considered age factors as well as sexual maturity, which is particularly relevant to the boys' performance.

Kinetic Energy

The kinetic energy (KE) developed in a limb depends on its mass and the square of its velocity, or $KE = M \cdot v^2$; but $v = L \cdot t^{-1}$ and $t \propto L$; therefore, $KE \propto M \cdot L^2 \cdot L^{-2} \propto M$ (or L^3). It follows that the work performed during a single movement, calculated per unit of body weight, in producing and utilizing kinetic energy in the limbs should be the same in large and small animals (Hill, 1950).

External Work

Hill (1950) points out that the external work done in overcoming the resistance of air or water is proportional to the square of the linear dimensions, i.e., the surface area.

Therefore, the effect of this resistance should be the same in similar animals of different size. In running uphill at a given speed, the effect of the slope is inversely proportional to the linear size L. The smaller the animal, the faster it should be when running uphill.

We have concluded that maximal speed is independent of an animal's size. If one animal is 1,000 times as heavy as another, a movement will be 10 times greater than in the smaller animal. However, the larger animal has to take only one-tenth the number of steps, each step taking 10 times as long, in order to attain the same speed as the smaller animal. Since the work per movement and per unit of body weight is the same in the two animals, it follows that it will take roughly 10 times as long for the larger animal to become exhausted during a maximal run.

Energy Supply

It is obviously important that the power output permitted by the mechanical design be matched by an equivalent supply of chemical energy.

As mentioned above, work (being the product of force and distance moved) is proportional to $L^2 \cdot L = L^3$, or M. The total energy output should therefore be related to the mass of the muscles and the body weight in similar animals. The *power,* or work output per unit of time, must then be proportional to $L^3 \cdot t^{-1}$, that is, $L^3 \cdot L^{-1} \propto L^2$ ($\propto M^{2/3}$).

It is well established that the *basal metabolic rate* in animals with large differences in body size, from the mouse to the elephant, follows this prediction. The resting oxygen uptake is actually proportional to $M^{0.74}$ rather than to $M^{2/3}$, but this difference is surprisingly small considering the wide variation in size, shape, and other factors (Brody, 1945, Heusner, 1982). This relation tells us that smaller animals must be more metabolically active per unit of body weight than larger ones. In proportion to its weight, a mouse has to eat 50 times more than a horse in order to maintain its basal metabolism.

Theoretically speaking, it is expected that the *maximal oxygen uptake* should be proportional to L^2 or $M^{2/3}$. This holds extremely well for trained athletes, as is evident from Fig. 9-4. The lower panel of this figure illustrates that the maximal oxygen uptake, expressed as $ml \cdot min^{-1} \cdot kg^{-2/3}$, is not related to body weight, and may therefore be used as a meaningful fitness index instead of the conventional method of expressing maximal oxygen uptake as $ml \cdot min^{-1} \cdot kg^{-1}$, which penalizes heavy individuals. Since maximal *cardiac output* and *pulmonary ventilation* are also volumes per unit of time, they should be proportional to $L^3 \cdot L^{-1} = L^2$ or $M^{2/3}$. Weibel and Taylor (1981) noted that in animals ranging in body weight (W) from 7 g to 260 kg, the maximal oxygen uptake measured during treadmill running scaled roughly to $W^{0.75}$. The estimated diffusing capacity of the lung, however, scaled approximately to $W^{1.0}$.

It thus appears that the larger animals require a larger pulmonary diffusion capacity to transfer O_2 at a given rate from the air to the blood than do smaller animals. There is another approach with which to analyze this relationship, giving the same result. Cardiac output is the product of the frequency of heartbeats and stroke volume. Frequency is proportional to L^{-1} and stroke volume to L^3, and therefore $\dot{Q} \propto L^{-1} \cdot L^3 \propto L^2$.

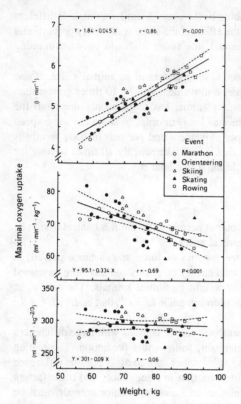

Figure 9-4
Maximal oxygen uptake in a group of Norwegian top athletes trained in different events, expressed as liters · min⁻¹; ml · min⁻¹ · kg⁻¹; and as ml · min⁻¹ · kg⁻²/₃ *(By courtesy of O. Vaage and L. Hermansen.)*

Similarly, pulmonary ventilation is the product of respiratory frequency and tidal air: $\dot{V}_E \propto L^{-1} \cdot L^3 \propto L^2$.

With pulmonary ventilation proportional to $\dot{V}_{O_2}$ and with the production of CO_2 proportional to $\dot{V}_{O_2}$, we should expect the alveolar P_{CO_2} to be the same in different mammals, which it is. Asmussen calculated that the vital capacity measured in subjects seven to thirty years of age (P.-O. Astrand, 1952) was proportional to $L^{3.1}$ in males and $L^{3.0}$ in females, and thus very close to the expected $L^{3.0}$. It may therefore be concluded that, children have lung volumes which are proportionally dimensional to their body size.

In heavy exercise, *heat production* is great and related to $\dot{V}_{O_2}$ or L^2, i.e., to the surface of the body, from which most of the excess heat is lost, at least in humans. There is also heat loss via the expired air, increasing within limits, in direct proportion to $\dot{V}_{O_2}$.

Döbeln (1956b) points out that a dimensional analysis should be based on body weight minus adipose tissue, since fat is metabolically inactive. For a group of 65 young female and male subjects, he calculated the value of the exponent b in the equation maximal oxygen uptake $= a \cdot$ (body weight $-$ adipose tissue)^b and found that $b = 0.71$ (maximal oxygen uptake predicted from the Åstrand and Åstrand nomogram; adipose tissue calculated by means of hydrostatic weighing).

Maximal Aerobic Power in Children

There are very few data available on maximal values for oxygen uptake and cardiac output in animals of different size to put these hypotheses to the test. In fully grown humans, the variations in body size are rather limited. For the male subjects eight to eighteen years of age, studied by P.-O. Åstrand (1952), Asmussen has calculated that the maximal attainable oxygen uptake is proportional to $L^{2.9}$ and not, as expected, to $L^{2.0}$.

Thus, it appears that children's maximal oxygen uptake is not as high as expected for their size, and compared with adults, they do not have the aerobic power to handle their weight. It is also significant that the eight-year-old boy could increase his basal metabolic rate only 9.4 times during maximal running for 5 min, but the seventeen-year-old boy could attain an aerobic power which was 13.5 times the basal power. Therefore, the child has less in the way of a power reserve than the adult. It should also be emphasized that the young subjects had a significantly higher oxygen uptake per kilogram of body weight than the older boys and adults when running at a given submaximal speed on a treadmill (P.-O. Åstrand, 1952). These two factors together may explain the fact that children have difficulty in following their parents' speed, even if maximal oxygen uptake per kilogram body weight may be the same. The children's lower efficiency can be explained partially by their high stride frequency, which is an expensive utilization of energy per unit of time.

In Chap. 7, we discussed the relatively low muscular strength of children. It is conceivable that their aerobic power may be adapted to their muscular machinery. In his analyses, Asmussen found that muscular strength in eight- to sixteen-year-old boys was proportional to $L^{2.89}$, or exactly the same as the maximal oxygen uptake ($\propto L^{2.90}$).

Since the oxygen is transported by the hemoglobin, it is of interest to compare the children's total amount of hemoglobin (Hb_T) with their body size and maximal oxygen uptake. In most of the subjects just discussed, the total hemoglobin was determined. In similar animals of different size, Hb_T should be proportional to M. However, per kilogram of body weight, the younger boys had only 78 percent of the amount of hemoglobin of the older boys. Thus, the amount of hemoglobin was definitely not proportional to body size. Assuming that the maximal oxygen uptake is proportional to Hb_T, we may calculate the maximal $\dot{V}_{O_2}$ in children from the equation $\dot{V}_{O_2} = a \cdot Hb_T^{2/3}$ and use a value for a calculated from the data on older boys. It is found that the child's maximal oxygen uptake should be 1.92 liters $\cdot$ min^{-1}, if the child's hemoglobin is as effective in transporting oxygen as the adult. This calculated value is not far from the determined 1.75 liters $\cdot$ min^{-1}. Using the exponent 0.74, the calculated maximal aerobic power in seven- to nine-year-old children will be 1.78 liters $\cdot$ min^{-1}, or very close to real maximum.

The sample of subjects selected for these analyses was limited to 21 individuals, but it was a homogeneous group. They were nonobese and in the same state of training. The results support the assumption that the maximal oxygen uptake in children and young adults is proportional to the muscular strength and to $Hb_T^{0.76}$, or roughly to $Hb_T^{2/3}$, but *not* to $M^{2/3}$.

It may be concluded that children are physically handicapped when compared with adults (and fully grown animals of similar size). When related to the child's dimensions, its muscular strength is low and so is its maximal oxygen uptake and other parameters of importance for the oxygen transport. Furthermore, the mechanical efficiency of children is often inferior to that of adults. The introduction of dimensions in the discussion of children's performance clearly indicates that they are not mature working machines.

Maximal Aerobic Power in Women

For female subjects, eight to sixteen years of age, the maximal $\dot{V}_{O_2}$ is proportional to $L^{2.5}$. In light of the previous discussion, the noted discrepancy from the expected L^2 is not surprising.

Women have approximately the same maximal oxygen uptake per kilogram of fat-free body mass as men. However, it should be higher in women because of their smaller size (Döbeln, 1956a and b). The lower Hb concentration in women may explain why they cannot fully utilize their cardiac output for oxygen transport.

Figure 9-5 presents data on 227 children and young adults, which stresses these points further. There is a very high correlation between maximal oxygen uptake and body weight for male, nonobese subjects (upper figure). The lower maximal oxygen uptake for female subjects above 40 kg of body weight (age about fourteen years) is largely explained by their higher content of adipose tissue. The lower concentration of hemoglobin in the women's blood also contributes to the observed difference between the sexes. When one relates the total amount of hemoglobin to the maximal oxygen uptake, the difference between regression lines for the female and male subjects is insignificant (lower figure). It should be noted that the exponent b in the equation $y = a \cdot x^b$ (i.e., maximal $\dot{V}_{O_2} = a \cdot Hb_T^b$) is 0.76.

Maximal Cardiac Output

From this analysis, it is evident that maximal $\dot{V}_{O_2}$ should be proportional to L^2 (or $M^{2/3}$). However, data on children and adults do not support this assumption, since the exponent is closer to 3. There are, nevertheless, reasons to believe that biological factors may have modified the aerobic power in the subjects studied. Since $\dot{V}_{O_2}$ must

---

Figure 9-5
Upper figure: Maximal oxygen uptake measured during cycling or running in 227 female and male subjects four to thirty-three years of age in relation to body weight. For male subjects the exponent $b = 0.76$ in the equation maximal $\dot{V}_{O_2} = a \cdot M$. (For male subjects, $y = -0.108 = 0.060\,x$; $r = 0.980 \pm 0.004$; deviation from regression line $= 7.5$ percent.)
Lower figure: Maximal oxygen uptake for 94 of the same subjects, age seven to thirty years, in relation to total amount of hemoglobin. In the equation $\dot{V}_{O_2} = a \cdot Hb^b$, the exponent $b = 0.76$. (For all subjects, $r = 0.970 \pm 0.006$; $2 \cdot SD$ within shadowed areas. For the determination of total hemoglobin, a CO method was used and the absolute values may be doubtful.) The subjects were all fairly well trained, and none of them was overweight. *(Modified from P.-O. Åstrand, 1952.)*

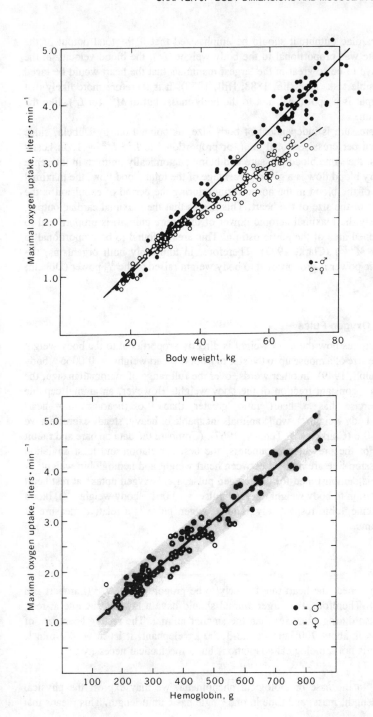

be related to cardiac output, it should be emphasized that if the total output of the heart per minute was proportional to the body weight (L^3), the blood velocity in the aorta would have to be so great in the largest mammals that the heart would be faced with an impossible task (Hoesslin, 1888; Hill, 1950). It is therefore more likely that the cardiac output is proportional, not to the body mass, but to $M^{2/3}$ (or L^2), like the basal metabolism.

The blood pressure is independent of body size, as pointed out by Döbeln, since pressure is force per cross-sectional area, or proportional to $L^2 \cdot L^{-2} = 1$. In hearts working against the same blood pressure and being anatomically uniform in the sense that the coronary blood flow is a given percentage of the total blood flow, the maximal linear velocity of the blood in the aortic ostium during the period of expulsion is the same regardless of the size of the heart. This means that the maximal cardiac output, and consequently the maximal aerobic power of the entire animal, is proportional to the cross-sectioned area of the aortic ostium. This area is found to be proportional to L^2 (or actually $M^{0.72}$) (Clark, 1927). Therefore, in uniformly built organisms, the maximal aerobic power is proportional to body weight raised to the $^2/_3$ power (Döbeln, 1956b).

Heart Weight, Oxygen Pulse

Figure 9-6 illustrates how the heart weight is directly proportional to the body weight in mammals the size of a mouse up to the size of a horse (heart weight $= 0.0066 \cdot$ body weight$^{0.98}$; Adolph, 1949). In other words, over the full range of mammalian size, the heart weight is a constant fraction of the body weight. However, an animal capable of severe exercise has a heart ratio greater than 0.6 (heart ratio = heart weight $\cdot$ 100 $\cdot$ body weight^{-1}), while animals incapable of heavy, steady exercise have ratios less than 0.6 (Clark, 1927; Tenney, 1967). (Compare the data on hare and rabbit in Fig. 9-6.) For the presented parameters, the best correlation and least deviation from the regression line are noticed between heart weight and hemoglobin weight. A similar picture is demonstrated for the oxygen pulse, i.e., oxygen uptake at rest/heart rate, and its relation to body weight (oxygen pulse $= 0.061 \cdot$ body weight$^{0.99}$), blood volume, and hemoglobin respectively. Thus, oxygen pulse is a relative measure of the stroke volume.

Heart Rate

As already mentioned, the heart rate is likely to be proportional to L^{-1} (Lambert and Teissier, 1927). Therefore, the larger animal should have a lower heart rate at rest, and possibly also during exercise, than the smaller animal. The resting heart rate of the 25-g mouse is about 700, and of the 3,000-kg elephant, it is 25 beats $\cdot$ min^{-1}. This difference is not a biological adaptation, but a mechanical necessity.

Summary In the case of biological phenomena, we may express the physical basic units of length, mass, and time in one single basic unit: length. This means that

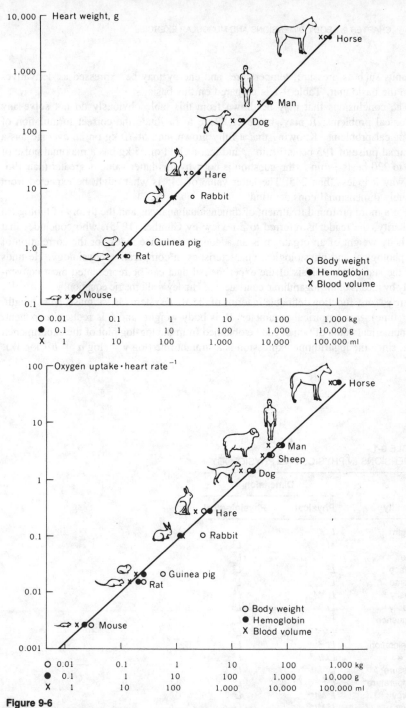

Figure 9-6
Heart weight (upper figure) and oxygen uptake per heartbeat (oxygen pulse) in relation to body weight, total hemoglobin weight, and blood volume, respectively, in various mammals. *(From various sources summarized by Sjöstrand, 1961.)*

all units such as pressure, temperature, and energy may be expressed as derivatives from the basic unit. Table 9-1 is prepared on this basis.

The conclusions that may be drawn from this table obviously do not solve any biological problems. It may, however, serve to facilitate the correct formulation of biological problems. Knowing that a fully grown man of 70 kg on an average has a maximal pulse of 195 beats $\cdot$ min^{-1}, and that a child of 35 kg has a maximal pulse of 210 to 220 beats $\cdot$ min^{-1}, the question is not why this latter value is greater than 195, but why it is less than 245. The latter value of 245 is what might be expected from a purely dimensional consideration.

For a more profound treatment of dimensional analysis and the theory of biological similarity, the reader is referred to a review by Günther (1975), who concludes that the body weight of an organism is an adequate reference index for the correlation of morphological and physiological characteristics in comparative physiology. He finds that the statistical analysis of the experimental data can be represented most conveniently by means of the logarithmic equivalent of Huxley's allometric equation: $y = a \cdot W^b$, where y is any function definable in terms of the MLT system (M = mass, L = length, T = time), a is empirical parameter, W is body weight, and b is reduced exponent, the numerical value of which can be obtained from the log-log plot of the experimental data, since the logarithmic expression is a straight line ($\log y = \log a + b \cdot \log W$).

TABLE 9-1
DIMENSIONS IN PHYSICS AND PHYSIOLOGY

	Dimension	
Quantity	Physical	Physiological
Length	L	L
Mass	M	L^3
Time	t	L
Surface	L^2	L^2
Volume	L^3	L^3
Density	$L^{-3}M$	L^0
Velocity	Lt^{-1}	L^0
Frequence	t^{-1}	L^{-1}
Flow	L^3t^{-1}	L^2
Acceleration	Lt^{-2}	L^{-1}
Force	LMt^{-2}	L^2
Pressure	$L^{-1}Mt^{-2}$	L^0
Temperature*	L^2t^{-2}	L^0
Energy	L^2Mt^{-2}	L^3
Power	L^2Mt^{-3}	L^2

*Physical dimension from J. C. Georgian, The Temperature Scale. *Nature*, 201:695, 1964.
Source: By courtesy of W. v. Döbeln.

Secular Increase in Dimension

We shall now consider a few additional applications of the effect of dimensions on human performance. Since, in many countries, man has been growing taller in recent generations, some improvement in athletic performance is to be expected. This steady secular increase in growth is typical of countries with a satisfactory nutritional status. Besides the increase in such bodily dimensions as height and weight at all ages from birth to adulthood, the maturation of certain physiological functions, notably those connected with sexual maturity, is also accelerated. There has been a steady decrease in age of menarche, from about seventeen years of age in 1840 to thirteen and a half years of age in 1960 (Tanner, 1962). A similar trend of earlier maturation of boys is also apparent from the available data; boys now reach their maximal height at an earlier age than they did a generation ago. The influence of sexual maturity on performance is evident from Fig. 9-2. The rapid increase in height and weight that is accompanied by puberty, generally occurs earlier in girls than in boys, but is at any rate subject to marked individual variation (Tanner, 1962). Thus, the peak height velocity may in extreme cases occur in a nine-year old girl and in a sixteen-year old boy (Lindgren, 1978; Fig. 7-12). Since this adolescent growth spurt has a profound effect on physical performance, it may be unfair, if not harmful, to group children athletically in classes according to age. Asmussen and his collaborators (1955, 1964, 1967) point out that although the height and weight for children of a given age have shifted upward during recent years, the weight-height curves have remained practically unaltered during the last couple of decades. It is also a fact that in spite of the general increase in height and weight of Olympic athletes during the last thirty years, their body proportions are remarkably constant. It may therefore be assumed that even with the increased dimensions, the present-day taller athletes are geometrically no different from those of earlier generations.

If the height of an athlete is 184 cm (the mean height of the participants in decathlon in Rome in 1960), and the height of an athlete thirty years before was 176 cm (mean height of a decathlon athlete in 1930), their heights will compare as 1.06:1, and their muscle strength as 1.13:1 (Asmussen, 1964). This means that, owing to different dimensions, the average top athlete now is 6 percent taller than the top athlete of thirty years ago, but his muscular strength should be 13 percent greater than thirty years ago. Therefore, the maximal work that the muscles of the athlete in decathlon could perform should be 20 percent greater than thirty years ago, for the maximal work a muscle can produce is the product of its maximal force and the distance it can shorten. When the size of the oxygen-transporting organs is the limiting factor, the taller athlete should consequently be able to deliver 13 percent more oxygen to his muscles per unit of time than the smaller athlete.

In such events as throwing the javelin or putting the shot, the increase in bodily dimensions may influence the achievements in two ways: In the first place, the strength of the athlete increases in proportion to the second power of the person's height. This will tend to improve the results, particularly since the weight of the equipment is constant and not varied with the weight of the thrower. Secondly, the greater height from which the javelin and shot start their flight will cause them to travel further.

These two factors, and the first one in particular, would result in better records, and may partially account for the improvements in records that have taken place. Anyway, there is clearly a good physiological basis for the selection of tall throwers.

This discussion is included to demonstrate that in some events, part (but only part) of the improvement in results over the years may be due to the athletes' dimensional change. Khosla (1968) finds that the winners in different throwing events in the Olympic Games in Rome and Tokyo were on the average definitely taller and heavier than their competitors. Similarly, the winners in jumping and running, with the exception of 10,000-m and marathon running, were taller. This, he claims, is unfair to people who are less tall, and suggests the classification of the competitors according to height and weight in events in which body size may influence the results.

Old Age

Figure 9-7(a) summarizes data from the literature on different static and dynamic dimensions in adults from twenty up to sixty years of age. It is evident that a strict interrelation of these functions based on dimensions alone does not occur. The body height is maintained constant, but body weight, heart weight, and heart volume increase with age. Blood volume and total amount of hemoglobin are not markedly changed. Heart rate at a given submaximal work load is the same in the old and the young, i.e., the oxygen transport per heartbeat (oxygen pulse) is constant. However, maximal oxygen uptake, heart rate, stroke volume, pulmonary ventilation, and muscular strength decrease significantly with age.

As is evident from Fig. 9-7(b), the urinary elimination of creatinine is reduced by about 30% from the age of thirty to age seventy five, which indicates a reduction of total muscle mass of about the same order. Concomitantly with this muscular atrophy, there is a corresponding decline in muscle strength, i.e., about a 30 percent drop in the strength of the back muscles, as indicated in Fig. 9-7(b).

SUMMARY

We have given a number of examples proving that static and dynamic functions in animals of different sizes in many cases have dimensions which are mechanically meaningful and desirable. The strength of a muscle is adjusted to the strength of bones, tendons, joints, connective tissue, and the muscle itself as a matter of safety. Furthermore, it provides a reasonable efficiency of the movements. There is, in general, a remarkable adjustment of the links involved in the chain of oxygen supply and energy output, so that none of the individual links is much stronger or weaker than necessary. On the other hand, there are examples showing that organisms of different size are not in all respects similar and uniform; there are deviations from the general trend. Deviations may sometimes be of a physical nature; in larger animals, for example, the weight of the skeleton is relatively high. It is also known that training may markedly improve physical performance. This improvement is sometimes accompanied by changes in organic dimensions, but this does not always occur (see Chap. 10). The performance of children is lower than expected from their dimensions, which clearly indicates that

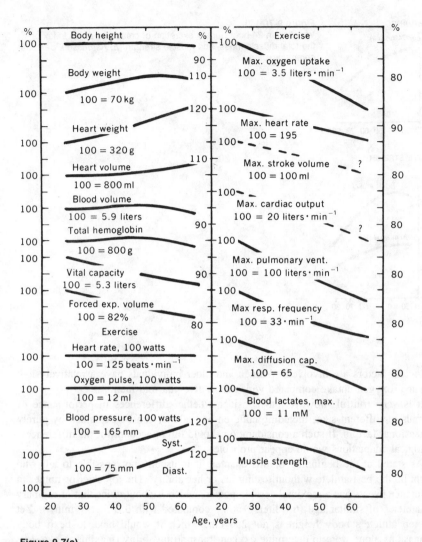

Figure 9-7(a)
Variation in some static and dynamic functions with age. Data have been collected from various studies, including healthy male individuals. For data on the same function, only one study was consulted. The values for the twenty-five-year-old subjects = 100 percent; for the older ages, the mean values are expressed in percentage of the twenty-five-year-old individuals' values. The mean values cannot be considered as "normal values," but their trends illustrate the effect of aging. Note that the heart rate and oxygen pulse at a given work load (100 watts or 600 kpm · min⁻¹, oxygen uptake about 1.5 liters · min⁻¹) are identical throughout the age covered, but the maximal oxygen uptake, heart rate, cardiac output, etc., decline with age. The data on cardiac output and stroke volume are based on relatively few observations and are therefore less certain.

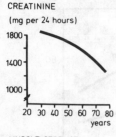

CREATININE
(mg per 24 hours)

Figure 9-7(b)
Variation in 24-hour urinary excretion of creatinine as an index of the total muscle mass, and muscle strength *(E. Asmussen, personal communication.)*

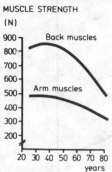

MUSCLE STRENGTH
(N)

biological factors are involved. Women and older individuals have a relatively low maximal oxygen uptake compared with twenty-five-year-old men.

It is very fruitful, however, to consider whether differences in performance of animals of different size, including children and adults, can be explained by purely dimensional factors. If such a consideration should fail to account for the differences, biological adaptations would appear probable.

As stated at the beginning of this chapter, it is usually not possible to tell who might be the best athlete without testing his or her ability. The top skier or runner in endurance events has a maximal aerobic power, which is about twice that of ordinary nonathletes of similar age (6.0 liters · min^{-1} compared with 3.0 liters · min^{-1}). Yet the top athlete's body height is not 250 cm, which it would have to be if body dimensions alone were to determine oxygen-transporting ability (maximal $\dot{V}_{O_2} \propto L^2$). In fact, these top athletes are of average height, and some of their dimensions are very similar to those of an average person, while others are different (see Chap. 10).

REFERENCES

Adolph, E. F.: Quantitative Relations in the Physiological Constitution of Mammals, *Science,* **109:**579, 1949.

Asmussen, E.: Growth and Athletic Performance, *FIEP,* **34**(4)**:**22, 1964.

Asmussen, E., and K. Heeböll-Nielsen: A Dimensional Analysis of Physical Performance and Growth in Boys, *J. Appl. Physiol.* **7:**593, 1955.

Asmussen, E., and E. H. Christensen: "Kompendium i Legemsövelsernes Specielle Teori," Köbenhavns Universitets Fond til Tilvejebringelse af Läremidler, Köbenhavn, 1967.

Åstrand, P.-O.: "Experimental Studies of Physical Working Capacity in Relation to Sex and Age," Ejnar Munksgaard, Copenhagen, 1952.

Borelli, J. A.: De Motu Animalium, Lugduni in Batavis, Apud Danielem à Guerbeeck . . . , 1685.

Brody, S.: "Bioenergetics and Growth," Reinhold Book Corporation, New York, 1945.

Clark, A. J.: "Comparative Physiology of the Heart," University Press, Cambridge, England, 1927.

Denny, M.: Locomotion: The Cost of Gastroped Crawling, Science, 208(4449):1288, 1980.

Diamond, J.: The Biology of the Wheel, Nature, 302(5909):572, 1983.

Döbeln, W. v.: Human Standard and Maximal Metabolic Rate in Relation to Fat-free Body Mass, Acta Physiol. Scand., 37(Suppl. 126):1956a.

Döbeln,, W. v.: Maximal Oxygen Intake, Body Size, and Total Hemoglobin in Normal Man, Acta Physiol. Scand., 38:193, 1956b.

Döbeln, W. v.: Kroppsstorlek, Energiomsättning och Kondition, in G. Luthman, U. Åberg, and N. Lundgren (eds.), "Handbok i Ergonomi," Almqvist & Wiksell, Stockholm, 1966.

Galilei, G.: Discorsi et Dimonstrazioni Matematiche, Interne à due Nueve Scienze, Elzévir, 1638.

Günther, B.: Dimensional Analysis and Theory of Biological Similarity, Phys. Rev. 55:659, 1975.

Heglund, N., T. A. McMahon, and C. R. Taylor: Scaling Stride Frequency and Gait to Animal Size: Mice to Horses, Science 186:1112, 1974.

Hill, A. V.: The Dimensions of Animals and Their Muscular Dynamics, Proc. Royal Inst. Great Britain, 34:450, 1950.

Hoesslin, H. v.: Ueber die Ursache der scheinbaren Abhängigkeit des Umsatzes von der Grösse der Körperoberfläche, Arch. F. Anat. Physiol., Physiol. Abs., p. 323, 1888.

Huesner, A. A.: Energy Metabolism and Body Size, I Is the 0.75 Mass Exponent of Kleiber's Equation a Statistical Artifact?, Resp. Physiol., 48:1, 1982.

Khosla, T.: Unfairness of Certain Events in the Olympic Games, Brit. Med. J., 4:111, 1968.

Lindgren, G.: Growth of Schoolchildren with Early, Average and Late Ages of Peak Height Velocity, Ann. Human Biol., 5(3):253, 1978.

Lambert, R., and G. Teissier: Théorie de la Similitude Biologique, Ann. Physiol., 3:212, 1927.

McMahon, T. A.: Using Body Size to Understand the Structural Design of Animals: Quadrupedal Locomotion, J. Appl. Physiol., 39:619, 1975.

Schmidt-Nielsen, K.: Energetic Cost of Locomotion: Swimming, Running and Flying, Science, 177:222, 1972.

Sjöstrand, T.: Relationen Zwischen Bau und Funktion des Kreislaufsystems Unter Patholo-gischen Bedingungen, Forum Cardiologicum, Boehringer & Soehne, Mannheim-Waldhof, Heft 3, 1961.

Tanner, J. M.: "Growth and Adolescence," Blackwell Scientific Publications, Ltd., Oxford, 1962.

Taylor, C. R., K. Schmidt-Nielsen, and J. L. Raab: Scaling and Energetic Cost of Running to Body Size in Mammals, Am. J. Physiol., 219:1104, 1970.

Tenney, S. M.: Some Aspects of the Comparative Physiology of Muscular Exercise in Mammals, Circulation Res., 20:1–7, 1967.

Thompson, D. W.: "On Growth and Form," Cambridge University Press, New York, 1943.

Weibel, E. R. and C. R. Taylor: Design of the Mammalian Respiratory System. Respir. Physiol. 44:1, 1981.

CHAPTER 10

PHYSICAL TRAINING

CONTENTS

INTRODUCTION

TRAINING PRINCIPLES
  Continuous Versus Intermittent Exercise

TRAINING METHODS AND BIOLOGICAL LONG-TERM EFFECTS OF
  TRAINING
  Locomotive Organs:
  Muscle Strength
  Muscle Mass
  Disuse
  The Role of the Central Nervous System
  Which is the Most Effective Strength Training Program?
  Isometric Strength Training,
  Dynamic Strength Training
  Do Anabolic-Androgenic Steroids Improve Strength?
  Training of Anaerobic Power
  "The Dallas Study"
  Training of Aerobic Power:
  Circuit Training
  Year-Round Training
  Tests
  Oxygen-Transporting System
  Peripheral Adaptation to Aerobic Training

Peripheral Circulation
How Does Endurance Training Affect the Muscle Fiber Composition and
   Metabolic Profile?
Endurance
Detraining
Specificity of Training
Recovery After Exercise
Mechanical Efficiency, Technique
Body Composition
Blood Lipids and Lipoproteins
Hormones
MISCELLANEOUS:
Does Hypoxia Enhance the Training Effect?
Training and Tolerance to Hot Environments
Stretching
Training and Oxygen Radicals
Training and Terminal Nerve Fiber Sprouting
Psychological Aspects
Muscular Soreness
ISCHEMIC HEART DISEASE
CONTRAINDICATIONS FOR PHYSICAL TRAINING

INTRODUCTION

In several sections of this book, it has been shown that different organs or organ systems may be affected by a variety of factors. The effect may be transitory or it may last for a considerable period of time; that is, an adaptation takes place. Figure 7-1, for example, illustrates that the aerobic muscular capacity is affected by training, deconditioning, and acclimatization (altitude, heat, cold). In this chapter, we shall discuss how physical training affects the body morphologically as well as functionally.

Conclusions concerning the effect of physical training have often been drawn from studies of well-trained persons, and the data obtained have been compared with similar data from studies on sedentary individuals. The disadvantage of such *cross-sectional studies* is that it may be impossible to determine whether any difference observed depends on constitutional dissimilarities or on the training as such. It is in any case quite obvious that the great maximal aerobic power which is characteristic for the top athlete in endurance largely depends on endowed organic advantages. Thus, a person with a maximal oxygen uptake of 45 ml $\cdot$ kg^{-1} $\cdot$ min^{-1} cannot, under any circumstance, no matter how well trained, attain a maximal oxygen uptake of 80 ml $\cdot$ kg^{-1} $\cdot$ min^{-1}, which is required for Olympic medals in certain sport events (Fig. 10-1). It may not be quite fair, but it is nevertheless a fact that the "choice of parents" is important for athletic achievement (see Klissouras, 1976; Bouchard and Lortre, 1984)

For these reasons *longitudinal studies* must be designed, in which the same individual is followed for shorter or longer periods of time. However, such studies are rare. Since they are difficult to perform, usually only a few subjects have been included in each study. Even this approach, with long observation periods of the same individuals, is not without objections from a scientific point of view:

1 If the objective is to examine such problems as the effect of physical training, the selection of subjects is critical. If the selection is not done on a strictly random basis, the material may not be representative. If a selection is done from a group of volunteers, it must be remembered that those who volunteer for such studies may do so for special reasons which may be medical, social, psychological, or personal. In any event, such volunteers may not be representative of the population as a whole.

2 Any such study may well represent an intervention of the subject's normal pattern of life. This may in itself bring about different effects: the subjects may perhaps change

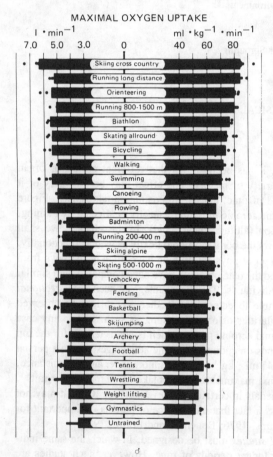

(a)

Figure 10-1
Average maximal oxygen uptake in liters per minute (left part) milliliters per kilogram body weight times minutes for male (a) and female (b) Swedish national teams in different sports. Dots indicate individual values higher than the mean value. *(Compiled from various sources by Ulf Bergh.)*

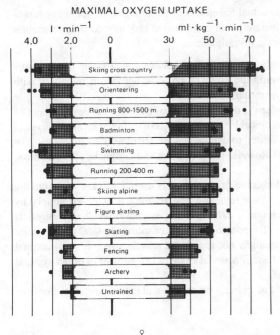

MAXIMAL OXYGEN UPTAKE

$l \cdot min^{-1}$ $ml \cdot kg^{-1} \cdot min^{-1}$

4.0   2.0   0   3U   50   70

Skiing cross country
Orienteering
Running 800-1500 m
Badminton
Swimming
Running 200-400 m
Skiing alpine
Figure skating
Skating
Fencing
Archery
Untrained

♀

(b)

Figure 10-1
(cont.)

their diet, smoking habits, or other habits which may elicit effects on the organism which, per se, have nothing to do with the training. It may also be difficult to avoid affecting the control group by the mere fact that the group is being studied and therefore the focus of special attention. In studying patients, arriving at a clear-cut experimental situation is especially complicated in that ethical considerations, in such cases especially, may be brought into the foreground.

3 Experience has shown that the drop-out frequency is rather high among subjects of prolonged training studies.

4 It is quite difficult to establish a person's physical condition objectively prior to the training, and any training effect obviously depends on the initial level at the start of the training. As mentioned on several occasions, at present no method of investigation is available which can definitely reveal and separate the influence of constitutional factors on the one hand, and the effects of training on the other. One is therefore largely forced to rely on the individual's own statements. However, what one person may characterize as a physically active life may, to another, represent a sedentary life. Statements as to occupational and recreational activities may in themselves represent only indications rather than precise, meaningful information regarding degree of physical activity, etc. The duration and intensity of such activities are important factors to be considered.

5 The intensity (tempo) of the actual training is difficult to assess, define, or reproduce. Moreover, it is a time-consuming and complicated task to record this variable, especially in the case of large-scale investigations.

On the basis of these considerations, it is understandable that different investigators arrive at different results concerning the effect of physical activity or inactivity with regard to qualitative as well as quantitative changes.

In this chapter we shall consider (1) the physiological basis for the development of a training program and (2) the biological long-term effects of different levels of physical activity. The discussion will be limited to data which are fairly well documented. (For further discussion, see Saltin and Rowell, 1980; Matoba and Gollnick, 1984; Strauss, 1984.)

When discussing training and training principles, it is essential to keep in mind the specific purpose of the training. As a rule, the best training is achieved simply by carrying out the activity for which one is training. As is evident from Chap. 7 physical training may influence a number of the factors which constitute physical performance capacity; that is, it may cause changes not merely in muscle strength and maximal oxygen uptake, but also structural and functional changes in a number of organ systems as well as psychological changes.

Table 10-1 summarizes effects of training on organs and organ functions. Some of

TABLE 10-1
EFFECTS OF TRAINING ON ORGANS AND ORGAN FUNCTIONS*

Organ or function	Increase	Decrease	No effect	References
Locomotive organs				
Strength of bones and ligaments	x			Viidik, 1966; Tipton et al., 1974; Booth and Gould, 1975; White et al., 1984
Thickness of articular cartilage	x			Holmdahl and Ingelmark, 1948
Muscle mass (hypertrophy)	x		x	Marpurgo, 1897; Saltin and Gollnick, 1983
Number of muscle cells			x	Gollnick et al., 1981
Fiber composition			?	Howald, 1982; Salmons and Henriksson, 1981; Schantz and Henriksson, 1983
Muscle strength	x			Clarke, 1973; Atha, 1981; Maughan, 1984
ATP, phosphocreatine muscle	x		x	Yakovlev, 1977; Saltin and Gollnick, 1983

TABLE 10-1
(cont.)

Organ or function	Increase	Decrease	No effect	References
"Anaerobic" enzyme activity in muscle	x		x	Roberts et al., 1982; Saltin and Gollnick, 1983
Oxidative enzyme activity in muscle	x			Newsholme and Leech, 1983; Saltin and Gollnick, 1983
Myoglobin	x			Whipple, 1926; Yakovlev, 1977; Hickson, 1981
Capillary density, muscle	x			Hudlická, 1982; Saltin and Gollnick, 1983; Fig. 10-13
Arterial collaterals, muscle	x			Schoop, 1966
Buffer capacity, muscle	x			Sahlin and Henriksson, 1984
Circulation				
Heart volume	x		x	Rost and Hollmann, 1983; Blomqvist and Saltin, 1983; Fig. 10-9
Heart weight	x		x	Siebert, 1929; Rost and Hollmann, 1983
Capillary density, heart	x			Petrén et al., 1936; Hudlická, 1982
Coronary collaterals	?		x	Blomqvist and Saltin, 1983; Haskell, 1984
Blood volume, total hemoglobin	x			Taylor et al., 1949; Greenleaf, 1984
Buffer capacity			x	Sharp et al., 1983
Hemoglobin concentration		x	x	Clement and Sawchuk, 1984; Hallberg and Magnusson, 1984
Plasma protein concentration			x	Sharp et al., 1983
Cardiac output:				
rest			x	Blomqvist and Saltin, 1983
submaximal exercise		?	x	Blomqvist and Saltin, 1983
maximal exercise	x			Blomqvist and Saltin, 1983; Fig. 10-8
Heart rate:				
rest		x		Steinhaus, 1933; Israel, 1982

(continued)

TABLE 10-1
(cont.)

Organ or function	Increase	Decrease	No effect	References
Heart rate (cont):				
submaximal exercise		x		Christensen, 1931; Israel, 1982; see "Cardiac Output"
maximal exercise		?	x	Robinson and Harmon, 1941a; Ekblom, 1969; Kilbom, 1971
Stroke volume:				
rest	x			see "Cardiac Output"
submaximal exercise	x			see "Cardiac Output"
maximal exercise	x			see "Cardiac Output"
a-$\bar{v}O_2$ difference:				
rest			x	see "Cardiac Output"
submaximal exercise	?		?	see "Cardiac Output"
maximal exercise	x		x	see "Cardiac Output" Fig. 10–8
Oxygen uptake:				
rest			x	P.-O. Åstrand, 1956
submaximal exercise		x	x	see text
maximal exercise	x			Robinson and Harmon, 1941b; P.-O. Åstrand, 1956; Lortie et al., 1984; Seals et al., 1984b; Fig. 10-8
Blood lactate concentration:				
rest			x	see text
submaximal exercise		x		P.-O. Åstrand, 1956; Denis et al., 1982; Hurley et al., 1984; Fig. 10-5
maximal exercise	x		x	P.-O. Åstrand, 1956; Kilbom, 1971
Local blood flow, muscle:				
submaximal exercise		x		Kiens and Saltin, 1985
maximal exercise	x			Klausen et al., 1982; Blomqvist and Saltin, 1983
Arterial blood pressure:				
rest		?	x	Seals and Hagberg, 1984

TABLE 10-1
(cont.)

Organ or function	Increase	Decrease	No effect	References
submaximal exercise		?	?	Seals and Hagberg, 1984
maximal exercise	?		?	Ekblom, 1969
Respiration				
Lung volumes:				
adults			x	P.-O. Åstrand, 1956
adolescents	?		?	P.-O. Åstrand, 1956; Hamilton and Andrew, 1976
Pulmonary ventilation:				
rest			?	see text
submaximal exercise		x	x	see text
maximal exercise	x†			Ekblom et al., 1968
Tidal air:				
rest	?		x	see text
submaximal exercise	?		x	see text
maximal exercise	?		?	see text
Respiratory rate:				
rest		?	x	see text
submaximal exercise		?	x	see text
maximal exercise	x†			
Diffusing capacity:				
rest			x	Anderson and Shephard, 1968; Saltin et al., 1968; Hamilton and Andrew, 1976
submaximal exercise			x	see "rest"
maximal exercise	x†			see "rest"
Miscellaneous				
Body density	x			Skinner et al., 1964; Pařizkova, 1977; Wilmore, 1982
Blood:				
serum cholesterol		x	x	Altekruse and Wilmore, 1973
serum triglycerides		x	x	Skinner et al., 1964; Higuchi et al., 1984; Gaesser and Rich, 1984
high density lipoproteins	x		?	Gaesser and Rich, 1984; Kiens et al., 1984

*The references quoted here primarily refer to recent papers or review articles. References to earlier studies are usually included in the above-quoted publications.
†Secondary to the increase in maximal oxygen uptake.

the data were obtained on animals, others on humans. There are still many open questions, as indicated in the table. At the end of this chapter, the table will be discussed in some detail.

TRAINING PRINCIPLES

Physical training entails exposing the organism to a training load or work stress of sufficient intensity, duration, and frequency to produce a noticeable or measurable training effect, i.e., an improvement of the functions for which one is training. In order to achieve such a training effect, it is necessary to expose the organism to an overload, that is, to a stress which is greater than the one regularly encountered during everyday life. Generally speaking, it appears that exposure to the training stress is associated with some catabolic processes, such as molecular breakdown of stored fuel and other cellular components, followed by an overshoot or anabolic response that causes an increased deposition of the molecules which were mobilized or broken down during exposure to the training load.

The intensity of the load required to produce an effect increases as the performance is improved in the course of training. The training load is therefore relative to the level of fitness of the individual. The fitter a person is, the more it will take to improve that fitness. Finally, it becomes a matter of time and motivation to continue when the elite athlete has to devote several hours a day to training. The need for a gradual increase in training load with improved performance, in the case of the effect of heart rate, was demonstrated as early as in 1931 by Christensen. He observed that regular training with a given standard exercise rate gradually lowered the heart rate. Further training did not modify this heart rate response. After a period of training on a heavier load, the original standard rate of exercise could then be performed with a still lower heart rate. This general principle is apparent during training of a number of functions: *An adaptation to a given load takes place; in order to achieve further improvement, the training intensity has to be increased.* This principle has been elucidated by several studies summarized in other sections of this chapter. However, there is no linear relationship between amount of training and the training effect. For instance, 2 hr of training per week may cause an increase in maximal oxygen uptake, say by 0.4 liter $\cdot$ min^{-1}. If the training is twice as much, that is 4 hr per week, the increase in O_2 uptake will not be twice as great (0.8 liter $\cdot$ min^{-1}) but possibly 0.5 to 0.6 liter/min (see Fig. 10-2).

The major effects in a strength training program are triggered by the first seconds of the first maximal contraction at each session. The gain is most pronounced when the subject is in a low state of training; most of the gain is noticed during the first five weeks of a daily training program (see Atha, 1981).

Obviously, there is a limit to the increase and the rate and magnitude of the increase vary from one individual to the next.

The exact magnitude of the training load that will produce an optimal training effect is not established in general terms. It varies not only from one individual to another,

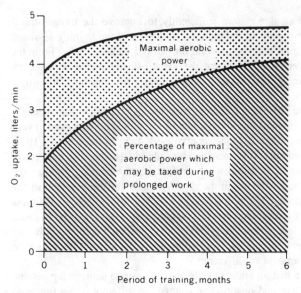

Figure 10-2
Training causes an increase in maximal oxygen uptake. With training a subject is also able to tax a greater percentage of his maximal oxygen uptake during prolonged work.

but also with age. On the whole, it appears that in athletic training, the greater the intensity the better, up to a certain limit.

For an improvement of the aerobic power in average sedentary persons, it appears that a training intensity in excess of approximately 50 percent of the individual's maximal aerobic power, which roughly corresponds to the rate of exercise causing the person to be slightly out of breath, may be sufficient to produce a significant improvement (see Kilbom, 1971; Pollock, 1973). It is commonly recommended that the exercise rate during training should gradually be increased to 70–80 percent of the individual's maximal aerobic power, i.e., approximately 75–85 percent of her or his maximal heart rate (see Fig. 8-10).

From these statements it is clear that physical activity is not synonymous with physical training, for the physical activity has to be maintained at a certain level of intensity in order to result in a training effect. In addition to a certain minimal level of intensity, the training stimulus must be of a certain duration in order to produce any training effect. Here again the relationship between training load and duration on the one hand, and training effect on the other, is not clearly established in general terms, and probably never will be, for it depends on which organ systems and what kind of training one is dealing with. Thus, while an increase in isometric muscle strength may be achieved by a few near-maximal contractions of a few seconds' duration once a day, such a brief exposure, even to a very intense physical exercise, may not affect the cardiovascular system sufficiently to improve the oxygen-transporting capacity. The same applies to the required frequency of the training sessions, i.e., how often one has to train in order to attain the desired training effect. From the available evidence, however, it is clear that the same effect may be obtained by relatively shorter daily training sessions as by much longer training sessions two or three times a week.

Older individuals (above fifty years of age) may be less trainable than younger ones. But it should be kept in mind that some effect of training is noticed even at very old age (See Liesen and Hollman, 1976; Seals et al., 1984a,b; Mazzeo et al., 1984; the latter group studying aged rats). It is important for individuals who wish to improve their general state of fitness to ascertain what amount of training may produce the most satisfactory result. One has to weigh the time available for training against the effects achieved by the training. If one has reached such a level of physical fitness that several additional hours of training per week would be necessary in order to attain a further improvement of a few percentage points or less, the small gain would hardly be worth the effort. One has to accept the fact that the daily fluctuations in one's state of fitness may exceed this amount of difference.

It is a general experience that an athlete needs several years to achieve top results. It is not clear what qualitative and quantitative changes in different organ functions may cause the slow, gradual improvement of the results. An analysis of the demands that a particular athletic event places on the body should form the basis for the training program, taking into consideration whatever deficiencies there may be in the athlete's resources or capabilities to meet these demands.

A certain amount of training of the oxygen-transporting organs is necessary for all categories of athletes, regardless of the nature of the athletic event. Thus, the individual will be better able to cope with the special training required for the event. Furthermore, even the warm-up prior to the event requires a certain amount of fitness. The general training for all individuals, irrespective of profession, age, and sex, should include: (1) training of the O_2-transporting function to improve maximal aerobic power and endurance; (2) improvement of muscle strength including the abdominal muscles (Chap. 6); (3) training aimed at maintaining joint mobility, the enhancement of the metabolism of the articular cartilage (Chap. 6), and the development of improved coordination (Chap. 3). In addition, (4) low-energy consumers should be stimulated to increase their metabolism through regular exercise, eventually becoming high-energy consumers (Chap. 12). As far as the O_2-transporting function is concerned, a distinction should be made between factors involved primarily in the heart and central circulation, and factors involved in the peripheral circulation. In regard to the *central circulation*, the training is effective and less strenuous if as large a muscle mass as possible is engaged in the training. In the case of *peripheral circulation*, it is a matter of training the muscles, which will be engaged in the performance of the type of event or activity, where an improvement is desired. There are many supporters of the assumption that the central circulation limits the maximal oxygen uptake in activities engaging large muscle groups. On the other hand, peripheral factors may be limiting factors in the case of endurance, defined as time to exhaustion at submaximal levels of oxygen uptake. (See Chaps. 4 and 7; Blomqvist and Saltin, 1983.)

Continuous Versus Intermittent Exercise

It has often been discussed whether physical training is most effective when the exercise is accomplished continuously or intermittently, i.e., with periods of more intensive muscular activity followed by periods of mild exercise or rest. In the following discussion, we shall present a summary of results from a few studies in which the physiological effects of these types of exercise are compared (I. Åstrand et al., 1960; Christensen et al., 1960).

One subject was made to accomplish a certain amount of work (635 kJ) in the course of 1 hr. This could be done either by uninterrupted exercise with a load of 175 watts, or by intermittent exercise with a heavier load, interrupted by rest periods at regular intervals. The double work rate (350 watts) was chosen; thus the required amount of work (635 kJ) could be accomplished by 30 min of exercise within the span of 1 hr. Exercising continuously without any rest periods, the subject could tolerate this high work rate for only 9 min, at the end of which he was completely exhausted. If, instead, he exercised for 30 s, rested for 30 s, exercise for 30 s, and so on, he could complete the work with moderate exertion. The longer the activity periods, the more exhausting the exercise appeared, even though the rest periods were correspondingly increased. Some of the results of these studies are summarized in Table 10-2. It appeared that with exercise periods of 3 min interrupted by 3-min rest periods, the load on the oxygen-transporting organs was maximal (oxygen uptake, 4.60 liters $\cdot$ min^{-1}, heart rate 188), and the degree of exertion was particularly high (blood lactate 13.2 mM).

At least one conclusion may be drawn from these experiments: For the purpose of taxing the oxygen-transporting organs maximally, exercise periods of a few minutes'

TABLE 10-2
DATA ON ONE SUBJECT PERFORMING 635 kJ (64,800 kpm) ON A CYCLE ERGOMETER WITHIN 1 HR WITH DIFFERENT PROCEDURES

Type of exercise		Oxygen uptake		Pulmonary ventilation	Heart rate	Blood lactic acid,
		liters $\cdot$ hr^{-1}	liters $\cdot$ min^{-1}	liters $\cdot$ min^{-1}	beats $\cdot$ min^{-1}	mM
Continuous						
175 watts		146	2.44	49	134	1.3
350 watts*			4.60	124	190	16.5
Intermittent						
350 watts						
Exercise	Rest					
½ min	½ min	154	2.90†	63†	150	2.2
1 min	1 min	152	2.93†	65†	167	5.0
2 min	2 min	160	4.40	95	178	10.5
3 min	3 min	163	4.60	107	188	13.2

*Could be performed for only 9 min.
†Measured during ½ min.
Source: I. Åstrand et al., 1960.

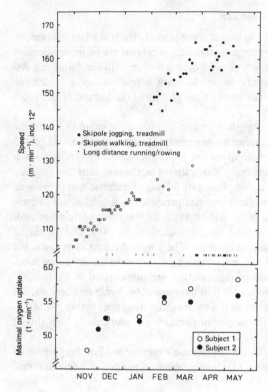

Figure 10-3
The effect of "ski-pole walking" and running on a treadmill (inclination, 12°) on the maximal oxygen uptake in two athletes. *(By courtesy of O. Vaage.)*

duration represent an effective type of activity. An example of the effect of intermittent exercise of this type is presented by Vaage (unpublished results, Fig. 10-3). Two athletes trained on the treadmill (inclination, 12°) almost daily for over 100 days. Walking with ski-poles, they started at a speed of about 100 m · min⁻¹, but later they ran as the treadmill speed was increased to over 140 m · min⁻¹ in order to attain maximal exertion. During the first period, they walked on the treadmill for 4 to 5 min three to five times in succession, with 5- to 9-min rest pauses between each exercise period. During the subsequent period, they alternated on different days between this very strenuous uphill treadmill walking and running in short bouts of about 20 s, interspersed with about 10-s rests for about 8 min. The whole 8-min sequence was repeated after 5- to 9-min rest periods, and altogether was repeated two to five times each day. This training was replaced by long-distance running or rowing for 40 to 150 min each time for about 30 days, as indicated by ′ in Fig. 10-3, showing the effect of this program on the maximal oxygen uptake of the two subjects.

Table 10-3 presents data on one of the subjects obtained in a study by Christensen et al. (1960) of intermittent treadmill running using shorter exercise periods. It is striking that when the subject ran for 10 s followed by a 5-s pause, he could run for 20 min during the 30-min period at a high speed without undue fatigue and with a

TABLE 10-3
DATA ON ONE SUBJECT DURING INTERMITTENT RUNNING FOR 30 MIN AT 20 KM/HR ON A
TREADMILL. BETWEEN THE RUNNING PERIODS, WHICH WERE VARIED, THE SUBJECT WAS
STANDING BESIDE THE TREADMILL. DURING CONTINUOUS RUNNING, HE COULD PROCEED
FOR 4.0 MIN, COVERING A DISTANCE OF ABOUT 1,300 M. OXYGEN UPTAKE: 5.6
LITERS · MIN^{-1}: PULMONARY VENTILATION: 158 LITERS · MIN^{-1}; BLOOD LACTIC ACID
CONCENTRATION 16.5 mM

Periods exercise-rest, s	Distance, m	Oxygen uptake, liters · min^{-1} Exercise Highest	Average	Rest	Pulmonary ventilation liters · min^{-1} Exercise Highest	Average	Rest	Blood lactate, mM
5–5	5,000	. . .	4.3	4.5	. . .	101	101	2.5
5–10	3,330	. . .	3.4	3.0	. . .	81	77	1.8
10–5	6,670	5.6	5.1	4.9	157	142	140	4.8
10–10	5,000	4.7	4.4	3.8	109	104	95	2.2
15–10	6,000	5.3	5.0	4.5	140	130	144	5.6
15–15	5,000	5.3	4.6	3.8	110	90	95	2.3
15–30	3,330	3.9	3.6	2.8	96	79	64	1.8

Source: Data from Christensen et al., 1960.

low blood lactic acid concentration. At the end of each running period, the load on
the oxygen-transporting system was maximal, or 5.6 liters · min^{-1}. On the average,
the oxygen uptake when the subject was running was 5.1 liters, but the oxygen
requirement can be calculated to be 7.3 liters per exercise minute. How the deficit of
about 46 kJ, $(7.3\text{-}5.1) \cdot 21$, is covered is still an open question. The low blood lactate
concentration indicates that anaerobic glycogenolysis was not the important energy
supplier. The high energy-containing phosphate compounds may have served as a
buffer, supported by aerobic processes utilizing oxygen bound to the myoglobin in
addition to the amount of oxygen transported during the running. It should be noted
that in all the experiments, the oxygen uptake and pulmonary ventilation were also
high during the interspersed resting periods. However, the results indicate that the
duration and spacing of exercise and resting periods are rather critical with respect to
the peak load on the oxygen-transport system. If the resting period of 5 s is prolonged
to 10 s (running for 10 s), the peak oxygen uptake observed will be reduced from 5.6
to 4.7 liters · min^{-1}. Running at the same speed for 15 s, then resting for 15 s, did
not bring the oxygen uptake (5.3 liters · min^{-1}) to a maximum.

Figure 10-4 presents another example of how critical the exercise intensity is for
the load on the oxygen-transporting organs in intermittent exercise with activity periods
of short duration. During running at a speed of 22.75 km · hr^{-1}, for 20 s, 10-s rest,
followed by running, and so on, the oxygen uptake becomes maximal. If the speed
is reduced to 22.0 km · hr^{-1}, the oxygen uptake is reduced to about 90 percent of the
maximum. On the other hand, the subject can continue for about 60 min at the lower
speed as against only 25 min at the higher speed of 22.75 km · hr^{-1}. It is an important
but unsolved question which type of training is most effective: to maintain a level

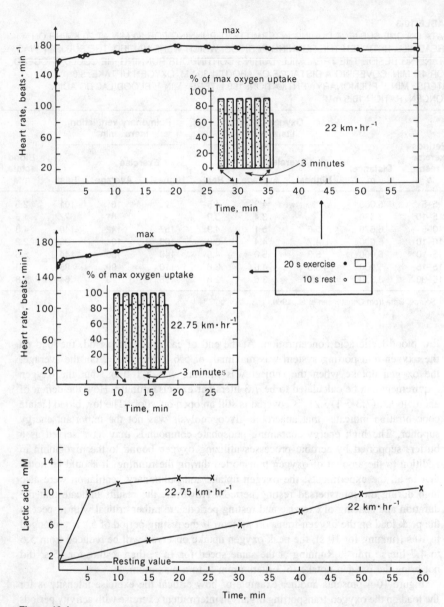

Figure 10-4
Oxygen uptake, heart rate, and blood lactate concentration during a training program, running on the treadmill, with short exercise and rest periods (20 and 10 s respectively) at speeds of 22.0 km · hr⁻¹ (upper panel) and 22.75 km · hr⁻¹ (lower panel). Note that the heart rate and oxygen uptake are not maximal in the first case (22.0 km · hr⁻¹) but that they are in the second case (22.75 km · hr⁻¹). It should also be noted that the total exercise time is reduced to half in the latter case. The lactic acid concentration in the blood had reached about the same level in both cases when the exercise had to be discontinued. *(From Karlsson et al., 1967b.)*

representing 90 percent of the maximal oxygen uptake for 40 min, or to tax 100 percent of the oxygen uptake capacity for about 16 min.
Other examples are presented in Chap. 7; e.g., see Fig. 7-5.

Summary A series of studies have shown that maximal oxygen uptake (and cardiac output) may be attained in connection with repeated periods of exercise of very high intensity of as short duration as 10 to 15 s, provided the rest periods between each burst of activity are very short (of equal or shorter duration than the exercise periods). In more prolonged exercise of several minutes duration, the duration of the rest periods is less critical. If the activity periods exceed about 10 min or so, a high level of motivation is required in order to attain maximal oxygen uptakes. In the case of continuous activity, the high tempo required for the severe taxation of the oxygen-transporting system must alternate with periods of reduced tempo.

With very short exercise periods of about 30 s or less, a very severe load may be imposed upon both muscles and oxygen-transporting organs without the engagement of anaerobic processes leading to any significant elevation of the blood lactate. It is thus possible to select the proper load and exercise and rest periods in such a manner that the main demand is centered on (1) muscle strength without a major increase in the total oxygen uptake; (2) aerobic processes without significantly mobilizing anaerobic processes; (3) anaerobic processes without maximal taxation of the oxygen-transporting organs; and (4) both aerobic and anaerobic processes simultaneously. The alternatives 2 and 4 do not entail maximal taxation of muscle strength; alternative 3 does not necessarily require maximal strength. In the following we shall consider how these principles may be applied.

TRAINING METHODS AND BIOLOGICAL LONG-TERM EFFECTS OF TRAINING

In this section, we shall discuss in some detail the various long-term training effects, indicated in Table 10-1. It should be emphasized that by long-term effects, we mean changes that will require a certain amount of time (weeks, months, or years) to be developed. Furthermore, available training methods, based on scientific data or experience, for achieving such effects, are discussed.

Locomotive Organs

In Chap. 6 it was pointed out that *bone, ligament,* and *joint cartilage* are affected by use as well as by disuse. Bone structures that are not stressed may disappear and new bone trabeculae may be created where altered mechanical forces increase the demand for sturdiness. The hard interstitial substance consists of carbonates and phosphates of calcium, and may constitute up to about 60 percent of the dry, fat-free weight in adult life. After a long period of inactivity, it may be as low as about 40 percent (Ingelmark, 1957). The increased urinary calcium and phosphorus elimination during prolonged bed rest depends on a demineralization of the bones (Deitrick et al., 1948; Rodahl et al., 1967; Greenleaf, 1984).

A reduction in bone density is found not only as a consequence of prolonged bed rest, but also in connection with immersion in water or weightlessness during space flights. The common denomination is a reduced compression force on weight-bearing bones (see Greenleaf, 1984). It is, therefore, not surprising that an increased compression force, applied with springs along the longitudinal axis of the long bones during bed rest, can markedly reduce the urinary calcium output (Issekutz et al., 1966). One problem for the aging person is often a reduced bone density, i.e., osteoporosis; postmenopausal women are particularly vulnerable. There are data available indicating that habitual physical exercise can reduce the degree of osteoporosis (see White et al., 1984; see also Chap. 6).

The thickness of the articular cartilage is greater in trained animals than in untrained ones. The compressability of the cartilage will therefore increase with training, providing greater possibilities of compensating for any incongruence of the articular surfaces of the cartilages. The contact surface will increase, and when stressed, the force per unit of surface will thus decrease (Holmdahl and Ingelmark, 1948; Ingelmark, 1957). Training causes a hypertrophy of the intercellular substance of connective tissue, increasing the volume of tendons and ligaments, and enhances their tensile strength. The hypertrophy of the tendons is accompanied by a hypertrophy of the muscles attached to them (Ingelmark, 1948). In the bone-ligament-muscle system, the attachment of the ligament to the bone is the weakest point, although it tends to become stronger in trained animals (Viidik, 1966; Tipton et al., 1974). (For details see Booth and Gould, 1975.)

Muscle strength   was discussed in Chaps. 2 and 3. It was pointed out that there is a high correlation between the cross-sectional area of a muscle and its potential to produce strength, first documented by Ikai and Fukunaga (1970). However, there are factors other than the cross-sectional area of a muscle that are of significance for its maximal strength. Muscle fibers may be attached to the tendons in a manner representing a mechanical disadvantage, e.g., those with pennate fiber arrangements. Furthermore, the possibility of engaging the muscle by impulses from the central nervous system, is of importance. This leads us to the question: how much of the training effect on muscle strength is due to neural adaptations, and how much is due to adaptive changes in the skeletal muscle. From the following discussion it is evident that we do not know.

There are some recent reviews discussing various aspects of strength training (Atha, 1981; Sale and MacDougall, 1981; Maughan, 1984; Matoba and Gollnick, 1984). A high tension must be produced for strength development. It is, therefore, of interest to recall the physiological events illustrated by Fig. 2-13. The maximal tension that a muscle can develop is reached in a fast eccentric contraction (lengthening the muscle). An intermediate position is held by the isometric contraction, and the maximal tension becomes progressively reduced the faster a maximal concentric contraction is performed. For reasons given in Chap. 2, the terminology will be: *eccentric dynamic, isometric,* and *concentric dynamic contractions,* respectively. Under normal conditions, the muscles do not perform isotonic ("same tension") contractions, because the demand on the muscles involved when moving a given external resistance varies with the mechanical design. For example, when lifting a weight kept in the hand by an elbow flexion, the muscle's lever changes during the movement and the demand on the flexor muscles varies. With reference to Fig. 2-13, it is evident that in tests of the dynamic strength before and after a period of training, it is of vital importance to control the speed of contraction. For this purpose, a machine was constructed to control the speed of shortening of the activated muscle over a chosen range of movement (Thistle et al., 1967). (It should be pointed out that such an *isokinetic* muscle activation with a constant angular velocity of a limb segment, does not lead to a constant velocity of shortening of the muscles involved. This is particularly evident when it is a question of two-joint muscles and muscles that follow twisted paths. Furthermore, the centers of rotation of joint and apparatus cannot be kept coincidental.) The isokinetic exercise does not simulate natural movements, with the possible exception of swimming, and therefore the value of isokinetic training is difficult to evaluate. For the testing of maximal dynamic strength and endurance, the isokinetic testing apparatus is very useful.

A comparison of different protocols for *muscle strengthening,* in terms of effectiveness and efficiency, is complicated by the fact that it is difficult to compare the total load on the muscles involved, e.g., how to quantify the demands imposed by isometric and dynamic training programs, respectively. Furthermore, individuals respond differently when exposed to a given program, and when entering such a program they may be in different states of muscular fitness. In many studies, the number of subjects has been small, and this inevitably complicates a statistical evaluation of the results. Therefore, at present, discussions of optimal training regimes are based more on experience and observations than on hard scientific data. It is also important to analyze whether the goal of a strength training program is to become just physically fit, a champion in weight lifting, a good track and field athlete, a swimmer or to achieve bodybuilding. The following text will be limited to a general discussion of physiological effects of strength training.

Strength can be defined as the ability to develop force against a resistance in a single contraction of restricted duration (Atha, 1981). When training with weights in dynamic contractions, one talks about "nRM load" = number of repetitions maximal load. The weight is so chosen that it can be lifted "n" times in good style, but it is too heavy to lift "n + 1" times. (If 1RM is 100 percent, 3RM are 90–95 percent, 6RM approximately 85 percent, 10RM 75 percent, and 15 RM about 65 percent of the 1RM load.) With interspaced intervals, such a set is eventually repeated a number of times.

It should be pointed out that *all* fiber types may be recruited in a maximal effort in a slow contraction against high resistance as well as in a fast contraction, resistance being less. With the force-velocity curve in mind, it should also be pointed out that it may be misleading to express the demand on maximal dynamic strength in running, swimming, jumping etc., as a percentage of maximal isometric strength. At a high velocity of contraction, "submaximal" force may, in this respect, be maximum or at least close to maximum.

Muscle mass When starting a strength training program there may, during the first weeks, be a 20 to 40 percent increase in strength without a noticeable increase in the cross-sectional area of the muscle involved. In the classical study by Ikai and Fukunaga (1970), they noticed that three maximal 10s isometric contractions repeated daily for about 100 days increased the isometric strength approximately 90 percent with only about a 25 percent increase in the cross-sectional area of the muscles involved. One explanation for this finding is a more efficient activation of the motor units, illustrated by an increase in the electromyographic activity during a maximal effort. The gradually occurring muscle hypertrophy is not achieved by increasing the number of muscle fibers (Gollnick et al., 1981), but by an increase in the cross-sectional area of the muscle fiber area, which may include both the fast-twitch and the slow-twitch fibers (See Timson et al., 1985.)

Actually, in its ability to adapt its size to physiological demands, the skeletal muscle is a very plastic tissue. This is demonstrated by the shape of bodybuilders and weight lifters. High resistance, slow velocity training, can thus induce a hypertrophy of both type I and type II fibers because both fiber types are engaged (see Sale and MacDougall, 1981). In cross-sectional studies, the ratio between the area of type II and type I fibers in weight and power lifters, and bodybuilders is about 1.6 and 1.1 to 1.2 in "normal," active people, and about 1.0 in sprinters (see Saltin and Gollnick, 1983). Endurance training, on the other hand, may selectively enlarge the type I fibers. However, there are reports that muscle groups that have been subjected to intensive endurance training have "normal" or relatively small type I fiber areas (Jansson and Kaijser, 1977; Nygaard et al., 1977). A typical strength training *will not* stimulate capillary growth, nor any development of mitochondrial enzyme systems. With an increase in the muscle fiber size there is actually an increase in the fiber area that each capillary must supply with oxygen and substrates. These factors may explain the relatively poor maximal aerobic power and endurance in muscles exclusively trained for strength. In contrast, endurance training will induce an increase in capillary density and mitochondrial volume, which certainly can enhance endurance (see below under "aerobic training"). The energy demand of muscle activation at maximal or near maximal force is mainly covered by a breakdown of ATP and PCr (Phosphocreatine). Neither strength training nor endurance training affects the ATP and PCr concentrations significantly.

Repeated overloading of a muscle, as well as chronic stretch of either a normal or denervated muscle, are factors that can increase the muscle mass. Very little is known concerning the regulatory mechanisms behind the "disturbed" balance between protein synthesis and protein breakdown, favoring the synthesis (see Goldberg et al., 1975; Goldspink, 1980; Saltin and Gollnick, 1983). At any rate, muscle hypertrophy and strength are related to an increase in myofibrillar protein (actin and myosin) and, therefore, increased contractile filaments per muscle fiber. When overstretched, more sarcomeres may develop, making the muscle fiber longer.

Disuse of a skeletal muscle, e.g., by immobilization or joint fixation, will result in muscle atrophy, and a reduced cross-sectional area of both type I and type II fibers is noticed. Actually, an immobilization of a leg following injury could, after some months, reduce both fiber types by an average of some 40 to 45 percent (Sargeant et al., 1977). There is also reduced activity of the oxidative enzymes (see Booth and Gollnick, 1983). It is not surprising that prolonged exposure to a weightless environment in space will also result in a loss of muscle mass. It was mentioned that a stretch of a muscle could initiate a hypertrophy. It is, therefore, logical that the fixation of a joint in such a manner that the muscles are shortened below their resting length, may enhance the atrophy more than expected by inactivity alone.

A stimulation of a denervated muscle will not lead to a hypertrophy, on the contrary, it cannot prevent an atrophy (Gutmann et al., 1961; see Henriksson et al., 1982). Apparently the nerve cell has a trophic function essential for the protein synthesis.

Patients confined to bed rest or with a joint immobilized by a cast can counteract muscular atrophy by subjecting the muscles to submaximal isometric contractions of a few seconds duration, repeated a couple of times per day. The same applies to astronauts during prolonged space flights.

One study by MacDougall et al. (1980), may serve to illustrate the magnitude of the effects of use and disuse on the skeletal muscle mass. Seven healthy male subjects were observed under controlled conditions, following 5 to 6 months of heavy resistance training, and then again after 5 to 6 weeks of immobilization in elbow casts. Cross-sectional fiber areas were calculated from sections of needle biopsies taken from triceps brachii. Training resulted roughly in a 100 percent increase in maximal elbow extension strength. Both fast-twitch and slow-twitch fiber areas increased significantly (with about 40 and 30 percent, respectively). Immobilization resulted in a 40 percent reduction in strength; the reduction in fiber areas was slightly above 30 percent for fast-twitch and 25 percent for slow-twitch fibers.

Strength training does not modify the fiber composition in the skeletal muscle. Bodybuilders and power/weight lifters as well as athletes in track events demanding strength and power have fiber distributions in their muscles that are within the range of nonathletes.

The role of the central nervous system (CNS) in all aspects of muscle strength is quite obvious, since the muscles are the slaves of their motoneurons. They can do nothing unless manipulated by electrical stimulation, either directly or via their motor nerves.

It was mentioned earlier in this chapter that during a period of strength training there is an increase in strength without any noticeable muscle hypertrophy, particularly in the initial phase of the training. And in Chap. 2, it was pointed out that the measured muscle strength may vary greatly from test to test in the same subjects. An explanation for these two observations may be a graded inhibition on the motoneurons innervating the skeletal muscles. The inhibition on these interneurons may in turn be inhibited (a so-called disinhibition, see Fig. 3-2), e.g., as a result of training. The less the inhibition on these motoneurons, the more motor units may contract at tetanus frequency, eventually becoming synchronized. In certain situations (danger, competition, etc.), this inhibition may also be inhibited, resulting in the development of larger muscular strength (see Ikai and Steinhaus, 1961). "Gates" may open up, or close, allowing for an increased, or decreased nerve impulse traffic to a given muscle group.

There is a third example, which supports the role of neural factors as a moderator of strength, even if the subject's effort is maximal: the effect of a strength training is related primarily to the practiced activity, particularly in events involving isometric training. Training the elbow flexors with the elbow joint at a 90° angle will increase the strength at that joint position but the effect is less pronounced when the strength is measured in a 45° or 135° angle (see Clarke, 1973).

How can this particular specificity in training be explained? As mentioned, joint angles determine the muscle length, alter the mechanical advantages of the system through which the muscles act, modify the synergic support that muscles receive, and in addition, they produce changes in the neural traffic to and from muscles and joints that affect the responsiveness of the muscles to central commands (Atha, 1981). All aspects considered, both in the case of muscle, and in the case of CNS, the activation pattern is not identical when the joint is at a 90° angle, and when it is at a 45° or 135° position.

Another example: The effect of a strength-training program carried out with the subjects in an upright position, was evaluated with the subjects in the upright as well as in a supine position. The improvement was much less pronounced when the test was performed in the unfamiliar position (Rasch and Morehouse, 1957).

Actually, there is a special nervous circuit for every single muscle activity and movement. Therefore, it is not surprising that some of the effects of training and practice will modify the "behavior" of the CNS. The technique that is used may be improved, and this learning process involves the CNS. Many strength tests and training activities include factors involving technique. A considerable confusion has emerged from the fact that some investigators have used the same activity both for testing and for training, while others have applied a test situation that did not correspond to the actual training.

It should also be mentioned that training the muscle strength of one arm or leg often results in a much smaller but significant increase in a similar exercise performed with the untrained arm or leg. It is thought that neural factors are behind this transfer effect.

Which is the Most Effective Strength Training Program?

From the preceding discussion, it should be evident that this question is unsolvable from a physiological point of view. The important point is the purpose of the training, and the person's potential to respond to the training. Many studies have been designed to compare the effect of isometric to dynamic muscle training, with different resistances and a different number of repetitions per session, per day, or per week. In dynamic activities, variations in the speed of contractions in eccentric or concentric contractions have been compared. Lastly, mixed programs have been evaluated. The reader is referred to the reviews mentioned. Atha (1981) in his extensive analysis of the literature concludes that a high tension, not necessarily fatiguing, is the stimulus that leads to increased strength, but it is not clear which of the three contractions: eccentric, isometric, or concentric is the most effective. Clarke, in his 1973 review, came to a similar conclusion. There are specific programs that have produced a strength improvement with individual variations from 0 up to 100 percent. At present, this is an area in which we must base our comments on a mixture of scientific evidence and belief.

Isometric Strength Training Isometric strength training can be conducted with simple equipment. It takes approximately 4 s to reach maximal tension (the time differs between different muscle groups). It is a common recommendation that the contraction should be maintained for about 6 s. Increasing the number of repetitions is advantageous for increasing strength, but the advantage is not proportional to the number of repetitions. Isometric contractions repeated 5 times, 3 times a week is an efficient program because it achieves significant gains at a minimal cost. As mentioned previously, isometric training improving the strength in one joint position does not always mean improved strength in other joint positions. Thus, for the development of a more general strength, the training should include a variety of joint positions. The superiority of daily training can be illustrated by Hettinger's observation (Hollman and Hettinger, 1976) that the effect of every-second-day training was 80 percent of that of daily training. This type of training does not enhance the aerobic power and endurance, but there are some transfer effects to dynamic strength tests. Isometric contractions are effective in preventing substantial loss of muscle mass, and muscle function, during periods of recovery from injury with joint immobilization.

Dynamic Strength Training When testing the effect on muscle strength of the number of repetitions in one training session, the majority of the experts are of the opinion that the 5RM to 6RM loads probably represent the optimum, and are more effective than 2RM and 10RM. At any rate, since the 6RM load is about 85 percent of the 1RM, it shows that the load does not have to be maximal. Individual athletes and coaches have their favorite programs. Presently, the trend is that weight-and power lifters include fewer repetitions, from 1 to 5RM, and bodybuilders prefer 8 to 12 repetitions with 3 to 5 and approximately 10 sets, respectively for each activity. Alternatively upper limb, trunk, and lower limb muscles are exercised 2 to 3 times per week. For dynamic strength training, special equipment and apparatus are available, such as barbells, dumbbells, and pulleys. Most sophisticated are the *isokinetic* machines, which can secure a maximal demand on torque over the entire functional range of joint movements. So far, there is no scientific evidence showing that isokinetic training is superior to dynamic or isometric strength-training methods. Most authorities are in favor of slow-speed isokinetic training rather than fast-speed training for the development of strength and muscle hypertrophy. Possible exceptions are training programs to enhance fast and powerful muscle contractions, in which cases such exercises are preferable. It should be kept in mind that a high muscular tension is important for the development of strength, and with a high speed of contraction the tension drops. From this viewpoint, one should expect that a program concentrating on *eccentric muscle exercises* should be superior to other programs. Actually, eccentric training is effective in increasing muscle strength, but it is no more effective than either isometric or concentric training (see Atha, 1981). Again, *maximal* loading does not necessarily elicit maximal strength gains. It is a disadvantage with eccentric exercises in that they are apt to produce sore muscles in untrained persons. Secondly, with heavy loads exceeding 1RM, these exercises cannot be handled in an isometric or concentric muscle contraction, which may create a potentially risky situation.

A combination of different training methods seems to have some advantage over single exercise methods in developing strength and muscle hypertrophy. On the basis of their studies, Hakkinen and Komi (1981) suggest a combination of concentric and eccentric exercises for an optimal training of maximal force. To go one step further, it has been suggested that a complete strength training program should devote some 15 percent of the entire training period to eccentric training, 10 percent to isometric, and 75 percent to concentric training.

The specificity of training complicates, as already mentioned, a comparison of different methods, particularly isometric versus dynamic training programs (see Duchateau and Hainout, 1984). It is understandable that the coach or athlete tries to develop strength-training exercises, which in their pattern of movement come as close as possible to the strength demands of the event in question. In many ways, the best form of training for any given activity is to perform that activity. Track and field athletes, soccer and football players, volleyball and basketball players often train repetitive jumps with a bounce-loading of the leg muscles, including jumps down from a bench before "take-off" to increase the load; swimmers try to swim at high speed with their normal stroke against a pull of weights or other forces attached to her or his waist arranged in such a way that the drag is exerted horizontally on the swimmer, just to mention a few examples.

Experiments have been made with *electrical stimulation* of the muscles indirectly via their motor nerves, or directly on the muscles, which has resulted in some positive reports (see Hollman and Hettinger, 1976, p. 245). The drawback of these experiments is that coordination is not properly trained. Dooley et al (1984), observed no increase in maximal voluntary strength after training using involuntary electrically evoked contractions. They interpret this result as support for the contention that the increase in maximal voluntary strength as the result of voluntary muscle training is due to changes in the neural drive to the muscle, rather than to an increase in the intrinsic force generating capacity of the muscle fibers. These authors do not report any data on eventual changes in the muscle mass as a consequence of repeated electrical stimulation.

Do Anabolic-Androgenic Steroids Improve Strength?

For years it has been a controversial question as to whether or not the effect of high-intensity strength training could be enhanced by the use of anabolic-androgenic steroids. The American College of Sports Medicine (1984) has revised its 1977 position stand; the new statement reads as follows:

"1. Anabolic-androgenic steroids in the presence of an adequate diet can contribute to increase in body weight, often in the lean mass compartment.

2. The gains in muscular strength achieved through high-intensity exercise and proper diet can be increased by the use of anabolic-androgenic steroids in some individuals.

3. Anabolic-androgenic steroids do not increase aerobic power or capacity for muscular exercise.

4. Anabolic-androgenic steroids have been associated with adverse effects on the liver, cardiovascular system, reproductive system, and psychological status in therapeutic trials and in limited research on athletes. Until further research is completed, the potential hazards of the use of the anabolic-androgenic steroids in athletes must include those found in therapeutic trials.

5. The use of anabolic-androgenic steroids by athletes is contrary to the rules and ethical principles of athletic competition as set forth by many of the sports governing bodies. The American College of Sports Medicine supports these ethical principles and deplores the use of anabolic-androgenic steroids by athletes."

Summary The major determinant of muscle strength, is the cross-sectional area of the muscle. Other factors will, however, modify the strength, such as neural factors, the internal muscle architecture, limb length, and joint structure. The muscle fiber composition is not critical, with the possible exception when contractions are performed at high velocities. In such cases, a high percentage of fast-twitch fibers may be advantageous. In the initial stages of a training program, strength increases more rapidly than the muscle mass. Gradually, there is a muscle hypertrophy, more so if the training load is at maximum or close to maximum. The increase in muscle mass is due to an increase in the cross-sectional area of the individual muscle fibers by the formation of more myofibrils. The number of muscle fibers and the fiber composition are maintained constant.

It is not clear which type of strength training is most effective: isometric or dynamic contractions, eccentric or concentric activation with speed control (isokinetic), or not. There is a specificity in the training effects and therefore it is an advantage if strength training activities can be designed to fit the specific needs of the individual. For obvious reasons, the load should be greater than the demands of "normal" life, i.e., an overload, and the resistance should be progressively increased.

There is a low correlation between strength, speed, and endurance. The highest loads do not produce the greatest gains; 5RM to 6RM have been suggested as optimum in dynamic exercises. Weight lifters usually include few repetitions in their programs, bodybuilders often favor 8 to 12 RM, with relatively slow contractions and with up to 10 sets being completed in each session. In an alternative manner, different exercises and muscle groups are trained 2 to 4 times per week. It is recommended that eccentric, isometric, and concentric exercises are included, for instance in the following proportions: 15 percent, 10 percent, and 75 percent, respectively of the time spent on each mode of exercise. Fast contractions can enhance performance in powerful and fast activities.

For endurance training, more repetitions are necessary to elicit favorable effects on local enzyme systems and capillary density.

For the ordinary "fitness seeker" a few isometric contractions at near maximal effort, lasting at least some seconds to allow all muscle fibers in the active muscle to be recruited, is a simple and effective program if repeated about 3 times per week. It should be pointed out that isometric training has position-specific effects. For example, training with the arm bent at 90° at the elbow, gives a strength gain most pronounced at that angle with much less improvement at a less than 60° and more than the 120° position.

With reference to Chap. 4, it should be emphasized that maximal or near maximal efforts in isometric and slow dynamic muscle contractions are associated with a relatively high blood pressure and heart rate, particularly if a large muscle mass is involved. This extra load on the cardiovascular system can be hazardous for patients with cardiovascular diseases. For the recovering patient or recreational athlete, it is not of vital importance to include sophisticated programs and isometric exercises, since submaximal loads can be quite effective in developing strength.

Training of Anaerobic Power

In Chap. 7, it was emphasized that there are no methods available for an accurate measurement of the anaerobic power or capacity. Therefore, it is not possible to evaluate objectively whether or not a specific anaerobic training program is effective or not. It is a common finding that the peak blood lactate concentration is higher in a track athlete after that athlete has finished an important competition than after an "all out" test on the treadmill. What then, is the limiting factor in efforts with maximal demands on the anaerobic energy yielding system? Saltin and Gollnick (1983) conclude that lactate per se is not responsible for the termination of exercise, nor are the enzymes involved in the anaerobic metabolism significantly inhibited.

However, if one can become adapted to a low pH and increase one's tolerance to it by exposing the tissue to the anaerobic endproducts, the following program should be effective: maximal effort for about 1 min, followed by 4- to 5-min rest, then a further period of 1 min maximal effort, followed by 4- to 5-min rest, and so on. At the end of four or five such exercise periods, a highly motivated runner may gradually attain lactate concentrations in the skeletal muscle in excess of 30 mM, and 20 mM or more in the blood, and an arterial pH approaching 7.0 or lower. In order to produce major changes in the cellular environment, e.g., affecting the respiratory function and causing dyspnea, large muscle groups should be engaged in activities such as running.

Training of the anaerobic motor power is important for many groups of athletes. Since this form of training is psychologically exhaustive, it should not be introduced until a month or two prior to the competitive season. Such strenuous training is not recommended for average persons.

At a given submaximal O_2 uptake, the content of *lactic acid* in the blood is lower in a trained subject than in an untrained one, as shown by a series of studies (Fig. 10-5). This may be interpreted as an expression of a more effective oxygen transport during the beginning of exercise as well as throughout the entire period of exercise, leading to a diminished anaerobic energy yield. While an onset of blood lactate accumulation may be observed at an O_2 uptake corresponding to 50 to 60 percent of the maximal oxygen uptake in an untrained individual, this percentage may be elevated to 70 to 80 percent or even higher in a well-trained individual (see Chap. 7, and Denis et al., 1982; Hurley et al., 1984). Coyle et al. (1983), observed that their well-trained ischemic heart disease patients could run at 100 percent of their maximal oxygen uptake before the blood lactate increased 1 mM above the base line. A similar increase in lactate concentration was already noticed in healthy, well-trained subjects when they were running at about 85 percent of their maximal oxygen uptake.

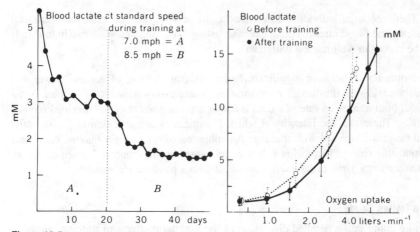

Figure 10-5
To the left: Progressive decrease of blood lactate for a standard amount of exercise: running on a treadmill at 7 mph for 10 min. During the first 20 days (*A*), training consisted of running daily on the treadmill for 20 min at 7 mph. A steady level of blood lactic acid is reached around 3 mmoles. During the following 30 days (*B*), training is increased to running at 8.5 mph for 15 min daily. Blood lactic acid decreases further, and a new steady level is reached around 1.5 mM after the standard test. *(From Edwards et al., 1939.)*
To the right: Peak concentration of lactic acid in blood in relation to oxygen uptake up to maximum before (dotted line) and after (solid line) 16 weeks of physical training. Mean values on eight subjects exercising on a cycle ergometer. Vertical bars indicate ±1 standard deviation. *(From Ekblom et al., 1968.)*

As mentioned, strength training with maximal or near maximal efforts for approximately 5 s periods will not significantly affect the resting level of ATP and PCr concentration, nor the activity of the enzymes involved in the anaerobic and aerobic energy yield. Sprint-type interval training (e.g., repeated 200 m dashes) will increase the activity in "anaerobic" enzymes, including PFK and LDH (see Roberts et al., 1982). During maximal effort, children usually attain relatively low muscle lactate concentrations (e.g., approximately 9 mM · kg^{-1} wet muscle in Eriksson's eleven to thirteen year-old boys; 1972). They have also much lower PFK activity as compared with adults. After a period of training, including sprint and endurance training, the PFK activity as well as the lactate concentration may increase (e.g., to about 14 mM in the aforementioned study). It is assumed that PFK participates in the regulation of the flow of substrate through the glycolytic pathways. However, it is estimated that the potential of the enzymes in this pathway far exceeds the demands during maximal exercise. It is an open question why the activity of most, if not all these enzymes, is so luxuriously high in a sedentary person's skeletal muscles and yet increases further during training (see Saltin and Gollnick, 1983).

It has been reported that elite athletes in sports with great demands on anaerobic power and capacity had a significantly higher buffer capacity in the skeletal muscles than the more sedentary subjects (Sahlin and Henriksson, 1984).

There are large individual differences in the peak lactate concentration in skeletal muscles and blood after maximal exercise lasting some minutes. This maximum tends to be higher in well-trained individuals.

Summary The most significant finding is that training of aerobic power and endurance lowers the lactate concentration in exercising muscles as well as in the blood, both at a given rate of exercise and at a given percent of the maximal oxygen uptake. Therefore, the intensity at which a continuous accumulation of lactate starts will be shifted upward with training. An enhanced oxygen supply may be one factor behind this change (see below), but an inhibition by the improved potential for an aerobic energy yield in the mitochondria, is also a possible mechanism.

The Dallas Study

Before going into a detailed discussion of the long-term effects of training, aimed at improving the maximal oxygen uptake and endurance, it might be useful to present one study that has become classical in its design.

Saltin et al. (1968), carried out extensive studies on the effect of a 50-day period of physical training following a 20-day period of bed rest in five male subjects, aged nineteen to twenty-one. Three of the subjects had previously been sedentary and two

Figure 10-6
Heart rate and oxygen uptake recorded in two subjects during training with alternating 3-min running and 3-min rest. The efforts were not maximal, but the oxygen uptake reached maximal values, as did the heart rate. *(From Saltin et al., 1968.)*

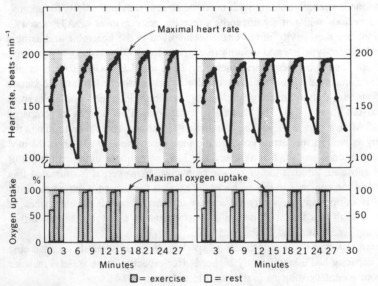

of them had been physically active. The training program was rather intensive and continuously supervised. The weekly schedule included two workouts daily, Monday through Friday; on Saturdays there was only one workout, and Sundays were free. The workouts consisted of both interval and continuous exercise, mainly outdoor running of from 4 up to 11 km (2.5 up to 7 miles). In the interval exercise, the speed was chosen so that the oxygen demand during 2- to 5-min periods of running was at or near the individual's maximal oxygen uptake (example see Fig. 10-6). During the continuous running, usually for more than 20 min, oxygen uptake varied from 65 to 90 percent of the subject's maximal value. One of the subjects, who had not trained before, was covering an average of 64 km per week (40 miles), and one of the previously trained subjects covered about 80 km per week (50 miles). The maximal oxygen uptake dropped from an average of 3.3 in the control study (before bed rest) to 2.4 liters $\cdot$ min^{-1} after bed rest, a 27 percent decrease (see Fig. 10-7). The stroke volume during supine

Figure 10-7
Changes in maximal oxygen uptake, measured during running on a motor-driven treadmill, before and after bed rest and at various intervals during training; individual data on five subjects. Arrows indicate circulatory studies. Heavy bars mark the time during the training period at which the maximal oxygen uptake had returned to the control value before bed rest. *(From Saltin et al., 1968.)*

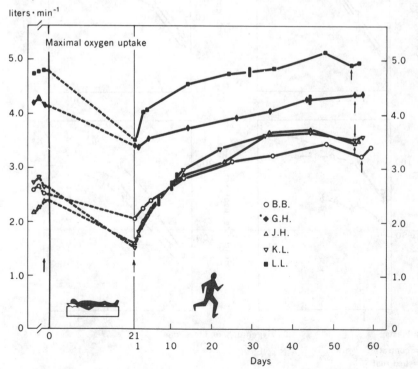

exercise on a cycle ergometer at 100 watts (oxygen uptake, 1.5 liters · min⁻¹) decreased from 116 to 88 ml, or about 25 percent. The heart rate increased from 129 to 154 beats · min⁻¹. The cardiac output at this standard load fell from 14.4 to 12.4 liters · ⁻¹. Thus, the arteriovenous O_2 difference became somewhat increased. Also, during upright exercise at submaximal loads, there was a reduction in cardiac output (15 percent) and stroke volume (30 percent) (Fig. 10-8). An oxygen uptake that could normally be attained with a heart rate of 145 required a heart rate of 180 beats · min⁻¹ after bed

Figure 10-8
Mean values of cardiac output, arteriovenous oxygen difference, stroke volume, and heart rate in relation to oxygen uptake during running at submaximal and maximal intensity before (c), after (b) a 20-day period of bed rest, and again after a 50-day training period (t). (Modified from Saltin et al., 1968.)

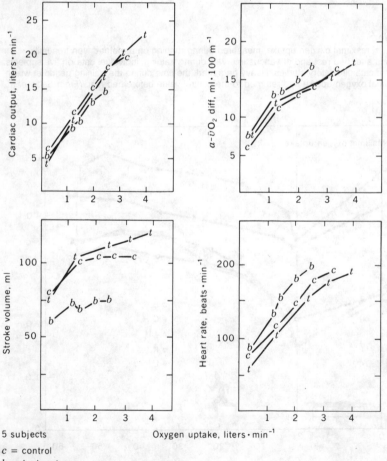

5 subjects

c = control
b = bed rest
t = training study

rest. During maximal treadmill exercise, the cardiac output fell from 20.0 to 14.8 liters · min^{-1} (a 26 percent reduction). It should be recalled that the maximal oxygen uptake was now 27 percent less. Since the oxygen content of arterial blood did not change, the maximal arteriovenous O_2 difference was, therefore, not modified by the bed rest. These findings are illustrated in Fig. 10-8. The maximal heart rate was not altered, so the fall in maximal cardiac output was due to a reduction of stroke volume.

The physical training produced an increase in maximal oxygen uptake from 2.52 to 3.41 liters · min^{-1} in the previously sedentary subjects (an increase of 33 percent), and from 4.48 to 4.65 liters · min^{-1} in the previously active subjects (Fig. 10-7). The most dramatic improvement in maximal aerobic power was noted with the three usually sedentary subjects when comparing the "after bed rest" values with the posttraining ones. They actually increased their maximal oxygen uptake by 100 percent, or from 1.74 to an average of 3.41 liters · min^{-1} as mentioned. This study illustrates how critical the level of physical activity is before a training regime for the evaluation of the effectiveness of a training program to improve the maximal aerobic power. In the previously physically active subjects, the improvement was only 4 percent. Starting with the "after bed rest" level, the increase was, however, 34 percent (from 3.48 to 4.65 liters · min^{-1}). For the three previously sedentary subjects, the improvement was 33 and 100 percent, respectively. In other words, the training program applied may be said to have caused the maximal oxygen uptake of the participants to increase from 4 to 100 percent, depending on how the initial level is defined. From Fig. 10-7 it is evident that the three sedentary subjects exceeded their control values as early as about 10 days after the commencement of the training. The two previously active subjects required 30 to 40 days to achieve the noted improvement.

Figure 10-9 illustrates the variation in maximal oxygen uptake for the three normally inactive subjects. The higher maximum noticed during their normal sedentary life, compared with the more extreme inactivity when immobilized in bed, was due to a higher cardiac output. The further improvement in maximal oxygen uptake with intensive training was partially due to a still higher cardiac output and partly to an increased arteriovenous O_2 difference by a more complete extraction of oxygen from the blood in the tissues.

Figure 10-8 summarizes the circulatory data for the five subjects. The training did increase the stroke volume and decrease the heart rate at submaximal work rates. The maximal heart rate was, however, not modified by the training. The maximal cardiac output in the sedentary subjects decreased from 17.2 to 12.3 liters · min^{-1} during bed rest; after the training it rose to 20.2 liters · min^{-1}. The arteriovenous oxygen difference, in the same experiments, was 147, 149, and 170 ml·l^{-1} blood, respectively. The stroke volume fell from 90 to 62 ml during bed rest, but after training it increased to 105 ml. The changes in heart rate at various levels of oxygen uptake due to inactivity and caused by activity are obvious. However, when the heart rate was related to the oxygen uptake in percentage of the maximum, the heart rate response remained the same in the different conditions.

The mean value for the heart volume in the three sedentary subjects was 740 ml. After bed rest, it was 690 ml; it increased to 810 ml at the end of the training, an increase of 17 percent. The increase in stroke volume was 69 percent, however.

Blood volume decreased significantly during bed rest (from 5.0 to 4.7 liters in the

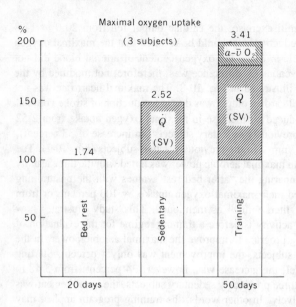

Figure 10-9
Maximal oxygen uptake during treadmill running for three subjects after bed rest ($= 100$ percent), when they are habitually sedentary, and after intensive training, respectively. The higher oxygen uptake under sedentary conditions compared with bed rest is due to an increased maximal cardiac output ($\dot{Q}$). The further increase after training is possible because of a further increase in maximal cardiac output and arteriovenous O_2 difference ($a\text{-}\bar{v}O_2$). The maximal heart rate was the same throughout the experiment; therefore the increased cardiac output was due to a larger stroke volume (SV). *(From data obtained by Saltin et al., 1968.)*

5 subjects, i.e., a 7 percent reduction). The fall in plasma volume was slightly more pronounced than in the red cell mass. During training, the plasma volume and red cell mass increased again, in most subjects above the control values.

The lean body mass changed from 66.3 to 65.3 kg after bed rest and increased to 67.0 after training. There were no significant changes in ultrastructural morphology of voluntary muscle (quadriceps) during the experimental period. The basal heart rate recorded during sleep was on the average 51.3 beats/min after bed rest, and it decreased to 39.7 at the end of the training.

The unique feature of this study is its broad illustration of the effects of prolonged bed rest on the oxygen-transporting system, followed by a specific training program in the same subjects. It is, therefore, not surprising that data from this study are almost always quoted when subjects concerning use or disuse, rehabilitation or exercise physiology in general, are discussed.

Let us now discuss in more detail (1) training principles to improve aerobic performance, and (2) training effects on the oxygen-transporting system.

Training of Aerobic Power

It has already been pointed out that physical activity ranging from repeated exercise periods of a few seconds' duration up to hours of continuous activity may involve a major load on the oxygen-transporting organs and thereby induce a training effect, provided the exercise load is sufficiently high. Practical experience has shown that exercise with large muscle groups for 3 to 5 min, followed by rest or light physical activity for an equal length of time, then a further exercise period, etc., as required by the individual's ambition and the objective of the training, is an effective method of training. The tempo does not have to be maximal during the activity periods. It is not necessary to be exhausted when the exercise is discontinued (for an explanation, see Fig. 7-3, showing that maximal oxygen uptake may be reached without the subject's being exhausted). Mild exercise, such as jogging, between the heavier bursts of activity may be advantageous, since the elimination of lactic acid is faster than at complete rest (see Chap. 7). It has been shown experimentally that the cardiac output and the blood pressure attain their highest values at a load that produces the maximal oxygen uptake. During "supermaximal" exercise, the oxygen uptake as well as the cardiac output and stroke volume may even attain lower values than they do at a slightly lower work rate. There is no evidence to support the assumption that it is important to engage the anaerobic processes to any extreme degree in order to train the aerobic motor power.

The justification for a submaximal tempo in the optimal training of the oxygen-transporting system may be further supported by Fig. 10-10. For six subjects, an individual speed was determined, which brought them to complete exhaustion at the end of the fourth minute of running. On other days, the speed of the treadmill was decreased stepwise by 0.5 to 1 km $\cdot$ hr^{-1} without changing the total distance of the run. A reduction in speed by as much as 3 km $\cdot$ hr^{-1} for some of the subjects did not reduce the oxygen uptake. Therefore, since maximal oxygen uptake can be attained at a submaximal speed, this lower speed may be sufficient and probably optimal as a training stimulus. A highly motivated individual and someone with a high anaerobic capacity will show a wide plateau; others may just be able, or willing, to push themselves to the point where the maximal oxygen uptake is reached (for example, 250

Figure 10-10
Individual data on oxygen uptake in relation to speed for two female (triangles) and four male subjects. Treadmill was set at an angle of 3°. The highest speed for each subject could be maintained for just 4.0 min. *(From Karlsson et al., 1967a.)*

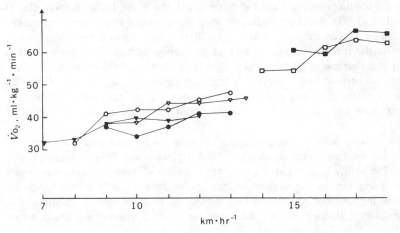

watts in Fig. 7-3*b*), or not even that far. It is thus impossible to delineate a border value where the greatest load on the oxygen-transporting organs is attained. In the case of healthy young persons, the speed of running may be reduced to about 80 percent of the maximum, which may be maintained for a period of 3 to 5 min. If, in other words, the distance that may be covered by running, swimming, or bicycling in a matter of, say, 3.0 min is covered, instead, in about 3.5 min, the demand on the aerobic processes remains the same. Therefore, the stopwatch, in such cases, should be used to maintain a reduced tempo, not to stimulate the trainee to attain a better achievement in terms of better timing. Figure 10-6 illustrates how this type of training may be arranged, and what effect it has on heart rate and oxygen uptake. (It should be emphasized that in exercise in the upright position the maximal stroke volume is attained *during* and not after the exercise. It is a misconception to believe that the advantage of interval training is that frequent recovery periods as such should elicit an effective training of the central circulation.)

As an aid to determine whether the load has been maximal or nearly maximal, the heart rate during the exercise may be used. It should not differ more than 10 beats $\cdot$ min^{-1} from the individual's maximal heart rate, assessed during controlled laboratory experiments. A person accustomed to heavy exercise can also sense when the pulmonary ventilation has reached the steep part of the slope on the curve relating pulmonary ventilation to oxygen uptake (see Fig. 5-9). This type of training is certainly far more pleasant than when the tempo is higher. Thus, one should definitely distinguish between training of the aerobic and the anaerobic processes. In competitive events, in which a superior effect is required in both these processes, the training obviously should also include this combination. This aspect, however, should be postponed until a few months prior to the start of the season; otherwise, the athlete may be unable to keep up the training program for psychological reasons.

The type of training described for the oxygen-transporting system, with a submaximal tempo for periods of 3 to 5 min, may indeed increase the maximal oxygen uptake. It is probably also effective in eliciting many of the side effects that will be discussed later in this book. In most of the earlier studies of the effects of training, the intensity during the training sessions included maximal efforts. In the 1970s, Kilbom (1971) and Pollock (1973) conducted studies showing that a demand exceeding 50 percent of the maximal oxygen uptake, repeated some 30 min three times a week was enough to elevate the maximal oxygen uptake in previously untrained individuals. When training at 50 percent of the maximal oxygen uptake a 5 to 10 percent increase in this $\dot{V}_{O_2}$ max has been noticed in previously sedentary subjects. When training at 70 to 80 percent of maximal oxygen uptake, the improvement is on the average 15 percent, however, with large individual variations at least partly due to the initial level of fitness.

Brisk walking, rope skipping, and stair climbing can produce significant improvement in aerobic power (see Jones et al., 1962; Ogawa et al., 1974; Ilmarinen et al., 1979). The ability to exercise for prolonged periods of time, utilizing the largest possible percentage of the maximal oxygen uptake, may probably be developed just by exercising continuously during long periods of time (endurance training). The capacity to store glycogen in the muscles and the ability to mobilize and to utilize free fatty acids play a major role in prolonged exercise. Endurance athletes may devote a great deal of time to "distance training." They may run or ski up to 250 to 350 km (150 to 220 miles) weekly. At present, there are no studies available for an objective evaluation of the effects of such training quantities. It should be emphasized that many top athletes of today are professional in the sense that they are "employed" full time in their sports and are thus free to devote a great deal of time to training.

However, it should again be emphasized that the recruitment of motor units is dependent on the rate of exercise. At low intensities, slow-twitch fibers (type I) are activated, and with increased load more and more units become recruited and the rate of fiber contractions increases. At high intensities the fast-twitch fibers (type II) are also thrown into action (see Chap. 3). Analyses of the exercising muscle groups have revealed that the slow-twitch fibers are the first to lose their glycogen content in prolonged submaximal activity. However, when these fibers are depleted of their glycogen stores, fast-twitch fibers are apparently recruited, for a reduction of glycogen content has been observed in these fast fibers under such conditions. When the subjects are exercising at rates exceeding 100 percent of maximal V_{O_2}, both fiber types appear to be continuously involved (Gollnick et al., 1974). Thus, in heavy interval training all fibers in the activated muscle groups may be active simultaneously. As pointed out in Chap. 7, the fast-twitch fibers may be forced to depend primarily on an anaerobic energy yield since the oxygen supply is not sufficient to cover the demand of all the fibers for an aerobic metabolism. In "distance training" the oxygen is, most of the time, below the maximum. With the depletion of the glycogen in the slow-twitch fibers, a central nervous system mechanism, eventually assisted by an afferent impulse traffic from muscle spindles, will cause the recruitment of new motor units, the high-threshold fast-twitch fibers being the last reserve. The timing of these events depends on initial glycogen stores, rate of work, and on the exercising person's will power. At any rate, the biochemical and morphological adaptations to chronic exercise will occur locally in the training fibers. One can speculate that an intensive interval training provides a balanced adaptation of both slow- and fast-twitch fibers to a combined aerobic and anaerobic stress situation, as in a 5,000 or 10,000 m race. An exclusive "distance training" will promote the fast-twitch fibers' aerobic metabolism. Actually the aerobic potential including the enzymes of the citric acid cycle and the electron-transport chain of both type IIa and type IIb fibers can attain the same level as in the type I fibers (see Saltin and Gollnick, 1983).

The question is whether this submaximal exercise is very effective in improving the maximal aerobic power. If, during training, there is no marked increase in maximal oxygen uptake, it is of questionable value to devote much time to increasing the aerobic potential of *all* muscle fibers. If the objective of the training is to improve the ability to perform prolonged exercise, it may be advisable to concentrate on a "distance training," for it may be effective in stimulating an increase in capillary density, myoglobin content, and activity in enzymes specialized on the metabolism of free fatty acids. However, for an endurance athlete, the optimum must be to combine submaximal, maximal, and "supermaximal" training. (There are experiments on rats showing that running durations of moderately intense exercise longer than 2 hours $\cdot$ day^{-1} do not further increase the oxidative capacity of the exercising muscles; Terjung, 1976.)

It is interesting to note that today's top athletes are not superior in their maximal oxygen uptake in comparison to the middle distance runners of the 1930–1950 era (Åstrand, 1956; Fig. 10-1). The extreme "distance training" typical for many training programs today has not been very effective in improving the maximal aerobic power considering the relatively modest amount of time devoted to training in earlier days.

It should be emphasized that it takes up to about 48 hours to refill emptied glycogen stores (Piehl, 1974). This must be an additional complication for those who have the ambition to train daily for several hours at high rates of exercise.

Whatever "special effects" these two types of training may elicit in order to improve the maximal aerobic power and the maximal aerobic capacity, a certain overlapping in the elicited effect is highly probable.

The athlete often needs a certain amount of variation in the training. According to the above, there are considerable possibilities for variation, even though certain programs are more critical than others concerning exercise time, rest periods, and intensity.

In the case of patients and completely untrained individuals, it is out of the question to prescribe an accelerated tempo in connection with the training of circulation. A previously bedridden patient should be satisfied with a training load that commences by elevating the heart rate by about 30 beats $\cdot$ min^{-1} above the resting value (to about 100 beats $\cdot$ min^{-1}). For the habitually sedentary individual, an elevation of the heart rate by about 60 beats $\cdot$ min^{-1} may be a suitable initial intensity. The principle of intermittent exercise is also valid for these categories of individuals. With daily training periods of from 15- to 30-min duration, or even with only a few training periods per week, the tempo may gradually be increased. The individual's health, age, and interest may determine how strenuous the training should be. It should be pointed out that in many cases it is not necessary, or in some cases even desirable, to attain maximal oxygen uptake and cardiac output during the training. In working with patients and average individuals, the "endurance training" may conveniently be accomplished by such activities as walking or bicycling. We can conclude that for an untrained individual an exercise that demands an oxygen uptake exceeding 50 percent of his or her maximal will, when repeated two to three half hours a week, gradually increase the maximal oxygen uptake (see Kilbom, 1971; Kearney et al., 1976). Training at an 80 percent level of maximal oxygen uptake may elicit a good effect (e.g., an increase in maximal oxygen uptake of about 15 percent; see Pollock, 1973; Hollman and Hettinger, 1976).

The problem is to teach a person how to find the exercise intensity that demands 70 to 80 percent of her or his maximal aerobic power, with no sophisticated technical aids available. If very breathless, the intensity is too high, if a fluent conversation is possible the intensity is too low. Chow and Wilmore (1984) noticed that about 50 percent of their subjects learned to pace themselves quite accurately by applying "ratings of perceived exertion" (see Borg, 1982). A classical rule of thumb has been to train at a heart rate of 195 minus the age in years. However, due to the large individual variations in maximal heart rate, this advice is quite useless (see Chap. 4).

It has repeatedly been emphasized that large muscle groups must be engaged when training the central circulation. Figure 10-1 may be of interest in this connection. In a number of top Swedish athletes belonging to the National Team, the maximal oxygen uptake was determined by laboratory experiments. In most cases the treadmill was used. It is evident that the athletes who had the highest maximal oxygen uptakes had selected events that placed heavy demands on their aerobic power, which means that these events also represent an excellent form of training of the oxygen-transporting system. Ball games fall rather low on the scale (data for handball, basketball, and soccer in Chap. 14 support this statement). This may be explained by the fact that each period of activity with a high tempo is frequently interrupted by periods of reduced tempo (Table 10-2). Most competitive calisthenics entail exercise periods up to 1 min. Needless to say, calisthenics may be performed in such a manner that they entail an excellent training effect of the aerobic power (straddle jump, sequences of movements engaging large muscle groups).

"Circuit training" (Morgan and Adamson, 1962) entails a series of activities performed one after the other. At the end of the last activity, one starts from the beginning again and carries on until the entire series has been repeated several times. By a preliminary test, in which many of the activities may be performed at maximal exertion, the number of repetitions of each activity is determined. The advantage with this circuit training is that every individual undergoes a program adjusted to his or her level of fitness. Persons may follow their own improvement by recording the time required for the series of repetitions, and endeavor to shorten the time required. At the end of a few weeks' training, a new test is performed on the number of repetitions of each activity the individual can manage (for example, push-ups). This training produces a high degree of motivation. The program may be accomplished in a limited space. The disadvantage is that an untrained individual is exposed to tests requiring maximal exertion. It has been found that a correctly devised program varying the involvement of large and small muscle groups, mixing static and dynamic exercise, does not produce maximal oxygen uptake measured on the cycle ergometer, but only about 80 percent of the maximal O_2 uptake (Hedman, 1960). In spite of this, the heart rate is almost maximal, the lactic acid concentration in the blood is very high, and the degree of exertion is considerable. Circuit training may be included in a training program, not only for athletes (especially those who fall within the lower part of Fig. 10-1) but also for school children for the sake of variation and for experience. It may also be applied very effectively in a strength training program.

Year-round Training In all cases, regularity in the performance of training is important. Within a month it is possible to develop a reasonable level of fitness, strength, and so on, but these qualities are lost when the training is discontinued. As already pointed out, less effort is required to maintain a certain level of fitness than to develop this level in the first place. With nonathletes, it may therefore be worthwhile, when time permits, to devote more time to training, and then, later on, to endeavor to maintain the acquired level of fitness by only a few training periods per week.

The athlete often has to train many different functions (aerobic and anaerobic power, strength, endurance, technique). It may be practical and sometimes necessary, because of time limitations, to concentrate on some of these functions at certain periods. This particular type of training has to be continued, however, even though it may be at a reduced intensity, when the athlete concentrates on the next function. Otherwise, the individual will lose the improvement already attained. The problem is to decide how intensively the different functions have to be trained in order to maintain a satisfactory level.

The keen competition of today necessitates year-round training. For a variety of reasons this is necessary, especially with regard to the oxygen-transporting system. This is particularly important for the aging athlete. The reason why many thirty- to thirty-five-year-old athletes may continue to rank among the world elite in endurance events is often due to relatively hard training during all seasons of the year. The older athlete may be what is known as "hard to train," compared with younger athletes, and the person whose fitness has been allowed to deteriorate may have considerable difficulty in regaining the former level of training. The seasonal variations in maximal oxygen uptake may be pronounced. Figure 10-11 presents data on three top cross-country skiers. It should be noted that they have apparently more or less reached a ceiling, and the intensive daily training they undergo over the years will not markedly affect the maximum. Included are data on two subjects who began an intensive training in 1969. The improvement in maximal oxygen uptake is quite remarkable. On the

Figure 10-11
Data on maximal oxygen uptake of three internationally successful cross-country skiers, and two "normal" subjects who started an intensive physical training in 1969. *(By courtesy of U. Bergh and B. Ekblom.)*

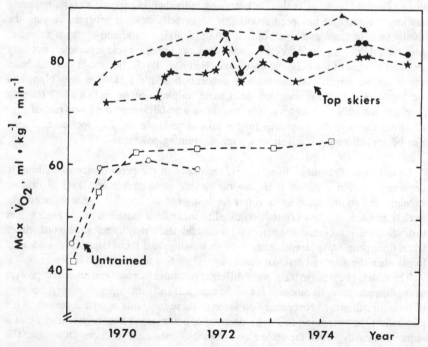

other hand, they could not compete with the athletes who apparently had a more favorable genetic background.

Tests Objective tests to determine the effect of training on different functions are both important and desirable. Such tests may be an aid in the development of the program and they encourage the individual to continue the training. Apart from the cycle ergometer test, a simple step test may be used (Chap. 8). Fig. 8-5 presents an example of the heart rate during a standardized work rate during a period of training.

Oxygen-transporting System

Like the skeletal muscle, the heart is adaptable to variations in the individual's physical activity. It is commonplace for athletes in endurance events to have a large *heart volume*. In older studies, including the results presented in Fig. 10-12, the heart volume was estimated from biplane radiographs. Recent studies applying echocardiographic and radionuclear techniques support the older observations. Endurance trained people usually have an increase in left ventricular end-diastolic volume without changes in wall thickness, which means an increase in ventricular mass. In their summary of eight longitudinal studies of endurance training in sedentary individuals Péronnet et al. (1981) report a 2.5 percent increase in left ventricular end-diastolic volume at rest. It is a very modest increase, but expressed in volume changes it is a 16 percent

Figure 10-12
The heart volume per kilogram body weight for members of different national teams of Germany and a group of untrained individuals of the same age. Each dot represents one subject. Note the large individual variations within a group. *(From Roskamm, 1967.)*

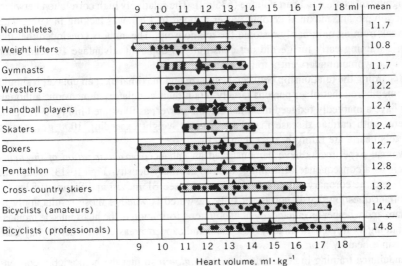

	9 10 11 12 13 14 15 16 17 18 ml	mean
Nonathletes		11.7
Weight lifters		10.8
Gymnasts		11.7
Wrestlers		12.2
Handball players		12.4
Skaters		12.4
Boxers		12.7
Pentathlon		12.8
Cross-country skiers		13.2
Bicyclists (amateurs)		14.4
Bicyclists (professionals)		14.8

9   10   11   12   13   14   15   16   17   18

Heart volume, $ml \cdot kg^{-1}$

increase. The average gain in maximal oxygen uptake was actually 17 percent. The ventricular volume in endurance trained athletes was 30 to 35 percent larger than in sedentary individuals. Blomqvist and Saltin (1983) conclude in their review that most studies have not demonstrated significant training effects, enhancing the contractile performance of the heart muscle, i.e., the quality of the myocardium does not change. The data from the study of Schaible et al. (1981), who found evidence for improved contractile performance in their running or swimming trained rats, remain to be confirmed. They also noted higher levels of actomyosin and myosin ATPase activity, and discovered that the hearts of conditioned rats were partially resistant to hypoxia.

In relation to body weight, the rabbit has a small heart compared to the hare (see Fig. 9-6). With training, the rabbit becomes more like the hare as far as its physical constitution is concerned (Arshavsky, personal communication). It is a general trend that physically active animals have a higher heart-weight/body-weight ratio than related inactive animals (e.g., wild animals compared with the same species kept in a zoo).

Many of the endurance trained athletes maintain their large heart size even after they have adopted a sedentary life style. In other cases, significant reductions have been noted with significant decreases in left ventricular wall thickness measured at least one-and-a-half year after cessation of training (Schumacher and Howald, 1984; see Chap. 4).

Fig. 4-28 illustrates the relation between heart volume and maximal oxygen uptake for 30 girl swimmers. What happened with these parameters after the termination of their intensive training? Sixteen of them were restudied 10 years later, in 1971; at that time the mean heart volume was the same as in 1961, or 625 ml (Eriksson et al., 1975). The maximal oxygen uptake had decreased from 2.80 to 2.17 liters · min⁻¹, and with a 6 kg gain in body weight, the weight corrected maximal oxygen uptake had declined from 51.4 to 36.1 ml · kg⁻¹ · min⁻¹. After a 12-week swim-training (in 1971) the maximal oxygen uptake increased by 14 percent, to 2.47 liters · min⁻¹, with no significant change in heart volume (635 ml). The relatively high correlation between heart volume and maximal aerobic power noticed in 1961 was lacking in 1971. The 12-week training only partly restored the relationship. The authors concluded that the intense training early in life did not constitute an apparent advantage 10 years later when the women underwent a training program.

Regarding the relationship between heart size and maximal oxygen uptake, it should be remembered that older persons largely retain their circulatory dimensions although they have a markedly reduced maximal aerobic power, as illustrated in Fig. 9-7. Also, patients with cardiac disease may have a heart weight exceeding 1000 g, whereas athlete's hearts are rarely heavier than 500 g.

Animal experiments have demonstrated an enhanced *vascularization of the heart muscle* as a consequence of training, both in capillary density (Fig. 10-13; Hudlická, 1982) and the coronary vascular bed as a whole (see Blomqvist and Saltin, 1983). An important question that is not as yet fully answered is whether those results apply to humans and whether an enhancement of collateral formation in the heart muscle induced by training is limited to already vascular handicapped areas or whether it may also occur in a healthy heart.

Endurance training increases the *total hemoglobin* so that the hemoglobin concen-

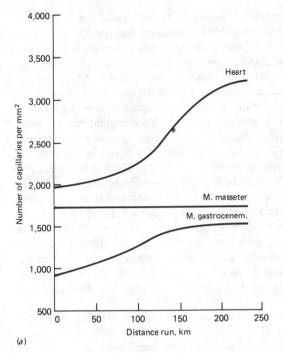

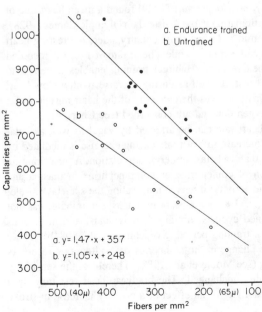

Figure 10-13
Increased vascularization of a muscle as the result of training. *(a)* Capillary density in three muscles of guinea pigs in the course of daily training on a treadmill at a speed that was gradually increased to 60 m · min⁻¹, running distance about 1,800 m each time. Animals were analyzed after different training times. Note that in the masseter muscle there was no change in capillary density in contrast with the systematically exercised muscles. *(Modified from Petrén, 1936.)* *(b)* Diagram showing the relation between the capillary density (capillaries per mm²) and the fiber area (fibers per mm²). Each point represents the values from one person. Two diameter values are indicated to show the relationship between the number of fibers per mm² and the lesser fiber diameter. It is seen that the number of capillaries decreases with increasing fiber area or fiber diameter. Increasing the fiber diameter with 10μm reduces the number of capillaries per mm² by about 150. It is evident that an endurance-trained person (with maximal oxygen uptake of more than 70 ml · kg⁻¹ · min⁻¹) may have fewer capillaries per mm² than an untrained person (with maximal oxygen uptake of 50 ml · kg⁻¹ · min⁻¹), if their fiber diameters are sufficiently different. *(By courtesy of P. Brodal, F. Ingjer, and L. Hermansen.)*

tration is usually maintained constant. However, it is not unusual for well-trained endurance athletes to have a lower hemoglobin concentration than nonathletes, despite an increase in their total amount of circulating hemoglobin (see Hunding et al., 1981). One possible explanation for this phenomenon is an increased mechanical destruction of red cells (hemolytic anemia). Why, then isn't it compensated for? Hallberg and Magnusson (1984) have launched an interesting hypothesis: the trained person has a higher concentration of 2,3-DPG shifting the oxygen dissociation curve of the hemoglobin to the right (see Chap. 5). At any level of blood oxygen tension, this shift leads to an increased delivery of oxygen to the tissue. The sensor responsible for the erythropoietin level, regulating the rate of red cell production, will receive the same information about the oxygen delivery potential of the arterial blood of the athletes as if, at a normal level of 2,3-DPG, their hemoglobin concentrations were normal. In other words, the regulatory mechanisms are cheated. The increased level of 2,3-DPG would therefore be associated with a reduction in the production of erythropoietin and thereby induce a lower hemoglobin concentration. These events would explain the development of a "sports anemia" even if the supply of iron is sufficient. Hallberg concludes that ". . . during the development of the human race—by selection of the fittest—the ability to run very long distances was not decisive for survival. Running short distances fast, to catch or escape wild animals, was probably far more important for our predecessors."

Prolonged bed rest, fluid immersion, and exposure to outer space reduce the plasma volume and red cell volume (see Miller et al., 1964; Greenleaf, 1984).

It has long been established that individuals known to have considerable endurance usually have a slow resting *heart rate*. Hoogerwerf (1929) found a mean heart rate of 50 beats $\cdot$ min^{-1} in 260 athletes participating in the Amsterdam Olympic Games (1928), the lowest value being 30 beats $\cdot$ min^{-1}. In one cross-country skier, the resting heart rate was repeatedly as low as 28 beats $\cdot$ min^{-1}, while a heart rate of 170 was recorded during heavy exercise (unpublished results). Habitual training enables a person to achieve a certain cardiac output at rest, as well as during exercise, with a slow heart rate and a large stroke volume. This improves the economy of the heart muscle as far as the energy requirement and oxygen demand are concerned (see Chap. 4).

It should be recalled that the heart rate can be affected by vagal as well as beta-adrenergic activities, and that the increase in heart rate during exercise is mediated by a combination of a vagal withdrawal and a beta-adrenergic stimulation. At rest, training does not modify the concentration of epinephrine or norepinephrine in plasma and myocardial tissue. The sympathetic activity in nerves innervating the skeletal muscle, expressed as sympathetic bursts per 100 heart beats or bursts per minute was not significantly different in well-trained cyclists, untrained individuals, and in persons studied before and after an 8-week training period (Svedenhag et al., 1984b). There are data indicating that there is no change in the number of beta-adrenergic receptors in the heart as an effect of training (see Moore et al., 1982). Therefore, the sensitivity of the S-A node apparently remains unchanged. The bradycardia at rest may be a consequence of an elevated vagal activity or an effect of a primary increase in stroke volume.

During exercise, the plasma concentration of epinephrine and norepinephrine is lower at any given oxygen uptake (and cardiac output) after training. At a given percentage of maximal oxygen uptake or given heart rate, the plasma concentration of those hormones is unchanged or slightly reduced (see Blomqvist and Saltin, 1983; Hollmann et al., 1981, respectively). Apparently the reduced heart rate in the trained person is associated with a reduced sympathetic activity, as well as an increase in vagal activity, at least at moderate rates of exercise. (It should be pointed out that the plasma concentration of norepinephrine is a relatively poor indicator of the sympathetic drive because the reuptake of this hormone at the point of release is very effective.)

Whether or not the stretch receptors in the dilated and hypertrophic atria might elicit a bradycardia has also been discussed. There is no evidence, however, in support of such a mechanism. A cardiac patient may also have a large heart, but this is not associated with any bradycardia. The increased blood volume caused by training causes an improved venous return and enhances the filling of the heart during diastole.

Many *patients* are treated with *beta-adrenergic blockers*. How do they respond to physical training? Many studies performed on patients, healthy individuals, as well as on animals have been devoted to this problem. Most studies report similar improvement in maximal oxygen uptake in subjects on chronic beta-adrenergic blockage as in nontreated controls. The hemodynamic responses to milder exercises were also similar in the two groups, but at higher rates of exercise the reduction in heart rate after training seems to be less pronounced in the subjects on medication. (See Vanhees et al., 1982; Svedenhag et al., 1984a.) Mullin and associates (1984), on the basis of experiments on rats, came to the same conclusion. In addition, they concluded that training with or without limited elevation of the exercise heart rate did not stimulate to an enhanced functional potential in the heart muscle, a question that was discussed above. Apparently, the stimulus of cardiac beta-receptors or the magnitude of the heart rate increase as such during the training sessions are not essential for the development of training bradycardia (see Nylander, 1985).

Summary Endurance training induces an increase in heart volume and heart muscle mass involving all chambers. There is a moderate increase in the plasma volume and total amount of hemoglobin with a maintained or in some cases a reduced hemoglobin concentration. Very noticeable is the reduction in heart rate at rest as well as during exercise. It may be concluded that the mechanisms underlying the effects of training on the heart function are by no means clear. The final result is a diminished demand on the oxygen consumption and thereby on the blood flow through the heart muscle at a given cardiac output. There are good indications to suggest that the coronary vascular bed is increased, at least at the capillary level.

Already, Robinson and Harmon (1941b) documented that a period of training could increase the subjects' *maximal oxygen uptake*. Since then, numerous studies have been devoted to the effect of training programs of different intensity, frequency, and duration on the oxygen transporting system. (For references see Saltin and Rowell, 1980; Blomqvist and Saltin, 1983, Table 10-1.) In sedentary people, low-intensity training with sessions lasting about 30 min, repeated 3 times per week, and demanding approximately 50 percent of the maximal oxygen uptake, can increase the maximal oxygen uptake 5 to 10 percent after 6 to 12 weeks. With more intensive training a 10 to 20 percent improvement is a common finding (training demanding 70–80 percent of the maximal aerobic power, or higher). As pointed out when discussing the "Dallas study", a given training program could increase the maximal oxygen uptake from 4 up to 100 percent, depending on the pretraining conditions. There are large individual differences in the response to training (see Pollock, 1973; Lortie et al., 1984). When expressing an improvement in percentage of the maximal oxygen uptake, it is not surprising that those individuals who start with a relatively low level of fitness improve the most. However, natural endowments set a final ceiling for the improvement.

Studies including individuals up to 70 years of age give similar figures for training-induced increases in maximal oxygen uptake as in young and middle-aged populations (see Liesen and Hollmann, 1976; Seals et al., 1984b). The normal decline in maximal oxygen uptake with age beyond twenty years can apparently be modified by regular training, corresponding to a rejuvenation of some 15 to 20 years (see Fig. 7-13). It should be pointed out that in the group of former physical education students, included in Fig. 7-13, restudied 20 years later, a decline in maximal oxygen uptake was noted without exception, although most of them were still very active physically. Preliminary data from repeated measurements 12 years later showed, surprisingly, no further decline in maximal oxygen uptake in the 27 women who took part in the study, and just a few percent reduction in the 26 men (mean values for oxygen uptake in liters $\cdot$ min^{-1}).

Few studies are available analyzing the effect of interval and distance training in *prepubescent children*. One problem is to separate the effects of growth and of training and control groups are important. When comparing young boys engaged in strenuous training programs (such as ice hockey), with groups who did not participate in regular training programs, the maximal oxygen uptake (liters $\cdot$ min^{-1}) was not significantly different (Hamilton and Andrew, 1976). This was also the case in a longitudinal study (see Lussier and Buskirk, 1977; Shepard, 1982). Kobayaski et al., (1978) note that: ". . . a remarkable increase in aerobic power was not observed in trained boys before the age of peak height growth velocity" (which is related to the onset of puberty, see Chap. 7 and Fig. 7-13). More studies are needed with different training intensity, duration, and frequency to provide a more comprehensive knowledge of the potential to improve aerobic power by training in children.

The maximal oxygen uptake during one-leg exercise on a cycle ergometer is normally 75 percent of the maximal oxygen uptake attained when cycling with both legs. Training one leg increases the maximal oxygen uptake in the one leg exercise some 15 to 20 percent, but there is no change with the untrained leg. The two-leg maximum is only 5 to 10 percent higher after the period of training. The maximal oxygen uptake during exercise with the trained leg is actually 85 to 90 percent of the two-leg maximum (see Davies and Sargeant, 1975; Blomqvist and Saltin, 1983). The conclusion is that the training effect of one-leg exercise is mainly peripheral and confined to that particular leg, but in two-leg exercise the central cardiovascular system is limiting the oxygen uptake (see below).

What *central circulatory adjustments* can explain the training-induced increase in *maximal oxygen uptake?* According to the Fick principle (Chap. 4), increased cardiac output and/or arteriovenous (a-$\bar{v}$) oxygen difference will increase the maximal oxygen uptake. In longitudinal studies of sedentary young men, an increase in maximal cardiac output and systemic a-$\bar{v}$O$_2$ difference both contribute to the increased $\dot{V}_{O_2}$ max, often at a 1:1 ratio. With prolonged training resulting in a further increase in the maximal aerobic power, an increment in the cardiac output is the cause of the increase. It should be recalled that there is no significant difference in the maximal a-$\bar{v}O_2$ difference between well-trained athletes with very high maximal aerobic power and trained subjects with much lower $\dot{V}_{O_2}$ max (Fig. 4-21). The individual variations in maximal stroke volume are much larger than differences in a-$\bar{v}O_2$ differences (within each sex).

There are few longitudinal studies of the effect of training on the central circulation in women and older man. Kilbom (1971) reported no increase in a-$\bar{v}$ O_2 difference in women. For older men, there are similar observations (see Blomqvist and Saltin, 1983), but an increase in a-$\bar{v}$ O_2 difference has also been noted (Seals et al., 1984b).

In most of the longitudinal studies reported, including the "Dallas study," training and bed rest did not modify the maximal heart rate; a variation in maximal cardiac output is therefore exclusively caused by changes in stroke volume.

The *cardiac output during submaximal exercise* at a given rate of work does not change as a result of training. Well-trained endurance athletes attain the same cardiac output as sedentary individuals. As mentioned, the heart rate, at a given oxygen uptake, becomes reduced. Whether the primary event is the increase in stroke volume or decrease in heart rate is at present impossible to say, as discussed above. The larger post-training left ventricular end-diastolic volume is a consequence of the prolonged diastole.

From a review of the literature Seals and Hagberg (1984) concluded that ". . . most of the studies reviewed reported modest reductions in *blood pressure* ($\bar{x} \leq 10$ mm Hg) at rest and during submaximal exercise after training. However, even the modest reductions in blood pressure reported in these studies must be interpreted with caution because of numerous methodological shortcomings and inadequate study design, most notably the omission of non-exercising hypertensive control groups."

Man-i and Imachi (1981), have developed a device that permits the continuous transmission of the systolic blood pressure telemetrically to a recorder from exercising subjects. They observed that the blood pressure of trained subjects tended to be more stable and remained lower during exercise than in the untrained individuals. While the blood pressure in the untrained rose or fell more or less in keeping with the heart rate, the blood pressure of the trained subjects changed less, and returned more rapidly to the original level after exercise.

Well-trained athletes with cardiac output around 40 liters $\cdot$ min^{-1} during maximal exercise have the same blood pressure as less active subjects with much lower maximal oxygen uptake. Since the mean blood pressure is the product of the cardiac output times the peripheral resistance, this resistance must be reduced in proportion to the increase in cardiac output. A reduced sympathetic drive and capillary growth in the post-training state will contribute to a reduction in the peripheral resistance.

Training has a relatively small effect on *pulmonary function*. In Chap. 5, it was mentioned that the capacity of the respiratory muscles to ventilate the lungs is not fully utilized in maximal exercise. Since the vital capacity does not change with training, at least not in adults, no major changes in the respiratory depth are to be expected during exercise. At any given level of pulmonary ventilation the mechanical work of breathing is the same for trained and untrained persons.

During submaximal exercise of relatively low intensity, the *pulmonary ventilation* *per liter O_2 consumed* does not change materially with training. Apparently, the depth of respiration is increased somewhat, associated with a corresponding reduction in the respiratory rate. During heavier exercise, the ventilation per liter O_2 uptake is reduced, but reaches a higher level during maximal exercise. A study of Fig. 5-9 facilitates the explanation of these findings. During light exercise, the level of the pulmonary ventilation is primarily determined by the CO_2 production, which is directly related to the O_2 utilization. During heavier exercise, the pH is also altered, primarily by the end-products from the glycogenolysis, which causes a relatively steeper increase in the pulmonary ventilation. Since the blood lactate concentration is generally lower during submaximal exercise following training, the respiratory drive is reduced, and the result is a lower pulmonary ventilation. The higher maximal ventilation is partially due to the increased maximal aerobic power, leading to an increased CO_2 production, and is also due in part to the higher maximal lactic acid level. Thus, the untrained individual and individuals with low maximal aerobic power fall within the left side of the shaded area in Fig. 5-9, while training moves the curve downward to the right.

Summary The magnitude of the increase in maximal oxygen uptake during 2 to 3 months of training, 30 min each section, 3 times per week, is on the order of 10 to 20 percent, but with large individual variations. Sedentary unfit persons will improve most. Depending on the initial level of fitness, duration and number of sessions per week, the increase in maximal oxygen uptake varies from 0 to almost 100 percent, according to the literature. The frequent recommendation that the training sessions should be three times per week can partly be explained by the fact that this regimen has been predominantly used in the research. Furthermore, as mentioned in the introduction, training six times per week is not twice as effective as training three times per week in improving aerobic power; there is an element of cost-benefit philosophy in the recommendation for the training of the nonathlete. Both interval training and more continuous exercise are effective.

There is a dual basis for an increased maximal oxygen uptake: (1) an increased maximal cardiac output; (2) an increased a-$\bar{v}$ O_2 difference. In younger subjects with initially low maximal oxygen uptake, there is an equal increase in cardiac output and a-$\bar{v}$ O_2 difference during the first 2 to 3 months. With continued intensive training a further increase in the maximal aerobic power is exclusively explained by an additional increased cardiac output. Since the maximal heart rate does not change significantly during training, the elevated maximal cardiac output is exclusively due to an increase in stroke volume.

During submaximal exercise at a given aerobic power, the cardiac output does not change, the heart rate is lowered and stroke volume increased. The systemic arterial blood pressure is moderately reduced or unaffected by the training.

Different stages of physical fitness have little effect on the pulmonary function.

Peripheral adaptation to aerobic training is particularly evident in the trained skeletal muscle. When discussing the effects of strength training it was pointed out that the muscle is a very plastic tissue. As we shall see, that statement also applies to its response to aerobic training.

As mentioned, a useful model for studies of the effect of physical training has been developed: the muscles of one limb are engaged in the training, the contralateral limb serving as a control. The advantage of this model is that both the trained and the "control" tissues in the post-training experiments are exposed to the same arterial supply of substrates, hormones, autonomic neuron messages, and central nervous system commands. From the discussion in Chap. 4, it is evident that in exercise involving a large muscle mass, the central circulation, i.e., the volume of oxygen offered to the skeletal muscles, is the limiting factor. Saltin et al. (see Saltin, 1985), in experiments including maximal dynamic exercise involving just the quadriceps muscles in knee extensions, concluded that the potential for oxygen uptake per kg muscle tissue was 0.3 liter · kg^{-1} tissue · min^{-1}, and that the capacity of the skeletal muscles to utilize the blood flow exceeded the capacity of the heart to provide that flow by a factor of two to three. Therefore, a locally improved aerobic performance would be of no advantage in a "whole body exercise" unless the central circulation was improved.

Another model is to train each leg separately and then to analyze the situation during maximal one-leg and two-leg exercise, respectively. In the first case, the maximal oxygen uptake increased by 19 percent and the cardiac output by 16 percent. In the two-leg exercise the improvement was not so large, or 11 and 11 percent, respectively (Klausen et al., 1982). (The systemic mean blood pressure decreased in all exercise situations by approximately 10 mm Hg, 1.3 kP.) One interesting finding was made: One leg, e.g., the left leg, exercised at a high work rate and continued to do so when the right leg started a similar exercise. Then, the blood flow in the left leg decreased due to an increased vascular resistance in the leg. As a consequence, a more pronounced metabolic acidosis developed. Evidently, there is a sympathetic vasoconstriction even in maximally contracting muscles. These data are important as a basis for the discussion of the effect of training on the regulation of regional blood flow. The study also confirms the common view that the central circulation is limiting the maximal oxygen uptake when large muscle groups are exercising. The central circulation would not have the capacity to provide an adequate perfusion pressure and to supply oxygen to the total mass activated if all the precapillary vessels were dilated. In the case of our ancestors, prolonged running at high speed was probably not necessary for survival. Consequently, there was no need to develop a central circulation capable of supporting a maximally dilated vascular bed in all muscles at the same time.

Before proceeding with a more detailed discussion, let us list some of the effects of endurance training on the skeletal muscle. There is an *increase* in:

- capillary density
- mitochondrial volume, with an increase of the enzymes involved in the citric acid cycle and the electron-transport chain
- utilization of free fatty acids at an energy demand related to the muscle's maximal aerobic power
- myoglobin concentration (still controversial)
- potential to store glycogen
- local a-$\bar{v}$ O_2 difference at submaximal rates of exercise
- maximal flow rate through the muscle

The subsequent discussion will also include comments on eventual changes in muscle fiber composition.

The *capillary density* in skeletal (and cardiac) muscle increases during periods of endurance training. Also long-term electrical stimulation for a sufficient length of time (minimum 4–7 days, 8 hours a day) can promote capillary sprouting and proliferation (see Hudlická, 1982). It is not known whether there are mechanical factors connected with increased blood flow, metabolic factors connected with a greater demand for oxygen, and/or accumulation of metabolites, which triggers the capillary growth. In contrast, training with the aim of improving maximal muscular strength does not stimulate capillary growth. Due to an increase in the muscle fiber area, this training actually may reduce the capillary density (see Saltin and Gollnick, 1983; Tesch et al., 1984).

An increase in capillary density may reduce the distance between the blood and the cell interior, which will enhance the exchange of gases, substrates, and metabolites. The surface area available for this exchange will increase. With more capillaries in a given tissue volume, more blood can be accommodated in this vascular bed per unit of time; the mean transit time (MTT) is reduced, which allows a more complete exchange of material. Saltin (1985) points out that the primary advantage of the high capillary density in highly trained skeletal muscle is probably that it allows for an adequate MTT at high flow rates, thereby promoting the exchange of material. (Capillary growth in the heart muscle would produce a similar advantage.)

An increase in the myoglobin contents of the trained muscle will enhance the diffusion of oxygen (Chap. 5).

There is actually a high correlation between the capillary density in the skeletal muscles and their maximal oxygen uptake. Two examples: Brodal et al. (1977), counted the number of capillaries per fiber and per mm^2 in biopsy samples from the vastus lateralis of the quadriceps muscle in untrained and trained individuals (figs. 10-13b, 10-14). The mean number of capillaries per fiber was 32 percent greater in the trained versus the untrained group. The difference was highly significant ($p < 0.001$), and was almost of the same order of magnitude as the difference in maximal O_2 uptake between the two groups, i.e., 40 percent. The number of capillaries per mm^2 was 39 percent higher in the trained group than in the untrained group, the difference being highly significant ($p < 0.001$).

In a subsequent longitudinal study, Ingjer (1979) obtained muscle biopsy specimens from 7 women before and after a 24-week systematic endurance training program, causing an average increase in $\dot{V}_{O_2}$ max of 25 percent. A corresponding increase in the number of capillaries occurred. The number of capillaries per muscle fiber increased on the average from 1.39 to 1.79 (29 percent), and the number of capillaries per mm^{-2} increased on the average from 348 to 438. Thus, it appears that a number of new capillaries had been formed in the muscle during the training period, evidently as the result of the training itself. Ingjer (1979) observed no increase in the cross-sectional area of the muscle fibers, indicating that endurance training of this kind did not stimulate to an increase thickness of the muscle fiber.

One of the most important *metabolic effects* of training of the oxygen transporting system is the increase in the capacity of the oxidative pathways. This is reflected by

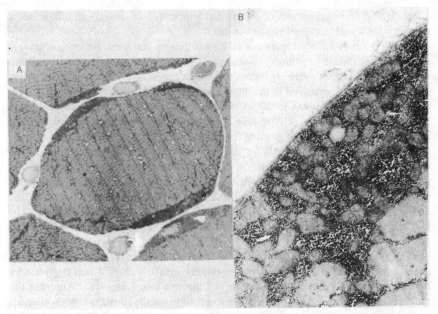

Figure 10-14
A: Electron microscopic picture ($\times 2000$) showing cross-section of a muscle fiber from one of the endurance-trained subjects studied by Ingjer (1979) after the 24-week training period. The muscle fiber is surrounded by four capillaries. The dark areas in the periphery of the fiber are mitochondria and granules of glycogen. A section of such an area is shown in B in greater magnification ($\times 20000$). The glycogen granules appear as black dots. The light spots in A are fat particles. *(By courtesy of Brodahl and Ingjer.)*

a large increase in the number and size of mitochondria in the trained skeletal muscles (see Howald, 1982; Saltin and Gollnick, 1983). There is a more or less parallel increase in the activities of the oxidative enzymes, including those of the citric acid cycle and electron transfer chain, and a slightly less increase in the enzyme involved in the beta-oxidation of the free fatty acids (FFA) beta-hydro-acyl-CoA-dehydrogenase, HAD. At one time, it was assumed that the enzyme activity limited the maximal aerobic power. It is now evident that, as was the case with the glycolytic enzymes, there seems to be an abundant supply of enzymes, considering the limited volume of oxygen which can be offered during maximal exercise. One study illustrates this fact: Henriksson and Reitman (1977) noticed a 19 percent increase in maximal aerobic power in their 13 subjects who had trained for endurance from 8 to 10 weeks. The activities of succinate dehydrogenase (SDH), and cytochrome oxidase had increased 32 and 35 percent, respectively (muscle samples from vastus lateralis). Within two weeks post-training, the cytochrome oxidase activity had returned to the pretraining level and after six weeks the SDH activity was back to the control level. However, the maximal oxygen uptake was still high, or 16 percent above the pretraining level. The authors conclude that an enhancement of the oxidative potential in skeletal muscle is not a necessity for a high maximal oxygen uptake.

Therefore, it is an open question as to whether or not there are one or several enzymes in the oxidative pathways that are rate limiting. One enzyme, which has been very popular in studies of the dynamics of chronic exposure to activity and inactivity, is the SDH activity, involved in the citric acid cycle. Apparently there is a more than ample supply of SDH, even in the sedentary muscle, but it is a good marker, and as mentioned, changes in its activity reflect the activity of other enzymes. An increase in its activity indicates that a training has been effective.

There is no good explanation for the regulatory mechanisms behind changes in enzyme activity. From experiments on rats, Henriksson et al. (1985) concluded that neither adrenomedullary hormones nor local sympathetic nerves are necessary for the post-training increase in muscle mitochondrial enzymes. Henriksson, Salmons, and Lowry (personal communication) found that 10 weeks of continuous stimulation (10 Hz) of a muscle triggered a much more dramatic increase in enzyme activity than did physical training.

It is an old observation that the trained individual can cover more of the energy demand during exercise at a given rate of oxygen uptake with *FFA and triglycerides as substrate* and thereby save glycogen and glucose (see Chap. 12). Applying the model to train one leg and use the other leg as a control has the advantage, as mentioned, that both legs are exposed to the same effects of diet, and concentration of substrates and hormones. With this model, Henriksson (1977) showed conclusively that the degree of FFA utilization was higher in the trained leg than in the untrained leg, indicating a difference in preference for substrate. This observation has been frequently confirmed (see Costill et al., 1979; Saltin and Gollnick, 1983; Kiens et al., 1985).

How, then, can the enhanced fat metabolism in a trained muscle be explained? It is not possible to answer this question completely, but there are some factors that may contribute. It was mentioned that both an increased capillary bed and a reduced mean transit time of blood favor a transport of substrates. The physiological site of the action of lipoprotein lipase (LPL) is in the luminar surface of the capillary epithelium. With an increased capillary bed, more binding sites for LPL will be available. This enzyme has a much higher activity in a trained muscle compared with an untrained muscle, in fact it may be as much as a 40 percent increase (Kiens et al., 1985). The degradation of triglyceride-rich particles is largely dependent on the activity of LPL. More FFA may be available for the exercising muscles, but in addition the LPL activity will transfer more surface material into high-density lipoprotein (HDL), and transfer it into the plasma compartment. This mechanism may explain the elevated plasma HDL concentration, noticed in endurance trained individuals (Kiens et al., 1984; see below).

Inside the cell, an increased activity of HAD may enhance a β-oxidation of FFA. Eventually, an increased concentration of acetyl-CoA from the FFA may compete with the formation of acetyl-CoA from pyruvate, thereby reducing the breakdown of glycogen. An increase in the concentration of citric acid may also inhibit glycogenolysis (see Chap. 12). It should be emphasized that more than enough of FFA is offered to the exercising muscle, and only a small percentage of the offered substrate (some 5 percent) is actually taken up by the muscle (Saltin and Gollnick, 1983). At any rate, in the endurance trained muscle, an increased FFA metabolism serves to save glycogen, and a given depot of glycogen will last longer. This will enhance performance in prolonged exercise.

The *glycogen content* of a muscle is very much affected by the diet. Endurance training may more or less empty the glycogen stores in the muscle fibers and the diet of hard training athletes is often rich in carbohydrate. These two factors alone might explain the slightly increased glycogen stores in trained muscle. In the trained muscle, there is an increase in the activity of glycogen synthetase and the glycogen branching enzyme. These events may explain an enhanced potential for storing glycogen in trained muscles (Holloszy and Booth, 1976).

Peripheral Circulation   With endurance training, a smaller fraction of the cardiac output is distributed to muscles activated *at a given rate of oxygen demand*. A plausible explanation is that an increased capillary bed allows a larger blood volume to be exposed to the skeletal muscles per unit of time, securing an exchange of gases, substrates, metabolites, and heat energy. Such a modification of the blood distribution would allow for a more liberal distribution of the blood flow to the liver, kidneys, skin, etc., which would be an advantage, particularly during prolonged exercise.

During *maximal exercise,* the trained muscle receives more blood per unit of time than untrained muscle. It was mentioned that there is a roughly parallel increase in maximal oxygen uptake and capillary density after some months of training. We can now add that the maximal blood flow through the muscles engaged in maximal exercise, follows the same pattern, quantitatively speaking. Due to the increased capillary bed, the mean transit time may be kept optimal, or close to it.

It is difficult to explain a gradually reduced blood flow through a muscle, active at a given rate of work, following a period of training. It was mentioned that the post-training sympathetic activity is diminished, which in itself should dilate the resistance vessels. Conceivably, the release of vasodilating mediators, produced by local factors, might be reduced by training. In the preceding text, it was emphasized that during maximal exercise with a large muscle mass involved, evidently some constriction of the precapillary resistance vessels must take place to secure an adequate perfusion pressure to vital organs (including the brain). Apparently part of this vasoconstriction is modified in the post-training situation.

How Does Endurance Training Affect the Muscle Fiber Composition and Metabolic Profile?

In Chap. 2, it was pointed out that continuous electrical stimulation of a nerve inner-vating an animal's fast-twitch muscle with a slow frequency (10 Hz), will gradually transform the muscle, so that all of its characteristics will become those typical for slow-twitch fibers (see Salmons and Henriksson, 1981; Pette, 1984). The capillary density and the activity of oxidative enzymes increase during the first week, and then follow a change in the contractile characteristics of the fibers. Some three to six weeks after the onset of nerve stimulation, the molecular structure of myosin and tropomyosin of the former type II fibers are now identical with those of the type I fibers. When the electrical stimulation is discontinued, there is a gradual retransformation of the fibers back to their original type II structure and behavior. This illustrates the plasticity of the skeletal muscle. A change of the type I fibers into type II fibers by a continuous stimulation with a high frequency, as typical for large motoneurons, is not as evident, particularly if the muscle's innervation is intact.

It is still an open question as to whether or not a large volume of endurance training can increase the number of slow-twitch fibers at the expense of fast-twitch fibers (see Schantz and Henriksson, 1983; Pette, 1984; Chap. 2).

Endurance Available data indicate that the maximal oxygen uptake in liters $\cdot$ min^{-1} or ml $\cdot$ kg^{-1} $\cdot$ min^{-1}, depending on the task, is of decisive importance in strenuous exercise lasting approximately 5 to 30 min. In prolonged energy demanding activities, the maximal aerobic power is less important for performance. Peripheral factors will be more important for success. When the energy demand exceeds approximately 70 percent of the maximal aerobic power, the enhanced FFA metabolism, due to training, seems to be important because a given glycogen depot then may last longer (Chap. 12).

Summary the reader is referred to the points listed previously. One effect of interval/endurance training is that the trained individual can tax a larger percentage of her or his maximal oxygen uptake. The reason for this is not obvious. A more effective O_2 supply to the exercising muscle due to an enlargement of the vascular bed, an enhanced diffusion, a higher enzymatic potential for an increased utilization of free fatty acids as a substrate, an increased glycogen content, and a higher psychological "fatigue threshold" may play a role. A higher oxygen uptake can be maintained without an accumulation of lactate in muscle, blood, or other tissues. With training, individuals become more accustomed to the exertion and are more willing to push themselves closer to their limit. It is important to keep in mind that achievements in endurance events may be improved beyond those that are mirrored in the observed changes in the maximal aerobic power. From a practical standpoint, it may be logical to list four components of a rational training program aimed at developing the different types of power:

1 Bursts of intense activity lasting only a few seconds may develop muscle strength and stronger tendons and ligaments.

2 Intense activity lasting for about 1 min, repeated after about 4 min of rest or mild exercise, may develop the anaerobic power.

3 Activity with large muscles involved, less than maximal intensity, for about 3 to 5 min, repeated after rest or mild exercise of similar duration may develop the aerobic power.

4 Activity of submaximal intensity lasting as long as 30 min or more may develop endurance, i.e., the ability to tax a larger percentage of the individual's maximal aerobic power.

Detraining

The adaptations in the oxygen-transporting system to regular exercise of various intensity, duration, and frequency are reversible, with the exception of heart size, which in many individuals may remain enlarged. Bed rest is an extreme form of inactivity, and the "Dallas study" provides a good illustration of its negative effects on maximal oxygen uptake (Fig. 10-7) and other functions.

It was mentioned earlier that the enzyme activity (in this case succinate dehydrogenase and cytochrome oxidase) fell markedly during the first weeks of detraining, while the maximal oxygen uptake was only slightly reduced (Henriksson and Reitman, 1977). The half-life of the oxidative enzymes seems to be approximately 1 week. (The half-life of the glycolytic enzymes is much shorter, or between about 30 min up to a few days; Illg and Pette, 1979).

Endurance seems to decline more rapidly than maximal aerobic power. This might be explained by the relatively fast reduction in the oxidative enzyme systems. Individuals who have been involved in intensive training for a long time are, on the other hand, reported to be more resistant to a loss in muscle mitochondria and capillary density than are subjects who have trained for only a few months (Coyle et al., 1984).

Specificity of Training

When discussing muscular strength it was pointed out that specific training may not improve the strength in other types of activities. In certain situations, there is a similar lack of transfer of the training effects of the oxygen transporting system. Above all, the adaptations in the skeletal muscles, just described, are limited to the muscles actually engaged in the training.

The most readily available indication of a training effect on the maximal aerobic power, is the heart rate response to a given submaximal rate of oxygen uptake. As mentioned, an increased maximal aerobic power is almost always associated with a reduced heart rate and *vice versa*. Three examples will be presented to illustrate the specificity of the heart rate response to training:

1 As mentioned, when training one leg, the increase in maximal oxygen uptake in that type of exercise is larger than when performing the exercise with both legs simultaneously. The drop in heart rate is also more pronounced in the one-leg than in the two-leg experiment.

2 With both legs trained separately, there is a significant decrease in the heart rate during the submaximal one-leg test (e.g. by 11 percent), but during the two-leg exercise the change is insignificant (e.g. a 2 percent decrease in the study by Klausen et al., 1982).

3 Clausen (1976) summarized data from studies on healthy, young subjects in whom the circulatory response to arm-and-leg exercise was assessed after a period of training of either arms or legs. Arm training caused a marked reduction in heart rate during exercise with the trained arm muscles (from 137 to 118 beats $\cdot$ min^{-1}). During exercise with the nontrained leg muscles, a much less pronounced decrease in heart

rate was seen (from 132 to 124 beats $\cdot$ min^{-1}). After leg training, however, the decrease in heart rate was about the same in the test with the trained leg muscles (from 135 to 122) as with the untrained arm muscles (from 127 to 112 beats $\cdot$ min^{-1}). It seems logical that there is more of a transfer effect on the central hemodynamics after training with large muscle groups involved, compared with training of small muscle groups. Also in these situations, the most prominent training effects are evident when the test is performed with the trained limbs.

Fig. 10-15 gives an example of data on maximal oxygen uptake, measured on a champion swimmer when swimming in a swimming flume and when running on a treadmill. For about four years, his maximal oxygen uptake was the same when running, but varying oxygen uptake values were measured during maximal swimming, due to variations in the training of the arms. Peaks were noted when the swimmer was most successful (winning two Olympic gold medals in 200 and 400 m medley, 1972).

Two identical twin sisters reached similar maximal oxygen uptake when running (3.6 liters $\cdot$ min^{-1}), but the sibling who had been training for competitive swimming attained the same maximal oxygen uptake of 3.6 liters $\cdot$ min^{-1} when swimming, which was a 30 percent higher maximal oxygen uptake than her twin sister who was not trained for competitive swimming. When swimming with arms only, the swim-trained sister could reach a 50 percent higher maximal oxygen uptake (I. Holmér, and P.-O. Åstrand, 1972; see also Magel et al., 1975).

The conclusion is: (1) a treadmill test is not a good predictor of performance in

Figure 10-15
Oxygen uptake during swimming and running at maximal intensities by one world-class swimmer over a six-year period. Note the unchanged oxygen uptake during running from 1970, whereas oxygen uptake during swimming with the whole stroke and with only the arms varied considerably during the seasons. *(Updated from Holmér, 1974.)*

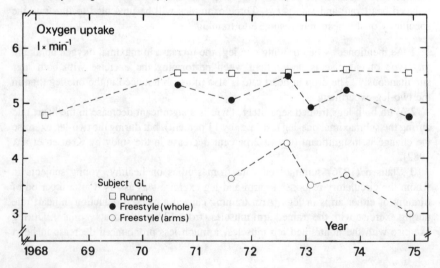

other types of activities; (2) in order to utilize the aerobic potential in an optimal way in a given activity, one must train in that activity.

Recovery after Exercise

There are very few longitudinal studies concerning the payment of oxygen debt, resting level of heart rate, cardiac output, temperature, etc., before and after training. The fast recovery of athletes after muscular exercise, in contrast, is well established.

Hartley and Saltin (1968) observed that untrained subjects who (1) exercised for 6 min at an O_2 uptake of about 40 percent of their maximal aerobic power, then (2) exercised for 6 min at a load of about 70 percent of their maximum, followed by (3) a 10-min rest, and then (4) repeated the 40 percent load, now have the same O_2 uptake and cardiac output as under (1), but they had a significantly higher heart rate, in many cases as much as 20 beats $\cdot$ min^{-1} higher, and consequently also had a smaller stroke volume. Following training, this increase in heart rate was much less, and well-trained athletes may actually accomplish the entire procedure with the results from (1) and (4) being almost identical. This type of testing appears quite promising as a method of assessing the level of physical fitness or the level of habitual physical exercise.

Mechanical Efficiency, Technique

In activities that are relatively uncomplicated technically, such as walking, running, or bicycling, there is a very slight increase in efficiency with training, but this increase is less than the variability among individuals. There is no definite difference in the consumption of energy per kilogram of body weight and per kilometer in differently trained runners, as groups (see Åstrand, 1956; Sjödin and Svedenhag, 1985). It is interesting to note that there is a wide variation in running economy even in elite marathon runners, with no significant correlation between running economy and performance time of marathon runners within a narrow racing time range. There are reports of world-class runners with maximal oxygen uptake below 70 ml $\cdot$ kg^{-1} $\cdot$ min^{-1}, but they are also characterized by an extremely efficient running economy. It is impossible to say whether this is a training effect or whether it is due to inherent qualities (see Sjödin and Svedenhag, 1985).

On a cycle ergometer, both Olympic medal bicyclists and untrained persons have the same mechanical efficiency at submaximal work rates. In Eskimos, totally unfamiliar with the use of a bicycle, however, Vokac and Rodahl (1976) found a distinctly higher oxygen uptake at a given work rate on the cycle ergometer than they found in whites. The more complicated the exercise, the greater the individual variations in mechanical efficiency and the greater the improvement with training. It should be pointed out that the mechanical efficiency of a person performing heavy exercise may be overestimated if the assessment is based on measurement of O_2 uptake, since anaerobic processes may have contributed to the energy yield.

In many achievements, the aim of the training is not primarily to reduce the energy expenditure during the event in question, but to attain an improvement "at all costs," for example by increasing the developed power. An example is given by Lauru (1957), who describes an athlete performing a broad jump from a force platform in which force phenomena during movements vertically, frontally, and transversely could be recorded and analyzed. Before training, the push against the platform during the jump reached 1000 N. During the training, the push increased to 1400 N. The performance was at least partially improved by an improved coordination and a smoother sequence of motions.

Motor learning is so specific in nature that one cannot speak of a general learning ability with reference to motor coordination in skilled movements (see Chap. 3).

Everyone has had personal experience with learning techniques, and has experienced how skill and dexterity is maintained for years without practice, e.g., the strokes in swimming, the tennis serve, the balance on the bicycle, the grip on a guitar etc. *But,* the level of endurance is lost. However, extreme skill can be maintained by training up to a very old age. André Segovia gave brilliant concerts at the age of 91 (1984), and the late Arthur Rubenstein played Chopin perfectly at the age of 88.

Body Composition

During inactivity or habitual training, the body weight may remain relatively constant although inactivity often produces a gradual weight gain. During intensive training, the density of the body increases while the skin-fold thickness decreases (see Pářizková, 1977; Wilmore, 1982).

An increased body density, i.e., with a 0.01 unit, indicates a reduction in fat content. This is also supported by the reduced skin-fold thickness observed. At the same time, there is a certain proportional increase in the muscle mass and the blood volume.

In the case of fat-free body mass, however, it is a matter of only a few kilograms variation in the body weight with different states of training. It is therefore evident that change in body weight alone is an inadequate index of possible alterations in body composition.

Blood Lipids and Lipoproteins

A number of studies during the last two to three decades have examined the possible relationship between physical activity and blood lipid levels. Many of them have investigated the effect of a specific training program on the blood cholesterol level. The results are conflicting: about half of the studies show a blood cholesterol-lowering effect of exercise; the other half show no effect (see Altekruse and Wilmore, 1973).

The question of blood lipoproteins has been the subject of renewed interest by the development of techniques that make it possible to fractionate cholesterol compounds into subfractions, notably high density lipoprotein (HDL), low density lipoprotein (LDL), and very low density lipoproteins (VLDL). LDL has been associated with the deposition of cholesterol in the blood vessels, while HDL has been claimed to have the opposite effect. High values of HDL are therefore considered to be desirable, and low levels of HDL are considered to be a risk factor in the development of coronary artery disease (for references see Paffenbarger and Hyde, 1980; Sutherland and Woodhouse, 1980). This question is still controversial, but a number of studies have concluded that physical activity above a certain level of intensity, is associated with high blood HDL levels, and a decrease in the LDL/HDL ratio (Wood et al., 1985). A possible mechanism for the observed elevation of HDL may be sought in the more extensive capillary bed as a result of training, leading to an increase in the lipoprotein lipase activity. This enzyme has the potential to transfer more surface material to HDL, and to release it into the plasma compartment (Kiens et al., 1984).

Hormones

It has already been mentioned that training will affect the balance between sympathetic and parasympathetic activities. At physical activities taxing more than about 30 percent of the individual's maximal oxygen uptake, increases in the plasma concentrations of both norepinephrine and epinephrine have been demonstrated. Norepinephrine levels are increased at exercise rates lower than those eliciting significant increases in epinephrine levels. The higher the work rate, the higher the increase in catecholamine concentration for a given increase in exercise intensity; but the increment is less following physical training. This is the case both at a given absolute submaximal exercise load and a given relative exercise load, expressed in percentage of maximal oxygen uptake (see Galbo, 1983). Winder et al. (1978), have shown that a significant reduction in catecholamine response to heavy exercise may occur after one week of vigorous training. The full effect is achieved at the end of 3 weeks, after which no further decrement in catecholamine response to heavy exercise occurs. However, the heart rate response to exercise continues to fall after the catecholamine response has leveled off. It is also well-documented that physical exertion results in elevated levels of cortisol (Favier et al., 1983).

It appears well established that the release of β-endorphins is elevated after prolonged, strenuous exercise (see Dearman and Francis, 1983). These neurohormones (β-endorphin, encephalin, and dynorphin), which have an effect resembling that of certain opiates, play an important role in general physiological stress reactions. They serve to reduce pain and enhance a feeling of well being. The increased release of endorphins during strenuous exercise may thus, at least in part, explain the feeling of well being commonly experienced at the end of a training session. As a matter of fact, attachment of endorphins to specific receptors in the brain, has been demonstrated in individuals after long distance running. It is also of interest to note that the endorphin antagonist nalaxon has the opposite effect (Janal et al., 1984; Surbey et al., 1984).

Physical training may cause an increased glucose tolerance. As a consequence, the individual may get along with less insulin (James et al., 1984). Evidently, this is due to an increased insulin sensitivity, as a consequence of training (for references, see Wallberg-Henriksson et al., 1982).

There is evidence that increased physical activity and training will increase the *thyroxine turnover* (Terjung and Winder, 1975). Plasma concentrations of growth hormone increase during exercise, but there are controversial reports on the effects of training in this response.

We can conclude that physical training does affect many hormone producing systems, but the significance of these responses is far from evident. As mentioned, many of the adaptations on a cellular level triggered by training are unique for the trained muscles, but are not found in "idle" muscles that must be subjected to the same changes in hormone levels. (For a more general discussion of the effect of training on the endocrine system see Galbo, 1983.)

MISCELLANEOUS

Does Hypoxia Enhance the Training Effect?

Training under hypoxic conditions (at low barometric pressure, or inhaling air with a reduced O_2 concentration) does not improve maximal strength or maximal oxygen uptake beyond the level achieved when training under normoxic conditions. Athletes who prepare to compete in events demanding a high oxygen uptake at an altitude above approximately 1500 m should allow for 1 to 2 weeks of acclimatization. At the present time, there are no convincing data to support the common belief that hypoxic training will improve performance when exposed to normal oxygen pressures (see Chap. 15).

Training and Tolerance to Hot Environments

Prolonged submaximal exercise performance is generally impaired by high ambient temperature. Repeated training sessions with an intensity and duration that will elevate the body temperature and elicit sweating, promote an acclimatization to heat. However, for an optimal performance during prolonged exercise in hot environments, acclimatization programs that include heat exposure seem necessary (see Chap. 13; Fortney and Vroman, 1985). So far we do not know how effective training in a hot environment is. At a given oxygen uptake and cardiac output there is often a gradual decrease in stroke volume, and certainly an elevated heart rate. In prolonged exercise, a smaller fraction of the maximal oxygen uptake can be maintained compared to exercise under temperate conditions.

Stretching

There are different stretching techniques: ballistic stretch, slow-stretch, and contract-relax stretching. The contract-relax stretching seems to be the most effective form of stretching to increase the range of motion in a joint. A common technique is to start with a maximal isometric contraction of the involved muscle group for 4 to 6 s, followed by complete relaxation lasting at least 2 s and then to conclude with a slow passive extension of the joint as far as possible without causing pain, and to maintain this maximally extended position for 8 to 20 s. This sequence is then repeated 5 to 6 times for each muscle group.

Stretching does not improve muscle strength, but it can prevent the quite commonly occurring reduced range-of-motion in joints engaged in training programs, which may persist for at least 24 hours after the training session (see Wiktorsson-Möller et al.,

1983). There are also reports to the effect that stretching may prevent muscle and tendon injuries. (For more details see Anderson, 1980.)

Training and Oxygen Radicals

It is well established that mitochondrial respiration in vitro, and most likely also in vivo, generates free oxygen radicals as a byproduct of aerobic metabolism. Such radicals, especially hydroxyl radicals, may initiate lipid peroxidation and thus result in cell injuries. In the course of evolution, aerobic cells, such as muscle fibers, have been equipped with enzymatic (peroxidases) and nonenzymatic scavenger systems against these radicals (see Chance et al., 1979; Del Maestro, 1980).

It appears likely that the increased aerobic metabolism, due to exercise as a byproduct, may cause a proportional increase in the production of free radicals. It has been pointed out that this situation may represent a negative consequence of training. We do not know at present whether or not this is the case. There are recent observations, however, which indicate that endurance training actually may *increase* the resistance of skeletal muscle to injuries caused by lipid peroxidation (Gilbert, 1981; Salminen and Vihko, 1983).

Training and Terminal Nerve Fiber Sprouting

In Chap. 3, it was mentioned that a denervated muscle fiber could be reinnervated by the process of fiber sprouting. From neighboring nerve terminals there is a sprouting; new functioning neuromuscular contact may be established. It has also been emphasized that the muscle mass becomes reduced at old age, presumably due to the death of motoneurons. Some of the denervated muscle cells will regain their innervation by terminal sprouting. The question is whether or not a stimulus such as endurance training may serve as a positive intervention, which will result in a reduced rate of senile muscle atrophy. From experiments performed on aging rats, trained by running, Stebbins et al. (1985), concluded that exercise can in fact stimulate the production of growth configuration and terminal sprouting.

Psychological Aspects

"There is a general assumption that physical fitness has psychological correlates. The concept of body-mind relationships is age-old. Psychosomatic research has indicated that physical changes result from continued psychological states; it seems logical to assume the reverse: that psychological changes result from physical states, such as fitness. Although there is a general assumption that this is so and considerable claims rather vaguely documented, there are surprisingly few firmly validated data" (Hammett, 1967). There are several studies demonstrating positive correlations between athletic ability on the one side and the general level of social adjustment and many psychological factors on the other. However, as emphasized by Hammett, they cannot be regarded as valid indications of psychological change related to increasing physical activity since it is equally possible that they reflect predilections; that is, persons with certain psychological characteristics may gravitate to physical training programs. There are too few longitudinal studies performed to date to permit us to draw any firm conclusions. In Kilbom's study (1971), the female sales personnel were questioned five times a day about some emotional dimensions and about subjective well-being. The emotional variables (irritation, interest in work, joy, etc.) were not significantly influenced, either from the morning to the afternoon, or before as compared to after training. However, the physical variables showed that fatigue was significantly more pronounced in the afternoon than in the morning, but it was less pronounced after training.

It appears that humans have a tendency to become inactive after they have reached puberty. Unfortunately, physical activity without a definite purpose becomes, from then on, rather rare. In the past, and in some countries and occupations even today, physical activity was a part of one's daily life and work. Inasmuch as the need for physical activity in connection with most daily occupations is about to be abolished, it will be necessary to devote some of the leisure time to physical activity in order to maintain an optimal function of the body. In this connection, it should be emphasized that effective physical training does not have to be very stressful. Nevertheless, it is a common misconception that physical training is unpleasant and difficult to arrange. Often the ideal solution is to acquire a hobby that involves some degree of physical activity, even though the training may not be very intensive. As a rule, most people can be motivated to exercise, provided proper facilities are available. Simple training tracks can, for example, be constructed in parks. Psychologically speaking, it is far more attractive to walk or run a 5-km course than to cover a 1-km course 5 times, to say nothing of 100 laps on a 50-m course. It is more acceptable to cover a distance that offers a certain amount of variation than to spend an equal amount of energy indoors on a stationary cycle.

On the whole, it must be admitted that for nonathletes, physical training may not be sufficiently enjoyable to make them exercise merely for fun, without being motivated by the prospects of the benefits they may obtain from the activity. More or less one must accept the fact that one has to train and that exercise is a matter of necessity in order to become fit, whether one likes it or not. It may, therefore, be best to get the exercise program over with the first thing in the morning—the sooner the better; then it is done. Eventually, as the benefits become apparent, one will miss the program if one does not keep it up. As an alternative, one may join a fitness club to train in a group; it may be more fun that way, and perhaps easier to continue, as one will be missed if he or she should fail to show up.

The most typical physical activity of our ancestors was the brisk walking of rather long distances in their search for food. The driving force was hunger, not exercise. Most likely, none of them were joggers. It is an interesting observation that dog owners are quite convinced that the dog, which is also another mammal, needs exercise. It is tempting to suggest: Go out and walk your dog even if you do not have a dog.

Muscular Soreness

If an untrained individual suddenly performs vigorous exercise, the engaged muscles may become painful, especially after eccentric exercise. The muscular soreness is accompanied by stiffness, tenderness, and reduced muscle strength. The symptoms usually appear from a few hours to a day after the exercise, and may be most pronounced during the second day. The symptoms gradually fade away so that the muscles become free of symptoms after 4 to 6 days.

The symptoms are probably, at least in part, caused by mechanical damage of the myofibrillar structures (Fridén, 1983), and the connective tissues within the muscle and in its attachment to the tendon. Secondarily, histamine and other elements produced or mobilized as a consequence of the inflammatory process, may cause edema and pain (see Armstrong, 1984). The repair of the damaged tissue results in a stronger muscle, much less susceptible to further injuries, even if the subsequent exercise is much more severe.

The localized, sustained, and painful cramp that occasionally can throw a muscle into vigorous involuntary contraction, after prolonged monotonous exercise or sometimes even during sleep, cannot be explained at present. A stretch of the muscle usually relieves it from the cramp, probably due to inhibitory impulses from the stimulated Golgi tendon organs (Sherrington's lengthening reaction).

ISCHEMIC HEART DISEASE

The not uncommon belief that competitive athletic exertion may be harmful to the heart and frequently leads to permanent damage is an old misconception that dates back to Hippocrates; that this is not the case has been shown conclusively (Dublin et al., 1949; Rook, 1954; Montoye, 1962; Pyörälä et al., 1967). It is also clear that sudden, unexpected death is very rare in young top-ranking athletes (Zeppilli and Venerando, 1981).

It is now generally recognized that physical activity is one of the factors of importance in the prevention of ischemic heart disease with atherosclerosis and infarction, in addition to factors such as heredity, diet, cigarette smoking, and way of life in general (Paffenbarger and Hyde, 1984). Some of the numerous studies on this subject (Zukel et al., 1959; Fox and Skinner, 1964; Fox and Haskell, 1968; Shaper et al, 1971; Brunner and Manellis, 1971; Kannel et al., 1971; Salonen et al., 1982; Salonen, 1983) have shown that ischemic heart disease is less frequent among physically active than among physically inactive individuals. It appears from the available evidence that the mortality of ischemic heart disease is twice as high in physically inactive as in active individuals. The chance of surviving the first attack of cardiac infarction is 2–3 times greater in patients who previously have been physically active than in those who have led a sedentary life. However, Tibblin and Wilhelmsen (1975) have questioned a number of Swedish men born in 1913 and found that the occurrence of myocardial infarct was no less in men whose occupation involved heavy physical work, while those who were physically active during their leisure time had a lower incidence of myocardial infarction than those who were inactive. Essentially similar conclusions were made by Gyntelberg (1974) on the basis of his studies in Copenhagen. For a detailed review of epidemiological studies on the association of physical activity with coronary heart disease, see Leon and Blackburn (1981).

In a recent review, Blackburn (1983) discusses the possible role of physical activity in the prevention of coronary heart disease, based on observations from anthropology, physiology, clinical and population studies. He concludes that habitual physical activity is importantly involved in the control and the prevention of elevated cardiovascular risk factors. Yet, in 1982 Kannel made the statement that "there are no data yet available that prospectively test the relation of physical fitness to subsequent coronary heart disease (Kannel, 1982)." Evidently, this is not entirely the case.

In a study of 17,944 middle-aged male office workers, Morris et al., (1980) found that over an $8\frac{1}{2}$ year period, those who engaged in vigorous physical activity and kept fit had an incidence of coronary heart disease (CHD) that was less than half that of their colleagues who recorded no vigorous exercise. The CHD rate of the vigorously active men was especially lower in fatal clinical manifestations throughout the age-range studied. This was also the case in men with a family history of CHD, the obese, cigarette smokers, as well as men with severe hypertension and subclinical angina.

Erikssen et al., (1981) examined a group of 2014 apparently healthy men aged forty to fifty-nine years of age. All of them were subject to a graded symptom-limited cycle ergometer test (a near maximal cycle exercise ECG test). A closer examination revealed that 182 of them had symptoms or signs indicating possible coronary artery disease. The remaining 1832 men were free of coronary heart disease. It was found that the higher levels of physical fitness were associated in all age groups with lower cigarette consumption, lower blood pressure, and lower serum lipids.

A subsequent follow-up study of this group of subjects seven years later (Erikssen and Mundal, 1982), revealed that the number of deaths from infarction or ischemic heart disease was significantly less in those who had a high level of physical fitness, than in those who were physically unfit. Thus, a poor state of physical fitness has a negative impact on the risk of dying from coronary heart disease.

The same study also showed that a poor state of physical fitness has a negative impact on the risk of coronary heart disease progression. Of the 182 men who were found not to be quite healthy, 115 had signs strongly suggesting latent coronary heart disease. In 110 of them, coronary angiography was performed after informed consent. Of these, 69 were found definitely to have coronary artery disease. Thirty two of them belonged to the lowest physical fitness classification within their age group. This low level of fitness was not caused by angina pectoris, for a positive ECG was the sole coronary heart disease (CHD) indicator in almost all of these subjects.

During the subsequent follow-up study, 20 of the 32 men belonging to the lowest physical fitness group showed definite coronary heart disease progression; i.e. they had had infarction, died from CHD or had angiographic evidence of progression. In contrast to this, only 1 out of 11 in the upper physical fitness quartile showed CHD progression according to the same criteria at the 7-year follow-up.

It is thus clear, that even those persons who have a high level of physical fitness may have coronary artery disease as evidenced by work electrocardiography or coronary angiography, but fewer of them die from it than do physically unfit individuals.

Essentially similar conclusions may be drawn from the observations of Mundal et al. (1985), who studied the incidence of coronary heart disease, level of fitness, and cardiac death in a group of military officers. Lie and Erikssen (1985) conducted a 5- and 7-year follow-up study of a group of 122 fit middle-aged and older competitive cross-country skiers, and found a remarkably low death rate of coronary heart disease. After 7 years, one skier had died from coronary heart disease and three had developed angina pectoris, as against an expected number of about a dozen in the general population (Lie and Erikssen, 1985).

Thus, it appears that those who are fit are less likely to die from coronary artery disease. Fitness is the result of physical training, i.e., physical activity of sufficient intensity and duration to produce training effect. Whether or not such training may prevent persons otherwise prone to contracting coronary artery disease from developing this disease, is not clear. But at any rate a well-trained physically active person may probably be more alive while living, even though they may not live longer. (For a further discussion, see Rapaport, 1980.)

A number of studies have shown that carefully controlled, gradually increasing physical activity has a beneficial effect on the cardiac function in patients who have had a coronary infarction (Hollman et al., 1981). A gradually increasing training program, as a rule, improves the physical work capacity and the electrocardiogram, so that the patient may tolerate a greater work load without symptoms. There may be several reasons for the beneficial effect of physical activity on the ischemic symptoms. It is conceivable that physical activity increases vascularization of the heart muscle, as shown experimentally in dogs (Eckstein, 1957), although, Tharp and Wagner (1982) failed to find any evidence in rats that exercise training stimulated the multiplication of cardiac capillaries. They consider that the beneficial effect of exercise on the heart is probably a result of an enlargement of the coronary arteries or the collateral circulation. In addition, physical activity will, as in healthy individuals, lower the heart rate also in patients with ischemic heart disease, so that the same physical work rate may be endured with a slower heart rate. This will lengthen the diastole. Thus, the blood perfusion of the heart muscle itself is enhanced, since the perfusion of the myocardium is about $2\frac{1}{2}$ times greater during the diastole than during the systole. From all the evidence, it appears that it is this latter factor, which is of the greatest importance in the exercise treatment of myocardial ischemia. Systematic studies by Barry et al., (1966) showed a significant improvement of the cardiac function with improvement of the electrocardiographic changes after several weeks of physical training associated with a significant reduction in the heart rate at a given submaximal work rate. When the training was discontinued and the work pulse increased to the original higher value, the electrocardiographic improvements also disappeared. However, with a further training period, the same improvements were once more produced simultaneously with a lowering of the work pulse, which suggests that the nonpermanent beneficial effect has to be primarily ascribed to the lowering of the heart rate. If the beneficial effect were to be attributed to an increased vascularization of the myocardium as a consequence of training, one would have expected the improvement to be of a longer duration, or more or less a lasting improvement.

As already mentioned a number of studies have examined the relationship between physical activity, or particularly endurance training, plasma lipids and the risk of coronary heart disease. On the whole, they conclude that physical activity of the endurance type tends to produce higher levels of high density lipoproteins and lower plasma concentrations of low density lipoproteins than is found in physically inactive persons of the same age and sex, and that this may be of significance in the prevention of coronary heart disease. (For references and further discussions, see Paffenberger and Hyde, 1980; Francis, 1980; Keys, 1980; Höstmark, 1982; Cowan, 1983; and Roberts, 1984.) It appears that weight loss and exercise have independent and additive effects on increasing high density lipoprotein cholesterol levels (Sopko et al. 1982). In this connection, it should be mentioned that daily exercise significantly reduced coronary atherosclerosis in monkeys (Kramsch et al., 1982).

The major differences between persons engaged in regular endurance-type exercise and physically inactive controls appears to be a lower plasma triglyceride concentration and greater high density lipoprotein levels, due to increased levels of the subfraction HDL_2 and aproprotein A-I (Haskell, 1984). According to Haskell, these lipoprotein changes require exercise training of the endurance-type at moderate intensity, amounting to some 4 MJ (1000 kcal) per week. Above this level, there is a dose-response relationship, with greater changes occurring up to an expenditure of 19 MJ per week.

As is evident from the facts presented in this chapter, the effects of regular physical activity are indeed beneficial. This is certainly the case from a physiological point of view since these effects constitute a definite improvement in a number of basic physiological functions.

Nevertheless, there are clinicians concerned with their concepts of primary and secondary prevention, who seriously question the value of physical exercise in terms of preventing or treating disease. Their skepticism is based on the lack of clear-cut evidence from large-scale epidemiological studies, showing beyond a shadow of a doubt that physical activity can prevent or favorably modify the development of disease processes such as coronary heart disease. In all probability, it will never be possible to conduct such population studies in a manner that will satisfy those who are in doubt, or that will meet the most stringent of requirements for the validity of such studies. Nonetheless, this is a weakness that must be admitted. But granting the existence of these uncertainties, and the basis for the doubters' doubt, the clear-cut physiological evidence referred to above should at any rate not be overlooked. This fact alone, should be reason enough for the advocation in favor of habitual physical activity.

CONTRAINDICATIONS FOR PHYSICAL TRAINING

In some cases, physical activity may be contraindicated. More often than not, however, it may be a matter of reducing the level of activity or training rather than completely abolishing it.

The practicing physician is often faced with the question whether a patient may continue the training, or if the training should be discontinued, and if so, when can the patient begin again. At times it may be quite difficult for the physician to be certain as to what advice to offer the patient. A sensible recommendation would depend on, in addition to a careful medical examination, a comprehensive knowledge of the physiological demand that the training or activity in question places upon the patient, as well as some insight into the pathophysiology of the patient's condition. Because of uncertainty on the part of the physician, it is frequently observed that, in order to be on the safe side, the doctor recommends prolonged rest or inactivity, or may ask the patient to "take it easy," without specifying precisely what is meant by the term: taking it easy. In consequence, the conscientious patient may end up being more inactive than the physician had wished to suggest. In the case of an athlete engaged in active competitive sports, the consequence of such prolonged inactivity may be the loss of an entire season. In the case of nonathletes engaged in regular fitness training, such a prolonged interruption of the training may actually be the end of his or her training habit because the patient never gets around to taking it up again.

There are a few medical conditions, however, that do require an interruption of the training program, either partially or completely, depending on the nature of the ailment.

Upper respiratory infections with sore throat and nasal congestions are not infrequent in athletes. As long as the patient has no fever, it may not be necessary to discontinue the training program. A few days break in the training may be all that is necessary. If the patient, on the other hand, is febrile, one has to be more cautious. His or her resting pulse as well as the work pulse at a standard work load, is almost invariably elevated. As a general rule, the patient should refrain from all training as long as febrile. When afebrile, he or she should start gently and increase the intensity of training gradually, increasing to full intensity in about a week. One of the reasons why such great caution has been recommended in the case of flu-like symptoms in athletes is the fear of myocarditis as a possible complication (Jokl, 1964). The actual risk of developing myocarditis under such conditions, is still an open question, however. A few cases of sudden death in young athletes have been attributed to possible myocarditis, but few documented cases have actually been reported in the literature. Nonetheless it is advisable to have this possibility in mind, when treating young athletes.

Lower respiratory infections, such as acute bronchitis and pneumonia, require cessation of training until the principal signs and the sedimentation rate have returned to normal. Preferably, x-ray examination of the chest should be made before the training is started again.

Acute nephritis requires an immediate stop of all types of strenuous physical training. The training should be discontinued until urine analysis reveals normal findings and the blood pressure has returned to normal. It should also be kept in mind that the disease, as a rule, has seriously affected the fitness of the patient, and that it will take a long time for the patient to regain his or her original level of fitness. Therefore, the retraining should be slow and gradual.

Patients with chronic nephritis should stay away from competitive training. In particular, they should refrain from participating in prolonged strenuous physical exertion, such as cross-country ski races covering long distances. Anuria has been observed in presumably healthy individuals with no known history of kidney disease after such extremely strenuous athletic events (Refsum and Strømme, 1974). Moderate physical activity, on the other hand, may be beneficial for patients with chronic nephritis, like everyone else. Patients with urinary tract infections, such as cystitis or pyelitis should not participate in physical training during the acute phase of the illness.

Acute hepatitis is definitely a contraindication for physical training. Since this disease, in general, has a rather protracted course it is necessary to accept a prolonged break in the training program. The training should not be resumed until the hepatomegalia has disappeared and the liver function tests are normal. In these cases it is particularly important that the retraining starts very gently and that the patient is regularly checked by the physician during the retraining period. This also applies to patients with infectious mononucleosis. Such patients often have an enlarged spleen. Because of the danger of a possible rupture of the spleen the patient should not resume physical training, at any rate in the case of contact sports, until the enlarged spleen has returned to normal. Careful jogging or similar training may be permitted, however.

REFERENCES

Altekruse, E. B., and J. H. Wilmore: Changes in Blood Chemistries Following a Controlled Exercise Program, *J. Occup. Med.*, **15**:110, 1973.

American College of Sports Medicine: Position Stand on the Use of Anabolic-Adrogenic Steroids in Sports, *Am. Coll. Sports Med.*, Indianapolis, IN, 1984.

Anderson, B.: "Stretching," Shelter Publishing, Bolinas, Calif., 1980.

Anderson, T. W., and R. J. Shephard: Physical Training and Exercise Diffusing Capacity, *Intern. Z. Angew. Physiol.*, **25**:198, 1968.

Armstrong, R. B.: Mechanisms of Exercise-Induced Delayed Onset Muscular Soreness: A Brief Review. *Med. Sci. Sports Exerc.*, **16**(6):529, 1984.

Åstrand, I., P.-O. Åstrand, E. H. Christensen, and R. Hedman: Intermittent Muscular Work, *Acta Physiol. Scand.*, **48**:448, 1960.

Åstrand, P.-O.: Human Physical Fitness with Special Reference to Sex and Age, *Physiol. Rev.*, **36**:307, 1956.

Atha, J.: Strengthening Muscle, *Exerc. Sport Sci. Rev.*,**9**:1, 1981.

Barry, A. J., J. W. Daly, E. D. R. Pruett, J. R. Steinmetz, N. C. Birkhead, and K. Rodahl: Effects of Physical Training in Patients who Have had Myocardial Infarction, *Am. J. Cardiol.*, **17**:1, 1966.

Blackburn, H.: Physical Activity and Coronary Heart Disease: A Brief Update and Population View, *J. Cardiac Rehab.*, **3**:101, 1983.

Blomqvist, C. G., and B. Saltin: Cardiovascular Adaptations to Physical Training, *Ann. Rev. Physiol.*, **45**:169, 1983.

Booth, F. W., and P. D. Gollnick: Effects of Disuse on the Structure and Function of Skeletal Muscle, *Med. Sci. Sports Exerc.*, **15**:415, 1983.

Booth, F. W., and E. W. Gould: Effects of Training and Disuse on Connective Tissue, in J. H. Wilmore, and J. F. Keogh (eds.), "Exercise and Sport Sciences Reviews," vol. 3, p. 83, Academic Press, Inc., New York, 1975.

Borg, G. A. V.: Psychophysical Base of Perceived Exertion, *Med. Sci. Sports Exerc.*, **14**:377, 1982.

Bouchard, C., and G. Lortre: Heredity and Endurance Performance, *Sports Med.*, **1**:38, 1984.

Brodal, P., F. Ingjer, and L. Hermansen: Capillary Supply of Skeletal Muscle Fibers in Untrained and Endurance-Trained Men, *Am. J. Physiol.*, **232**:H705, 1977.

Brunner, D., and G. Manelis, Physical Activity at Work and Ischemic Heart Disease, in O. A. Larsen and P. O. Malmberg (eds): "Coronary Heart Disease and Physical Fitness," Munksgaard, Copenhagen, 1971.

Chance, B., H. Sies, and A. Boveris; Hydroperoxide Metabolism in Mammalian Organs, *Physiol. Rev.*, **59**:527, 1979.

Chow, R. J., and J. H. Wilmore: Continuing Medical Education: The Regulation of Exercise Intensity by Rating Perceived Exertion, *J. Cardiac Rehab.*, **4**:382, 1984.

Christensen, E. H.: Beiträge zur Physiologie schwerer körperlicher Arbeit, *Arbeitsphysiol.*, **4**:1, 1931.

Christensen, E. H. R. Hedman, and B. Saltin: Intermittent and Continuous Running, *Acta Physiol. Scand.*, **50**:269, 1960.

Clarke, D. H.: Adaptations in Strength and Muscular Endurance Resulting from Exercise, in J. H. Wilmore (ed.), "Exercise and Sport Sciences Reviews," vol. 1, p. 73, Academic Press, Inc., New York, 1973.

Clausen, J. P.: Circulatory Adjustments to Dynamic Exercise and Effect of Physical Training in Normal Subjects and in Patients with Coronary Disease, *Progr. Cardiovasc. Dis.*, **18**:459, 1976.

Clement, D. B., and L. L. Sawchuk: Iron Status and Sports Performance, *Sports Med.*, **1**:65, 1984.

Costill, D. L., W. J. Fink, L. H. Getchell, J. L. Ivy, and F. A. Witzmann: Lipid Metabolism in Muscle of Endurance-Trained Males and Females, *J. Appl. Physiol.: Resp. Environ. Exerc. Physiol.*, **47**:787, 1979.

Cowan, G. O.: Influence of Exercise on High-Density Lipoproteins, *Am. J. Cardiol.*, **52**:(4):13B, 1983.

Coyle, E. F., W. H. Martin, A. A. Ehsani, J. M., Hagberg, S. A. Bloomfield, D. R. Sinacore, and J. O. Holloszy: Blood Lactate Threshold in Some Well-Trained Ischemic Heart Disease Patients, *J. Appl. Physiol.: Resp. Environ. Exerc. Physiol.*, **54**:18, 1983.

Coyle, E. F., W. H. Martin III, D. R. Sinacore, M. J. Joyner, J. M. Hagberg, and J. O. Holloszy: Time Course of Loss of Adaptations after Stopping Prolonged Intense Endurance Training, *J. Appl. Physiol.*, **57**:1857, 1984.

Davies, C. T. M. and A. J. Sargeant: Effects of Training on the Physiological Responses to One- and Two-leg Work, *J. Appl. Physiol.*, **38**:377, 1975.

Dearman, J., and K. T. Francis: Plasma Levels of Catecholamines, Cortisol and Beta-endorphins in Male Athletes after Running 26.2, 6 and 2 miles, *J. Sports Med.*, **23**(1):30, 1983.

Deitrick, J. E., G. D. Whedon, and E. Shorr: Effects of Immobilization upon Various Metabolic and Physiologic Functions of Normal Men, *Am. J. Med.*, **4**:3, 1948.

Del Maestro, R. F.: An Approach to Free Radicals in Medicine and Biology, *Acta Physiol. Scand.*, (Suppl. 492):153, 1980.

Denis, C., R. Fouquet, P. Poty, A. Geyssant, and J. R. Lacour: Effect of 40 Weeks of Endurance Training on the Anaerobic Threshold, *Int. J. Sports Med.*, **3**:208, 1982.

Dooley, P., M. J. N. McDonagh, and M. J. White: Training using Involuntary Electrically Evoked Contractions Does not Increase Voluntary Strength, *J. Physiol.*, **346**:61P, 1984.

Dublin, L. I., A. J. Lotka, and M. Spiegelman: "Length of Life: A Study of the Life Table," The Ronald Press Company, New York, 1949.

Duchateau, J., and K. Hainaut: Isometric or Dynamic Training: Differential Effects on Mechanical Properties of a Human Muscle, *J. Appl. Physiol.: Resp. Environ. Exerc. Physiol.*, **56**:296, 1984.

Eckstein, R. W.: Effect of Exercise and Coronary Artery Narrowing on Coronary Collateral Circulation, *Circulation Res.*, **5**:230, 1957.

Edwards. H. T., L. Brouha, and R. T. Johnson: Effect de l'entrainement sur le taux de l'acide lactique au cours du travail musculaire, *Le Travail Humain*, **8**:1, 1939.

Ekblom, B.: Effect of Physical Training on Oxygen Transport System in Man, *Acta Physiol. Scand.* Suppl. 328, 1969.

Ekblom, B., P.-O. Åstrand, B. Saltin, J. Stenberg, and B. Wallström: Effect of Training on Circulatory Response to Exercise, *J. Appl. Physiol.*, **24**:518, 1968.

Erikssen, J., K. Forfang and J. Jervell: Coronary Risk Factors and Physical Fitness in Healthy Middle-aged men, *Acta Med. Scand.*, (Suppl. 645):57, 1981.

Erikssen, J., and R. Mundal: The Patient with Coronary Artery Disease without Infarction: Can a High-Risk Group be Identified?, *Ann. N. Y. Acad. Sci.*, **382**:438, 1982.

Eriksson, B. O.: Physical Training, Oxygen Supply and Muscle Metabolism in 11 to 13-year-old Boys, *Acta Physiol. Scand.*, (Supl. 384), 1972.

Eriksson, B. O., A. Lundin, and B. Saltin: Cardiopulmonary Function in Former Girl Swimmers and the Effects of Physical Training, *Scand. J. Clin. Lab. Invest.*, **35**:135, 1975.

Favier, R., J. M. Pequignot, D. Desplanches, M. H. Mayet, J. R. Lacour, L. Peyrin and R. Flandrois: Catecholamines and Metabolic Responses to Submaximal Exercise in Untrained Men and Women, *Eur. J. Appl. Physiol.*, **50**:393, 1983.

Fortney, S., and N. B. Vroman: Exercise Performance and Temperature Control: Temperature Regulation during Exercise and Implications for Sports Performance and Training, *Sports Med.*, **2**:8, 1985.

Fox, S. M., and J. S. Skinner: Physical Activity and Cardiovascular Health, *Am. J. Cardiol.*, **14**:731, 1964.

Fox, S. M., and W. L. Haskell: Physical Activity and the Prevention of Coronary Heart Disease, *Bull. N. Y. Acad. Med.*, **44**:950, 1968.

Francis, K. T.: HDL Cholesterol and Coronary Heart Disease, *Southern Med. J.*, **73**(2):169, 1980.

Fridén, J.: Exercise-Induced Muscle Soreness, Umeå Univ. Med. Dissertations, new series, No. 105, Umeå, 1983.

Gaesser, G. A., and R. G. Rich: Effects of High- and Low-Intensity Exercise Training on Aerobic Capacity and Blood Lipids, *Med. Sci. Sports Exerc.*, **16**:269, 1984.

Galbo, H.: Hormonal and Metabolic Adaptation to Exercise, Georg Thieme Verlag, Stuttgart, New York, 1983.

Gilbert, D.L. (ed.): "Oxygen and Living Processes. An Interdisciplinary Approach," Springer-Verlag, New York, 1981).

Goldberg, A. L., J. D. Etlinger, D. F. Goldspink and C. Jeblecki: Mechanism of Work-Induced Hypertrophy of Skeletal Muscle, *Med. Sci. Sports*, **7**:185, 1975.

Goldspink, G.: Growth of Muscle, in D. F. Goldspink (ed.), "Development and Specialization of Skeletal Muscle," Cambridge University Press, Cambridge, 1980.

Gollnick, P. D., K. Piehl, and B. Saltin: Selective Glycogen Depletion Pattern in Human Muscle Fibres after Exercise of Varying Intensity and at Varying Pedalling Rates, *J. Physiol.*, **241**:45, 1974.

Gollnick, P. D., B. F. Timson, R. L. Moore, and M. Riedy: Muscular Enlargement and Number of Fibers in Skeletal Muscles of Rat, *J. Appl. Physiol.: Resp. Env. Exerc. Physiol.*, **50**:936, 1981.

Greenleaf, J. E.: Physiological Responses to Prolonged Bed Rest and Fluid Immersion in Humans, *J. Appl. Physiol.: Resp. Environ., Exerc. Physiol.*, **57**:619, 1984.

Gutmann, E., R. Beranek, P. Hnik, and J. Zelena: Physiology of Neurotrophic Relations. *Proc. 5th Nat. Congr. Czech. Physiol. Soc., 1961.*

Gyntelberg, F.: Physical Fitness and Coronary Heart Disease in Copenhagen Males, Aged 40–49, III., *Dan. Med. Bull.*, **21**:49, 1974.

Hakkinen, K., and P. V. Komi: Effect of Different Combined Concentric and Eccentric Muscle Work Regimes on Maximal Strength Development, *J. Human Movement Studies*, **7**:33, 1981.

Hallberg, L., and B. Magnusson: The Etiology of "Sports Anemia," *Acta Med. Scand.*, **216**:145, 1984.

Hamilton, P., and G. M. Andrew: Influence of Growth and Athletic Training on Heart and Lung Functions, *Europ. J. Appl. Physiol.*, **36**:27, 1976.

Hammett, V. B. O.: Psychological Changes with Physical Fitness Training, *Can. Med. Ass. J.*, **96**:764, 1967.

Hartley, L. H., and B. Saltin: Reduction of Stroke Volume and Increase in Heart Rate after a Previous Heavier Submaximal Work Load, *Scand. J. Clin. Lab. Invest.* **22:**217, 1968.

Haskell, W. L.: Exercise-Induced Changes in Plasma Lipids and Lipoproteins, *Preventive Med.,* **13:**23, 1984.

Haskell, W. L.: Cardiovascular Benefits and Risks of Exercise: The Scientific Evidence, in R. H. Strauss (ed.): "Sports Medicine," W.B. Saunders Company, Philadelphia, p. 57, 1984.

Hedman, R.: Fysiologiska Synpunkter på Cirkelträning, *Tidskrift i Gymnastik,* **87:**1, 1960.

Henriksson, J.: Training Induced Adaptation of Skeletal Muscle and Metabolism during Submaximal Exercise, *J. Physiol.,* **270:**661, 1977.

Henriksson, J., and J. S. Reitman: Time Course of Changes in Human Skeletal Muscle Succinate Dehydrogenase and Cytochrome Oxidase Activities and Maximal Oxygen Uptake with Physical Activity and Inactivity, *Acta Physiol. Scand.,* **99:**91, 1977.

Henriksson, J., H. Galbo, and E. Blomstrand: Role of the Motor Nerve in Activity-induced Enzymatic adaptation in Skeletal Muscle, *Am. J. Physiol.,* **242** (Cell Physiol. 23):C272, 1982.

Henriksson, J., J. Svedenhag, E. A. Richter, N. J. Christensen, and H. Galbo: Skeletal Muscle and Hormonal Adaptation to Physical Training in the Rat: Role of the Sympatho-Adrenal System, *Acta Physiol. Scand.,* **123:**127, 1985.

Hickson, R. C.: Skeletal Muscle Cytochrome and Myoglobin, Endurance, and Frequency of Training, *J. Appl. Physiol.: Resp. Environ. Exerc. Physiol.,* **51:**746, 1981.

Higuchi, M., I. Hashimoto, K. Yamakawa, E. Tsuji, M. Nishimuta, and S. Suzuki: Effect of Exercise Training on Plasma High-density Lipoprotein Cholesterol Level at Constant Weight, *Clin. Physiol.,* **4:**125, 1984.

Hollmann, W., and Th. Hettinger: "Sportmedizin—Arbeits-und Trainingsgrundlagen," F. K. Schattauer Verlag, Stuttgart, 1976.

Hollmann, W., R. Rost, H. Liesen, B. Dufaux, H. Heck, and A. Mader: Assessment of Different Forms of Physical Activity with Respect to Preventive and Rehabilitative Cardiology, *Int. J. Sports Med.,* **2:**67, 1981.

Holloszy, J. O., and F. W. Booth: Biochemical Adaptations to Endurance Exercise in Muscle, *Ann. Rev. Physiol.,* **18:**273, 1976.

Holmdahl, D. E., and B. E. Ingelmark: Der Bau des Gelenkknorpels unter verschiedenen Funktionellen Verhältnissen, *Acta Anat.,* **6:**309, 1948.

Holmér, I.: Physiology of Swimming Man, *Acta Physiol. Scand.,* (Suppl. 407), 1974.

Holmér, I., and P.-O. Åstrand: Swimming Training and Maximal Oxygen Uptake, *J. Appl. Physiol.* **33:**510, 1972.

Hoogerwerf, S.: Elektrokardiographische Untersuchungen der Amsterdamer Olympiakämpfer, *Arbeitsphysiol.,* **2:**61, 1929.

Höstmark, A. T.: Physical Activity and Plasma Lipids, *Scand. J. Soc. Med.,* (Suppl. 29):83, 1982.

Howald, H.: Training-induced Morphological and Functional Changes in Skeletal Muscle, *Int. J. Sports Med.,* **3:**1, 1982.

Hudlická, O.: Growth of Capillaries in Skeletal and Cardiac Muscle, *Circ. Research,* **50:**451, 1982.

Hunding, A., R. Jordal, and P. E. Pauley: Runners Anaemia and Iron Deficiency, *Acta Med. Scand.,* **209:**315, 1981.

Hurley, B. F., J. M. Hagberg, W. K. Allen, D. R. Seals, J. C. Young, R. T. Cuddihee, and J. O. Holloszy: Effect of Training on Blood Lactate Levels during Submaximal Exercise, *J. Appl. Physiol.: Resp. Environ. Exerc. Physiol.,* **56:**1260, 1984.

Ikai, M., and T. Fukunaga: A Study on Training Effect on Strength per Unit Cross-Sectional Area of Muscle by Means of Ultrasonic Measurement, *Int. Z. Angew. Physiol.*, **28:**172, 1970.

Ikai, M., and A. H. Steinhaus: Some Factors Modifying the Expression of Human Strength, *J. Appl. Physiol.*, **16:**157, 1961.

Illg, D., and D. Pette: Turnover Rates of Hemokinase I, Phosphofructokinase, Pyruvate Kinase, and Creatinine Kinase in Slow-twitch Soleus Muscle and Heart of the Rabbit, *Eur. J. Biochem.*, **97:**267, 1979.

Ilmarinen, J., R. Ilmarinen, A. Koskela, O. Korhonen, F. Fardy, T. Partanen and J. Rutenfranz: Training Effects of Stair-Climbing during Office Hours in Female Employees, *Ergonomics*, **22**(5):507, 1979.

Ingelmark, B. E.: Der Bau der Sehnen während verschiedener Altersperioden und unter wechselnden funktionellen Bedingungen I, *Acta Anat.*, **6:**113, 1948.

Ingelmark, B. E.: Morpho-physiological Aspects of Gymnastic Exercises, *FIEP-Bull.*, **27:**37, 1957.

Ingjer, F.: Effects of Endurance Training on Muscle Fibre ATP-ase Activity, Capillary Supply and Mitochondral Content in Man, *J. Physiol.*, **294:**419, 1979.

Israel, S: "Sport und Herzschlagfrequenz," Sportmedizinische Schriftenreihe 21, Johann Ambrosius Barth, Leipzig, 1982.

Issekutz, B., Jr., J. J. Blizzard, N. C. Birkhead, and K. Rodahl: Effect of Prolonged Bed Rest on Urinary Calcium Output, *J. Appl. Physiol.*, **21:**1013, 1966.

James, D. E., E. W. Kraegen, and D. J. Chisholm: Effect of Exercise Training on Whole-Body Insulin Sensitivity and Responsiveness, *J. Appl. Physiol.: Resp. Environ. Exerc. Physiol.*, **56**(5):1217, 1984.

Janal, M. N., E. W. D. Colt, W. C. Lark, and M. Glusman: Pain Sensitivity, Mood and Plasma Endocrine Levels in Man Following Long-Distance Running: Effects of Naloxone, *Pain*, **19:**13, 1984.

Jansson, E., and L. Kaijser: Muscle Adaptation to Extreme Endurance Training in Man, *Acta Physiol. Scand.*, **100:**315, 1977.

Jokl, E.: "Heart and Sport," Charles C. Thomas, Springfield, Ill., 1964.

Jones, D. M., C. Squires, and K. Rodahl: Effect of Rope Skipping on Physical Work Capacity, *Res. Quart.*, **33:**236, 1962.

Kannel, W. B.: Exercise and Sudden Death, *J. Am. Med. Assoc.*, **248**(23):3143, 1982.

Kannel, W. B., T. Gordon, and M. J. Schwartz: Systolic Versus Diastolic Blood Pressure and Risk of Coronary Heart Disease. The Framingham Study. *Am. J. Cardiol.*, **27:**335, 1971.

Karlsson, J., P.-O. Åstrand, and B. Ekblom: Training of the Oxygen Transport System in Man, *J. Appl. Physiol.*, **22:**1061, 1967a.

Karlsson, J., L. Hermansen, G. Agnevik, and B. Saltin: Energikraven vid löpning, *Idrottsfysiologi*, Rapport 4, Framtiden, Stockholm, 1967b.

Kearney, J. T., G. A. Stull, J. L. Ewing, Jr., and J. W. Strein: Cardiorespiratory Responses of Sedentary College Women as a Function of Training Intensity, *J. Appl. Physiol.*, **41:**822, 1976.

Keys, A.: Alpha Lipoprotein (HDL) Cholesterol in the Serum and the Risk of Coronary Heart Disease and Death, *Lancet*, Sept.: 603, 1980.

Kiens, B., H. L. Lithell, and B. Vessby: Further Increase in High Density Lipoprotein in Trained Males after Endurance Training, *Eur. J. Appl. Physiol.*, **52:**426, 1984.

Kiens, B., B. Essén-Gustavsson, N. J. Christensen, and B. Saltin: Skeletal Muscle Substrate Utilization with Exercise. Effect of Endurance Training, unpublished, 1985.

Kilbom, Å: Physical Training in Women, *Scand. J. Clin. Lab. Invest.*, **28**(Suppl. 119), 1971.

Klausen, K., N. H. Secher, J. P. Clausen, J. Hartling, and J. Trap-Jensen: Central and Regional Circulatory Adaptation to One-leg Training. *J. Appl. Physiol.: Resp. Environ. Exerc. Physiol.*, **52**:976, 1982.

Klissouras, V.: Prediction of Athletic Performance: Genetic Considerations, *Can. J. Appl. Sport Sci.*, **1**:195, 1976.

Knuttgen, H. G., J. A. Vogel and J. Poortmans: "Biochemistry of Exercise" in International Series on Sport Sciences, Vol. 13. Hitman Kinetics Publishers, Inc. Champaign, Ill., 1983.

Kobayaski, K., K. Kitamura, M. Miura, H. Sodeyama, Y. Murase, M. Miyashita, and H. Matsui: Aerobic Power as Related to Body Growth and Training in Japanese Boys: A Longitudinal Study, *J. Appl. Physiol.: Resp. Environ. Exerc. Physiol.*, **44**:666, 1978.

Kramsch, D. M., A. J. Aspen, B. M. Abramowitz, T. Kreimendahl, and W. H. Hood, Jr.: Reduction of Coronary Atherosclerosis by Moderate Conditioning Program in Monkeys on Atherogenic Diet, *New Engl. J. Med.*, **305**:1483, 1982.

Lauru, L.: Physiological Study of Motion, *Advanced Management*, **22**:17, 1957.

Leon, A. S., and H. Blackburn: Physical Activity in the Prevention of Coronary Heart Disease: An Update 1981, in C. B. Arnold (ed.): "Advances in Disease Prevention." Springer-Verlag Publishing Company, N. Y., **1**:97, 1981.

Lie H., and J. Erikssen: "Five Year Follow-up Study of ECG-Aberations, Latent Coronary Heart Disease and Cardiopulmonary Fitness in Various Age Groups of Norwegian Cross-Country Skiers, unpublished, 1985.

Liesen, H., and W. Hollmann: Leistungsverbesserung und Muskelstoffwechsel-adaptionen durch Ausdauertraining im Alter, *Geriatric*, **6**:150, 1976.

Lortie, G., J. A. Simoneau, P. Hamel, M. R. Boulay, and F. Landry: Responses of Maximal Aerobic Power and Capacity to Aerobic Training, *Int. J. Sports Med.*, **5**:232, 1984.

Lussier, L., and E. R. Buskirk: Effects of an Endurance Training Regimen on Assessment of Work Capacity in Prepubertal Children, *Ann. N. Y. Acad. Sci.*, **301**:734, 1977.

MacDougall, J. D., G. C. B. Elder, D. G. Sale, J. R. Maroz, and J. R. Sutton: Effects of Strength Training and Immobilization on Human Muscle Fibres, *Eur. J. Appl. Physiol.*, **43**:25, 1980.

Magel, J. R., G. F. Foglia, W. D. McArdle, B. Gutin, G. S. Pechar, and F. I. Katch: Specificity of Swim Training on Maximum Oxygen Uptake, *J. Appl. Physiol.* **38**:151, 1975.

Man-i, M., and Y. Imachi: Effect of Training on Blood Pressure and Heart Rate Measured Continuously during Exercise, *J. Sports Med.*, **21**(2):97, 1981.

Marpurgo, P.: Über Aktivitäts-Hypertrophie der wilkürlichen Muskeln. *Virchows Arch.*, **150**:522, 1897.

Matoba, H., and P. D. Gollnick: Response of Skeletal Muscle to Training, *Sports Med.*, **1**:240, 1984.

Maughan, R. J.: Relationship between Muscle Strength and Muscle Cross-sectional Area; Implications for Training, *Sports Med.*, **1**:263, 1984.

Mazzeo, R. S., G. A. Brooks, and S. M. Horvath: Effects of Age on Metabolic Responses to Endurance Training in Rats, *J. Appl. Physiol.: Resp. Environ. Exerc. Physiol.*, **57**:1369, 1984.

Miller, P. B., R. L. Johnson, and L. E. Lamb: Effects of Four Weeks of Absolute Bed Rest on Circulatory Functions in Man, *Aerospace Med.*, **35**:1194, 1964.

Montoye, H. J.: Physiology of Training Including Age and Sex Differences, *J. Sport Med.*, **2**:35, 1962.

Moore, R. L., M. Riedy, and P. D. Gollnick: Effect of Training on Beta-Adrenergic Receptor Number in Rat Heart, *J. Appl. Physiol.: Resp. Environ. Exerc. Physiol.*, **52**:1133, 1982.

Morgan, R. E., and G. T. Adamson: "Circuit Training," G. Bell and Sons, Ltd., London, 1962.

Morris, J. N., M. G. Everitt, R. Pollard, S. P. W. Chave, and A. M. Semmence: Vigorous Exercise in Leisure-Time: Protection against Coronary Heart Disease, *Lancet*, **2:**1207, 1980.

Mullin, W. J., R. E. Herrick, V. Valdez, and K. M. Baldwin: Adaptive Responses of Rats Trained with Reductions in Exercise Heart Rate, *J. Appl. Physiol.: Resp. Environ. Exerc. Physiol.*, **56:**1378, 1984.

Mundal, R., J. Erikssen, and K. Rodahl: Assessment of Physical Activity as Judged by Questionnaire and Personal Interview with Particular Reference to Fitness and Coronary Mortality, unpublished, 1985.

Newsholme, E. A., and A. R. Leech: "Biochemistry for the Medical Sciences," John Wiley & Sons, Chichester, U.K., 1983.

Nygaard, E., H. Bentzen, M. Houston, H. Larsen, E. Nielsen, and B. Saltin: Capillary Supply and Morphology of Trained Human Skeletal Muscle, paper presented at the 27th International Congress of Physiological Sciences, Paris, 1977.

Nylander, E.: Training-induced Bradycardia in Rats on Cardioselective and Non-selective Beta Receptor Blockade, *Acta Physiol. Scand.*, **123:**147, 1985.

Ogawa, S., K. Asano, Y. Furuta, T. Obara, and T. Fujimaki: The Effect of Intermittent Rope Skipping on Aerobic Work Capacity, *Bull. Inst. Sport Sci., Faculty Phys. Educ., Tokyo U. Educ.*, **23:**1, 1974.

Paffenbarger, R. S., and R. T. Hyde: Exercise and Protection against Heart Attack, *New Engl. J. Med.*, **302**(18):1026, 1980.

Paffenbarger, R. S., and R. T. Hyde: Exercise in the Prevention of Coronary Heart Disease, *Preventive Med.*, **13:**3, 1984.

Pařízková, J.: "Body Fat and Physical Fitness," Martinus Nijhoff B. V. Medical Division, The Hague, 1977.

Péronnet, F., R. J. Ferquson, H. Perrault, G. Ricci, and D. Lajoie: Echocardiography and the Athlete's Heart, *Phys. Sports Med.*, **9:**102, 1981.

Petrén, T.: Die totale Anzahl der Blutkapillaren im Herzen und Skelettmuskulatur bei Ruhe und nach langer Muskelübung, p. 165, *Verhandl. Anatom. Gesellsch* (Suppl. *Anat. Anz., 81*), 1936.

Petrén, T., T. Sjöstrand, and B. Sylvén: Der Einfluss der Trainings auf die Heufigkeit der Capillaren in Herz- und Skelettmuskulatur, *Arbeitsphysiol.*, *9:*376, 1936.

Pette, D.: Activity Induced Fast to Slow Transitions in Mammalian Muscle, *Med. Sci. Sports Exerc.*, **16:**517, 1984.

Piehl, K.: Glycogen Storage and Depletion in Human Skeletal Muscle Fibres, *Acta Physiol. Scand., (Suppl. 402), 1974.*

Pollock, M. L.: The Quantification of Endurance Training Program, in J. H. Wilmore (ed.), "Exercise and Sport Sciences Review," vol. 1, p. 155, Academic Press, Inc., New York, 1973.

Pyörälä, K. M., I. Karvonen, T. Taskinen, J. Takkunen, H. Kyrönseppä, and P. Peltokallio: Cardiovascular Studies on Former Endurance Athletes, *Am. J. Cardiol.*, **20:**191, 1967.

Rapaport, E.: "Current Controversies in Cardiovascular Disease," W. B. Saunders Company, Philadelphia, 1980.

Rasch, P. J., and L. E. Morehouse: Effect of Static and Dynamic Exercise on Muscular Strength and Hypertrophy, *J. Appl. Physiol.*, **11:**29, 1957.

Refsum, J. E., and S. B. Strømme: "Urea and Creatinine Production and Excretion in Urine during and after Prolonged Heavy Exercise, *Scand. J. Clin. Lab. Invest.*, **33:**247, 1974.

Roberts, A. D., R. Billeter, and H. Howald: Anaerobic Muscle Enzyme Changes after Interval Training, *Int. J. Sports Med.*, **3:**18, 1982.

Roberts, W. C.: An Agent with Lipid-Lowering Antihypertensive, Positive Inotropic, Negative Chronotropic, Vasodilating, Diuretic, Anorexigenic, Weight-Reducing, Cathartic, Hypoglycemic Tranquilizing, Hypnotic and Antidepressive Qualities, *Am. J. Cardiol.*, **53**(1):261, 1984.

Robinson, S., and P. M. Harmon: The Lactic Acid Mechanism and Certain Properties of the Blood in Relation to Training, *Am. J. Physiol.*, **132**:757, 1941a.

Robinson, S., and P. M. Harmon: The Effects of Training and of Gelatin upon Certain Factors Which Limit Muscular Work, *Am. J. Physiol.*, **133**:161, 1941b.

Rodahl, K., N. C. Birkhead, J. J. Blizzard, B. Issekutz, Jr., and E. D. R. Pruett: Physiological Changes during Prolonged Bed Rest, in G. Blix (ed.), "Nutrition and Physical Activity," p. 107, Almqvist & Wiksell, Stockholm, 1967.

Rook, A.: An Investigation into the Longevity of Cambridge Sportsmen, *Br. Med.J.*, **1**:773, 1954.

Roskamm, H.: Optimum Patterns of Exercise for Healthy Adult, *Can. Med. Ass. J.*, **22**:895, 1967.

Rost, R., and W. Hollmann: Athlete's Heart—A Review of its Historical Assesment and New Aspects, *Int. J. Sports Med.*, **4**:147, 1983.

Sahlin, K., and J. Henriksson: Buffer Capacity and Lactate Accumulation in Skeletal Muscle of Trained and Untrained Men, *Acta Physiol. Scand.* **122**:331, 1984.

Sale, D., and D. MacDougall: Specificity in Strength Training: A Review for the Coach and Athlete, *Can. J. Appl. Sport Sci.*, **6**:87, 1981.

Salminen, A., and V. Vihko: Endurance Training Reduces the Susceptibility of Mouse Skeletal Muscle to Lipid Peroxidation in vitro, *Acta Physiol. Scand.*, **117**:109, 1983.

Salmons, S., and J. Henriksson: The Adaptive Response of Skeletal Muscle to Increased Use, *Muscle & Nerve*, **4**:94, 1981.

Salonen, J. T.: Physical Activity and Risk of Myocardial Infarction, Stroke and Death, *Suomen Liikuntalääketiede*, (Finnish Sports Exer. Med.) **2**:28, 1983.

Salonen, J. T., P. Puska, and J. Tuomilehto: Physical Activity and Risk of Myocardial Infarction, Cerebral Stroke and Death, *Am. J. Epidemiol.*, **115**(4):526, 1982.

Saltin, B.: Hemodynamic Adaptations to Exercise, *Am. J. Cardiol.*, **55**(10):42D, 1985.

Saltin, B., B. Blomqvist, J. H. Mitchell, R. L. Johnson, Jr., K. Wildenthal, and C. B. Chapman: Response to Submaximal and Maximal Exercise after Bed Rest and Training, *Circulation*, **38**(Suppl. 7): 1968.

Saltin, B., and L. B. Rowell: Functional Adaptations to Physical Activity and Inactivity, *Fed. Proc.*, **39**:1506, 1980.

Saltin, B., and P. D. Gollnick: Skeletal Muscle Adaptability: Significance for Metabolism and Performance, in L. D. Peachey, R. H. Adrian, and S. R. Geiger (eds.): "Handbook of Physiology," section 10, chap. 9, p. 555, Williams and Wilkins Company, Baltimore, 1983.

Sargeant, A. J., A. Young, C. T. M. Davies, C. Maunder, and R. H. T. Edwards: Functional And Structural Changes Following Disease of Human Muscle, *Clin. Sci. Mol. Med.*, **52**:337, 1977.

Schaible, T. F., S. Pehpargkul, and J. Scheuer: Cardiac Responses to Exercise Training in Male and Female Rats, *J. Appl. Physiol.: Resp. Environ. Exer. Physiol.*, **50**:112, 1981.

Schantz, P., and J. Henriksson: Increases in Myofibrillar ATPase Intermediate Human Skeletal Muscle Fibers in Response to Endurance Training, *Muscle & Nerve*, **6**:553, 1983.

Schoop, W.: Auswirkungen gesteigerter körperlicher Aktivität auf gesunde und krankhaft veränderte Extremitätsarterien, in Roskamm et al. (eds.), "Körperliche Aktivität und Herz und Kreislauferkrankungen," p. 33, Johann Ambrosius Barth, Munich, 1966.

Schumacher, R., and H. Howald: Echocardiographische Nachuntersuchungen von ehemaligen Spitzensportlern mit typischen Sportherz, *Schweiz. Ztschr. Sportmed.*, **32**:117, 1984.

Seals, D. R., and J. M. Hagberg: The Effect of Exercise Training on Human Hypertension: A Review, *Med. Sci. Sports Exerc.*, **16**:207, 1984.

Seals, D. R., J. M. Hagberg, B. F. Hurley, A. A. Ehsani, and J. O. Holloszy: Effects of Endurance Training on Glucose Tolerance and Plasma Lipid Levels in Older Men and Women, *JAMA*, **252**:645, 1984a.

Seals, D. R., J. M. Hagberg, B. F. Hurley, A. A. Ehsani, and J. O. Holloszy: Endurance Training in Older Men and Women, *J. Appl. Physiol.: Resp. Environ. Exerc. Physiol.*, **57**:1024, 1984b.

Shaper, A. G., J. N. Morris, and T. W. Meade: The London Busmen, in O. A. Larsen, and P. O. Malmberg (eds.): "Coronary Heart Disease and Physical Fitness," Munksgaard, Copenhagen, 1971.

Sharp, R. L., L. E. Armstrong, D. S. King, and D. L. Costill: Buffer Capacity of Blood in Trained and Untrained Males, in H. G. Knuttgen, J. A. Vogel and J. Poortmans: "Biochemistry of Exercise," in International Series on Sport Sciences, Vol. 13, Hūman Kinetics Publishers, Inc., Champaign, Ill., p. 595, 1983.

Shephard, R. L.: "Physical Activity and Growth," Year Book Medical Publishers, Inc., Chicago, 1982.

Siebert, W. W.: Untersuchungen über Hypertrophie des Skelettmuskels, *Z. Klin. Med.*, **109**:350, 1929.

Sjödin, R., and J. Svedenhag: Applied Physiology of Marathon Running, *Sports Med.*, **2**:83, 1985.

Skinner, J. S., K. O. Holloszy, and T. K. Cureton: Effects of a Program of Endurance Exercises on Physical Work, *Am. J. Cardiol.*, **14**:747, 1964.

Sopko, G., A. S. Leon, D. R. Jacobs, N. Foster, J. Moy, D. Casal, J. Anderson, and C. McNally: Effect of Exercise and/or Weight Loss on Blood Lipids in Obese Men on Controlled Diets, *Circulation*, **66**(Suppl. 2):285, 1982.

Stebbins, C. L., E. Schultz, R. T. Smith, and E. L. Smith: Effects of Chronic Exercise during Aging on Muscle and End-Plate Morphology in Rats, *J. Appl. Physiol.*, **58**:45, 1985.

Steinhaus, A. H.: Chronic Effects of Exercise, *Physiol. Rev.*, **13**:103, 1933.

Strauss, R. H. (ed.): "Sports Medicine", W. B. Saunders Company, Philadelphia, 1984.

Surbey, G. D., G. M. Andrew, F. W. Cervenko, and P. P. Hamilton: Effects of Naloxon on Exercise Performance, *J. Appl. Physiol.: Resp. Environ. Exerc. Physiol.*, **57**:674, 1984.

Sutherland, W. H. F., and S. P. Woodhouse: Physical Activity and Plasma Lipoprotien Lipid Concentrations in Man, *Atherosclerosis*, **37**:285, 1980.

Svedenhag, J., J. Henriksson, A. Juhlin-Dannfelt, and K. Asano: Beta-Adrenergic Blockade and Training in Healthy Men—Effects on Central Circulation, *Acta Physiol. Scand.*, **120**:77, 1984a.

Svedenhag, J., B. G. Wallin, G. Sundlöf, and J. Henrikkson: Skeletal Muscle Sympathetic Activity at Rest in Trained and Untrained Subjects, *Acta Physiol. Scand.*, **120**:499, 1984b.

Taylor, H. L., A. Henschel, J. Brozek, and A. Keys: Effects of Bed Rest on Cardiovascular Function and Work Perormance, *J. Appl. Physiol.*,**2**:223, 1949.

Terjung, R. L.: Muscle Fiber Involvement during Training of Different Intensities and Durations, *Am. J. Physiol.*,**230**:946, 1976.

Terjung, R. L., and W. W. Winder: Exercise and Thyroid Function, *Med. Sci, Sports*,**7**:20, 1975.

Tesch, P. A., A. Thorsson, and P. Kaiser: Muscle Capillary Supply and Fiber Type Characteristics in Weight and Power Lifters, *J. Appl. Physiol: Resp. Environ, Exerc. Physiol.*, **56**:35, 1984.

Tharp, G. D., and C. T. Wagner: Chronic Exercise and Cardiac Vascularization, *Eur. J. Appl. Physiol.*, **48**:97, 1982.

Thistle, H. G., H. J. Hislop, M. Moffroid, and E. W. Lowman: Isokinetic Contraction: A New Concept of Resistive Exercise, *Arch. Phys. Med. Rehabil.*, **48:**279, 1967.

Tibblin, G., and L. Wilhelmsen: Physical Activity and Risk of Myocardial Infarction, *Läkartidningen*, **72:**343, 1975.

Timson, B. F., B. K. Bowlin, G. A. Dudenhoeffer, and J. B. George: Fiber Number, Area, and Composition of Mouse Soleus Muscle Following Enlargement, *J. Appl. Physiol.*, **58:**619, 1985.

Tipton, C. M., R. D. Matthes, and D. S. Sandage: In Situ Measurement of Junction Strength and Ligament Elongation in Rats, *J. Appl. Physiol.*, **37:**758, 1974.

Vanhees, L., R. Fagard, and A. Amerg: Influence of Beta-Adrenergic Blockade on Effects of Physical Training in Patients with Ischemic Heart Disease, *Br. Heart J.*, **48:**33, 1982.

Viidik, A.: Biomechanics and Functional Adaptation of Tendons and Joint Ligaments, in F. G. Evans (ed.), "Studies on the Anatomy and Function of Bone and Joints," p. 17, Springer-Verlag OHG, Heidelberg, 1966.

Vokac, Z., and K. Rodahl: Maximal Aerobic Power and Circulatory Strain in Eskimo Hunters in Greenland, Nordic Council for Arctic Medical Research Report no. **16:**16, 1976.

Wallberg-Henriksson, H., R. Gunnarsson, J. Henriksson, R. Defronzo, P. Felig, J. Østman, and J. Wahren: Increased Peripheral Insulin Sensitivity and Muscle Mitochrondrial Enzymes but Unchanged Blood Glucose Control in Type I Diabetics After Physical Training, *Diabetes*, **31:**1044, 1982.

Whipple, G. H.: The Hemoglobin of Striated Muscle, 1, Variations Due to Age and Exercise, *Am. J. Physiol.*, **76:**693, 1926.

White, M. K., R. B. Martin, R. A. Yeater, R. L. Butcher, and E. L. Radin: The Effects of Exercise on the Bone of Postmenopausal Women, *Int. Orthopaedics*, (Sicot), **7**(4):209, 1984.

Wiktorsson-Möller, M., B. Öberg, J. Ekstrand, and J. Gilquist: Effects of Warming-Up, Massage, and Stretching on Range of Motion and Muscle Strength in the Lower Extremity, *Am. J. Sports Med.*, **11:**249, 1983.

Wilmore, J. H.: "Training for Sport and Activity," 2nd ed., Allyn and Bacon, Inc., Boston, 1982.

Winder, W. W., J. M. Hagberg, R. C. Hickson, A. A. Ehsani, and J. A. McLane: Time Course of Sympatho-adrenal Adaptation to Endurance Exercise Training in Man, *J. Appl. Physiol.*, **45:**370, 1978.

Wood, P. D., R. B. Terry, and W. L. Haskell: Metabolism of Substrates: Diet, Lipoprotein Metabolism, and Exercise, *Federation Proc.*, **44**(2):358, 1985.

Yakovlev, N. N.: "*Sportbiochemie,*" Sportmedizinische Schriftenreihe Nr. 14, Joh. Ambrosius Barth, Leipzig, 1977.

Zeppilli, P., and A. Venerando: Sudden Death and Physical Exertion, *J. Sports Med.*, **21:**299, 1981.

Zukel, W. J., R. H. Lewis, P. E. Enterline, R. C. Painter, L. L. Ralston, R. M. Fawcett, A. P. Meredith and B. Peterson: A Short-Term Community Study of The Epidemiology of Coronary Heart Disease, *Am. J. Public Health*, **49:**1630, 1959.

CHAPTER 11

APPLIED WORK PHYSIOLOGY*

CONTENTS

INTRODUCTION

FACTORS AFFECTING THE ABILITY TO PERFORM SUSTAINED
   PHYSICAL WORK

ASSESSMENT OF WORK LOAD IN RELATION TO WORK CAPACITY
   Assessment of the Maximal Aerobic Power
   Assessment of the Physical Work Load
   Assessment of the Load Exerted on Specific Muscles
   Assessment of the Organism's Response to the Total Stress of Work

ENERGY EXPENDITURE OF WORK, REST, AND LEISURE
   Classification of Work
   Daily Rates of Energy Expenditure
   Energy Expenditure During Specific Activities

FATIGUE
   General Physical Fatigue
   Local Muscular Fatigue

VIBRATION

CIRCADIAN RHYTHMS AND PERFORMANCE
   Circadian Rhythms in the Human
   Shift Work

EFFECTS OF MENSTRUATION

INTRODUCTION

The practical application of the principles of work physiology involves the study of the functions of the organism subjected to the many stresses of muscular work. Since some degree of muscular activity is required in all kinds of work, even in most intellectual occupations, and in all expressions of life, work physiology is of interest not only to those concerned with manual labor.

The main objective of the work physiologist is to make it possible for individuals to accomplish their tasks without undue fatigue so that at the end of the working day, they are left with sufficient vigor to enjoy their leisure time.

With the development of mechanization, automation, and many work-saving devices, modern technology has contributed greatly to the elimination of much heavy physical work. Nevertheless, some heavy physical work, at least occasionally, is still obligatory in a number of occupations, such as commercial fishing, agriculture, forestry, construction, transportation, and many service occupations. In some cases, when the work load appears excessively high, it is evident that the only reason why the task can be accomplished is that it is performed by intermittent work, i.e., brief work periods interspaced with short periods of rest. However, the tendency is for the elimination of physical strain, while at the same time the need for increased output and greater efficiency through rationalization and automation has resulted in an accelerated tempo in most industrial operations. The short working week has resulted, on the whole, in a greater work intensity. The outcome is increasing nervous tension and mounting emotional stress. Consequently, the greatest problem in many industrial operations today is not the physical load, but rather, the mental stress and unfavorable working environment.

However, mechanization does not always lead to reduced energy expenditure. Thus Oberoi et al. (1983), measured the energy expended by hand washing clothes and by machine washing in 15 Indian women. The mean energy expenditure in washing clothes by hand while sitting on a 6.5 cm stool, and while standing at sink level were not significantly different from the mean energy expenditure found during washing clothes by machine.

The task facing the work physiologist in the field is to assess the strain imposed on the working organism by the total stress of the work and the working environment. Since practical experience has shown that one cannot tax more than some 30 to 40 percent of one's maximal aerobic power during an 8-hour working day without developing subjective or objective symptoms of fatigue, one of the most obvious problems is to determine the ratio between work load and work capacity. If the burden placed upon the worker is too high in relation to the person's capacity for sustained physical work, fatigue invariably will develop. This is true whether the work in question involves the entire body (large muscle groups) or only part of it (small muscle groups). A basic task for the work physiologist must therefore be that of measuring the rate at which

---

*In this chapter dealing with occupational work we have deliberately used the term "work" meaning job, manual labor, mental pursuit, etc., instead of the term "exercise" used throughout the rest of the book.

the work is being done, i.e., the work load, and of matching this rate with the worker's ability to perform the work.

FACTORS AFFECTING THE ABILITY TO PERFORM SUSTAINED PHYSICAL WORK*

The relationship between the work load and work capacity is affected, however, by a complicated interplay of many factors, internal as well as external, which must be taken into consideration (Fig. 11-1).

Ability to perform physical work basically depends on the ability of the muscle cell to transform chemically bound energy in the food into mechanical energy for muscular work, that is, into the energy-yielding processes in the muscle cell (see Chap. 2). This in turn depends on the capacity of the service functions that deliver fuel and oxygen to the working muscle fiber, i.e., on the nutritional state, nature, and quality of the food ingested, frequency of meals (see Chap. 12), oxygen uptake including pulmonary ventilation (see Chap. 5), cardiac output and oxygen extraction (see Chap. 4), and the nervous and hormonal mechanisms that regulate these functions.

Many of these functions depend on somatic factors, which may be partially genetically endowed; others may depend on sex, age, body dimensions, and state of health. In addition, physical performance is to a significant extent a function of psychological factors, notably motivation, attitude to work, and the will to mobilize one's resources for the accomplishment of the task in question. Several of these factors may be affected by training and adaptation.

Physical performance may also, directly or indirectly, be greatly influenced by factors in the external environment. Thus, air pollution may affect physical performance directly by increasing air-way resistance and thereby pulmonary ventilation, and indirectly, by causing ill health. Noise is a stress, which may not only damage hearing,

---

Definition of terms and units: Power = ability to work; the faculty of performing work; capacity for performance; the rate of transfer of energy. *Capacity* = ability; maximum power output. *Load* = the burden placed upon the worker; the rate at which work is being done at any time.

Units: In order to facilitate the transition from old units to the new SI system (Le Systéme International d'Unités), both the conventional and the new units will be used.

The new unit for *force* (F) is newton (N). F = m · a, where m = mass; a = acceleration. 1 N = 1 kg · m · s^{-2} = the force which gives the mass of 1 kg an acceleration of 1 m · s^{-2}. In the old system, the commonly used unit for force is kp (1 kp is the force acting on the mass of 1 kg at normal acceleration of gravity): 1 kp = 9.80665 N, or approximately 10 N.

The unit for *work* or energy is derived from the equation: W = F · L = force times distance. The unit for force is N and the unit for distance is m (meter); therefore the unit for work or energy is N · m = *joule* (J). 1000 J = 1 kJ : 1000 kJ = 1 MJ; 1 cal = 4.1868 J; 1 kcal = 4.1868 kJ.

For power (work/time) the following units apply:

1 watt = 1 joule · s^{-1} = 6.12 kpm · min^{-1}
9.81 watts = 1 kpm · s^{-1}
16.35 watts = 100 kpm · min^{-1}
1 hp (horsepower) = 736 watts = 75 kpm · s^{-1} = 4500 kpm · min^{-1}

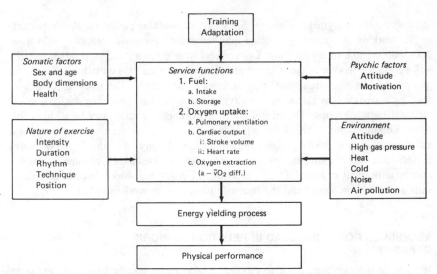

Figure 11-1
Factors affecting physical performance.

but also causes an elevation of heart rate and affects other physiological parameters that reduce physical performance. Cold weather, if severe, may in itself reduce physical performance because of numbness of the hands or lowered body temperature (the opposite effect of a warm-up prior to an athletic competition). But it may also involve the hobbling effect of bulky clothing and the slowing down of ordinary simple functions because of snow and ice. Heat, if intense may greatly reduce endurance because of the need for more of the circulating blood volume to be devoted to the transportation of heat rather than to the transportation of oxygen (see Chap. 13), and because of the effect of dehydration often accompanying heat exposure as a result of loss of body fluids (sweating). Although high gas pressures, encountered in underwater operations in connection with modern offshore oil exploration, present new and rather unique problems for the work physiologist, the drastic reduction in physical work capacity at high altitudes is one of the best studied problems concerning the environmental effects on physical work capacity.

Finally, the nature of the work to be performed, apart from work intensity and duration, is of decisive importance when considering an individual's capacity to endure prolonged work stress. Since all life functions generally consist of rhythmic, dynamic muscular work in which work and rest, muscle contraction and relaxation, are interspersed at more or less regular, fairly short intervals, the ideal way to perform physical work is to perform it dynamically, with brief work periods interrupted by brief pauses. This routine will provide some rest during the actual work period, so the worker may avoid fatigue and exhaustion and be able to leave the workplace with sufficient vigor left over for the enjoyment of leisure. Similarly, the working position is also important in that working in a standing position may represent a greater circulatory strain than does working in a sitting position. Conversely, working in a standing position, which will permit the worker to move about and thereby vary the load on individual muscle groups and facilitate circulation, may at times be preferable. The working technique may be of major importance in conserving energy and in providing varied use of different muscle groups. The monotony of a working operation may be a stress for some individuals but a relief for others who can carry on the work more or less automatically while thinking about something else. In any case, the tempo of work performance may be extremely important and impose stresses which, in some instances, may be unbearable or harmful to the individual. Finally, the work schedule, including shift work, is a problem requiring increasing attention in modern industry.

ASSESSMENT OF WORK LOAD IN RELATION TO WORK CAPACITY

In view of the fact that maximal oxygen uptake varies greatly from one person to another, a work load that is fairly easy for one worker may be quite exhausting for another. Suppose two men are to perform the same task, such as carrying a heavy load uphill, requiring an energy expenditure of about 2 liters of oxygen per min. One of the individuals has a maximal oxygen uptake of 6.0 liters $\cdot$ min^{-1}, the other 2.0 liters $\cdot$ min^{-1}. In the first case, the individual is merely taxing 30 percent of his aerobic power. Consequently, he can carry on all day without fatigue, as is normally true when the work load is less than about 40 percent of the individual's maximal aerobic power. Furthermore, he can continue to cover more than half his energy expenditure from the oxidation of fat. The second man, on the other hand, is taxing his aerobic power maximally and can carry on for only a few minutes, during which time he is compelled to rely on his carbohydrate stores as a major source of metabolic fuel (see Chap. 12).

Expression of the work load, as such, in absolute values (liter oxygen uptake per min) may therefore be quite meaningless. Instead, it should be expressed in percentage of the individual's maximal aerobic power. This means that the ratio between load and power should be assessed individually; that is, the individual's maximal oxygen uptake has to be determined and his or her rate of work has to be assessed. In general, the same principle also applies to the muscle groups which are engaged in the performance of the work in question, since only a certain percentage of the maximal muscle strength can be taxed without developing muscular fatigue (see Chap. 3).

Assessment of the Maximal Aerobic Power

A person's maximal aerobic power may be determined by direct measurement of the individual's maximal oxygen uptake, or estimated on the basis of data obtained from submaximal tests (see Chap. 8).

Assessment of the Physical Work Load

The physical work load may be assessed either by measurement of the oxygen uptake during the actual work operation or by indirect estimation of the oxygen uptake on the basis of the heart rate recorded during the performance of the work.

Measuring the Oxygen Uptake in a Typical Work Situation   Since the validity of using oxygen uptake as a basis for measuring energy expenditure has been established, this indirect calorimetry has been used to determined the energy cost of a great variety of human activities. Figure 11-2 presents oxygen uptakes based on average values for various loads on a cycle ergometer, compared with equivalent activities. With the development of highly portable devices for collecting expired air under field conditions and rapid methods of analyzing the oxygen and carbon dioxide content of the air samples, a vast body of knowledge of the energy cost of physical work has been accumulated.

In field studies, the classical method is to collect the expired air in Douglas bags carried on the subject's back. At present, other methods are available by which the volume of the expired air is measured with flow meters, and aliquot samples of the expired air are collected in a small rubber bladder (the Max Planck respirometer,

Figure 11-2
Cycle ergometer work and equivalent physical activity.

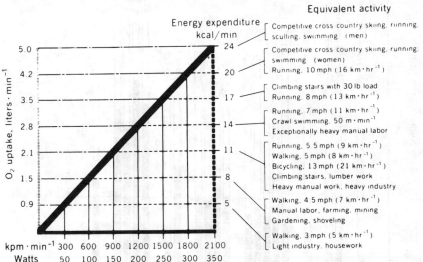

Energy expenditure for "equivalent activity" is only approximate. It is meant merely as a general guide. It depends, among other things, on the weight of the subject. The examples listed are based on the average person with a body weight of 160 lb (70.75 kg)

Müller and Franz, 1952; the integrating motor pneumotachygraph, or IMP, Wolff, 1958). The expired air is then analyzed for O_2 and CO_2, using a conventional gas-analysis technique (Haldane or Micro-Scholander) or electronic O_2 and CO_2 analyzers. If accuracy is not too critical (± 10 percent), it may be sufficient to analyze only the O_2 content by a portable O_2 analyzer. Various methods are described and discussed by Consolazio et al. (1963), Durnin and Passmore (1967), Banister and Brown (1968), Morehouse (1972), Kamon (1974), and Harrison et al. (1982). Wilson and Sklenka (1983) have designed a modified portable system for measuring energy cost during particularly intense, dynamic activity.

An example of measured oxygen uptake in typical commercial fishing operations is given in Fig. 11-3. It is evident that the measured oxygen uptake during the work performance represents the energy expenditure only at the time when the expired air sample is collected, and may not be representative for the work performed during the whole working day. Furthermore, it is a common experience that the test subject tends to be affected by the investigation, causing the test situation to be atypical. The test

Figure 11-3
Measured oxygen uptake in typical fishing operations. *(From I. Åstrand et al., 1973.)*

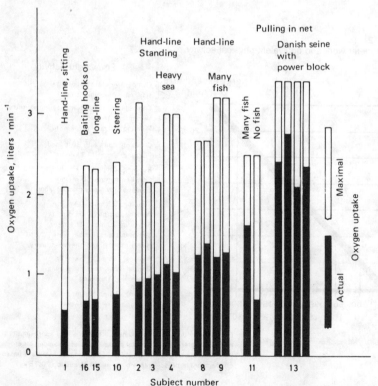

equipment is apt to affect heart rate, pulmonary ventilation, and oxygen uptake, and may hamper the actual work operation.

In contrast to this, the indirect assessment of the work load on the basis of the continuously recorded heart rate reveals a general picture of the overall activity level during the entire working day. Moreover, on the basis of time-activity records for each subject, collected by an observer during the whole working day, it is possible to separate the different activities with respect to heart rate. Thus, the indirect assessment of work load based on the recorded work pulse may be preferable in many work situations.

Indirect Assessment by Recording the Heart Rate during Work   In a given person, there is generally a linear relationship between O_2 uptake and heart rate. Therefore, the heart rate, under certain standardized conditions, may be used to estimate work load, if the work load-heart rate relationship has been established for the individual in question, if roughly the same large muscle groups are engaged in the work in both cases, and if environmental temperature, emotional stress, etc., are the same (see Chap. 4).

The individual's circulatory response to work engaging large muscle groups may be measured on a cycle ergometer. Starting at a low load such as 50 watts, the load is increased stepwise every 6 min, usually by 50 watts, until a heart rate of about 150 beats $\cdot$ min^{-1} is reached (Fig. 11-4). On the basis of the resulting line representing the relationship between the individual's work load and heart rate, it is possible to estimate the work load from the heart rate recorded during a specific work situation in the field (Fig. 11-5, 11-6, showing typical examples).

The recording of the heart rate in the field is most conveniently accomplished with the aid of a portable miniature battery-operated tape recorder (such as those produced by Avionics, Philadelphia, U.S.A., or Hellige, Freiburg/Breisgau, Germany, or Oxford Instrument Co., England) monitoring the heart rate continuously in the form of a simplified electrocardiogram. This can also be accomplished by the use of a portable miniature computer, such as the Vitalog (produced by the Vitalog Corporation, Palo

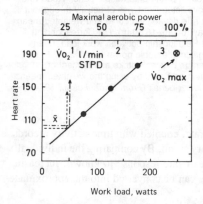

Figure 11-4
The individual relationship between the heart rate at different submaximal work loads and the predicted corresponding oxygen uptake (Åstrand, 1967) is established graphically (scale $\dot{V}_{O_2}$). The measured maximal oxygen uptake ($\dot{V}_{O_2}$ max) is used to construct another parallel scale which shows the load expressed in percentage of the individual's maximal aerobic power. The weighted mean ($\bar{x}$) of the continuous recording of the heart rate is then used to asess the approximate average oxygen uptake during work as well as the load expressed as percentage of the maximal aerobic power.

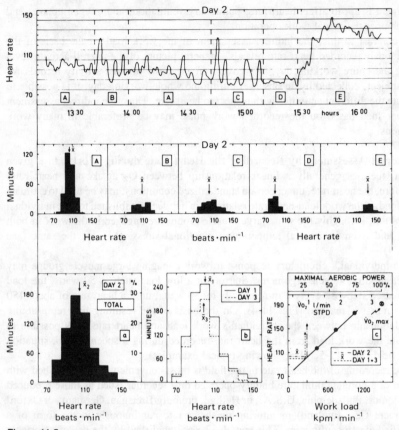

Figure 11-5
Example of analysis of the heart rate recorded at sea in a twenty-one-year-old net-fisherman (T.F.). Upper panel: Heart rate curve in the latter part of the second day observation period. Activities: A—Arranging and putting out net. B—Bleeding and cleaning the fish. C—Other unspecified activities. D—Resting. E—Unloading the catch at the pier. Middle panel: Heart rate distribution curves and weighted means ($\bar{x}$) of the above listed five types of activity throughout the whole observation period. Lower panel: (a) Heart rate distribution curve and weighted mean ($\bar{x}_2$) for the whole observation period of the second day (06.45–16.50 hr). (b) Corresponding results from the first and third day of observation. (c) Relationship between work load, heart rate, estimated oxygen uptake ($\dot{V}_{O_2}$) and percent of maximal aerobic power in subject T.F. as assessed in the ergometer test. Estimation of the average oxygen uptake and percent of maximal aerobic power during the 3 days of observation on board by using the weighted means ($\bar{x}$) of the heart rate from sections (a) and (b) of the lower panel. *(From Rodahl et al., 1974a.)*

Alto, California, U.S.A.). The recorded heart rate, coupled with time-activity records, shows the degree and variations of the circulatory strain. By comparing the individual's heart rate during work in the field with the heart rate response to known, increasing work loads on a cycle ergometer, the heart rate can be converted into the approximate oxygen uptake (Fig. 11-5).

The heart rate curves thus obtained may be replayed and transcribed on recording

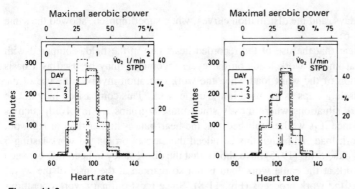

Figure 11-6
Weighted means ($\bar{x}$) and distribution curves of the heart rate and corresponding work load in two subjects during consecutive days of observation.
Left panel: forty-one-year-old long-line fisherman (K.S.)
Right panel: fifty-six-year-old catch handler (M.E.) on land. *(From Rodahl et al. 1974a.)*

paper and graphically evaluated (I. Åstrand et al., 1973), or evaluated by computer analysis (Rodahl et al., 1974a).

It should be borne in mind that the estimation of oxygen uptake from recorded heart rate may be subject to considerable inaccuracy. However, in a field study by Rodahl et al. (1974a), 24 direct measurements of oxygen uptake with the Douglas bag method were compared with the oxygen uptake calculated from the simultaneously recorded heart rate in six fishermen. The calculated values deviated from the measured ones in both directions by no more than ± 15 percent (Fig. 11-7).

The reproducibility of the results from day to day in such field studies over a 3-day period was examined in connection with a study of fishermen by Rodahl et al. (1974a). They found a remarkable reproducibility of the day-to-day results in the same individual doing the same work. The weighted mean heart rates were practically the

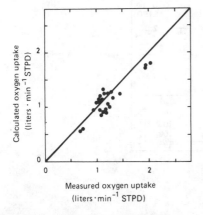

Figure 11-7
Relationship between oxygen uptake measured by the Douglas bag method and calculated from the simultaneously recorded heart rate. Twenty-four observations in six subjects during the various fishing operations. *(From Rodahl et at., 1974a.)*

same for all 3 days, and the distribution curves, when superimposed, showed the same shape (Fig. 11-6).

It is thus clear that the use of the recorded heart rate in the field, compared with the heart rate at known work loads on the cycle ergometer, may be used as a basis for the estimation of the work load when the work operation involves the use of the same large muscle groups as are used in the cycle work. This comparison may not be feasible in work situations where mostly small muscle groups are involved, such as in arm work, since it is well established that the heart rate is higher than in leg work at the same work load. Although this is indeed the case in prolonged work lasting 5 min or more, Vokac et al. (1975) have shown that the discrepancy in heart rate between arm and leg work at the same work load is not so marked at the onset of the work, but increases as the work proceeds (Fig. 11-8). Since most ordinary work operations involve a dynamic type of work with a rhythmic alteration between muscular contraction and relaxation, in which each period of work effort is rather brief, it appears that the use of the recorded heart rate as a basis for the estimation of work load may be acceptable even in many work situations involving arm work or the use of small muscle groups.

Since the heart rate response to one and the same work load varies individually, the circulatory strain is best expressed as percentage of the heart rate reserve of the subject, the heart rate reserve being the difference between maximal heart rate and resting heart rate. In order to overcome the time-dependent variability, the heart rate

Figure 11-8
Heart rates in the first 6 min of cycling at 900 kpm · min⁻¹ (150 watts) (A) and of arm cranking at 600 kpm · min⁻¹ (100 watts) in sitting (B) and standing (C) positions. *(From Vokac et al., 1975.)*

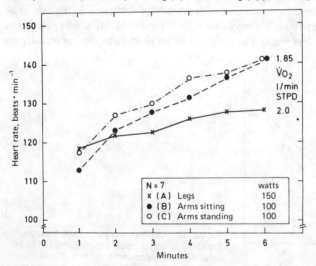

results may be presented as weighted means ($\bar{x}_w$), which are calculated from the individual's values multiplied by the fraction of the sum of their durations (Rodahl et al., 1974a).

Verma et al. (1979), have suggested that a combination of minute ventilation and heart rate is a better predictor of energy expenditure than either of the two variables used singly.

Summary The continuous recording of the heart rate permits an uninterrupted collection of data which reflect the work load during the entire day. A quantitative numerical analysis of the recorded data, supplemented by visual analysis of the replayed heart rate curves (Fig. 11-5), permits a comprehensive and dynamic evaluation of the circulatory strain imposed by work loads of varying intensity. The use of a computer makes it possible to analyze large series of observations with respect to the mean values, peak values, time distribution, and the occurrence and duration of excessively high heart rates, for example, over 50 percent of the heart rate reserve (the maximal heart rate minus the heart rate at rest). Since the heart rate response to one and the same work load varies with individuals, the circulatory strain is best expressed as a percentage of the subject's heart rate reserve (I. Åstrand, 1960). The results can be presented more conveniently and more graphically by converting the recorded heart rate individually into the corresponding estimated oxygen uptake (I. Åstrand, 1967). Estimated oxygen uptake serves, then, as the measure of the work load, and, expressed in percentage of the subject's maximal aerobic power, it indicates the relative degree of the exertion in the same way as the percentage of the heart rate reserve. In most cases, the reliability of the conversion is adequate for all practical purposes of field investigation (Rodahl et al., 1974a).

Assessment of the Load Exerted on Specific Muscles

While oxygen uptake and heart rate may be used to assess the magnitude of muscular work in general, these parameters may not be suitable for the monitoring of localized muscular loads (Malmqvist et al., 1981). Under certain conditions, however, the recording of heart rate is a quite sensitive index of muscular force, and may even respond significantly to minor changes in posture (Hanson and Jones, 1970).

Electromyographic recordings, on the other hand, will permit load measurements on single muscles, as well as groups of muscles. This can be accomplished by using surface electrodes. It is well established that the electromyogram (EMG) does reflect the magnitude of the muscle engagement (Fig. 11-9), and may be used to measure the exerted force in percent of the maximal voluntary muscle strength. For a more detailed discussion of the principles underlying the use of electromyography in the evaluation of local muscular load, see Hagberg (1981a). Bobet and Norman (1982) have shown that the average electromyogram may provide useful information even in the investigation of whole-body tasks, but that their use is less straightforward in this respect than in the investigation of more constrained or simplified tasks.

EMG amplitude

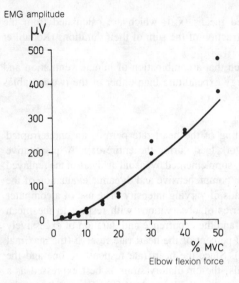

Figure 11-9
Relation between EMG-activity from the biceps brachii and elbow flexion force as a percentage of the maximal voluntary contraction (% MVC). *(From Hagberg, 1981a.)*

Assessment of the Organism's Response to the Total Stress of Work

The total stress imposed on the organism by a given work situation (physical as well as psychological) is generally reflected by nervous and hormonal stimulation, more or less proportional to the degree of the stress.

Hormonal Response   It is well known that the total stress response of the organism is reflected by the sympatheticoadrenomedullary activity. This may be roughly assessed by measuring urinary excretion of epinephrine and norepinephrine by the method described by von Euler and Lishajko (1961), as modified by Andersson et al. (1974), using the resting night urine as base value. Expressed in ng · min^{-1}, it may serve as a measure of occupational stress. A nearly tenfold increase in the epinephrine excretion and about a fourfold increase in norepinephrine excretion were observed during the workday, as compared with the excretion of coastal fishermen during the night (I. Åstrand et al., 1973). An even greater increase in norepinephrine excretion was observed in war college cadets during a strenuous battle course including marked sleep deprivation (Holmboe et al., 1975).

An increase both in epinephrine and norepinephrine may be due to a number of single factors or a combination of them. Generally, both the circadian rhythm (Fröberg et al., 1972) and the change of body posture from the recumbent to the standing position (Sundin, 1956) increase the catecholamine excretion during the day. This increase may be markedly enhanced by the effect of physical exertion (von Euler and Hellner, 1952), cold (Lamke et al., 1972), and emotional factors (von Euler, 1964; Levi, 1967). The level of plasma catecholamines increases with both the duration and the severity of the muscular exertion (Banister and Griffiths, 1972).

It is generally agreed that urinary catecholamine excretion may be used to quantitate stress response in a laboratory situation, and in simulated work situations (Cox et al., 1982). It may also be used in field studies when the collection of urinary samples is carefully supervised and controlled. Jenner et al. (1980), have reported the analysis of catecholamine excretion rates of men living in 12 villages north of Oxford. Their results showed differences according to day of week and occupation. Work-day adrenaline excretion rates were higher than rest-day rates, rates for nonmanual occupations were higher than for manual occupations and rates for professional and managerial occupations were higher than for other occupations.

The fact remains, however, that it is at best, very difficult to obtain complete urinary collections; the volume of residual urine is a source of error, and it is difficult to obtain samples more frequently than every two hours or so. This means that acute stress situations of brief duration may not be reflected in the urinary sample, which shows the average value for as much as a couple of hours. In addition, there are considerable individual differences in urinary catecholamine elimination, even among persons exposed to the same level of work stress (Fig. 11-10). Finally, extensive studies of urinary catecholamine elimination in a large number of workers at a variety of work places, which were carried out at the Norwegian Institute of Work Physiology, have failed to show any significant elevation of excretion rates of most industries. Similarly, Follenius et al. (1980), have shown that although it is generally recognized that exposure to high levels of noise in the working environment may be a stressful experience for those who are exposed, high levels of noise do not necessarily induce

Figure 11-10
Urinary excretion of norepinephrine in war academy cadets during simulated battle course.
(From Holmboe et al., 1975.)

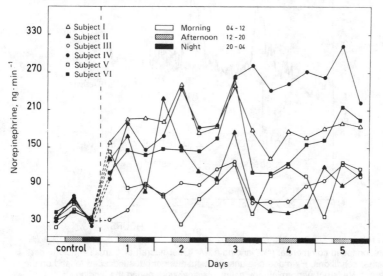

any significant endochrinological changes. For these reasons, this method is not to be recommended in order to assess levels of work stress. Instead, continuously recorded heart rate may be routinely employed as an indication of autonomous nervous system response to stress, both physical and mental.

Nervous Response An increased sympathicotonus, brought about by emotional as well as physical stress, will give rise to an accelerated heart rate, which consequently may serve as an index of stress response. It is thus well established that the heart rate increases linearly with increasing physical work load, provided there is no major change in the subject's emotional state Fig. 11-4. Also, the heart rate may vary markedly as a consequence of emotional stress in an individual who is at rest or subject to a constant light work load (Fig. 11-11).

The fact that the mean heart rate increases during demanding mental load was shown by Hitchen et al. (1980), who observed that the heart rate increased by 15 percent during a demanding mental task.

Combined with a detailed activity log kept by a trained observer, the continuously recorded heart rate may indeed provide useful information concerning a subject's response to stressful situations.

Heart rate variability has been suggested as an indicator for nonphysical stress. The basis for this is that heart rate variability is known to decrease under a mental load (Hitchen et al., 1980). In a critical evaluation of this approach, Luczak (1979) concludes

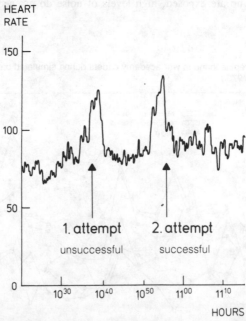

Figure 11-11
Heart rate response in commercial pilot landing a twin-Otter aircraft at a small airfield under difficult weather conditions. He made one unsuccessful attempt but succeeded to land on the second attempt. He is not performing any actual physical work during the landing.

that this method is a good indicator of nonphysical strain in the laboratory with steady state and well-defined conditions. However, it is very difficult to handle, and has restricted value for the evaluation of strain in practical situations with varying environmental and physical as well as emotional conditions. Specifically, Luczak (1979) points out that heart rate variability cannot be used if the subject is required to speak, since this affects the pattern of respiration and hence the heart rate. Nor can it be used if the test situation involves muscular activity, since this too affects heart rate. The same applies to the environmental temperature.

ENERGY EXPENDITURE OF WORK, REST, AND LEISURE

Classification of Work

Evidently, the human being is not ideally suited to be a source of mechanical power, and in this respect cannot compete with modern mechanical devices, such as a bulldozer or a truck. The power output of an average man engaged in prolonged work over an 8-hour working day may amount to little more than 0.1 horsepower (1 horsepower ~ 750 watts). A horse may yield at least 7 times that amount, and an ordinary farm tractor, 70 horsepower. Yet humans must be physically active, or they will deteriorate. The solution, therefore, is to include some physical activity in the daily work, to provide the opportunity for varied use of the locomotive system, and to select a proper ratio between work and rest in order to supply adequate recuperation *during* work.

In most instances, at least in the Western world with its advanced technology, excessively heavy work can easily be eliminated with technical aids; it is merely a matter of cost and priority. Establishing limits for permissible physical work loads is therefore of limited practical value. Of far greater importance to the worker today is the manner in which the work is being performed, the opportunity to influence the working situation and to govern one's own rate of work, the safety and the general atmosphere of the working environment, the arrangement of work shifts, etc. In most jobs in modern industry, the worker or operator is able to adjust the rate of work according to one's personal capacity. However, there are some exceptions, as when the work is performed by a team. Here, the weak have to keep up with the strong. In such teamwork, older workers, who are generally slower and who have a reduced physical working capacity, may be hard-pressed to keep pace with the younger members of the team. In any event, great individual differences do exist in physical working capacity, and practical experience has indicated that a work load taxing 30 to 40 percent of the individual's maximal oxygen uptake is a reasonable average upper limit for physical work performed regularly over an 8-hour working day. Similarly, no more than 40 percent of maximal muscle strength should be applied in repetitive muscular work in which the time of each muscular contraction is about one-half the time of each period of relaxation (Banister and Brown, 1968).

The physiological and psychological effects of a given energy output (per min, per 8 hr, per day) are determined by the individual's maximal aerobic power, size of the engaged muscle mass, working position, whether work is intermittent at a high rate or continuous at a lower intensity, and environmental conditions. In general, a person's subjective experience of a particular work load or rate of work is more closely related to heart rate than to oxygen uptake during the performance of the work, since the work pulse, in addition to the actual work load, also reflects emotional factors, heat, the size of the engaged muscle groups, etc.

Bearing these reservations in mind, the following identification of prolonged physical work, classified as to severity of work load and to cardiovascular response, may be of some use. These figures refer to average individuals twenty to thirty years of age, and can be used only as general guidelines in view of the vast individual variations in ability to perform physical work.

In terms of oxygen uptake:

Light work	up to 0.5 liter $\cdot$ min^{-1}
Moderate work	0.5–1.0 liter $\cdot$ min^{-1}
Heavy work	1.0–1.5 liter $\cdot$ min^{-1}
Very heavy work	1.5–2.0 liter $\cdot$ min^{-1}
Extremely heavy work	over 2.0 liter $\cdot$ min^{-1}

In terms of heart rate response:

Light work	up to 90 beats $\cdot$ min^{-1}
Moderate work	90–110 beats $\cdot$ min^{-1}
Heavy work	110–130 beats $\cdot$ min^{-1}
Very heavy work	130–150 beats $\cdot$ min^{-1}
Extremely heavy work	150–170 beats $\cdot$ min^{-1}

Daily Rates of Energy Expenditure

An estimate of daily energy expenditure is important not only for the calculation of energy need. It is also necessary in order to determine fairly accurately the level of physical activity of an individual or a group of individuals, in connection with attempts to ascertain the role of physical activity in health and disease, for example, in the prevention and treatment of ischemic heart disease. Such an estimate of energy output can be made by several methods: (1) the 24-hr recording of heart rate by portable miniature recorders, as described earlier in this chapter; (2) estimation based on time-activity data and measurements of the energy cost of all pertinent activity; and (3) assessment (by food weighing) of daily food intake required to maintain body weight. All three of these methods are about equally accurate and reliable, with an error of no more than about 15 percent (Rodahl, 1960).

As expected, there is a wide individual variation in energy output depending on occupation, leisure activity, and attitude or individual proneness to physical activity in general. The range of daily rates of energy expenditure is from 1340 up to 5000 kcal (5.63 to 21.00 MJ), (Durnin and Passmore, 1967). About 2900 kcal $\cdot$ day^{-1} (12 MJ $\cdot$ day^{-1}) is a reasonable expenditure for a man who is not engaged in heavy manual

TABLE 11-1

Activity	Time, hr	Man Rate kcal · min⁻¹	kJ	Man Total kcal · min⁻¹	MJ	Woman Rate kcal · min⁻¹	kJ	Woman Total kcal · min⁻¹	MJ
Sleeping and lying*	8	1.1	4.6	540	2.3	1.0	4.2	480	2.0
Sitting†	6	1.5	6.3	540	2.3	1.1	4.6	420	1.7
Standing‡	6	2.5	10.5	900	3.8	1.5	6.3	540	2.3
Walking§	2	3.0	12.5	360	1.5	2.5	10.5	300	1.3
Other¶	2	4.5	18.8	540	2.3	3.0	12.5	360	1.5
Total	24			2880	12.2			2100	8.8

*Essentially basal metabolic rate plus some allowance for turning over or getting up or lying down.
†Includes normal activity carried on while sitting, e.g., reading, driving an automobile, eating, playing cards, and desk or bench work.
‡Includes normal indoor activities while standing and walking spasmodically in limited areas, e.g., performing personal toilet, moving from one room to another.
§Includes purposeful walking, largely outdoors, e.g., home to commuting station to work site, and other comparable activities.
¶Includes spasmodic activities in occasional sport exercises, limited stair climbing, or occupational activities involving light physical work. This category may include weekend swimming, golf, tennis, or picnic using 20 to 85 kJ · min⁻¹ (5 to 20 kcal) for limited time.

labor but who is regularly active during leisure time. A reasonable figure for his wife would be about 2100 kcal · day^{-1} (9 MJ · day^{-1}) (Table 11-1), representing energy expenditure by reference man and woman, according to the Food and Nutrition Board (1964). It is possible, however, that these figures may be slightly too high because of the increasing inactivity and the sedentary life led by large segments of the population.

Available evidence shows that the effect of environmental temperature, as such, on the metabolic work is very small (Consolazio, et al., 1963). Gray et al. (1951) reported variations in metabolism up to 4 percent in men working at a fixed cycle ergometer work load at ambient temperatures of $-15°C$ and $32°C$. This variation, however, is probably within the experimental error. It appears that any increase in the energy cost of work in a cold environment is primarily due to the additional energy cost of bodily movement in bulky clothing through difficult terrain, snow, etc. Nelson et al. (1948) found that metabolic heat production for a given amount of work remained unchanged in three men walking in seven hot environments between $32°C$ and $49°C$.

In the majority of professional activities (including office work), housework, light industry, laboratory and hospital work, and retail and distribution trade, the energy output is less than 5 kcal · min^{-1} (less than 20 kJ · min^{-1}, or less than 1 liter O_2 · min^{-1}) (Fig. 11-12). In the building industry, agriculture, the iron and steel industries, and the armed services, there are many jobs which occasionally demand a caloric expenditure of up to 7.5 kcal · min^{-1} (or 30 kJ · min^{-1}) or even higher, particularly if mechanical aids are few and prefabricated materials are utilized to only a small extent. Still higher energy demands are found in fishing, forestry, mining, and dock labor, where figures up to or exceeding 10 kcal · min^{-1} (40 kJ · min^{-1}) have been reported.

Even if her energy output may be relatively low, a housewife, because of her lower maximal aerobic power, may strain herself as much doing domestic work as a lumberjack, farmer, commercial fisherman, or miner whose work requires a higher energy output, but who also has a higher maximal aerobic power.

The energy expenditure in recreational activity naturally covers the whole range from near resting values up to utilization of the full power of aerobic and anaerobic processes, depending on the type of activity and the degree of vigor with which it is pursued.

Various attempts have been made to establish maximal permissible limits for daily energy output for men working at the same task on a year-round basis (Banister and Brown, 1968). Lehmann (1953) suggested 4800 kcal · day^{-1} (20 MJ · day^{-1}) as the limit. Subtracting about 2300 kcal (9.66 MJ) for basal metabolism, eating and various basic necessities, leisure activity and travel to and from work, 2500 kcal (10.34 MJ) is left for the actual 8-hour work. Banister and Brown (1968) consider 2000 kcal (8.40 MJ) a more suitable load for heavy workers, giving an average rate of energy expenditure of 4.2 kcal · min^{-1} (18 kJ · min^{-1}).

Establishing such norms may be quite meaningless, however, in view of the great individual differences in physical work capacity or fitness. Furthermore, the level of activity in many industrial tasks is actually self-regulatory in that the rate of work and the spacing of rest pauses are set by the individual's level of physical fitness. In fact, in some cases, such as the older commercial fishermen studied by Rodahl et al. (1974a), the only way it is possible for a person to endure work loads close to the permissible

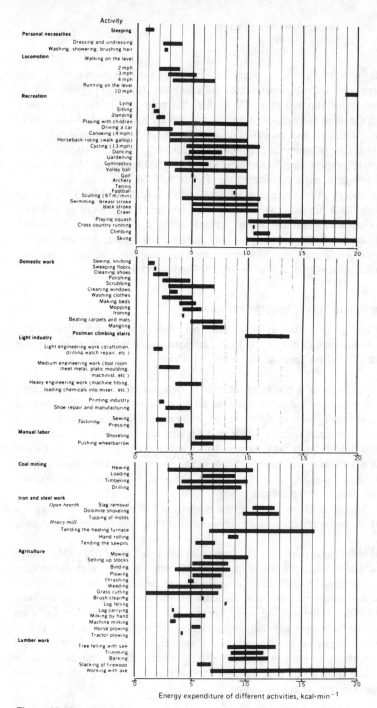

Figure 11-12

Energy expenditures of different activities. Bars denote range of data presented in literature (See also Karvonen, 1974.)

physiological limits, day after day, year after year, is by working intermittently, with periods of high work intensity interspersed with frequent, brief rest periods.

In terms of strain imposed upon the worker, the peak load of the task is more important than the mean energy expenditure. A steel worker may expend 10 kcal · min^{-1} (40 kJ · min^{-1}) during 1 hour of shoveling gravel or dolomite, but during the rest of the 8-hour shift, energy output may be only 2 to 2.5 kcal · min^{-1} (8–10 kJ · min^{-1}). The 8-hour energy expenditure is then 600 + 900 = 1500 kcal (6.30 MJ). A worker with a job which demands a consistent rate of work with no peak loads may attain a higher 8-hour energy expenditure (e.g., 480 · 4 = 1920 kcal, (or 8.00 MJ), without requiring as strong a physique as does the steel worker. Heavy work or awkward working positions may often hamper the recruitment for certain types of work, even though these factors may be operating for only short periods of time. The same applies to many types of automated industrial operations where monotony or lack of personal influence on the process may appear boring, or even harmful.

Energy Expenditure during Specific Activities

Data for O_2 uptake, expressed as kilocalories per minute, of a variety of activities taken from various sources but mostly from Passmore and Durnin (1955) and Spitzer and Hettinger (1958), are summarized in Fig. 11-12.

Sleeping   In the past it has been difficult to obtain accurate values for energy expenditure in a sleeping subject because of the discomfort and restriction imposed by noseclip and mouthpiece, or the technical problem of leaks when a face mask is used. The use of the open-circuit method with a plastic hood covering the entire head of the patient and the Noyons basal metabolism apparatus (Issekutz et al., 1963) offers many distinct advantages when making metabolic measurements in a resting or sleeping individual. With this method, repeated samples can be taken throughout the observation period without disturbing the subject in any way.

Generally, one-third of the 24-hr period is spent in bed sleeping or resting. The energy spent in this way accounts for about one-tenth to one-quarter of the daily expenditure.

The metabolism, as a rule, may fall below the basal metabolic rate (BMR) when the fasting subject is asleep. The basal metabolic rate is the rate of energy metabolism in a resting individual 14 to 18 hours after eating. At the start of the night's sleep, however, the metabolism may be above the basal level because of the specific dynamic action (SDA) effect of the evening meal. These two factors thus tend to cancel each other out, so that the energy expenditure throughout the night is not far from that of the BMR value. In any case, a deviation of 10 percent above or below the normal basal level represents an error of less than about 3 percent of the total 24-hr energy expenditure. The BMR value may therefore be taken as a measure of the metabolic rate of a subject in bed, asleep or awake.

Personal Necessities   Under normal circumstances, an individual spends not more than 1 hr of the day in carrying out personal necessities, such as washing, shaving, brushing hair, cleaning teeth, dressing, and undressing. Energy expenditure values for such activities are summarized in Fig. 11-12.

Sedentary Work   For all practical purposes, the energy expenditure of mental work, including office work, etc., is not materially different from that of sitting or standing unless such occupations involve a great deal of physical activity, such as walking, bending, and opening drawers. Passmore and Durnin (1955) list a mean value of 1.6 kcal · min^{-1} (7 kJ · min^{-1}) for miscellaneous office work while sitting, and 1.8 kcal · min^{-1} (8 kJ · min^{-1}) while standing.

Although the brain utilizes a substantial part of the total O_2 uptake of the body at rest, it is well established that mental work requires only an insignificant rise in O_2 uptake, at least as long as the mental effort is not associated with markedly increased muscular tension or emotional stress (Benedict and Carpenter, 1909). Benedict and Benedict (1933) observed no substantial difference in metabolism at rest and during mental effort in six subjects engaged in 15-min periods of arithmetic exercises. Others (Eiff and Göpfert, 1952) have found an average rise in metabolism during mental efforts of as much as 11 percent, but this difference was apparently due to a concomitant rise in muscle tone. However, it is indeed interesting to note that although the central nervous system is extremely sensitive to lack of oxygen and lowered oxygen tension, an increase in intellectual functions is not associated with a significant increase in the overall O_2 uptake.

Housework   From Fig. 11-12 it is apparent that domestic work involves many tasks which may be classified as fairly heavy physical work (Oberai et al. 1983), although modern equipment has contributed greatly to making life somewhat easier for the person keeping house today. This is all relative, however, since performance of physical work depends not only on the severity of the work load, but also on the physical work capacity of the individual. For this reason, the lighter work load of present-day domestic occupations, due to modernizations, may represent, relatively speaking, as heavy a load on present-day housekeepers as that of their grandmothers, if the latter were more physically fit.

Figure 11-13 presents an example of a continuous recording of heart rate during a day's work in a Swedish housewife, showing that a great deal of physical activity is involved even in modern housekeeping.

Light Industry   A fair amount of data is available regarding the energy cost of different kinds of manual labor and industrial tasks. With the development of automation, the physical work load of the industrial worker on the whole has been greatly reduced. This is evident from an early study by Kagan et al. (1928), who compared energy expenditure by men assembling machinery entirely by hand with those using a conveyer system. In the former case, the energy expenditure varied from 5.2 to 6.4 kcal · min^{-1} (22 to 27 kJ · min^{-1}); in the latter case, it varied between 1.8 and 4.7 kcal · min^{-1} (8 to 20 kJ · min^{-1}). According to Passmore and Durnin (1955) and Durnin and Passmore (1967), a wide variety of industrial activities, classified by them as light industry, demanded energy expenditure rates between 2 to 5 kcal · min^{-1} (8 to 20

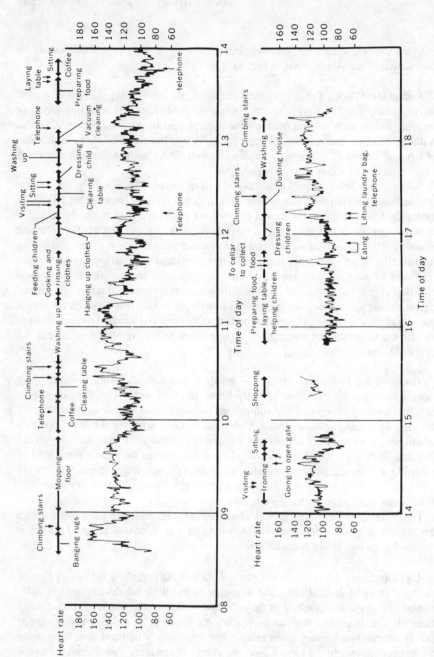

Figure 11-13
Heart rate in a housewife during an average day of housework.

kJ · min^{-1}) for men and 1.5 to 4 kcal · min^{-1} (6 to 17 kJ · min^{-1}) for women. The authors pointed out that further subdivision or classification is quite impracticable.

Recent unpublished studies carried out by the Institute of Work Physiology in Norway show that the work load in most light industries is well below 30 percent of the worker's maximal oxygen uptake.

Manual Labor   A number of studies have shown that the energy expended during the performance of similar types of work may vary greatly, depending on the technique used in accomplishing the work. This is even true for relatively simple activities, such as carrying a load, depending on how the activity is performed. Bedale (1924) showed that when a person carried a load, the energy expenditure was minimal when a yoke across the shoulder was used; it was maximal when the load was carried on the hip under the arm. The energy cost is less when the load is kept close to the center of the body (Soule et al., 1978). When carrying a load, the energy expenditure rises markedly with the increased speed of walking (Brezina and Kolmer, 1912; Cathcart et al., 1923).

Garg and Ayoub (1980) have shown that the recommended maximal loads to be lifted by humans, vary with the different criteria used for the assessment (biomechanical, psychosocial, metabolic, or circulatory). At any rate, handles have a profound effect on maximal acceptable weights to be lifted (Garg and Saxena, 1980). The subject's own weight in addition to that of the load, should also be taken into account when assessing the optimal load to be carried (Pierrynowski et al., 1981). It seems heavier to walk with a load than to simply hold it standing still. Evans et al., (1983), recorded heart rate and EMG in seven young men who held and carried loads of 15, 20, 25, 30, and 40 kg until exhaustion. The mean maximum time for load holding was found to be hyperbolically related to the mass of the load, the mean maximum time for load carrying being shorter than that for load holding. The heart rate at exhaustion was linearly related to the load, and greater when the load was carried than when held. Electromyographic activity in the forearm flexor muscle increased when the load was carried and fluctuated synchronously with the stepping frequency.

The physiological response to lifting a load depends on the manner in which it is being lifted. Kumar (1980) measured intra-abdominal pressure by telemetry and the activity of the erector spinal and external oblique muscles by electromyography in male volunteers lifting 10 kg from ground to knee, hip and shoulder levels in the sagittal, lateral, and oblique planes. He found a significant difference between the responses in these three planes; the saggital plane activities evoking least response. The intra-abdominal pressures and the electromyographic activity of the erector spinal and the external oblique activity were significantly correlated in each of the three planes.

The importance of using proper technique when lifting is evident from the studies of Ekholm et al. (1982), who showed that the compressive forces on the lower back (the L5-S1 disc) were about 4400 N when lifting 12.8 kg with straight knees, as against 3400 N when lifting with bent knees, straight back and the burden moved between the knees and close to the pelvis. It should be emphasized, however, that this only applies as long as the load is kept close to the body. Furthermore, the inertial forces due to accelerations of the body should also be considered (Leskinen et al., 1983).

In horizontal lifting, the preferred weight is lower than in vertical lifting (Ljungberg et al., 1982), a fact that should be kept in mind when considering recommendations for tolerable weights to be handled manually.

The energy cost of climbing up and going downstairs with a load is according to Crowden (1941), 11 times that of walking on the level bearing the same load (see also Pimental and Pandolf, 1979). In a study of the energy cost of transporting a load with the aid of a wheelbarrow, Hansson (1968) showed that the energy expenditure is higher the smaller and softer the wheel, and that a two-wheel cart is more efficient than a wheelbarrow with a single wheel.

There are few data available to assess the overall energy expenditure of building work. Accurate assessment of the activity is particularly difficult because of the variety of individual operations involved.

A rough estimate of the efficiency of a particular work operation may be achieved by measuring the energy cost based on oxygen uptake or heart rate, or by recording output at a constant O_2 uptake or heart rate. Thus the influence of the choice of tools on the rate of work and work effort was studied in a group of men engaged in the nailing of impregnated paper under standardized conditions (Hansson, 1968). The rate attained in nailing when using a stapler was 3 to 4 times higher than when using an ordinary hammer. There were no significant differences, however, in oxygen uptake per minute, quality of work, or estimated degree of fatigue when using the different tools. I. Åstrand et al. (1968) have further studied the energy cost of hammer nailing at three different heights: at bench level, into a wall at head level, and into a ceiling above the head. The number of hammer strokes did not differ significantly in the three situations. However, the number of nails driven per minute was lower when nailing into the wall (10.6 nails · min^{-1}) than when nailing into the bench (14.6), and still lower for nailing into the ceiling (4.5 nails · min^{-1}), indicating that the strokes became less powerful or were less well aimed when nailing into the wall and ceiling than when nailing into the bench. Nailing in the three positions resulted in an oxygen uptake of about 1.0 liter · min^{-1} in each case. The 11 subjects, all of whom were skilled carpenters, also performed leg work on a cycle ergometer with the same oxygen uptake. Nailing into the wall or ceiling resulted in a greater elevation of the heart rate than nailing at bench level; the heart rate during cycle exercise at a comparable oxygen uptake was lower (or 102 beats · min^{-1}) than for all types of nailing (130 beats · min^{-1} when nailing into the ceiling). It is interesting to note that leg exercise with an oxygen uptake of 1.4 liters · min^{-1} gave approximately the same rise in heart rate as nailing with an oxygen uptake of only 1.0 liter · min^{-1}. It should be noted that this finding refers to prolonged continuous work in a more or less steady-state situation, and not to intermittent work as described elsewhere in this chapter. The intra-arterially measured blood pressures during nailing were higher than during leg exercise at a given oxygen uptake (and probably cardiac output), the difference being most pronounced between nailing into the ceiling and cycle work ($P < 0.01$).

These examples illustrate that (1) the work output may vary with the tools used and with the working positions, even though the energy expenditure is the same; and (2) the physiological effects of a given energy demand may vary considerably, depending on tools and techniques.

Studies of coal miners in different countries have shown a good general agreement for the work with pick and shovel. It appears that the energy expenditure of shoveling ranges, as a rule, from 6 to 7 kcal · min^{-1} (25 to 29 kJ · min^{-1}). In a German study (Lehmann et al., 1950), the mean gross energy expenditure during actual coal mining was 5 kcal · min^{-1} (21kJ · min^{-1}), and the mean expenditure of energy per minute for the total time spent underground was 3.5 (15 kJ). Walking to and from the coal face in a stooping position may, according to Passmore and Durnin (1955), require as much as 10 kcal · min^{-1} (42 kJ · min^{-1}). It appears that in spite of the increased mechanization, mining is still hard physical work. Studies in the iron and steel industry (Lehmann et al., 1950; Christensen, 1953) show wide variations in energy expenditure, so that generalizations are not justified. Although some of the workers periodically may work very hard, many are doing only light work in most modern plants.

Several studies have verified the general impression that farming is hard work, at least in the busy season, especially where the advantage of mechanization is not available (Durnin and Passmore, 1967). Milking by hand requires about 4.5 kcal · min^{-1} (18 kJ · min^{-1}) as against 3.5 kcal · min^{-1} (15 kJ · min^{-1}) for machine milking. Horse ploughing entails 6 kcal · min^{-1} (25 kJ · min^{-1}), whereas tractor ploughing requires less than 5 kcal · min^{-1} (21 kJ · min^{-1}) (Hettinger and Wirths, 1953a and b).

In Norwegian coastal fishermen, the average energy expenditure of all activities on board during a whole day at sea amounted from 34 to 39 percent of the fisherman's maximal aerobic power, with occasional peaks of up to 80 percent (Rodahl, et al., 1974a). The heart rate exceeded 50 percent of the fisherman's heart rate reserve for 9 to 23 percent of the time. The most strenuous activities were pulling in the seine with a power block (oxygen uptake up to 2.7 liters min^{-1}) and unloading the catch, taxing the subjects by more than 50 percent of their maximal aerobic power for two-thirds of the duration of these activities. In general, it appears that the energy expenditure of commercial fishermen during the active fishing season may reach levels as high as about 5000 kcal · day^{-1} (21 MJ · day^{-1}).

It is generally recognized that lumber work involves heavy expenditure of energy. In fact, lumbering is probably the hardest form of physical work, requiring sometimes as much as 6000 kcal · day^{-1} (25 MJ · day^{-1}) (Lundgren, 1946). Hansson (1965) noted that lumberjacks with very high earnings were characterized by a particularly high maximal aerobic power compared with that of the average earner. With regard to muscle strength and precision in a variety of standardized work operations, there was no significant difference between the two categories. Because of his higher aerobic power, the top worker could attain a higher work output, he did not take as long breaks, and he became less tired at a given energy output, compared with his less productive colleague. It appears that workers involved in manual labor, who are more or less free to set their own pace, normally accept working with an energy output which is less than about 40 percent of their individual maximal aerobic power (I. Åstrand, 1967; Chap. 7).

Recreational Activities A general summary of measurements of energy expenditure of a variety of common recreational activities is given in Fig. 11-12, ranging from light indoor recreation and games to strenuous outdoor sports (see also Chap.

14). It should be pointed out that the listed figures can be taken as approximate values only, in view of the many factors which may have a profound influence upon the energy expenditure of such activities. Individuals pursue their recreations with very different degrees of vigor. Swimming, ball games, dancing, gardening, etc., can be enjoyed with an energy expenditure which ranges from low to very high. Playing volleyball or taking a swim does not involve the same degree of physical activity for all persons. Anyone, even the partially incapacitated person, can find a suitable recreation.

Data for energy expenditure of playing children have been published by Taylor et al. (1948, 1949) and Cullumbine (1950).The reported values range from 1 to 3 kcal $\cdot$ min^{-1} (4 to 13 kJ $\cdot$ min^{-1}). (Walking and running are discussed in Chap. 14). Data for energy expenditure during different sports activities are given by Banister and Brown (1968) and show variations from 2 kcal $\cdot$ min^{-1} (playing pool) to 23 kcal $\cdot$ min^{-1} (8 to 100 kJ $\cdot$ min^{-1}) (sprinting).

Military Activities   Data representing the energy cost of various military activities (Goldman, 1965; Rodahl et al, 1974b) are mostly gathered during training or maneuvers and not under realistic combat conditions. In general, it appears that the energy expenditure under these conditions very seldom exceeds 10 kcal $\cdot$ min^{-1} (40 kJ $\cdot$ min^{-1}), and that energy expenditure in excess of 7.0 kcal (30 kJ) seldom lasts more than 10 min.

FATIGUE

General fatigue may be a symptom of disease. It may also be psychological in nature, often associated with lack of motivation, lack of interest, low reserve capacity, etc. None of these problems will be dealt with here since they fall ouside the scope of this book. We shall deal only with fatigue that has a functional physiological basis. Such physiological fatigue, thought to be a warning mechanism preventing overstraining the organism or part of it, may be general and systematic, or it may be local and, as a rule, muscular in nature.

General Physical Fatigue

Christensen (1960) defines physical fatigue as a state of disturbed homeostasis due to work and to work environment. This may give rise to subjective as well as objective symptoms.

A more comprehensive discussion of the general aspects of fatigue, both physical and mental, will be found in a collection of papers that appeared in *Ergonomics, 14*(I):1–187, 1971.

Grandjean (1979) has developed a neurophysiological model of fatigue, involving an activating and inhibitory system, which he maintains is confirmed to some extent by field studies.

It should be emphasized, however, that, so far, very little is known about the nature of this disturbed homeostasis. Thus, all that can be said with certainty at present is that the fall in the blood sugar observed in a fasting subject engaged in prolonged submaximal work lasting several hours causes disturbed homeostasis in the central nervous system leading to a feeling of fatigue as one of the symptoms of hypoglycemia. This hardly ever occurs in regular work. It is also clear that the accumulation of lactic acid in muscles engaged in intense work involving anaerobic metabolic conditions is a sign of disturbed homeostasis leading to symptoms of local fatigue. But this is not likely to occur in ordinary manual labor.

Subjective Symptoms of Fatigue These symptoms may range from a slight feeling of tiredness to complete exhaustion. Attempts have been made to relate these subjective feelings to objective physiological criteria such as the accumulation of lactate in the blood. While such a relationship is often observed in connection with prolonged strenuous physical efforts such as an athletic event, this relationship is not usually present in prolonged light or moderate work. Subjective feelings of fatigue usually occur at the end of an 8-hr workday when the average work load exceeds 30 to 40 percent of the individual's maximal aerobic power, and certainly when the load exceeds 50 percent of the maximal aerobic power.

Objective Symptoms of Fatigue I. Åstrand (1960) observed a rise in heart rate in subjects working at a load corresponding to about 50 percent of the individual's maximal oxygen uptake during a period of about 8 hours. However, since these experiments were carried out during the day (morning and afternoon), it is still an open question as to whether these changes are in fact due to, or partially due to, the development of fatigue, or may be the result of the normal circadian rhythms, causing both a rise in heart rate and rectal temperature. In normal environment a rise in rectal temperature above 38°C is noted when the oxygen uptake exceeds 50 percent of the individual's maximal oxygen uptake (I. Åstrand, 1960; Chap. 13).

Volle et al. (1979), examined the possible fatigue induced in workers submitted to a compressed work week, i.e., 40 hours in 4 days, as compared with the usual 40 hours in 5 days. Two groups of workers from two different factories manufacturing similar products, one group practicing the 4-day week, the second the 5-day week, participated in the study. The data did not reveal any significant difference between the two groups in terms of reaction time, heart rate, blood pressure, body temperature, O_2 uptake, CO_2 output, etc. measured before and after the first and the last day of the week. The only difference observed was that the critical flicker fusion frequency and the right-hand strength showed a significantly greater impairment in the 4-day week group. However, to what extent these parameters do in fact represent a greater level of fatigue may be questioned.

Local Muscular Fatigue

This subject is discussed in considerable detail in Chap. 3. The practical application of this discussion may be summarized as follows: Both in static and dynamic muscular work, endurance is related to the developed force expressed in percentage of the maximal force which the muscle can develop. Thus, in isometric muscular contraction (see Fig. 3-25), a 50 percent load can be maintained for about 1 minute, while the contraction may be maintained almost indefinitely as long as the load is significantly less than 15 percent of the maximal force of the muscle. Similar relationships exist in the case of rhythmic contractions (see Fig. 3-26). Thus, if the load corresponds to about 80 percent of maximal strength, only 10 contractions per min can be maintained in a steady state, but if the load is reduced to about 60 percent of maximal strength, a contraction rate of 30 · min^{-1} can be maintained. Therefore the stronger the muscles, the greater load they can endure without developing muscular fatigue. But, in any situation, the load and the rate of contraction have to be adjusted according to the strength of the muscle in order to avoid muscular fatigue.

Hagberg (1981b) observed a rapid decrease of the endurance time at contraction levels above 15 to 20 percent of maximum voluntary contraction for both the sustained isometric and dynamic exercise involving elbow flexions. This value may well be too high for prolonged work, for other field studies have indicated that loads below 10 percent of the maximal voluntary strength of a muscle may produce symptoms of local muscular fatigue in the course of a working day. Malmqvist et al. (1981), have shown that fatigue may occur even when the force exerted is small and that a stereotyped task has a greater tendency to produce localized muscle fatigue than a more varied one, even if the latter is heavier.

Fatigue not only modifies the characteristics of the force developed during isometric contractions. For the same value of force, the fatigued muscle is also more compliant than the nonfatigued muscle (Vigreux et al., 1980).

Kvarnstrøm (1983) has used the EMG approach to study the relationship between work load on local muscle groups and the occurrence of musculoskeletal disorders in a manufacturing industry, with special attention to occupational shoulder disorders. In light manufacturing industry, neck-shoulder problems dominated, and were more common among women and immigrants of either sex compared with local males.

Fitts et al. (1982), in studies performed on rats, have provided evidence that prolonged exercise produces alterations in contractile and biochemical properties of type I and IIa, but not type IIb fibers, and that muscle fatigue is not necessarily correlated with glycogen depletion.

Tesch (1980) studied muscle fatigue with special reference to lactate accumulation, during short-term, intense exercise. The results seemed to indicate a higher rate of lactate production in fast-twitch fibers than in slow-twitch fibers.

A shift in the EMG power spectra toward lower frequencies has been demonstrated during fatiguing contractions (Lindström et al., 1977; Petrofsky and Lind, 1980), and is being used as an index of local muscular fatigue (Hagberg, 1981a,b; Petrovsky et al., 1982).

Sadoyama and Miyano (1981) have introduced a conduction velocity measure, which is derived from a surface EMG model, as an index of muscle fatigue. They suggest that the spectral shift to a lower frequency is concerned with conduction velocity of action potential along the muscle fibers. This view appears to be supported by Sadoyama et al. (1983), and Komi (1983). However, Naeije and Zorn (1982) conclude, on the basis of their experiments in which 8 subjects were made to contract their m. biceps brachii as long as possible at 50% of maximum voluntary strength, that great

EMG power spectral shifts to lower frequencies during fatigue may occur without a concomitant change in muscle fiber action potential conduction velocity. It is evident that both electrical, mechanical, and metabolic changes in the muscle are interrelated during fatigue contractions (Komi, 1983). At any rate, however, the EMG may provide useful information when assessing fatigue during specific work operations.

VIBRATION

The pathophysiological consequences of working with vibrating tools such as pneumatic drills and motor-driven hand saws, are well known. The effect of whole-body vibration on the worker is far less understood (Seidel et al., 1980; Ullsperger and Seidel, 1980).

CIRCADIAN RHYTHMS AND PERFORMANCE

Circadian Rhythms in the Human

In human beings, a variety of physiological functions, such as heart rate, oxygen uptake, rectal temperature, and urinary excretion of potassium and catecholamines, show distinct rhythmic changes in the course of a 24-hr period, with the values falling to their lowest during the night (low dip around 4 A.M.) and rising during the day, reaching their peak in the afternoon. This phenomenon is known as circadian rhythms, and is thought to be regulated by several separately operating biological clocks. It occurs in most individuals, although there are apparently a few exceptions; some individuals show reversed rhythms, the rectal temperature, for example, being highest at night (Folk, 1974).

These rhythmic changes in physiological functions have been found to be associated with changes in performance. This relationship appears to exist especially in the case of rectal temperatures and performance. In general, the lowest performance is observed early in the morning (about 4 A.M.). Thus, the delay in answering calls by switchboard operators on night shift was twice as long between 2 and 4 A.M. as during the daytime (Colquhoun, 1971). A similar relationship may exist also in the case of athletic performance. Thus, A. Rodahl et al. (1976), studying the performance of top swimmers who competed under comparable conditions early in the morning and late in the evening, found that the swimmers performed significantly better in the evening than in the morning ($P < .001$).

These findings show that circadian rhythms must be considered when interpreting the results from prolonged physiological experiments and when performing fitness tests in athletes at different times of the day.

Shift Work

The fact that human beings are "day-animals" and that some of their basic physiological functions which are associated with their performance capacities are subject to circadian rhythmic changes, suggests that humans may not ideally be suited for night work.

Nonetheless, shift work has been practiced for generations in one form or another. Yet, little precise information is available as to what effects shift work has on physiological functions or physical performance, and there is no general agreement as to what type of shift work or work schedule is to be preferred. Most of the available information refers to clinical, social, or psychological aspects of shift work (Aanonsen, 1964). A review of the literature indicates that the health of shift workers in general is good in spite of such complaints as loss of sleep, disturbance of appetite and digestion, and a high rate of ulcers. The social and domestic effects of shift work represent greater problems than do the physiological effects. The results of studies pertaining to the effects on productivity are conflicting, as are results concerning accident rates. Absenteeism because of illness appears to be lower among shift workers than among day workers. It has been suggested that the physiological and biological effects are probably related to circadian rhythms rather than to work schedule. To what extent such circadian rhythms are related to health, performance, and a feeling of well-being is still undetermined.

Systematic studies of men engaged in rotating shift work and in continuous night work (Vokac and Rodahl, 1974, 1975) indicate that shift work does represent a physiological strain on the organism. It causes a desynchronization between functions such as body temperature and the biological clocks governing these functions. These studies show that there are considerable individual differences in the reaction to shift work, supporting the general experience that not everyone is equally suited for such work (some individuals consistently show relatively high values for urinary catecholamine elimination during shift work, whereas others have consistently low values). As judged by the catecholamine excretion, the greatest strain occurs when the worker, after several free days, starts work on night shift. The results of this study indicate that it is preferable, from a physiological standpoint, to distribute the free days more evenly throughout the entire shift cycle; that is, to alternate between work and free time regularly instead of assigning several consecutive free days.

The study of continuous night work shows that at the onset, body temperature and work pulse fell in the course of the night as if the subject were sleeping, although he was working (Fig. 11-14). It takes several weeks for this normal rhythm to be reverted (i.e., for an increase in body temperature in the course of the night work). In view of this, it would appear unrealistic to keep shift workers on continuous night work for prolonged periods in order to obtain the benefit of the reverted physiological reactions, since such a reversion takes too long to occur and is lost when interrupted by a single day.

Disturbances in circadian rhythms may give rise to considerable problems for those who have to travel by air from one continent to another in order to conduct business, to take part in political negotiations, or to participate in athletic competitions. It is an open question whether the indisposition or functional disturbances experienced after such intercontinental flights are in fact due to disturbed circadian rhythms, to loss of sleep, or to both. It is a common experience, however, that by being able to sleep during such travel, if necessary by using sleep-producing drugs, the individual can maintain a reasonable functional capacity in spite of the rapid shift from one time to another. (For further references see Rutenfranz and Colquhoun, 1978 and 1979).

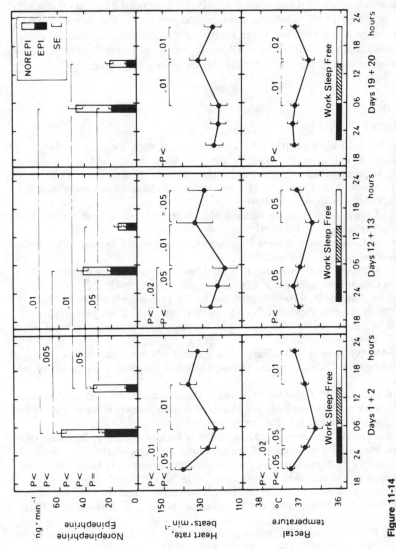

Figure 11-14
Mean urinary catecholamine excretion, heart rate at a fixed submaximal work load, and rectal temperature in four steel mill workers during 3 weeks continuous night shifts. *(From Vokac and Rodahl, 1975.)*

EFFECTS OF MENSTRUATION

Observations concerning the effect of menstruation on physical work capacity have yielded conflicting results (Gamberale et al., 1975). In general it appears that a subjectively perceived exertion may be more pronounced during the menstrual cycle, while no change in heart rate or oxygen uptake is apparent during the different phases of the menstrual cycle. Furthermore, Olympic gold medals have been attained by menstruating women athletes. (For further references, see Drinkwater, 1973.)

REFERENCES

Aanonsen, A.: "Shift Work and Health," Universitetsforlaget, Oslo, 1964.

Andersson, B., S. Houmøller, C-G, Karlson, and S. Svensson: Analysis of Urinary Catecholamines: An Improved Auto-Analyzer Fluorescence Method, *Clin. Chim. Acta,* **51:**13, 1974.

Åstrand, I.: Aerobic Work Capacity in Men and Women with Special Reference to Age, *Acta Physiol. Scand.,* **49**(Suppl. 169): 1960.

Åstrand, I.: Degree of Strain during Building Work as Related to Individual Aerobic Work Capacity, *Ergonomics,* **10:**293, 1967.

Åstrand, I., A. Guharay, and J. Wahren: Circulatory Responses to Arm Exercise with Different Arm Positions, *J. Appl. Physiol.,* **25:**528, 1968.

Åstrand, I., P. Fugelli, C. G. Karlsson, K. Rodahl, and Z. Vokac: Energy Output and Work Stress in Coastal Fishing, *Scand. J. Clin. Lab. Invest.,* **31:**105, 1973.

Banister, E. W., and S. R. Brown: The Relative Energy Requirements of Physical Activity, Chap. 10, in "Exercise Physiology," Academic Press, Inc., New York, 1968.

Banister, E. W., and J. Griffiths: Blood Levels of Adrenergic Amines during Exercise, *J. Appl. Physiol.,* **33:**674–676, 1972.

Bedale, E. M.: "Comparison of the Energy Expenditure of a Woman Carrying Loads in Eight Different Positions," Medical Research Council Industrial Fatigue Research Board, no. 29, 1924.

Benedict, F. G., and T. M. Carpenter: "Influence of Muscular and Mental Work on Metabolism and Efficiency of the Human Body as a Machine," U.S. Department of Agriculture, Office of Experimental Stations, Bul. 208, 1909.

Benedict, F. G., and C. G. Benedict: "Mental Effort in Relation to Gaseous Exchange, Heart Rate and Mechanics of Respiration," Carnegie Institute, Washington, D.C., no. 446, 1933.

Bobet, J., and R. W. Norman: Use of the Average Electromyogram in Design Evaluation. Investigation of a Whole-Body Task, *Ergonomics,* **25**(12):1155, 1982.

Brezina, E., and W. Kolmer: Über den Energieverbrauch bei der Geharbeit unter dem Einfluss verschiedener Geschwindigkeiten und Verschiedener Belastungen, *Biochem. Z.,* **38:**129, 1912.

Cathcart, E. P., D. T. Richardson, and W. Campbell: Maximum Load to be Carried by the Soldier, *J. Roy. Army Med. Corps.,* **40:**435; **41:**12; **87:**161, 1923.

Christensen, E. H.,: "Physiological Valuation of Work in the Nykoppa Iron Works," in W. F. Floyd and A. T. Welford (eds.), Ergonomic Society Symposium on Fatigue, p. 93,Lewis, London, 1953.

Christensen, E. H.: Muscular Work and Fatigue, Chap. 9 in K. Rodahl and S. M. Horvath (eds.), "Muscle as a Tissue," McGraw-Hill Book Company, New York, 1960.

Colquhoun, W. P.: Circadian Variations in Mental Efficiency, "Biological Rhythms and Human Performance," pp. 39–108, Academic Press, Inc., London, 1971.

Consolazio, C. F., R. E. Johnson, and L. J. Pecora: "Physiological Measurements of Metabolic Functions in Man," McGraw-Hill Book Company, New York, 1963.

Cox, S., T. Cox, M. Thirlaway, and C. Mackay: Effects of Simulated Repetitive Work on Urinary Catecholamine Excretion, *Ergonomics*, **25**(12):1129, 1982.

Crowden, G. P.: Stair Climbing by Postmen, *The Post* (London), p. 10, July 26, 1941.

Cullumbine, H.: Heat Production and Energy Requirements of Tropical People, *J. Appl. Physiol.*, **2**:640, 1950.

Drinkwater, B. L.: Physiological Responses of Women to Exercise in J. H. Wilmore (ed.), "Exercise and Sport Sciences Reviews," p. 125, Academic Press, Inc., New York, 1973.

Durnin, J. V. G. A., and R. Passmore: "Energy, Work and Leisure," William Heinemann, Ltd., London, 1967.

Eiff, A. W., and H. Göpfert: Ausmass und Ursachen der Energieumsatz-Veränderungen bei geistiger Arbeit, *Z. Ges. Exp. Med.*, **120**:72, 1952.

Ekholm, J., U. P. Arborelius, and G. Németh: The Load on the Lumbo-Sacral Joint and Trunk Muscle Activity During Lifting, *Ergonomics*, **25**(2):145, 1982.

Euler, U. S. von,: Quantification of Stress by Catecholamine Analysis, *Clinical Pharmacology and Therapeutics*, **5**:398–404, 1964.

Euler, U. S. von, and S. Hellner: Excretion of Noradrenaline and Adrenaline in Muscular Work, *Acta Physiol. Scand.*, **26**:183–191, 1952.

Euler, U. S. von, and F. Lishajko: Improved Technique for the Fluorimetric Estimation of Catecholamines, *Acta Physiol. Scand.*, **51**:348–356, 1961.

Evans, O. M., Y. Zerbib, M. H. Faria, and H. Monod: Physiological responses to load holding and load carriage, *Ergonomics*, **26**(2):161, 1983.

Fitts, R. H., J. B. Courtright, D. H. Kim and F. A. Witzmann: Muscle Fatigue with Prolonged Exercise: Contractile and Biochemical Alterations, *Am. J. Physiol.*, **242**:C65, 1982.

Folk, G. Edgar, Jr.: "Textbook of Environmental Physiology," 2nd ed., Lea & Febiger, Philadelphia, 1974.

Follenius, M., G. Brandenberger, C. Lecornu, M. Simeoni, and B. Reinhardt: Plasma Catecholamines and Pituitary Adrenal Hormones in Response to Noise Exposure, *Eur. J. Appl. Physiol.*, **43**:253, 1980.

Food and Nutrition Board: "Recommended Dietary Allowances," Nat. Acad. Sci., Nat. Res. Coun., pub. 1146, Washington, D.C., 1964.

Fröberg, J., C. G. Karlsson, L. Levi, and L. Lidberg: Circadian Variations in Performance, Psychological Ratings, Catecholamine Excretion, and Diuresis during Prolonged Sleep Deprivation, *Intern. J. of Psychobiol.*, **2**:23–36, 1972.

Gamberale, F., L. Strindberg, and I. Wahlberg: Female Work Capacity during the Menstrual Cycle: Physiological and Psychological Reactions, *Scand J. Work Environ. & Health*, **1**:120, 1975.

Garg, A., and M. M. Ayoub: What Criteria Exist for Determining How Much Load Can Be Lifted Safely?, *Human Factors*, **22**(4):475, 1980

Garg, A., and U. Saxena: Container Characteristics and Maximum Acceptable Weight of Lift, *Human Factors*, **22**(4):487, 1980.

Goldman, R. F.: Energy Expenditure of Soldiers Performing Combat Type Activities, *Ergonomics*, **8**:322, 1965.

Grandjean, E.: Fatigue in Industry, *Br. J. Industr. Med.*, **36**:175, 1979.

Gray, E. L., C. F. Consolazio, and R. M. Karl: Nutritional Requirements for Men at Work in Cold, Temperate and Hot Environments, *J. Appl. Physiol.*, **4**:270, 1951.

Hagberg, M.: On Evaluation of Local Muscular Load and Fatigue by Electromyography, Arbete och Hälsa, Vetenskaplig Skriftserie 1981: 24, Arbetarskyddsverket, Stockholm, Sweden, 1981a.

Hagberg, M.: Muscular Endurance and Surface Electromyogram in Isometric and Dynamic Exercise, *J. Appl. Physiol.: Respirat. Environ. Exercise Physiol.*, **51**:(1):1, 1981b.

Hanson, J. A., and F. P. Jones: Heart Rate and Small Postural Changes in Man, *Ergonomics*, **13**(4), 483, 1970.

Hansson, J. E.: The Relationship between Individual Characteristics of the Worker and Output of Work in Logging Operations, *Studia Forestalia Suecia*, no. 29, Skogshögskolan, Stockholm, 1965.

Hansson, J. -E.: Work Physiology as a Tool in Ergonomics and Production Engineering, Al-Rapport 2, *Ergonomi och Produktionsteknik*, National Institute of Occupational Health, Stockholm, 1968.

Harrison, M. H., G. A. Brown, and A. J. Belyavin: The "Oxylog": An Evaluation, *Ergonomics*, **25**(9):809, 1982.

Hettinger, T., and W. Wirths: Der Energieverbrauch beim Hand und Motorpflügen, *Arbeitsphysiol.*, **15**:41, 1953a.

Hettinger, T., and W. Wirths: Über die körperliche Beanspruchung beim Hand und Maschinemelken, *Arbeitsphysiol.*, **15**:103, 1953b.

Hitchen, M., D. A. Brodie, and J. B. Harness: Cardiac Responses to Demanding Mental Load, *Ergonomics*, **23**(4), 379, 1980.

Holmboe, J., H. Bell and N. Norman: Urinary Excretion of Catecholamines and Steroids in Military Cadets Exposed to Prolonged Stress, *Försvarsmedicin*, **11**:183, 1975.

Issekutz, B., Jr., N. C. Birkhead, and K. Rodahl: Effect of Diet on Work Metabolism, *J. Nutr.*, **79**:109, 1963.

Jenner, D. A., V. Reynolds, and G. A. Harrison: Catecholamine Excretion Rates and Occupation, *Ergonomics*, **23**(3)237, 1980.

Kagan, E. M., P. Dolgin, P. M. Kaplan, C. O. Linetzkaja, J. L. Lubarsky, M. F. Neumann, J. J. Semernin, J. S. Starch, and P. Spilberg: Physiologische Vergleichsuntersuchung der Hand-und Fleiss-(Conveyor) Arbeit, *Arch. Hyg. (Berlin)*, **100**:335, 1928.

Kamon, E.: Instrumentation for Work Physiology, *Trans. N.Y. Acad. Sci.*, no. 7, **36**:625, 1974.

Karvonen, M. J.: Work and Activity Classifications, in L. A. Larson (ed.): "Fitness, Health, and Work Capacity," p. 38, Macmillan Publishing Company, Inc., New York, 1974.

Komi, P. V.: "Electromyographic, Mechanical and Metabolic Changes during Static and Dynamic Fatigue. Biochemistry of Exercise, International Series on Sports Sciences, vol. 13, p. 197, Human Kinetics Publishers Inc., Champaign, Ill., 1983.

Kumar, S.: Physiological Responses to Weight Lifting in Different Planes, *Ergonomics*, **23**(10): 987, 1980.

Kvarnstrøm, S.: Occurrence of Musculoskeletal Disorders in a Manufacturing Industry, with Special Attention to Occupational Shoulder Disorders, *Scand. J. Rehab. Med. (Suppl. 8)*, 1983.

Lamke, L. O., S. Lennquist, S. O. Liljedahl, and B. Wedin: The Influence of Cold Stress on Catecholamine Excretion and Oxygen Uptake of Normal Persons. *Scand J. Clin. Lab.Invest.*, **30**:57–62, 1972.

Lehmann, G., E. A. Müller, and H. Spitzer: Der Kalorienbedarf bei gewerblicher Arbeit. *Arbeitsphysiol*, **14**:166, 1950.

Lehmann, G., "Praktische Arbeitsphysiolgie," Thieme, Stuttgart, 1953.

Leskinen, T. P. J., I. Stålhammar, and A. A. Kuorinka: A Dynamic Analysis of Spinal Compression with Different Lifting Techniques, *Ergonomics*, **26**(6):595, 1983.

Levi, L: Sympatho-adrenomedullary Responses to Emotional Stimuli: Methodologic, Physiologic and Pathologic Considerations, in E. Bajusz (ed.), "An Introduction to Clinical Neuroendocrinology," pp. 78–105, S. Karger, New York, 1967.

Lindström, L., R. Kadefors, and I. Petersén: An Electromyographic Index for Localized Muscle Fatigue, *J. Appl. Physiol.: Resp. Environ. Exerc. Physiol.*, **43**:750, 1977.

Ljungberg, A-S., E. Gamberale, and Å. Kilbom: Horizontal lifting—Physiological and Psychological Responses, *Ergonomics*, **25**(8):741, 1982.

Luczak, H.: Fractioned Heart Rate Variability. Part II. Experiments on Superimposition of Components of Stress, *Ergonomics*, **22**(12):1315, 1979.

Lundgren, N.: Physiological Effects of Time Schedule Work on Lumbar Workers, *Acta Physiol. Scand.*, **13**(Suppl. 41): 1946.

Malmqvist, R., I. Ekblom, L. Lindström, I. Petersén and R. Örtengren: Measurement of Localized Muscle Fatigue in Building Work, *Ergonomics*, **24**(9):695, 1981.

Morehouse, L. E.: "Laboratory Manual for Physiology of Exercise," The C. V. Mosby Company, St. Louis, 1972.

Müller, E. A., and H. Franz: Energieverbrauchsmessungen bei beruflicher Arbeit mit einer verbesserten Respirations-Gasuhr, *Arbeitsphysiol*, **14**:499, 1952.

Naeije, M., and H. Zorn: Relation between EMG Power Spectrum Shifts and Muscle Fibre Action Potential Conduction Velocity Changes during Local Muscular Fatigue in Man, *Eur. J. Appl. Physiol.*, **50**:23, 1982.

Nelson, N. A., W. B. Shelley, S. M. Horvath, L. W. Eichna, and T. F. Hatch: Influence of Clothing, Work and Air Movement on the Thermal Exchanges of Acclimatized Men in Various Hot Environments, *J. Clin. Invest.*, **27**:209, 1948.

Oberoi, K., M. K. Dhillon, and S. S. Miglani: A Study of Energy Expenditure during Manual and Machine Washing of Clothes in India, *Ergonomics*, **26**(4):375, 1983.

Passmore, R., and J. V. G. A. Durnin: Human Energy Expenditure, *Physiol. Rev.*, **35**:801, 1955.

Petrofsky, J. S., and A. R. Lind: The Influence of Temperature on the Amplitude and Frequency Components of the EMG during Brief and Sustained Isometric Contractions, *Eur. J. Appl. Physiol.*, **44**:189, 1980.

Petrofsky, J. S., R. G. M. Glaser, and C. A. Phillips: Evaluation of the Amplitude and Frequency Components of the Surface EMG as an Index of Muscle Fatigue, *Ergonomics*, **25**(3):213, 1982.

Pierrynowski, M. R., D. A. Winter, and R. W. Norman: Metabolic Measures to Ascertain the Optimal Load to be Carried by Man, *Ergonomics*, **24**(5):393, 1981.

Pimental, N., and K. B. Pandolf: Energy Expenditure while Standing or Walking Slowly Uphill or Downhill with Loads, *Ergonomics*, **22**(8):963, 1979.

Rodahl, A., M. O'Brien, and R. G. R. Firth: Diurnal Variation in Performance of Competitive Swimmers, *J. Sports Med. and Physical Fitness*, **16**:72, 1976.

Rodahl, K.: "Nutritional Requirements under Arctic Conditions," Norsk Polarinstitutt Skrifter No. 118, Oslo University Press, Oslo, 1960.

Rodahl, K., Z. Vokac, P. Fugelli, O. Vaage and S. Maehlum: Circulatory Strain, Estimated Energy Output and Catecholamine Excretion in Norwegian Coastal Fisherman, *Ergonomics*, **17**:585–602, 1974a.

Rodahl, K., T. Wessel-Aas, P. O. Huser and T. S. Nilsen: A Physiological Evaluation of a Waterproof, Partially Permeable Protective Suit against Chemical and Bacteriological Warfare, *Försvarsmedicin*, **10**:24, 1974b.

Rutenfranz, J., and W. P. Colquhoun (eds.): Shiftwork: Theoretical Issues and Practical Problems, *Ergonomics*, **21**(10):737, 1978.

Rutenfranz, J., and W. P. Colquhoun: Circadian Rhythms in Human Performance, *Scand. J. Work Environ. Health*, **5**:167, 1979.

Sadoyama, T., and H. Miyano: Frequency Analysis of Surface EMG to Evaluation of Muscle Fatigue, *Eur. J. Appl. Physiol.*, **47**:239, 1981.

Sadoyama, T., T. Masuda, and H. Miyano: Relationship between Muscle Fibre Conduction Velocity and Frequency Parameters of Surface EMG during Sustained Contraction, *Eur. J. Appl. Physiol.*, **51**:247, 1983.

Seidel, H., R. Bastek, D. Bräuer, C. Buchholz, A. Meister, A-M, Metz, and R. Rothe: On Human Response to Prolonged Repeated Whole-Body Vibration, *Ergonomics*, **23**(3):191, 1980.

Soule, R. G., K. B. Pandolf, and R. F. Goldman: Energy Expenditure of Heavy Load Carriage, *Ergonomics*, **21**(5):373, 1978.

Spitzer, H., and Th. Hettinger: "Tafeln für Kalorienumsatz bei körperlicher Arbeit," REFA publication, Darmstadt, 1958.

Sundin, I.: The Influence of Body Posture on the Urinary Excretion of Adrenalin and Nora-drenalin, *Acta Med. Scand.*, **154**:(Suppl. 313): 1956.

Taylor, C. M., M. W. Lamb, M. E. Robertson, and G. MacLeod: Energy Expenditure for Quiet Playing and Cycling of Boys 7 to 15 Years of Age, *J. Nutr.*, **35**:511, 1948.

Taylor, C. M., O. F. Pye, A. B. Caldwell, and E. R. Sostman: Energy Expenditure of Boys and Girls 9 to 11 Years of Age (1) Sitting Listening to the Radio (Phonograph), (2) Sitting Singing and (3) Standing Singing, *J. Nutr.*, **38**:1, 1949.

Tesch, P.: Muscle Fatigue in Man with Special Reference to Lactate Accumulation during Short-Term Intense Exercise, *Acta Physiol. Scand.* (Suppl. 480): 1980.

Ullsperger, P., and H. Seidel: On Auditory Evoked Potentials and Heart Rate in Man During Whole-Body Vibration, *Eur. J. Appl. Physiol.*, **43**:183, 1980.

Verma, S. S., M. S. Malhotra, and J. Sen Gupta: Indirect Assessment of Energy Expenditure at Different Work Rates, *Ergonomics*, **22**(9):1039, 1979.

Vigreux, B., J. C. Cnockaert, and E. Pertuzon: Effects of Fatigue on the Series Elastic Component of Human Muscle, *Eur. J. Appl. Physiol.*, **45**:11, 1980.

Vokac, Z., and K. Rodahl: "A Study of Continuous Night Work at the Norwegian Street Mill at Mo i Rana," Nordic Council for Arctic Medical Research Report, no. 10, 1974.

Vokac, Z., H. Bell, E. Bautz-Holter, and K. Rodahl: Oxygen Uptake/Heart Rate Relationship in Leg and Arm Exercise, Sitting and Standing, *J. Appl. Physiol.*, **39**:54, 1975.

Vokac, Z., and K. Rodahl: Field Study of Rotating and Continuous Night Shifts in a Steel Mill, Proceedings of the Third International Symposium on Night- and Shiftwork in Dortmund, Oct. 28–31, 1974, in P. Colquhoun, S. Fokard, P. Knaut, and R. Rutenfranz (eds.), "Experimental Studies in Shiftwork," Westdeutscher Verlag, Opladen, 1975.

Volle, M., G. R. Brisson, M. Pérusse, M. Tanaka, and J. Doyan: Compressed Work-Week: Psycho-Physiological and Physiological Repercussions, *Ergonomics*, **22**(9):1001, 1979.

Wilson, G. D., and M. P. Sklenka: A System of Measuring Energy Cost during Highly Dynamic Activities, *J. Sports Med.*, **23**:155, 1983.

Wolff, H. S.: The Integrating Motor Pneumotachograph: A New Instrument for the Measurement of the Energy Expenditure by Indirect Calorimetry, *Quart. J. Exper. Physiol.*, **43**:270, 1958.

CHAPTER 12

NUTRITION AND PHYSICAL PERFORMANCE

CONTENTS

ENERGY LIBERATION AND TRANSFER
ENERGY TRANSFORMATION
High-Energy Phosphates
THE OXIDATION OF FUEL
ATP Formation
CONTROL SYSTEMS
INTERPLAY BETWEEN ANAEROBIC AND AEROBIC ENERGY YIELD
RELATIVE IMPORTANCE OF THE DIFFERENT ENERGY STORES
SUMMARY
NUTRITION AND PHYSICAL PERFORMANCE
NUTRITION IN GENERAL
DIGESTION
ENERGY METABOLISM AND THE FACTORS GOVERNING THE
    SELECTION OF FUEL FOR MUSCULAR EXERCISE
REGULATORY MECHANISMS
FOOD FOR THE ATHLETE
Events Lasting Less Than 1 Hour
Events Lasting Between 1 and 2 Hours
Events Lasting for Several Hours
PHYSICAL ACTIVITY, FOOD INTAKE, AND BODY WEIGHT
Energy Balance
Regulation of Food Intake

"Ideal" Body Weight
Obesity
Slimming Diets
Optimal Supply of Nutrients

ENERGY LIBERATION AND TRANSFER

The physiology of muscular work and exercise is basically a matter of transforming bound energy into mechanical energy.

Many similarities exist between the "human engine" and the combustion engine constructed by human beings. In the combustion engine, gasoline and air are introduced into the cylinder. The spark from the spark plug initiates the explosive combustion of the gas mixture. Chemical energy is transformed into kinetic energy and heat. The expansion of the gas forces the piston to move, and a system of mechanical devices can transfer this motion to the wheels. The motor is cooled by fluid or air to prevent overheating. The waste products are expelled with the exhaust. As this motor can work only in the presence of oxygen, its function is aerobic. When the gasoline tank is empty, the engine can no longer continue to run, since the operation of the combustion engine is dependent upon a continuous supply of fuel. In an automobile, the self-starter provides the energy for the first movements of the pistons. This energy comes from the electrical accumulator (battery); the starter can thus work in the absence of oxygen, or anaerobically. The stored energy of the battery is quite limited, however, so that the battery must be frequently recharged.

"Living organisms, like machines, conform to the law of the conservation of energy, and must pay for all their activities in the currency of metabolism" (Baldwin, 1967). In the human machine, the muscle fibers are the pistons. When fuel is available and a spark is introduced to start the breaking down of the fuel, part of the energy, which is thus liberated, can cause movement of the pistons. Heat and various waste products are produced.

In the following paragraphs we shall summarize the chemical processes involved in the human machine, omitting, for the sake of simplicity, the more complicated steps of the reactions and placing less emphasis on the actual chemistry involved than on the account of where and how the energy is released. For a more complete study, the reader is directed to more detailed textbooks and reviews (Needham, 1971; Conn and Stumpf, 1972, Lehninger, 1982; Newsholme and Leech, 1983).

ENERGY TRANSFORMATION

Energy is required for the various kinds of biological activities performed by living organisms. There is a demand for energy, for we have: (1) a synthesis of new cell materials from simpler precursors so that, in the mature organism, the rate of formation of new molecules balances the rate of degradation of the "old" ones; (2) a transport of materials against gradients of concentration, and this active movement of molecules and ions is often called osmotic work; (3) a mechanical work particularly evident in contracting muscles; and (4) production of heat to maintain a body temperature of about 37°C. This heat is in many situations just a by-product of the above mentioned processes.

Originally, the energy from the sun is trapped by the green plants and conserved in the form of foodstuffs. In the course of this process, carbon dioxide and water are consumed by the plant and oxygen is liberated into the atmosphere. When the plant is consumed by an animal, the energy it contains may be utilized by the animal cells where the foodstuffs, through combination with oxygen, are once again brought back to carbon dioxide and water, while part of the energy is utilized for the types of biological activities described in (1) through (3), and the rest is liberated as heat (4).

Chemical reactions involving transfer of energy may be divided into two kinds:

1 Reactions that liberate energy in some form (most often as heat). These reactions may take place spontaneously and are called *exergonic*.

2 Reactions that cannot take place unless energy in some form is added to the system; these are *endergonic* reactions.

In biological chemistry we are dealing with more or less complex organic molecules where the atoms are bound together with covalent bonds. Each bond has a certain energy level. As a rule, bond breakage is exergonic, whereas bond formation (synthesis) is an endergonic process. Different molecules will yield different amounts of released energy as they are broken down to simpler molecules: they have a different *chemical potential*.

The chemical potential may be thought of as an equivalent of the more familiar potential energy of a waterfall. The water in a lake located 1,600 m above sea level has a potential energy of 16 kJ (1600 kpm) per liter, which may be utilized as electrical power as the water flows down to the sea. Analogously, glucose has a chemical potential of roughly 16 kJ (some 4 kcal) per gram that may be used for work as it is broken down into carbon dioxide and water. A resynthesis needs added external energy, just as energy has to be added in order to get the water back from sea level into the lake.

This waterfall model may be useful to clarify other points concerning chemical energy and transformation. Let us assume that water from a top reservoir A flows down to a bottom reservoir B, thereby losing potential energy; but some of this energy will be used to drive a turbine. This A → B process is exergonic, very much like a chemical reaction where reactant A gives product B.

Without a stopper of some kind, the water from the lake would keep falling at maximal speed and at a maximal rate of energy liberation until the lake would be completely empty. However, in the operation of an electric power plant, as in living organisms, there are times, such as during the night, when comparatively little energy is needed, and when, therefore, a full-speed operation of the power plant would represent an enormous waste of potential energy. Both systems (power plant and living organisms) have therefore built in a valve device that will automatically regulate the flow of water or molecules according to the energy need at any given moment. Water from the bottom reservoir may be lifted to another reservoir with higher potential energy, i.e., from B to C. This B → C process is endergonic, and obviously does not happen by itself. Now the energy released in the exergonic A → B process may be used to drive the endergonic B → C process. The same can be accomplished in biochemical reactions, but in both cases some type of coupling device is needed. In our waterfall model, we can couple the turbine directly to the pump shafts by some mechanical device. Alternatively, we may let the turbine drive a generator; the electric power from this generator can then directly drive an electric pump motor or charge accumulators which later may be used for driving the pump motor.

Regardless of what type of coupling we select, it is impossible to pump the water to an energy level of A or higher with the energy obtained from the waterfall alone. If the system is to work, the water level C has to be somewhat lower than A, unless we could construct a *perpetuum mobile,* which is impossible as it would violate the second law of thermodynamics. There is always a loss of energy during transformation, a frictional loss mostly coming out as heat. This is the case with waterfalls as it is in biochemical reactions. An exergonic reaction may drive an endergonic reaction through some type of coupling, but *all* the energy in the exergonic reaction cannot be regained as potential energy. There will always be a heat loss.

In biological chemistry one will find several directly coupled reactions. In our exergonic A → B and endergonic B → C processes, the following combination or coupling may well work:

$$A + B \rightarrow B + C \ (+ \text{ heat})$$

The coupling between two processes or reactions may be a common intermediate (I) (some type of carrier, like the molecular complex RH_2), as in the following scheme:

$$A + B \rightarrow I \rightarrow B + C \ (+ \text{ heat})$$

This would be analogous to a gearbox placed between the turbine and the pump in our waterfall model.

However, nature has chosen more flexible solutions to the problem of how to get an endergonic reaction going. Instead of direct coupling, we find systems analogous to our power plants: the energy is transformed and stored in another and much more readily available form. It is not very practical to have our vacuum cleaner directly coupled to a water turbine, but it works nicely via electric power!

A classical equation for an exergonic oxidative metabolic process is the oxidation of glucose:

$$C_6H_{12}O_6 + 6O_2 \underset{\text{photosynthesis}}{\overset{\text{oxidation}}{\rightleftharpoons}} 6CO_2 + 6H_2O + \text{energy}$$

The glucose has a much higher energy level than CO_2 and H_2O, and this energy is liberated by the oxidation and made available for processes that require energy. (One mole of glucose can yield a maximum of 2870 kJ, or 686 kcal, of chemical energy.) The reaction is reversible. With chlorophyll as a catalyst, CO_2 and H_2O can combine again, the necessary energy for this endergonic reaction being provided by light. Thus the circle is completed: Animals dissimilate or break down carbohydrates (catabolism), and plants can also assimilate or synthesize carbohydrates (anabolism).

The animals require the energy-rich and complex products of photosynthesis to pay the costs of the biological work mentioned. They can do this by degrading the structures of such molecules as glucose. Secondly, they also need complex carbon compounds, such as glucose, as building blocks for the synthesis of their own cellular materials since they cannot, like plants, use CO_2 for this purpose.

Oxidation is defined chemically as the loss of electrons from an atom or molecule; *reduction* is defined as a gain of electrons. In many reactions the electrons are carried in the form of hydrogen atoms, and the oxidized compound is thereby dehydrogenated. In animals, organic fuels, such as carbohydrates, lipids, and proteins, constitute the major electron donors, and sooner or later oxygen is the final electron acceptor or oxidant of the fuel; that is, it is an *aerobic* oxidation and is also called *respiration*. As an alternative pathway, the glucose and glycogen molecules can be broken down into two or more fragments, and one of these fragments then becomes oxidized by another. This energy yield is *anaerobic*, and the processes are named glycolysis and glycogenolysis, respectively.

In biological oxidation-reduction reactions, an intermediary carrier of electrons usually acts together with catalysts (enzymes and coenzymes). In the *mitochondria* the aerobic energy-transducing system is bound into its structures. The mitochondria has two membranes, of which the inner one has inward folds, or invaginations, called cristae, which greatly increase the surface area. (It has been estimated that the mitochondria of blowfly flight muscle has 400 m^2 of inner membrane surface per gram of mitochondria protein.) This inner membrane surface contains enzymes (e.g., the cytochromes) that act in a chain to carry electrons from nutrient molecules to oxygen. Each one of the enzymes specializes in carrying atoms or electrons in just one range of the molecular chain in the substrate. In other words, there is a lock-and-key fit of the substrate molecule to a small patch on the surface of the enzyme molecule.

High-Energy Phosphates

Electric power may be stored in accumulators or batteries. Our living cells have solved the problem of storing energy in a similar way. The most abundant "battery packs" used in the cell are the compound called adenosinetriphosphate (ATP). This compound belongs to a class of molecules called high-energy phosphates. This term is somewhat imprecise, but the crucial point is that in such molecules a great deal of the total energy content is concentrated in one (or two) phosphate bonds. (We have already mentioned that the energy in a molecule lies in the bonds between the atoms.) There exist other energetically equivalent compounds of this type (GTP, UTP, etc.), but ATP is the one most often used when energy is called for in biological systems. (See Table 12-1 for these and other abbreviations and their meanings.) Furthermore, such phosphate bonds donate their energy by very simple reactions, i.e., hydrolysis or phosphate transfer, thereby providing energy for driving endergonic processes.

Almost every energy-demanding process in a cell is associated with the use of the energy from ATP stored in a rapidly usable form such as ATP.

TABLE 12-1
ABBREVIATIONS

ATP	adenosine triphosphate
GTP	guanosine triphosphate
UTP	uridine triphosphate
ADP	adenosine diphosphate
AMP	adenosine monophosphate
P_i	orthophosphate (PO_4^{3-})
PP_i	pyrophosphate
G-I-P	glucose-l-phosphate
G-6-P	glucose-6-phosphate
F-6-P	fructose-6-phosphate
F-1, 6-P_2	fructose-1, 6-diphosphate
PEP	phosphoenol pyruvate
SDH	succinate dehydrogenase
SuccCoA	succinylcoenzyme A
OAA	oxaloacetate
NAD	nicotinamide-adenine-dinucleotide
FAD	flavin-adenine-dinucleotide
GS	glycogen synthetase
GP	glycogen phosphorylase
HK	hexokinase
TCA	tricarboxylic acid
PFK	phosphofructokinase
FDPase	fructosediphosphatase
LDH	lactate dehydrogenase
PDH	pyruvate dehydrogenase
PCr	phosphocreatine

A battery contains a limited amount of energy; it needs continuous recharging to be constantly operative. The same holds for ATP. Here the charging consists of combining ADP and phosphate in an endergonic reaction to ATP, and the thus-stored energy is liberated by the typical exergonic reaction:

$$ATP + H_2O \rightarrow ADP + P_i + energy$$

ATP contains, in fact, one more energy-rich bond, and in a separate chemical reaction, the transformation may proceed further to adenosine monophosphate (AMP):

$$ATP + H_2O \rightarrow AMP + PP_i + energy$$

We can conclude that ATP is one universal intracellular vehicle of chemical energy; it saves or conserves some of the energy yielded in the degradation of fuel molecules and can transfer its energy, by donation of its terminal high-energy phosphate group, to processes requiring energy within the cell: biosynthetic processes, active transport of material against gradients, and muscle contraction. Without the ATP-ADP system, no cell could function or survive!

ADP is like a discharged battery. In order for it to be used again, it has to be recharged. ATP, then, operates as the immediately usable "energy pack" for driving endergonic processes, but it is certainly no energy *store* in the usual sense of the term. The muscle cell contains a type of "accumulator" with a somewhat greater capacity than the ATP "energy pack," namely, phosphocreatine (PCr). This is also a high-energy phosphate compound, and is in equilibrium with ATP through the following reactions:

$$ATP + C \rightleftarrows ADP + PCr$$

PCr operates as an immediate store for ATP regeneration, but also this energy store is fairly rapidly depleted (within a matter of seconds or minutes in strenuous muscular work). The resynthesis of ATP (reversal of the reaction $ATP + H_2O \rightarrow ADP + P_i + $ energy) therefore has to go on continuously. The energy source for this process is the breakdown of more complex molecules (foodstuffs) to simpler ones, and ultimately to CO_2 and H_2O. Organic compounds containing carbon and hydrogen may be totally combusted to CO_2 and H_2O by oxygen (Fig. 12-1). This can be done either in a bomb calorimeter or by catalytic oxidation with the aid of enzymes in a living cell. It should be noted that nitrogen is never oxidized in animal cells (although much energy lies in such a process!). It is mostly excreted as urea ($NH_2 CO NH_2$), in itself an energy-demanding process.

A total combustion of a compound in a bomb calorimeter gives off all the energy in one burst. In our waterfall model, for example, all the water in the lake would suddenly fall down into the sea. Obviously, such a burst would then make it technically very difficult to catch all the available energy by a turbine. On the other hand, it is possible to build several power plants in a river, taking out energy in parts as the water flows downward and gradually loses its initial potential energy. In fact, the cell uses a similar technique. Combustible material (foodstuffs, fuel) is broken down in discrete steps, thereby giving the system an opportunity to catch much of the released energy step by step.

A combustion is an oxidation, but an oxidation does not necessarily involve the direct participation of oxygen. Most of the single-step oxidations in biological pathways are, in fact, dehydrogenations. The hydrogen atoms are later carried through different reactions and ultimately combined with oxygen to give water. This is of particular importance for our exercising muscle cell, for several energy-yielding breakdown reactions of our fuel do take place anaerobically, that is, without the presence of oxygen.

The main stored energy-yielding fuels are carbohydrate in the form of glycogen, which is a glucose polymer, and fatty acids in the form of acylglycerols (triglycerides).

The third class of foodstuffs, proteins, are metabolized in a rather complex manner. The carbon skeletons of the different amino acids will, after deamination and transport of nitrogenous compounds to the urea cycle, enter the glycolysis-Krebs cycle scheme

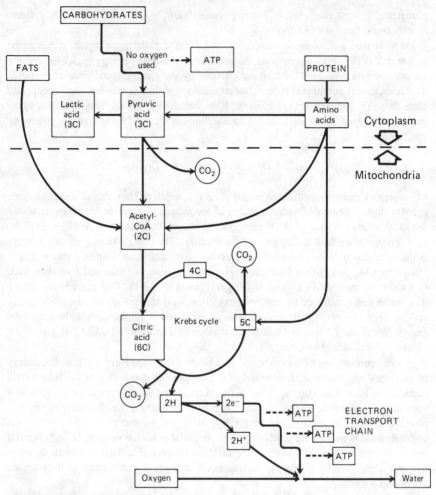

Figure 12-1
Organic compounds (fats, carbohydrates, proteins) containing carbon and hydrogen may be totally combusted to CO_2 and H_2O by oxygen, that is, by the process of a catalytic oxidation with the aid of enzymes in the living cell. For some of the steps the number of carbon atoms involved are given.

at different points. Normally, the contribution from these compounds to the total energy expenditure in skeletal muscles during exercise is very small and may be ignored.

THE OXIDATION OF FUEL

In the total carbohydrate combustion, we have an initial anaerobic sequence, glycolysis, followed by an aerobic phase. The glycolytic enzymes are located in the cytoplasm of the cells, whereas the aerobic oxidation of carbohydrates and also fatty acids takes place inside the mitochondria. The last step of the entire mitochondrial oxidation is oxygen-dependent, while the cytoplasmic glycolysis is not. Therefore, whereas the breakdown of fatty acids is strictly aerobic, the breakdown of glucose may proceed either aerobically or anaerobically.

In the reactions by which carbohydrates are broken down to lactates, ATP serves as a phosphate carrier and NAD as an electron carrier. All the intermediates of the glycolysis between glucose and pyruvate are phosphorylated compounds. The phosphorylation makes these intermediates unable to penetrate through the cell membrane; they are trapped inside the cell and cannot diffuse out. Ultimately the phosphates become the terminal phosphate group of ATP as the glycolysis and glycogenolysis proceed. ATP can then transform some of the yielded energy to various cell activities. Figure 12-2 gives, in a schematic way, the ATP cost and the ATP yield for an anaerobic breakdown of one 6-carbon unit into pyruvate. Starting with glycogen, the net energy yield covers the resynthesis of three ATP, but with glucose as the substrate, the net ATP formation is limited to two molecules per molecule of glucose.

The aerobic oxidation of glucose or glycogen is identical down to pyruvate. If the supply of oxygen is sufficient, only unimportant quantities of pyruvate will be reduced to lactate, if anything at all. Instead pyruvate will enter the mitochondria through a very complex enzyme system (pyruvate dehydrogenase, PDH). This system catalyzes a reaction known as oxidative decarboxylation, using NAD as H-acceptor and, after CO_2 release, attaching the rest of the pyruvate molecule to coenzyme A, whereby the compound acetyl-coenzyme A is formed. This reaction is essentially irreversible, and no other backward reaction involving the same reactants and products is known. Acetyl-CoA is also the product of the initial fatty acid β-oxidation (Fig. 12-1), and combines with oxaloacetate (OAA) to form citrate, thereby entering the *tricarboxylic acid cycle* (also named the *Krebs citric acid cycle*). In this cycle a cyclic sequence of reactions catalyzed by a multienzyme system accepts the acetyl-CoA as a fuel and dismembers it to yield CO_2 and hydrogen atoms.

In this cycle we find another irreversible oxidative decarboxylation: from α-keto-

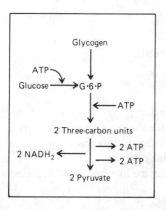

Figure 12-2
A summary of the demand and yield respectively of ATP molecules in the anaerobic breakdown of carbohydrates to pyruvate in the cell cytoplasm. Note the formation of two $NADH_2$.

glutarate (KG) to succinyl-CoA, i.e., a one-way traffic around the cycle. The α-KG formation is also associated with a decarboxylation: The two carbons that enter the cycle as acetyl will leave it as two CO_2's. These facts have rather important implications: Fatty acids, which inevitably are broken down to acetyl-CoA, can never give rise to a net synthesis of carbohydrate. On the other hand, since acetyl-CoA is also the precursor for fatty acid synthesis, carbohydrates can be broken down to this compound and act as precursors for fat.

We can identify three main steps in the oxidation of fatty acids: (1) the formation of acetyl-CoA from fatty acids; (2) the degradation of the acetyl residue of acetyl-CoA by the tricarboxylic acid cycle to yield CO_2 and hydrogen atoms; and (3) the transport of the electrons from the corresponding hydrogen atoms to molecular oxygen in the respiratory chain. Some of the yielded energy is caught by the formation of ATP. The enzyme complex catalyzing the formation of acetyl-CoA, is inhibited by a high concentration of ATP in the mitochondria. However, whenever the ADP concentration is high and ample pyruvate is available, the enzyme complex is turned on and more acetyl-CoA, which is the fuel for the cycle, is formed. This reaction is also enhanced by Ca^{2+}. The formation of acetyl-CoA from pyruvate is an irreversible step; in other words; pyruvate cannot be synthesized from acetyl-CoA.

ATP Formation

We get 3 ATP from the glycogenolysis (whether pyruvate is reduced to lactate or oxidized in the mitochondria), while we get 36 ATP from the aerobic mitochondrial oxidation, resulting in a total of 39 ATP. For palmitate, the most abundant fatty acid in the body, we will obtain 129 ATP when this C_{16} acid is metabolized to CO_2 and H_2O.

$$C_{15}H_{31}COOH + 129\ P_i + 129\ ADP + 23\ O_2 \rightarrow 16\ CO_2 + 16\ H_2O + 129\ ATP$$

It is estimated that the average yield is 138 moles of ATP per mole of a mixture of fatty acids resembling the composition of human adipose tissue.

CONTROL SYSTEMS

Biochemical reactions may be controlled by several means:

1 By controlling the effective concentrations of the reactants and/or products in a particular reaction.

2 By allosteric feedback control of the enzyme catalyzing the reaction. That is a mechanism by which some particular compound will bind to the enzyme protein at a site other than the substrate binding site, thereby changing the catalytic power. Such modulators may be positive (activators) or negative (inhibitors).

3 By differences in the amount of the particular enzyme involved in the reaction.

In order to appreciate the magnitude of the energy yield involved in the different metabolic reactions, it may be useful to take a closer look at the first two of these control mechanisms:

A reversible reaction

$$A + C \rightleftharpoons B + D$$

will come to an equilibrium when the reaction rate in the forward direction equals that of the backward direction, whether we start with A + C or with B + D. From the law of mass action, we can formulate an equilibrium constant

$$K_{eq} = \frac{[D]\,[B]}{[A]\,[C]}$$

If we now either add more reactant (A or C) or remove some of the products (B or D), the reaction in either case will proceed forward. Even if the equilibrium lies far to the left (having small concentrations of B and D), the reaction may proceed until all the reactants A and C have been converted, provided that the products are continuously removed. This is indeed what happens in several biological reaction sequences. Products in one reaction are entering as reactants in another new reaction, and we get a flow through several steps (i.e., glycolysis), even if some of the steps have an equilibrium which favors a backward direction of the reaction.

The enzymes catalyzing these steps may be controlled, both externally (by hormonal control, often via cyclic AMP and protein kinases) and internally (by allosteric feedback). As all of these reactions are designed primarily to meet the energy demand of the cell, it is not surprising that the key energy-exchanging nucleotides ATP, ADP, and AMP, and also P_i, are directly involved in the feedback regulation.

Thus, the electron-transport/oxidative-phosphorylation device is a very tightly coupled one. However, several compounds may act as uncouplers; that is, they cause a reoxidation of the entering $NADH_2$ or $FADH_2$ without ATP resynthesis. This, incidentally, is what happens physiologically in the brown fat in hibernators and other animals. Here the reoxidation energy is directly utilized as heat, and not used for ATP resynthesis. Normally, however, the ADP concentration will directly control the O_2 consumption; there will be no reoxidation as long as the ADP concentration is low. Low ADP normally means high ATP, and the control depends not so much on the concentration of one of these nucleotides as on the ratio between them. It is the ATP/ADP ratio that primarily determines how the flow through glycolysis and mitochondrial oxidation will take place. This ratio is also of the utmost importance in controlling the glycolysis in the phosphofructokinase (PFKase)-fructose-diphosphatase (FDPase) step. These two reactions represent most of the control of the glycolysis and the gluconeogenesis (the reverse of glycolysis, i.e., resynthesis of carbohydrate).

INTERPLAY BETWEEN ANAEROBIC AND AEROBIC ENERGY YIELD

It should be emphasized that we have at our disposal a very accurate method for measuring the body's total aerobic metabolic rate in the determination of the oxygen uptake. On the other hand, good methods for a quantification of the anaerobic energy yield are presently not available.

It is the availability of oxygen in the cell that determines the extent to which the metabolic processes can proceed aerobically and anaerobically. In other words, the procedure depends on whether the cytoplasmic $NADH_2$ can be reoxidized in the mitochondria at the same pace as it is formed. At rest and during moderate exercise, the oxygen supply is sufficient and the energy metabolism is essentially aerobic The ATP concentration is high and the ADP concentration is low. With increasing severity of exercise, ADP accumulates, the breakdown of glycogen speeds up, and the reduction of NAD is correspondingly faster. At some critical intensity, the oxygen-transporting system cannot provide enough oxygen to the cells, and part of the pyruvic acid which is formed must act, in addition to oxygen, as a hydrogen acceptor. Under these conditions, some of the cytoplasmic coenzyme $NADH_2$ is reoxidized anaerobically by pyruvate, which is transformed to lactic acid, and the rest of the cytoplasmatically formed $NADH_2$ is reoxidized in the mitochondria in the aerobic process. When the exercise is intensified further, an increasingly greater proportion of the $NADH_2$ (formed in the cytoplasm) will have to be reoxidized by pyruvate, and consequently a larger part of the exercise is done anaerobically. When the intensity of the exercise exceeds 100 percent of the maximal oxygen uptake, all further increase in intensity will have to be done anaerobically. Since anaerobic glycolysis furnishes one-thirteenth as much ATP per mole glucose-unit from glycogen as does aerobic oxidation, the glycolysis will have to be speeded up 13 times more for each percentage point of increase in exercise intensity at work rates above 100 percent of maximal oxygen uptake than at work rates below about 60 percent of maximal oxygen uptake.

At the very start of the exercise, even if the exercise intensity is below 60 to 70 percent of the maximal oxygen uptake, some anaerobic energy metabolism does take place while the oxygen supply to the active muscle cells is being adjusted to meet the actual demand, that is, during the first 45 to 90 seconds when the cardiac output is changing from a resting condition to the level required by the work rate. After the cessation of exercise, the metabolic "generators" continue to run for a while to replenish the energy in an immediately accessible form and because there is an increased post-exercise metabolic rate (see p. 301).

It should be noted that one molecule of extra oxygen is consumed in the formation of about 6.5 ATP when glycogen is completely oxidized (6 O_2 will produce 39 ATP, or 6.5 ATP per O_2), but only 5.6 ATP when a fatty acid is oxidized (23 O_2 will produce 129 ATP, or only 5.6 ATP per O_2). When the oxygen supply to a muscle becomes limited during heavy exercise, glycogen contributes relatively more to the energy yield than does fat. This obviously gives a better utilization of the oxygen transported to the muscle.

Although it is the availability of oxygen in the cell that determines the extent to which the metabolic processes can proceed aerobically or anaerobically, the exact regulatory mechanism is unknown. However, the intracellular enzymes engaged in the aerobic metabolism can still operate maximally at oxygen tensions in the order of about 1 mm Hg (0.1 kPa) if this is a critical value, the capillary oxygen pressure must be higher than 1 mm Hg to secure an effective diffusion gradient in order to supply the cell with the oxygen needed for the aerobic processes to proceed. An increase in concentration of the phosphate ADP, and AMP, which may also be formed, will enhance the oxidation. The positive effects in various enzyme systems may not be linearly related to the ADP concentration. McGilvery (1975) has suggested that the rate of conversion of triosephosphates to pyruvate may vary as the square, and the rate of oxidative phosphorylation as the cube of the ADP concentrations in its low ranges. It means that high concentrations of ADP are needed for a strong stimulation of the glycogenolysis. It should be emphasized that the variations in ADP concentrations are modest in absolute terms, but with a reduction in the concentration of phosphocreatine to one-tenth the initial level (the ATP concentration may be reduced below 0.5 of its resting level), there is a 20-fold change in the concentration of ADP and a nearly 700-fold increase in the concentration of AMP at a constant pH (McGilvery, 1975). The effects of such dramatic changes are difficult to predict.

Newsholme concludes that oxoglutarate dehydrogenase activities, indicating the maximum flux through the citric acid cycle, are probably the only ones that can provide quantitative information on the aerobic capacity of a muscle (see Newsholme and Leech, 1983).

When oxygen is not available in sufficient amounts, which is particularly true at the beginning of heavy exercise, the glycogenolysis is, so to speak, the only alternative to provide energy for the regeneration of ATP. It probably takes seconds before the oxidative system in the mitochondria can be thrown into full swing; it takes time for the ADP that is accumulating in the cytoplasm to diffuse into the matrix of the mitochondria.

RELATIVE IMPORTANCE OF THE DIFFERENT ENERGY STORES

A given weight of an organic compound contains a fixed amount of potential energy locked up in the bonds between the atoms of its molecules. Knowing the amount of such available compounds within the body and their energy content gives us the energy stores of the human machine. Table 12-2 summarizes data which, however, are approximate and subjected to large individual fluctuations, particularly the content of fat. In spite of these individual differences, some general statement can be made about the relative importance of the different compounds. Their specific importance depends heavily on both the intensity and the duration of the exercise. During maximal exercise the energy demand may exceed 200 kJ (50 kcal) min^{-1}. The supply from a breakdown of all the ATP available would cover only a maximal effort of about one second, and the generation of ATP from the breakdown of all the phosphocreatine would cover another few seconds of maximal effort (Bergström et al., 1971). It is well established

TABLE 12-2
THE FIGURES ARE VERY APPROXIMATE. THE GLYCOGEN CONCENTRATION
CAN BE ANYTHING FROM ALMOST ZERO UP TO 250 MMOL KG^{-1}, AND
CERTAINLY THE FAT CONTENT IS SUBJECT TO LARGE VARIATIONS. IT IS
ASSUMED THAT ONLY PART OF THE MUSCLE MASS IS ACTIVATED.

	Energy mole^{-1}		Concentration	Total energy in humans (body weight 75 kg, muscle weight 20 kg)	
	kJ	kcal	mmol kg^{-1} wet muscle	kJ	kcal
ATP	42	10	5	4	1
Phosphocreatine	44	10.5	17	15	3.6
Glycogen	2900	700	80	4600	1100
Fat	10,000	2400	—	300,000	75,000

that maximal speed can be maintained for less than 10 seconds, that is, for less time than it takes to run the 100-m dash, and the explanation may be that "rapid energy" is no longer available because of an exhaustion of phosphocreatine and eventually also of ATP. In addition to the energy coming directly from the ATP and phosphocreatine stores, some of the energy in a 100-m dash will come from the glycogenolysis which is rapidly speeding up during the effort, giving rise to increasing amounts of lactate.

Lactate is not normally considered a form of stored energy. However, when a certain amount of anaerobic exercise has been done, the concomitant production of lactate is by no means wasted. If the intensity of the exercise is reduced to aerobic conditions, lactate is readily converted back to pyruvate in the exercising muscles and can be oxidized in the mitochondria, replacing glycogen as fuel. If, on the other hand, the anaerobic exercise is followed by rest, the lactate-via-pyruvate is converted back to glycogen in the liver (the Cori cycle), and probably also in the muscles themselves (see Chap. 7; Fig. 12-3).

The total amount of energy that can be derived from the ATP, phosphocreatine, and lactate stores is of limited importance when the activity period exceeds 15 to 30 min. In this situation the energy demand may be in the order of 20 to 80 kJ · min^{-1}, and it is observed that both the ATP and the phosphocreatine concentrations are only moderately lowered. The levels of lactate are also modest compared with what is found in maximal exercise. It is therefore evident that during prolonged exercise, the rapid and continuous energy production from the oxidation of glycogen and fatty acids is extremely important. The rather complicated regulation of the ratio between the utilization of these two main fuels will be discussed later in this chapter. For a further discussion of the limiting factors in aerobic and anaerobic energy yield, see Saltin and Gollnick (1983).

Whereas maximal exercise of short duration in essence depends only on the ATP and phosphocreatine stores, and whereas prolonged exercise depends only on the oxidation of glycogen and fat (free fatty acids), exercise with duration from 1 to 10

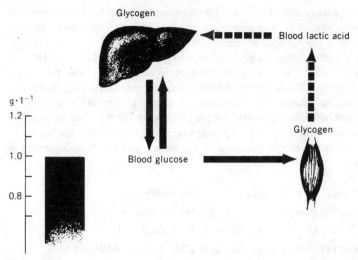

Figure 12-3
Blood glucose can form glycogen in the liver and the muscles (enzyme hexokinase at the first step) but can be released only from the liver into the blood because of a lack of the proper enzyme for this formation in the muscles (glucose-6-phosphatase). The formation of glycogen is induced by insulin and by a rise in blood sugar. The formation of glucose is induced by epinephrine and glucagon, and a formation of glucose from glycogen is stimulated by a fall in blood sugar level.

min is much more complex from the viewpoint of fuel utilization. When exercise is performed to exhaustion within this time interval, probably all fuel stores are utilized at the same time, but the relative amount of each fuel changes from second to second. At the start of vigorous exercise, utilization of ATP and phosphocreatine is predominant; then the anaerobic conversion of glycogen to lactate takes over more and more, and toward the end of the exercise, the oxidation of glycogen and eventually of fat will predominate. (For the total daily ATP-ADP turnover, see p. 2.)

SUMMARY

High-energy phosphate compounds represent the common currency for the transfer of energy within the living organism. The ATP-ADP system is the primary carrier of chemical energy in each and every cellular reaction. As quickly as ADP is formed, it is rephosphorylated, and the metabolic generation of ATP provides a general mechanism for the coupling of energy-yielding and energy-requiring processes. If the process is anaerobic, ADP is rephosphorylated by phosphocreatine, or, at the expense of glycogenolysis or glycolysis, lactate is formed. In aerobic conditions, ADP is rephosphorylated during oxidative phosphorylation by the mitochondria using glycogen, glucose, or free fatty acids as fuel.

In the absence of oxygen the skeletal muscles can work only for short periods of time, and the total energy available is very limited compared with the aerobic exercise situation. Theoretically, ATP and phosphocreatine could cover the energy demand alone, but only for a few seconds of heavy exercise.

Glucose has a slightly lower energy content than glycogen; in addition, ATP must provide energy to "lift" it into the glycogen-pyruvic acid system. The fatty acids contribute significantly in the aerobic oxidation by entering the tricarboxylic acid cycle. The amino acids, after deamination, can enter this cycle via pyruvate or acetyl-CoA and can be completely oxidized or synthesized to glycogen or fat. As a fuel for muscular contractions, however, the oxidation of proteins normally plays a very limited role.

The main steps in the energy exchange in the muscle cell can be summarized in the following way:

$$(1)\ \text{ATP} \rightleftharpoons \text{ADP} + P_i + \text{free energy}$$

$$\text{Anaerobic}\ (2)\ \text{Phosphocreatine} + \text{ADP} \rightleftharpoons \text{creatine} + \text{ATP}$$

$$(3)\ \text{Glycogen or glucose} + P_i + \text{ADP} \rightleftharpoons \text{lactate} + \text{ATP}$$

$$\text{Aerobic}\ (4)\ \text{Glycogen and free fatty acids} + P_i + \text{ADP} + O_2 \rightarrow$$

$$CO_2 + H_2O + \text{ATP}$$

In this schematic presentation, all quantitative aspects and the efficiency of the processes are disregarded.

NUTRITION AND PHYSICAL PERFORMANCE

To a certain extent we are a product of what we eat. Molecules and atoms in the food we ingest are used to build and maintain the different cells, tissues, and organs in the body. It is therefore fairly essential that the necessary elements of nutrition be included in our daily diet in order to provide the proper building material and supplies for our tissues.

Food is also fuel for the biological machinery of the body. The energy-containing foodstuffs we consume are oxidized in the cells with the aid of the oxygen we inhale. As a result of this process, all the energy we need for our existence and accomplishments is liberated.

It therefore appears reasonable to expect that nutrition may well play a role in physical performance. It is well known that undernutrition definitely does impair performance, but it is still an open question whether more than enough of an essential element of nutrition is better than just enough.

The question of what an athlete should eat in order to achieve superior performance is as old as the recorded history of organized sports. The practice of consuming large quantities of meat to replenish the supposed loss of muscular substances during heavy muscular exercise was first recorded in Greece during the fifth century B.C. (Christophe and Mayer, 1958). Two athletes, instead of eating the predominantly vegetarian diet of the time, adopted a regimen consisting of large quantities of meat, resulting in increased body bulk and weight. Many of the food taboos of primitive tribes are related to this question. The discovery of vitamins offered a new and promising area of experimentation by the more imaginative minds on the basis of the assumption that if a little of something is good, a great deal of it must be much better. More recently, protein tablets have figured prominently in dietary discussion, especially among weight lifters. Since the muscles consist mainly of protein, it was assumed that ingestion of excess protein might enhance muscle growth and improve strength, as the Greek pioneers believed some 2,000 years ago.

Evidently, human beings are genetically endowed with the potential to eat, digest, and metabolize both plant and animal foods in a variety of combinations; they can live on an almost exclusive animal diet, vegetable diet, or a combination of the two. The earliest humans consumed a considerable amount of meat. This is evident from the findings of large accumulations of animal remains where they lived, and from the tools they used, mainly designed for the processing of game. Furthermore, their living sites were selectively located in areas where there was likely to be an abundance of large grazing animals. Widespread use of aquatic food appears to be a recent phenomenon, since shells and fish bones are infrequently found in archeological material older than 20,000 years. From all indications, big game hunting attained increased significance with the appearance of truly modern human beings. However, during the period shortly before the introduction of agriculture and animal husbandry, there was a shift away from big game hunting toward a broader spectrum of food sources. This is apparent from the remains of fish, shellfish, and small game, as well as tools suitable for processing plant foods, such as grindstones and mortars. With the introduction of agriculture, the proportion of meat in the diet declined drastically, while vegetable foods eventually accounted for some 90 percent of the diet. However, since the industrial revolution, the animal protein content of Western diets has once more increased (see Eaton and Konner, 1985).

On the basis of the available evidence, Eaton and Konner (1985) have estimated the probable daily nutrition of paleolithic human beings. They conclude that the paleolithic people generally ate much more protein and far less fat than we do. Their diet contained more essential fatty acids, and much higher ratios of polyunsaturated to saturated fats than our diet, although their cholesterol intake was high. Evidently their calcium intake far exceeded even the highest estimates of minimal daily requirement. Meat and entrails from game animals provided them with high iron intakes. Their intake of dietary fiber was much higher than ours, while their sodium intake was remarkably low. By all indications, their vitamin intake greatly exceeded ours.

On the whole, one is left with the impression that the ancient diet of our paleolithic ancestors was superior to ours in terms of promoting health. This impression is further strengthened by the observation that coronary heart disease, hypertension, and diabetes are relatively unknown among the few surviving hunter-gatherer populations whose way of life and eating habits most closely resemble those of preagricultural human beings (Eaton and Konner, 1985). Even among the more isolated Eskimos living in Alaska 35 years ago, hypertension, and coronary heart disease were extemely rare (Rodahl, 1953).

Repeated claims have been made by different workers over the last several decades that physical exercise performance of average persons can be significantly improved by special diets or dietary supplements (Simonson, 1951; Fenn et al., 1983), and even greater improvements can thus be made in athletic performance (Mayer and Bullen, 1960). Yet, Mayer and Bullen conclude: "The concept that any well-balanced diet is all that athletes actually require for peak performance has not been superseded." It is the purpose of this chapter to examine whether this, in the light of more recent evidence, is still true.

NUTRITION IN GENERAL

Today, our knowledge of the number of essential nutrients is rather comprehensive. Patients have been living and maintaining good nutritional conditions and health for years, exclusively on intravenous administration of approximately 45 different nutrients. Actually, one patient has been living on total parenteral nutrition since September 1970 (Hallberg, et al., 1982) as is still in good health in 1985.

We need food as building blocks for our tissues, and we need food for energy. The most important building material is protein. It is essential for building new cells and tissues in growth and development, and for replacing parts of old cells which constantly are being broken down. There are different kinds of proteins, depending on their amino acid composition. Plant proteins are not identical with proteins of animal origin. The tissues of our body need some amino acids which we cannot live without because the body is unable to synthesize them. They are therefore vital, and are known as the essential amino acids. Meat, fish, eggs, milk, and cheese contain proteins with about the proper composition of amino acids. They are therefore the most suitable sources of protein.

An adult needs about 1 g protein of proper composition per kg body weight per day. This corresponds to about 70 g for a full-grown person. Athletes may perhaps have a somewhat greater protein requirement during periods of intense training, especially that involving muscle strength. However, the exact protein requirements of different categories of athletes in training have, so far, not been established by scientifically controlled metabolic balance studies. (FAO and WHO have in their recommendations of 1973 suggested that an intake of 0.57 g egg protein per kg body weight should maintain nitrogen balance in young, healthy individuals. See "Energy and Protein Requirements" World Health Organization Technical Report Series No. 522, 1973. However, this recommended protein intake requires a relatively high-energy intake; see Garza et al., 1976.)

Minerals are also necessary for the maintenance of body structures, and we need a number of vitamins for a variety of catalytic processes in our biological machinery. However, food is also the fuel for the body machinery. Some of the ingested food is metabolized and used as soon as it is resorbed. But most of it is stored temporarily, mostly in the form of fat. This stored energy may then be mobilized as needed.

In the development of the ideal energy stores for the mobile animal, such as the human, nature has to meet a number of requirements. For reasons of portability, each molecule should carry a large amount of energy per unit weight. The material should be fitted into various oddly shaped spaces and compartments of the body. It should possess a great storage stability and, at the same time, be readily available and capable of being rapidly converted into oxidizable substrate when needed, without being spontaneously explosive. As in all other efficient operations, the overhead cost should be low, that is, the handling expenses, including costs of storage and transport, should be minimal.

As pointed out by Dole (1964), fat or triacylglycerol (triglycerides) meet these requirements remarkably well. They are high in energy content, and they are stable, yet readily mobilized. The amount of energy held per unit weight of any molecule depends on its content of oxidizable carbon and hydrogen. Oxygen in a molecule of stored energy merely adds dead weight, since oxygen atoms can be obtained from the air as needed. Fat, therefore, approaches maximal storage efficiency since it contains as much as 90 percent carbon and hydrogen and has an energy density of 39 kJ · g^{-1} (9.3 kcal). It is much superior in storage efficiency to carbohydrates, which contain only 49 percent carbon and hydrogen and have an energy density of only 17 kJ · g^{-1} (4.1 kcal). The difference is even greater when one considers that fat is deposited in droplets, while carbohydrate is deposited together with an appreciable amount of water, in mammals 2.7 g water per g dried glycogen (Weis-Fogh, 1967). In other words, hydration dilutes the energy density of glycogen to about 4 kJ · g^{-1} (1 kcal). Thus, carbohydrate is a rather inferior material as an energy store in terms of portability. The energy value of 1 g of adipose tissue, which does not consist of pure fat, is about 25 to 29 kJ (6 to 7 kcal). On the other hand, it should be pointed out that there is an almost 10 percent higher energy yield per liter of oxygen used when carbohydrate is combusted than when protein and fat are burned.

The different animal species have, through adaptation, met their own peculiar needs. In migrating fishes like the salmon and the eel, and in birds, fat constitutes the main source of energy. Weis-Fogh (1967) points out that owing to impaired weight economy, carbohydrate cannot sustain a bird in flight for more than a few hours, and the recorded endurance of 1 to 3 days observed in some typical migrants therefore depends almost exclusively on the utilization of fat mobilized from stored triglycerides.

DIGESTION

The digestion of food starts in the oral cavity where the food is broken up into smaller particles by the teeth during chewing, and is mixed with saliva. The secretion of saliva is brought about by reflexes. It is greatest when food is dry and gritty. The saliva contains enzymes which split starch into maltose. This process is more complete the longer the food is chewed.

When the chewed food has attained the proper consistency (bolus), it is gathered by the tongue and pushed backward toward the pharynx. The swallowing reflex automatically provides for the swallowing of the bolus, whereby the food is moved by peristalsis down the esophagus into the stomach. The stomach is not a relaxed bag but a slightly constricted muscular container. It encompasses the stomach content by an even, gentle pressure and is expanded as the stomach is filled. This tension of the stomach wall apparently is connected with the sensation of satiety.

As the food is dissolved by the gastric juice which is secreted from numerous glands in the stomach mucosa, it is ejected into the duodenum. The gastric secretion is regulated by both neurogenic and hormonal mechanisms. The secretory glands may be excited by smell and taste, and possibly also by the mere thought of food. Fatty food, on the other hand, suppresses the production. The gastric juice causes the breakdown of proteins and the emulsification of fat.

The digestion of the food (chyme) in the small intestine is accomplished by the intestinal juice, which contains several digestive enzymes capable of breaking down carbohydrates, fat, and proteins into molecules which can be transported through the intestinal wall into the blood. The remainder is moved along by the intestinal peristalsis down into the large intestine and eventually expelled as feces.

The nutrient molecules which are resorbed through the intestinal wall are partially used at once as building blocks for the cells and tissues of the body, and as fuel. The rest is stored, partly in the form of glycogen, but for the greater part as fat in the adipose tissues around the guts, under the skin, and for a small part also in the muscles.

For a more comprehensive review, the reader is referred to the standard textbooks of physiology and nutrition. A simplified schematic presentation of the actions of the different digestive enzymes may be summarized as follows:

Sequence of protein digestion:

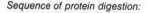

Protein		*Active enzyme*	*Site of action*
↓	←	pepsin (+ acid)	stomach
Proteoses and peptones			
↓	←	trypsin chymotrypsin	small intestine
Polypeptides and dipeptides			
↓	←	peptidases	small intestine
Amino acids			

Sequence of fat digestion:

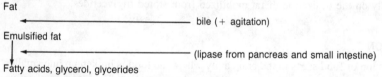

Fat

⟵ bile (+ agitation)

Emulsified fat

↓ ⟵ (lipase from pancreas and small intestine)

Fatty acids, glycerol, glycerides

Sequence of carbohydrate digestion:

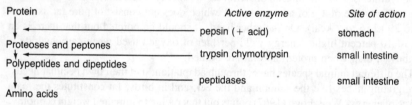

STARCHES

γ - amylase (saliva)

Amylases (pancreatic, intestinal)

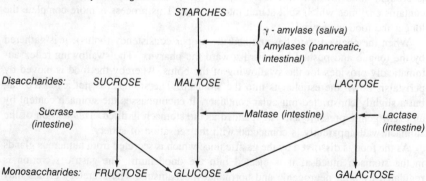

Disaccharides:	SUCROSE	MALTOSE		LACTOSE
	Sucrase (intestine) →	← Maltase (intestine)		← Lactase (intestine)
Monosaccharides:	FRUCTOSE	GLUCOSE		GALACTOSE

ENERGY METABOLISM AND THE FACTORS GOVERNING THE SELECTION OF FUEL FOR MUSCULAR EXERCISE

Of the various nutrients in the food we eat, it is only the carbohydrates, fat, and proteins that can yield energy for muscular exercise. Alcohol (ethanol), which contains 29 kJ (7 kcal) per g, cannot be utilized either directly or indirectly by the muscle (Schürch et al., 1982). The three energy sources however, do not contribute equally to the energy-yielding processes in the muscle cell. The fact that nitrogen excretion is not significantly increased during muscular exercise in the fed individual (Crittenden, 1904; Cathcart and Burnett, 1926; Hedman, 1957) was taken as an indication that protein is not used as a fuel to any appreciable extent as long as the energy supply is adequate (Pettenkofer and Voit, 1866; Chauveau, 1896; Krogh and Lindhard, 1920). Under such conditions, proteins are used almost exclusively to replace those parts of the cells which are being broken down. This is particularly true when an individual is undernourished or fasting, and during prolonged moderately heavy exercise lasting 4–6 hours or more (Refsum et al., 1979; Dohm et al., 1985). Normally, however, protein is only a minor source of energy during exercise (see Calles-Eskandon et al., 1984; Wilmore and Freund, 1984; Wolfe et al., 1984). From a teleological point of view it would indeed be a catastrophe if the skeletal muscle fibers, like cannibals, consumed themselves; thus an inhibition of protein breakdown during exercise is a necessary provision for the maintenance of an intact muscle machinery. It is known that some nitrogen is eliminated during exercise in the sweat. For this reason nitrogen excretion by the kidneys may give a false impression of the total rate of protein deamination. During dynamic exercise there may be a reduction in protein synthesis and an increase in protein catabolism, and the branched chain amino acids may in particular contribute to the energy liberation (in the glucose-alanine cycle).

From these facts, it is clear that the choice of fuel for the exercising muscle is mainly limited to carbohydrate and fat. The percent participation of these two fuels in the energy metabolism is usually assessed by the determination of the nonprotein respiratory quotient (R), which is the ratio CO_2 volume produced/O_2 volume utilized. An estimation of the amount of O_2 (ml · min^{-1}) used for the oxidation of fat can be obtained from the formula: $(1 - R) \cdot 0.03^{-1} \cdot O_2$. In this calculation, the R is handled as a "nonprotein R." The fact that the energy value of O_2 varies with R is not taken into consideration, but this may represent, at the most, a 7 percent difference.

The percentage participation of the two major fuels in the energy metabolism depends on a variety of factors:

1 Type of muscular exercise: whether it is (*a*) continuous or intermittent; (*b*) brief or prolonged; (*c*) light or heavy in relation to the maximal aerobic power of the engaged muscle groups

2 State of physical training: whether the individual is untrained or well-trained

3 The diet: whether it is high or low in carbohydrates

4 State of health: certain pathological conditions, such as diabetes, affecting the organism's choice of fuel

With increasing intensity of the exercise, there is a gradual change toward a proportionally greater share of energy yield from carbohydrates (see Fig. 12-4). This has

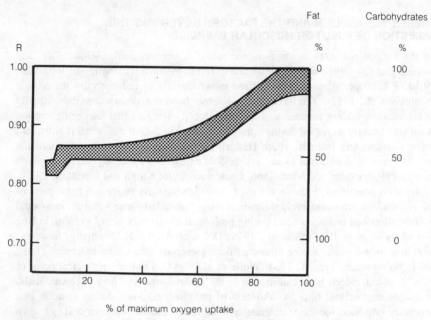

Figure 12-4
Nonprotein respiratory quotient at rest and during exercise related to the oxygen uptake in percent of the subject's maximal oxygen uptake. To the right, the percentage contribution to the energy yield of fat and carbohydrate. Prolonged exercise, endurance training, and diet can markedly modify the metabolic response.

certain advantages in terms of energy yield. One liter of oxygen can, in the respiration chain, oxidize glycogen and yield energy for a regeneration of about 6.5 moles of adenosine triphosphate (ATP). When fatty acids are oxidized, the ATP formation is reduced to 5.6 mole/l of oxygen consumed.

On the other hand, the adequacy of the oxygen supply to the exercising muscle cell is of prime importance because the oxidation of free fatty acids (FFA) depends on oxygen as the hydrogen acceptor. Thus, an inadequate oxygen supply more or less restricts the usable fuel to carbohydrate, of which there are limited stores. In addition, anaerobic oxidation of carbohydrate utilizes only about 8 percent of its energy as compared to an aerobic oxidation of this fuel. When the oxygen supply is lacking, the carbohydrate oxidation proceeds only as far as the formation of lactate. The accumulation of lactic acid in the muscle may impair the function of the muscle cells. Furthermore, the increased lactate in the blood may inhibit the mobilization of FFA (Fredholm, 1969). This in turn suppresses fat metabolism further by limiting the supply of FFA substrate to the muscle cell, although a study by Ahlborg et al. (1976) suggests an augmented removal of FFA from the plasma pool in humans following lactate infusion.

Since the ability to utilize fat as a fuel depends on the oxygen-transporting capacity, the choice of fuel for the exercising muscle depends on the work rate in relation to the individual's maximal oxygen uptake. The greater the maximal oxygen uptake, the greater the percentage contribution of fat to the energy metabolism at a given work rate. Since training increases the maximal oxygen uptake, it also increases the facility for utilizing fat as a source of muscular energy during certain types of activity (Costill et al., 1979; Rahkila et al., 1980).

In prolonged exercise, it is a distinct advantage for the fasting individual to be able to utilize fat as a source of muscular energy, since the fat stores are infinitely larger than those of carbohydrates. Stored fat amounts perhaps to some 400 MJ (10^5 kcal) or more in a well-fed individual of average size. The available energy in the form of stored ATP and phosphocreatine is only a few kJ, sufficient for only seconds of strenuous physical effort. Stored glycogen amounts to some 8MJ (2000 kcal); see below.

The fact that physical training which increases the maximal oxygen uptake also increases the individual's facility for fat utilization was shown by Issekutz et al. (1965) in dogs. A trained and an untrained dog performed the same work rate on the treadmill, which the untrained dog could endure for only 30 min (Fig. 12-5). In the untrained dog, the blood lactate rose to 8mM, whereas, in the trained dog, it was only 3 mM. In the trained dog, the utilization of FFA rose during the experiment, while in the untrained dog it declined. One reason for this difference is that training enhances the O_2 supply to the active muscle cells, so that the exercise, in the case of the trained dog, may be performed to a greater extent aerobically. Consequently, less lactic acid is formed in the trained dog. The unfit dog, at the same work rate, produces more lactic acid, which inhibits the FFA release from the adipose tissue (for reasons which will be explained later). The result is that the plasma FFA drops. A decreased plasma FFA always means decreased turnover rate or decreased rate of utilization of FFA (see below).

At rest, fat and carbohydrate contribute about equally to the energy supply, as they also do during light or moderate exercise in the fasting individual (Christensen and Hansen, 1939). As the exercise progresses, fat contributes in an increasing amount to the energy yield (Fig. 12-6). During moderately heavy exercise (in fasting subjects) which may be endured for 4 to 6 hours (including rest pauses), as much as 60 to 70 percent of the energy may be derived from fat at the end of the exercise period (Pruett, 1971; Table 12-3). Thus, the longer the exercise lasts, the smaller the percentage contribution of carbohydrate to energy metabolism in the fasting subject.

Christensen and Hansen (1939), in their classical experiments, examined the participation of fat and carbohydrate in energy metabolism on the basis of the respiratory quotient during physical exercise of different intensities. In subjects on a *normal diet*, engaged in exercise of such an intensity that the metabolic processes were essentially aerobic, they found that about 50 to 60 percent of the energy was supplied by fat. In prolonged, standardized aerobic exercise of up to 3-hr duration, an increased participation of fat was observed, supplying up to 70 percent of the energy. In heavy exercise, on the other hand, where anaerobic metabolic processes were involved, their findings indicated a major participation of carbohydrates.

Subjects on an extremely *high-fat diet* for several days, in which less than 5 percent

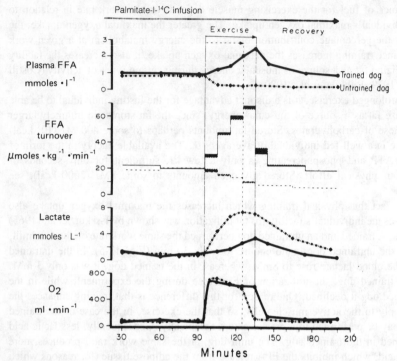

Figure 12-5
Effect of exercise on the FFA (free fatty acid) turnover in a trained dog (body weight 11.4 kg) and in an untrained dog (body weight 10.5 kg). At zero time, a priming dose of radiopalmitate was given intravenously, 220 and 276 mμc · kg^{-1}, respectively, followed by an infusion from 0 to 200 min of 12 and 15.3 mμc · kg^{-1} · min^{-1}. respectively. The trained dog ran on the treadmill for 40 min (from 95th to 135th min), and the untrained dog for 30 min (from 95th to 125th min). Shaded area represents differences between the rate of release and rate of uptake of FFA. *(From Issekutz et al., 1965.)*

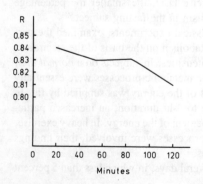

Figure 12-6
Respiratory quotient (R) in a subject exercising at 175 watts for 120 min while living on a normal diet *(Modified from Christensen and Hansen, 1939.)*

TABLE 12-3
AVERAGE RESPIRATORY QUOTIENTS (R) AT THE BEGINNING AND END OF $4\frac{1}{2}$-HR (WITH 15-MIN REST EVERY HOUR), AND THE PERCENTAGE OF THE ENERGY UTILIZED, WHICH WAS DERIVED FROM CARBOHYDRATE AT 50 PERCENT AND 70 PERCENT MAX $\dot{V}_{O_2}$ ON A STANDARD DIET. (*FROM PRUETT, 1971.*)

50% max $\dot{V}_{O_2}$					70% max $\dot{V}_{O_2}$				
R		% from carbohydrate			R		% from carbohydrate		
First hr	Last hr	First hr	Last hr	Average	First hr	Last hr	First hr	Last hr	Average
0.84	0.81	47	37	40	0.89	0.84	64	47	53

of the energy intake was derived from carbohydrates, were only able to exercise at a given intensity for a period of about 1 hr. Throughout the entire exercise period, 70 to 99 percent of the energy was obtained from fat combustion. Although there was a significantly reduced capacity for prolonged heavy exercise, it is noteworthy that the subjects were nevertheless able to carry on this exercise for 1 hr while utilizing fat almost exclusively as a source of fuel (Fig. 12-7).

In subjects on a very *high-carbohydrate diet,* where 90 percent of the food energy were derived from carbohydrates, the standard load could be performed for a much longer time, up to 4 hr. Initially, fat contributed only 25 to 30 percent to the metabolic fuel, compared with more than 70 percent when the diet was rich in fat. Gradually, the contribution from fat combustion increased and it was about 60 percent at the end of the exercise.

The subjects were able to exercise about 3 times as long on this very high carbohydrate diet than on the very high fat diet. In spite of the reduced carbohydrate utilization during prolonged moderate exercise, the availability of glycogen may nevertheless be a factor limiting endurance because of the limited stores.

At work rates up to some 75 percent of the individual's maximal oxygen uptake, performed with 15-min rest during each hour of exercise, and tolerated for perhaps as long as 3 to 6 hours, the hepatic glucose supply may be the limiting factor causing a drop in the blood sugar. This leads to central nervous system symptoms typical of hypoglycemia (dizziness, partial blackout, nausea, confusion, etc.) in fasting subjects (Pruett, 1971; Fig. 12-8) without depletion of the muscle glycogen depots. The fact that the hepatic sugar output and the blood sugar level indeed are the limiting factor under these conditions was clearly demonstrated by Christensen and Hansen (1939). Their experiments were carried out until the subjects became exhausted. At the point of exhaustion, both subjective symptoms and laboratory findings indicated that the subjects had hypoglycemia. At this point, 200 g of glucose was ingested. Within 15 min the subjective symptoms disappeared. The blood sugar level rose and the subjects could continue the exercise for another hour.

Similarly, studies at Lankenau Hospital (Rodahl et al., 1964) on the effect of feeding on blood glucose levels in prolonged exercise (15-min rest per hour of exercise) showed

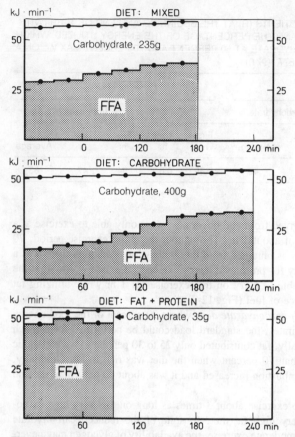

Figure 12-7
Increase in free fatty acid metabolism in prolonged exercise. One well-trained subject exercised on a cycle ergometer at 183 watt after a mixed diet, then at 176 watt after a carbohydrate-rich diet for 3 days. In another experiment 176 watt was preceded by a 3-day period on fat and protein, excluding the carbohydrates from the diet. The subject exercised until exhausted. The total energy output was calculated from the measured oxygen uptake and respiratory quotient (R) during 15 minute periods; the energy yield from carbohydrate and free fatty acids (FFA), respectively, was estimated from the (R) values. The calculated total carbohydrate consumption (g) is presented. Note how exercise time and the diet affects the choice of substrate. At a given rate of exercise the endurance time varied from 93 to 240 minutes depending on the diet. (The subject's maximal oxygen uptake was not determined.) *(Data from Christensen and Hansen, 1939.)*

that fed subjects could complete 6 hours of exercise at a load corresponding to approximately 60 percent of their maximal oxygen uptake without difficulty, their blood sugar level remaining more or less unchanged; the same subjects, when fasting, could barely complete the exercise period, and had a marked drop in their blood sugar level (Fig. 12-9). Later studies have shown that oral administration of 225 ml of a 5 percent glucose solution, given every 15 min, is sufficient to prevent hypoglycemia in subjects

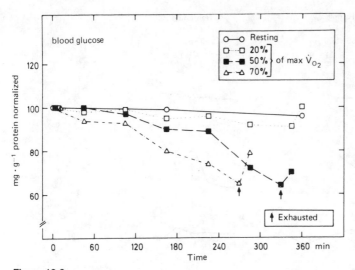

Figure 12-8
Effect of rest and three levels of exercise on blood glucose concentrations in seven subjects living on the standard diet. *(From Pruett, 1971.)*

exercising at 70 percent of their maximal oxygen uptake for 3 hours (Staff and Nilsson, 1971).

Pirnay et al. (1977), have demonstrated that by giving naturally labelled [^{13}C] glucose to subjects exercising at 50 percent V_{O_2} max that carbon-13 may be detected in the expired CO_2 as early as 15 min after the oral intake, indicating a rapid resorption and utilization of exogenous glucose during such levels of exercise. Even at higher exercise intensity (80 percent V_{O_2} mean) Bonen et al. (1981), observed that a considerable portion of glucose ingested during exercise was metabolized.

The central nervous system with its low glycogen content depends to a very great extent on the blood sugar. It has been calculated that in humans, approximately 60 percent of the hepatic sugar output serves the brain metabolism (Shreeve et al., 1956; Reichard et al., 1961). It would therefore seem essential that some sort of barrier should exist to prevent the blood glucose from freely entering the muscle cells and being used in their metabolism, since this might lead to a too rapid fall in blood glucose levels resulting in severe symptoms of hypoglycemia. In this respect, it should be pointed out that the permeability of the cell membrane for glucose depends on the plasma insulin concentration which falls parallel with the blood sugar during prolonged moderately heavy exercise (Pruett, 1971). This may serve to reduce glucose uptake by the exercising muscle cell. Furthermore, enzymes are necessary for the uptake of glucose across the membrane, and at least one such enzyme, hexokinase, is inhibited by products from the breakdown of glycogen (Hultman, 1967). Therefore, stored muscle glycogen is a more readily available substrate for the energy metabolism in the exercising muscle cell than exogenous glucose. This fact is an advantage for the central nervous system, which might otherwise be competing with the muscles for

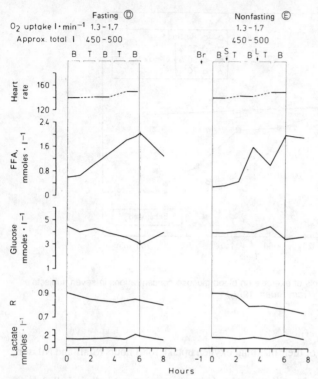

Figure 12-9
Effect of prolonged exercise. D: fasting subjects; B: exercise on the cycle ergometer at 100 watts for 60 min; T: walk on the treadmill with a slope 8.6 percent at a speed 5.6 km · hr⁻¹ (3.5 mph) for 60 min; E: nonfasting subjects; Br: breakfast; S: snack; L: lunch. *(From Rodahl et al., 1964.)*

blood glucose and suffer from hypoglycemia as a consequence. However, during prolonged exercise there is a significant utilization of glucose in the exercising muscles. This is partly balanced by an enhanced hepatic gluconeogenesis from various glucose precursors e.g., alanine and glycerol. Eventually glucagon may have a stimulatory effect on the hepatic uptake of glucose precursors and on the gluconeogenesis. Wahren et al. (1975) estimated the total glucose output from the liver during 4 hours of exercise (at 30 percent of the subject's maximal oxygen uptake) to be about 75 g. The glucose production in the liver was estimated to 15 to 20 g. With a total glycogen content in the liver amounting to 75 to 90 g, it is evident that the situation may be critical in prolonged heavy exercise (see Hultman and Nilsson, 1973). According to Issekutz (1981), the relatively small decline in plasma glucose and insulin in the exercising organism seems to play a major role in increasing hepatic glucose production and in limiting the glucose uptake of exercising muscle during prolonged activity. This mechanism maintains glucose homeostasis as long as possible and guarantees that both liver and muscle glycogen participate in about equal proportions in the elevated glycolysis of the exercising muscle.

At heavier work rates that are above 75 to 80 percent of the individual's aerobic power, and that can be tolerated for only about an hour and a half at the most, a significant depletion of the muscle glycogen stores may be the factor limiting endurance. Under these conditions, there is usually no fall in blood sugar, because the duration of the exercise period is too short to cause a depletion of hepatic glycogen stores to any significant extent, while the muscle glycogen stores may indeed be markedly reduced. Normally, some glucose from the blood enters the muscle cell at all times. However, when the muscle glycogen store is nearly depleted, as it is toward the end of heavy prolonged exercise, evidently an increasing amount of glucose enters the muscle cells from the blood.

By the needle biopsy technique, small pieces of muscle tissue (10 to 20 mg) can be sampled in the human, and their content of glycogen and other substances analyzed (Bergström, 1962). By obtaining samples during various stages of different dietary regimens, during exercise, etc., the variation in glycogen content of the muscles can be followed.

On a normal mixed diet, the glycogen content in the quadriceps femoris muscle in the individual ranges from about 50 to 100 mmol $\cdot$ kg^{-1} wet muscle (Hultman, 1967). In the deltoid muscle, the glycogen content is significantly lower, or about 50 mmol $\cdot$ kg^{-1} g^{-1} wet muscle. Figure 12-10 presents data from experiments in which biopsies were taken every twentieth min during exercise on a cycle ergometer, with an average oxygen uptake of 77 percent of the individual's maximal aerobic power. Two groups of subjects participated in the study; 10 were trained, 10 were untrained and had a lower maximal oxygen uptake. Therefore, the "77 percent load" corresponds to a mean oxygen uptake of 3.4 liters $\cdot$ min^{-1} for the trained and only 2.8 liters $\cdot$ min^{-1} for the

Figure 12-10
Average values for glycogen content, expressed in glucose units, in needle-biopsy specimens from the lateral portion of the quadriceps muscle taken before and at intervals during exercise until exhaustion, in a group of 20 subjects. Oxygen uptake averaged 77 percent of the maximal aerobic power. *(Modified from Hermansen et al., 1967.)*

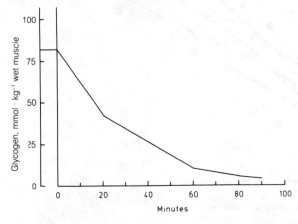

untrained subjects. After about 90 min the exercise had to be terminated because of the subjects' exhaustion. The glycogen content was then in the order of 5 mmol · kg⁻¹ wet muscle. It should be emphasized, however, that the biopsies were taken only from the lateral and superficial portion of the quadriceps femoris muscle, and the analyses do not necessarily reveal events in other portions of the active muscles. The respiratory quotient was higher in the untrained group (about 0.95) than in the trained subjects (about 0.90), indicating that FFA metabolism plays a larger role in the trained men. The actual combustion of glycogen was about 155 mmol · min⁻¹ in both groups, despite the difference in energy expenditure (3.4 versus 2.8 liters O_2 per min). At the end of exercise, the blood sugar level was still 4.4 mM.

The available evidence suggests that at work rates exceeding about 75 percent of the individual's maximal oxygen uptake, the initial glycogen content in the skeletal muscles is decisive for the individual's ability to sustain such exercise for more than an hour. Thus, Bergström et al. (1967) showed that at a work rate of about 75 percent of the maximal oxygen uptake, the larger the initial muscle glycogen stores, the longer

Figure 12-11
Relation between initial glycogen content in the quadriceps muscle in nine subjects who had been on different diets, and maximal exercise time when exercising at a given work rate demanding 75 percent of maximal aerobic power. Broken lines denote subjects who had been on a carbohydrate diet prior to the fat-plus-protein diet. *(From Bergström et al., 1967.)*

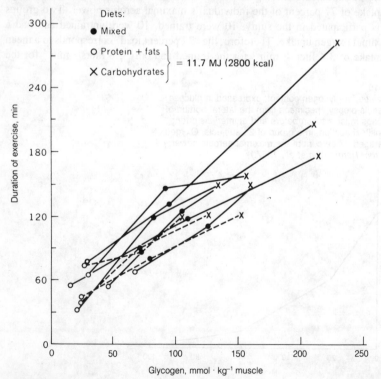

the subject could continue to exercise at this load (Fig. 12-11). After a normal mixed diet giving an initial glycogen content of about 100 mmol · kg^{-1} wet muscle, the subject could tolerate the 75 percent work rate for 115 min. After the subject spent 3 days on an extreme fat and protein diet, the glycogen concentration was reduced to about 35 mmol · kg^{-1} wet muscle and the standard load could be performed for only about 60 min. After 3 days on a carbohydrate-rich diet, the subject's glycogen content became higher, 200 mmol · kg^{-1} wet muscle, and the time on the 75 percent work rate could now be prolonged to about 170 min on the average. It was further observed that the most pronounced effect was obtained if the glycogen depots were first emptied by heavy prolonged exercise and then maintained low by giving the subject a diet low in carbohydrate, followed by a few days with a diet rich in carbohydrates (see Fig. 12-12). With this procedure, the glycogen content could exceed 200 mmol · kg^{-1} wet muscle and the heavy load could be tolerated for longer periods, in some subjects for more than 4 hr. The total muscle glycogen content under these conditions could exceed 700 g.

Bergström and Hultman (1966) also performed an experiment in which one subject exercised with his left leg and the other subject simultaneously exercised with his right

Figure 12-12
Different possibilities of increasing the muscle glycogen content. For further explanation, see text. *(From Saltin and Hermansen, 1967.)*

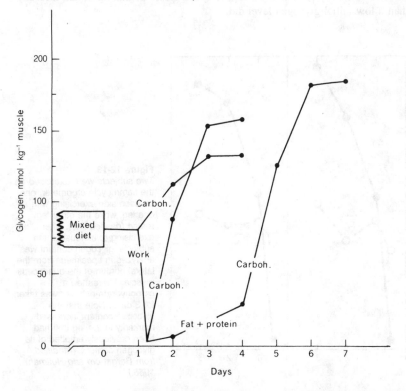

leg on the same cycle ergometer. After several hours' exercise, the exercising leg of each individual was almost depleted of glycogen while the resting leg still had a normal glycogen content. Feeding the subjects a carbohydrate-rich diet on the following days did not markedly influence the depots of the resting limb, but in the previously exercised leg the glycogen content increased rapidly until the values were about twice as high as those in the nonexercised leg (Fig. 12-13). This experiment shows that exercise with glycogen depletion enhances the resynthesis of glycogen. It also shows that the factor must be operating locally in the exercising muscle. (See Bergström et al., 1972).

In prolonged athletic events in which the work rates exceed about 75 percent of the maximal oxygen uptake, not only endurance, but also speed, is affected by the initial muscle glycogen content. This is schematically illustrated in Fig. 12-14. A group of subjects participated in two 30-km cross-country running races, on the first occasion after their normal mixed diet, and on the second occasion after a few days on an extremely high carbohydrate diet after previously emptying the glycogen depots. At several points on the track, the running time was recorded. It was observed that the lower the initial muscle glycogen content, the lower the ability to maintain a high running speed toward the end of the race. It should be noted, however, that even in the subjects with the lowest initial muscle glycogen content, the speed was maintained during the first hour of the race. This shows that a high glycogen content in the muscles did not enable the subject to attain a higher speed at the beginning of the race any more than a low initial glycogen level did.

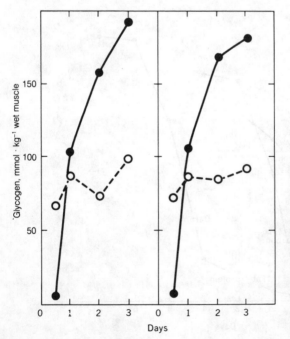

Figure 12-13
Two subjects were exercised on the same cycle ergometer, one on each side exercising with one leg, while the other leg rested (dashed line). After exercising to exhaustion the subjects' glycogen content was analyzed in specimens from the lateral portion of the quadriceps muscle. Thereafter, a carbohydrate-rich diet was taken for 3 days. Note that the glycogen content increased markedly in the leg that had been previously emptied of its glycogen content. *(Modified from Bergström and Hultman, 1966.)*

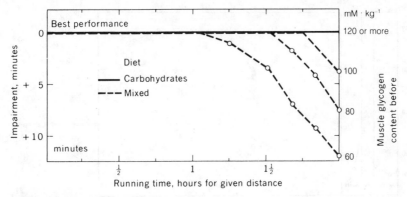

Figure 12-14
Schematic illustration of the importance of a high glycogen content in the muscle before a 30-km race (running). The lower the initial glycogen store, the slower became the speed at the end of the race compared with the race performed when the muscle glycogen content was 120 mmol · kg^{-1} muscle or more at the start of the race. For the first hour, however, no difference in speed was observed. *(By courtesy of B. Saltin.)*

Under such conditions of very intense or near maximal effort of sufficient duration, the draining of the muscle glycogen apparently is high enough to cause a significant depletion of the muscle glycogen stores. In this event, then, the cause of exhaustion is located in the exercising muscle tissue. However, even at the point of such exhaustion, the muscle is still utilizing significant amounts of glycogen or glucose as judged by the respiratory quotient.

It should be pointed out that there is one drawback with the high glycogen storage: it was mentioned that each gram of glycogen is stored together with about 2.7 g of water. With a glycogen storage of 700 g, there is then an increase in body water amounting to about 2 kg. In activities in which the body weight has to be lifted, an excessive glycogen store should therefore be avoided.

It is thus clear that under certain conditions, i.e., during heavy exercise equivalent to about 75 percent or more of the maximal oxygen uptake, the level of the muscle glycogen stores may significantly affect performance after about 1 hour. This may be influenced by the diet (Fig. 12-12).

At extremely heavy or near maximal rates of exercise, glycogen is the major source of energy for muscular exercise (see Gollnick, 1985). Under such conditions, the energy metabolism is largely anaerobic, the fuel being carbohydrate with the accumulation of lactic acid. The lactic acid suppresses the FFA mobilization, which further reduces the available FFA substrate in the muscle cell.

Following cessation of such intense or near maximal exercise, however, there is a marked rise in plasma FFA levels lasting for several hours, or actually until the next meal is taken (Fig. 12-15). Such a postexercise rise in FFA levels may also occur following intermittent exercise (Fig. 12-16).

There are other fuels which may participate as sources of energy, such as amino acids, acetoacetate, and 3-hydroxy-butyrate, but which play an insignificant role, at

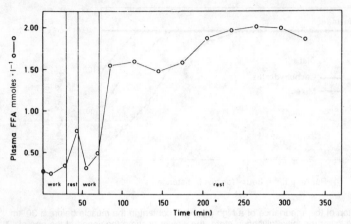

Figure 12-15
The plasma FFA (free fatty acid) response to two exhausting exercise bouts on the cycle ergometer of 87 percent max V_{O_2}. *(From Pruett, 1971.)*

any rate in well-fed individuals engaged in intense muscular exercise at high levels of energy expenditure (McGilvery, 1975). During prolonged fasting, however, Owen et al. (1967) have shown that ketone bodies may replace glucose oxidation in the brain.

The metabolic rate of the nervous system is exceedingly high even at rest. Although intellectual efforts do not seem to cause an appreciable elevation of the overall oxygen uptake of the individual, it appears likely that the rate of energy utilization of the nervous tissues is markedly increased during physical exertion, owing to the vastly increased nervous impulse traffic, the increased rate of the sodium pump, etc. The energy supply to the central nervous system during strenuous physical exercise should therefore be subject to more attention, especially since some unpublished observations have indicated reduced CNS function in athletes during strenuous physical exertion (such as orienteering competitions).

The question as to how long it takes for the depleted glycogen stores to be restored is of considerable practical importance. A complex starch-based diet will build up the glycogen stores in the skeletal muscles within one to two days. The interval between meals seems to be unimportant and the adherence to a high sugar-glucose diet does not appear to offer any advantage (see Costill et al., 1981; Wilmore and Freund, 1984).

REGULATORY MECHANISMS

The utilization of metabolic fuel is regulated by the interplay of a number of factors, both physiological and biochemical in nature (Fig. 12-17).

First of all, a high carbohydrate diet favors a higher participation of carbohydrate in energy metabolism. This fact was first described by Christensen and Hansen in 1939. It was confirmed by Issekutz et al. in 1963 and by Bergström et al., 1967, and Pruett in 1971. Issekutz et al. (1963) found that it is the carbohydrate intake rather

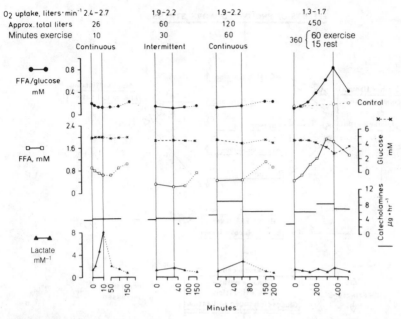

Figure 12-16
Effects of exercise of different intensity and duration (heavy exercise on the cycle ergometer for 10 min; intermittent run: 5 s exercise, 5 s rest, on the treadmill, slope 8.6 percent, speed 12 km · hr⁻¹, 7.5 mph for 30 min; continuous exercise on the cycle ergometer at 150 watts for 60 min, and alternating exercise on the cycle ergometer at 100 watts for 60 min and exercise on the treadmill with a slope 8.6 percent at a speed 5.6 km · hr⁻¹, 3.5 mph for 60 min for a total period of 360 min or until exhaustion), in normal fasting subjects in relation to blood lactate levels, urinary catecholamine excretion, plasma FFA, and blood glucose. *(From Rodahl et al., 1964.)* Note the rise in plasma FFA and the drop in blood glucose during prolonged exercise and the marked rise in blood lactate level during heavy short exercise, while it is essentially unchanged during the prolonged exercise. Note also the rise in plasma FFA following the cessation of short, heavy exercise.

than the amount of fat in the diet which determines whether the preferred fuel is FFA or carbohydrate. Ingestion of 100 g glucose immediately before exercise causes a shift of exercise metabolism toward carbohydrate and a corresponding reduction in FFA metabolism. The reason probably is that the carbohydrate intake causes an increased production of insulin which inhibits FFA mobilization, and therefore suppresses FFA oxidation. In this connection, it may be of interest to note that Wirth et al. (1981) have shown that basal plasma insulin concentrations appear to be lower in athletes than in nonathletes, due to reduced insulin secretion and increased sensitivity. During exercise, however, insulin secretion is diminished independent of the training state.

Secondly, the utilization of FFA during muscular exercise is partly determined by the level of plasma FFA. An increased plasma FFA concentration always means an increased rate of FFA utilization. At any given plasma FFA level, the FFA turnover rate is about twice as high during exercise as during rest (Fig. 12-18) (Issekutz et al.,

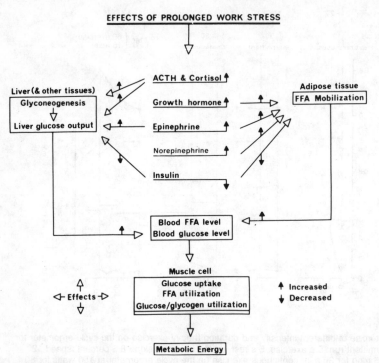

EFFECTS OF PROLONGED WORK STRESS

Figure 12-17
Schematic representation of the effect of exercise on hormone secretion and the concomitant effect of those hormones on fat and carbohydrate metabolism *(From Pruett, 1971.)*

1964; Paul, 1975). As pointed out in Chap. 10, one-leg training induced during a standardized two-leg exercise increased utilization of FFA in the trained leg compared with the untrained leg. Certainly the arterial concentration of FFA was identical in the two legs. Apparently local factors affected by training will also influence the FFA utilization.

The plasma FFA level during exercise is in turn determined by the combined effects of several factors:

1 Norepinephrine is a most powerful stimulator of FFA mobilization. Even small increases in norepinephrine cause a marked rise in plasma FFA levels associated with a corresponding increase in FFA turnover (Issekutz, 1964; Rodahl and Issekutz, 1965) (Fig. 12-19). Both during short, intense, and prolonged, moderately heavy exercise, there is a considerable increase in catecholamine production (Fig. 12-16) which would increase FFA mobilization and utilization.

2 The lactate accumulated during strenuous muscular exercise will, on the other hand, suppress the mobilization of FFA from the adipose tissues. The possibility that lactate might function as a physiological inhibitor of FFA mobilization in severe

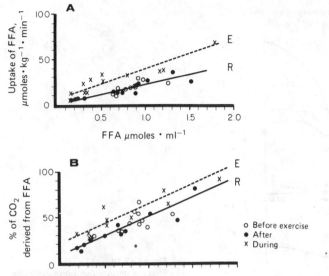

Figure 12-18
A: interrelationship between plasma FFA concentraton and uptake of FFA, at rest (R) and during the last 10 min of exercise (E). $Y_R = 23.8 \times -1.5$; SE of regression coefficient; ± 2.5 ($P < .001$), $Y_E = 32.8 \times +6.1$. SE of the regression coefficient: ± 3.1 ($P < .001$). **B:** interrelationship between plasma FFA level and the contribution of FFA to the fat oxidation at rest (continuous lines), and in the 30th min of exercise (broken line). $Y_R = 41.9 \times +9.3$; SE of the regression coefficient: ± 6.3 ($P < .001$), $Y_E = 42.2 \times +19.8$; SE of the regression coefficient: ± 9.2 ($P < .01$). *(From Issekutz et al., 1964.)*

exercise was first suggested by Issekutz and Miller (1962), who observed an inverse correlation between FFA and lactate levels in the blood (Fig. 12-20). Miller et al. (1964) showed that lactate infusion decreased the inflow of FFA into the plasma in depancreatized dogs. Fredholm (1969) induced FFA mobilization in isolated adipose tissue preparations by sympathetic nerve stimulation, which could be counteracted by lactate infusion in physiological concentrations. From his experiments, it appears that an increased re-esterification of FFA is the major mechanism underlying the lactate effect. This view is supported by the findings of Issekutz et al. (1975).

In prolonged, moderately heavy exercise, the lactate level is not significantly increased, but the catecholamine level is (Fig. 12-16). This may explain the observed rise of FFA mobilization under such conditions. In contrast, the accumulation of lactate during short, intense exercise may prevent a rise in plasma FFA during the exercise period itself, but permits a subsequent rise after exercise when the lactate level has dropped (Fig. 12-16) (Rodahl et al., 1964).

3 A number of hormones other than catecholamines are affected by physical exercise. As already mentioned, insulin suppresses FFA mobilization. However, insulin is decreased during prolonged exercise (Fig. 12-21; Pruett, 1971). Insulin, therefore, cannot prevent the observed rise in FFA mobilization during prolonged exercise. It is

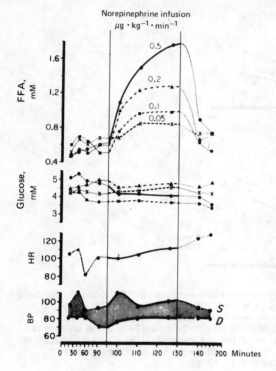

Figure 12-19
Effect of intravenous norepinephrine infusion on the plasma FFA and blood glucose of a resting nonanesthetized dog. Heart rate (HR) and blood pressure (BP) values obtained with the highest infusion rate ($0.5 \ \mu g \cdot kg^{-1} \cdot min^{-1}$) are shown on the figure. *(From Issekutz, 1964.)*

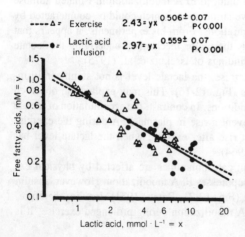

Figure 12-20
Treadmill experiment on a dog (9.5 kg). Interrelationship between plasma FFA level and blood lactic acid concentration. *(From Issekutz and Miller, 1962.)*

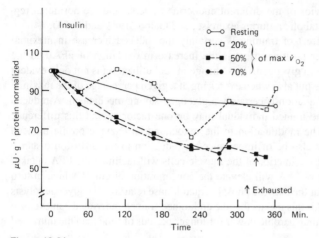

Figure 12-21
The effect of rest and three levels of exercise on plasma immunoreactive insulin concentrations in six subjects living on a standard diet. *(From Pruett, 1971.)*

known that growth hormone is increased during exercise (for reference see Hartley, 1975). This increase may affect both glucose and FFA metabolism. The administration of growth hormone is reported to cause an increase in plasma FFA lasting several hours (Grunt et al., 1967). It therefore appears likely that the long-lasting increase in FFA plasma levels following the cessation of short, intense exercise is due to a growth hormone effect rather than to a norepinephrine effect, since this effect probably is of a fairly short duration.

It is well known that cortisol, which is produced during both physical and mental stress, has a profound influence on carbohydrate metabolism. It increases the catabolic breakdown of proteins and the formation of carbohydrates from protein, which in turn may enter the metabolic pool. Issekutz and Allen (1971) have shown in experiments on dogs that even short-term treatment with metylprednisolone increases the glycogen depots of the body, causing a corresponding increase in endurance. This process, however, takes place at the expense of the protein content of the tissues and may therefore be harmful in the long run. As the result of prolonged cortisol administration, muscular wastage occurs, causing reduced muscle strength and endurance.

It has been noted that an intake of caffeine in the form of regular coffee will enhance the FFA metabolism. This will have a glycogen sparing effect, which may improve endurance (see Essig et al., 1980). It should be pointed out, however, that a period of exercise as in the case of a regular warm-up also will elevate the FFA concentration in the blood, and thus cause an increased FFA utilization.

From this brief review it is thus clear that the hormonal regulation of fat and carbohydrate metabolism during exercise depends on a very complicated balance between many factors, representing a series of unsolved problems for future research into the regulation of energy supply during heavy exercise. For a more comprehensive discussion of the behavior of the different endocrine systems and the hormonal regulation of substrate metabolism during exercise, see Galbo (1981 and 1983).

A very important effect of training is certainly the induced increase in maximal oxygen uptake. Another positive effect is the increase in oxidation of FFA and the reduced energy yield from glycogen. This change in fuel utilization is not only evident at a given metabolic rate but also when exercising at a given percentage of the maximal aerobic power. This adaptation is certainly a glycogen-saving mechanism. A reduced lactate production in the trained individual may be one factor behind this difference in metabolic pattern. The modification in the mitochondrial enzyme profile induced in trained muscles may also be of importance. A reduction in the diffusion distance between capillaries and the interior of the muscle cells will facilitate the FFA uptake. An increased oxidation of FFA will elevate the concentration of citrate which in turn will inhibit the activity of the enzyme phosphofructokinase retarding the glycogenolysis and lactate formation. All this means that the endurance-trained individual can exercise closer to his or her maximal aerobic power for longer periods of time than the untrained person.

FOOD FOR THE ATHLETE

The nutritional requirements in general have been briefly reviewed earlier in this chapter. On the whole, these requirements also hold for the athlete. A varied, well-balanced diet in adequate amounts is all that is necessary from a nutritional point of view for the body to function optimally, and for providing a biological basis for top performance. It is true that the requirements for protein and certain minerals may be somewhat increased in athletes during training in events which require muscular strength. However, the total food intake in athletes undergoing such training is also increased, often by as much as $4MJ \cdot day^{-1}$ (1000 kcal) or more in connection with the regular training and participation in competitions. If their diet is balanced, they will then automatically take in more of these nutrients as they increase their food intake. This is true provided that their diet is not too high in fat, i.e., that no more than 35 to 40 percent of the energy is derived from fat, and that the content of refined sugar is relatively low, with the consumption of milk, fish, meat, vegetables, fruit, berries, and grain products being comparatively high. In this way, an adequate intake of proteins, iron, calcium, and vitamins is secured. This diet holds for athletes in general, whether they are engaged in the training of strength or endurance, whether they are engaged in competitions or not. However, the purpose of the training must also be kept in mind. The nutritional body building should be geared to that for which the body is being trained. If we train for participation in long-distance running where the body weight must be carried long distances, the body weight and muscle mass should be no greater than is necessary in order to support the functional needs. All body weight in excess of this requirement is dead weight. The same holds for high jumpers who have to lift the body against the force of gravity. In weight lifting and throwing, on the other hand, these considerations are unimportant.

Nutritional surveys in Norway (K. Solvoll, unpublished results) have verified the general impression that throwers eat more than runners. The throwers are heavier because of their larger muscle mass. Therefore their overall energy requirement is greater. However, the energy intake per kg body weight is about the same. The percentage of the energy from proteins, fat, and carbohydrates is about the same in throwers and runners. Fourteen percent of the energy is taken in the form of proteins, close to 40 percent in the form of fat, and about 46 percent in the form of carbohydrates. The surveys showed that both groups of athletes consumed, as a rule, much more meat and milk than the population as a whole.

When discussing the dietary requirements of athletes, it is necessary to distinguish between events of very short duration, which mainly involve technique and muscular strength and which last for only seconds or a minute or two at the most, and events which last for a long period lasting up to several hours, and which therefore require endurance. In the case of the endurance events, it is necessary from a nutritional standpoint to distinguish between events lasting less than an hour and events of significantly longer duration.

Events Lasting Less Than 1 Hour

In very intense physical exertion or athletic events lasting less than 1 hour, the available supply of stored energy fuel is generally ample to cover the need. Under such conditions a special diet is unnecessary. Because the digestion of a meal causes a redistribution of blood from the muscles to the guts, physical exercise shortly after a meal will result in a competition between the guts and the exercising muscles for the available blood supply. Heavy muscular exercise should therefore be avoided immediately following a heavy meal. As a general rule, a meal should not be ingested later than $2\frac{1}{2}$ hr prior to an athletic event. Furthermore, the last meal prior to the event should be light. It should consist only of ingredients which the individual knows from experience can be tolerated well. Excessive quantities of carbohydrates should not be ingested prior to the event because of a possible subsequent insulin effect, previously explained.

Events Lasting Between 1 and 2 Hours

In the case of athletic events involving large muscle groups at very high work rates for periods exceeding an hour or so, the available evidence suggests that it would be advisable for the competitor to ingest ample quantities of carbohydrates several days preceding the events in order to fill the muscle glycogen depots (Fig. 12-12). The individual should avoid heavy muscular activity which might deplete the existing glycogen depots prior to the event. On the other hand, it is not advisable to live on a high carbohydrate diet regularly, since this would condition the metabolic processes to a high utilization of carbohydrate fuel, rather than FFA, as already explained (see Durnin, 1982; Wilmore and Freund, 1984).

Events Lasting for Several Hours

In the case of very prolonged physical exertions lasting 3 to 4 hours or more, it might also be an advantage to ingest ample quantities of carbohydrates several days preceding the event, as mentioned above. Here, however, it is more important, relatively speaking, to ingest carbohydrates during the actual event in order to supplement the hepatic glucose output.

If the period of exertion during the event is very long, it might be advantageous to consume moderate amounts of sugar, preferably in the form of a flavored glucose solution, just before warm-up. During a 50-km race, cross-country skiers may ingest as much as 1 liter of various sugar solutions corresponding to about 50 to 400 g sugar, divided into seven to eight portions, taken at the various control posts some 5 to 6 km apart. The ideal arrangement would be to issue the sugar solution just before a slack downhill slope so that the skier could drink the solution while gliding downhill. The sweet sugar solution should be rinsed down with water, which would also serve to replace some of the fluid loss. In view of the fact that the sugar solution is rather concentrated, it is important that the competitor become accustomed to it during training in order to avoid any unpleasant reaction during the event itself. This regimen has been well tried and has survived the test of time. It may therefore be recommended for other competitive events such as long-distance bicycle racing, walking, and marathon running.

It should be emphasized that the rate of absorption of glucose, water, and various minerals in the gastrointestinal tract is not affected by exercise, at least not by loads demanding less than about 70 percent of the maximal oxygen uptake. In Fordtran and Saltin's experiments (1967), at least 50 g glucose was emptied from the stomach during 1 hr of heavy exercise. This amount corresponded to one-fourth to one-half the carbohydrates required by the body during this period. The data of Fordtran and Saltin (1967) suggest that gastric emptying and intestinal absorption of saline solutions could be rapid enough to replace all the losses of sweat incurred during heavy exercise, even in hot environments. An addition of glucose to water may cause inhibition of gastric emptying, and high concentration of sugar in orally ingested water may cause large amounts of fluid to be retained in the stomach. This may produce abdominal discomfort during exercise. In our experience with cross-country skiers, we have noticed that some of them prefer a weak (about 10 percent) glucose solution, whereas others tolerate and really prefer a 30 to 40 percent glucose solution.

With the exception of the days preceding the event and on the day itself, the athlete should consume a regular, well-balanced diet. The greatly increased energy expenditure automatically increases the appetite, with the result that the competitor ingests more food. If the diet is well balanced to start with, the increased intake will also automatically cover the athlete's increased requirements for protein, vitamins, and minerals. It has been shown that during a 10-day period, maximal oxygen uptake is not affected by a reduction of the protein intake to 4 g $\cdot$ day^{-1} (Rodahl et al., 1962), nor is the performance improved by an increased ingestion of protein, up to 160 g $\cdot$ day^{-1} (Darling et al., 1944). If an extra supply of protein seems advisable in connection with muscle training or "muscle building" during convalescence or during growth, there is no reason to resort to costly preparations, since all the required protein and all the essential amino acids can be obtained from meat, fish, and milk. The old rule of thumb that 1 g of protein $\cdot$ kg body weight^{-1} $\cdot$ day^{-1} covers the demand may well be valid also for the hard-training athlete. However, properly controlled metabolic balance studies to determine the exact protein requirement of the athlete have not as yet been carried out.

With regard to the distribution and the frequency of meals throughout the day, it appears that rather frequent, moderate meals are more effective in yielding maximal performance than fewer, larger meals. Hutchinson (1952) and Mayer and Bullen (1960) recommend that athletes should have at least three meals a day. It is a common experience, however, that athletes in active training often are unable to maintain regular meal schedules, and may therefore incur detrimental effects.

It is well known that vitamin deficiency causes an impairment of the performance capacity, but it takes a very long time to develop such vitamin deficiency in an individual living on a deficient diet (Rodahl, 1960). For this reason, an individual may be without any vitamin intake for a week or so without any detectable detrimental effect on the work capacity. As long as the usual vitamin intake is adequate, additional vitamin intake does not improve the performance (Keys, 1943; Simonson, 1951; Mayer and Bullen, 1960). Particular attention has been focused on the possible beneficial effect of the B-vitamins on performance capacity, but the studies, almost without exception, have been negative. In the case of the water-soluble vitamins, the ingestion of large quantities of vitamin pills is a rather expensive way of increasing the vitamin content of the urine, which serves no useful purpose in the first place.

The iron intake is at present subject to considerable interest because low serum-iron values have been observed in many top athletes. It has been pointed out that inasmuch as the oxygen is transported by hemoglobin, which contains iron, an iron deficiency may cause a reduced capacity for oxygen transport. It is hard to conceive how reduced serum-iron levels may affect the oxygen-transporting system as long as the hemoglobin content of the blood is normal, as it usually is in the athletes in question. However, it is conceivable that the observed reduction in the serum iron may be interpreted as a forerunner for a reduction in the hemoglobin content of the blood. On the basis of the available evidence, it appears that there is no obvious justification for excessive iron intake in the athlete. However, in a recent review of the iron status in regard to sports performance, Clement and Sawchuk (1984) recommend that the iron status of athletes be monitored regularly, including routine examination of serum ferritin and hemoglobin levels, (see also Chap. 10).

In certain sports, such as boxing, wrestling, and horse-racing, athletes at times subject themselves to stringent dietary regimens combined with dehydration in an endeavor to lose weight and thus to obtain a lower weight classification, for which they do not properly qualify. This allows them the advantage of competing with contestants who weigh less than they themselves normally do. Such weight classifications have been established to provide competition on an equitable basis. The violation of these standards by sudden, self-inflicted starvation and dehydration not only defies good sports ethics, but may also perhaps harm the health of the individual concerned. Such practices have been condemned by the American Medical Association (1959, 1967) and the American College of Sports Medicine (1976).

During competitive cross-country orientation, a mean fluid loss of 3.0 liters during the 90-min duration of the competition has been observed (Saltin, 1964). Even greater fluid loss of up to 5 or 6 liters has been observed during ski racing and in bicycle racing in a hot environment. Sweat loss of a similar magnitude has been reported for many industrial operations.

In the army it used to be advocated that during strenuous field maneuvers, the soldier should drink as little as possible in order to reduce the amount of sweating. This advice is in total violation of the physiological fact that sweating is a vital process which, during heavy muscular activity or in the heat, cools the body and prevents overheating. It is now generally recognized that the lost water has to be replaced, preferably at the same rate at which it is lost (Adolph, 1947; see Chap. 13).

As part of the dietary advice for the athlete, emphasis should be placed on the fact that weight loss through dehydration should be avoided. If the training is intense, and especially if it occurs in a hot climate, ample fluid should be taken, even on the day preceding the event. It is advisable to drink water a few hours prior to the event. In the case of prolonged efforts in a hot climate, adequate fluid intake is essential. Apparently the death of an athlete during the bicycle race in Rome in 1960 may have been due, at least in part, to extreme dehydration. A dehydration corresponding to a loss of body water in excess of 1 to 2 percent of the body weight should in any case be avoided (Adolph, 1947; Ladell, 1955; Saltin, 1964).

The sweat is hypotonic compared with the body fluid (Robinson and Robinson, 1954), so that relatively more fluid than salt is lost from the body during sweating. Sweating, therefore, causes an increase in the NaCl concentration in the body. Under these conditions, ingestion of extra NaCl is contraindicated (Ladell, 1955). Only in prolonged activity associated with intense sweating and complete rehydration is the ingestion of additional salt indicated, provided that the daily diet contains sufficient amounts of salt.

It should be kept in mind that whatever the physiological principles for an optimal diet, the practical considerations dictate that the diet has to be acceptable to the individual. If an athlete believes in a food fad or in a miracle pill, the fad or the pill may result in victory, provided, of course, that the diet otherwise is fully adequate.

PHYSICAL ACTIVITY, FOOD INTAKE, AND BODY WEIGHT

Energy Balance

The energy requirement is essentially a question of energy balance or energy intake versus energy expenditure. Any excess intake of food energy over and above the daily need will be stored as fat. The result is weight gain.

The energy requirement is directly proportional to body size and degree of physical activity. As a rough guide, sedentary, middle-aged people need about 150 kJ (35 kcal) per kg body weight per day. This corresponds to about 10.5 MJ (2500 kcal) per day for a 70-kg person.

As long as such a person adheres to his or her general activity pattern in energy balance, there will be only minor fluctuations in energy expenditure. If the individual regularly takes in roughly this amount of energy, the body weight will remain the same. On growing older, however, the person is apt to be less active and therefore to experience a gradual decline in energy expenditure and a decline in the resting metabolic rate. Consequently, if food intake continues to be about the same as before, the person will no longer spend what is taken in. The difference will be deposited as stored energy in the fat depots. Thus, if the person takes in some 1.5 MJ (350 kcal) more than is spent each day (the amount of energy contained in a piece of apple pie), 10.5 MJ (3500 kcal) will be stored in 10 days. This means that the individual will have deposited about half a kg of fat in the body, for there are approximately 29 MJ (7000 kcal) in each kg of stored fat. This uptake, if continued day after day, will result in a gain of about 18 kg in body weight in a year. Simply omitting a single piece of apple pie from the daily diet could have prevented this weight gain. Or, if the person had kept up his or her former level of physical activity, the slice of apple pie could have been enjoyed every day without gain in weight.

It is a common observation that athletes, such as runners and javelin throwers engaged in intense training, need about 200 kJ (50 kcal) per kg body weight per day. Thus, a javelin thrower weighing about 90 kg consumes about 19 MJ (4500 kcal) per day. A 70-kg runner consumes between 12.5 and 14.5 MJ (3000 and 3500 kcal) daily. The higher energy intake, compared with that of most sedentary individuals, is caused by the fact that athletes, owing to their daily 1- to 3-hour training activities, expend at least 4 MJ (1000 kcal) more per day than normal persons. Jogging, for example, involves an energy expenditure of more than 1.7 MJ (400 kcal) per hour.

As mentioned previously, about one-third to one-half the total daily energy expenditure is used to maintain the basal metabolic rate (BMR). This BMR may vary greatly from one individual to another. A BMR 10 percent below or above the normal average is not uncommon. In sedentary individuals with a daily total energy expenditure of about 10.5 MJ (2500 kcal), a difference in BMR of 20 percent may mean a difference of 2.1 MJ (500 kcal) per day in overall energy metabolism merely due to a basic difference in BMR. This may well explain why some persons remain slim although they eat more than some heavy persons who are equally active. It should be emphasized that such small differences in BMR, 10 percent below and 10 percent above the average, are clinically classified as normal and that a BMR 10 percent higher than the average will mean an additional energy expenditure of 375 MJ (90,000 kcal), or the equivalent of more than 10 kg body fat, in a year.

On a short-term basis, on the other hand, only muscular exercise can cause major differences in energy expenditure, in that such exercise easily may cause a marked increase in energy expenditure over the resting value. However, physical activity of a more moderate degree, spread out over the whole day, may also count in the long run. There are great individual differences in an individual's attitude to activity. Some instinctively seek it, whereas others avoid it. Some remain seated while others get up and move about. These are indeed small differences but in the course of the day, they may add up to several hundred kilojoules, for it takes only some 15 steps to expend 4 kJ (1 kcal). It should be pointed out, however, that a great deal of uncertainty is associated in the prediction of changes in body weight from studies of energy intake and energy output. The available methods are not accurate enough when applied over long periods of time.

There is no real basis for the attitude that one should refrain from unnecessary activity that spends extra energy when large segments of the world population are starving. Physical exercise increases the utilization only of fat and carbohydrate, not of protein, which is the critical nutrient for poorly fed people. Therefore, exercise does not deprive the needy people of their essential proteins.

The relationship between food intake, body weight, and physical activity has been systematically studied in laboratory animals. Mayer et al. (1954) exercised mature rats, accustomed to a "caged" sedentary existence, on a treadmill for increasing daily periods. They observed that for moderate exercise of 20- to 60-min duration, there was no corresponding increase in food intake. In fact, there was a significant decline in food intake, and consequently body weight also decreased. When the exercise was extended from 1 to 6 hr, food intake increased linearly with energy expenditure, and body weight was maintained. When the daily exercise was extended beyond 6 hr, the animals lost weight, their food intake decreased, and their appearance deteriorated. It has also been shown that rats that are overfed voluntarily reduce their activity. This, in turn, increases the weight gain, and a vicious circle is created. On the other hand, when fat rats are fed a calorically restricted diet, their spontaneous activity increases as the weight loss progresses. Thus, both slight and exhausting activities do not appear to be directly related to corresponding changes in food intake.

It has been observed that animals whose activities are restricted by confinement in small cages consume more food than they require and therefore accumulate fat (Gasnier and Mayer, 1939; Ingle and Nezamis, 1947). This is the basis for the practice of fattening cattle, pigs, and geese by restricting their activity.

Mayer et al. (1956) made a study in an industrial population in India engaged in a very wide range of physical activity, from tailors and clerks to coolies carrying heavy loads for several hours a day. The diet was quite uniform for each individual and showed little variety within groups and from group to group. It was observed that the sedentary individuals had the highest energy intakes and the highest body weight. The light workers had lower energy intakes, but had a low body weight. The groups engaged in medium-heavy and very heavy work had increasing energy intakes, but their body weights were the same. It thus appears that sedentary individuals, like the animals, are apt to eat more than they need and to become obese. The remarkable thing is that, similar to the rats on light treadmill exercise, the group engaged in light exercise did not eat more than the sedentary workers; they did, in fact, eat less. Essentially similar findings were made by Woo et al. (1982).

Felig et al. (1983), have compared resting metabolic rates in obese and nonobese subjects in the fasting state and after meal ingestion. Their data indicate that absolute energy expenditure is greater in the obese, than in the nonobese, in the fasting as well as in the postprandial state and that no defect in the thermogenetic response to a mixed meal is demonstrable in obesity. They conclude that the achievement of body weight maintenance in obese subjects, therefore, requires a greater energy intake than in nonobese subjects.

Summary   Within a wide range of energy expenditures through various degrees of physical activity, there is an accurate balance between energy output and intake so that the body weight is maintained constant. If the daily activity is very intensive and prolonged, the spontaneous energy intake is often less than the output, with a reduction in body weight as a result. A daily energy expenditure below a threshold level often leads to an energy surplus and consequent obesity. In this case, satiety is not reached until more energy has been taken in than has been expended.

Regulation of Food Intake

The exact mechanism by which persons regulate the amount and kind of food they need is not completely understood. As in many control systems, one has to extrapolate from animal studies (Hamilton, 1965; Mayer and Thomas, 1967). The concept that the hypothalamus contains a feeding center responsible for the urge to eat or the initiation of feeding, as well as a satiety center capable of exerting inhibitory control over this feeding center is now considered to be faulty (Kandel and Schwartz, 1981). It appears that the brain is not organized into discrete centers that control specific functions, but that these functions are performed by neural circuits distributed among several structures in the brain.

Mayer and Thomas (1967) point out that there is some sort of short-term feedback: (1) Information concerning the nutritive value of ingested foods is relayed from gastric sensors to the hypothalamic regulation system by way of neural or humoral pathways, or both. (2) Food intake is also intimately related to the regulation of blood glucose. The information concerning the availability of glucose to the cells mediated via glucose receptors is more important than the actual blood glucose concentration. Hypoglycemia is, however, almost always followed by hunger sensations. (3) The control of heat exchange is connected in a complex manner with the center regulating food intake. An elevation of the body temperature inhibits the sensation of hunger (Andersson, 1967). This may explain the poor appetite experienced during periods of intense physical activity and the poor appetite often noticed in patients with fever. All this information is integrated with myriads of other exteroceptive and interoceptive inputs at a particular moment, determining the balance of appetite and satiety and the initiation, continuance, or termination of the feeding response.

Cumulative errors in this system could, in turn, be corrected through a long-term feedback, or a lipostatic regulation, i.e., an inhibition of food intake whenever sufficient energy is derived from the mobilization of surplus body fat. How the adjustment of feeding is modified by the state of the fat stores is not clear. It is noticed, however, that after a period of weight gain, food intake is often reduced and the body weight becomes stabilized at a new level.

As pointed out by Garrow (1978) at a symposium on energy balance, humans are not rats. He observed, however, that the rat has an astonishing ability to regulate its energy intake over long periods of time, but it applies only if the food available is monotonous, if the rat has been fed *ad libitum* with this diet, and if flavoring agents, especially sweeteners, are excluded. These conditions are not relevant for the human species. For primitive man there was available a relatively small selection of naturally occurring animal and plant foods. Appetite, which may at one time have been a reliable guide to a correctly balanced diet, is now merely a sensation which can be manipulated in many ways by food manufacturers. In experiments, subjects have been over- or underfed by various methods (e.g., by supplement of food by an intragastric tube, or by meals with differing energy density but indistinguishable in taste and volume). The correlation between the state of energy balance and the voluntary energy intake is very low, and the individual's ability to regulate the food intake, at least over a short term, is very poor. For a more complete discussion of body energy balance and food intake, see Le Magnen (1983); and Wood et al. (1985).

"Ideal" Body Weight

The body size and shape are largely determined by the skeletal size, for a certain amount of muscle and other tissues usually go along with a certain amount of bone. Therefore, the "ideal" body weight, the body weight which includes only a minimal amount of body fat, depends largely on the skeletal size. The "ideal" body weight may be modified to some extent by enlargement of the muscles by training, especially such training as weight lifting; a person may therefore be overweight without being obese. However, for all practical purposes the excess weight, or the weight over and above the ideal body weight, represents accumulated body fat.

There are graphs and tables, usually based on height and sex, giving the so-called ideal body weight. One of the best known of these norms is the Metropolitan Life Insurance table (1963). Such norms are, however, rather inaccurate and meant only as a general guide. Thus the need for a relatively simple but meaningful method of assessing body composition, including the amount of adipose tissue, is apparent. There are several methods for a quantiative classification of body build based on a series of anthropometric measurements, including various diameters (e.g., the radial-ulnar diameter), skin-fold thickness, body densitometry, soft tissue radiography, methods using fat-soluble gases, and total body potassium determination (von Döbeln, 1959; Mayer and Thomas, 1967; Wilmore and Behnke, 1969; Björntorp, 1974; Garrow, 1978; Wilmore, 1983). Body composition, particularly in athletes, is a better guide for determining the desirable weight than the standard height-weight-age tables because of the high proportion of muscular content in their total body composition. However, for practical purposes, obesity is most conveniently diagnosed and judged by the application of simple height-weight methods. These measurements are at least quite accurate.

There are observations of animals as well as of humans showing that an increase in physical activity may have a profound effect on the body composition, increasing the protein/fat ratio, but the body weight may remain the same (Larsson, 1967). However, most studies have a weakness concerning the relationship between physical activity and body composition in that they do not include records of food intake. It has been well established that lean body mass and body fat content are altered by both undernutrition and overeating, and that these alterations are greater in magnitude than those produced by changes in physical activity (Forbes, 1985).

As a general guide, the body weight at the age of about twenty is usually not far from the "ideal" body weight for most adults later on in life.

Obesity

If energy intake exceeds energy output, the excess energy will be stored mainly as adipose tissue. If this state of affairs is maintained over a period of time, it will lead to obesity. As Mayer and Thomas (1967) have pointed out, obesity is often the result of too little physical activity rather than of overeating.

Greene (1939) studied more than 200 overweight adult patients in whom the onset of obesity could be traced to a sudden decrease in activity. In a study concerning the relationship of body weight and physical activity in children, Bruch (1940) found that inactivity was characteristic of the majority of 160 obese children examined; 76 percent of the boys and 88 percent of the girls were physically inactive. It has been suggested by Rony (1940) that laziness or a decreased tendency to muscular activity is a primary characteristic of obese subjects, and Bronstein et al. (1942) found that the majority of the obese children they studied spent most of their leisure time in sedentary activities. Similar observations have been made by Simonson (1951), Juel-Nielsen (1953), and others. Other studies (Peckos, 1953; Fry, 1953) indicate that obese children do not have higher average energy intakes than do control children of the same height and age.

In a study by Johnson et al. (1956), energy intake and activity were systematically compared in paired groups of obese and normal-weight school girls. Their findings were that suburban high school girls were generally not very active, but nevertheless there was a marked difference between the groups in that the obese groups were much less active than the nonobese. Generally speaking, the time spent by the obese groups in sports or any other sort of exercise was less than half that spent by the lean girls. Energy intakes were generally larger in the nonobese girls than in the obese, and it was concluded that inactivity was more important than overeating in the development of obesity. It is interesting to note that when these school girls attended summer camp, they all, both obese and nonobese, almost without exception lost weight under a program of enforced strenuous activity in spite of simultaneous increased food intake.

Stefanik et al. (1959), in a summer camp study, found that obese boys had significantly smaller energy intakes both during the school year and at the summer camp than the nonobese controls. Similar observations have been made by Bullen et al. (1964).

Chirico and Stunkard (1960) made a rough estimate of the degree of physical activity of obese and normal subjects of similar occupation and social status. They were asked to wear a pedometer for recording the number of steps throughout the day. The nonobese subjects were about twice as active as the obese ones. Durnin (1967) has also noticed that overweight individuals are less physically active than those who are nonobese.

Mayer and Thomas (1967) remark that comparison of the hunger and satiety pictures in obese and nonobese individuals suggests that abnormalities of satiety may be more prevalent in the obese than abnormalities of hunger. Nisbett (1968) made observations which fit this assumption. He told his subjects, normal as well as overweight, to eat as many sandwiches as they wanted, but a limited number were served at the table, the rest being left in a refrigerator in the same room. He found that the overweight individuals confronted with three sandwiches ate 57 percent more than those confronted with only one sandwich. In contrast, normal and underweight subjects were completely unaffected by the difference between experimental conditions; both groups ate as many sandwiches whether they were initially offered one or three. Obese individuals would habitually eat everything they were served in a typical meal, but they might not necessarily refill from the refrigerator. Schachter et al. (quoted from Nisbett, 1968) found that obese subjects ate no more food after being deprived for several hours than they did after being recently fed, while normal subjects ate much more food after they had been deprived.

Recent studies have distinguished between two types of obesity: One type in which there is an increased number of fat cells (hyperplasia-obesity), and another in which there is a normal number of fat cells, but each fat cell has an increased content of triglycerides (hypertrophy-obesity). (For a review see Björntorp, 1974.) The total number of fat cells in adipose tissue in women is greater than in men. In transectional studies, the cell size seems to increase with an increase in body fat, but levels off at about 30 kg of body fat. The fat cell number, on the other hand, increases with the increase in body fat over the whole range studied (up to 90 kg body fat). Children already obese before the age of one year have a larger number of fat cells than children with a similar degree of obesity of later onset. This supports the hypothesis that during the first year of life, the adipose cells are particularly susceptible to multiplication (hyperplasia) (see Brook, 1972). (There are similar observations from animal experiments indicating that obesity at young age increases the number of fat cells.) According to Salans et al. (1973), there may be a second critical period for such a hyperplasia between nine and thirteen years of age. Otherwise, short-term longitudinal studies and transectional data on obese subjects of different ages, do not indicate that the adipose cell number increases with age and obesity. Therefore, subjects with adult-onset obesity seem to have a normal number of fat cells, but an enlarged fat cell size. When fat is lost there is apparently no loss of fat cells.

Thus, it may be argued that the small fat cells of the reduced obese patients are particularly avid to replenish their fat content, and as such the individual is forever prey to rapid weight gain (Garrow, 1978, p. 244). Björntorp (1974) also emphasizes that patients with hyperplasia obesity of early debut and an increased body cell mass seem to be difficult to treat by ordinary dietary and exercise procedures. When treated with a connventional low-energy diet they seem to fail to lose weight after reaching a certain fat cell size. Obese patients who have lost weight by a restricted energy intake are very prone to regain weight. Therefore, one must try at the earliest stage of life to prevent obesity. In patients with hypertrophy-obesity, the prognosis for weight reduction is better than in the case of hyperplasia-obesity (Rognum and Kindt, 1973). Vinten et al. (1983), have shown that the total number of fat cells in the organism is not changed by training. (For additional reading on the subject of obesity and diet, see Mayer, 1968; Garrow, 1978.) Rognum et al. (1982), observed a 2.7 kg mean body fat loss and a marked reduction in average fat cell size (from 0.34 μg to 0.24 μg) in 12 well-trained men after 5 days of almost continuous, strenuous combat exercise combined with marked energy deficiency. No significant changes were found in the total number of fat cells. The decrease in fat cell size was most pronounced in the gluteal subcutaneous region, followed by the abdominal region, but no significant decrease in fat cell size was observed in samples from the subcutaneous fat tissue of the thigh. This finding may indicate that gluteal fat depots are most important for energy supply.

Summary It appears conclusively that obesity is to a large extent the result of reduced activity with a maintenance of an "old-fashioned" appetite center set for an energy expenditure well above the one typical for a sedentary individual. This is true for children as well as for adults. The reason for their different attitudes toward physical exertion is not clear. When studying very obese individuals on the cycle ergometer, the authors noted that obese subjects complained of fatigue and felt exhausted at low work loads, when the blood lactic acid levels were not significantly elevated and would be easily tolerated by normal individuals. Obese individuals are characterized by their response to external rather than to internal cues in their eating.

It is particularly important to stimulate young individuals to regular physical activity, for such activity will in the long run effectively counteract obesity by keeping the individual within the range where spontaneous energy intake is properly regulated by the energy output. Obesity in infancy may increase the number of fat cells, causing predisposition for subsequent overweight. The treatment of obesity is particularly difficult in patients with an increased number of fat cells.

Slimming Diets

There are numerous slimming diets available, but most of them are not based on physiological principles. It should be emphasized that the most lenient way to reduce weight involves allowing adequate time for the measures to take effect. A weight loss of more than 0.5 to 1.0 kg per week is not to be recommended. The obese person's diet should be critically examined and an attempt made to eliminate about a thousand kilojoules per day, for example, by substituting artificial sweetenings for sugar and low-fat milk for whole milk, and by eliminating butter and all visible fat. Furthermore, a 2-km walk per day would add 400 kJ (100 kcal) to the expenditure, the result being a total net reduction in fatty tissue equivalent to about 1.5 MJ $\cdot$ day^{-1}. A combination of a small dietary restriction and an increased energy expenditure equivalent to a total of some 2 MJ (500 kcal) per day will mean the loss of half a kilo in body weight per week.

It is a common experience that when a slimming diet is instituted, there is often a sudden drop in body weight, followed by a more gradual decline. This initial drop, which may indeed have a gratifying and encouraging effect, may be due largely to loss of body water, which is regained later on. During the first day or two of energy restriction, stored glycogen is mobilized to cover the energy requirement. Since each gram of glycogen is stored together with almost 3 g of water, the mobilized glycogen liberates this water, which is eliminated and may thus account for the initial drop in body weight. When returning to an adequate diet, glycogen is again stored, together with the required amount of water. This, then, would account for the rapid transient weight gain when changing from energy restriction to adequate diet.

It is a mistake to avoid carbohydrates completely, as muscle and nerve cells need them in their metabolism. They are particularly important for anyone who is physically active.

Optimal Supply of Nutrients

With the exception of children during growth, of women during pregnancy, and sometimes of convalescents, the energy intake should not exceed the energy expenditure. The energy requirements vary naturally with the individual's physical activity. On the other hand, the need for most of the nutrients is comparatively independent of the individual's activity level; therefore the less active the individual, the higher is the content of the essential nutrients required per energy unit in order to obtain the desired optimal nutritional level (Wretlind, 1967). In a homogeneous population, a dietary tradition is usually quite similar. Blix (1965) found that a linear relation exists between daily energy supply and supply of many nutrients (protein, calcium, vitamin A, thiamin, iron). This is illustrated in Fig. 12-22. Through the centuries people obtained their choice of food geared to an energy output of 12.5 MJ (3000 kcal) or more, which also gave all the nutrients needed. For many of the nutrients mentioned, the energy intake should actually exceed about 10.5 MJ (2500 kcal) to ensure an adequate supply. In other words, the diet in Sweden, and no doubt in many other countries, seems to be adjusted to persons with an energy requirement of at least 10.5 to 12.5 MJ (2500 to 3000 kcal) (Wretlind, 1967). This diet, however, is not suitable for the large number of "low energy consumers" who actually do exist, as exemplified in Fig. 12-22. This unsuitability may explain the rather common disturbances in the state of health and well-being associated with malnutrition even in countries with plenty of food available. It has been noted that in Sweden, as well as in many other countries, the percentage of energy from protein has remained remarkably constant from the end of the nineteenth century to the present day, or between 11 and 12 percent; the proportion of fat in the energy supply, however, has greatly increased from about 20 to 40 percent. Per capita, the amounts of energy of protein and carbohydrates have decreased, but the amount of fat has increased during this period of time (Wretlind, 1967).

There are two general ways of improving the nutritional conditions of the low energy consumers: (1) Change their food habits so that their diet consists of a higher content of essential nutrients per energy unit than it now does. A dietary habit should be developed so that the requirements of those who consume only 6.3 to 8.4 MJ (1500 to 2000 kcal) per day can be satisfied. Wretlind (1967) has pointed out, as an example, that a 20 percent increase in the iron content of the diet (from 10 to 12 mg $\cdot$ day^{-1}) would reduce the risk of anemia from 25 to 3 percent. (2) Low energy consumers should be stimulated to become high energy consumers by taking part in regular physical activity in one form or another. By an increase in energy output, they can, without the risk of obesity, eat more and automatically get a greater supply of essential nutrients.

Summary An improved diet in a modern society should include the following: (1) The fat content should be between 25 and no more than 35 percent of the energy

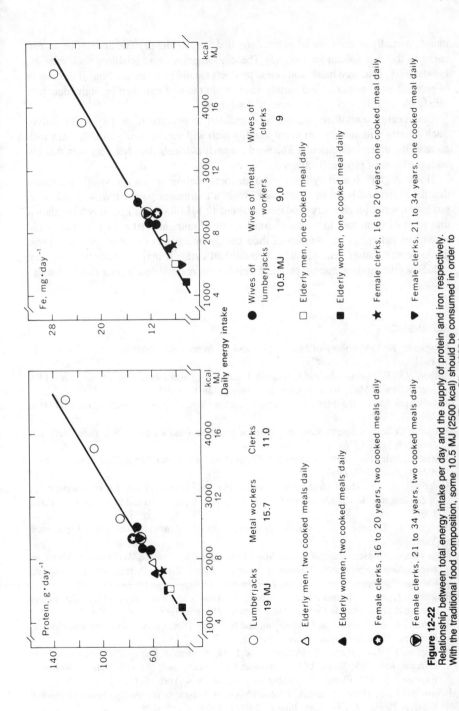

Figure 12-22
Relationship between total energy intake per day and the supply of protein and iron respectively. With the traditional food composition, some 10.5 MJ (2500 kcal) should be consumed in order to ensure a sufficient supply of these and other nutrients. *(From Blix, 1965.)*

○ Lumberjacks 19 MJ Metal workers 15.7 Clerks 11.0

△ Elderly men, two cooked meals daily

▲ Elderly women, two cooked meals daily

✪ Female clerks, 16 to 20 years, two cooked meals daily

◑ Female clerks, 21 to 34 years, two cooked meals daily

● Wives of lumberjacks 10.5 MJ Wives of metal workers 9.0 Wives of clerks 9

□ Elderly men, one cooked meal daily

■ Elderly women, one cooked meal daily

★ Female clerks, 16 to 20 years, one cooked meal daily

▼ Female clerks, 21 to 34 years, one cooked meal daily

intake, partially in the form of polyunsaturated fatty acids. (2) The amount of refined sugar in the diet should be low. (3) The consumption of vegetables, fruit, berries, low-fat milk, fish, lean meat, and cereal products should be relatively high. The content of protein, iron, calcium, and certain vitamins in the diet will then be high enough to satisfy the requirements of low energy consumers.

Low energy consumers should be stimulated to become more physically active. Such activity will allow them to eat more, which will automatically furnish them with more of the essential nutrients. The need for such nutrients does not vary significantly with the level of physical activity.

However, a high activity level and high energy intake do not necessarily guarantee that the individual is on the "safe side" from a nutritional point of view. Studies on girl swimmers with an energy balance of about 17 MJ (4000 kcal) per day have shown that some of them had an intake of some essential nutrients just at or even below the recommended values. Too much of their energy intake was based on sweetened soft drinks, cakes, candy, etc. (Hultén, unpublished data). It may be devastating if they suddenly give up their intense physical training but maintain their poor dietary habits.

REFERENCES

Adolph, E. F.: "Physiology of Man in the Desert," Interscience Publishers, Inc., New York, 1947.

Ahlborg, G., L. Hagenfeldt, and J. Wahren: Influence of Lactate Infusion on Glucose and FFA Metabolism in Man, *Scan. J. Clin. Lab. Invest.*, **36**:193, 1976.

AMA committee on the Medical Aspects of Sports, Wrestling and Weight Control, *JAMA*, **201**:541, 1967.

American College of Sports Medicine: Position Stand on Weight Loss in Wrestlers, *Med. Sci. Sports*, **8**(2):xi, 1976.

American Medical Association (editorial): Crash Diets for Athletes Termed Dangerous, Unfair, *A.M.A. News*, **2**:2, 1959.

Andersson, B.: The Thirst Mechanism as a Link in the Regulation of the "Milieu Intérieur," in "Les Concepts de Claude Bernard sur le Milieu Intérieur," p. 13, Masson et Cie, Paris, 1967.

Baldwin, E.: "Dynamic Aspects of Biochemistry," 5th ed., Cambridge University Press, New York, 1967.

Bergström, J.: Muscle Electrolytes in Man, Determined by Neutron Activation Analysis on Needle Biopsy Specimens, A Study on Normal Subjects, Kidney Patients, and Patients with Chronic Diarrhoea, *Scand. J. Clin. Lab. Invest.*, **14**(Suppl. 68):1962.

Bergström, J., and E. Hultman: Muscle Glycogen Synthesis after Exercise: An Enhancing Factor Localized to the Muscle Cells in Man, *Nature*, **210**:309, 1966.

Bergström, J., L. Hermansen, E. Hultman, and B. Saltin: Diet, Muscle Glycogen and Physical Performance, *Acta Physiol. Scand.*, **71**:140, 1967.

Bergström, J., R. C. Harris, E. Hultman, and L. O. Nordensjö: Energy Rich Phosphagens in Dynamic and Static Work, in B. Pernow and B. Saltin (eds.), "Muscle Metabolism during Exercise," p. 341, Plenum Publishing Corporation, New York, 1971.

Bergström, J., E. Hultman, and A. F. Roch-Norlund: Muscle Glycogen Synthetase in Normal Subjects, *Scand. J. Clin. Lab. Invest.*, **29**:231, 1972.

Björntorp, P.: Effects of Age, Sex and Clinical Conditions on Adipose Tissue Cellularity in Man, *Metabolism,* **23:**1091, 1974.

Blix, G.: A Study on the Relation between Total Calories and Single Nutrients in Swedish Food, *Acta Soc. Med. Upsal.,* **70:**117, 1965.

Bonen, A., S. A. Malcolm, R. D. Kilgour, K. P. MacIntyre, and A. N. Belcastro: Glucose Ingestion Before and During Intense Exercise, *J. Appl. Physiol.,* **50**(4):766, 1981.

Bronstein, I. P., S. Wexler, A. W. Brown, and L. J. Halpern: Obesity in Childhood, *Am. J. Diseases of Children,* **63:**238, 1942.

Brook, C. G. D.: Evidence for a Sensitive Period in Adipose-cell Replication in Man, *Lancet,* **2:**624, 1972.

Bruch, H.: Energy Expenditure of Obese Children, *Am. J. Diseases of Children,* **60:**1082, 1940.

Bullen, B. A., R. B. Reed, and J. Mayer: Physical Activity of Obese and Nonobese Adolescent Girls Appraised by Motion Picture Sampling, *Amer. J. Clin, Nutr.,* **14:**211, 1964.

Calles-Eskandon, J., J. J. Cunningham, P. Snyder, R. Jacob, G. Huszar, J. Loke, and P. Felig: Influence of Exercise on Urea, Creatinine, and 3-methyl Histidine Excretion in Normal Human Subjects, *Am. J. Physiol.* **246:**E 334, 1984.

Cathcart, E. P., and W. A. Burnett: Influence of Muscle Work on Metabolism in Varying Conditions of Diet, *Proc. Roy. Soc. (Biol.),* **99:**405, 1926.

Chauveau, A.: Source et Nature du Potentiel Directement Utilisé dans le Travail Musculaire d'après les Exchanges Respiratoires, chez l'homme en Etat d'Abstinence, *C.R.A. Sci. (Paris),* **122:**1163, 1896.

Chirico, A. M., and A. J. Stunkard: Physical Activity and Human Obesity, *New Engl. J. Med.,* **263:**935, 1960.

Christensen, E. H., and O. Hansen: Arbeitsfähigkeit und Ehrnährung, *Skand, Arch. Physiol.* **81:**160, 1939.

Christophe, J., and J. Mayer: Effect of Exercise on Glucose Uptake in Rats and Men, *J. Appl. Physiol.,* **13:**269, 1958.

Conn, E. E., and P. K. Stumpf: "Outlines of Biochemistry," 3d ed., John Wiley and Sons, New York, 1972.

Clement, D. B., and L. L. Sawchuk: Iron Status and Sports Performance, *Sports Medicine,* **1**(1):65, 1984.

Costill, D. L., W. J. Fink, L. H. Getchell, J. L. Ivy, and F. A. Witzmann: Lipid Metabolism in Skeletal Muscle of Endurance-Trained Males and Females, *J. Appl. Physiol.,* **47**(4):787, 1979.

Costill, D. L., W. M. Sherman, W. J. Fink, C. Maresh, M. Witten, and J. M. Miller: The Role of Dietary Carbohydrates in Muscle Glycogen Synthesis after Strenuous Running, *Am. J. Clin. Nutr.,* **34:**1831, 1981.

Crittenden, R. H.: "Physiological Economy in Nutrition," Frederick A. Stokes Company, New York, 1904.

Darling, R. C., R. E. Johnson, G. C. Pitts, R. C. Consolazio, and P. F. Robinson: Effects of Variations in Dietary Protein on the Physical Well-being of Men Doing Manual Work, *J. Nutr.,* **28:**273, 1944.

Döbeln, W. von: Anthropometric Determination of Fat-free Body Weight, *Acta Med. Scand.,* **165:**37, 1959.

Dohm, G. L., G. J. Kasperels, E. B. Tapscott, and H. A. Barakat: Protein Metabolism during Endurance Exercise, *Fed. Proc.,* **44**(2):348, 1985.

Dole, V. P.: Fat as an Energy Source, in K. Rodahl and B. Issekutz, Jr. (eds.), "Fat as a Tissue," McGraw-Hill Book Company, New York, 1964.

Durnin, J. V. G. A.: Activity Patterns in the Community, *Can. Med. Ass. J.*, **96**:882, 1967.

Durnin, J. V. G. A.: Muscle in Sports Medicine—Nutrition and Muscular Performance, *Int. J. Sports. Med.* **3**:52, 1982.

Eaton, S. B., and M. Konner: Paleolithic Nutrition, *New Engl. J. Med.*, **312**(5):283, 1985.

Essig, D., D. L. Costill, and P. J. Van Handel: Effects of Caffeine Ingestion on Utilization of Muscle Glycogen and Lipid during Leg Ergometer Cycling, *Int. J. Sports Med.*, **1**:86, 1980.

Felig, P., J. Cunningham, M. Levitt, R. Hendler and E. Nadel: Energy Expenditure in Obesity in Fasting and Postprandial state, *Am. J. Physiol.*, **244**(1):E45, 1983.

Fenn, C. E., J. B. Leiper, I. M. Light, and K. O. Maugnan: Effects of the Oral Administration of Fluid, Electrolytes and Substrate on Endurance Capacity in Man, *J. Physiol.* **341**:66P, 1983.

Forbes, G. B.: Body Composition as Affected by Physical Activity and Nutrition, *Fed. Proc.*, **44**(2):343, 1985.

Fordtran, J. S., and B. Saltin: Gastric Emptying and Intestinal Absorption during Prolonged Severe Exercise, *J. Appl. Physiol.*, **23**:331, 1967.

Fredholm, B. B.: Inhibition of Fatty Acid Release from Adipose Tissue by High Arterial Lactate Concentrations, *Acta Physiol. Scand.*, **77**(Suppl. 330):1969.

Fry, R. C.: A Comparative Study of "Obese" Children Selected on the Basis of Fat Pads, *Am. J. Clin. Nutr.*, **1**:453, 1953.

Galbo, H.: Endochrinology and Metabolism in Exercise, *Int. J. Sports Med.*, **2**(4):203, 1981.

Galbo, H.: "Hormonal and Metabolic Adaptation to Exercise," Georg Thieme, Stuttgart, 1983.

Garrow, J. S.: "Energy Balance and Obesity in Man," North-Holland Publishing Company, Amsterdam/American Elsevier Publ. Inc., New York, 2nd. ed. 1978.

Garza, C., N. S. Scrimshaw, and V. R. Young: Human Protein Requirements: the Effect of Variations in Energy Intake within the Maintenance Range, *Am. J. Clin. Nutr.*, **29**:280, 1976.

Gasnier, A., and A. Mayer: Recherches sur la Régulation de la Nutrition. I, Qualités et Côtes des Méchanismes Régulateurs Généraux; III, Méchanismes Régulateurs de la Nutrition et Intensité du Métabolisme, *Ann. Physiol.*, **15**:145, 1939.

Gollnick, P. D.: Metabolism of Substrates: Energy Substrate Metabolism During Exercise and as Modified by Training, *Fed. Proc.*, **44**(2):353, 1985.

Greene, J. A.: Clinical Study of the Etiology of Obesity, *Ann. Intern. Med.*, **12**:1797, 1939.

Grunt, J. A., J. F. Crigler, Jr., D. Slone and J. S. Soeldner: Changes in Serum Insulin, Blood Sugar and Free Fatty Acid Levels Four Hours after Administration of Human Growth Hormone to Fasting Children with Short Stature, *Yale J. Biol. and Med.*, **40**:68, 1967.

Hallberg, D., B. Hallgren, O. Schuberth, and A. Wretlind: Parenteral Nutritional Goals and Achievements, Part II, *Nutritional Support Services*, **2**:35, 1982.

Hamilton, C. L.: Control of Food Intake, in W. S. Yamamoto and J. R. Brobeck (eds.), "Physiological Controls and Regulations," p. 274, W. B. Saunders Company, Philadelphia, 1965.

Hartley, L. H.: Growth Hormone and Catecholamine Response to Exercise in Relation to Physical Training, *Med. Sci. Sports*, **7**:34, 1975.

Hedman, R.: The Available Glycogen in Man and the Connection between Rate of Oxygen Intake and Carbohydrate Usage, *Acta Physiol. Scand.*, **40**:305, 1957.

Hermansen, L., E. Hultman, and B. Saltin: Muscle Glycogen during Prolonged Severe Exercise, *Acta Physiol. Scand.*, **71**:129, 1967.

Hultman, E.: Studies on Muscle Metabolism of Glycogen and Active Phosphate in Man with Special Reference to Exercise and Diet, *Scand. J. Clin. Lab. Invest.*, **19**(Suppl. 94): 1967.

Hultman, E., and L. Nilsson: Liver Glycogen as Glucose-supplying Source during Exercise, in J. Keul (ed.): "Limiting Factors of Physical Performance," p. 179, Georg Thieme, Stuttgart, 1973.

Hutchinson, R. C.: Meal Habits and Their Effects on Performance, *Nutr. Abstr. Rev.*, **22:**283, 1952.

Ingle, D. J., and J. E. Nezamis: The Effect of Insulin on the Tolerance of Normal Male Rats to the Overfeeding of a High Carbohydrate Diet, *Endocrinology*, **40:**353, 1947.

Issekutz, B., Jr.: Effect of Exercise on the Metabolism of Plasma Free Fatty Acids, in K. Rodahl and B. Issekutz, Jr. (eds.), "Fat as a Tissue," chap. 11, McGraw-Hill Book Company, New York, 1964.

Issekutz, B., Jr.: Effects of Glucose Infusion on Hepatic and Muscle Glycogenolysis in Exercising Dogs, *Am. J. Physiol.*, **240**(5):E451, 1981.

Issekutz, B., Jr., and H. Miller: Plasma Free Fatty Acids during Exercise and the Effect of Lactic Acid, *Proc. Soc. Exp. Biol. Med.*, **110:**237, 1962.

Issekutz, B., Jr., N. C. Birkhead, and K. Rodahl: Effect of Diet on Work Metabolism, *J. Nutr.*, **79:**109, 1963.

Issekutz, B., Jr., H. I. Miller, P. Paul, and K. Rodahl: Source of Fat Oxidation in Exercising Dogs, *Am. J. Phys.*, **207:**583, 1964.

Issekutz, B., Jr., H. I. Miller, P. Paul, and K. Rodahl: Aerobic Work Capacity and Plasma FFA Turnover, *J. Appl. Physiol.*, **20:**293, 1965.

Issekutz, B., Jr., and M. Allen: Effect of Metylprednisolone on Carbohydrate Metabolism of Exercising Dogs, *J. Appl. Physiol.*, **31:**813, 1971.

Issekutz, B., Jr., W. A. S. Shaw, and T. B. Issekutz: Effect of Lactate on FFA and Glycerol Turnover in Resting and Exercising Dogs, *J. Appl. Physiol.*, **39**(3):349, 1975.

Johnson, M. L., B. S. Burke, and J. Mayer: Relative Importance of Inactivity and Overeating in the Energy Balance of Obese High School Girls, *Am. J. Clin. Nutr.*, **4:**37, 1956.

Juel-Nielsen, N.: On Psychogenic Obesity in Children, *Acta Paediat. (Stockholm)*, **42:**130, 1953.

Kandel, E. R., and J. H. Schwartz: "Principles of Neural Science," Edward Arnold, Ltd., London, 1981.

Keys, A.: Physical Performance in Relation to Diet, *Fed. Proc.*, **2:**164, 1943.

Krogh, A., and J. Lindhard: Relative Value of Fat and Carbohydrate as Source of Muscular Energy, *Biochem. J.*, **14:**290, 1920.

Ladell, W. S. S.: Effects of Water and Salt Intake upon Performance of Men Working in Hot and Humid Environments, *J. Physiol. (London)*, **127:**11, 1955.

Larsson, S.: Diet, Exercise, and Body Composition, in G. Blix (ed.), "Nutrition and Physical Activity," p. 132, Almqvist & Wiksell, Uppsala, 1967.

Lehninger, A. L.: "Bioenergetics," W. A. Benjamin, Menlo Park, Calif., 1973.

Lehninger, A. L.: "Principles of Biochemistry," Worth Publisher, New York, 1982.

Le Magnen, J.: Body Energy Balance and Food Intake: A Neuroendocrine Regulatory Mechanism, *Physiol. Rev.*, **63**(1):314, 1983.

Mayer, J.: "Overweight Causes, Cost and Control," Prentice-Hall, Inc., Englewood Cliffs, N.J., 1968.

Mayer, J., N. B. Marshall, J. J. Vitale, J. H. Christensen, M. B. Mashayekhi, and F. J. Stare: Exercise, Food Intake and Body Weight in Normal Rats and Genetically Obese Adult Mice, *Am. J. Physiol.*, **177:**544, 1954.

Mayer, J., P. Roy, and K. P. Mitra: Relation between Caloric Intake, Body Weight and Physical Work in an Industrial Male Population in West Bengal, *Am. J. Clin. Nutr.*, **4:**169, 1956.

Mayer, J., and B. Bullen: Nutrition and Athletic Performance, *Physiol. Rev.*, **40:**369, 1960.

Mayer, J., and D. W. Thomas: Regulation of Food Intake and Obesity, *Science,* **156:**328, 1967.

Metropolitan Life Insurance Company: "How to Control Your Diet," p. 4, 1963.

McGilvery, R. W.: The Use of Fuels for Muscular Work, in H. Howald and J. R. Poortmans (eds.), "Metabolic Adaptation to Prolonged Physical Exercise," p. 12, Birkhäuser Verlag, Basel, 1975.

Miller, H. I., B. Issekutz, Jr., P. Paul, and K. Rodahl: Effect of Lactic Acid on Plasma Free Fatty Acids in Pancreatectomized Dogs, *Am. J. Physiol.,* **207:**1226, 1964.

Needham, D. M.: "Machina Carnis. The Biochemistry of Muscular Contraction in Its Historical Development," Cambridge University Press, Cambridge, 1971.

Newsholme, E. A. and A. R. Leech: "Biochemistry for the Medical Sciences," John Wiley & Sons, Chichester, 1983.

Nisbett, R. E.: Determinants of Food Intake in Obesity, *Science,* **159:**1254, 1968.

Owen, O. E., A. P. Morgan, H. G. Kemp, J. M. Sullivan, M. G. Herrera, and G. F. Cahill, Jr.: Brain Metabolism during Fasting, *J. Clin. Invest.,* **46:**1589, 1967.

Paul, P.: Effects of Long Lasting Physical Exercise and Training on Lipid Metabolism in H. Howald and J. R. Poortmans (eds.), "Metabolic Adaptation to Prolonged Physical Exercise," p. 156, Birkhäuser Verlag, Basel, 1975.

Peckos, P. S.: Caloric Intake in Relation to Physique in Children, *Science,* **117:**631, 1953.

Pettenkofer, M. von, and C. Voit: Üntersuchungen über dem Stoffverbrauch des normalen Menschen, *Z. Biol.,* **2:**459, 1866.

Pirnay, F., M. Lacroix, F. Mosora, A. Luyckx, and P. Lefebre: Glucose Oxidation during Prolonged Exercise Evaluated with Naturally Labeled ^{13}C Glucose, *J. Appl. Physiol.,* **43:**258, 1977.

Pruett, E. D. R.: "Fat and Carbohydrate Metabolism in Exercise and Recovery, and Its Dependence upon Work Load Severity," Institute of Work Physiology,, Oslo, 1971.

Rahkila, P., J. Soimajärvi, E. Karvinen, and V. Vittko: Lipid Metabolism During Exercise: II Respiratory Exchange, Ratio and Muscle Glycogen Content During 4 hr Bicycle Ergometer in Two Groups of Healthy Men, *Eur. J. Appl. Physiol.,* **44:**245, 1980.

Refsum, H. E., L. R. Gjessing, and S. B. Strømme: Changes in Plasma Amino Acid Distribution and Urine Amino Acids Excretion During Prolonged Heavy Exercise, *Scand. J. Clin. Lab. Invest.,* **39:**407, 1979.

Reichard, G. A., B. Issekutz, Jr., P. Kimbel, R. C. Putnam, N. J. Hochella, and S. Weinhouse: Blood Glucose Metabolism in Man during Muscular Work, *J. Appl. Physiol.,* **16:**1001, 1961.

Robinson, S., and A. H. Robinson: Chemical Composition of Sweat, *Physiol. Rev.,* **34:**202, 1954.

Rodahl, K.: "North," Harper & Bros., New York, 1953.

Rodahl, K.: "Nutritional Requirements under Arctic Conditions," Norsk Polarinstitutts Skrifter no. 118, Oslo University Press, 1960.

Rodahl, K., S. M. Horvath, N. C. Birkhead, and B. Issekutz, Jr.,: Effects of Dietary Protein on Physical Work Capacity during Severe Cold Stress, *J. Appl. Physiol.,* **17:**763, 1962.

Rodahl, K., H. I. Miller, and B. Issekutz, Jr.,: Plasma Free Fatty Acids in Exercise, *J. Appl. Physiol.,* **19:**489, 1964.

Rodahl, K., and B. Issekutz, Jr.,: Nutritional Effects on Human Performance in the Cold, in L. Vaughan (ed.), "Nutritional Requirements for Survival in the Cold and at Altitude," p. 7, Arctic Aeromedical Laboratory, Fort Wainwright, Alaska, 1965.

Rognum, T. O., and E. Kindt: Fettcellestörrelse hos 14 kvinnelige overvektige pasienter, *T. Norske Laegeforen.,* **93:**1737, 1973.

Rognum, T. O., K. Rodahl, and P. K. Opstad: Regional Differences in the Lipolytic Response of the Subcutaneous Fat Depots to Prolonged Exercise and Severe Energy Deficiency, *Eur. J. Appl. Physiol.,* **49:**401, 1982.

Rony, H. R.: "Obesity and Leanness," Lea & Febiger, Philadelphia, 1940.

Salans, L. B., S. W. Cushman, and R. E. Weismann: Studies of Human Adipose Tissue, Adipose Cell Size and Number in Non-obese and Obese Patients, *J. Clin. Invest.,* **52:**929, 1973.

Saltin, B.: Aerobic Work Capacity and Circulation at Exercise in Man: With Special Reference to the Effect of Prolonged Exercise and/or Heat Exposure, *Acta Physiol. Scand.,* **62**(Suppl. 230):1964.

Saltin, B., and L. Hermansen: Glycogen Stores and Prolonged Severe Exercise, in G. Blix (ed.): "Nutrition and Physical Activity", p. 32, Almqvist & Wiksell, Uppsala, 1967.

Saltin, B., and P. D. Gollnick: Skeletal Muscle Adaptability: Significance for Metabolism and Performance, Handbook of Physiology, chap. 19, American Physiological Society, Washington,, D.C., 1983.

Schürch, P. M., J. Radimsky, R. Iffland, and W. Hollman: The Influence of Moderate Prolonged Exercise and a Low Carbohydrate Diet on Ethanol Elimination and on Metabolism, *Eur. J. Appl. Physiol.,* **48:**407, 1982.

Shreeve, W. W., N. Baker, M. Miller, R. A. Shipley, G. E. Ingefy, and J. W. Craig: C^{14} Studies in Carbohydrate Metabolism: Oxidation of Glucose in Diabetic Human Subjects, *Metabolism,* **5:**22, 1956.

Simonson, E.: Influence of Nutrition on Work Performance, Nutrition Fronts in Public Health, Nutrition Symposium Series no. 3, p. 72, National Vitamin Foundation, New York, 1951.

Staff, P. H., and S. Nilsson: Fluid and Glucose Ingestion during Prolonged Severe Physical Activity, *Tidsskrift for Den Norske Laegeforening,* **16:**1235, 1971.

Stefanik, P. A., F. P. Heald, Jr., and J. Mayer: Caloric Intake in Relation to Energy Output of Obese and Non-obese Adolescent Boys, *Am. J. Clin. Nutr.,* **7:**55, 1959.

Van Handel, P. J., W. J. Fink, G. Branam, and D. L. Costill: Fate of ^{14}C Glucose Ingested During Prolonged Exercise, *Int. J. Sports Med.,* **1:**127, 1980.

Vinten, J., and H. Galbo: Effect of Physical Training on Transport and Metabolism of Glucose in Adipocytes, *Am. J. Physiol.,* **244**(2):E129, 1983.

Wahren, J., P. Felig. L. Hagenfeldt, R. Hendler, and G. Ahlborg: Splanchnic and Leg Metabolism of Glucose, Free Fatty Acids and Amino Acids during Prolonged Exercise in Man, in H. Howald and J. R. Poortmans (eds.), "Metabolic Adaptation to Prolonged Physical Exercise," p. 144, Birkhäuser Verlag, Basel, 1975.

Weis-Fogh, T.: Metabolism and Weight Economy in Migrating Animals, Particularly Birds and Insects, in G. Blix (ed.), "Nutrition and Physical Activity," p. 84, Almqvist & Wiksell, Uppsala, 1967.

Wilmore, J. H.: Body Composition in Sport and Exercise: Directions for Future Research, *Med. Sci. Sports and Exercise,* **15:**21, 1983.

Wilmore, J. H., and A. Behnke: An Anthropometric Estimation of Body Density and Lean Body Weight in Young Men, *J. Appl. Physiol.,* **27:**25, 1969.

Wilmore, J. H., and B. J. Freund: Nutritional Enhancement of Athletic Performance, *Nutr. Abstract and Reviews in Clin. Nutr.* Series A, **54:**1, 1984.

Wirth, A., C. Diehm, H. Mayer, H. Mörl, I. Vogel, P. Björntorp, and G. Schlierf: Plasma C-peptide and Insulin in Trained and Untrained Subjects, *J. Appl. Physiol.,* **50**(1):71, 1981.

Wolfe, R. R., M. H. Wolfe, E. R. Nadel, and J. H. F. Shaw: Isotopic Determination of Amino Acid-urea Interactions in Exercise in Humans, *J. Appl. Physiol.: Respirat. Environ., Exercise*

Woo, R., Y. S. Garrow, F. X. Pi-Sunyer: Effect of Exercise on Spontaneous Calorie Intake in Obesity, *Am. J. Clin. Nutr.*, **36**:470, 1982.

Wood, P. D., R. B. Terry, and W. L. Haskell: Metabolism of Substrates: Diet, Lipoprotein Metabolism, and Exercise, *Fed. Proc.*, **44**(2):358, 1985.

Wretlind, A.: Nutrition Problems in Healthy Adults with Low Activity and Low Caloric Consumption, in G. Blix (ed.), "Nutrition and Physical Activity," p. 114, Almqvist & Wiksell, Uppsala, 1967.

CHAPTER 13

TEMPERATURE REGULATION

CONTENTS

HEAT BALANCE

METHODS OF ASSESSING HEAT BALANCE

MAGNITUDE OF METABOLIC RATE

EFFECT OF CLIMATE

    Cold

    Heat

EFFECT OF EXERCISE

TEMPERATURE REGULATION

ACCLIMATIZATION

    Heat

    Cold

LIMITS OF TOLERANCE

    Normal Climate

    Cold Injury

    Hypothermia

    Failure to Tolerate Heat

    Upper Limit of Temperature Tolerance

    Age

    Sex

    State of Training

COORDINATED MOVEMENTS

MENTAL WORK CAPACITY

WATER BALANCE
Normal Water Loss
Thirst
Water Deficit
PRACTICAL APPLICATION
Physical Work
Warm-Up
Radiation
Air Motion
Clothing
Microclimate

The protected human being may well tolerate variations in environmental temperature between −50°C and +100°C. But a person can tolerate a variation of only about 4°C in deep body temperature without impairment of optimal physical and mental performance. Changes in body temperature affect cellular structures, enzyme systems, and numerous temperature-dependent chemical reactions and physical processes that take place in the body. The maximal limits which the living cell can tolerate range from about −1°C at one end of the scale, when the ice crystals formed during freezing break the cell apart, to thermal heat coagulation of vital proteins in the cell at about 45°C at the other end of the scale. Only for shorter periods of time can they tolerate an internal temperature exceeding 41°C. In fact, many animals, including humans, live their entire lives only a few degrees removed from their thermal death point.

The hot end of the scale is more problematic than the cold end, for people can protect themselves more easily against overcooling than overheating. Consequently, the controlling mechanism for temperature regulation is particularly geared to protect the body tissues against overheating (Hardy, 1967).

HEAT BALANCE

If the heat content of the body is to remain constant, heat production and heat gain must equal heat loss, according to the equation: $M \pm R \pm C - E = 0$ where M = metabolic heat production, R = radiant heat exchange (positive if the environment is hotter than the skin temperature, but negative if the temperature of the environment is lower than that of the skin), C = convective heat exchange (positive if the air temperature is higher than that of the skin, negative if the reverse), E = evaporative heat loss. This equation is valid only for conditions when the body temperature is constant. If the body temperature varies, a correction must be introduced, and the following equation is applicable: $M \pm S \pm R \pm C - E = 0$, where S = storage of body heat (Winslow et al., 1939). S is positive if the body heat content is falling, negative if the heat content increases. The specific heat of most tissues is about 0.83. Conductive heat exchange (K) is in most conditions negligible but increases in importance during such activities as swimming, since water has a heat-removing capacity that is some 20 times that of air.

One very important function of the blood circulation is to transport heat: to cool or to heat various tissues as may be needed, and to carry excess body heat from the interior of the body to the body surface; i.e., skin. The blood is very effective in this function, for it has a high heat capacity (0.9), which means that the blood may carry a great deal of heat with only a moderate increase in temperature. Conductance of the tissue (kJ · m^{-2} · hr^{-1} · °C^{-1}) is the amount of heat given off per square meter body surface per hour and per degree temperature difference between the interior of the body and its surroundings. When the skin blood flow increases, there is a rise in skin temperature, and the conductance increases. When the skin blood flow is reduced, there is a drop in skin temperature, and the conductance is reduced, i.e., the insulating value of the skin is increased.

This control of body temperature, the balance between overcooling and overheating, is the role of temperature regulation. This regulation endeavors to keep the temperature of certain tissues, such as the brain, heart, and guts, relatively constant. Within the body, the temperature is by no means uniform. The greatest gradient is found between the "shell" (the skin) and the "core" (deep central areas including heart, lungs, abdominal organs, and brain). The temperature of the core may be as much as 20°C higher than that of the shell, but the ideal difference between shell and core is about 4°C at rest. Even within the core the temperature varies from one place to another. This complicates the calculation of the heat content of the body and makes it difficult to study temperature regulation. Evidently the term *body temperature* is a misnomer. The maintenance of a normal body temperature is actually quite compatible with considerable gains or losses of heat. The problem is, which temperatures are being regulated.

METHODS OF ASSESSING HEAT BALANCE

Measurements of the *deep body temperature* (core temperature) may be accomplished with the aid of mercury thermometers, thermocouples, or thermistors. The classical site of measurement is the rectum. Since the temperature in the rectum (T_r) varies with distance from the anus, it is customarily measured at a depth of 5 to 8 cm. This rectal temperature is, in a resting individual, slightly higher than the temperature of the arterial blood; it is about the same as liver temperature, but slightly lower (0.2 to 0.5°C) than the part of the brain where the thermal regulatory center is located. During physical exertion or exposure to heat, the temperature of this part of the brain increases more rapidly than does the rectal temperature, and the time interval until a new temperature equilibrium is established has been found to be about 30 min (Nielsen, 1938). The temperature increase or decline in the brain and in the rectum is of the same magnitude, however. The rectal temperature is therefore a representative indicator for the purpose of assessing changes in the deep body temperature, provided the measurement is made under steady-state conditions, i.e., after some 30 to 40 min. It has been found that the eardrum temperature is a fairly good indication of the actual brain temperature. This may be obtained by placing a thermocouple, introduced through the ear, against the eardrum (Benzinger and Taylor, 1963). However, the eardrum temperature is not quite identical with the temperature in the thermoregulatory center. Furthermore, it varies with the ambient air temperature (Houdas and Ring, 1982). Another alternative is to measure the temperature in the esophagus, which is relatively accessible for such measurements. Although the temperature of the esophagus is not identical with any of the above-mentioned core temperatures, it generally changes parallel to those temperatures (Saltin and Hermansen, 1966; Nielsen, 1969). In exercise and heat studies, the oral temperature measurement has its limitations (Strydom et al., 1966), but otherwise may be used as a practical index of deep body temperature. Mairiaux et al., (1983) have shown that sublingual oral temperature variations were highly correlated with esophagal temperature variations under steady state conditions and under air temperature variations; under these conditions oral temperature represented a better estimate of esophagal temperature than did rectal temperature. The reverse was the case under work-rate variations, however.

The skin temperature is measured with a radiometer, or by placing thermocouples or thermistors on the skin at certain locations. For clinical purposes, infrared thermography may be used, measuring the infrared radiation emitted by the skin surface. (For further details see Houdas and Ring, 1982.) The mean skin temperature ($\overline{T}_S$) is calculated by assigning certain factors to each of the measurements in proportion to the fraction of the body's total surface area represented by each specific area, as follows (Hardy and Dubois, 1938):

Head	0.07
Arms	0.14
Hands	0.05
Feet	0.07
Legs	0.13
Thighs	0.19
Trunk	0.35
	1.00

A simple weighing formula for the computation of the mean skin temperature from observations of four areas of the body has been developed (Ramanathan, 1964). The formula, which may give values identical to the elaborate Hardy-Dubois formula, is:

$$\overline{T}_s = 0.3\, T_{chest} + 0.3\, T_{arm} + 0.2\, T_{thigh} + 0.2\, T_{leg}$$

Furthermore, Ramanathan (1964) confirms Teichner's previous observation (Teichner, 1958) that the temperature on the surface of the medial thigh is a satisfactory measure of the mean skin temperature under most conditions. In fact, the mean difference of the temperature values for the medial thigh temperature and the mean skin temperature calculated according to the Hardy-Dubois formula in 39 observations, was only 0.17°C. It should be pointed out, however, that this may not be the case in a very cold environment.

For the calculation of the heat content of the body, the following equation may be applied:

$$\text{Heat content} = 0.83\ W\ (0.65T_r + 0.35\ \overline{T}_S)$$

where W represents body weight, 0.83 is the specific heat of the body, and 0.65 and 0.35 are the factors assigned to the rectal and the mean skin temperatures, respectively (Burton, 1935). The specific heat of the body has been assumed to be 0.83 for nearly a century, but it varies with the individual's body composition and may range from 0.70 up to 0.85 (see Hardy et al., 1970, p. 345.) It is also clear that the heat content of the body during exercise in a hot environment cannot be determined from any fixed ratio of $\overline{T}_S$ and T_r (see Wyndham, 1973.)

The *metabolic rate,* or the magnitude of heat production, is assessed by the measurement of oxygen uptake. The volume of 1 liter oxygen consumed corresponds to approximately 20 kJ (4.8 kcal). Human calorimeters, large enough to measure the rate of heat production by direct calorimetry, have been constructed and are in use in certain laboratories.

The *evaporative heat loss (E)* plays a major role in the cooling of the skin and the blood during exposure to heat. At normal skin temperature, the evaporation of 1 liter of sweat requires 2.4 MJ (580 kcal). The magnitude of the sweat loss may be estimated simply by weighing the subject nude or dressed in dry clothing, before and after the experiment, and by weighing food and fluid ingested and stools and urine voided during the period of observation. Furthermore, weight loss due to respiratory gas exchange should be included in the calculation, and may be accounted for as follows (Snellen, 1966):

$$C_{g_e} = \dot{V}_{O_2}\ (1.977 \cdot R - 1.429)$$

where C_{g_e} = weight loss in g $\cdot$ min^{-1} due to respiratory gas exchange

$\dot{V}_{O_2}$ = oxygen uptake in liters $\cdot$ min^{-1} STPD

R = respiratory quotient

1.977 = weight in g of 1 liter CO_2 STPD

1.429 = weight in g of 1 liter O_2 STPD

Such measurements are valid for the calculation of heat balance only as long as all the sweat produced during the experiment is actually evaporated. On the other hand, whether the produced sweat is evaporated or part of it has run off the body, the sweat rate is an indication of the magnitude of the heat stress.

The *air temperature* affects convective heat loss or gain *(C)* and is most conveniently measured with the aid of the usual mercury thermometer. If the thermometer is exposed to radiation, it should be shielded. A piece of tinfoil may be used, but care should be taken to allow free air passage around the thermometer.

The *humidity of the air* may be measured with the aid of a sling psychrometer or an electronic device for measuring humidity. The rate of evaporation of the produced sweat is greatly dependent on the humidity of the air.

The *air movement*, which may be measured by a hot-wire animometer, affects both convective heat exchange *(C)* and evaporative heat loss *(E)*.

The *radiant heat exchange (R)* depends on the temperature difference between the individual and the surroundings. This may be assessed by the values obtained from a mercury thermometer placed in a hollow, spheric black copper container with a diameter of 15 cm (a globe thermometer). Because of rapidly changing radiation intensities typical for many industrial operations, it is often almost impossible to obtain a true picture of the intensity of the radiation.

Afferent impulses from thermoreceptors in the skin, which respond to very rapid temperature changes, may signal peripheral thermal disturbances long before the central core temperature has been affected. Such impulses, integrated with various sensory input from the rest of the body, result in feelings of thermal comfort or discomfort (Hardy et al., 1971). More specifically, the sensation of thermal comfort appears to be the result of the interaction between signals evoking temperature sensation, input signals for temperature regulation, and sensations arising from thermoregulatory activities associated with skin blood flow, sweating, and shivering (Hardy et al., 1971). Generally speaking, the state of thermal neutrality or thermal comfort is characterized by core temperature of 36.6 to 37.1°C, and skin temperatures between 32° and 35.5°C (Precht et al., 1973). Apparently, comfort temperature may be subject to some degree of adaptation or accustomization. Thus, according to Burton and Edholm (1969), the preferred indoor temperature in Britain is about 18°C versus 24°C in the United States. However, at least part of the difference may be due to variation in the amount of clothing.

MAGNITUDE OF METABOLIC RATE

Human beings may be considered tropical animals inasmuch as they require an ambient temperature of 28 to 30°C in order to maintain thermal balance at rest, in the nude. That is, without clothing and shelter man could only survive in a narrow zone along the equator. The oxygen uptake under these conditions is about 0.20 to 0.30 liter · min^{-1}. It is slightly higher when the body size is larger. This corresponds to a production of 60 to 90 kcal · hr^{-1}, or 70 to 100 watts. This energy is the by-product of metabolic processes, which are essential for the maintenance of life. This produced heat makes up for the heat lost through convection *(C)*, radiation *(R)*, and evaporation *(E)*. Under these conditions, $C + R$ accounts for about 75 percent of the heat loss, and E accounts for only 25 percent. Heat loss through the lungs, through the saturation of the air with water vapor during respiration, accounts for about two-fifths of E. Not all the rest of E is due to the evaporation of sweat, for part of the water loss through the skin occurs without the involvement of the sweat glands, the so-called insensible perspiration. The total water loss through the skin amounts to a minimum of 0.5 liter · day^{-1}.

Muscular exercise is associated with an increase in metabolic rate. Since the mechanical efficiency (the ratio of external work to the extra energy used) may vary from

0 to 50 percent depending on the kind of exercise, at least 50 percent of the energy used is converted into heat. Well-trained athletes may, during short exercise periods of 5- to 10-min duration, attain an oxygen uptake of up to about 6 liters · min^{-1} (2000 W), and during more prolonged exercise as much as 4 to 5 liters · min^{-1}. The amount of heat thus produced during 1 hr could theoretically increase the body temperature of a 70-kg individual from 37°C to about 60°C if the excess heat were not dissipated. With fever, or during shivering due to cold exposure, the heat production may be increased two- to fourfold.

EFFECT OF CLIMATE

Cold

For a nude resting individual, the ideal ambient temperature is about 28 to 30°C. Under such conditions the mean skin temperature is about 33°C, and the temperature of the core is about 37°C. The temperature gradient from core to skin is then adequate to facilitate the transfer of the excess heat from the metabolically active tissues to the surroundings. Of the total amount of blood (the circulating blood volume) pumped by the heart, about 5 percent flows through the blood vessels of the skin. If the ambient temperature drops, the temperature difference between the skin and the environment is increased; this causes an increased heat loss through convection and radiation. A reduced heat flow to the skin would result in a gradual lowering of the skin temperature. This would produce a reduced temperature gradient between the skin and the environment. Actually, a reduction in the "conductance of the tissue" occurs, partly because of *vasoconstriction of the skin's blood vessels* causing a reduction in blood flow, partly because the blood in the veins of the extremities is deviated from the superficial to the deep veins. This vasoconstriction is most pronounced in the extremity (Vangaard, 1975). Because of the proximity of the deep veins to the arteries, a heat exchange occurs (Fig. 13-1). Owing to this system of countercurrent, in a subject exposed to an ambient temperature of 9°C, the blood leaving the heart will have a temperature of about 37°C. As it is flowing through the arm, it will be gradually cooled so that by the time it reaches the hand, it may have dropped to about 21°C. The returning venous blood absorbs a considerable part of the heat as the blood flows through the

Figure 13-1
Schematic illustration of the two possible ways that venous blood flow from the hand, the superficial veins, or the deeper ones anatomically close to the arteries can make a heat exchange possible.

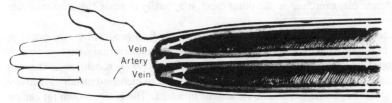

arm. In other words, cooling of arterial blood flowing through the arteries of the limbs depends on the rewarming of cold blood returning in adjacent veins from more distal areas. Thus, a cooling of the body core is prevented (Bazett et al., 1948; Schmidt-Nielsen, 1963). In a hot environment, on the other hand, the blood from the limbs returns primarily through superficial veins, thus facilitating further cooling of the blood.

The major effect of the vasoconstriction of the skin is a sudden displacement of the blood volume from the skin to the central circulation, as evidenced by a sudden rise in central blood volume, and a redistribution of blood from superficial to deep veins (Rowell, 1974).

The heat exchange between arterial and venous blood is still more important for arctic animals. Sea birds swim in the icy sea, and the large surfaces of their bare feet are exposed to the cooling effect of the water. However, little body heat is lost because their feet cool down to the temperature of the water, thus reducing the loss of metabolic heat (Irving, 1967). Warm feet of a gull or a duck standing on snow or ice would cause it to melt, and soon the feet would be frozen solid to the ground where they stood! Hogs, naked as a man in a cold environment, and narwhal, walrus, or seals in arctic waters can prevent the loss of heat by having a very cold skin. These animals have a considerable layer of subcutaneous fatty tissue as an effective insulation when their blood vessels constrict. It is an interesting observation that fats in the peripheral regions have a lower melting point than those in the warmer internal tissues; if they did not, the peripheral tissues and legs would become too inflexible in cold weather (Irving, 1967).

Thickly furred animals can use their bare extremities to release excess heat from the body (heat can also be dissipated by evaporation from the mouth and tongue during panting).

Through peripheral vasoconstriction, a sixfold increase in the insulating capacity of the skin and subcutaneous tissues is possible (Burton, 1963). This vascular constriction is particularly active in the fingers and toes. It has been estimated that the blood flow through the fingers may vary a 100-fold or more, from 0.2 to 120 ml blood $\cdot$ min^{-1} $\cdot$ 100 g^{-1} of tissue (Robinson, 1963). The disadvantage of this vasoconstriction is that the temperature in the peripheral tissues may approach that of the environment (Fig. 13-2). For this reason, one is apt to suffer cold fingers and toes. The blood vessels of the head are far less subject to active vasoconstriction.

Another protective mechanism to maintain heat balance is an *increase of the metabolic rate,* mediated through muscle activity in the form of shivering by a reflex mechanism. Shivering consists of a synchronous activation of practically all muscle groups; antagonists are made to contract against one another. Since the mechanical efficiency of shivering is 0 percent, the heat production is relatively high, and the metabolic rate may increase to 2 to 4 times that of resting metabolic rates. Ordinary dynamic muscular exercise, on the other hand, may easily increase the metabolic rate tenfold or more.

Whether or not a significant increase in metabolic heat production may occur in humans in response to cold exposure by the so-called nonshivering thermogenesis is still an open question. According to Houdas and Ring (1982), nonshivering thermogenesis may play a role in human neonates, possibly due to increased metabolic activity of their stores of brown fat, which are absent in adults. Therefore, brown fat cannot

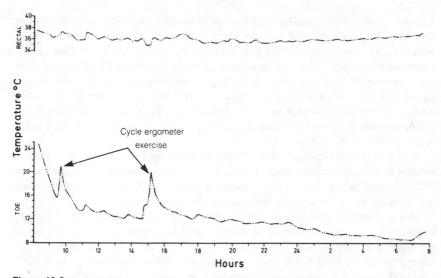

Figure 13-2
Rectal and toe temperatures in a nude subject exposed to 8°C continuously for 3 days. By the end of 24 hours, the skin temperature of the big toe had dropped to about 8°C, the temperature of the ambient air.

play any role in adult thermogenesis. Nor does it appear likely that catecholamines or thyroid hormones play a significant thermogenic role in man exposed to cold under ordinary circumstances, which leaves shivering or physical activity as the only ways of increasing metabolic heat production in the unprotected individual exposed to cold.

In male and female subjects submerged to the neck in swirling water of about 12°C, there was a wide individual variation in shivering intensity. The mean $\dot{V}_{O_2}$ during shivering was 1.39 1 · min⁻¹ (range 0.87 to 2.16). $\dot{V}_{O_2}$ during shivering was significantly correlated with max. $\dot{V}_{O_2}$ during exercise, $\dot{V}_{O_2}$ during maximum shivering being approximately 46 percent of max. $\dot{V}_{O_2}$ (Golden et al., 1979a). A similar value (50 percent) was observed by Golden et al. (1979b) in one swimmer not accustomed to swimming in water at 17.4°C, while shivering only taxed 22 to 26 percent of the max. $\dot{V}_{O_2}$ in two cold-accustomed swimmers, suggesting that habituation may be involved.

Even though the resources for maintaining the core temperature may be quite effective, this maintenance does to some extent take place at the expense of the peripheral tissues, the shell. Local cold injury may be the result in extreme conditions. During prolonged severe cold exposure, even the core temperature may drop. Under certain circumstances, especially when exposed to cold water, obese individuals may be better off in the cold than lean individuals because of the insulating value of the adipose tissue (see Keatinge, 1978; Holmér and Bergh, 1974; Golden et al., 1979a). An unclothed individual of average body build will be helpless from hypothermia after approximately 20 to 30 min in water at 5°C and after 1½ to 2 hrs in water at 15°C. With thick conventional clothing, these times can be substantially prolonged (see

Keatinge, 1969). Keatinge points out that one should not exercise in cold water in an attempt to keep warm; it will have the reverse effect. In one study on swimming in water at 18°C the oxygen uptake was elevated by approximately 0.5 liter · min^{-1} compared with swimming at the same speed in warmer water. For lean subjects there was, however, a significant drop in esophageal temperature. The maximal oxygen uptake and heart rate were markedly reduced. The effect of exposure to moderately cold water for 20 to 30 min will thus seriously impair physical performance and swimming may be hazardous (Holmér and Bergh, 1974; Davies et al., 1975; Golden et al., 1979b; Bergh, 1980).

Vangaard (1975) measured the skin temperature of the hand in a subject suddenly brought from a room temperature of 28°C into a cold chamber with a temperature of 9°C. He found that the drop in skin temperature as a result of the sudden cold exposure was identical to that observed in the same subject when the arterial blood flow to the hand was arrested by an inflated cuff placed around the arm. From this, it appears that sudden cold exposure causes an almost complete shutdown of the circulation to the exposed hand. Evidently every effort is made to maintain the core temperature at the expense of the shell temperature, and especially that of the hands and feet. Since manual dexterity is impaired with falling hand temperature, unprotected exposure to cold may lead to reduced performance, and increased risk of accidents.

An impairment of the performance of serial choice reaction time has been observed in subjects exposed to cold. Wyon (1982) has shown that lowering of the ambient temperature from 24° to 18°C or colder caused significant impairment of manual dexterity. This may be partly explained by changes in nervous conduction velocity. Vangaard (1982) investigated the relationship between local temperature and nervous conduction velocity in a peripheral motor nerve in subjects exposed to a minor cold stress. The decrease in conduction velocity was 15 m · s^{-1} for each 10°C fall in temperature (Fig. 13-3). At a local temperature of 8° to 10°C, a complete nervous block was established. This may explain the common finding that a local cooling in the extremities may be accompanied by a rapid onset of physical impairment. Bergh (1980), found that physical performance is reduced with reducing core temperature. The temperature effect amounted to 4 to 6 percent · °C^{-1} in maximal exercise of less than 3 min duration, and 8 percent · °C^{-1} at 3 to 8 min exercise. Maximal dynamic muscle strength is also reduced with reducing muscle temperature.

Davies and Young (1982) studied the effect of heating and cooling by immersion in water on the electrically evoked mechanical and contractile properties of the triceps sural in five healthy male subjects. An enhanced muscle temperature was associated with a decrease in time to peak tension and half relaxation time, but was without effect on supramaximal twitch and tetanic tension, and maximal voluntary contraction. Cooling increased the time to peak tension. Furthermore, cooling of the leg muscles reduced the peak power output during cycling and jumping markedly. (For a more detailed discussion, see LeBlanc, 1975.)

Exposure of the face to a cold stimulus results in a slowing down of the heart rate and an increased peripheral vasoconstriction (LeBlanc et al., 1978). Similarly, directing cold wind at the faces of subjects during exercise caused a significant lowering of heart rate (Riggs et al., 1981, 1983). The heart rate response to facial cooling during exercise is probably mediated via a central thermoregulatory response.

CHANGES IN NERVE CONDUCTANCE
WITH DECLINE IN TEMPERATURE

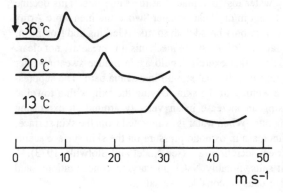

Figure 13-3
Relationship between local temperature and nervous conduction velocity in a peripheral motor nerve in subjects exposed to a minor cold stress *(From Vanguard, 1982.)*

Summary   When a resting person is exposed to a cold environment, there are two main mechanisms by which a lowering of the body temperature can be prevented: (1) a reduction in the peripheral blood flow with a secondary drop in the skin temperature (reducing the heat loss by radiation and convection); and (2) an increase in the metabolic heat production by shivering. In a hypothermic person, the combined effect of reduced efficiency and reduced maximal aerobic power will impair physical performance.

Heat

When the nude, resting individual is exposed to heat (when the ambient temperature exceeds 28°C), or during muscular activity, the heat content of the body tends to increase. Under such conditions, the blood vessels of the skin dilate, venous return in the extremities takes place through superficial veins, and the conductance of the tissue increases. In the comfort zone, the skin blood flow, as mentioned, amounts to about 5 percent of the cardiac minute volume; in extreme heat, it may increase to 20 percent or more. The increased heat flow to the skin increases the skin temperature. If the temperature of the surroundings is lower than that of the skin, heat loss is facilitated through $C + R$. If the heat load is sufficiently large, the sweat glands are activated, and as the produced sweat is evaporated, the skin is cooled. It has been calculated that there are at least 2 million sweat glands in the skin of an average individual. Recruitment of sweat glands from different areas of the body is not consistent between individuals (see Nadel et al., 1971). The activity of individual sweat glands follows a cyclic pattern. The sweat starts to drop off the skin when the sweat intensity has reached about one-third the maximal evaporative capacity (Kerslake, 1963).

The individual difference in the capacity for sweating is considerable; some people have no sweat glands at all. As a person becomes accustomed to heat, the amount of sweat produced in response to a standard heat stress increases. A person may produce several liters of sweat per hour. Workers exposed to intense heat may lose as much as 6 to 7 liters of sweat in the course of the working day. Sweat loss up to 10 to 12 liters in 24 hr has been reported (Leithead and Lind, 1964).

During prolonged exposure to a hot environment, there is a gradual reduction in the sweat rate, even if the body water loss is replaced at the same rate. This decline in sweat rate is greater in humid than in dry heat, greater "when the men wore Army tropical uniforms than when they wore only broadcloth shorts" (Gerking and Robinson, 1946). The explanation of this "fatigue" of the sweat mechanism is presently not clear. Ahlman and Karvonen (1961) report that exercise could again induce sweating after the sweating had ceased during repeated thermal stimuli in a sauna bath. The suppression in sweating is related to the wetting of the skin. Drying the skin with a towel at regular intervals or superimposing an increase in air velocity around an exercising subject will enhance the sweating rate. When water is evaporated from the skin surface, the solutes are left behind. An increase in osmotic pressure on the skin surface seems to produce an increase in the sweat secretion rate (see Nadel and Stolwijk, 1973).

The sweat contains different salts, notably NaCl, in varying concentrations, and excessive sweating may therefore cause a considerable salt loss.

Exposure of resting subjects to direct whole-body heating causes a rise in skin temperature that is accompanied by an almost immediate rise in heart rate and cardiac output, and a drop in total peripheral resistance and splanchnic blood flow. Cardiac output commonly increases 50 to 75 percent (Rowell, 1983). The excess cardiac output is diverted to the skin. This increased skin blood flow is further supplemented by a reduction in splanchnic and renal blood flow. Evidently, muscle blood flow is not increased (Rowell, 1983). As the skin temperature rises, cutaneous resistance vessels relax and cutaneous venous pressure and volume increase until a new level of wall tension is reached. This volume displacement is augmented by splanchnic vasoconstriction, which reduces distending pressure in the splanchnic veins, allowing them to empty passively. In this way, blood may be displaced from central to cutaneous venous beds (Rowell, 1983). The skin blood flow will increase by local heating. This blood flow can be further elevated if the whole body skin temperature and/or core temperature rise. Keeping the local skin temperature high will not abolish skin vasoconstriction response to lower body negative pressure (simulating the effect of a mild hemorrhage by pooling blood in the leg veins). Thus, local factors and reflex influences to a skin area can interact so as to modify the degree but not the pattern of the skin vasomotor response (Johnson et al., 1976).

Exposure to dry, hot air in the form of the Finnish sauna bath has been studied by Eisalo (1956). Both in healthy and in hypertensive subjects, he found that the cardiac output increased by an average of 73 percent (65 percent in the hypertensive group), the mean circulation time decreased by almost 60 percent, and the pulse rate increased by more than 60 percent. There was a slight but significant decrease in systolic blood pressure in healthy subjects, but in the hypertensive subjects it decreased by 29 mm Hg (3.9 kPa) on the average. Twenty minutes after the sauna, the mean decrease was 54 mm Hg (7.2 kPa). The diastolic blood pressure remained practically unchanged in the healthy subjects, while it decreased significantly in the hypertensive subjects. There was a statistically highly significant decrease in peripheral resistance in both groups of subjects.

Summary The resting person exposed to a hot environment experiences (1) a vasodilation in the skin, making an increased heat transfer from "core" to "shell" possible; and perhaps (2) an activation of the sweat glands, with the evaporation of the sweat taking heat from the body and causing an evaporative heat loss.

EFFECT OF EXERCISE

Since the mechanical efficiency of the human body is mostly below 25 percent, more than 75 percent of the total energy utilized is converted into heat. The greater the exercise intensity, the greater the total amount of heat produced. This excess heat has to be removed and dissipated in order to prevent overheating and hyperthermia.

Figure 13-4 shows an example of how the thermal balance is maintained during muscular exercise of different intensity over a 1-hr period. The body temperature increases during work, and this temperature elevation may be interpreted as the result of an active regulation (Christensen, 1931; Nielsen, 1938; Berggren and Christensen, 1950). Apparently, the core temperature rises in order to establish a gradient for heat flow from core to shell, and to stimulate sweating (Nielsen, 1980). The difference between "energy output" and "heat production" in Fig. 13-4 is an expression of the

Figure 13-4
Heat exchange at rest and during increasing exercise intensities (expressed in kilopond meters per minute along the abscissa) in a nude subject at a room temperature of 21°C. Further explanation in text. *(From M. Nielsen, 1938.)*

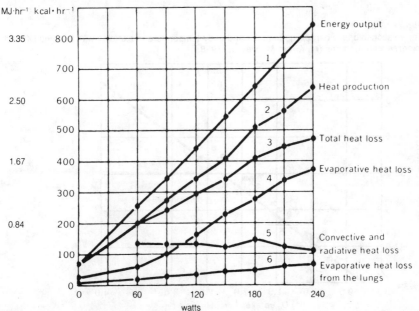

mechanical efficiency (about 23 percent), and the difference between "heat production" and "total heat loss" is a consequence of the elevated body temperature. Note that the convective and radiative heat losses are almost constant despite the large variations in heat production. Evaporation takes care of the extra heat loss as the rate of exercise increases.

Figure 13-5 shows how the rectal temperature, measured after about 45 min of exercise on the cycle ergometer, increases linearly with the O_2 uptake, at least up to an energy demand of about 75 percent of the individual's maximal aerobic power. At higher metabolic rates, the increase in core temperature seems to be curvilinearly related to the oxygen uptake (Davies et al., 1976). This end temperature does not depend on the absolute magnitude of the energy output but on the level of metabolism relative to the individual's maximal aerobic power. A subject with a maximal O_2 uptake of 2.0 liters $\cdot$ min^{-1} attains a body temperature of about 38°C during an exercise load, which demands an O_2 uptake of 1.0 liter $\cdot$ min^{-1}, i.e., 50 percent of the person's maximal aerobic power. A subject with a maximum of 5.0 liters $O_2 \cdot$ min^{-1} may expend $2\frac{1}{2}$ times more energy (O_2 uptake of 2.5 liters $\cdot$ min^{-1}) without the body temperature exceeding 38°C (I. Åstrand, 1960).

Figure 13-6 shows the relationship between muscle, rectal, and esophageal temperatures, measured simultaneously, and the oxygen uptake as a percentage of the individual's maximal oxygen uptake. As expected, the highest temperatures are seen in the exercising muscles where most of the heat is produced.

Nielsen (1938) had a subject perform a certain amount of work, 150 watts, at ambient temperatures varying between 5°C and 36°C. After 30 to 40 min of exercise,

Figure 13-5
The relationship between oxygen uptake and body temperature in exercise with the legs (X) and exercise with the arms (●). *(From M. Nielsen, 1938.)*

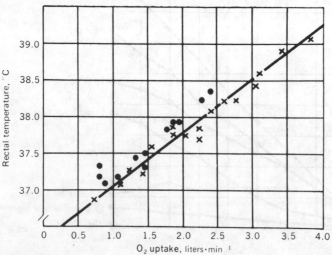

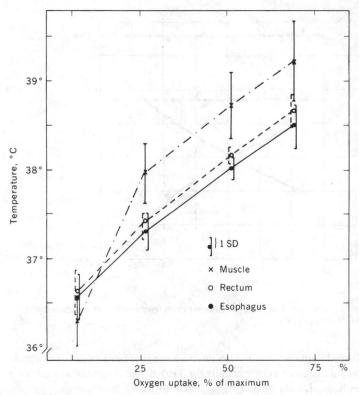

Figure 13-6
Average temperature measured simultaneously in the esophagus, the rectum, and the exercising muscle in relation to the oxygen uptake in percent of the individual's maximal oxygen uptake. Seven subjects were exercising for 60 min on a cycle ergometer. To the left, data obtained at rest. (SD = standard deviation.) *(From Saltin and Hermansen, 1966.)*

the rectal temperature was the same, regardless of the room temperature. Since the mechanical efficiency was constant, the heat dissipation had to be the same in all these experiments. Figure 13-7 shows how radiation and convection *(R + C)* accounted for about 70 percent of the heat loss at the lower ambient temperatures. The skin temperature was then 21°C. In the experiment carried out at the highest ambient temperature, the skin temperature was 35°C, and the body absorbed heat from the environment (i.e., 36°C air temperature). This was completely counteracted by an increasing evaporative heat loss. In the cold, the subject evaporated 150 g sweat; in the heat he evaporated 700 g. The greater variation in the rectal temperature at the end of the work period in all these experiments was only 0.11°C. This difference in body temperature may be brought about merely by evaporating about 11 g sweat. These findings have been confirmed by Nielsen (1969) and Stolwijk et al. (1968).

During maximal exercise the rectal temperature may exceed 40°C and the muscle temperature 41°C without causing any discomfort for the exercising person.

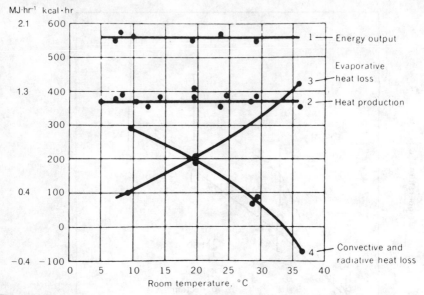

Figure 13-7
Heat exchange during exercise (150 watts) at different room temperatures in a nude subject. *(Modified from M. Nielsen, 1938.)*

During prolonged light exercise in a hot environment, the heart rate rises markedly while cardiac output increases more gradually for 30 to 40 min in spite of a progressive fall in stroke volume. During prolonged moderate to heavy muscular exercise, the heart rate does not reflect changes in cardiac output (CO) or in skin blood flow (SBF). Actually, stroke volume (SV) decreases while cardiac output is maintained by increased heart rate. During graded exercise, submaximal cardiac output is maintained by increased heart rate in spite of reduced stroke volume (Fig. 13-8) (Rowell, 1974, 1983). Suzuki (1980) concludes that any changes in cardio-circulatory function, which limit performance in hot environments, are due to the decreased stroke volume (i.e., heart filling) caused by alterations in the peripheral blood flow. In 20 to 25 min heavy exercise in the heat, the relative vasoconstriction contributes to the maintenance of an adequate stroke volume preventing a fall in cardiac output (Nadel et al., 1979). Thus, in this case, circulatory regulation appears to have precedence over temperature regulation.

The control of peripheral circulation in dynamic exercise in the heat may be complicated by changes in body position and especially by the upright position (see Roberts and Wenger, 1980; Rowell, 1983; Harrison, 1985).

The circulatory changes associated with heat exposure are exaggerated by hypohydration. Nadel et al. (1980), studied the influence of hydration state upon circulatory controls at an ambient temperature of 35°C in 4 relatively fit subjects during ergometer exercise (55 percent max. $\dot{V}_{O_2}$). They found that hypohydration resulted in a significantly reduced cardiac output during exercise, due to a further reduction in stroke

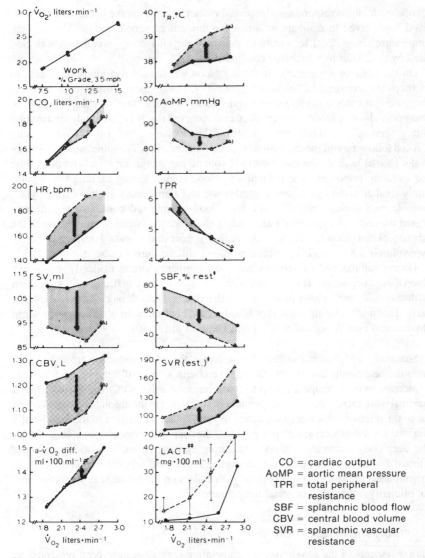

Figure 13-8
Cardiovascular responses to graded exercise in hot (43.3°C △—△) and neutral (25.6°C●——●) environments. In upper left-hand graph, $\dot{V}_{O_2}$ is plotted against work rate to show absence of temperature effects on $\dot{V}_{O_2}$. All other data are plotted against $\dot{V}_{O_2}$ during each of four levels of exercise. Arrows show direction of temperature-induced change in each variable. Largest reductions in CO and SV and increase in a-$\bar{v}$ O_2 difference were seen at highest work rate at 43.3°C. Two of six subjects were unable to reach this level in the heat; as a result, this effect is not seen in averaged responses. Data for SBF (*) were from 11 different subjects in an experiment of identical design. SVR (‡) was estimated from aortic MP (AoMP) values taken from six subjects in another study. Average blood lactate (lact. ‡‡) concentrations were higher at 43.3°C but not significantly so, as indicated by large standard deviation. *(From Rowell, 1974.)*

volume without corresponding elevation in heart rate. A marked reduction in skin blood flow served to maintain an already compromised venous return. The internal temperature (esophageal temperature) threshold for cutaneous vasodilation was elevated by 0.42°C in hypohydrated conditions.

On the basis of experiments involving blood withdrawal and blood reinfusion in six subjects exercising at 60 percent $\dot{V}_{O_2}$ in a 35°C environment, Fortney et al., (1981a) conclude that cardiac stroke volume and cutaneous blood flow vary in proportion to changes in absolute blood volume. They also observed that the rise in body temperature during exercise was significantly greater in hypovolumia, but was not significantly reduced following volume expansion, i.e., blood reinfusion. The same authors (Fortney et al., 1981b) have shown that the blood volume has significant effects on body fluid and sweating responses. The reduction in blood volume during exercise was significantly smaller under hypovolumic conditions, and significantly greater during hypervolumic than during control conditions. Blood volume reduction also significantly altered the control of sweating rate, tending to reduce the sweating rate relative to the esophageal temperature. Thus, both the body fluid and sweating responses during hypovolumia act to conserve circulating blood volume during exercise.

During submaximal exercise in a hot environment there is evidently a different blood flow distribution as compared with exposure to a normal climate. At a given cardiac output, more blood flows through the skin and less through the active muscle mass. This might explain the higher blood lactate concentration observed in the warm environment (see Windham et al., 1973; Dimri et al., 1980).

Summary   Muscular exercise may increase the heat production from 10 to 20 times the heat production at rest. During exercise in a "neutral" environment, there is an increase in body temperature up to a maximum of about 40°C or slightly higher at maximal work rates. The body temperature is not related to the absolute heat production but to the relative work rate, i.e., actual oxygen uptake in relation to the individual's maximal aerobic power; at a 50 percent load, the deep body temperature is about 38°C. The deep body temperature at rest and during exercise is, within a wide range, not affected by the environmental temperature; but the skin temperature is. In a given environment, the sweating rate is mainly dependent on the actual heat production and not primarily on the skin or rectal temperature.

TEMPERATURE REGULATION

Various aspects of the physiology of temperature regulation have been presented by Hardy et al., 1970; Bligh and Moore, 1972; Wyndham, 1973; Cabanac, 1975; Greenleaf, 1979; Nielsen, 1980; Kandel and Schwartz, 1981; Houdas and Ring, 1982. In this section we shall attempt to summarize some of the current concepts.

In the hypothalamus and the adjacent preoptic region, as shown in animal experiments, there are nerve cells, which by local heating and cooling may elicit the same reactions that occur during exposure to heat or cold (Hammel, 1965; Hardy, 1967). These cells belong to the temperature regulatory center, which is connected via nervous pathways with receptors in the skin, the central nervous system, and possibly elsewhere in the body, such as in the deep leg veins, the muscles, the abdomen, and the spinal cord (Hensel, 1974). These receptors consist of a net of fine nerve endings that are specifically activated by heat or cold stimuli (Zotterman, 1959; Hensel, 1981). They are especially sensitive to rapid changes in temperature and are highly susceptible to adaptation. In the heat receptors, the maximal frequency of the impulses occurs in a steady-state condition between 38 to 43°C; in the cold receptors, the maximal impulse frequency occurs at 15 to 34°C (Zotterman, 1959). At temperatures above 45°C, the cold receptors may again be activated. This may explain the paradoxical cold sensation experienced at the first contact with very hot water. The receptors register not only temperature changes but also temperature levels, particularly if the skin temperature is below 32°C in the case of the cold receptors, and above 37°C in the case of the heat receptors (Kenshalo et al., 1961). The number of active receptors determines to some extent the sensation of temperature. The fact that temperature sensation is a relative matter is best illustrated by the simple experiment of putting one finger in warm water, another finger in cold water. When both fingers are then simultaneously put into lukewarm water, the finger which previously was exposed to cold water will sense the lukewarm water as warm; the other finger will interpret it to be cold.

For the regulation of body temperature, the hypothalamic temperature regulatory center and the temperature-sensitive receptors in the skin play a dominating role (Smiles et al., 1976). Precisely how the temperature regulatory mechanism works is unknown. It appears that several temperature sensors are capable of triggering defense reactions independently in order to maintain thermal balance, and that the response is a function of several inputs combined at the same time. According to Cabanac (1975), the temperature sensors in the spinal cord may be at least one-quarter to one-half as sensitive as the hypothalamic temperature sensors. Warm stimulation of the spinal cord in animals is followed by all of the warmth defense reactions (skin vasodilation, reduced heat production, increase in evaporative heat loss) proportional to the spinal cord temperature. Cooling of the spinal cord in unanesthetized dogs induces shivering, peripheral vasoconstriction, and piloerection proportional to the magnitude of the stimulus (for references, see Cabanac, 1975). Nevertheless, it appears that the hypothalamus remains the main center of temperature regulation, for studies of the effects of spinal cord lesions have shown that the spinal network does not possess complete thermoregulatory capability (Walther et al., 1971). On the basis of the available evidence, Cabanac (1975) concludes that it is probably more accurate to consider the control system for the regulation of body temperature as consisting of a number of networks operating independently of one another.

As is evident from recent symposia (Poulos, 1981a and b; Blatteis, 1981), great uncertainty still prevails as to the neurophysiology of temperature regulation and the role of the newer putative central neurotransmitters in thermoregulation. Greenleaf (1979) has discussed the hypothesis that fluid and electrolyte changes influence thermal regulation within the finer control boundaries, assuming that the control of mammalian thermoregulation may have two separate components: a broad ban control that operates when core temperature deviates widely from its normal range, and a fine control that operates between these two limits. Evidently much research is still needed in order to clarify the postulated role which fluid-electrolyte composition and concentration in body cells may play in temperature regulation. For a discussion of neural and hormonal control of metabolic aspects of thermogenesis, see Federation Proceedings, vol. 38, no. 8, pp. 2147–2181, 1979, B.A. Horwitz, ed.

The *anterior hypothalamus* and the preoptic region are sensitive to changes in the local temperature. Many neurons have been observed to increase their discharge rate when heated; only a few cells increase their discharge frequency when the local tissue temperature is lowered (Hardy, 1967). Preoptic heating in animals exposed to a neutral environment can induce vasodilation in the skin and, eventually, panting and sweating. With the animal in a cool environment, such a local heating can inhibit the normal response of shivering and vasoconstriction of peripheral blood vessels. The depression of the metabolic rate causes the body temperature to drop.

An intact *posterior hypothalamus* is required to induce the reactions to a cold environment, namely shivering and an increase of metabolic rate, and to restrict flow of heat to the skin by a vasoconstriction of skin blood vessels. This posterior center, however, is temperature blind; it is essentially insensitive to local temperature changes. The function of this area is largely coordinating and it receives, rather than generates, temperature signals. Afferent impulses from cold receptors in the skin seem to be the main drive for this center.

The two thermoregulatory centers are in a way connected so that the response to stimulation of the anterior center includes stimulation of sweating, but inhibition of shivering and vasoconstriction. Conversely, the action of a stimulation of the posterior center involves a stimulation of vasoconstriction and an increase of heat production, but a simultaneous inhibition of responses to heat. The final common pathways include not only the motor pathways of the synaptic and somatic systems, but also blood humoral transmissions. Cold may, via the thermoregulatory center in the hypothalamus, affect the pituitary gland and the release of hormones, which in turn act on their target organs to release thyrotropic and adrenal hormones, increasing the heat production in the tissues. According to Hardy (1967), many large animals, however, show relatively little of the neuroendocrine response to cold. There are functional connections between the central control of body temperature and the areas regulating water and food intake (Andersson, 1967).

Under steady state conditions, McCaffrey et al. (1979), have demonstrated that cutaneous thermoreceptors influence sweating rate. During exercise, Davies (1980) observed that the integrating and modulating effect of skin temperature from different regions of the body is responsible for the control of sweat loss, under conditions of constant central thermal drive. That peripheral inputs are a major factor in thermo-regulatory processes also under nonsteady state conditions, was demonstrated by Libert et al. (1979), in subjects exposed to rapid or slow alterations in air and wall temperatures (28° to 45°C). They conclude that cutaneous receptors produce a positive and a negative rate component within the central thermal integrator. In nude, resting men exposed to environmental temperatures changing from neutral (28°C) to warm (50°C), Libert et al. (1978), observed that sweating started before appreciable variation in rectal temperature, suggesting a peripheral control of the onset of sweating.

Under steady state conditions, the central drive for sweating in humans has generally been described as a function of internal and mean skin temperature signals (Stolwijk,

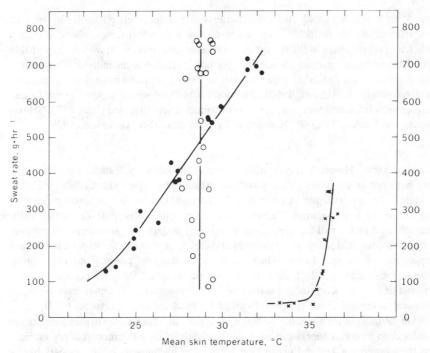

Figure 13-9
Steady-state values of sweat rate plotted against the corresponding values of mean skin temperature: (O) exercise intensity from 90 watts to 235 watts at constant environmental temperature of 20°C; (●) constant exercise intensity (150 watts) at environmental temperatures from 5 to 30°C; (X) experiments at rest at environmental temperatures from 25 to 44°C. *(From B. Nielsen, 1969.)*

1970) (Figure 13-9). The experiments of Libert et al. (1982) during thermal transients, however, show the insufficiency of explanations based on a simple additive function of core and skin temperature. During such thermal transients there appears to be an interplay of additive and multiplicative controls. The multiplier effect appears to be exhibited not only in the control of local sweat rate by local skin temperature, but also in making up the central drive for sweating.

Nielsen (1969) noticed a remarkable similarity in the thermoregulatory responses to passive heating by diathermia and to active heating by muscular exercise. Thus, at the same level of heat production, rectal and skin temperatures and estimated skin blood flow were increased to the same level in the two kinds of experiments. Therefore, she concludes, a work factor of nervous origin (from mechanoreceptors or cortical irradiation) may operate on the thermoregulatory centers at the start of exercise but hardly in the later phase of exercise. In intermittent exercise with the same heat production over a given period of time as in continuous exercise (with different intensity during the periods of activity), Nielsen (1969) found the same body temperature and sweat rate, which should also exclude nervous impulses related to the severity of exercise as the "work factor." She proposes that a chemical factor, liberated during exercise proportionate to the engagement of the aerobic processes, may be responsible for the temperature resetting. Since the work rate in relation to the individual's maximal aerobic power dominates the adjustment of the body thermostat more than the absolute aerobic power (I. Åstrand, 1960), attention should be focused at some stress factor, e.g., epinephrine and norepinephrine. (For further discussions see Gale, 1973; Schönbaum and Lomax, 1973; B. Nielsen, 1976; Sobocinska and Greenleaf, 1979.)

Summary Homeothermy is a fairly recent evolutionary feature. Consequently, we have not as yet developed any specific organs for temperature regulation. Physiological thermoregulation is controlled by nervous centers in the anterior preoptic region of the hypothalamus. The central thermoregulation is quite neurochemically complicated and is still far from being completely understood. A decrease in temperature activates cold receptors in the skin and certain neurons in the hypothalamus and upper part of the spinal cord. These further stimulate heat production by activating the nervous pathway, which contains cholinergic synapses. Body temperature may be maintained by vasoconstriction and shivering. Dopaminergic synapses have a role in the central control of the peripheral vasomotor mechanism (Lagerspetz, 1982).

On the basis of the available evidence, it appears that both a drop in skin temperature and a drop in core temperature may independently elicit a thermoregulatory increase in the oxygen uptake, and hence heat production in humans. It still appears likely, however, that both core and skin temperature have a combined influence on our subjective and physiological reactions to temperature changes in our immediate environment.

When exposed to heat, sweating and the dilation of the skin blood vessels do not run parallel. The produced sweat may in itself cause a vasodilation; on the other hand, the local skin temperature may affect the diameter of the blood vessels in that area. At high evaporative sweat rates, the skin may cool, causing vasoconstriction. It is known that heating or cooling of the skin of the head does lead to increased or decreased sweating over the rest of the body (McCaffrey et al., 1975).

It is probable that other impulses to the temperature regulatory center exist. Fever is caused when the "thermostats" are set at a higher level. The reasons for the diurnal variations in body temperature are unknown.

A great deal is known about the various thermosensitive elements such as cold and hot receptors and deep receptors, which sense temperature changes in different parts of the body core and shell. Similarly, a considerable amount of knowledge is also available concerning the effector elements that may contribute to heat production or heat loss in the body as required. What is still missing, however, is the understanding of the integrative mechanisms in the human central nervous system proper, which transform afferent inputs into efferent action resulting in mandatory adjustive measures such as piloerection, shivering, sweat secretion, vasoconstriction or vasodilation, endocrine gland stimulation, or changes in cardiopulmonary function.

ACCLIMATIZATION

Continuous or repeated exposure to heat, and possibly also to cold, causes a gradual adjustment or acclimatization resulting in a better tolerance of the temperature stress in question. Many plants prepare for the winter by increasing their carbohydrate content, and certain types of apple trees may, during the winter, tolerate air temperatures below $-40°C$, but during the month of July, they fail to survive air temperatures below $-3°C$. Certain insects accumulate the "antifreeze" glycerol in the fall, which enables them to survive cold. (For literature, see Dill, 1964.)

Heat

After a few days' exposure to a hot environment, the individual is able to tolerate the heat much better than when first exposed. This improvement in heat tolerance is associated with increased sweat production, a lowered skin and body temperature, and a reduced heart rate (Robinson et al., 1943; Kuno, 1956; Bass, 1963; Leithead and Lind, 1964; Wyndham, 1967; 1973; Rowell, 1974; Libert et al., 1983). An example is illustrated in Fig. 13-10. Usually the skin blood flow is reduced; in one experiment it declined from 2.6 to 1.5 liters $\cdot$ m^{-2} $\cdot$ min^{-1}, i.e., to about 60 percent of the original value (Bass, 1963). The increased sweat rate provides the possibility for a more effective cooling of the skin through the evaporative heat loss, and the resultant lowered skin temperature provides for a better cooling of the blood flowing through the skin. Thus, the body can afford to cut down on the skin blood flow. In acute experiments, the sweat glands have the capacity to produce more sweat than they do under ordinary circumstances. The reason why this capacity is not fully utilized until after several days' exposure to a hot environment is not known. The increase in sweat production may go as high as 100 percent (Leithead and Lind, 1964).

It is possible, then, to demonstrate objectively physiological alterations in response to prolonged exposure to heat. It is also possible to explain why the heat tolerance gradually increases. It is found that the skin blood flow in acute experiments increases at the expense of the blood flow through other tissues (Rowell et al., 1965; Rowell, 1974), although more recent findings suggest that this may not always be the case (see Rowell, 1983). As the individual becomes acclimatized, normal distribution of the blood flow is once more established. Within 4 to 7 days' exposure to a hot environment, most of the changes have taken place, and at the end of 12 to 14 days, the acclimatization is complete. Even a relatively short daily heat exposure will have some effect.

The effect of heat acclimatization persists several weeks following heat exposure, although some impairment in heat tolerance may be detected after a few days following cessation of exposure, such as after a long weekend, especially if the individual is fatigued and alcohol has been consumed (Bass, 1963; Williams et al., 1967).

During acute exposure to heat and the first phase of acclimatization, the heart rate during a standardized prolonged exercise is increased and the stroke volume reduced. Gradually the heart rate decreases and the stroke volume increases but seems to remain depressed throughout the acclimatization. The cardiac output remains relatively con-

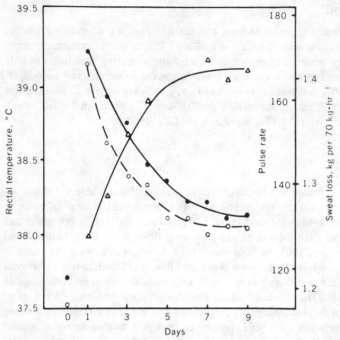

Figure 13-10
Mean rectal temperature (●), heart rate (O), and sweat loss (△) in a group of men during a 9-day acclimatization to heat. On day 0 they exercised for 100 min at a rate of energy expenditure of 1.2 MJ (300 kcal) · hr⁻¹ in a cool climate. On the following day they exercised in a hot climate (48.9°C dry-bulb and 26.7°C wet-bulb temperature). *(Modified from Lind and Bass, 1963.)*

stant. Thus, the major cardiovascular adjustments during heat acclimatization are reciprocal changes in heart rate and stroke volume while cardiac output and also arterial blood pressure remain essentially unaltered (Rowell, 1983; Wyndham et al., 1976). The mechanisms for these adjustments are far from revealed, however.

Some authors (Edholm and Weiner, 1981; Avellini et al., 1982) maintain that physical activity, as such, may not be necessary to achieve heat acclimatization, but may contribute to the extent that it causes a rise in body temperature. On the other hand, Inbar et al. (1981), have reported physiological changes compatible with heat acclimatization by mere physical conditioning in a neutral climate.

Measurements of total blood volume during acclimatization to heat have produced conflicting results, but most studies seem to indicate an increase (see Rowell, 1983). Senay et al. (1976), suggested that the most critical event in the first phase of heat acclimatization is an expansion of the plasma volume. However, Fortney and Senay (1979) studying the responses of 9 females to moderate exercise in a cool and a hot environment, during a 4-week training program and during heat acclimatization, found that training caused a significant expansion of resting plasma volume and total protein content, while acclimatization caused increased sweat rates and decreased skin temperature. Similarly, Harrison et al. (1981), suggest that the expansion of the intravascular volume with heat acclimatization should be regarded merely as a secondary effect of exercise and not as a primary adaptive response to heat. They maintain that hemoconcentration is the usual response to cycling exercise in the heat, while Senay (1978) observed a significant hemodilution in acclimatized exercising men (30 percent max $\dot{V}_{O_2}$) during exposure to heat (33.8°C).

Seasonal differences in acclimatization to heat have been reported in humans (Shapiro et al., 1981a). Acclimatized humans in the winter season exhibited lower sweat rates than in the summer. During winter acclimatization, the blood volume expansion was due to expansion both in the plasma and cellular compartments, whereas in the summer the acclimatization resulted in only plasma volume expansion.

Oddershede and Elizondo (1980) performed a complete body fluid compartment analysis during resting heat acclimatization in rhesus monkeys. Their results suggest that heat acclimatization in primates is characterized by a protein and fluid shift from the interstitial fluid compartment to the cardiovascular and the intracellular compartments.

Cold

Acclimatization to heat may be conveniently studied in subjects living in a hot climate or in laboratory experiments. Studies of adaptation to cold, on the other hand, require climatic experiments inasmuch as individuals, when moving to a cold climate, normally bring their semitropical climate with them. They have the ability to protect themselves against the cold with clothing or adequately insulated dwellings, even in arctic regions. Animals habitually exposed to cold develop an effective protection in the form of fur, and an effective heat exchange system in their peripheral blood vessels. In certain types of seals, the skin may be cooled to 0°C without an increase in oxygen uptake. In the human, a lowering of the skin temperature by a few degrees may cause a doubling of the metabolic rate. The human has then, with respect to reaction to cold, taken a path somewhat different from that followed by arctic animals.

Radomski and Boutelier (1982), on the basis of observed hormonal and thermal responses, have suggested that a degree of cold resistance can be induced rapidly in humans by short intermittent exposures to an intense cold stress. This cold resistance persists for a significant period of time after the last exposure. Budd (1962) has claimed convincing evidence of general acclimatization to cold in men, wintering in Antarctica.

According to some investigators, the metabolic rate of "cold-acclimatized" individuals is often elevated when they are exposed nude to a standardized cold stress, even though shivering is said to be less pronounced. Nevertheless, the metabolic rate is unchanged at normal room temperature. It is possible that hormones, notably norepinephrine, play a role in the elevation of the metabolic rate, but the mechanism is unclear. The aborigines of Central Australia and the bushmen of the Kalahari Desert, wearing little or no clothing, may be exposed to night temperatures of about 0°C or below. They do not shiver even though their core and skin temperatures keep falling throughout the night, and they can sleep. Control subjects were "fighting" against the cold environment by shivering, and they could not sleep. There are similar observations on other primitive populations and also on women divers of the Korean peninsula who do not increase their heat production by shivering to the same extent as individuals who have not been exposed to cold environment for longer periods of time (for references, see LeBlanc, 1975, p. 116.) The essential feature of cold adaptation seems to be reduced shivering. The energy saved is not very important, but this effect has other advantages. Shivering is disturbing and uncomfortable. "Shivering is the first line of defense but nobody likes it" (LeBlanc, 1975).

While it is difficult to demonstrate definite evidence of general physiological acclimatization to cold in humans, such acclimatization can be produced in animals exposed to severe cold. If rats are placed in a refrigerator kept at 5°C for as long as 6 weeks, the metabolic rate of the animals will increase twofold. They will double their food intake, and their thyroid function will increase. The main feature of the cold-acclimatized rat is its ability to maintain a high rate of heat production. This ability is absent in the nonacclimatized rat, which is unable to survive in the cold. If these findings in the cold-acclimatized rats are compared with findings in human beings, such as Eskimos or whites, who are habitually exposed to cold, it is observed that the changes found in the rat do not necessarily occur in the human. A person's food intake is not materially increased in a cold environment (Rodahl, 1963), nor is the rate of heat production. It is true that the Eskimo's rate of heat production is higher than that of whites, but much of this is due to the specific dynamic effect of diet, for Eskimo living on the white person's diet do not, as a rule, have any higher rate of heat production than the white person (Rodahl, 1952). When whites move to the Arctic, their metabolic rate is no higher than it was at home. Normal Eskimos do not show increased thyroid function compared with normal whites (Rodahl and Bang, 1957). The most probable reason is that in clothed individuals habitually exposed to cold environments, the degree of cold exposure is not severe enough to cause an appreciable increase in metabolism. Without increased metabolism, there is no need for increased food intake.

On the other hand, when the human is exposed to cold stress that is far more severe than that normally encountered by the clothed individual living in the Arctic, certain physiological changes do occur, which may be interpreted as hormonally induced adjustments to cold (Rodahl et al., 1962). When young men, dressed in shorts and sneakers only, were confined continuously to cold-chamber temperatures of 8°C for 3 to 10 days, they responded by violent shivering, which lasted more or less continuously night and day throughout the experiment, even at night when they slept under a blanket. They soon learned to continue to sleep in spite of the shivering. As a result, the increased heat production due to the shivering was maintained even during sleep, so that the body temperature remained normal and the subjects slept relatively comfortably. Occasionally, however, the same subjects, for some reason, failed to keep up the vigorous shivering during the night while they slept. Consequently their body temperature continued to drop and approached dangerously low levels. Under these conditions, the subjects occasionally objected to being disturbed, and wanted to be left in peace to continue to sleep. If this were allowed, the subject might continue to cool and, conceivably, might eventually die from hypothermia. It thus appears that an exposed person sleeping in the cold may actually freeze to death if the body's rate of cooling is slow and gradual, so that violent shivering is not produced, which would wake the person up.

These subjects in the climatic chamber doubled their metabolic rates because of their constant shivering. Their resting heart rate was markedly elevated, and the nitrogen loss in the urine was greatly increased. These changes were interpreted as being the result of hormonal changes brought about by the cold stress (Issekutz et al., 1962). It should be borne in mind that the cold exposure of these subjects in the climatic chamber (8°C) was far greater than the cold exposure of the Eskimo or any clothed group of individuals living in the Arctic. Consequently, all the cold-room subjects suffered ischemic cold injury of their feet, although the temperature never approached the freezing temperature of the tissue.

When nutritional deficiency or nutritional stress was superimposed on the cold stress in these subjects, there was a marked impairment in physical performance. After 9 days' constant exposure to 8°C in the nude, maximal oxygen uptake showed no deterioration when the individuals were fed an adequate diet consisting of 12.5 MJ (3000 kcal) and 70 g protein. When the energy intake was reduced to 6.3 MJ (1500 kcal) or the protein intake was reduced to 4 g · day^{-1}, or when both energy and protein intakes were reduced, a marked deterioration in physical performance capacity occurred within 3 to 5 days. The effect was most pronounced in the treadmill running time; maximal oxygen uptake in some of the subjects was less affected.

However, any physiological adaptation to cold in humans is of little practical value compared with the importance of know-how, experience, and state of physical fitness. There is definitely a limited capacity of the automatic thermostatic system. The success of the Eskimo in getting along in the cold depends on their ability to avoid the extreme cold. Since most arctic clothing (except fur clothing) is inadequate in terms of insulation to maintain thermal balance when an individual is exposed without shelter to extreme prolonged cold, the only way to survive is to remain active in order to increase the heat production. The fitter the Eskimo are, the longer they can do this. Thus, survival in the Arctic is a matter of survival of the fittest.

When a person, whether an Eskimo, a white, or a black person, allows his or her hands to be repeatedly exposed to cold for about $^1/_2$ hr daily for a few weeks, this cold stress will cause an increased blood flow through the hands, so that they will remain warmer and are not so apt to become numb when exposed to cold. Similar findings have been made in fish filleters, who habitually expose their hands to cold water (Hellström, 1965). This may be termed local acclimatization to cold. While it inevitably will cause a greater amount of heat to be lost from the hands, it will improve the ability of the hand and fingers to perform work of a precise nature in the cold. To this extent, local cold acclimatization is beneficial (LeBlanc, 1975; Nelms and Soper, 1962; Strömme et al., 1963).

It is still unknown as to what extent the reported differences in finger temperature response to local finger cooling reflect different psychological states such as apprehension, rather than reflecting racial differences (for reference see Provins and Clarke, 1960). One of us (K. R.) recorded finger skin temperatures in well adjusted and relaxed blacks familiar with the laboratory procedures, and in whites during hand emersion in ice water, and found no difference (unpublished results).

LeBlanc et al. (1975) found that repeated exposure of the hands and face to severe cold activates some adaptive mechanisms characterized by a diminution of the sympathetic response and a concomitant enhancement of the vagal activation normally observed when the extremities and the face are exposed to cold.

Enander et al. (1980) studied the reactions to hand cooling, including rewarming, in workers occupationally exposed to cold (slaughterhouse workers) compared with office workers. Considerable inter-individual differences in hand skin temperature reaction were found in both groups. They found no statistically significant difference in the hand skin temperature in the two groups, while the office workers rated significantly greater cold sensation as a result of the cold immersion, as well as a higher frequency of pain ratings. The authors suggested that the occupational cooling may not have been severe enough to produce physiological adaptation, although some psychological adaptation, i.e., accustomization, was indicated.

Summary   Resting humans acclimatize to heat by adjustments in sweating rate and core temperature threshold for sweating, and by increasing blood volume. Changes accompanying acclimatization to exercise in hot, dry conditions, include reduction in heart rate, skin, and core temperatures. Sweating begins at a lower body temperature and may increase at any given core temperature. Most of the changes occur during the first days of exposure. Apparently, cardiac output is not changed, but plasma volume may be increased (see Rowell, 1983).

Whether or not, or to what extent a real acclimatization to cold may occur in clothed humans exposed to cold climate, is still an open question. Local acclimatization, however, may occur in the hands of individuals regularly exposed to cold. Such localized cold stress may cause an increased blood flow through the hands.

LIMITS OF TOLERANCE
Normal Climate

Optimal function requires that the body temperature be maintained between 36.5 and 40.0°C. The ideal room temperature is about 20°C for clothed individuals, who are sitting or standing still. The more active the individual, the lower the room temperature should be. Thus, when performing heavy physical work, a person may prefer a room temperature of 15°C or even lower. Not only may acclimatization play a role, but habits and established traditions may affect the so-called comfort temperature. This may explain the fact that the preferred room temperature is higher in the United States than in England and higher in the summer than in winter in both the United States and England. The preferred room temperature in Singapore is higher than in the United States (Pepler, 1963).

Cold Injury

Cold injury may be inflicted in common winter sports such as skiing and skating, and in long distance runners competing in cold, windy conditions. Local cold injury may occur in the exposed parts of the body such as the face, hands, and feet, either due to the freezing of tissue and formation of ice crystals, i.e., frostbite, or by vasoconstriction causing deprivation of the blood circulation to the exposed parts, leading to ischemic cold injury. In the case of the former, it has been customary to distinguish between first degree frostbite, involving merely the superficial layer of the skin; second degree frostbite, with the formation of exudate containing blisters in the skin; and third degree frostbite, involving the freezing of the deeper tissues as well, including subcutaneous and muscle tissues. The pathological consequences of the freezing may, to some extent, depend on the speed of the cooling. In rapid cooling, the ice crystals formed may break the cells and cause tissue destruction and necrosis. In slow cooling ice crystals are formed in the tissues, but since solutes are excluded in the freezing process, the osmotic pressure in the extracellular fluid increases, pulling fluid out of the cell, with exudate formation as a result. This in itself will cause cell damage, augmented by the concomitant vasoconstriction with increased venous pressure, reduced capillary blood flow, blood cell aggregation, thrombosis, and necrosis.

The treatment of local cold injury consists of local rewarming in the case of first degree frostbite, blisters (second degree frostbite) should be left intact, third degree frostbite requires hospitalization. At any rate, the patient should be brought into shelter, tight clothing should be loosened, the patient should be kept warm, given hot drinks, and if possible made to be active in order to produce internal heat. The treatment should be kept up until the return of normal color and feeling in the affected parts, and the patient should be kept under surveillance. If normal functions are not restored within 30 min the patient should be hospitalized.

The danger of contracting cold injury of the lungs, especially during prolonged athletic competitions in the cold, involving high speed (high wind velocity), and high pulmonary ventilation, has been discussed a great deal over the years. Very few well-documented cases have been reported, however. Furthermore, it should be kept in mind that the alveolar temperature is nearly constant and independent of ambient temperature variations (Houdas and Ring, 1982).

Local cold injury of the eyes, i.e., transitory epithelial damage of the cornea with the formation of corneal edema and blurred vision, has been observed in cross-country skiers competing at very low temperatures (Kolstad and Opsahl, 1979). Similar afflictions have been reported among the early aviators, racing cyclists, speed skaters, and natives of the Arctic (for references, see Kolstad and Opsahl, 1979). In skiers, this particular type of injury is usually seen only during competitive ski races, not during training. It is especially apt to occur during long distance races at temperatures below − 15°C combined with wind. The symptoms usually develop toward the end of the race and consist of impaired or blurred vision evidently caused by pathological changes in the lower segment of the cornea. According to Kolstad and Opsahl (1979), who first described this phenomenon, the cause may be an impaired blinking reflex and an incomplete closure of the eyelids during blinking due to a lowering of the surface temperature of the cornea (Kolstad, 1969). As a result of the reduced blinking, the thin tear film covering the cornea and nourishing it, will not be maintained. This may explain the observed degeneration of the epithelium of the unprotected part of the cornea. The damage is transitory, however, and will usually heal completely within 24 hours. A possible prevention may be achieved by protecting the eyelids and the cornea by wearing suitable headgear, perhaps in combination with the use of contact lenses.

Hypothermia

Clinically speaking, hypothermia is a condition usually characterized by body temperatures below 35°C.

During the initial stages of hypothermia, the patient is shivering, gradually he or she becomes disorientated, apathetic, hallucinatory, or may become aggressive, excited, or even euphoric. As the rectal temperature gradually drops below 34°C, the patient may appear distant and stuperous, he or she may be unconscious, respiration is shallow, the pulse is weak. Cardiac arythmia may develop. There is loss of reflexes, and the pupils are dilated. Finally, the patient reaches the paralytic stage as the rectal temperature drops below 30°C. The skin is cold, no pulse can be detected, the pupils are dilated, there are no reflexes, and no heart sounds.

If the patient cannot be brought to hospital safely, and the treatment must be performed under field conditions, slow rewarming should be applied (0.5°C per hour). The danger of rapid rewarming in the field is a further drop in deep body temperature due to the return of cool venous blood from the skin and extremities to the core. This temperature drop may be fatal due to cardiac arythmia (ventricular fibrillation), which requires defibrillation.

It should be kept in mind, that even at more moderate degrees of body cooling, the muscle temperature may be significantly lowered, causing muscular weakness, impaired neuromuscular function, and reduced endurance. This may be the underlying cause of some of the fatal accidents among climbers and cross-country skiers.

Failure to Tolerate Heat

The vastly increasing number of participants in long distance running races under hot conditions has led to a marked increase in heat casualties and heat illnesses, such as heat exhaustion, heat syncope, and heat stroke; often in combination with dehydration, especially in the case of marathon runners (Sutton, 1984).

The most serious consequence of exposure to intense heat is heat stroke, which may be fatal. It is caused by a sudden collapse of temperature regulation leading to a marked rise in body heat content. The rectal temperature may be 41°C or higher. The skin is hot and dry. There are tachycardia and hypotension, metabolic acidosis, disseminated intravascular coagulation, and occasionally renal failure. The victim is confused or unconscious. This form of temperature-regulatory failure is rare. The risk is higher in nonacclimatized than in acclimatized individuals. Obese persons and older individuals are most susceptible. The treatment is rapid cooling (for instance by pouring cold water over the victim, the application of ice packs, etc.) until the rectal temperature has dropped below 39°C.

Since heat stroke often is associated with peripheral circulatory collapse, the oral temperature of the victim may not necessarily be very high, while the rectal temperature always is: This emphasizes the importance of measuring rectal temperature in long distance runners who collapse (Sutton, 1984).

Another type of temperature-regulation failure is the so-called anhidrotic heat exhaustion. The victim may have a body temperature of 38 to 40°C, and may sweat very little or not at all. He or she feels very tired, may be out of breath, and has tachycardia. The main trouble is reduced sweat production. When the patient stops exercising and is removed to a cool place, this condition rapidly improves.

A third type of serious disturbance due to heat exposure is excessive loss of fluid and salt, usually because of failure to replace fluid and salts lost through sweating. After several weeks' exposure, the patient may eventually experience cramps, the so-called miner's cramps, which in rare cases may be fatal. Intravenous administration of NaCl will promptly relieve the cramps.

Heat syncope is a less serious affliction due to heat exposure. This is primarily caused by an unfavorable blood distribution. A large proportion of the blood volume is distributed to the peripheral vessels, especially in the lower extremities as the result of prolonged standing, or by a reduction in blood volume due to dehydration. The result is a fall in blood pressure and inadequate oxygen supply to the brain, which may lead to unconsciousness. If the victim is placed in the horizontal position, preferably with the legs elevated, he or she will quickly regain consciousness. This type of heat collapse is one of the body's built-in safety mechanisms.

Certain individuals exhibit an untoward reaction to heat in the form of heat rash. This condition may make the individual unsuited for work in a hot environment (Minard and Copman, 1963).

Recent literature on acute and chronic occupational heat illness, and the usefulness of the various heat stress standards recommended by occupational health authorities, have been reviewed by Dukes-Dobos (1981).

Brain function is particularly vulnerable to heat (Baker, 1982). Tolerance to elevated deep body temperature is extended, if the brain is kept cool (Carithers and Seagrave, 1976). According to Baker (1982), studies in humans show that changes in heat loss from the face have a powerful effect on thermoregulation, which may be caused by a change in hypothalamic temperature.

Kloetzel et al. (1973) have indicated a relationship between prolonged heat exposure and hypertension. This has never been confirmed. In extensive studies of workers at a ferro-alloy foundry in Norway, carried out by the Norwegian Institute of Work Physiology (unpublished results), there was no significant difference in the blood pressure between heat exposed and nonheat exposed workers of similar age and duration of employment. Plasma renin concentrations were elevated, however, during the work shift in the heat exposed foundry workers, but this was also the case in workers in the same company engaged in prolonged heavy physical work in a cool environment. The cause of elevated renin concentration in both cases may be a reduced kidney blood flow.

Upper Limit of Temperature Tolerance

It is not feasible to quote exact permissible temperature limitations for the working environment. These limitations depend on the combination of the different climatic factors, on the nature and type of work in question, on the severity of the work rate, and on the duration of the work apart from clothing and other measures of protection.

Finally, there are wide individual variations in the tolerance of climatic stress, beside the effect of acclimatization.

Nevertheless, it is desirable at times to be able to objectively assess the environmental conditions at individual work places, especially as a point of reference when changes are being made, in order to determine to what extent such changes actually represent improvement.

For such assessment of the thermal heat load in industrial situations, several so-called heat stress indices have been introduced during the past 50 years. By such indices one tries to predict the heat load on the individual. They are usually based on data from physical measurements of air temperature (dry bulb), radiant temperature (globe), relative humidity (wet bulb), and air velocity. A measurement of the individual's metabolic rate is also often included, i.e., the energy demand of the work is measured or estimated (Kerslake, 1972; I. Åstrand et al., 1975). The American Conference of Governmental Industrial Hygienists (ACGIH, 1976) has recommended limit values for industrial application, based on the so-called Wet Bulb Globe Temperature index (WBGT). The limit values are based on the assumption that almost all acclimatized and fully-clothed workers with an appropriate fluid and salt intake should be able to work under the given conditions without their body temperatures rising above 38°C. There appears to be no valid physiological basis for this assumption, however, and there is no reason to believe that a rectal temperature above 38°C, in healthy people, is associated with any health hazard.

Furthermore, physical performance improves with elevated body temperature, and this is the reason for the warming-up procedures practiced by athletes prior to athletic competition. Besides, rectal temperatures around 38°C have been recorded in workers at a Norwegian aluminum plant upon arrival at the plant in the morning. They had bicycled to work. And finally, continuous recordings of body temperatures with the aid of miniature recorders in industrial workers exposed to heat, such as workers in a Norwegian aluminum plant, at times show rectal temperatures in excess of 38°C for extended periods (Fig. 13-11). Similar findings have been made in workers in other types of electrochemical and melting industry.

The WBGT index referred to above is rather complicated, and the recording of it is time-consuming. In the case of a predominantly dry, radiant heat environment, the much simpler and faster reacting Wet Globe thermometer or Botsball, devised by Botsford (1971) records similar values and the results are interchangeable with a simple formula (Ciriello and Snook, 1977).

At any rate, the recording of environmental temperatures in hot industrial operations in fractions of a degree, as commonly done, is not warranted and may even be misleading, in view of the fact that the temperature in the same spot in a pot room may fluctuate as much as 10°C or more in less than an hour.

The complexity of the assessment of heat stress in industrial operations is evident from the fact that in actual field studies no direct relationship can be found in many cases between the environmental temperature on the one hand and the different physiological responses to this temperature on the other. Thus, in a study carried out in an aluminum plant, one tapper was exposed to a Botsball heat stress of 30°C, without his rectal temperature exceeding 38°C, and his evaporative sweat loss was 344 ml · hour^{-1}.

BODY TEMPERATURE AND HEART RATE IN A POT WORKER IN AN ALUMINUM PLANT

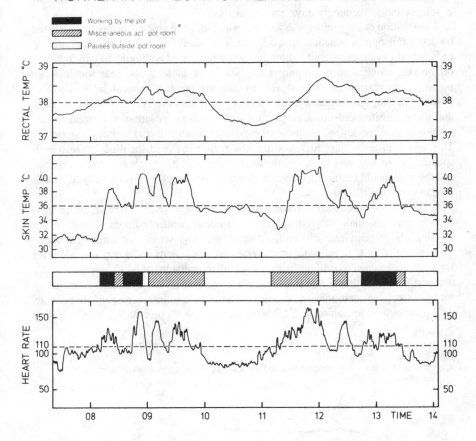

Figure 13-11
Body temperature and heart rate in a pot worker in an aluminum plant.

Another tapper, who was exposed to the same heat stress index, had a rectal temperature in excess of 38°C for 131 min, while his evaporative sweat loss was only 239 ml · hour⁻¹. One jacker, who was exposed to a heat stress index of about 38°C, had a rectal temperature higher than 38°C for 33 min, as compared with 117 min in the case of another jacker, who was exposed to a heat stress index of only 22°C. Similar observations have been made in other types of hot industries.

Ramsey and Chai (1983) have critically examined the inherent variability in heat stress decision rules. Their approach consisted of pairing WBGT values with the corresponding values of human heat exchange calculated from the heat stress index equation (see Belding, 1973). They suggest that simplified rules, when combined with appropriate administrative and work practices, may yield protection to the worker that is comparable to seemingly more precise methods.

The solution of the problems associated with industrial heat exposure may be based on a combination of practical measures such as: shielding of heat source, reducing each period of heat exposure to about 20 min, interspersed with brief 10 min, cooling-off breaks, elimination of strenuous physical work close to the heat source when possible, introducing mechanical aids to take the place of manual labor wherever possible, nursing the industrial process properly at all times so as to avoid complications that will necessitate drastic measures causing excessive exposure to heat stress. A key point is to avoid prolonged intense heat exposure, which will lead to profuse sweating. This will reduce the fluid loss, and thereby reduce the need for fluid intake, correspondingly. At any rate, however, the fluid loss should be replaced as it is being lost, and the aim should be for the worker to leave his or her place of work fully hydrated, able to enjoy their leisure time.

In laboratory experiments, Kamon (1979) studied different exercise cycles in 6 heat-acclimatized male subjects exercising in hot ambient conditions. Based on the leveling off of heart rate and rectal temperature, and setting the limit of a rectal temperature of 38°C, a schedule of exercise and rest periods of 20 min was adequate for a hot-dry ambient condition, with a dry bulb temperature of 50°C, wet bulb 25°C, irrespective of whether the resting took place under the same ambient conditions as for the exercising conditions, or at a neutral ambient condition (dry bulb temperature 23°C, wet bulb temperature 16°C). The work rate was 40 percent $\dot{V}_{O_2}$ max.

Figure 13-12
Worker in a Mg plant, slag removal, revised work schedule; approximately 20 min work, 10 min cooling off periods.

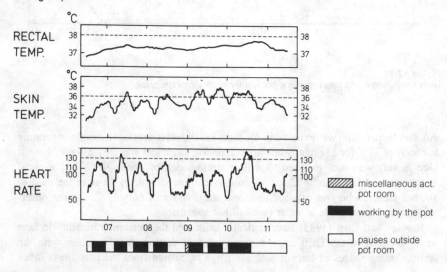

Essentially similar observations were made in real life in a Norwegian magnesium plant with a work schedule of 20 min work interspaced with a 10 min cooling-off period at prevailing outside ambient temperatures (see Fig. 13-12).

Age

Although the experimental data are still limited, the available evidence suggests that heat tolerance is reduced in older individuals (Robinson, 1963; Leithead and Lind, 1964; Lind et al., 1970). They start to sweat later than do young individuals. Following heat exposure, it takes longer for their body temperature to return to normal levels. Older people react with a higher peripheral blood flow, but their maximal capacity is probably lower. In one study, it was found that 70 percent of all individuals who suffered heat stroke were over sixty years of age (Minard and Copman, 1963). On the other hand, Davies (1979) observed no evidence for differences in thermoregulatory function, which could be ascribed to sex or age, in his one-hour treadmill exercise experiments on subjects eighteen to sixty-five years of age in a moderate environment.

However, Davies (1981) has shown that thermal responses of children are quantitatively different from young adults, evaporative sweat loss being lower in children and skin temperature higher for the same environmental conditions than in adults. (For a review of the literature concerning climate and the exercising child see Bar-Or, 1980.)

Sex

The available experimental evidence shows that women require lower evaporative cooling in both hot-wet and hot-dry environments (Shapiro et al., 1981b). Women have a lower tissue conductance in cold and a higher tissue conductance in heat than do men. This fact indicates a greater variation in the peripheral reaction to climatic stress in women. It appears that this fact is of no importance for the performance of work, however. Studies by Drinkwater et al. (1982), reveal that in healthy older women, aging does not diminish the functional capacity of the sweating mechanism to cope with heat stress while resting. From studies of active men versus active women during acclimatization to dry heat, Horstman and Christensen (1982) concluded that active women performed exercise of equal relative intensity in dry heat as well as active men. Ventilatory, metabolic, and cardiovascular differences between sexes were minimal.

Frye and Kamon (1983) observed no differences in sweating efficiency between sexes in the dry heat, but the women maintained a significantly higher sweating efficiency than the men in the humid heat. In both environments, the men recruited a significantly lower percentage of their available sweat glands than did the women.

Aerobic capacity is an important factor to be considered when men and women are compared in the heat. When fitness levels are similar, the previously reported sex related differences in response to an acute heat exposure seem to disappear (Avellini et al., 1980; Nielsen, 1980).

State of Training

It appears that a trained individual is better able to adjust to heat than one who is untrained. Physical training will enhance the sweating mechanism at a given level of central sweating drive. The increased metabolic rate during training raises a high thermoregulatory demand, and apparently this demand will induce an increased peripheral sensitivity of the sweat glands to the central sweating drive. It has been mentioned that for an individual who increases his maximal aerobic power by training, a given rate of exercise will require a lower percentage of his maximal oxygen uptake. Concomitantly, the core temperature will decrease during a standardized exercise. The increased capability for heat dissipation behind this adaptation is due to an enhanced sweating response (Shvartz et al., 1979). In addition, heat acclimatization results in a further enhancement of the sweating response at a given level of central sweating drive by the lowering of the zero point of the central nervous system drive for sweating. In other words, physical training seems to increase the slope of the sweating rate-core temperature curves, i.e., the activity of the sweat glands increases at a given core temperature. In contrast, acclimatization to heat seems to lower the threshold core temperature at which sweating starts (it moves the curves "to the left" without changing the slope. For further discussion, see Nadel et al., 1974). For an optimal heat acclimatization, simultaneous exposure to both heat and exercise is recommended. Physical training will also improve the circulatory potential. Thereby the trained individual may be able to maintain a cardiac output sufficient to meet metabolic requirements and the demand for peripheral bloodflow for a longer period of time than untrained people (Drinkwater et al., 1976.)

COORDINATED MOVEMENTS

The speed of nervous impulses and the sensitivity of the receptors are affected by the temperature of the tissues. At about 5°C, the skin receptors for pressure and touch do not react on stimulation. The execution of coordinated motions depends upon the inflow from these receptors to the central nervous system. The numbness in the cold is the result of this lack of sensitivity of the skin receptors. Irving (1966) reports that the skin at a temperature of 20°C was only one-sixth as sensitive as at 35°C; i.e, an impact on the skin had to be six times greater to be felt at the lower skin temperature. The muscle spindles show an increased sensitivity at moderately lowered muscle temperatures, but at 27°C the activity in response to a standardized stimulus is reduced to 50 percent; at a temperature of 15 to 20°C, it is completely abolished (Stuart et al., 1963). This phenomenon also contributes to the difficulty of performing fine coordinated movements in the cold. It may be partially responsible for an increased accident rate in certain types of manual work operations in cold environments.

For a more comprehensive review of the literature dealing with human manual performance in the cold, see Fox (1967), who concludes that performance decrements when exposed to low ambient temperature appear to stem from the physical lowering of the surface temperature of the hand. (See also Ellis, 1982; Tanaka et al., 1983.)

MENTAL WORK CAPACITY

An evaluation of the mental or intellectual performance during exposure to heat or cold is hampered by subjective variations and lack of suitable objective testing methods (Pepler, 1963). As a rule, a deterioration is observed when the room temperature exceeds 30 to 35°C if the individual is acclimatized to heat. For the unacclimatized, clothed individual, the upper limit for optimal function is about 25°C.

The observed deterioration in performance capacity refers to precise manipulation requiring dexterity and coordination, ability to observe irregular, faint optical signs, the ability to remain alert during prolonged, monotonous tasks, and the ability to make quick decisions. During a 3-hr drilling operation, the best results were achieved at 29°C, but at a room temperature of 33°C, the performance was reduced to 75 percent; at 35.5°C, to 50 percent; and at 37°C, to 25 percent. A high level of motivation may to some extent counteract the detrimental effect of the climate.

Wyon et al. (1979), examined the effect of moderate heat stress (up to 29°C) on mental performance in seventeen-year-old boys and girls. They were subject to rising air temperature conditions, typical of occupied classrooms in the range of 20° to 29°C. Sentence comprehension was significantly reduced by intermediate levels of heat stress in the third hour. A multiplication task was performed significantly more slowly in the heat by male subjects, showing a minimum at 28°C. Recognition memory showed a maximum at 26°C, decreasing significantly at temperatures below and above. The authors consider these results to support the hypothesis of reduced arousal in moderate heat stress in the absence of conscious effort.

Colquhoun (1970) observed that in the normal range of rectal temperatures, performance in a calculation test rose and fell with the rise and fall in body temperature.

WATER BALANCE
Normal Water Loss

Reasonable figures for the daily water loss are as follows: from gastrointestinal tract, 200 ml; respiratory tract, 400 ml; skin, 500 ml; kidneys, 1500 ml = 2600 ml. This loss is balanced by an intake as follows: as fluid, 1300 ml; as water in the food, 1000 ml; water liberated during the oxidation in the cells, 300 ml = 2600 ml. However, the water loss can increase considerably when the individual exercises or is exposed to a hot environment.

Water loss through the respiratory tract varies roughly with the pulmonary ventilation (dryness and temperature of the inspired air have some influence). The ventilation varies within a wide range directly with the production of CO_2, and this production is, in turn, proportional to the metabolic rate. Therefore, the water volume from oxidation, proportional to the metabolic rate, equals, by coincidence, roughly the water loss through the respiratory tract.

During very heavy exercise, glycogen is the preferred fuel. About 2.7 g of water is stored together with each gram of glycogen (Chap. 12), and this water becomes free as the glycogen is combusted. If during such heavy exercise, 5 MJ (1200 kcal) is totally metabolized, 80 percent, or 4 MJ (960 kcal), may be derived from glycogen. The liberated volume of water (including the water of oxidation) will be close to 800 ml. Assuming a mechanical efficiency of about 25 percent, 3.8 MJ (900 kcal) of the 5 MJ (1200 kcal) should be dissipated as heat if the body temperature should be maintained unchanged. An exclusive evaporative heat loss demands the evaporation of about 1500 ml of water to eliminate 3.8 MJ (900 kcal). Under these conditions, only approximately half the necessary water volume must be taken from body "stores," for the rest is apparently liberated in the processes producing the heat. It should be emphasized that more sweat may be secreted than is evaporated from the skin. On the other hand, radiative and convective heat exchange may reduce the demand on the evaporative heat loss. When the glycogen depots are again restored, extra water is certainly needed.

Thirst

In adults about 70 percent of the lean body weight is water, so there is a substantial buffer to cover water losses over limited periods of time. However, in the long run, water intake must balance water loss by the several routes mentioned. Hypothalamus and adjacent preoptic regions play the essential role in the thirst mechanism (Stevenson, 1965; Andersson, 1967). There are some sort of osmoreceptors reacting on an increase in the osmolarity of the intracellular fluid. Any change in the internal environment leading to cellular hypohydration (dehydration) will elicit thirst. A second effect of a rise in body fluid osmolarity is an increased secretion of antidiuretic hormone (ADH) from the neurohypophysis, an effect mediated from the same center (Verney, 1947). The kidneys must excrete a minimal amount of water as a vehicle for the elimination of solids. When water is in excess in the body, little or no ADH is brought to the kidneys and more water is excreted. A water deficit will, as stated, stimulate ADH secretion, causing an increased reabsorption of water by increasing the water permeability of the wall of the collecting ducts in the kidneys. It is a common observation that the volume of urine is reduced when sweating is profuse.

It has been shown that injection of minute amounts of hypertonic saline into, and electrical stimulation within, the anterior parts of the hypothalamus may elicit excessive drinking in the goat (Andersson, 1967). Similar stimulations will not only induce drinking but inhibit feeding. The intracellular fluid volume in the specific cells of the hypothalamus may be the crucial factor in thirst. The osmotic pressure across the cell membrane will of course influence this volume. There are volume receptors (stretch receptors) in the walls of the atria and great veins which can detect and reflexively adjust variations in the volume of intravascular fluid (see Gauer et al., 1970; Goetz et al., 1975; Zimmerman and Blaine, 1981).

Satiety is to a large extent a matter of behavior. Stevenson (1965) points out that a person usually waits until taking food to replace the last part of a water deficit.

Sensation of oral-pharyngeal dryness can elicit the urge to drink, but this reflex is not essential for the maintenance of a normal water intake. When one drinks, there is a temporary relief of thirst. This negative feedback operates partly from the oral-pharyngeal level, but it is also induced from stomach distension. The rapid relief of thirst after water intake is also explained by a normalization of the osmolarity of the blood. Not only does water move out of the gastrointestinal tract, but salts diffuse in the opposite direction along a concentration gradient.

The salt content of the sweat is less than that of the blood, and sweat loss therefore causes an increase in the salt concentration of the blood. As discussed earlier, increased salt concentration of the body fluids leads to the sensation of thirst and reduced urine volume. Certain studies have indicated that the salt content of sweat may be reduced as the result of acclimatization to heat, which should result in increased osmolarity and increased thirst at a given sweat loss (Robinson, 1963). The acclimatized individual is better able to maintain the fluid balance than one who is not acclimatized. Here again, experience may be an important factor, for a water deficit will negatively influence the physical condition.

It is a common observation that a voluntary water intake does not necessarily cover the water loss induced by excessive sweating (Pitts et al., 1944; Adolph, 1947; Leithead and Lind, 1964). The risk of a voluntary hypohydration is greatest in an individual unaccustomed to heat. The risk is also greater when a large portion of the food consists of dried or dehydrated rations, since a considerable volume of the daily water intake comes normally with the regular meals.

Summary Osmometric, volumetric, and thermal excitations appear to be nature's signals feeding information into the control system for water intake, located mainly in the hypothalamus. Sweat loss increases the osmotic pressure of the body fluids and thereby elicits the urge to drink. However, the sensation of thirst does not always "force" the individual to cover the water loss, particularly not when this loss is pronounced because of profuse sweating or because the individual does not eat normal meals containing a large amount of water.

Water Deficit

We have concluded that high sweat rates with excessive loss of body fluids may cause a deficit of body water (hypohydration or dehydration). The regulation of body temperature has priority over the regulation of body water. Therefore, a hypohydration can be driven very far, and may in fact be a threat to life if the environment is very hot and water is not available.

Prolonged exposure to heat and/or prolonged exercise certainly causes a hypohydration. In both situations a decrease in plasma volume has been noticed. Costill and Fink (1974) report a 16 to 18 percent reduction in plasma volume at a hypohydration equivalent to a 4 percent decrease in body weight. There is a shrinkage of the red cells during hypohydration (therefore, changes in hematocrit are not a reliable measure of changes in plasma volume; see Costill et al., 1974; Harrison et al., 1975). During exercise and/or exposure to heat stress there is a movement of protein from the interstitial spaces, particularly in skeletal muscles, to the vascular volume. The increase in plasma protein concentration will raise plasma oncotic (osmotic) pressure thereby helping to maintain blood volume by reducing water loss and enhancing water gain (Senay, 1972; Harrison et al., 1975). As suggested by Senay (1972), heat acclimatization may increase the ability to shift protein and fluid from the interstitial to the intravascular volume, which could improve the efficiency of the cardiovascular system. (The effect of exercise and heat stress on the protein translocation also creates methodological problems when plasma volume is evaluated from plasma protein concentrations.) Irrespective of the cause of sweating, hypohydration is associated with a decrease in stroke volume during exercise and a concomitant increase in heart rate during submaximal exercise (Fig. 13-13). It is remarkable, however, that during maximal exercise, oxygen uptake, cardiac output, and stroke volume are not modified by a sweat loss of up to 5 percent of body weight. However, endurance, i.e., the time that a standardized maximal exercise load can be tolerated, is definitely reduced after dehydration (Fig. 13-14). It has also been found that rectal temperatures are significantly higher in the dehydrated subject (Fig. 13-15), and the rise is related to the weight loss incurred (Gisolfi and Copping, 1974). The excessive rise in core temperature with hypohydration is probably due to inadequate sweating (Greenleaf and Castle, 1971).

Unpublished results from a study in a Norwegian cement factory, where the ambient temperature was fairly high even in winter, showed that the heart rate of the workers at a standardized submaximal work rate was significantly lower (more than 10 beats · min^{-1}) at the end of the work shift when 2 liters of fluid was taken in the course of the shift, compared with the days when no fluid was taken between meals.

The explanation for the gradual decrease in physical performance as a hypohydration develops is presently not available. It apparently cannot be primarily a modification of the aerobic energy yield, because the maximal aerobic power was not impaired in Saltin's experiments, despite a pronounced hypohydration. The explanation should be sought at the cellular level, where changes may occur during a hypohydration. The maximal isometric strength after a water deficit is reported to be unaffected (Saltin, 1964) or slightly decreased in connection with progressive hypohydration (Bosco et

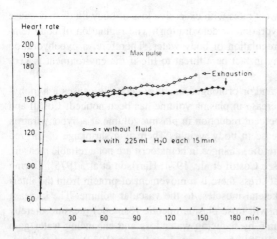

Figure 13-13
Mean heart rates in subjects running on the treadmill at 70 percent of the maximal O$_2$ uptake until exhaustion, with and without fluid. *(From Staff and Nilsson, 1971.)*

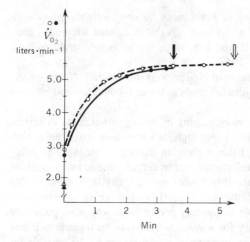

Figure 13-14
Oxygen uptake (liters · min⁻¹) at an exercise load which could be tolerated for 5¹/₂ min during normal conditions (unfilled symbols) but only for 3¹/₂ min after hypohydration (filled symbols). Arrows indicate maximal exercise time. *(From Saltin, 1964.)*

al., 1968). In any case, a reduced water content within the muscle cell and a disturbed electrolyte balance can easily influence the muscle cell's ability to contract and its susceptibility to metabolites. The reduction in work performance is more marked if a water deficit is caused by extended heavy exercise than after exposure to hot environment without exercise being involved.

Dehydration causes a reduced tilt-table tolerance. A person who normally could tolerate prolonged 45° head-up tilt with a heart rate of 90 beats · min⁻¹ fainted within

Figure 13-15
Mean rectal temperature in subjects running on the treadmill at 70 percent of their maximal oxygen uptake, with and without fluid. When no fluid was taken, rectal temperature rose significantly from the 45th to the 165th min. *(From Staff and Nilsson, 1971.)*

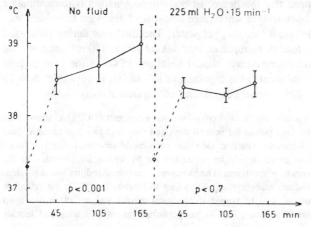

7.5 min after a fluid loss corresponding to 3 percent of his body weight. Following a fluid loss corresponding to 6 percent of his body weight, he fainted within 1.5 min. His heart rate before fainting was 115 and 135, respectively. A low stroke volume was a characteristic finding (Adolph, 1947).

Dehydration also effects crew members of high performance aircraft. Nunneley and Stribley (1979) have shown that dehydration tends to lower G-tolerance and increase the variability of response to heat.

During prolonged and physically heavy training or during participation in certain competitive sports, the sweat rate may be very high. In some cases it may be as high as 2 liters · hr^{-1}. An adequate water balance plays an important role in maintaining optimal performance capacity. It is unfortunate that in certain athletic events such as marathon running or walking, the established rules may actually limit the available fluid supply to the athletes. Similarly, the current practice of weighing in wrestlers at least 2 hr before the start of the day's match may permit those who have purposely become dehydrated in order to qualify for a lower weight class to replace their lost body water before the match if they are to compete late; this may not be possible for those who are to compete first. In any case, such dehydration is not only harmful to the individual, but also poor sportsmanship. (See the American College of Sports Medicine: Position Stand on Weight Loss in Wrestlers, *Med. Sci. Sports,* **8**(2): xi, 1976.)

Well-trained subjects are less affected in their performance by a hypohydration than untrained subjects (Buskirk et al., 1958; Saltin, 1964). Acclimatization to heat does not seem to protect from the deteriorating effect of a hypohydration.

The simplest method of determining whether the fluid intake has been adequate is by weighing the individual under standard conditions. Even a reduction in body weight of 1 to 2 percent may represent a deterioration in physical performance (Pitts et al., 1944; Adolph, 1947; Ladell, 1955; Saltin, 1964; Gisolfi and Copping, 1974).

According to Bar-Or et al. (1980), exercising children become progressively dehydrated when not forced to drink.

It should be emphasized that the degree of body hypohydration is overestimated from measurements of body weight when large amounts of glycogen have been metabolized, which will deliver a "surplus" of water. The fluid loss during prolonged heat exposure should preferably be replaced by drinking 100 to 150 ml water several times per hour. The water temperature should be about 15°C. In the case of heat exposure lasting for several weeks, the ingestion of salt tablets is advisable, 5 to 15 g per day depending on diet, climate, and degree of physical activity.

Excessive fluid loss can also occur in cold environments. Lennquist (1972) has shown that the cold-induced diuresis may persist for several days and may lead to a considerable fluid deficit, accompanied by hemoconcentration and reduction in blood volume. The cold-induced rise in osmolar excretion could largely be accounted for by significant increases in the excretion of sodium, chloride, and calcium. The increased excretion of sodium was dependent on a reduced tubular sodium reabsorption. According to Lennquist (1972), the releasing mechanism of cold diuresis is to be sought in the renal tubules and not in an increased glomerular filtration rate. Wallenberg (1974) has published results indicating that tubular sodium reabsorption in a cold-exposed person is influenced by plasma osmotic pressure and changes in arterial blood pressure. Cold-induced suppression of distal tubular sodium reabsorption could be almost completely abolished by an albumin infusion of 0.4 to 0.5 g per kg.

Summary An individual tolerates heavy physical exercise less well if subjected to a water deficit, even if the water loss is only about 1 percent of the body weight. At a submaximal work rate, the heart rate is increased, the stroke volume is reduced, and the body temperature is higher than normal. Drinking water to satiety may not fully compensate for a water loss.

PRACTICAL APPLICATION

Physical Work

It should be borne in mind that heat exposure in itself represents an extra load on the blood circulation. Exhaustion occurs much sooner during heavy physical exercise in the heat because the blood, in addition to carrying oxygen to the exercising muscle, also has to carry heat from the interior of the body to the skin. This represents an extra burden on the heart, which has to pump that much harder. This is convincingly demonstrated in Table 13-1, which shows the difference in heart rate in a subject performing the same work in a hot environment and in a cool one. The stress of heat and the hydrostatic factors in prolonged standing work may be added to the stress of work itself.

Williams et al. (1967) observed no difference in maximal O_2 uptake in subjects working in the heat and at comfort temperature. At submaximal work rates, however, they found that the major change in hemodynamics in the heat was an increase in heart rate and a fall in stroke volume. Neither cardiac output nor arteriovenous difference was significantly altered compared with comfortable conditions (Rowell, 1974). Williams et al. (1967) also demonstrated a larger lactate production in the subject who exercised in the heat as compared with one exercising in a neutral environment. This finding can be explained as a result of the reduced muscle blood flow.

Heat stress can also be a significant problem for pilots during low-level flights in hot climates, especially in fighter aircraft that impose high task loads and repetitive maneuvering forces (Nunneley and Flick, 1981). Pilots noted lowered G-tolerance and increased general fatigue on the hotter flights. Weight losses up to 2.3 percent were observed.

In the armed forces and in certain industries, the problem of an efficient and rapid method of acclimatizing a large number of people is often of practical importance. Daily exposure to a hot environment for about an hour will, after a week, result in some acclimatization. However, studies by Wyndham and co-workers (1973) to establish the minimal number of days required for acclimatization showed that a person cannot be acclimatized adequately for a normal shift of 6 to 8 hours in less than 4 hours per day and in less than 8 to 9 days. They apply a step test with a gradual increase in rate of exercise up to an oxygen uptake of 1.4 liters · min^{-1} combined with

TABLE 13-1
EFFECT OF ENVIRONMENTAL TEMPERATURE ON HUMAN RESPONSE TO STANDARD EXERCISE ON A CYCLE ERGOMETER FOR 45 MIN

Environmental temp.	Heart rate beats · min⁻¹	Rectal temp., °C T_r	Mean skin temp., °C T_s	O_2 uptake, liters · min⁻¹	Weight loss	
					kg	% of body weight
Cool	104	37.7	32.8	1.5	0.25	0.3
Hot steel mill, air temp. 40 to 50°C + radiation	166	38.8	37.6	1.5	1.15	1.6

At the same exercise rate, the temperature difference between core and shell is 4.9°C in the cool environment, but only 0.9°C in the heat. This necessitates a much greater skin blood flow in the heat. Hence, the markedly elevated heart rate in the heat.

heat stress conditions (air temperature 31.7°C, air saturated with water vapor). This program is applied to recruits for the gold mines in South Africa.

Warm-up

The benefit of the higher temperature during exercise lies in the fact that the metabolic processes in the cell can proceed at a higher rate, since these processes are temperature-dependent. For each degree of temperature increase, the metabolic rate of the cell increases by about 13 percent. At the higher temperature, the exchange of oxygen from the blood to the tissues is also much more rapid. Physical performance is improved following warm-up (Simonson et al., 1936; Asmussen and Böje, 1945). Furthermore, the nerve messages travel faster at higher temperatures. At the temperature of the human body, which is much higher than that of a frog, our nerve messages go up to 8 times as fast as those of the frog (Hill, 1927). Thus, there is a very good reason for a person to keep the body temperature up as he or she does, even at considerable expense, in order to be able to move more quickly. This is also the reason why athletes have discovered that it pays to warm up before an athletic event. This warming up may make a difference of 3 s in a 400-yard dash. The warming up may profitably consist of rather vigorous exercise, such as running at a rate of about 12 km/hr for 15 to 30 min just before the event (Högberg and Ljunggren, 1947). In the case of ordinary exercise, a 5-min warm-up consisting of light to moderate exercise is usually adequate.

Högberg and Ljunggren (1947) examined the effect of warm-up in the form of running at moderate speed combined with calisthenics on the speed of running 100, 400, or 800 m in well-trained athletes. They compared this effect with the effect of heating the body passively in a sauna bath for a period of 20 min prior to the race and found that the beneficial effect of passively elevating the body temperature by such a bath was much less than that of elevating the body temperature by a warm-up through physical exercise. In the 100-m dash, the improvement after a proper warm-up was on the order of 0.5 to 0.6 s, corresponding to 3 to 4 percent, compared with the results without any warm-up. In the 400-m race, the improvement amounted to 1.5 to 3.0 s, corresponding to 3 to 6 percent. In the 800-m race, the improvement was 4 to 6 s, or 2.5 to 5.0 percent. Thus the percentage improvement was roughly the same at all distances examined. Similar results have been obtained in swimming (Muido, 1946).

With regard to the duration of the warm-up, Högberg and Ljunggren (1947) observed better results after a 15-min warm-up than after a 5-min one, but no further significant improvement occurred in the 100-m race when the warm-up was extended from 15 to 30 min. The authors observed no deterioration in performance attributable to fatigue as a consequence of rather vigorous warm-up. They recommend a warm-up period of 15 to 30 min at a relatively high rate of energy expenditure (in their experiments about 3.0 to 3.4 liter O_2 uptake $\cdot$ min^{-1}, equivalent to running at a speed of 12 to 14 km $\cdot$ hr^{-1}). The duration and intensity of warm-up should be adjusted according to the environmental temperature and amount of clothing. The higher the environmental temperature and the greater the amount of clothing, the sooner the desired body temperature of about 38.5°C is attained (muscle temperature 39°C or higher). Ideally, the rest period between warm-up and the start of the race should be no more than a few minutes, in any case no more than 15 min. After 45 min rest, the beneficial effect of the warm-up is abolished, at which time the muscle temperature has also returned to pre-warm-up levels. The use of warm clothing is recommended during warm-up, and this clothing should be worn until the athlete is ready to start the race.

Barnard et al. (1973) report that strenuous exercise, without prior warm-up, induced abnormal ECG changes in 70 percent of their 44 normally asymptomatic subjects (aged from twenty-one to fifty-two years). Two minutes of jogging-in-place as a warm-up just prior to the exercise eliminated or reduced these abnormal ECG responses.

Figure 13-16 presents a summary of this discussion of the beneficial effect of warm-up for the physical performance. The improvement is particularly related to the increase in muscle temperature. The higher the muscle temperature (i.e., the heavier the preceding warm-up exercise), the better the performance. In accordance with this, Davies and Young (1982), observed that the peak power output during cycling and jumping was entirely related to muscle temperature, while heating of the leg muscles was without effect on supermaximal twitch and tetanic tension and maximal voluntary contraction.

Radiation

Figure 13-17 illustrates how the radiant heat may be reduced from 5,4 MJ (1300 kcal) · hr⁻¹ to about 54 kJ (13 kcal) · hr⁻¹ by placing an aluminum shield between the worker and the heat source, which in this particular case had a temperature of 188°C. An inexpensive protection against radiation may be provided by a sheet of masonite covered with tinfoil. It is important that the surface be kept clean. If the shield has to be transparent, substances that will reflect infrared light, such as glass, should be used.

It should be emphasized that behavior is an important aspect of the protection of the body in hot or cold environments. One can modify radiation to or from the body by changing the posture. The surface area of the human silhouette will increase by a factor of 3 when a person changes from crouching to an expanded body position. Just a change in orientation to the direction of the wind or sun can alter the heat exchange markedly.

Air Motion

Air motion increases the evaporation of sweat. However, if the air temperature is higher than that of the skin, the air motion may serve to increase the heat load in that it will cause the skin to pick up more heat through convection.

According to Edholm and Weiner (1981), the convective envelope of warmed air surrounding the body of a standing person may extend more than one meter above the head. In the supine position, the convective heat flow passes upward, and heat loss by convection is about 30 percent higher than in the upright position.

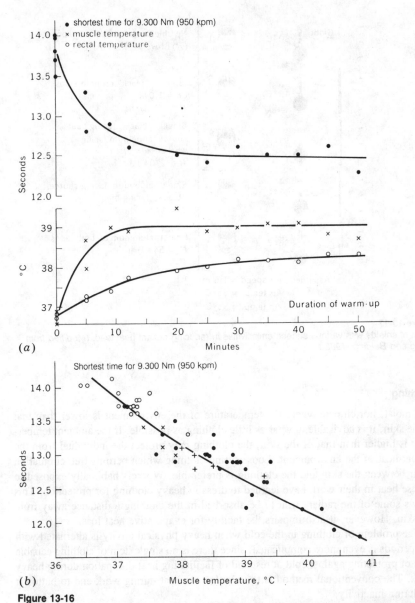

Figure 13-16
(a) Two lower curves show the temperature in the lateral vastus muscle (upper curve) and in the rectum (lower curve) after warm-up at a work rate of 160 watts of different durations (abscissa). The top curve shows the shortest period of time required for the completion of an energy output corresponding to 9.300 Nm on a cycle ergometer following warm-up of different durations as shown on the abscissa.
(b) The relationship between the time required for a spurt on the cycle ergometer (9.300 Nm) and the temperature measured in the lateral vastus muscle immediately prior to the spurt.
O = no warm-up; ● = 30-min warm-up of different intensity; X = warm-up in the form of warm showers; + = warm-up with the aid of diathermy. *(From Asmussan and Böye, 1945.)*

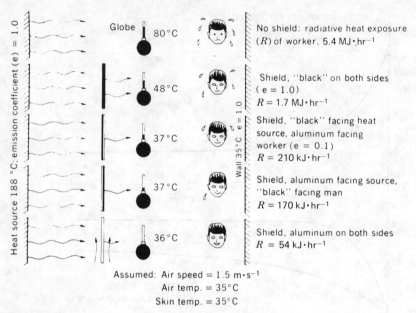

Figure 13-17
Effect of shields with various surface emissivities in reducing radiant heat load. *(Modified from Hertig and Belding, 1963.)*

Clothing

In a moist, hot climate where the temperature of the environment is lower than that of the skin, it is advisable to wear as little clothing as possible. If the ambient temperature is higher than that of the skin, the clothing may protect the individual from the radiant heat of the environment. Loose-fitting clothing which permits free circulation of air between the skin and the clothing is preferable. Workers habitually exposed to intense heat in their work have learned to dress in heavy clothing for protection. This allows some of the radiant heat to be absorbed in the clothing a distance away from the skin. However, it also impairs the facility for evaporative heat loss.

The problem of clothing in the cold when heavy physical ativity is alternated with rest periods is even more complicated, since there is no single item of clothing capable both of protecting against cold at rest and of facilitating heat dissipation during heavy work. The conventional method is to unbutton the coat during work and to button it up during inactivity.

It should also be pointed out that the surface of the hands represents about 5 percent of the total surface area of the body. In a nude individual, about 10 percent of the heat produced may be eliminated through the hands. In a clothed individual, up to 20 percent of the heat produced may be eliminated through the hands (Day, 1949).

In order to remain in heat balance, a person sleeping outdoors at −40°C needs protective clothing with an insulation value of about 12 Clo units. (A "Clo" unit equals the amount of insulation provided by the clothing a person usually wears at room temperature.) However, when the same individual is physically active, moving about or walking along, only the equivalent of 4 Clo units will be needed because the person's body heat production is now at least 3 times greater than it was when sleeping because of the increased metabolic rate associated with the increased physical activity (Fig. 13-18). This requirement is adequately met by the original double-layer caribou clothing of the Eskimo. Two layers of caribou fur, amounting to a thickness of 75 mm., have a total insulation value of about 12 Clo units (Fig. 13-19). This is adequate to maintain heat balance under practically any condition likely to be encountered by the Eskimo. Temperature measurements inside the Eskimo clothing, taken in the field, have confirmed that the Eskimo's body inside the clothing is indeed comfortably warm. There is, therefore, some truth to the old statement that the Eskimo, by virtue of their clothing, are really surrounded by a tropical climate. The ordinary uniform usually worn by airmen and soldiers in the Arctic, on the other hand, has an insulation value of only 4 Clo units, which is only one-third that of the Eskimo's clothing. The arctic uniform offers adequate protection for an active person at temperatures as low as $-40°C$, but the person would be in negative heat balance if inactive (Fig. 13-18). Temperatures below $-40°C$ occur on an average of about 2 days per month in the winter in the interior of Alaska.

The insulation value of most materials is proportional to the amount of air that is trapped within the material itself, since air is such a superb insulator. In the case of fur, air is trapped in the space between each hair, but the superior insulation quality of caribou fur, over and above other fur, lies in the fact that the caribou hair is hollow and contains trapped air inside each hair as well as in the spaces between them.

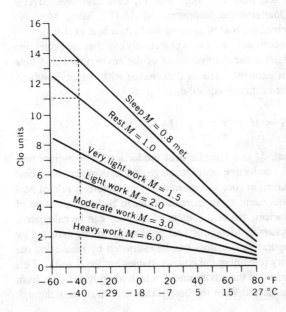

Figure 13-18
Insulating requirements in the cold when protected from the wind during different rates of heat production. A Clo unit is the thermal insulation which will maintain a resting man indefinitely comfortable in an environment of 21°C, relative humidity less than 50 percent, and air movement 6 $m \cdot min^{-1}$. The unit referred to as "met" is the metabolic rate (M) of a resting man, 3.5 ml $O_2 \cdot kg^{-1}$. *(From Burton and Edholm, 1969.)*

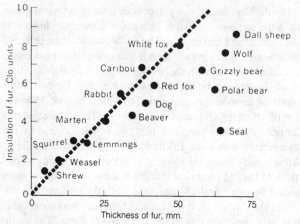

Figure 13-19
Insulating value of different furs. *(Redrawn from Scholander et al., 1950.)*

As a rough estimate, the insulating value of most clothing is approximately 1.6 Clo · cm^{-1}. As the insulating value is primarily a function of the amount of trapped air in the clothing, the type of fiber used is of less importance.

The Clo unit is a measure of the resistance of clothing to sensible heat transfer (Gagge, et al. 1941). Further systematic attempts to develop concepts and methods for the quantitative analyses of heat transfer through clothing have been made over a number of years at the U.S. Quartermaster laboratories at NATIC, Mass. More recently, Holmér (1983) has reviewed the role of clothing for human heat exchange with the environment, with particular emphasis on how to objectively evaluate the protection offered by a clothing system against the thermal stress of the environment and how to compare differences between garments. The heat exchange with the environment is expressed in terms of the heat exchange equation (see p. 582):

$$S = M \pm R \pm C - E$$

where S is storage of body heat, M is metabolic heat production, R is radiant heat exchange, C is convective and conductive heat exchange and E is evaporative heat loss. S can be calculated by measuring core and mean skin temperature (a rate of heat storage of about 40 W/m^2 is equivalent to an increase in mean body temperature of about 1°C per hour). Of these factors, the heat content of the body can be calculated, as described previously in this chapter. Metabolic heat production can be assessed by measuring oxygen uptake. Evaporative heat loss can be estimated by measuring the changes in body weight. The only remaining unknown variables are the R and C, i.e., heat transmitted via the skin surface to the environment, which roughly amounts from 80 to 90 percent of all heat transferred from the body. This heat must pass through the clothing layers, covering up to 95 percent of the body surface. This transfer takes place as dry (sensible) heat transfer, and as humid (latent) heat transfer.

The dry heat transfer, H_{dry}, accounts for the heat transported by convection, radiation, and conduction and may be defined by the followed equation:

$$H_{dry} = \frac{(T_s - T_a)}{R_c}$$

where T_s = mean skin temperature

T_a = ambient air temperature

R_c = resistance of clothing and air layers

to dry heat transfer, expressed in Clo units

The humid heat loss (H_{humid}) accounts for the latent heat of sweat evaporated at the skin surface and transported as vapor through clothing and air layers. The humid heat loss =

$$\frac{P_s - P_a}{R_c}$$

where P_s = the actual, average water vapor pressure at the skin surface

P_a = the ambient water vapor pressure, expressed in Pascal

R_c = the resistance to evaporative heat transfer by clothing and air layers.

Thermal properties of clothing systems may be determined either by the guarded hot plate technique, the thermal manikin or "copper man," or by means of direct or partitional calorimetry (Holmér and Elnäs, 1981).

A physiological comparison of the insulating value of clothing may be attained simply by measuring heat production (oxygen uptake) and heat loss as evidenced by changes in stored body heat (by applying the formula and procedures for obtaining rectal and mean skin temperatures described earlier in this chapter) in normal subjects under controlled climatic-chamber conditions. An example of such a study is presented in Fig. 13-20, in which similar garments made of nylon pile and wool pile were compared in paired experiments at rest for 1 hr and during 2 hr of fairly strenuous physical activity (treadmill walking at 100 m/min, 5° incline) followed by a 2-hr rest in a climatic chamber at $-20°C$ (the values represent the means of five subjects). Evaporative weight loss was determined by weighing the subjects in the nude before and after the experiment. The accumulation of moisture in the experimental clothing was assessed by weighing the garments before and after the experiment on a scale with an accuracy of ± 10 g. In this study, no significant difference could be detected between the two types of garments in terms of thermal insulation, nor in the ability of the two types of fabric to allow free escape of moisture produced by sweating during the physical activity (Rodahl et al., 1974).

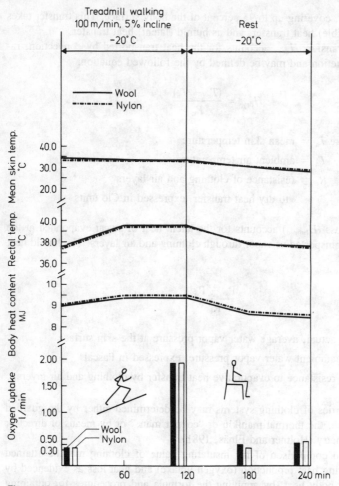

Figure 13-20
A comparison between similar garments made of nylon pile and of wool pile during 2-hr treadmill walking and 2-hr rest in a climatic chamber at −20°C. There was no statistically significant difference in O_2 uptake, skin or rectal temperature, or stored heat resulting from the two types of garments worn. Figures represent means of five subjects. *(From Rodahl et al., 1974.)*

Microclimate

Preferred environmental temperatures range from 17°C up to 31°C, depending on the climate and the clothing worn (Edholm and Weiner, 1981). However, 21° to 24°C represents the comfort zone for a large majority of individuals. Edholm and Weiner (1981) conclude that the most effective environmental temperature for mental effort is 28°C in terms of number of signals missed in a particular set of experiments.

Evidently the temperature of the face may be critical, for according to Crawshaw et al., (1975), the skin of the face is four times as sensitive to the sensation of heat as is the skin of the thigh, per unit surface area.

The solution to the problem of providing optimal working environment may be to create local microclimates by cooling or heating the clothing, by providing environmental suits to be worn under special circumstances, or by enclosing the work area in a suitable artificially made environment, which will facilitate an effective climatic control. Radiant heaters may be used, and exposure suits may be applicable. With any solution, there will be certain complications. In the ideal climate, the temperature of the skin is about 33°C, but not uniformly so. The feet are normally colder than the trunk, and a person may accept a greater lowering of the temperature of the feet without feeling cold. A bath, with a water temperature of 33°C, feels cold. In order for the bather to feel comfortable, the water temperature has to be about 35°C. However, such a water temperature does not produce temperature equilibrium; it will cause the body temperature to rise. As Burton (1963) puts it: man is not constructed to spend much time in water.

A local heating of the floor may represent an unphysiological manner of regulating room temperature (Burton, 1963) due to the fact that the receptors in the skin of the feet exert a relatively dominating influence upon the temperature-regulating center. An induced vasodilation of the feet is effective in increasing heat loss. In spite of warm feet, the subjects of one study eventually became cold. With this background, it might appear unphysiological to heat our dwellings, factories, etc., by keeping the floor hotter than the room air.

It is thus possible, by improper clothing or by local heating or cooling of limited skin areas, to upset the normal physiological temperature regulation. Local heating of hands and feet may, for example, bring about shivering and sweating at the same time. Even if the air temperature is high, heat loss by outgoing radiation to cold surfaces, like a cold window, may cause a most unpleasant sensation, commonly referred to as draft. Because of radiative heat loss to the night sky, a person exposed, unshielded, in the arctic environment may actually be exposed to a cold stress 10 to 20° colder than that which the air thermometer might indicate. In a small enclosed area, such as the cabin of an aircraft or a car, it is difficult to satisfy the requirement for adequate ventilation, hot or cold, without causing certain areas of the body to be too hot or too cold. It is evident that much money, as well as many heart beats and much sweating, could be saved by proper planning of lecture halls, office and factory buildings, and machinery, taking into account all of the factors that constitute the optimal climate. Even so, however, it is not possible at all times to satisfy the individual comfort requirements of all of those who work in large open rooms or offices, due to individual differences in temperature sensation, metabolic rate, clothing, etc.

Rohles and Munson (1981) recorded EEC and skin temperature in six men and six women while sleeping in environmental temperatures 10°C, 21.1°C, and 32.2°C. They found that the proportion of time in each sleep stage was not affected by the temperature of the sleep environment.

Vokac and Hjeltnes (1981) have shown that the fall in core temperature when going to bed may not be due to the onset of sleep, but to the thermal resistance of the bedclothes causing an elevation of the temperature of the air trapped under the bedclothes, resulting in a redistribution of body heat, increasing the peripheral temperature and lowering the core temperature.

The secret of success of an experienced arctic traveler or hunter lies in the ability to avoid the extreme cold. The Eskimo's dwelling, whether it be a skin tent, a peat-covered house, or a log cabin, is comfortably warm at all times. The temperature is kept around 21°C (70°F) in the day and may drop to about 10°C (50°F) during the night, when the sleeping Eskimo is well covered by fur. Although, in the summer, the Eskimo may spend as much as 9 hr or more out of doors, the average amount of time spent outside in the winter is only 1 to 4 hr. Furthermore, experience has taught arctic dwellers to take advantage of the characteristic temperature distribution in their environment, produced by the so-called temperature inversion during the winter. The coldest spot is at the surface of the snow, and especially in a depression in the terrain, such as a riverbed where cold, heavy air is trapped. A few feet up the hillside, the temperature is usually many degrees warmer. While the temperature at the actual snow surface may be as cold as $-50°C$, the temperature under the snow cover, in the narrow air space between the ground and the snow crust, is usually maintained at -9 to $-6°C$ throughout the winter. It is here that the field mice and other arctic rodents, which do not hibernate, run around perfectly comfortable all winter long. Actually, smaller animals, like weasels and mice, are not able to carry a fur thick enough for insulation and must therefore spend the winter mostly underneath the snow. By taking advantage of this characteristic temperature distribution, the experienced traveler may successfully escape the extreme degrees of cold stress.

REFERENCES

ACGIH (American Conference of Governmental Industrial Hygienists): "Threshold Limit Values for Chemical Substances and Physical Agents in the Workroom Environment with Intended Changes for 1976," Cincinnati, Ohio, 1976.

Adolph, E. F., and Members of the Rochester Desert Unit: "Physiology of Man in the Desert," Interscience Publishers, Inc., New York, 1947.

Ahlman, K., and M. J. Karvonen: Stimulating of Sweating by Exercise after Heat Induced "Fatigue" of the Sweating Mechanism, *Acta Physiol. Scand.*, **53**:381, 1961.

Andersson, B.: The Thirst Mechanism as a Link in the Regulation of the "Milieu Intérieur," in "Les Concepts de Claude Bernard sur le Milieu Intérieur," p. 13, Masson et Cie, Paris, 1967.

Asmussen, E., and O. Böje: Body Temperature and Capacity for Work, *Acta Physiol. Scand.*, **10**:1, 1945.

Åstrand, I.: Aerobic Work Capacity in Men and Women with Special Reference to Age, *Acta Physiol. Scand.*, **49**(Suppl. 169):67, 1960.

Åstrand, I., O. Axelson, U. Eriksson, and L. Olander: Heat Stress in Occupational Work, *AMBIO* **3**:37, 1975.

Avellini, B. A., E. Kamon, and J. T. Krajewski: Physiological Responses of Physically Fit Men and Women to Acclimatization to Heat, *J. Appl. Physiol.*, **49**(2):254, 1980.

Avellini, B. A., Y. Shapiro, S. M. Fortney, C. B. Wenger, and K. B. Pandolf: Effects on Heat Tolerance of Physical Training in Water and on Land, *J. Appl. Physiol.*, **53**:1291, 1982.

Baker, M. A.: Brain Cooling in Endotherms in Heat and Exercise, *Ann. Rev. Physiol.*, **44**:85, 1982.

Barnard, R. J., G. W. Gardner, N. V. Diaco, R. N. MacAlpin, and A. A. Kattus: Cardiovascular Responses to Sudden Strenuous Exercise—Heart Rate, Blood Pressure, and ECG, *J. Appl. Physiol.*, **34**:833, 1973.

Bar-Or, O.: Climate and the Exercising Child—A Review, *Int. J. Sports Med.*, **1**(2):53, 1980.

Bar-Or, O., R. Dotan, O. Inbar, A. Rothstein, and H. Zonde: Voluntary Hypohydration in 10- to 12-year-old boys, *J. Appl. Physiol.*, **48**(1):104, 1980.

Bass, E. E.: Thermoregulatory and Circulatory Adjustments during Acclimatization to Heat in Man, in J. D. Hardy (ed.), "Temperature: Its Measurement and Control in Science and Industry," vol. 3, part 3, p. 299, Reinhold Book Corporation, New York, 1963.

Bazett, H. C., L. Love, M. Newton, L. Eisenberg, R. Day, and R. Forster: Temperature Changes in Blood Flowing in Arteries and Veins in Man, *J. Appl. Physiol.*, **1**:3, 1948.

Belding, H. S.: "The Industrial Environment: Its Evaluation and Control", Dept. Health, Education and Welfare Publ. No. NIOSH (Wash. Govt. Printing Office), p. 563, 1973.

Benzinger, T. H., and G. W. Taylor: Cranial Measurements of Internal Temperature in Man, in J. D. Hardy (ed.), "Temperature: Its Measurement and Control in Science and Industry," vol. 3, part 3, p. 111, Reinhold Book Corporation, New York, 1963.

Berggren, G., and E. H. Christensen: Heart Rate and Body Temperature as Indices of Metabolic Rate during Work, *Arbeitsphysiol.*, **14**:255, 1950.

Bergh, U.: Human Power at Subnormal Body Temperatures, *Acta Physiol. Scand.*, (Suppl. 478):1980.

Blatteis, C. M.: The Newer Putative Central Neurotransmitters: Roles in Thermoregulation, *Fed. Proc.*, **40**(13):2735, 1981.

Bligh, J., and R. E. Moore (eds.): "Essays on Temperature Regulation," North-Holland, Amsterdam, 1972.

Bosco, J. S., R. L. Terjung, and J. E. Greenleaf: Effects of Progressive Hypohydration on Maximal Isometric Muscular Strength, *J. Sports Med.*, **8**:81, 1968.

Botsford, J. H.: A Wet Globe Thermometer for Environmental Heat Measurement, *Amer. Ind. Hyg. Assoc. J*, **32**:1, 1971.

Budd, G. M.: Acclimatization to Cold in Antarctica as Shown by Rectal Temperature Response to a Standard Cold Stress, *Nature (London)*, **193**:886, 1962.

Burton, A. C.: Human Calorimetry, II—The Average Temperature of the Tissues of the Body, *J. Nutr.*, **9**:261, 1935.

Burton, A. C.: The Pattern of Response to Cold in Animals and the Evolution of Homeothermy, in J. D. Hardy (ed.), "Temperature: Its Measurement and Control in Science and Industry," vol. 3, part 3, p. 363, Reinhold Book Corporation, New York, 1963.

Burton, A. C., and O. G. Edholm: "Man in a Cold Environment," Hafner Publishing Company, Inc., New York-London, 1969.

Buskirk, E. R., P. F. Iampietro, and D. E. Bass: Work Performance after Dehydration: Effects of Physical Conditioning and Heat Acclimatization, *J. Appl. Physiol.* **12**:189, 1958.

Cabanac, M.: Temperature Regulation, *Ann. Rev. of Physiol.* **37**:415, 1975.

Carithers, R. W., and R. C. Seagrave: Canine Hyperthermia with Cerebral Protection, *J. Appl. Physiol.*, **40**(4):543, 1976.

Christensen, E. H.: Beiträge zur Physiologie schwerer Körperlicher Arbeit, Die Körpertemperatur während und unmittelbar nach schwerer körperlicher Arbeit, *Arbeitsphysiol.*, **4**:154, 1931.

Ciriello, V. M., and S. H. Snook: The Prediction of WBGT from the Botsball, *Amer. Ind. Hyg. Assoc. J.*, **38**:264, 1977.

Colquhoun, W. P.: Circadian Rhythms Mental Efficiency and Shift Work, *Ergonomics*, **13**(5):558, 1970.

Costill, D. L., L. Branam, D. Eddy, and W. Fink: Alterations in Red Cell Volume Following Exercise and Dehydration, *J. Appl. Physiol.*, **37**:912, 1974.

Costill, D. L., and W. J. Fink: Plasma Volume Changes Following Exercise and Thermal Dehydration, *J. Appl. Physiol.*, **37**:521, 1974.

Crawshaw, L. I., E. R. Nadel, J. A. J. Stolwijk, and B. A. Stanford: Effect of Local Cooling on Sweating Rate and Cold Sensation, *Pflügers Arch.*, **354**:19, 1975.

Davies, C. T. M.: Thermoregulation during Exercise in Relation to Sex and Age, *Eur. J. Appl. Physiol.*, **42**:71, 1979.

Davies, C. T. M.: Influence of Air Flow and Skin Temperature on Sweating at the Onset, During and Following Exercise, *Ergonomics*, **23**(6):559, 1980.

Davies, C. T. M.: Thermal Responses to Exercise in Children, *Ergonomics*, **24**(1):55, 1981.

Davies, C. T. M., J. R. Brotherhood, and E. Zeidifard: Temperature Regulation during Severe Exercise with Some Observations on Effects of Skin Wetting, *J. Appl. Physiol.*, **41**:772, 1976.

Davies, C. T. M., and K. Young: Maximal Power Output in Relation to the Contractile Properties of the Triceps Surae in Man, *J. Physiol.*, **325**:51, 1982.

Davies, M., B. Ekholm, U. Bergh, and I-L. Kanstrup-Jensen: The Effect of Hypothermia on Submaximal and Maximal Work Performance, *Acta Physiol. Scand.*, **95**:201, 1975.

Day, R.: Regional Heat Loss, in L. W. Newburgh (ed.), "1, Physiology of Heat Regulation," p. 240, W. B. Saunders Company, Philadelphia, 1949.

Dill, D. B. (ed.): "Handbook of Physiology," sec 4, Adaptation to the Environment, American Physiological Society, Washington, D. C., 1964.

Dimri, G. P., M. S. Malhotra, J. Sen Gupta, T. Sampath Kumar, and B. S. Arora: Alterations in Aerobic-Anaerobic Proportions of Metabolism during Work in Heat, *Eur. J. Appl. Physiol.*, **45**:43, 1980.

Drinkwater, B. L., J. E. Denton, I. C. Kupprat, T. S. Talag, and S. M. Horvath: Aerobic Power as a Factor in Women's Response to Work in Hot Environment, *J. Appl. Physiol.*, **41**:815, 1976.

Drinkwater, B. L., J. F. Bedi, A. B. Loucks, S. Roche, and S. M. Horvath: Sweating Sensitivity and Capacity of Women in Relation to Age, *J. Appl. Physiol.*, **53**(3):671, 1982.

Dukes-Dobos, F. N.: Hazards of Heat Exposure, *Scand. J. Work Environ. Health*, **7**:73, 1981.

Edholm, O. G., and Y. S. Weiner: "The Principles and Practice of Human Physiology", Academic Press, London, 1981.

Eisalo, A.: Effects of the Finnish Sauna on Circulation, *Annals Medicinae Experimentalis et Biologicae Finniae, vol. 34*, (Suppl. 4), 1956.

Ellis, H. D.: The Effects of Cold on the Performance of Serial Choice Reaction Time and Various Discrete Tasks, *Human Factors*, **24**(5):589, 1982.

Enander, A., B. Skøldstrøm, and I. Holmér: Reaction to Hand-Cooling in Workers Occupationally Exposed to Cold, *Scand J. Work Environ. Health*, **6**:58, 1980.

Fortney, S. M., and L. C. Senay, Jr.: Effect of Training and Heat Acclimation on Exercise Responses of Sedentary Females, *J. Appl. Physiol.*, **47**(5):978, 1979.

Fortney, S. M., E. R. Nadel, C. B. Wenger, and J. R. Bove: Effect of Acute Alterations of Blood Volume on Circulatory Performance in Humans, *J. Appl. Physiol.*, **50**(2):292, 1981a.

Fortney, S. M., E. R. Nadel, C. B. Wenger, and J. R. Bove: Effect of Blood Volume on Sweating Rate and Body Fluids in Exercising Humans, *J. Appl. Physiol.*, **51**(6):1594, 1981b.

Fox, W. F.: Human Performance in the Cold, *Human Factors*, **9**(3):203, 1967.

Frye, A. J., and E. Kamon: Sweating Efficiency in Acclimated Men and Women Exercising in Humid and Dry Heat, *J. Appl. Physiol.*, **54**(4):972, 1983.

Gagge, A. P., A. C. Burton, and H. C. Bazett: A Practical System of Units for Description of Heat Exchange of Man with his Environment, *Science*, **94**:428, 1941.

Gale, C. C.: Neuroendocrine Aspects of Thermoregulation, *Ann. Rev. Physiol.*, **35**:391, 1973.

Gauer, O. H., J. P. Henry, and C. Behn: The Regulation of Extracellular Fluid Volume, *Ann. Rev. Physiol.*, **32**:547, 1970.

Gerking, S. D., and S. Robinson: Decline in the Rates of Sweating of Men Working in Severe Heat, *Am. J. Physiol.*, **147**:370, 1946.

Gisolfi, C. V., and J. R. Copping: Thermal Effects of Prolonged Treadmill Exercise in the Heat, *Med. Sci. Sports*, **6**:108, 1974.

Goetz, K. L., G. C. Bond, and D. D. Bloxham: Atrial Receptors and Renal Function, *Physiol. Rev.*, **55**:157, 1975.

Golden, F. St. C., I. F. G. Hampton, G. R. Harvey, and A. V. Knibbs: Shivering Intensity in Humans during Immersion in Cold Water, *J. Physiol. (London)*, **290**(2):48, 1979a.

Golden, F. St. C., I. F. G. Hampton, and D. Smith: Cold Tolerance in Long-distance Swimmers, *J. Physiol. (London)*, **290**(2):49, 1979b.

Greenleaf, J. E.: Hyperthermia and Exercise in D. Robertshaw (ed.), "International Review of Physiology, Environmental Physiology III," vol. 20, University Park Press, Baltimore, 1979.

Greenleaf, J. E., and B. I. Castle: Exercise Temperature Regulation in Man during Hypohydration and Hyperhydration, *J. Appl. Physiol.*, **30**:847, 1971.

Hammel, H. T.: Neurons and Temperature Regulation, in W. S. Yamamoto and J. R. Brobeck (eds.), "Physiological Controls and Regulations," p. 71, W. B. Saunders Company, Philadelphia, 1965.

Hardy, J. D.: Central and Peripheral Factors in Physiological Temperature Regulation, in "Les Concepts de Claude Bernard sur le Milieu Intérieur," p. 247, Masson et Cie, Paris, 1967.

Hardy, J. D., and E. F. DuBois: "The Technique of Measuring Radiation and Convection," *J. Nutr.*, **15**:461, 1938.

Hardy, J. D., A. P. Gagge, J. A. J. Stolwijk (eds.): "Physiological and Behavioral Temperature Regulation," Charles C. Thomas, Springfield, Ill., 1970.

Hardy, J. D., J. A. J. Stolwijk, and A. P. Gagge: in G. C. Whittow (ed.), "Comparative Physiology of Thermoregulation," vol. II, p. 327, Academic Press, Inc., New York, 1971.

Harrison, M. H.: Effects of Thermal Stress and Exercise on Blood Volume in Humans, *Physiol. Rev.*, **65**(1):149, 1985.

Harrison, M. H., R. J. Edwards, and D. R. Leitch: Effect of Exercise and Thermal Stress on Plasma Volume, *J. Appl. Physiol.*, **39**:925, 1975.

Harrison, M. H., R. J. Edwards, M. J. Graveney, L. A. Cochrane, and J. A. Davies: Blood Volume and Plasma Protein Responses to Heat Acclimatization in Humans, *J. Appl. Physiol.*, **50**(3):597, 1981.

Hellström, B.: "Local Effects of Acclimatization to Cold in Man," Universitetetsforlag, Oslo, 1965.

Hensel, H.: Thermoreceptors, *Ann. Rev. Physiol.*, **36**:233, 1974.

Hensel, H.: Thermoreception and Temperature Regulation, *Monographs of the Physiological Society*, no. 38, Academic Press, London, 1981.

Hertig, B. A., and H. S. Belding: Evaluation and Control of Heat Hazards, in J. D. Hardy (ed.), "Temperature: Its Measurements and Control in Science and Industry," vol. 3, part 3, p. 347, Reinhold Book Corporation, New York, 1963.

Hill, A. V.: "Living Machinery," Harcourt, Brace & World, Inc., New York, 1927.

Högberg, P., and O. Ljunggren: Uppvärmningens inverkan på löpprestationerna, *Svensk Idrott*, **40**, 1947.

Holmér, I: Role of Clothing for Man's Heat Exchange with the Environment, *Proceedings of the International Conference on Protective Clothing Systems*, Aug. 23–27, 1981, pp. 31–50, Stockholm, Sweden, 1983.

Holmér, I., and U. Bergh: Metabolic and Thermal Response to Swimming in Water at Varying Temperatures, *J. Appl. Physiol.*, **37**:702, 1974.

Holmér, I., and S. Elnäs: Physiological Evaluation of the Resistance to Evaporative Heat Transfer by Clothing, *Ergonomics*, **24**(1):63, 1981.

Horstman, D. H., and E. Christensen: Acclimatization to Dry Heat: Active Men Versus Active Women, *J. Appl. Physiol.*, **52**(4):825, 1982.

Horwitz, B. A.: Metabolic Aspects of Thermogenesis: Neural and Hormonal Control, *Fed. Proc.*, **38**(8):2147, 1979.

Houdas, Y., and E. F. J. Ring: "Human Body Temperature," Plenum Publishing Corporation, New York and London, 1982.

Inbar, O., O. Bar-Or, R. Dotan, and B. Gutin: Conditioning Versus Exercise in Heat as Methods for Acclimatizing 8-to 10-year-old Boys to Dry Heat, *J. Appl. Physiol.*, **50**(2):406, 1981.

Irving, L.: Adaptations to Cold, *Sci. Am.*, **214**(1):94, 1966.

Irving, L.: Ecology and Thermoregulation, in "Les Concepts de Claude Bernard sur le Milieu Intérieur," p. 381, Masson et Cie, Paris, 1967.

Issekutz, B., Jr., K. Rodahl, and N. C. Birkhead: Effect of Severe Cold Stress on the Nitrogen Balance of Men under Different Dietary Conditions, *J. Nutr.*, **78**:189, 1962.

Johnson, J. M., G. L., Brengelmann, and L. B. Rowell: Interactions between Local and Reflex Influences on Human Forearm Skin Blood Flow, *J. Appl. Physiol.*, **41**:826, 1976.

Kamon, E.: Scheduling Cycles of Work for Hot Ambient Conditions, *Ergonomics*, **22**(4):427, 1979.

Kandel, E. R., and J. H. Schwartz: "Principles of Neural Science," Edward Arnold Ltd., London, 1981.

Keatinge, W. R.: "Survival in Cold Water," Blackwell, Oxford, 2nd. ed., 1978.

Kenshalo, D. R., J. P. Nafe, and B. Brooks: Variations in Thermal Sensitivity, *Science*, **134**:104, 1961.

Kerslake, D. McK.: Errors Arising from the Use of Mean Heat Exchange Coefficients in Calculation of the Heat Exchanges of a Cylindrical Body in a Transverse Wind, in J. D. Hardy (ed.), "Temperature: Its Measurement and Control in Science and Industry," vol. 3, part 3, p. 183, Reinhold Book Corporation, New York, 1963.

Kerslake, D. McK.: "Monographs of the Physiological Society: The Stress of Hot Environment," Cambridge University Press, Cambridge, 1972.

Kloetzel, K., A. Etelvino de Andrade, J. Falleiros and J. Cota Pacheco: Relationship between Hypertension and Prolonged Exposure to Heat, *J. Occup. Med.*, **15**(11):878, 1973.

Kolstad, A.: Corneal Sensitivity by Low Temperatures, *Acta Ophthalmol.*, **47**:656, 1969.

Kolstad, A., and R. Opsahl: Cold Injury to Corneal Epethelium, A Cause of Blurred Vision in Cross-Country Skiers, *Acta Ophthalmol.*, **48**:789, 1979.

Kuno, Y.: "Human Perspiration," Charles C. Thomas, Springfield, Ill., 1956.

Ladell, W. S. S.: The Effects of Water and Salt Intake upon the Performance of Men Working in Hot and Humid Environments, *J. Physiol.*, **127**:11, 1955.

Lagerspetz, K. Y. H.: Reactions and Adaptions of Organisms to Cold., *Nordic Council Arct. Med. Res. Rep.*, **30**:7, 1982.

LeBlanc, J.: "Man in the Cold," American Lecture Series, Charles C. Thomas, Springfield, Ill., 1975.

LeBlanc, J., S. Dulac, J. Côté, and B. Girard: Autonomic Nervous System and Adaptation to Cold in Man, *J. Appl. Physiol.*, **39**:181, 1975.

LeBlanc, J., J. Côté, S. Dulac, and F. Dulon-Turcot: Effect of Age, Sex and Physical Fitness on Responses to Local Cooling, *J. Appl. Physiol.*, **44**:813, 1978.

Leithead, C. S., and A. R. Lind: "Heat Stress and Heat Disorders," Cassell & Company, Ltd., London, 1964.

Lennquist, S.: Cold Induced Diuresis, *Scand. J. Urology and Nephrology*, (Suppl. 9), 1972.

Libert, J. P., V. Candas, and J. J. Vogt: Sweating Response in Man During Transient Rises of Air Temperature, *J. Appl. Physiol.*, **44**(2):284, 1978.

Libert, J. P., V. Candas, and J. J. Vogt: Effect of Rate of Change in Skin Temperature on Local Sweating Rate, *J. Appl. Physiol.*, **47**(2):306, 1979.

Libert, J. P., V. Candas, J. J. Vogt, and P. Mairiaux: Central and Peripheral Inputs in Sweating Regulation during Thermal Transients, *J. Appl. Physiol.*, **52**(5):1147, 1982.

Libert, J. P., V. Candas, and J. J. Vogt: Modifications of Sweating Responses to Thermal Transients following Heat Acclimation, *Eur. J. Appl. Physiol.*, **50**:235, 1983.

Lind, A. R., and D. E. Bass: Optimal Exposure Time for Development of Acclimatization to Heat, *Fed. Proc.*, **22**:704, 1963.

Lind, A. R., P. W. Humphreys, K. J. Collins, K. Foster, and K. F. Sweetland: Influence of Age and Daily Duration of Exposure on Responses of Men to Work in Heat, *J. Appl. Physiol.*, **28**:50, 1970.

Mairiaux, P., J. C. Sagot, and V. Candas: Oral Temperature as an Index of Core Temperature during Heat Transients, *Eur. J. Appl. Physiol.*, **50**:331, 1983.

McCaffrey, T. V., G. S. Geis, J. M. Chung, and R. D. Wurster: Effect of Isolated Head Heating and Cooling on Sweating in Man, *Aviat. Space Environ. Med.*, **46**(11):1353, 1975.

McCaffrey, T. V., R. D. Wurster, H. K. Jacobs, D. E. Euler, and G. S. Geis: Role of Skin Temperature in the Control of Sweating, *J. Appl. Physiol.*, **47**(3):591, 1979.

Minard, D., and L. Copman: Elevation of Body Temperature in Disease, in J. D. Hardy (ed.), "Temperature: Its Measurement and Control in Science and Industry," vol. 3, part 3, p. 253, Reinhold Book Corporation, New York, 1963.

Muido L.: The Influence of Body Temperature on Performances in Swimming, *Acta Physiol. Scand.*, **12**:102, 1946.

Nadel, E. R., J. W. Mitchell, B. Saltin, and J. A. J. Stolwijk: Peripheral Modifications to the Central Drive for Sweating, *J. Appl. Physiol.*, **31**:828, 1971.

Nadel, E. R., and J. A. J. Stolwijk: Effect of Skin Wettedness on Sweat Gland Response, *J. Appl. Physiol.*, **35**:689, 1973.

Nadel, E. R., K. B. Pandolf, M. F. Roberts, and J. A. J. Stolwijk: Mechanisms of Thermal Acclimation to Exercise and Heat, *J. Appl. Physiol.*, **37**:515, 1974.

Nadel, E. R., E. Cafarelli, M. F. Roberts, and C. B. Wenger: Circulatory Regulation during Exercise in Different Ambient Temperatures, *J. Appl. Physiol.*, **46**(3):430, 1979.

Nadel, E. R., S. M. Fortney, and C. B. Wenger: Effect of Hydration State on Circulatory and Thermal Regulations, *J. Appl. Physiol.*, **49**(4):715, 1980.

Nelms, J. D., and J. G. Soper: Cold Vasodilatation and Cold Acclimatization in the Hands of British Fish Filleters, *J. Appl. Physiol.*, **17**:444, 1962.

Nielson, M.: Die Regulation der Körpertemperatur bei Muskelarbeit, *Skand. Arch. Physiol.*, **79**:193, 1938.

Nielsen, B.: Thermoregulation in Rest and Exercise, *Acta Physiol. Scand.* (Suppl. 323): 1969.

Nielsen, B.: Physical Effort and Thermoregulation in Man, *Israel J. Med., Sci.*, **12**:974, 1976.

Nielsen, B.: Exercise and Temperature Regulation in Z. Szelényi and M. Székely (eds.), in Satellite of the 28th International Congress of Physiological Sciences, Pécs, p. 537, 1980.

Nunneley, S. A., and F. Stribley: Heat and Acute Dehydration Effects on Acceleration Response in Man, *J. Appl. Physiol.,* **47**(1): 197, 1979.

Nunneley, S. A., and C. F. Flick: Heat Stress in the A-10 Cockpit: Flights over Desert, *Aviat. Space Environ. Med.,* **52**(9):513, 1981.

Oddershede, I. R., and R. S. Elizondo: Body Fluid and Hematologic Adjustments during Resting Heat Acclimation in Rhesus Monkey, *J. Appl. Physiol.,* **49**(3):431, 1980.

Pepler, R. D.: Performance and Well-being in Heat, in J. D. Hardy (ed.), "Temperature: Its Measurement and Control in Science and Industry," vol. 3, part 3, p. 319, Reinhold Book Corporation, New York, 1963.

Pitts, G. C., R. E. Johnson, and F. C. Consolazio: Work in the Heat as Affected by Intake of Water, Salt and Glucose, *Amer. J. Physiol.,* **142**:253, 1944.

Poulos, D. A.: Neurophysiology of Temperature Regulation: Introductory Remarks, *Fed. Proc.,* **40**(14):2803, 1981a.

Poulous, D. A.: Central Processing of Cutaneous Temperature Information, *Fed Proc.,* **40**(14), 2825, 1981b.

Precht, H., J. Christophersen, H. Hensel, and W. Larcker: "Temperature and Life," Springer-Verlag, Berlin, 1973.

Provins, K. A., and R. S. Clarke: The Effect of Cold on Mammal Performance, *J. Occup. Health,* **2**:169, 1960.

Radomski, M. W., and C. Boutelier: Hormone Response of Normal and Intermittent Cold-Preadapted Humans to Continuous Cold, *J. Appl. Physiol.,* **53**(3):610, 1982.

Ramanathan, L. N.: A New Weighing System for Mean Surface Temperature of the Human Body, *J. Appl. Physiol.,* **19**(3):531, 1964.

Ramsay, J. E., and C. P. Chai: Inherent Variability in Heat-Stress Decision Rules, *Ergonomics,* **26**:495, 1983.

Riggs, C. E., D. J. Johnson, B. J. Konopka, and R. D. Kilgour: Exercise Heart Rate Response to Facial Cooling, *Eur. J. Appl. Physiol.,* **47**(4):323, 1981.

Riggs, C. E., D. J. Johnson, R. D. Kilgour, and B. J. Konopka: Metabolic Effects of Facial Cooling in Exercise, *Aviat. Space Environ. Med.* **54**(1):22, 1983.

Roberts, M. F., and S. B. Wenger: Control of Skin Blood Flow during Exercise by Thermal Reflexes and Baroreflexes, *J. Appl. Physiol.,* **48**(4):717, 1980.

Robinson, S.: Circulatory Adjustments of Men in Hot Environments, in J. D. Hardy (ed.), "Temperature: Its Measurement and Control in Science and Industry," vol. 3, part 3, p. 287, Reinhold Book Corporation, New York, 1963.

Robinson, S., E. S. Turrell, H. S. Belding, and S. M. Horvath: Rapid Acclimatization to Work in Hot Climates, *Am. J. Physiol.,* **140**:168, 1943.

Rodahl, K.: Basal Metabolism of the Eskimo, *J. Nutr.,* **48**:359, 1952.

Rodahl, K.: Nutritional Requirements in the Polar Regions, U.N. Symposium on Health in the Polar Regions, Geneva, 1962, *WHO Public Health Paper,* **18**:97, 1963.

Rodahl, K., and G. Bang: "Thyroid Activity in Men Exposed to Cold," *Arctic Aeromed. Lab. Tech. Report* **57-36,** 1957.

Rodahl, K., S. M. Horvath, N. C. Birkhead, and B. Issekutz, Jr.: Effects of Dietary Protein on Physical Work Capacity during Severe Cold Stress, *J. Appl. Physiol.,* **17**:763, 1962.

Rodahl, K., F. A. Giere, P. H. Staff, and B. Wedin: A Physiological Comparison of the Protective Value of Nylon and Wool in a Cold Environment in A. Borg and J. H. Veghte (eds.), AGARD Report, no. 620, 1974.

Rohles, F. H., and D. M. Munson: Sleep and the Sleep Environment Temperature, *J. Environ. Psychol.,* **1**:207, 1981.

Rowell, L. B.: Human Cardiovascular Adjustment to Exercise and Thermal Stress, *Physiol., Rev.*, **54**:75, 1974.

Rowell, L. B., H. J. Marx, R. A. Bruce, R. D. Conn, and F. Kusumi: Reduction in Cardiac Output, Central Blood Volume and Stroke Volume with Thermal Skin in Normal Man during Exercise, *J. Clin. Invest.*, **45**:1801, 1965.

Rowell, L. B., Cardiovascular Adjustments to Thermal Stress, in "Handbook of Physiology, sect. 2: The Cardiovascular System," vol. III: Peripheral Circulation and Organ Blood Flow, part 2, chap. 25, p. 967, American Physiological Society, Bethesda, MD., 1983.

Saltin, B.: Aerobic Work Capacity and Circulation at Exercise in Man, *Acta. Physiol. Scand.*, **62**(Suppl. 230):1964.

Saltin, B., and L. Hermansen: Esophageal, Rectal and Muscle Temperature during Exercise, *J. Appl. Physiol.*, **21**:1757, 1966.

Schmidt-Nielsen, K.: Heat Conservation in Counter-current Systems, in J. D. Hardy (ed.), "Temperature: Its Measurement and Control in Science and Industry," vol. 3, part, 3, p. 143, Reinhold Book Corporation, New York, 1963.

Scholander, P. F., V. Walters, R. Hook, and L. Irving: Body Insulation of Some Arctic and Tropical Mammals and Birds, *Biol. Bull.*, **99**:225, 1950.

Schönbaum, E., and P. Lomax (eds.): "The Pharmacology of Thermoregulation," Karger, Basel, 1973.

Senay, L. C., Jr.: "Changes in Plasma Volume and Protein Content during Exposures of Working Men to Various Temperatures before and after Acclimatization to Heat: Separation of the Roles of Cutaneous and Skeletal Muscle Circulation, *J. Physiol.*, **224**:61, 1972.

Senay, L. C., Jr.: Early Response of Plasma Contents on Exposure of Working Men to Heat, *J. Appl. Physiol.*, **44**(2):166, 1978.

Senay, L. C., D. Mitchell, and C. H. Wyndham: Acclimatization in a Hot, Humid Environment: Body Fluid Adjustments, *J. Appl. Physiol.*, **40**:786, 1976.

Shapiro, Y., R. W. Hubbard, C. M. Kimbrough, and K. B. Pandolf: Physiological and Hematologic Responses to Summer and Winter Dry-heat Acclimation, *J. Appl. Physiol.*, **50**(4):792, 1981a.

Shapiro, Y., K. B. Pandolf, B. A. Avellini, N. A. Pimental, and R. F. Goldman: Heat Balance and Transfer in Men and Women Exercising in Hot-Dry and Hot-Wet Conditions, *Ergonomics*, **24**(5):375, 1981b.

Shvartz, E., S. Shibolet, A. Meroz, A. Magazanik, and Y. Shapiro: Prediction of Heat Tolerance from Heart Rate and Rectal Temperature in a Temperate Environment, *J. Appl. Physiol.*, **43**(4):684, 1977.

Shvartz, E., A. Bhattacharya, S. J. Sperinde, P. J. Brock, D. Sciaraffa, and W. van Beaumont: Sweating Responses during Heat Acclimation and Moderate Conditioning, *J. Appl. Physiol.*, **46**(4):675, 1979.

Simonson, E., N. Teslenko, and M. Gorkin: Einfluss von Vorübungen auf die Liestung beim 100 m. Lauf, *Arbeitsphysiol.*, **9**:152, 1936.

Smiles, K., R. S. Elizondo, and C. C. Barney: Sweating Responses during Changes of Hypothalamic Temperatures in the Rhesus Monkey, *J. Appl. Physiol.*, **40**:653, 1976.

Snellen, J. W.: Mean Body Temperature and the Control of Thermal Sweating, *Acta Physiol. Pharmacol. Neerl*, **14**:99, 1966.

Sobocinska, J., and J. E. Greenleaf: Cerebrospinal Fluid [Ca^{2+}] and Rectal Temperature Response during Exercise in Dogs, *Am. J. Physiol.*, **230**:1416, 1976.

Staff, P. H., and S. Nilsson: Vaeske og Sukkertilförsel under Langvarig Intens Fysisk Aktivitet, *Tidsskrift for Den Norske Laegeforening*, **16**:1235, 1971.

Stevenson, J. A. F.: Control of Water Exchange, in W. S. Yamamoto and J. R. Brobeck (eds.),

"Physiological Controls and Regulations," p. 253, W. B. Saunders Company, Philadelphia, 1965.

Stolwijk, J. A. J.: Mathematical Model of Thermoregulation, in J. D. Hardy, A. P. Gagge, and J. A. J. Stolwijk (eds.), "Physiological and Behavioral Temperature Regulation", chap. 48, p. 703; Charles C. Thomas, Springfield, Ill., 1970.

Stolwijk, J. A. J., B. Saltin, and A. P. Gagge: Physiological Factors Associated with Sweating during Exercise, *J. Aerospace Med.*, **39:**1101, 1968.

Strömme, S., K. L. Andersen, and R. W. Elsner: Metabolic and Thermal Responses to Muscular Exertion in the Cold, *J. Appl. Physiol.*, **18:**756, 1963.

Strydom, N. B., C. H. Wyndham, C. G. Williams, J. F. Morrison, G. A. G. Bredell, A. J. S. Benade, and M. von Rahden: Acclimatization to Humid Heat and the Role of Physical Conditioning, *J. Appl. Physiol.*, **21:**636, 1966.

Stuart, D. G., E. Eldred, A. Hemingway, and Y. Kawamura: Neural Regulation of the Rhythm of Shivering, in J. D. Hardy (ed.), "Temperature: Its Measurement and Control in Science and Industry," vol. 3, part 3, p. 545, Reinhold Book Corporation, New York, 1963.

Sutton, J. R.: Heat Illness in R. H. Strauss (ed.), "Sports Medicine," p. 307, W. B. Saunders Company, Philadelphia, PA, 1984.

Suzuki, Y.: Human Physical Performance and Cardio-circulatory Responses to Hot Environments during Sub-Maximal Upright Cycling, *Ergonomics,* **23**(6):527, 1980.

Tanaka, M., Y. Tochihara, S. Yamagaki, T. Ohnaka, and K. Yoshida: Thermal Reactions and Manual Performance during Cold Exposure while Wearing Cold-protective Clothing, *Ergonomics,* **26**(2):141, 1983.

Teichner, W. H.: Assessment of Mean Body Surface Temperature, *J. Appl. Physiol.*, **12:**169, 1958.

Vangaard, L.: Physiological Reactions to Wet-Cold, *Aviation Space and Environ. Med.*, **46:**33, 1975.

Vangaard, L.: Perifer Varmebalanse, in I. Holmér and J. Sundell (eds.): "Arbete i kallt klimat," Arbete och Hälsa, Arbetarskyddsverket, vol. 1, Stockholm, 1982.

Verney, E. B.: The Antidiuretic Hormone and the Factors Which Determine Its Release, *Proc. Roy. Soc.*, (ser B) **135:**25, 1947.

Vokac, Z., and N. Hjeltnes: Core-Peripheral Heat Redistribution during Sleep and Its Effect on Rectal Temperature, in Reinberg et al. (eds.): "Night and Shift Work Biological and Social Aspects," Advances in the Biosciences, vol. 30, Pergamon Press, Oxford and New York, 1981.

Wallenberg, L. R.: Reduction in Cold-induced Natriuresis Following Hyperoncotic Albumin Infusion in Man Undergoing Water Diuresis, *Scand. J. Clin. Lab. Invest,* **34:**233, 1974.

Walther, E. E., E. Simon, and C. Jessen: Thermoregulatory Adjustments of Skin Blood Flow in Chronically Spinalized Dogs, *Pflüger Arch.*, **332:**323, 1971.

Williams, C. G., C. H. Wyndham, and J. F. Morrison: Rate of Loss of Acclimatization in Summer and Winter, *J. Appl. Physiol.*, **22:**21, 1967.

Winslow, C.-E. A., A. P. Gagge, and L. P. Herrington: Influence of Air Movement upon Heat Losses from Clothed Human Body, *Am. J. Physiol.*, **127:**505, 1939.

Wyndham, C. H.: Effect of Acclimatization on the Sweat Rate/Rectal Temperature Relationship, *J. Appl. Physiol.*, **22:**27, 1967.

Wyndham, C.: The Physiology of Exercise under Heat Stress, *Ann. Rev. Physiol.*, **35:**193, 1973.

Wyndham, C. H., G. G. Rogers, L. C. Senay, and D. Mitchell: Acclimatization in a Hot Humid Environment: Cardiovascular Adjustments, *J. Appl. Physiol.*, **40**:779, 1976.

Wyon, D.: "Kyla Och Prestation" in I. Holmér and J. Sundell (eds.): "Arbete i Kallt Klimat", Arbete och Hälsa, Arbetarskyddsverket, vol. 1, Stockholm, 1982.

Wyon, D. P., I. Andersen, and G. R. Lundqvist: The Effects of Moderate Heat Stress on Mental Performance, *Scand. J. Work Environ. Health*, **5**:352, 1979.

Zimmerman, M. B., and E. H. Blaine: Water Intake in Hypovolemic Sheep: Effects of Crushing the Left Atrial Appendage, *Science*, **211**: 489, 1981.

Zotterman, Y.: Thermal Sensations, in J. Field (ed.), "1, Handbook of Physiology: Neurophysiology," vol. 1, p. 431, American Physiological Society, Washington, D.C., 1959.

14

APPLIED SPORTS PHYSIOLOGY

CONTENTS

INTRODUCTION
  Energy Yield
  Neuromuscular Function
ANALYSIS OF SPECIFIC ATHLETIC EVENTS
  Walking
  Running
  Bicycling
  Swimming
  Speed Skating
  Cross-Country Skiing
  Alpine Skiing
  Canoeing
  Rowing
  Ball Games
SEX DIFFERENCES

INTRODUCTION

The development of lightweight electronic instruments and devices capable of recording and transmitting impulses by telemetry, or by direct recording with the aid of portable miniature recorders, has made it possible to study a variety of physiological functions in the person exposed to different types of work stress, including athletic events. These studies have in recent years led to the accumulation of considerable data concerning physiological characteristics of the individual athlete, as well as physiological requirements of the specific athletic event. Such information may provide a foundation for the selection of athletes, for the analytical evaluation of technique and methods of training, and for the evaluation of training progress.

An attempt to schematically present the major components of physical or athletic performance in general is shown in Fig. 14-1.

Energy Yield

Activities which engage large muscle groups continuously during 1 min or more may often tax the *aerobic power* to a maximal degree and thereby also impose a maximal load on the circulation (see Fig. 7-4). In many types of exercise (walking, running, swimming, rowing, cycling, cross-country skiing, skating, or calisthenics), the individual may set a pace which calls for an aerobic power that may vary from low to maximal (Fig. 7-3b). These types of activities are therefore excellent examples of exercises suitable for developing general physical fitness.

However, the same types of activities may also tax the *anaerobic processes* to a great extent when performed at maximal intensity. The feeling of exertion depends largely on the rate at which glycogen is broken down to lactic acid. Naturally, maximal exercise in which small muscle groups are engaged, brings about the accumulation of lactic acid in the muscles involved and a feeling of exertion, although the total amount of energy applied may be quite small.

In exercise involving great intensity lasting for a few seconds, interrupted by periods of rest or light exercise, the peak load is moderate with respect to both aerobic processes and anaerobic energy yield, as evidenced by the formation of small amounts of lactic acid (Fig. 7-5; Tables 10-2, 10-3). Various kinds of ball games are typical examples of this type of activity.

In athletic events that require heavy exercise lasting more than 1 hr, the availability of muscle glycogen will eventually determine the level of achievement (Chap. 12). Special measures may therefore be necessary in order to attain improvement. Such events carried out in a hot climate place an additional demand on the heat-dissipating systems and on continuous replacement of lost water (Chap. 13).

Neuromuscular Function

Any evaluation of the engagement of individual muscle groups is impossible without the use of *electromyography*. Electromyography furnishes information pertaining to (1) which muscle or which parts of a muscle are activated; (2) the chronological order of the participation of the respective muscles in the activity; (3) the degree and duration of the contraction of the respective muscles in each movement. Such studies may also facilitate the development of an individual muscle training program. As a general rule, strength and technique are best trained through the utilization of the respective athletic event.

The *kinematic* analysis describes the geometrical form of a movement. In order to

PHYSICAL PERFORMANCE

Function	Structural Basis	Biochemical Processes Involved	Modifying Factors
1. Muscular strength	Motor unit:	a) Muscle cell contractile elements	Contraction-tension (static, dynamic) — Genetic endowment; sex; age — Training; use-disuse.
		Energy metabolism: chemical energy mech. work Aerobic-anaerobic processes, enzymat. react. Fuels: carbohydrate, fat; Nutritional intake, storage	— Training — Diet, training.
		b) Motoneuron, synapses, endplate	Excitation, impulse propagation membrane depolarization. — Psychic factors, CNS training
2. Joint mobility	Skeleton:	a) Muscular attachment b) Skeletal levers c) Joints & ligaments, articul. cartilage, synovia, bursae	— Training of joint mobility
3. Coordination	Neuromuscular apparatus:	Afferent efferent pathways, senses	Propagation of nerve impulses, facilitation, inhibition, regulation — Training, practice Psychic factors Drugs
4. Endurance	Oxygen transport organs:	a) Pulmonary ventilation b) O₂ binding capacity of the blood (Hb, blood vol., etc.) c) Cardiac output: Stroke volume Heart rate Cardiac pump: Myocardium Valves Pacemaker d) a-v̄O₂ diff.: local milieu in muscle cell (Shift of the dissociation curve, etc.) Venous return (muscle pump) Negative pressure in the thorax Redistribution of the blood volume to the muscle e) Fluid balance	— Training, altitude, air pollution, smoking, iron intake — Genetic endowment, state of health, training, environment (heat) — Fluid intake, fluid loss, heat, etc.
5. Will to win	Reticular formation, etc.		— Psychic factors, attitude

Locomotor Organs {1, 2, 3}
Service Organs {4}
Central Nervous System {5}

Figure 14-1
Physical performance.

arrive at an idea of the forces that produce movement, i.e, a *kinetic* analysis, force-platforms with strain gauges as force-sensitive devices may be used. It is typical for an elite athlete, such as a champion golf player, to be able to repeat precisely, again and again, a certain motion or force, while the path of movement, the force developed, and the electromyogram are practically identical each time (Carlsöö, 1967). A discussion of the neuromuscular function as applied to sport activities falls primarily within the area of kinesiology and is therefore beyond the scope of this book.

ANALYSIS OF SPECIFIC ATHLETIC EVENTS

Walking

All active individuals have to walk, and occasionally even run, in order to move about in the course of their normal daily life and activity, whether they otherwise engage in any recreational physical activity or not. The energy cost of walking may vary within wide limits, not only among individuals but also in the same individual, depending on the circumstances. It certainly depends on total body weight, including clothing, speed of walking, type of surface, and gradient (Fig. 14-2), and on whether the person limps or not (Molbech, 1966). The freely chosen step rate requires the least oxygen uptake at any given speed (Zarrugh and Radcliffe, 1978). If the subject is forced to use any other step rate at the same speed, the oxygen cost will be higher than that required to maintain that speed at their own freely chosen step rate.

Taylor and Heglund (1982) have emphasized the importance of the energy storage and recovery in the elastic elements of the extremities during locomotion. Their observations demonstrate that the storage of energy in elastic elements occurs in one part of the stride, and that this energy is recovered as useful work in another.

There is a fair amount of data accumulated from different countries on the energy cost of walking on the level, and generally, these data are in good agreement. Figure

Figure 14-2
The energy expenditure of walking under different conditions. (Data from Pugh, 1971; 1976.)

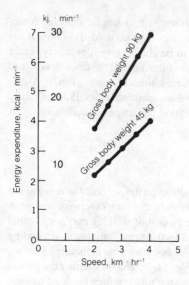

Figure 14-3
Effect of speed (km · hr⁻¹) and gross body weight (kg) on energy expenditure (kcal · min⁻¹ or kJ · min⁻¹) of walking. (Data from Passmore and Durnin, 1955)

14-3, based on data from Passmore and Durnin (1955), shows the combined effects of varying speed and varying body weight on the energy expenditure of walking.

A comprehensive treadmill-grade walking study was carried out by Margaria (1938), who found that going down a slope of 1 in 10 at varying speeds involved an energy expenditure of up to 25 percent less than walking on the level. However, on very steep declines, particularly at low speeds, energy expenditure may be considerably higher than when walking on the level.

Gordon et al. (1983) compared load carrying with just walking on a treadmill. They showed that both heart rate and the rating of perceived exertion increased linearly with increases in power output, but added load carriage brought about substantially larger increases in heart rate and rating of perceived exertion than did unloaded walking for equivalent increases in power. They suggest that these differential responses may be related to differences in muscular fatigue and biomechanical action.

The type of surface may affect the energy cost of walking from 23 kJ (5.5 kcal) · min⁻¹ on an asphalt road to 31 kJ (7.5 kcal) · min⁻¹ on a ploughed field for a 70-kg man walking at a speed of about 5.5 km · hr⁻¹ (Granati and Busca, 1945). Walking up and down stairs may represent an energy expenditure of as much as 42 kJ (10 kcal) · min⁻¹ for a 75-kg person (Passmore et al., 1952).

Going down stairs involves only about one-third the energy used in going up stairs. Going, and especially running, up stairs thus represents fairly heavy exercise, and it is therefore not surprising that so many housewives find this activity very tiring, especially when their care of sick children on an upper floor necessitates frequent ascents from the kitchen or elsewhere on the first floor. On the other hand, stair-climbing may be effective in improving physical fitness.

After the Olympic Games in London, 1948, the winner of the 10-km walking event, John Mikaelsson, was studied when walking on a treadmill. When simulating the race by adjusting the speed of the treadmill so that the walking speed during the race was attained (13.3 km · hr⁻¹ = 10 km in 45 min 13.2s), his measured oxygen uptake was 4.0 liters · min⁻¹, or 58 ml · kg⁻¹ · min⁻¹ (unpublished results). The oxygen uptake during competitive walking seems to be approximately 75 percent of the maximum; uphill it even exceeds that level. In 1972, the maximal oxygen uptake in Swedish male competitive walkers was around 72 ml · kg⁻¹ · min⁻¹ (see Fig. 10-1).

Running

The energy expenditure of running varies tremendously, as is to be expected. In adults, individual variations are fairly small at submaximal speeds. Under these conditions, O_2 uptake per kilogram body weight is the same regardless of sex or athletic rating (P.-O. Åstrand, 1956). On the other hand, the oxygen uptake per kilogram body weight is higher for children than for adults when they both run at a certain speed (P.-O. Åstrand, 1952). The reason for this is not quite clear. One possibility is that they do not utilize the elastic properties of the skeletal muscles, see p. 46.

There is very little difference in the measured values for energy expenditure per meter in elite runners (Margaria et al., 1975), indicating that any influence of running technique on energy expenditure must be small. Thus, in well-trained elite runners running at a speed of 20 km · hr⁻¹ on the level, the oxygen uptake varied only between 67 and 71 ml · kg⁻¹ · min⁻¹ (Karlsson et al., 1972). The energy expenditure is, on the other hand, greatly increased when running against the wind (Fig. 14-4). According to Davies (1980), the energy cost of overcoming air resistance on a calm day outdoors is calculated to be about 8 percent for sprinting (10 m · s⁻¹), 4 percent for middle-distance (6 m · s⁻¹), and 2 percent for marathon (5 m · s⁻¹) running.

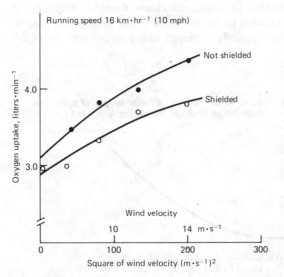

Figure 14-4
Showing the effect of increased air resistance due to wind on the energy expenditure of running. The curve "Not shielded" was obtained when the subject was running alone on the treadmill, and "Shielded" when he was about 1 m behind another runner. The wind was achieved by a big fan. An extrapolation of the data indicates that running 1500 m at about 4-min speed on a track in calm air close behind a pace-maker or a faster competitor may save up to 6 percent of the energy cost. Running behind and to one side of another runner gives a gain of about 1 s per lap. *(From Pugh, 1971.)*

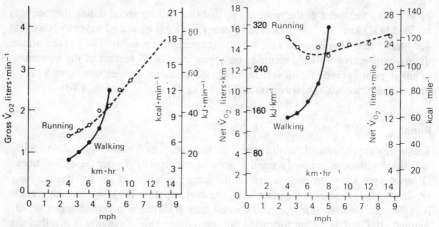

Figure 14-5
Energy cost of walking and running at different speeds. For calculation of the net energy cost (right panel) the oxygen uptake on the sitting subject was subtracted from the oxygen uptake measured during walking and running respectively. The body weight of the subject was 75 kg.

As long as the speed is kept below a certain level, which varies with the individual, it is more economical to walk than to run (Fig. 14-5). In elite walkers, this level is higher than in less expert walkers. It may cost the same amount of energy to run at a rate of 14 km · hr⁻¹ (8.7 mph) as it does to walk at a rate of only 10 km · hr⁻¹ (6.2 mph).

The O_2 uptake depends upon the stride length, as is evident from Fig. 14-6. The subject ran at a given speed paced by a metronome, and the oygen uptake was measured at steady state. In some experiments, he was free to choose the stride frequency. In general, the stride length that is natural for the individual is also the most economical one. The energy cost of running is greatly increased with a further increase of the

Figure 14-6
Oxygen uptake during running at a speed of 16 km · hr⁻¹ with different lengths of stride. The encircled cross represents the freely chosen length of stride. *(From Högberg, 1952.)*

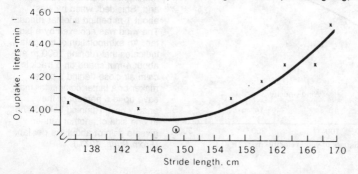

stride lengths. This fact has considerable practical significance. In long-distance running, the factor of economy and energy efficiency is important, so that it becomes essential to maintain the stride length most efficient for the individual. In short-distance running, such as the 100-m dash, on the other hand, where speed is more important than economy, the man who can run with rapid, long strides has an advantage, and he can afford to disregard energy cost for the limited period of time involved.

The well-known runner Nurmi had an unusually long stride length. It was thought that this was the key to his success, and efforts were made by others to copy his style. This, however, resulted in reduced efficiency and inferior results. It appears that the long stride was natural for Nurmi, and for him this represented the most economical style. This example clearly shows the danger of generalization on the basis of single cases and emphasizes the need for objective physiological studies.

Otherwise, an increase in the speed of running is brought about primarily by an increase in the stride length. An experienced 800-m runner ran on the treadmill at different speeds from 8 up to 30 km $\cdot$ hr^{-1}. The length of the stride increased more or less rectilinearly from about 80 to 220 cm, while the stride frequency increased from only about 170 to 230 steps $\cdot$ min^{-1} (Högberg, 1952). When increasing the running speed, the time spent in contact with the ground decreases markedly, whereas the time spent in the air first increases and then remains relatively constant. The increase in step frequency with the speed of running is therefore mainly due to a decrease in the time spent in contact with the ground (Cavagna et al., 1976). During the time of contact the body "bounces" on the supporting limb.

It should be emphasized that when running within a wide range of speeds the energy demand per kilometer is practically the same (jogging at lower speeds, running at higher speeds; see Fig. 14-6). Walking at lower speeds is less costly, but at higher speeds the energy cost approaches and even exceeds that of running (Zunts and Schumburg, 1901; Böje, 1944; Margaria et al., 1963). As a rule of thumb the energy cost of jogging or running is approximately 4 kJ $\cdot$ kg^{-1} $\cdot$ km^{-1}(1 kcal) but for walking at 4 to 5 km $\cdot$ hr^{-1} the energy demand is only half the figure, or 2 kj $\cdot$ kg^{-1} $\cdot$ min^{-1}(0.5 kcal). As pointed out, the surface, wind, and gradient will modify these figures. At about 4 km $\cdot$ hr^{-1} the work done at each step to lift the center of mass of the body equals the work done to increase its forward speed. The total mechanical energy involved (potential plus kinetic) is at this speed at a minimum, as is the energy cost (Cavagna et al., 1976).

In order to attain a level of achievement equivalent to the world elite in middle- and long-distance running, a maximal oxygen uptake close to, or preferably above, 80 ml $\cdot$ kg^{-1} $\cdot$ min^{-1} is necessary in the case of men (Saltin and P.-O. Åstrand, 1967). In women runners of 400 and 800 m, a maximal aerobic power of 65 ml $\cdot$ kg^{-1} $\cdot$ min^{-1} or higher is necessary in order to attain results qualifying for the elite class. An athlete with a high maximal oxygen uptake has the advantage of being better able to tolerate bouts of forced tempo than the competitor with a lower aerobic power in that he or she does not have to utilize the anaerobic energy yield to the same extent during the bouts of increased tempo. Ideally, the pace should be selected corresponding to the intensity when the oxygen uptake, following a linear increase with the increasing intensity, begins to level off (Fig. 7-3b).

On the basis of our present state of knowledge, it appears that a top runner who has had a proper warm-up attains his or her maximal oxygen uptake after about 45 s. Accordingly, it may be assumed that in a 400-m dash lasting 40 to 50 s, the runner taxes his or her anaerobic metabolic capacity maximally. In these cases, blood lactate concentrations of 20 to 25 mM have been measured. At shorter distances, i.e., in 100- to 200-m races, it is estimated that some 90 percent of the energy is derived from anaerobic metabolic processes. At distances exceeding 400 m, aerobic metabolic processes assume an increasingly important role. In an 800-m race, aerobic metabolic processes account for about 40 percent of the total energy utilized, and in a 1,500-m race, for 65 percent.

Some of the physiological changes observed in runners may be manifestations of stress reactions. A pronounced leucocytosis has been observed in blood samples collected immediately after the race in marathon runners (Wells et al., 1982).

Bicycling

Bicycling deserves our attention, not only as a popular sport, but also because of the extensive use of cycle ergometers as instruments for the study of the physiology of exercise. It has been the subject of extensive studies and a considerable body of information is available (see Whitt and Wilson, 1974; Faria, 1984; Sjøgaard, 1984).

Basically, the bicycle is a very efficient method of locomotion. The energy cost of bicycling is only about 1/5 that of walking, yet the speed may be 5 times greater. Improvements have been made aerodynamically by more streamlined construction, by the use of racing suits, and streamlined helmets to reduce drag. But above all, the training of the bicyclist has been vastly improved to meet the requirements of this demanding sport, which may include races ranging from a 200 m sprint to the 5000 km Tour de France.

The average aerobic power of most elite cyclists ranges between 70 and 74 ml · kg^{-1} · min^{-1}. The quadriceps muscle is recruited proportionally to the changes in oxygen uptake during cycling. The saddle height may significantly affect muscle involvement. Mechanical efficiency ranging from 19.6 to 28.8 percent for cycling at pedal frequencies between 40 and 100 rpm have been reported. Data on the effect on the efficiency of changing the pedal rate are conflicting, but there appears to be a significant advantage in employing a high pedaling rate at high power output. Evidently, the body posture during cycling is crucial for optimal performance. (For references, see Faria, 1984). Åstrand (1953) studied the effect of different pedal crank lengths (16–20 cm) on mechanical efficiency, riding a bicycle placed on a horizontal treadmill (speed 20 km · hr^{-1}, slope 2°). He found no difference in efficiency, while Faria (1984) suggests that the crank length on most commercially available bicycles and ergometers is probably too long for most cyclists.

Swimming

Swimming engages practically all muscle groups of the body. It is therefore not surprising that very high oxygen uptakes have been obtained on swimmers (P.-O. Åstrand et al., 1963; Holmér, 1974a). A maximal oxygen uptake of 3.75 liters · min^{-1} was attained by the female silver-medal winner in the 400-m freestyle in the Olympic

Games in Rome, 1960. In male swimmers of good world standard, a maximum of around 6 liters $\cdot$ min^{-1} has been measured.

Since the specific gravity of the body is not much different from that of water, the weight of the body submerged in water is reduced to a few kilograms. The obese individual especially may keep afloat with very little energy expenditure. Swimming may therefore be an easy task when performed at a low level of intensity. For this reason, swimming and various exercises performed in water are a very common form of training for physically handicapped individuals.

The functional demands of competitive swimming were evaluated in 22 girl swimmers on the basis of the relationship between oxygen uptake during swimming at competitive speed and maximal oxygen uptake during work on a cycle ergometer (P.-O. Åstrand et al., 1963). A very high correlation was observed, but the oxygen uptake during swimming averaged only 92.5 percent of the maximal oxygen uptake reached during cycling. However, five girls reached a higher value in the former case. The high correlation is also evident from the fact that the blood lactate concentration after swimming was of the same order of magnitude as after maximal cycling (10.3 and 10.5 mM respectively). Similar results were reported by P.-O. Åstrand and Saltin (1961) and Holmér (1974a); 12.8 mM after maximal swimming, and 12.9 mM after running for male swimmers. The quotient of pulmonary ventilation to oxygen uptake ($\dot{V}_E/\dot{V}_{O_2}$) was significantly lower during maximal swimming than during cycling (27.7 and 35.5, respectively). The reason for this relative hypoventilation may be the different mechanical conditions of breathing. The water pressure on the thorax makes the respiration more difficult. Furthermore, the breathing is not as free during swimming as in most other types of exercise, in that the respiration during competitive swimming is synchronized with the swimming strokes. Holmér et al. (1974) also noted a similar difference in the ratio between maximal pulmonary ventilation and maximal oxygen uptake when comparing the data obtained during swimming and running (29.8 and 37.4, respectively for 5 subjects). However, despite the relative hypoventilation in the swimming subjects compared with running, the arterial oxygen pressure and content were the same in the two types of exercise.

In recent years, world swimming records have been attained by girls at an increasingly younger age. In Chap. 7, an analysis of their physiological possibilities was presented. It was concluded that the girls, by the age of thirteen to fourteen years, had almost reached the maximal power of their aerobic processes. During puberty, the organism may respond more strongly to the training. In the aforementioned study of girl swimmers (P.-O. Åstrand et al., 1963), it was found that they had significantly greater functional dimensions than girls who had not taken part in competitive sport or undergone any special physical training. Vital capacity and heart volume were highly correlated to the maximal oxygen uptake.

Thus, young girls may exhibit a very high energy power. It has been shown that women during breaststroke swimming at a certain speed have a lower oxygen uptake (greater mechanical efficiency) than men. This may be explained by the fact that the lower specific gravity in women, due to their greater fat content, reduces the effort required to keep the body floating. However, considerable individual variations in technique are typical for swimming (see Holmér, 1974b).

Metabolic and cardiorespiratory responses of swimmers were compared by Holmér and Gullstrand (1980) during a so-called "hypoxic swim" training session, in which the swimmer breathes less often than usual, with a training session performing normal breathing. Their results do not support the view that "hypoxic swim" training has an advantage over normal swim training.

The energy requirement of different types of swimming is best mirrored in the record tables, which show that crawling is the most economical type of swimming as long as the swimmer masters the proper technique. It is possible to study the different swimming techniques in detail in a specially constructed swimming flume, such as that installed at the College of Physical Education in Stockholm. In this swimming flume, the water can be made to flow at different speeds through a channel, thus providing opportunity for studies similar to those on walking and running on a treadmill.

Holmér (1974a) studied the physiological responses to swimming in 87 subjects of varying states of skill and training in the Swedish swimming flume, compared with running and cycling. He found that oxygen uptake during swimming at given submaximal speed depended on the degree of swimming training, body dimensions, swimming technique, and swimming style. Thus, oxygen uptake at a given submaximal speed was higher for untrained than for trained swimmers, and for tall subjects than for short subjects. It is the arm stroke that has the highest efficiency, not the leg kick. In fact, the maximal speed with arm strokes was almost the same in freestyle (1.31 $m \cdot s^{-1}$) as for the whole stroke (1.34 $m \cdot s^{-1}$), but at a significantly lower oxygen uptake. One might speculate that over longer distances the leg kicks should be deemphasized because they may waste oxygen and blood flow. In breaststroke, the leg kicks are probably as important or more important than the arm strokes. The mechanical efficiency was 6 to 7 percent in freestyle and 4 to 6 per cent in breaststroke in elite swimmers. Most costly from the standpoint of energy expenditure is the butterfly stroke. (For details, see Holmér, 1974b; Di Prampero et al., 1974.) The increase in oxygen uptake with increasing swimming speed was linear or slightly exponential. Maximal oxygen uptake during swimming was, for elite swimmers, 6 to 7 percent lower than during running and approximately the same as during cycling. For subjects untrained in swimming, their maximal oxygen uptake during swimming was on the average 80 percent of the running maximum. When the body was submerged, vital capacity was reduced by 10 percent and the expiratory reserve volume was less than 1 liter as compared with 2.5 liters in air. The increase in tidal volume in water was exclusively achieved by the use of the inspiratory reserve volume. Heart rate, cardiac output, and stroke volume during submaximal swimming were of the same magnitude and increased with increasing speed in approximately the same way as during running. Heart rate was significantly lower in maximal swimming than in maximal running. The mean intra-arterial blood pressure at submaximal as well as maximal rates of work was higher in swimming than in running. (The circulatory data were observed on five subjects, who were studied when swimming and running respectively, see Holmér et al., 1974.) Breaststroke and butterfly swimming required 1 to 2 liters $\cdot$ min^{-1} higher oxygen uptake at a given submaximal speed than did freestyle and backstroke. As mentioned, the mechanical efficiency was 4 to 6 percent in breaststroke and 6 to 7 percent in freestyle in elite swimmers. Maximal oxygen uptake measured during swimming varied as a consequence of swimming training, whereas the maximal oxygen uptake measured during treadmill running remained relatively unchanged. Therefore, maximal oxygen uptake measured during running is not representative for performance in swimming, and training for the purpose of increasing a swimmer's maximal oxygen uptake should, to the greatest amount possible be done by swimming. (See Fig. 10-15).

Bonen et al., (1980) compared the maximal oxygen uptake obtained in swimmers during tethered swimming, free swimming, and flume swimming and found essentially identical results. They also compared maximal oxygen uptake obtained during arm ergometer exercise and during swimming and found large differences. Thus, predictions of a swimming maximal oxygen uptake from arm-ergometer data would yield considerable errors, as in the case of treadmill data (Fig. 10-15).

An analysis of world records (1985), shows that the average speed in swimming 1500 m is 82 percent of the best time at 100 m (about 14 min 55 s and 49 s respectively). In running 5000 m, the average speed is only 70 percent of the world record in 400 m (time abbout 13 min and 44 s respectively). The reason for the greater decline in speed in distance running compared with swimming could be that when running at a high speed the skilled runner may possibly utilize the potential to store kinetic energy in eccentrically activated extensor muscles in the subsequent concentric contraction during a step. At lower speeds this factor becomes less evident. In swimming there is no similar pattern of eccentric muscular activity at any speed.

When discussing the physiology of swimming, the effect of the temperature of the water on the swimmer should be considered. Bergh (1980) observed rectal temperatures below 35°C in 10 out of 49 swimmers in a 3.2 km. race, water temperature 19° C. Such low body temperature will impair performance. .

Speed Skating

Ekblom et al. (1967) have made a study of Swedish speed-skating athletes including the 1968 Olympic champion in 10,000 m, J. Höglin, the 1968 bronze medal winner, Ö. Sandler, and the 1964 World champion and Olympic champion in 10,000 m, J. Nilsson. They were studied when running on a treadmill as well as when skating at submaximal and maximal speeds.

Table 14-1 summarizes some of the maximal data attained in the two types of activities. The average value for maximal oxygen uptake was 5.49 liters $\cdot$ min^{-1} during treadmill running, as against 4.85 liters $\cdot$ min^{-1} when skating, showing a difference of about 12 percent. Otherwise, the achieved maximal values were rather uniform. It should be pointed out that the extra burden imposed by the equipment that the skaters had to carry for the collection of the expired air might have caused the oxygen uptake values to be higher than they would have been when skating without equipment. Most of the determinations were made on Sandler who, relatively speaking, had the highest oxygen uptake when skating (95 percent of that attained during treadmill running). The values for blood lactate and heart rate, which were attained in connection with skating competitions were similar to those obtained during the determination of maximal oxygen uptake.

Table 14-1

		|								
					Maximal values					
		Oxygen uptake				Pulmo-nary ventila-tion, l · min⁻¹		Blood lactates, mM		Heart rate, beats · min⁻¹
Subject	Best time 10,000 m	l · min⁻¹		ml · kg⁻¹ · min⁻¹						
		R*	S*	R	S	R	S	R	S	R	S
Sandler, Ö.	15,20.6	5.77	5.48	79.0	75.1	184	183	20.3	19.9	186	186
Nilsson, J.	15,47.0	5.70	4.80	79.2	66.7	172	139	17.0	18.0	188	186
Höglin, J.	15,23.6	5.39	4.69	71.9	62.5	137	159	15.4	15.4	185	185
Claesson	17,08.0	5.39	4.89	64.9	58.9	141	141	17.4	17.4	194	190
Nilsson, I.	16,19.4	5.20	4.38	76.5	64.4	138	121	16.8	16.6	171	171
Mean value		5.49	4.85	74.3	65.5	154	149	17.4	17.4	185	184

*R = Running; S = Speed skating

When the ice condition is excellent, the maximal oxygen uptake expressed in liters per min is probably of greater practical importance than the maximal oxygen uptake expressed as ml O_2 per kg body weight per min. During speed skating the center of gravity of the body is moving relatively parallel to the ice surface without the marked vertical movements made during running. The potential to perform work is related from a dimensional point of view to the body mass (L^3), but the air resistance is proportional to the body surface area (L^2) (Chap. 9). Therefore, a better performance can be expected from the larger subject, and more so at higher speeds (see Di Prampero et al., 1976). When the ice surface is soft, it is a drawback for the skater to be heavy, since weight increases the friction. The results from the study mentioned show that the elite long-distance skaters have a maximal aerobic power of about 5.5 liters · min⁻¹. The 500-m race and especially the 1,500-m race (skating time, slightly under 2.0 min) impose particularly high demands on the anaerobic power (the peak blood lactate concentration for three of the skaters at the end of the race at the Swedish championship competition in 1965 averaged as follows: 500 m, 13.6 mM; 1,500 m, 17.3 mM; 5,000 m, 15.1 mM; 10,000 m, 13.3 mM). The technique is also different in the shorter distances compared with the longer ones. The skaters who come first in the 500-m race, and to some extent also in the 1,500-m race, infrequently win the 10,000-m race, thus illustrating that the physiological requirements are different in the two types of races. The skaters referred to in Table 14-1 are all typical long-distance skaters.

Figure 14-7 presents oxygen uptakes (Douglas bag method) during skating at different speeds at two different skating rinks. As is evident, the curve is not linear. An increase in the speed from 4 to 6 m · s⁻¹ requires an additional 0.7 liter · min⁻¹, while the increase from 8 to 10 m · s⁻¹ necessitates an increase in the aerobic power of 2.0

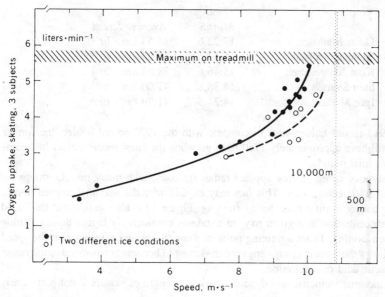

Figure 14-7
Oxygen uptake during speed skating at different speeds. The different symbols represent
experiments performed at different speed-skating rinks. In excellent ice conditions as well as at
high altitude (reduced atmospheric pressure = reduced air resistance), the curve is shifted to
the right; i.e., the same expenditure of energy gives greater speed. Vertical pillars indicate speed
during the skaters' best performances on 500 and 10,000 m respectively. Subjects J. Höglin, J.
Nilsson, and Ö. Sandler. *(From Ekblom et al., 1967.)*

liters · min⁻¹. (A possible contribution from anaerobic processes was not taken into
consideration in this study.) The explanation must be sought primarily in the increased
air resistance at the higher speed. It increases with the speed raised to the second
power, and accounts for a large part of the energy expenditure at high speeds. Thoman
(personal communication) has calculated, from experiments with Sandler in a wind
tunnel, that at a speed of 10 m · s⁻¹, 70 percent of the external work is devoted to
overcoming the air resistance, while the remaining 30 percent is required to overcome
the friction of the ice. Thus, a more ideal aerodynamic profile would to a large extent
reduce the air resistance. The speed-skating style, with the arms held on the back, is
undoubtedly the result of experience. It is also known that the type of clothing is very
important in regard to the air resistance. If the conventional speed-skating suit of a
competitive speed skater were replaced with a suit made of a material similar to that
used by many skin divers, the time on a 5,000-m speed-skating race might, theoretically
speaking, be improved by 10 to 20 s.

The increase in speed of the skater is perhaps best illustrated by comparing the
world records for the 10,000-m speed skating events from 1913 through 1984–85:

		Time *Min: S*	*Average Speed*
1913	Oscar Mathisen	17:22,6	34.53 km · hr⁻¹
1952	Hjalmar Andersen	16:32,6	36.27 km · hr⁻¹
1960	Knut Johannessen	15:46,6	38.03 km · hr⁻¹
1976	Sten Stensen	14:38,8	41.00 km · hr⁻¹
1984	Igor Malkov	14:21,5	41.79 hm · hr⁻¹

If the 1913 record holder were to compete with the 1976 record holder, the former would still have approximately 1750 m to go when the latest record holder had completed his final round.

The style of speed skating appears rather rigorous, with many muscle groups involved in static contraction. This fact may explain why the maximal oxygen uptake is lower during skating than during running. Figure 14-8 also shows that the blood lactate concentration at a given oxygen uptake is considerably higher during skating than when cycling. From a training point of view, it appears essential that the speed skater allows the muscle groups engaged in skating to become accustomed to tolerating a high lactic acid concentration.

The maximal isometric and dynamic muscle strength of skaters if not particularly

Figure 14-8
Blood lactate concentration at different work rates expressed in percentage of maximal oxygen uptake during speed skating and bicycling. The bicycling data are from well-trained noncyclists. The speed-skating data are obtained from well-trained speed skaters. At corresponding oxygen uptakes, more lactic acid is produced during speed skating than during bicycling. *(From Ekblom et al., 1967.)*

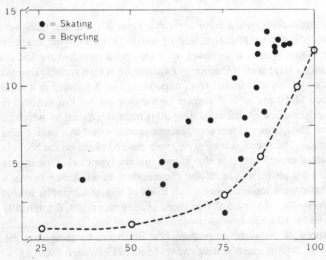

great. It has been observed, however, that the leg muscles of the skaters who are the best sprinters and specialize in 500-m skating are much stronger than those of long-distance skaters.

Cross-country Skiing

The use of skis to transport the body across the snow has several advantages: It enables a person to move comparatively easily across loose, deep snow where walking on foot may be extremely difficult or almost impossible (Ramaswamy et al., 1966); it is possible to move much faster downhill on skis than on foot; finally, it allows the skier to move into the wilderness independently of roads.

Since more muscle groups generally are engaged in skiing than in walking (the use of the arms to pull and push on the ski-poles), the overall energy expenditure involved in transporting the body on skis from one place to another may be as high as, or higher than, the energy expenditure when moving the body the same distance on foot. In fact, the energy expenditure of skiing, especially uphill, may occasionally be surprisingly high. Meen et al. (1972) measured the oxygen uptake in the same subjects when skiing uphill at maximal speed and when running at maximal effort on the treadmill. They found slightly higher oxygen uptakes during skiing than during treadmill running. The mean difference was about 2.5 to 3.0 percent. However, Åstrand and Saltin (1961) did not find any significant difference in maximal oxygen uptake in the two activities.

Because skiing engages about all of the major muscle groups in the body, cross-country skiing is an excellent method of training for physical fitness and dynamic muscular endurance. For the same reason, top cross-country skiers generally have exceedingly high maximal oxygen uptakes. A maximal oxygen uptake of 7.4 liters $\cdot$ min^{-1} has been reported in the Finnish cross-country skier Mieto (Bergh, 1974). Evidently, cross-country skiers have the highest maximal aerobic power ever recorded: 94 ml oxygen uptake $\cdot$ kg^{-1} $\cdot$ min^{-1} for a male Olympic champion; 75 ml $\cdot$ kg^{-1} $\cdot$ min^{-1} for a female skier (unpublished results). Data on maximal oxygen uptake collected on the best Swedish and Norwegian skiers from 1955 to 1985 are roughly at the same level, indicating no major improvement in maximal aerobic power over the last decades.

An additional advantage of cross-country skiing is that the work rate varies greatly with the changing features of the terrain, with very high loads when climbing uphill and lighter loads when sliding downhill (see Fig. 14-9).

Rönningen (1976) compared the energy cost of walking on foot on a hard, snow-covered level road with skiing on a level trail alongside the road. He used six young war college cadets, dressed in military arctic uniforms and carrying 22.5 kg on their backs. In both cases, the subjects moved at a speed of about 5 km $\cdot$ hr^{-1}. He obtained identical values for oxygen uptake (about 1.5 liters O_2 $\cdot$ min^{-1}) although the subjects stated that skiing felt easier than walking, probably because, in skiing, the work is accomplished by a larger muscle mass, since the arm muscles are being used to aid locomotion through the ski-poles. At higher speeds on hard snow the skiing is more economical. On loose snow, the energy cost when skiing at 3.5 to 4.0 km $\cdot$ hr^{-1} was about 70 percent of the demand during walking. For walking with snow shoes, the energy cost was somewhere between these extremes (Christensen and Högberg, 1950).

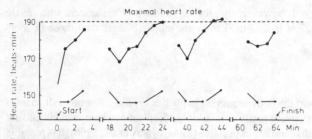

Figure 14-9
Upper panel: Heart rate of the winner in a 21-km cross-country ski race. *Lower panel:* Arrows indicate terrain profile: → = level ground; ↗ = uphill, ↓ = downhill. *(From Bergh, 1974.)*

Skiing uphill at an incline of 7 to 8 percent on a cross-country trail at a speed of about 4 km · hr⁻¹ required an oxygen uptake of about 2.3 liters · min⁻¹ on the average. One of the cadets was studied during a winter maneuver covering 120 km across the mountains in Norway in 5 days. Moving on skis with a load of 22.5 kg on the back, at a speed slightly less than 4 km · hr⁻¹, required an average oxygen uptake of 1.8 liters · min⁻¹, corresponding to roughly 40 percent of the individual's maximal oxygen uptake. When two men pulled a 72-kg sled while on skis, the oxygen uptake increased to 2.3 liters · min⁻¹ on the average, corresponding to more than 50 percent of the individual's maximal oxygen uptake. Christensen and Högberg (1950) compared the energy cost of carrying an extra load of 30 kg in a rucksack and the transportation of the same load on a sled. The oxygen uptake in the rucksack experiment was 70 to 80 percent of the figures when the sled was pulled (at 4.0 km⁻¹ · hr⁻¹).

The efficiency of skiing is illustrated by the following examples: A good skier covered 30 km on a snow-covered lake in 1 hr 20 min (speed 6.25 m · s⁻¹). The best time in the Swedish Vasa race, 86 km (with some 10,000 participants), is 4 hrs 10 min (average speed 5.75 m · s⁻¹). The terrain is relatively flat. In the Olympic Games in 1984 the 50 km winner finished after 1 hr 50 min 55s (average speed 7.51 m · s⁻¹). For comparison, the best time in track running 10,000 m gives a speed of 6.13 m · s⁻¹, and for marathon about 42.2 km, 5.53 m · s⁻¹.

It is thus clear that skiing, even at submaximal speeds, requires a high aerobic work capacity. Accordingly, most elite cross-country skiers have maximal oxygen uptakes of 5.5 liters · min⁻¹ or more (in excess of 80 ml · kg⁻¹ · min⁻¹), with a maximum of 94 ml obtained in an Olympic Champion (15 km race), an extremely high aerobic power. The corresponding figures for the top Swedish female cross-country skiers are 3.5 to 4.4 liters · min⁻¹ or 70 to 75 ml · kg⁻¹ · min⁻¹. Recording of the heart rate during competitive skiing (Fig. 14-9) shows that it reaches maximal levels during uphill skiing, and drops, at the most, to only some 20 beats · min⁻¹ below maximal values during downhill skiing and only slightly below maximal values during skiing on fairly long stretches of level ground. It appears that top competitive cross-country skiers may need to tax some 85 percent or more of their maximal oxygen uptake during a race. Millerhagen et al. (1983), on the basis of a study of combined arm and leg exercise, suggest that skiing could be made more economical if effective use was made of a well-developed upper body.

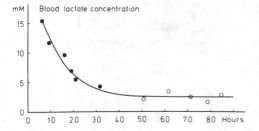

Figure 14-10
Blood lactate concentration at the end
of races of distances from 10 to 85 km.
Filled circles denote mean values, open
circles represent individual values
(From P.-O. Åstrand et al., 1963.)

From Fig. 14-10 it is evident that the longer the duration of the race, the lower the blood lactate concentration measured at the end of the race (Åstrand et al., 1963). It is thus evident that a high anaerobic capacity is important in the shorter-distance, 5- to 10-km races, and particularly in the relay races on skis where the tempo may be very uneven. This reduced ability to liberate energy by glycogenolysis with a lactate production after prolonged exercise may be related to reduced glycogen stores (Asmussen et al., 1974), inhibitory effects on key enzymes or other factors presently unknown (Costill et al., 1971).

Measurements of the maximal oxygen uptake in Swedish elite skiers throughout the year consistently show the highest values at the end of January, and the lowest values in May and June. The difference may be in excess of 5 percent (see Fig. 10-11) (Bergh, 1974).

In recent years, roller-skis have been widely used by competitive skiers as a training device during the seasons of the year when there is no snow on the ground. It appears, on the basis of a comparison between the oxygen uptake during maximal roller-skiing and maximal treadmill running, that roller-skiing may tax the oxygen-transporting system during prolonged roller-skiing as much as, or more than, during running. Thus roller-skiing and running may conceivably be of equal value as a method of endurance training (Fig. 14-11).

Figure 14-11
Heart rate during 90 min roller-skating on a hilly road. In this subject a heart rate of 180
beats · min⁻¹ corresponds to 90 to 93 percent of his maximal oxygen uptake. *(From Bergh, 1974.)*

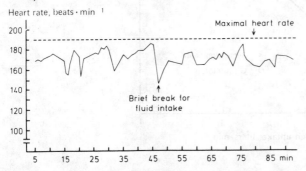

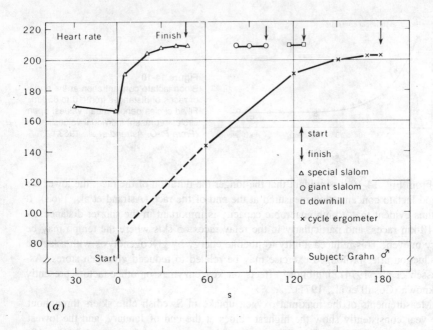

(a)

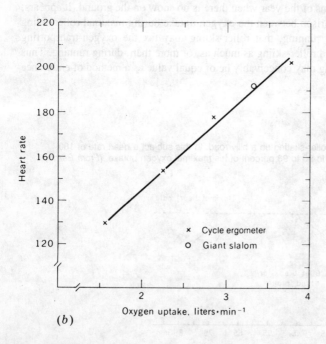

(b)

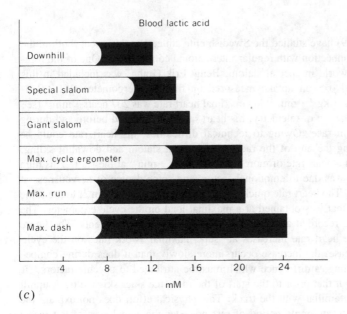

(c)

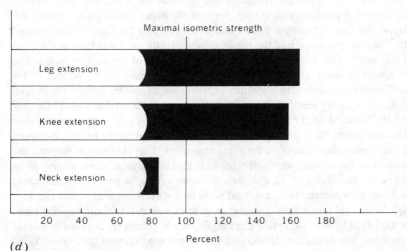

(d)

Figure 14-12
Data on male top athletes in Alpine skiing.
(a) Heart rate before and during competitions in Alpine skiing and during maximal exercise on a cycle ergometer. Note the high heart rate before start and the rapid increase in heart rate after start.
(b) Heart rate in relation to oxygen uptake during cycling and giant slalom the day after actual race on the same course.
(c) Peak blood lactate concentration after various activities; mean values of three to five determinations. "Max. run" = 3 × 1,000 m at maximum speed with a few minutes' rest in between; "Max. dash" = 5 × 50 s in a similar manner.
(d) Maximal isometric muscle strength; 100 percent = data obtained on a group of recruits.
(From Agnevik et al., 1969.)

Alpine Skiing

Agnevik et al. (1969) have studied the Swedish elite athletes in this event, both in the laboratory and in connection with regular international competitions. One of the foremost skiers in the world in special slalom, Bengt-Erik Grahn, was included in this study. His maximal oxygen uptake measured on the cycle ergometer was 3.9 liters $\cdot$ min^{-1} or 66 ml $\cdot$ kg^{-1} $\cdot$ min^{-1}; his maximal heart rate was 207 beats $\cdot$ min^{-1} (see Fig. 14-12). With the aid of telemetry, his heart rate was recorded before and during a competitive slalom race. Owing to technical difficulties, his heart rate could be followed only during the end of the race in both giant slalom and downhill skiing. Figure 14-12 shows a heart rate of more than 160 beats $\cdot$ min^{-1} at the start. This high heart rate no doubt was due to emotional factors and some degree of nervousness at the start of the race. The heart rate quickly rose to over 200 beats $\cdot$ min^{-1}, to the same maximal heart rate that was obtained at a maximal load on the cycle ergometer. The same heart rate was recorded at the end of the other competitive events. The figure also shows how the heart rate increases at "supermaximal" work rates on the cycle ergometer. In this case, the increase occurs more slowly than it does during competitions in Alpine skiing, a difference which must be attributed to psychic factors. (It should be pointed out that prior to the start of the race, the skier skis the track uphill in order to become familiar with the track. This physical effort does not explain the high heart rate, since an ample period of rest precedes the actual start of the race.) The day following the giant slalom race, some of the participants covered exactly the same track, but then the oxygen uptake was determined by the Douglas bag method as well. The subjects completed the run in almost exactly the same time as they had during the actual race. Figure 14-12 presents an example of the data collected. The oxygen uptake in this particular subject was 3.3 liters $\cdot$ min^{-1} or 87 percent of his maximal O_2 uptake measured in laboratory experiments. In another subject, the oxygen uptake was 3.9 liters $\cdot$ min^{-1} or 78 percent of his maximal aerobic power. On this occasion the recorded heart rates were submaximal. Figure 14-12 shows that the heart rate measured lies on the regression line, which was obtained for the heart rate-oxygen uptake relationship during tests on the cycle ergometer. This observation supports the assumption that the competition itself represents an extra psychic stress leading to an elevated heart rate. Figure 14-12 also presents mean values for peak blood lactic acid in connection with competitions, maximal work on the cycle ergometer, and running. The high lactic acid levels during skiing, in spite of a relatively short exercise time, may be explained by the assumption that certain muscle groups are engaged in intense static exercise. The best skier, Grahn, attained in his special event the highest lactic acid levels of the entire group. Finally, it is evident from Fig. 14-12 that the Alpine skiers are characterized by a great isometric strength in the stretch muscles of the legs. A group of military recruits is shown for comparison (values taken as 100 percent). The skiers had even greater muscle strength than a group of weight lifters. Various studies (Eriksson et al., 1977) have confirmed these results. The winner of the World Cup in alpine skiing in 1976 and 1977 (Ingemar Stenmark) had a maximal oxygen uptake of 5.2 liters $\cdot$ min^{-1}, or 70 ml $\cdot$ kg^{-1} $\cdot$ min^{-1} when running on the treadmill. The static muscular strength of the legs' extensor muscles was very high on the skiers, on an average 2900 N (Stenmark leading with 3430 N). Athletes from other sports events studied had mean values at or below 2500 N. It can be calculated that the force the alpine skier must develop during actual skiing can reach several thousand Newtons. Muscle lactate concentrations up to 24 mmole $\cdot$ kg^{-1} wet muscle have been measured, and concentrations around 15 mM in the blood are common findings after competitions. A relatively wide scatter in muscle fiber composition of the quadriceps muscle has been noted, with a mean value of 43 percent slow-twitch fibers (type I) (Eriksson *et al.*, 1977).

It is thus clear that competitive Alpine skiing places heavy demands on both aerobic and anaerobic motor power. Great strength in the muscle groups involved is required. In addition to these basic requirements, the technique will obviously determine the level of achievement. From the training standpoint, it is important to learn to master a good technique in spite of a high lactic acid concentration in the muscles.

Watanabe and Ohtsuki (1977) performed wind-tunnel experiments in order to measure aerodynamic forces acting on an alpine skier running down a slope at top speed, and to clarify their relation to the skier's postural changes. The aerodynamic advantage of the so-called "egg-shaped" posture in alpine skiing was confirmed. However, even at the lowest posture, lateral extension of the arm caused a substantial increase in drag, which was compatible with the increase caused by raising the trunk.

Canoeing

Canoeing is rather unique in that it is about the only competitive sport in which mainly the arm and trunk muscles are engaged in endurance efforts. In addition, the same individual competes in 500-, 1,000-, and 10,000-m races, lasting from about 1.45 to 45 minutes.

For the success of the canoeist, maximal aerobic power is more critical than body weight, for the slightly greater resistance caused by the friction of the canoe in the water by a few kilos in extra body weight is insignificant. It is therefore more meaningful to express the canoeist's aerobic power in terms of liters $O_2 \cdot$ min^{-1} than in ml $O_2 \cdot$ kg$^{-1} \cdot$ min.$^{-1}$ According to Tesch et al. (1974, 1976) the mean maximal oxygen uptake of the present Swedish elite group of canoeists (including two world champions) is 5.40 liters $\cdot$ min^{-1} (range 4.7 to 6.1) or 68 ml $\cdot$ kg$^{-1} \cdot$ min^{-1} (64 to 75); see Fig. 10-1. From Fig. 14-13 it appears that some of the canoeists utilize all their aerobic power when paddling the canoe.

Although back, abdominal, chest, and shoulder muscles are also engaged in canoeing, the major part of the work is performed by the arms. A high maximal oxygen uptake in arm exercise must therefore be a major requirement for the canoeist. Tesch et al. (1974) compared the maximal oxygen uptake during leg exercise (cycle ergometer exercise) with the maximal oxygen uptake during arm cranking in Swedish elite canoeists and found very high values for arm exercise, both in absolute terms (liters $\cdot$ min^{-1}) and in percentage of the maximal oxygen uptake measured during leg exercise. Swedish male junior and senior canoeists combined had a mean maximal oxygen uptake in exercise with the legs of 5.13 liters $\cdot$ min^{-1} as against 4.45 li-

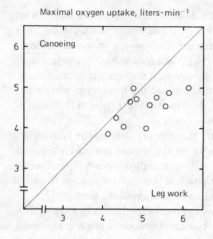

Maximal oxygen uptake, liters·min⁻¹

Figure 14-13
Relationship between maximal oxygen uptake during canoeing and during exercise with the legs in elite canoeists. *(From Tesch et al., 1974.)*

ters · min⁻¹ in arm exercise (87 percent of that in leg exercise), and 4.48 liters · min⁻¹ during canoeing. In elite women canoeists, the maximal oxygen uptake during arm exercise was 90 percent of that obtained during exercise with the legs. The corresponding figures for a group of Swedish weight lifters was 78 percent, and 71 percent for physical education students. A comparison between maximal oxygen uptake in elite Swedish canoeists during leg exercise (cycle ergometer) and arm cranking, and during 500-m, 1,000-m, and 10,000-m canoe racing, is presented in Fig. 14-14. One of the canoeists, who had started systematic canoe training when very young, showed during repeated tests, consistently maximal oxygen uptakes in arm exercise that were as high as, or higher than, in leg exercise. The fact that he started canoe training at an early age might explain, at least in part, his exceptionally high aerobic work capacity with the arms.

In a more recent review of the physiological characteristics of elite kayak paddlers, Tesch (1983) points out that in comparison with other groups of athletes known to

Oxygen uptake in junior and senior canoeists

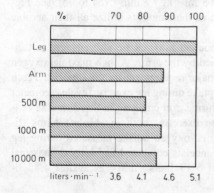

Figure 14-14
Oxygen uptake (liters · min⁻¹ and in percentage of the maximal oxygen uptake measured during leg exercise) in Swedish elite canoeists (junior and senior) during the final minutes of 500-m, 1,000-m, and 10,000-m races. The maximal oxygen uptake in arm exercise is included for comparison. *(From Tesch et al., 1974.)*

exhibit great upper-body muscle strength, kayakers were found to possess high values for shoulder strength, endurance, and anaerobic capacity. Total body maximal oxygen uptake averaged 5.36 l · min⁻¹. The values for arm cranking and paddling were 4.30 and 4.67 l · min⁻¹. High blood lactate levels were observed under training conditions and at the end of a race.

For comparison, it may be pointed out that in a group of Greenland Eskimo kayak-hunters studied by Vokac and Rodahl (1977); the maximal oxygen uptake in arm cranking averaged 89 percent (81 to 104 percent) of that of cycling on a cycle ergometer. This is approximately the same rate between maximal oxygen uptakes in arm and leg exercise as found by Vrijens et al. (1975) in various white subjects, including highly trained athletes specialized in kayak paddling.

Studies of Swedish elite canoeists (Tesch et al., 1976) indicate that canoeists generally are unable to tax their maximal oxygen uptake fully during the 500-m race. In 1,000-m canoe races, almost maximal heart rates have been recorded (189 beats per min compared to a maximum of 195) during the last half of the race. In two Swedish canoeists studied during a 10,000-m race lasting 45 min, the mean heart rate during the entire race was kept at a level corresponding to 96 and 98 percent of the maximal heart rate. In one of them, the heart rate was maximal during the last 2,000 m of the race.

According to Tesch et al. (1976), the higher the maintained speed during the race, the greater the lactate levels in the blood, the highest values being obtained at the end of the 500-m race (Fig. 14-15). Under comparable circumstances, blood lactate values are higher in senior than in junior elite canoeists, and higher in the final heat than in the same canoeist's semifinal or trial heat, perhaps indicating the positive relationship between degree of exertion and blood lactate levels. The metabolic demands for kayak competitions at 500 and 10,000 m are apparently different, in that the energy contribution from glycolytic precursors and the anaerobic component is of greater relative importance in short distances than in exercise of long duration (Tesch and Karlsson, 1984).

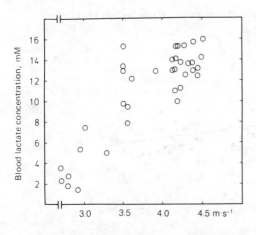

Figure 14-15
Blood lactate levels (mM) in elite canoeists, related to mean speed (m · sec⁻¹). *(From Tesch et al., 1974.)*

While the maximal oxygen uptake measured during maximal leg exercise tends to remain more or less unchanged throughout the year in elite canoeists, the maximal oxygen uptake measured during arm cranking was, on the average, 8 percent higher during the active canoeing season than during the winter (Tesch et al., 1976).

Rowing

Secher (1973, 1983) has compiled the winning times in international rowing championships during the period from 1893 to 1971. The results for the eight are presented in Fig. 14-16. As is evident from the figure, there is a considerable scatter of the data owing to varying wind and water conditions. Because these external conditions may change very quickly (in a matter of minutes), it is quite meaningless to compare individual results from one race to another even when the time interval is as short as 1 hr. Mean results of races from several regattas, on the other hand, may be an indication of the general level of performance for any one year. It is observed that, on this basis, the mean improvement in rowing performance is 0.66 s per year, corresponding to a 1.6 percent increase in speed. Secher (1973) explains this improvement as the result of a combination of factors: the increasing body size and consequently an increase in the maximal aerobic power of the general population; better selection and training of rowers; better rigging of the boats, allowing better technique and greater mechanical efficiency; and better boats, causing less water resistance.

The water resistance against the boat is equal to the force (F) that moves the boat, and it increases proportionally to the velocity raised to the second power ($F \propto v^2$). Since work (W) is force times distance ((L), i.e., $W = F \cdot L$) and power is force times distance divided by time (power $= F \cdot L \cdot t^{-1} = F \cdot v$) that is force times velocity, the energy expenditure (or power) of the rower increases proportionally to velocity raised to the third power (power $= F \cdot v \propto v^2 \cdot v \propto v^3$).

By actually measuring the oxygen uptake of the rower while rowing at submaximal

Figure 14-16
Improvements of results in international rowing championships. *(From Secher, 1973.)*

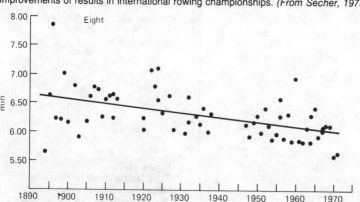

intensity during near steady-state conditions, Secher (1983) observed that the mean oxygen uptake during rowing increases with velocity to the power 2.4. Di Prampero et al. (1971) concluded from their data that "the mechanical power output necessary to maintain the boat progression as well as the energy expenditure appears to increase as the 3.2 power function of the average speed," which is not far from Secher's figure.

By extrapolating the lines of best fit in Fig. 14-16 to 1971, we arrive at a mean value for all boats of 4.9 m · s^{-1}. At this speed, the mean energy demand in the case of the single, double, and pair amount to 6.4 liters O_2 per min (Secher, 1983). One may therefore expect that the maximal aerobic power of the rower may be a limiting factor in rowing performance. This supposition is strengthened by the finding of a positive correlation between the average oxygen uptake of the crew and their placing in international championships (Fig. 14-17). The mean maximal oxygen uptake was 6.1 liters · min^{-1} for the crew taking first place (which is close to the figure 6.25 liters · min^{-1} referred to above), 5.7 liters · min^{-1} for the crew taking sixth place, and 5.1 liters · min^{-1} for the crew attaining the twelfth place. Furthermore, a comparison of the mean maximal oxgen uptakes of individual international champion oarsmen shows the following figures for the maximal oxygen uptakes in liter · min^{-1}: German: 5.9 ($n = 5$); Norwegian: 5.8 ($n = 6$); Danish: 5.7 ($n = 4$) (Vaage and Secher, personal communication). In a study by Secher et al. (1976a), the mean maximal oxygen uptake of a group of 20 top rowers (members of national teams) was 5.0 liters · min^{-1} as against 4.3 liters · min^{-1} in a group of 44 beginners. In a group of 21 highly conditioned oarsmen, Clark et al. (1983) observed an average maximal oxygen uptake of 6.6 1 · min^{-1}. These findings point to the maximal oxygen uptake as a limiting factor in rowing performance.

So far, we have considered the performance in rowing championships, expressed in terms of mean speed during the entire race. However, a closer analysis of the speed attained at different points during the race shows a general speed pattern starting with an initial very high speed followed by a decline to a steady level after the first quarter of the race, and then increasing again in a final spurt. Secher et al. (1976b) have measured oxygen uptake during simulated racing conditions on a cycle ergometer

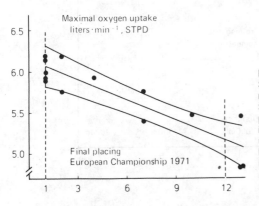

Figure 14-17
Showing the relationship between the maximal oxygen uptake of individual Scandinavian rowers and their final placing in the European championship in 1971; the higher the maximal oxygen uptake, expressed in liters · min^{-1}, the better the performance. The outer lines indicate 95 percent confidence limit. *(From Secher, Vaage, and Jackson, 1976.)*

leading to about the same state of exhaustion in all the subjects at the end of 6 min. Oxygen uptake increased more rapidly during the simulated racing experiment than during the control experiment when the speed was kept constant during the entire 6-min period. They conclude that the greater energy expenditure early in the simulated race was at least partially covered by the more rapid attainment of the maximal oxygen uptake. This appears to provide a physiological basis for the procedure generally practiced during racing of this kind. It may also supply the scientific basis for the development of an optimal speed profile for any type of race. In this connection it should be noted that the heart rate, generally speaking, attains its peak value more rapidly during a real race than during a training race (Fig. 14-18). It should also be kept in mind that the greater the degree of exertion in events lasting several minutes, the greater the anaerobic contribution to the overall energy metabolism, evidenced by a high blood lactate level (Table 14-2).

Ishiko (1971) measured the force acting on the oars of rowers of different international standing. He found that the force varied greatly, with peak values in the range of 800 N (Fig. 14-19). According to the force-velocity relationship, the greatest power is attained at 30 to 40 percent of the maximal voluntary isometric strength of the muscle (Monod and Scherrer, 1972; see Fig. 2-13). One would therefore expect that the rowing strength of a rower ought to be in excess of $800 \cdot 2.5 =$ about 2000 N. Secher (1975) finds an average rowing strength of 2000 N in seven international competitive rowers, 1800 N in national Danish oarsmen, and 1600 N in beginners. However, his results reveal that champion rowers do not show any greater muscle

Figure 14-18
Heart rate of a rower during a real race and during a training race. *(By courtesy of E. D. R. Pruett.)*

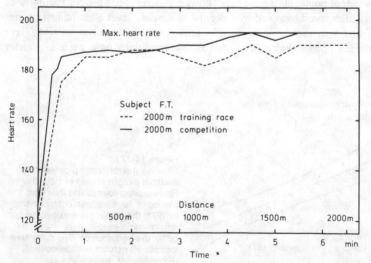

TABLE 14-2
PEAK BLOOD LACTATE LEVELS IN NORWEGIAN ELITE ROWERS DURING THE
SEMIFINAL IN THE EIGHT IN THE EUROPEAN CHAMPIONSHIP REGATTA IN 1969 AND
IN A NORWEGIAN NATIONAL REGATTA THE SAME YEAR. (BY COURTESY OF O.
VAAGE.)

	Peak blood lactate levels (mM)		
		Rowing	
Subjects	Maximum treadmill running	Norwegian national regatta	European championship regatta, Klagenfurt
O. N.	10.2	13.4	15.9
T. W.	6.4	10.1	12.0
S. N.	11.6	13.7	17.2
A. H.	11.8	17.6	18.6
T. T.	14.7	20.4	18.6
K. S. J.	10.2	16.7	18.8
E. H.	13.1	11.8	17.4
T. A.	7.1	13.2	14.6
Mean ± SD	10.6 ± 2.8	14.6 ± 3.4	16.6 ± 2.4

Figure 14-19
Examples of force-time curves, measured at the oar, of Japanese, German, and American
oarsmen. The horizontal line gives time course, and the vertical deviations are proportional to
the force developed during one stroke. *(From Ishiko, 1971.)*

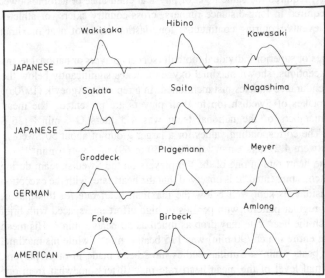

strength when their different muscle groups are tested by conventional tests of muscle strength (Secher, 1983).

The best rowers show a clear tendency to be heavy. This can be explained by the positive correlation between body weight and maximal oxygen uptake, discussed in Chap. 9, and the fact that there is no statistical significant correlation between maximal oxygen uptake in ml per kg body weight and rowing performance (Vaage, personal communication). In addition, Secher (1975) finds a positive correlation between rowing strength and body weight. This means that maximal oxygen uptake divided by body weight is not a good indicator of rowing performance (Secher et al., 1976a). It also appears that the larger body mass may be associated with a larger anaerobic capacity. It may therefore be concluded that the three attributes of greater maximal oxygen uptake, greater rowing strength, and greater anaerobic capacity together make a superior rower, and that these attributes more than compensate for the stronger water resistance caused by the greater body weight.

From their theoretical analysis of the oar movement and the forces involved, Celentano et al. (1974) estimated that at a given mean speed the energy required to cover a given distance decreased slightly when decreasing the speed oscillation at each stroke. Thus, it should be advantageous to increase the frequency of strokes, within the limits set by the efficiency of muscular contraction.

Ball Games

Football (Soccer) Generally speaking, most types of ball games represent more or less intermittent work with frequent interchanges of short bursts of physical effort interspaced with brief pauses. For this reason, various ball games, including football, have not, in the past, required the same level of physical endurance or aerobic power in the players as required in long-distance runners, cross-country skiers, or athletes engaged in similar events requiring continuous, long-lasting effort of near maximal intensity.

In the past, studies of superior individual football (soccer) players or national teams have, with some exceptions, shown maximal oxygen uptakes significantly below the levels found in the endurance athletes just mentioned. In a report by Agnevik (1970a), maximal oxygen uptakes of Swedish top football players are presented. The mean value for the 11 members of the national team was 4.3 liters $O_2 \cdot min^{-1}$ (56.5 ml $\cdot$ kg^{-1} $\cdot$ min^{-1}). The corresponding values for a larger group of about 50 Swedish top football players were 4.2 liters $\cdot$ min^{-1} or 58.6 (50 to 69) ml $\cdot$ kg^{-1} $\cdot$ min^{-1}.

In Fig. 14-20, the heart rate in one of the top players on the Swedish team during a major football match is presented. It is observed that the heart rate, with the exception of a few brief occasions, is kept well below the maximal heart rate for this player, and reflects a fairly regular pattern, with periods of high effort interspaced with brief pauses, during which the heart rate may drop as much as 50 beats $\cdot$ min^{-1}. His mean heart rate during the entire match (90 min) was 175 beats $\cdot$ min^{-1}, while his maximal heart rate was 189 beats $\cdot$ min^{-1}. Similar studies have been made in other players, and whereas the actual level of the mean heart rate may differ somewhat from one

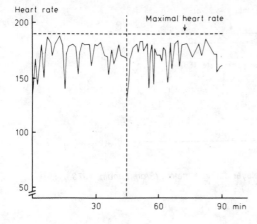

Figure 14-20
Heart rate in a top Swedish football (soccer) player on the national team during an important match. *(From Agnevik, 1970a.)*

player to another during the match, the alternating pattern with changing heart rate is a fairly common feature found in most football players studied.

Blood lactate values up to 16 mM have been measured, but 11 mM may be fairly common among football players at the end of a match.

Table Tennis Lundin (1973) has reported the results of measurements of the maximal oxygen uptakes in seven male Swedish elite table tennis players, including three world champions. The mean maximal oxygen uptake was 4.42 liters $\cdot$ min^{-1} (3.6 to 5.1) corresponding to 65.0 ml $\cdot$ kg^{-1} $\cdot$ min^{-1} on the average. It is interesting to note that, according to Lundin, the mean maximal oxygen uptake in Swedish elite table tennis players has been increasing in recent years, perhaps indicating more emphasis on physical fitness and endurance in this athletic event also. According to Lundin, there are small changes in maximal oxygen uptake in the same player from month to month, possibly because of the fact that the elite table tennis player is engaged in systematic training all year round.

Measurements of the actual oxygen uptake with the Douglas bag technique during a simulated table tennis match in seven elite players showed that, on the average, they taxed slightly more than 70 percent of their maximal aerobic power, corresponding to roughly 50 ml O_2 $\cdot$ kg^{-1} $\cdot$ min^{-1}.

The heart rate during important matches varied considerably. In some cases, the heart rate was maintained close to the maximal level (Fig. 14-21). In other cases, the heart rate dropped far below the maximal level, as in Fig. 14-22, when the losing player appeared to have lost interest in the game. In general, however, Lundin (1973) concludes that among top Swedish table tennis players, the heart rate during a match is, on the whole, 20 to 30 beats $\cdot$ min^{-1} below the maximal level.

During the actual playing, the blood lactate concentrations were around 2 to 3 mM with peak values up to 5mM. Similar concentrations were observed in the muscles. In four players, the proportion of slow- and fast-twitch fibers in the deltoid and

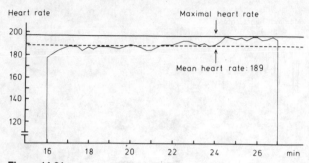

Figure 14-21
Heart rate in a Swedish elite table tennis player during a match. *(From Lundin, 1973.)*

quadriceps muscles varied markedly, with a mean value of approximately 45 percent fast-twitch fibers (type II).

Badminton A study of a group of Swedish top badminton players (Agnevik, 1970b) showed that the mean maximal oxygen uptake of the women players was 2.57 (2.3 to 2.7) liters · min^{-1} or 46.9 ml · kg^{-1} · min^{-1}. For the male players, the corresponding figures were 3.74 (3.0 to 4.4) liters · min^{-1}, or 55.6 ml · kg^{-1} · min^{-1}. This indicates that the maximal oxygen uptake of these badminton players is quite low compared with that of other elite athletes in endurance events.

Measurements of the oxygen uptake during a simulated badminton match revealed oxygen uptakes up to 3.9 liters · min^{-1} for the men (as against 3.9 liters · min^{-1} during maximal exercise on the cycle ergometer) and 2.6 liters · min^{-1} for the women (as against 2.6 liters · min^{-1} during maximal cycling) showing that during a hard game the players may actually tax all their aerobic power maximally.

Recordings of heart rates during important international matches showed that the players attained near maximal values in single matches, whereas doubles and mixed-

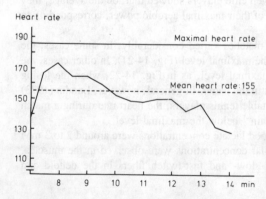

Figure 14-22
Heart rate in a Swedish elite table tennis player during a losing match. *(From Lundin, 1973.)*

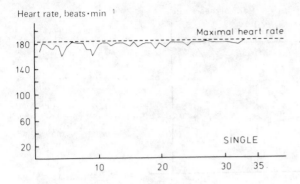

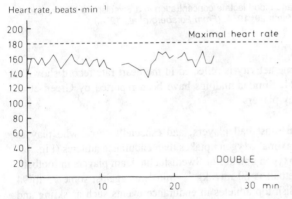

Figure 14-23
Heart rates recorded in an elite badminton player during a match. Upper panel: singles; lower panel: doubles. *(From Agnevik, 1970b.)*

doubles matches clearly showed lower heart rates (Fig. 14-23). Blood lactate values around 12 to 13 mM were obtained.

Ice Hockey According to Forsberg et al. (1974), the mean maximal oxygen uptakes of the Swedish elite ice hockey team during the 1973–1974 season was 4.9 liters · min⁻¹ (4.5 to 5.5) or 65 ml · kg⁻¹ · min⁻¹ (61 to 69). The values have shown a tendency to increase since 1960, probably because of more intense endurance training. The values are somewhat higher than for Swedish elite football or bandy players.

The oxygen uptake during simulated matches, measured by the Douglas bag technique, showed that the players at maximal intensity utilized at the most 85 to 90 percent of their maximal oxygen uptakes measured during maximal running on the treadmill.

Heart rates recorded during important international ice hockey matches showed mean heart rates of about 180 beats · min⁻¹, with peaks reaching maximal heart rate values (204 beats · min⁻¹). Heart rates were recorded and blood lactate concentrations were measured in a Swedish forward player during all three periods in the international championship match against Soviet Russia in 1974. As is evident from Fig. 14-24, the more intense the play, the higher the blood lactate level. Again, the intermittent

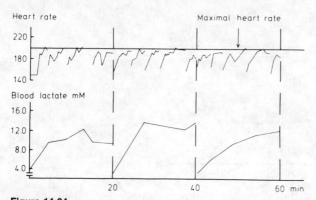

Figure 14-24
Relationship between heart rate and blood lactate concentration in a Swedish ice hockey player during a match against Russia, March 29, 1974. *(From Forsberg et al., 1974.)*

nature of this type of ball-game activity is reflected in the heart rate record shown in the upper panel of Fig. 14-24. Similar findings have been reported by Green et al. (1976) in Canadian ice hockey players.

Summary It appears that most ball players, and especially those who play as team members, have lower maximal oxygen uptakes than endurance athletes (Fig. 10-1). Thus, the mean maximal oxygen uptakes of Swedish elite team players in football, bandy, ice hockey, and basketball is 60 ml · kg^{-1} · min^{-1}, as against some 75 to 94 ml · kg^{-1} · min^{-1} in the Swedish top athletes in endurance events such as skiing and orienteering. The explanation may be sought in the fact that team ball games usually involve varying degrees of intermittent exercise, with brief pauses between short bouts of strenuous effort, while endurance events, such as long-distance running, represent a continuous effort of more or less high intensity. It thus appears that while ball games, and especially table tennis and badminton, may offer an excellent activity for the development of physical fitness for the ordinary person, they hardly represent a suitable training activity for the development of maximal aerobic power in top endurance athletes.

SEX DIFFERENCES

A comparison of world records attained by male and female elite athletes as of 1984 is presented in Table 14-3. On the whole, the difference is remarkably similar regardless of the athletic event.

Apart from the smaller body size and hence smaller strength and power of the average woman compared with the average man, women athletes differ from men in other ways; hormonal functions differ and menstruation may be a handicap for the female athlete. Menstruating women athletes have been reported to perform poorer

TABLE 14-3
SHOWING THE SEX DIFFERENCES IN ATTAINED
WORLD RECORDS IN 1984, EXPRESSED ACCORDING
TO THE FORMULA $\dfrac{♀ - ♂}{♀}$ · 100 (percent).

TRACK & FIELD		SWIMMING	
100 m	7.7	Free Style	
200 m	9.2	100 m	9.9
400 m	8.6	200 m	8.8
800 m	10.3	400 m	7.3
1500 m	9.3	800 m	6.4
3000 m	10.9	1500 m	7.3
5000 m	13.2	Breast Stroke	
10000 m	12.8	100 m	9.7
Marathon	9.9	200 m	10.1
		Butterfly	
HIGH JUMP	15.5	100 m	8.4
BROAD JUMP	19.8	200 m	7.1
SPEED SKATING		Backstroke	
500 m	7.9	100 m	8.9
1000 m	8.5	200 m	8.9
1500 m	7.9	Medley	
3000 m	6.7	200 m	7.8
BICYCLING		400 m	6.8
5 km	12.7		
20 km	10.4		

than usual in events requiring endurance, such as tennis and rowing (McNeill and Mozingo, 1981). This has led to the use of oral contraceptives in order to *avoid* menstruation during important athletic competitions. However, the use of these oral contraceptives may in itself elicit physiological alterations that may in turn affect athletic performance. Thus, it has been suggested that oral contraceptives may affect blood clotting characteristics, blood pressure, and water retention, as well as affecting carbohydrate and lipid metabolism. At any rate, specific studies of the metabolic cost of predetermined standardized submaximal exercise loads with and without the use of oral contraceptives have shown that there is a statistically significant increase in the oxygen uptake at submaximal exercise loads during all three phases of the menstrual cycle, (McNeill and Mozingo, 1981).

REFERENCES

Agnevik, G.: Fotboll, *Idrottsfysiologi,* rapport no. 7, Trygg-Hansa, Stockholm, 1970a.
Agnevik, G.: Badminton, *Idrottsfysiologi,* rapport no. 8, Trygg-Hansa, Stockholm, 1970b.
Agnevik, G., B. Wallström, and B. Saltin: A Physiological Analysis of Alpine Skiing, Unpublished Report, 1969.

Asmussen, E., K. Klausen, L. Egelund Nielsen, O. S. A. Techow, and P. J. Tönder: Lactate Production and Anaerobic Work Capacity after Prolonged Exercise, *Acta Physiol. Scand.*, 90:731, 1974.

Åstrand, P.-O.: "Experimental Studies of Physical Working Capacity in Relation to Sex and Age," Munksgaard, Copenhagen, 1952.

Åstrand, P. O.: Study of Bicycle Modifications using a Motor Driven Treadmill-Bicycle Ergometer, *Arbeitsphysiologie*, 15:23, 1953.

Åstrand, P.-O.: Human Physical Fitness with Special Reference to Sex and Age, *Physiol. Rev.*, 36:307, 1956.

Åstrand, P.-O., and B. Saltin: Maximal Oxygen Uptake and Heart Rate in Various Types of Muscular Activity, *J. Appl. Physiol.*, 16:977, 1961.

Åstrand, P.-O., L. Engström, B. Eriksson, P. Karlberg, I. Nylander, B. Saltin, and C. Thorén: Girl Swimmers, *Acta Paediat.* (Suppl. 147): 1963.

Åstrand, P.-O., I. Hallbäck, R. Hedman, and B. Saltin: Blood Lactates after Prolonged Severe Exercise, *J. Appl. Physiol.*, 18:619, 1963.

Bergh, U.: "Längdlöpning," *Idrottsfysiologi*, rapport no. 11, Trygg-Hansa, Stockholm, 1974.

Bergh, U.: Human Power at Subnormal Body Temperatures, *Acta Physiol. Scand.* (Suppl. 478), 1980.

Böje, O.: Energy Production, Pulmonary Ventilation, and Length of Steps in Well-trained Runners Working on a Treadmill, *Acta Physiol. Scand.*, 7:362, 1944.

Bonen, A., B. A. Wilson, M. Yarkony, and A. N. Belcastro: Maximal Oxygen Uptake during Free, Tethered, and Flume Swimming, *J. Appl. Physiol.*, 48(2):232, 1980.

Carlsöö, S.: A Kinetic Analysis of the Golf Swing, *J. Sports Med.*, 7:76, 1967.

Cavagna, G. A., H. Thys, and A. Zamboni: The Sources of External Work in Level Walking and Running, *J. Physiol.*, 262:639, 1976.

Celentano, F., G. Cortili, P. E. Di Prampero, and P. Cerretelli: Mechanical Aspects of Rowing, *J. Appl. Physiol.*, 36:642, 1974.

Christensen, E. H. and P. Högberg: Physiology of Skiing, *Arbeitsphysiologie*, 14:292, 1950.

Clark, J. M., F. C. Hagerman, and R. Gelfand: Breathing Patterns during Submaximal and Maximal Exercise in Elite Oarsmen, *J. Appl. Physiol.*, 55(2):440, 1983.

Costill, D. L., K. Sparks, R. Gregor, and C. Turner: Muscle Glycogen Utilization during Exhaustive Running, *J. Appl. Physiol*, 31:353, 1971.

Davies, C. T. M.: Effects of Wind Assistance and Resistance on the Forward Motion of a Runner, *J. Appl. Physiol.*, 48(4):702, 1980.

Di Prampero, P. E., G. Cortili, F. Celentano, and P. Cerretelli: Physiological Aspects of Rowing, *J. Appl. Physiol.*, 31:853, 1971.

Di Prampero, P. E., D. R. Pendergast, D. W. Wilson, and D. W. Rennie: Energetics of Swimming in Man, *J. Appl. Physiol.*, 37:1, 1974.

Di Prampero, P. E., G. Cortili, P. Mognoni, and F. Saibene: Energy Cost of Speed Skating and Efficiency of Work against Air Resistance, *J. Appl. Physiol.*, 40:584, 1976.

Ekblom, B., L. Hermansen, and B. Saltin: Hastighetsåkning på Skridsko, *Idrottsfysiologi*, rapport no. 5, Framtiden, Stockholm, 1967.

Eriksson, A., A. Forsberg, L. Källberg, P. Tesch, and J. Karlsson: Alpint, *Idrottsfysiologi*, rapport no. 17, Framtiden, Stockholm, 1977.

Faria, I. E.: Applied Physiology of Cycling, *Sports Med.*, 1:287, 1984.

Forsberg, A., B. Hultén, G. Wilson and J. Karlsson: "Ishockey," *Idrottsfysiologi*, rapport no. 14, Trygg-Hansa, Stockholm, 1974.

Gordon, M. J., B. R. Goslin, T. Graham, and J. Hoare: Comparison betwen Load Carriage and Grade Walking on a Treadmill, **Ergonomics, 26**(3):289, 1983.

Granati, A., and L. Busca: Il Lavoro della Tribiatura, *Boll. Soc. Ital. Biol. Sper.*, **20**:51, 1945.

Green, H., P. Bishop, M. Houston, R. McKillop, R. Norman, and P. Stothart: Time-motion and Physiological Assessments of Ice Hockey Performance, *J. Appl. Physiol.*, **40**:159, 1976.

Högberg, P.: How Do Stride Length and Stride Frequency Influence the Energy Output during Running? *Arbeitsphysiol.*, **14**:437, 1952.

Holmér, I.: Physiology of Swimming Man, *Acta Physiol. Scand.* (Suppl. 407): 1974a.

Holmér, I.: Energy Cost of Arm Stroke, Leg Kick, and the Whole Stroke in Competitive Swimming Styles, *Eur. J. Appl. Physiol.*, **33**:105, 1974b.

Holmér, I., E. M. Stein, B. Saltin, B. Ekblom, and P.-O. Åstrand: Hemodynamic and Respiratory Responses Compared in Swimming and Running, *J. Appl. Physiol.* **37**:49, 1974.

Holmér, I., and L. Gullstrand: Physiological Responses to Swimming with a Controlled Frequency of Breathing, *Scand. J. Sports Sci.*, **2**(1):1, 1980.

Ishiko, T.: "Biomechanics of Rowing," in *Medicine and Sport,* vol. 6, Biomechanics II, p. 249, Karger, Basel, 1971.

Jokl, E., P. Jokl, R. Green, and B. Reinhardt: Running and Swimming World Records, *Am. Corr. Ther. J.*, **30**(no. 5), 1976.

Karlsson, J., L. Hermansen, G. Agnevik, and B. Saltin: "Löping," *Idrottsfysiologi,* rapport no. 4, Trygg-Hansa, Stockholm, 1972.

Lundin, A.: "Bordtennis," *Idrottsfysiologi,* rapport no. 12, Trygg-Hansa, Stockholm, 1973.

Margaria, R.: Sulla Fisiologia e Specialmente sul Consumo Energetico, della Marcia e della Corsa a Varie Velocita ed Inclinazioni del Terreno, *Atti dei Lincei,* **7**:299, 1938.

Margaria, R., P. Cerretelli, P. Aghemo, and G. Sassi: Energy Cost of Running, *J. Appl. Physiol.*, **18**:367, 1963.

Margaria, R., P. Aghemo, and F. Piñera Limas: A Simple Relation between Performance in Running and Maximal Aerobic Power, *J. Appl. Physiol.*, **38**(2):351, 1975.

McNeill, A. W., and E. Mozingo: Changes in the Metabolic Cost of Standardized Work Associated with the Use of an Oral Contraceptive, *J. Sports Med.*, **21**:238, 1981.

Meen, H. D., R. Gullestad, and S. B. Strömme: En sammenligning av maksimalt oksygenopptak under skisprint i motbakke og löp på tredemölle, *Kroppsöving,* no. 7, 134, 1972.

Millerhagen, J. O., J. M. Kelly, and R. J. Murphy: A Study of Combined Arm and Leg Exercise with Application to Nordic Skiing, *Can. J. Appl. Sports Sci.* **8**(2), 92, 1983.

Molbech, S.: "Energy Cost of Level Walking in Subjects with an Abnormal Gait," in K. Evang and K. L. Andersen (eds.), "Physical Activity in Health and Disease," Universitetsforlaget, Oslo, 1966.

Monod, H., and I. Scherrer: "How Muscles Are Used in the Body," in C. H. Bourne (ed.), "The Structure and Function of Muscle," vol. 1, Academic Press, Inc., New York, 1972.

Passmore, R., J. G. Thomson, and G. M. Warnock: Balance Sheet of the Estimation of Energy Intake and Energy Expenditure as Measured by Indirect Calorimetry, *B. J. Nutr.*, **6**:253, 1952.

Passmore, R., and J. V. G. A. Durnin: Human Energy Expenditure, *Physiol. Rev.*, **35**:801, 1955.

Pugh, L. G. C. E.: The Influence of Wind Resistance in Running and Walking and the Mechanical Efficiency of Work Against Horizontal or Vertical Forces, *J. Physiol.* **213**:255, 1971.

Pugh, L. G. C. E.: Air Resistance in Sport, in E. Jokl, R. L. Anand, and H. Stoboy (eds.) "Advances in Exercise Physiology," S. Karger, Basel, 1976.

Ramaswamy, S. S., G. L. Dua, V. K. Raizada, G. P. Dimri, K. R. Viswanatan, J. Madhaviah, and T. N. Srivastava: Effect of Looseness of Snow on Energy Expenditure in Marching on Snow-covered Ground, *J. Appl. Physiol.*, **21**:1747, 1966.

Rönningen, H.: Unpublished results, 1976.

Saltin, B., and P.-O. Åstrand: Maximal Oxygen Uptake in Athletes, *J. Appl. Physiol.* **23**:353, 1967.

Secher, N.: Development of Results in International Rowing Championships 1893–1971, *Med. Sci. Sports*, **5**:195, 1973.

Secher, N.: Isometric Rowing Strength of Experienced and Inexperienced Oarsmen, *Med. Sci. Sports*, **7**:280, 1975.

Secher, N. H.: The Physiology of Rowing, *J. Sports Sci.* **1**:23, 1983.

Secher, N. H., O. Vaage, and R. C. Jackson: Rowing Performance and Maximal Oxygen Uptake, unpublished results, 1976a.

Secher, N. H., R. A. Binkhorst, and W. Ruberg-Larsen: Oxygen Uptake during Simulated Racing Conditions, unpublished results, 1976b.

Sjøgaard, G.: Muscle Morphology and Metabolic Potential in Elite Road Cyclists during a Season, *Int. J. Sports Med.*, **5**:250, 1984.

Taylor, C. R., and N. C. Heglund: Energetics and Mechanics of Terrestrial Locomotion, *Ann. Rev. Physiol.*, **44**:97, 1982.

Tesch, P. A.: Physiological characteristics of elite kayak paddlers, *Can. J. Appl. Sports Sci.* **8**(2):87, 1983.

Tesch, P., K. Piehl, G. Wilson, and J. Karlsson: Kanot, *Idrottsfysiologi*, rapport no. 13, Trygg-Hansa, Stockholm, 1974.

Tesch, P., K. Piehl, G. Wilson, and J. Karlsson: Physiological Investigations of Swedish Elite Canoe Competitors, *Med. Sci. Sports*, **8**:214, 1976.

Tesch, P. A., and J. Karlsson: Muscle Metabolite Accumulation following Maximal Exercise. A Comparison between Short-term and Prolonged Kayak Performance. *Eur. J. Appl. Physiol.* **52**:243, 1984.

Vokac, Z., and K. Rodahl: Maximal Aerobic Power and Circulatory Strain in Eskimo Hunters in Greenland, *Nordic Council for Artic Med. Res.*, Report no. 1, 1977, Oulu Univ. Press, Oulu, Finland.

Vrijens, J., J. Hoekstra, J. Bouckaert, and P. van Tyvanck: Effects of Training on Maximal Working Capacity and Haemodynamic Response during Arm and Leg Exercise in a Group of Paddlers. *Eur. J. Appl. Physiol.*, **34**:113, 1975.

Watanabe, K., and T. Ohtsuki: Postural Changes and Aerodynamic Forces in Alpine Skiing, *Ergonomics*, **20**(2):121, 1977.

Wells, C. L., J. R. Stern, and L. H. Hecht: Hematological Changes following a Marathon Race in Male and Female Runners, *Eur. J. Appl. Physiol.*, **48**:41, 1982.

Whitt, F. R., and D. G. Wilson: Bicycling Science, Ergonomics and Mechanics, The Massachusetts Institute of Technology Press, Cambridge, Mass., 1974.

Zarrugh, M. Y., and C. W. Radcliffe: Predicting Metabolic Cost of Level Walking, *Europ. J. Appl. Physiol.*, **38**:215, 1978.

Zuntz, L., and W. Schumburg: "Studien zu einer Physiologie des Marsches," Bibliothek v. Coler, Band 6, Verlag von A. Hirschwald, Berlin, 1901.

CHAPTER 15

FACTORS AFFECTING PERFORMANCE

CONTENTS

INTRODUCTION
HIGH ALTITUDE
    Physics
    Physical Performance
    Limiting Factors
    Oxygen Transport
    Adaptation to High Altitude
    Performance after Return to Sea Level
    Practical Applications
HIGH GAS PRESSURES
    Pressure Effects
    Nitrogen
    Oxygen
    Carbon Dioxide
    Oxygen Inhalation in Sports
TOBACCO SMOKING
    Circulatory Effects
    Respiratory Effects
    Metabolic Effects
    Smoking Habits among Athletes
ALCOHOL AND EXERCISE
    Neuromuscular Function
    Aerobic and Anaerobic Power
    Metabolic Effects

CAFFEINE
DOPING
THE WILL TO WIN

INTRODUCTION

Figure 7-1 is a schematic presentation of the components of the aerobic work capacity and the factors which affect this capacity. Most of these factors have been discussed earlier rather extensively from various viewpoints and in different connections. Thus the effect of the type of exercise was examined in Chap. 7, oxygen-transporting functions were discussed in Chaps. 4 and 7, metabolic aspects in Chap. 12 and training and deconditioning in Chap. 10. As far as the effect of the environment is concerned, the effect of temperature and water balance was explored in Chap. 13. The present chapter will be devoted to a combined discussion of the remaining factors mentioned in Fig. 7-1, including the effect of altitude.

HIGH ALTITUDE

The climbing of high mountains has always fascinated humans. The quest for new discoveries has taken men beyond the earth's atmosphere into outer space. Permanent human habitation is encountered up to an altitude of over 4,500 m. It is becoming increasingly popular to spend summer and winter vacations in high mountain areas undertaking the fairly heavy physical exercise involved in hiking, mountaineering, or skiing. In spite of pressure cabins, passenger flights at high altitudes involve a certain lack of oxygen. The air pressure in the airplane cabin usually corresponds to an altitude of 1,000 to 1,500 m. The decision to hold the 1968 Olympic Games in Mexico City at an altitude of 2,300 m created a special interest in the problems concerning the effects of altitude on physical performance. Several symposia have been held to discuss various problems related to physical performance at high altitude. (For reports, see Dill, 1964; *Schw. Zschr. Sportmed.*, **14**:1-329, 1966; Margaria, 1967; Jokl and Jokl, 1968; Roskamm et al., 1968; Loeppky and Riedesel, 1982). Luft (1964) has reviewed historic events in the exploration of altitude and the facilities available in various parts of the world for research in this field (see West, 1981).

Physics

In the nineteenth century, Bert (1878) recognized that the detrimental effects of high altitude were due to the *diminished partial pressure of oxygen* at reduced barometric pressure. The barometric pressure at a given altitude depends on the weight of the air column over the point in question. The atmosphere is compressed under its weight. Its pressure and density are therefore highest at the surface of the earth and decrease exponentially with altitude. Because of temperature differences and turbulence, any sedimentation of the gas molecules of different molecular weight is avoided, and the chemical composition of the atmosphere is practically uniform up to an altitude of more than 20,000 m.

Table 15-1 presents barometric pressure and the oxygen pressure of the inspired air (tracheal air) at various altitudes. With a constant oxygen concentration of 20.94 percent of the dry air, the oxygen pressure of the inspired air in the trachea, saturated with water vapor, can easily be calculated from the formula

$$P_{O_2} = (P_{Bar} - 47) \cdot 20.94 \cdot 100^{-1}.$$

This means that at an altitude of about 19,000 m, where the barometric pressure is 47 mm Hg (6.3 kPa), there should be nothing but water molecules in the trachea!

The oxygen tension of the alveolar air, and thereby also the oxygen tension of the arterial blood, is determined by the magnitude of the pulmonary ventilation in addition to the composition and pressure of the inspired air. The more frequently the air is exchanged, the more closely the composition of the pulmonary air resembles the inspired air (water vapor subtracted). This will be discussed below.

The reduced density of the air at high altitudes affects the mechanics of breathing. Part of the work of breathing is expended in moving the air against the resistance of the airways. The resistance is relatively high when the flow is turbulent, as it is during exercise. Therefore, the influence of reduced density is more noticeable at high rates of air flow as in hyperpnea, during heavy exercise, or in flow-dependent pulmonary function tests. The maximal breathing capacity is considerably higher at high altitude than at sea level (Cotes, 1954; Miles, 1957; Ulvedal et al., 1963). The respiratory muscles' ability to exert pressure seems to be reduced at low barometric pressure; however, the net effect of the reduced resistance to air flow at lowered barometric pressure is a *diminished respiratory work* to move a given volume of air in and out of the lungs (Fenn, 1954). On several occasions, pulmonary ventilations of 200 liters $\cdot$ min^{-1} have been measured during maximal exercise at high altitude (see below). Another effect of the reduced density of the air at a low barometric pressure is a diminished external air resistance. The air resistance changes with the wind speed raised to the second power. The external work is therefore reduced at high altitude in sprint-type activities, speed skating, cycling, and alpine skiing with high velocities. The *air temperature* is on the whole lower, the higher the altitude. With a mean annual temperature of 15°C at sea level, the air temperature decreases linearly by 6.5°C $\cdot$ 1,000 m^{-1} to about 11,000 m. The air also becomes increasingly *dry* with increasing altitude. Therefore the water loss via the respiratory tract is higher at high altitudes than at sea level. If much work is performed at high altitude, this loss may give rise to a hypohydration and a sensation of soreness and dryness in the throat.

The *solar radiation* is more intense at high altitude and the ultraviolet radiation may cause difficulties in the form of sunburn or snow blindness. Finally, *the force of gravity* is reduced with the distance from the earth's center. High altitudes may therefore have some favorable effect in the case of athletic events involving jumping or throwing.

Physical Performance

The reduced performance at high altitudes is well established; it is already evident at an altitude of about 1,200 m in the case of heavy exercise engaging large muscle

TABLE 15-1
BAROMETRIC PRESSURE (STANDARD ATMOSPHERE) AT VARIOUS ALTITUDES AND THE PRESSURE OF OXYGEN AFTER THE INSPIRED GAS HAS BEEN SATURATED WITH WATER VAPOR AT 37°C (TRACHEAL AIR).*

Altitude		Pressure,		P_{O_2} tracheal air,		Altitude		Pressure,		P_{O_2} tracheal air,	
m	ft	mm Hg	kPa	mm Hg	kPa	m	ft	mm Hg	kPa	mm Hg	kPa
0	0	760	101.3	149	19.9	5,500	18,050	379	50.5	69	9.2
500	1,640	716	95.5	140	18.7	6,000	19,690	354	47.2	64	8.5
1,000	3,280	674	89.9	131	17.5	6,500	21,330	330	44.0	59	7.9
1,500	4,920	634	84.5	123	16.4	7,000	22,970	308	41.1	55	7.3
2,000	6,560	596	79.5	115	15.3	7,500	24,610	287	38.3	50	6.7
2,500	8,200	560	74.7	107	14.3	8,000	26,250	267	35.6	46	6.1
3,000	9,840	526	70.1	100	13.3	8,500	27,890	248	33.1	42	5.6
3,500	11,840	493	65.7	93	12.4	9,000	29,530	230	30.7	38	5.1
4,000	13,120	462	61.6	87	11.6	9,500	31,170	214	28.5	35	4.7
4,500	14,650	433	57.7	81	10.8	10,000	32,800	198	26.4	32	4.3
5,000	16,400	405	54.0	75	10.0	19,215	63,000	47	6.3	0	0.0

*Based on dry conditions for average temperature at altitude when the temperature at sea level is 15°C and the barometric pressure is 760 mm Hg (101.3 kPa).

groups for about 2 min or longer. As an example of the physical strain experienced during the conquest of high mountain peaks, Somervell (1925) may be quoted: "It may be of interest to record one or two personal observations which I made while climbing in the neighborhood of 27,000 to 28,000 feet. *Pulse:* The heart rate during the actual motion upwards was found to be beating 160-180 per minute, sometimes even more, regular in rhythm and of good volume. . . . *Respiration:* About 50 to 55 per minute while climbing. Approaching 28,000 feet I found that for every single step forward and upward, seven to ten complete respirations were required." Norton (1925) writes that at an altitude of 8,500 m he required 1 hr to climb a distance of 35 m even though the climb was not particularly difficult.

Sutton et al. (1983), have made an interesting analysis of the remarkable ascent of Mount Everest by Habeler and Messner without the use of supplementary oxygen. It has been pointed out that although Habeler and Messner were exceptional athletes, they had almost reached their limits as they were climbing the last 48 m to the summit, collapsing into the snow every few steps, and barely crawling on. While they have proved that it is indeed possible to climb Mount Everest without supplementary oxygen under fortunate circumstances, only a drop in the ambient barometer pressure would perhaps have been enough to make it impossible. It can only be done at the expense of extreme hyperventilation and respiratory alkalosis (observed P_{ACO_2} 7.5 mm Hg − 1 kPa, P_{AO_2} 35 mm Hg − 4.6 kPa, pH_a 7.7) (West, 1983; Hackett et al., 1983; West and Lahiri, 1984).

Schoene (1982) reports that a group of climbers, who have successfully climbed to at least 7800 m had significantly higher ventilatory responses to hypoxia, and hypercapnia, than a group of outstanding long-distance runners. He suggests that this factor may be selective for this elite group of athletes.

Athletic competitions at different altitudes provide an experimental condition where highly motivated, well-trained athletes subject themselves to all-out tests. Leary and Wyndham (1966) report observations from South Africa, where important track meetings take place at altitudes of 1,500 m or above as well as at sea level. They conclude that the best performances recorded in middle- and long-distance events were on the coast, whereas sprinters recorded better times at medium altitude. South African championships held at medium altitude have been consistently dominated by athletes domiciled at such altitudes, and those who acclimatized themselves by training for medium altitude for 3 to 4 weeks before the championship competitions. In competitions in the mile event, five international class athletes all performed better at sea level. The average time was 4 min and 0.5 s in the best coastal performance and 4 min and 15.2 s in Johannesburg (altitude 1,760 m. or 5,780 ft).

In competitions in Mexico City (altitude 2,300 m), the same or better performance in running distance up to 400 m has been reported. In 1,500-m running, there is an impairment of about 3 percent, and in 5,000- and 10,000-m, an impairment of roughly 8 percent compared with sea level. In swimming, one finds an impairment of the time for 100 m of about 2 to 3 percent and an impairment of about 6 to 8 percent in the case of 400-m or longer distances when comparing Mexico City altitude with that of sea level (Goddard and Favour, 1967; Craig, 1969; Shephard, 1973). In jumping and throwing , some differences have been recorded among athletes competing at different altitudes. The world record in long jumping (8.90 m) by Bob Beaman in the Mexico

City Olympic Games should be particularly noted. It has frequently been stated that recovery time in Mexico City is considerably longer than at low altitudes. Various types of collapses occurred with relatively high frequency during the 1968 Olympic Games in Mexico City (see Jokl and Jokl, 1977).

This brief summary shows that in types of exercise requiring an intense activity of no more than 1-min duration, and especially in events of this nature where the technique is of primary importance, there is no noticeable difference in performance between sea level and high altitude, at least up to an altitude of 2,500 m. The capacity to perform heavy exercise for 2 min or more is, on the other hand, definitely reduced at high altitude, with the exception of types of activities in which the air resistance plays a major role.

What then, is the physiological explanation for this phenomenon?

Limiting Factors

We may base our analysis on the factors listed in the table of factors affecting physical performances at the beginning of Chap. 7. If one disregards altitudes above 3,000 m, which may also cause a disturbance of psychological functions, it is evident that it is the aerobic power which is affected by a reduced oxygen pressure in the inspiratory air. Studies have shown that exercise during acute exposure to high altitude causes a rise in the blood lactate concentration at lighter work rates than is the case at sea level. At a fixed rate of exercise, the lactate level is higher, but the maximal concentration attained during exhaustive exercise is roughly the same as at sea level (Edwards, 1936; Asmussen et al., 1948; P.-O. Åstrand, 1954; Stenberg et al., 1966; Buskirk et al., 1967; Hermansen and Saltin, 1967). Figure 15-1 gives an example of how the blood lactate concentration is affected during exercise at sea level, 2,300 m, and 4,000 m simulated altitude. In this subject, however, the lactate concentration was the same when it is related to the relative, instead of the absolute, oxygen uptake. This must be interpreted as indicating that the anaerobic processes are brought into play at a relatively lower work rate at high altitude. The maximal anaerobic power, at least the part of it which is determined by the glycogenolysis, on the other hand is not affected. The fact that the maximal oxygen debt is the same following maximal exercise at sea level and at altitude also points in the same direction (Saltin, 1966; Buskirk et al., 1967).

As an example, Saltin (1966) mentions that the sprinter Bodo Tümmler ran 1,500 m in Stockholm in 3.42 min and in Mexico City in 3.54 min. The O_2 uptake during the subsequent 60-min rest was 38 and 42 liters respectively; peak blood lactate concentration was 18.6 and 19.3 mM respectively. After a 3,000-m steeplechase, the values obtained for Bengt Persson in Stockholm were 28 liters O_2, blood lactates 18.2 mM, running time 8.34 min. Corresponding values in Mexico City were 33 liters O_2, 20.3 mM, and 9.32 min.

It is difficult to explain the higher lactic acid level in the blood at a standard submaximal work rate at high altitude in view of the fact that the oxygen uptake is the same as at sea level (Christensen, 1937; Asmussen and Chiodi, 1941; P.-O. Åstrand, 1954; Pugh et al., 1964). Recent studies have indicated a slower increase in oxygen

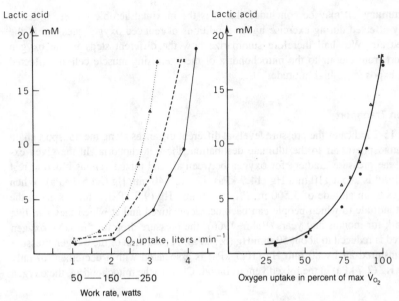

Figure 15-1
Blood lactate concentrations in one subject during exercise at sea level (filled dots) and acute exposure to 2,300 (crosses) and 4,000 m (triangles) (760,580, and 462 mm Hg or 101, 77, and 66 kPa, respectively). Exercise times for all submaximal work rates were 10 min, and all blood samples were taken between the ninth and tenth min. Exercise times for the maximal work rates were 4 to 5 min, and peak values for lactic acid are given.
In the left panel the absolute oxygen uptake is on the abscissa; on the right, oxygen uptake is in percentage of the maximum. *(From Hermansen and Saltin, 1967.)*

uptake at the beginning of exercise, i.e., the oxygen deficit is enlarged. Therefore, the demand of the anaerobic energy yield is correspondingly increased (see Chap. 7; Raynaud et al., 1974).

With regard to the neuromuscular function, Christensen and Nielsen (1936) have shown that speed and strength are not affected by moderate hypoxia. In a test on a Hill's wheel, contraction time being less than 6 s, their subjects attained the same maximum at sea level as at a barometric pressure of 440 (58.7 kPa) and 390 mm Hg (52.0 kPa), respectively.

From a psychological point of view, a stay at even moderate altitudes may represent a considerable stress, or in any case an unusual sensation. From experience one knows how a certain exercise intensity feels or affects the body under normal circumstances. The same physical effort at higher altitudes produces a higher pulmonary ventilation, a higher heart rate, and possibly other symptoms of fatigue, which under normal conditions would not be customary at this work rate. An adaptation to the new situation gradually takes place. For the athlete, the tactics may be different in training and competitions at high altitude compared with the situation at sea level. This will be further discussed at the end of this chapter.

Summary It may be concluded that it is the maximal aerobic power, which is directly affected during exercise under conditions of reduced oxygen pressure in the inspired air. We shall therefore summarize how the different steps in the oxygen transport from the air to the mitochondria of the exercising muscle cell are affected during exposure to high altitude.

Oxygen Transport

Figure 15-2 indicates the pressure levels at different distances along the transport chain from atmospheric air to the ultimate destination, the mitochondria. In the given example, the pressure gradient for oxygen between air and mixed venous blood at rest at sea level is about 110 mm Hg (14.7 kPa), i.e. 150-40 mm Hg (20-5.3 kPa). When one goes to an altitude of 5,500 m, $P_{O_2} = 70$ mm Hg (9.3 kPa), which is about the highest altitude to which people can become acclimatized and at which they can live and work for months and years (Rahn, 1966), the pressure-drop for the same oxygen delivered is reduced to about 50 mm Hg (6.7 kPa). A reduction of the oxygen pressure in the inspired air by 80 mm Hg (10.7 kPa) is associated with a decrement of only 20 mm Hg (2.7 kPa) in the mixed venous blood. Close to the mitochondria, the oxygen

Figure 15-2
A comparison of the oxygen cascade from inspired air to tissue in the human at sea level and at 5,500 m (18,000 ft). *(Modified from Rahn, 1966.)*

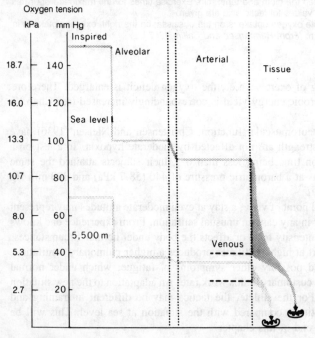

pressure may be 10 mm Hg (1.3 kPa) in the sea-level situation and about 5 mm Hg (0.7 kPa) at 5,400 m. This pressure is still adequate to provide optimal conditions for the oxidative enzyme reactions (Chap. 12). There are many factors explaining the increased "conductance" to keep the tissue oxygen pressure at this almost constant level. We shall first consider the *acute hypoxia*.

1 The *pulmonary ventilation* at a given oxygen uptake is markedly elevated (see Fig. 15-3). In this subject the ventilation at an oxygen uptake of 4.0 liters · min⁻¹ was 80 liters · min⁻¹ when pure oxygen was inhaled, 105 liters · min⁻¹ at sea level when the subject was breathing air, 140 liters · min⁻¹ at 2,000 m, and 160 liters · min⁻¹ at 3,000 m, or twice as high as when breathing oxygen. Even at sea-level conditions there exists a hypoxic drive which is evident when this subject's oxygen uptake exceeds 1.5 liters · min⁻¹. This hypoxic hyperpnea is elicited through the reflex pathway originating in the chemoreceptors of the carotid body and aortic body (see Chap. 5). Since the production of carbon dioxide is roughly the same at a given oxygen uptake, this hyperventilation will inevitably wash out CO_2 from the blood into the inspired air, the dissolved CO_2 of the blood being more affected than the bicarbonate. The secondary effect of the hyperpnea will therefore be a rise in the pH of the blood, i.e., an

Figure 15-3
Pulmonary ventilation (BTPS) in relation to oxygen uptake in one subject at different work rates when breathing oxygen or air during exposure to various simulated altitudes. Note the high pulmonary ventilation of 190 liters · min⁻¹ reached during exercise at 3,000-m altitude. *(From P.-O. Åstrand, 1954.)*

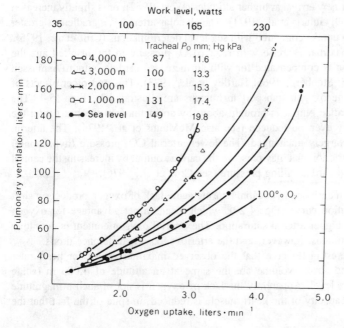

uncompensated respiratory alkalosis. The reduced P_{CO_2} and elevated pH of the arterial blood exert an inhibitory influence on the respiratory center. On the other hand, the earlier accumulation of lactic acid in the blood must be considered to cause the pH to fall.

At sea level, the alveolar P_{CO_2} was about 38 mm Hg (5.1 kPa) in the experiment, bringing the pulmonary ventilation to 80 liters $\cdot$ min^1 (Fig. 15-3). The reduction in ventilation when breathing oxygen brought the alvealor P_{CO_2} to a somewhat higher level; at a simulated altitude of 4,000 m. the alvealor P_{CO_2} became reduced to 28 mm Hg (3.7 kPa) during exercise, with a similar pulmonary ventilation of 80 liters $\cdot$ min^{-1} (see also Dejours et al., 1963; Dempsey et al., 1972). The effect is that the hypoxic drive during exercise at high altitude must be stronger than reflected in the magnitude of the pulmonary ventilation. If the alveolar and arterial P_{CO_2} are maintained at about 40 mm Hg (5.3 kPa) by the addition of a proper volume of CO_2 to the inspired air; the pulmonary ventilation during exercise in hypoxic conditions will be still higher than illustrated in Fig. 15-3. The ventilatory response must be viewed as a physiological compromise with the call for an adequate oxygen supply matched against the need to maintain the acid-base balance as normal as possible.

Owing to the great increase in ventilation during exercise at high altitude and a given oxygen uptake, the alveolar P_{O_2} is higher than what it normally would be. This obviously facilitates the diffusion of oxygen to the blood in the pulmonary capillaries.

The maximal pulmonary ventilation during exercise at high altitude is the same as, or higher than, at sea level (Stenberg et al., 1966; Saltin, 1966; Grover and Reeves, 1967; Roskamm et al., 1968).

2 The diffusing capacity is reported to be unchanged in people who have lived at sea level after their new arrival at higher altitudes (West. 1962), or is slightly increased (Reeves et al., 1969; Gularia et al., 1971). The alveolar-arterial P_{O_2} gradient is greater for a given oxygen uptake compared with sea-level determinations (Cruz et al., 1975).

3 During submaximal exercise with reduced O_2 pressure in the inspired air, the lower O_2 saturation is compensated for with an increased cardiac output (Asmussen and Nielsen, 1955; Stenberg, 1966; Hartley et al., 1973). This and other effects of high altitude on the oxygen transport in humans are illustrated by Fig. 15-4. The increase in the cardiac output is brought about by an increase in the heart rate; the stroke volume may even be reduced (see also McManus et al., 1974). The arterial blood pressure is largely unchanged. The lower arterial CO_2 pressure (hypocapnia) causes a venoconstriction that may preserve the cardiac output by increasing the central venous volume and cardiac filling pressure (see Cruz et al., 1976).

The previously mentioned hypocapnia during acute lack of oxygen produces a shift in the Hb dissociation curve (Fig. 4-20), which means a net advantage for oxygen transport due to a higher arterial saturation. The arterial oxygen content is definitely reduced at high altitudes, however, and the arteriovenous O_2 difference drops.

It is rather interesting to note that the observed maximal values for heart rate, cardiac output, and stroke volume are the same at an altitude of 4,000 m (acute exposure) as at sea level. Apparently the lack of oxygen is not of such a magnitude that the pumping capacity of the heart muscle is reduced, in spite of the fact that the

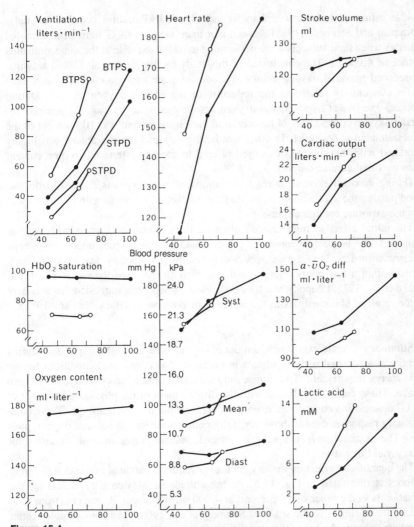

Figure 15-4
Mean values on six subjects studied at two submaximal work rates and one maximal work rate (cycle ergometer) at sea level (filled dots) and when acutely exposed to simulated altitude of 4,000 m in a decompression chamber (open dots). Note that the maximal oxygen uptake was reduced to 72 percent of the value at sea level. Abscissa = oxygen uptake in percentage of the maximum attained at sea level. *(From Stenberg et al., 1966.)*

Pa_{O_2} is estimated to be lower than 50 mm Hg (6.7 kPa). In a comparable study, Blomqvist and Stenberg (1965) showed that there were no ECG signs of myocardial ischemia when their subjects were performing maximal exercise at the same simulated altitude of 4,000 m. The conclusion of this study by Stenberg et al. (1966) was that the reduced maximal oxygen uptake in moderate acute hypoxia compared with normoxia was closely related to the reduction of the arterial oxygen content. During maximal exercise at hypoxia, the oxygen uptake was on an average 72 percent, the arterial oxygen content was 74 percent, and the cardiac output was 100 percent of the values attained at sea level. In other words, the maximal oxygen uptake was highly correlated with the volume of oxygen offered to the tissue (arterial oxygen content times maximal cardiac output).

During maximal exercise at sea level, almost all the oxygen is extracted from the blood passing the active muscles, so that there is nothing more to gain in this respect at acute exposure to high altitude.

The initial effect of high altitude often includes mountain sickness with various symptoms like headache, vomiting, physical and mental fatigue, interrupted sleep, digestive disorders. In rare cases, previously healthy persons may develop pulmonary edema within a few hours after exposure to high altitude (see Reeves et al., 1969; Clarke et al., 1975; Loeppky and Riedesel, 1982). It has been suggested that this may be the result of abnormalities in the handling of body water (Hackett et al., 1981).

Summary Acute exposure to a reduced oxygen pressure in the inspired air during exercise is associated with hyperpnea in excess of that at sea-level conditions for the same energy requirement. The cardiac output also rises out of proportion to the oxygen uptake. These factors, combined with the displacement of the physiological range of the O_2 dissociation curve to its steep part, enhance the oxygen transport. These compensatory responses cannot, however, fully compensate for the reduced oxygen pressure. The maximal oxygen uptake is reduced, and the importance of the anaerobic energy yield increases.

The quantitative effect on the oxygen transport during maximal exercise at different altitudes is illustrated in Fig. 15-5. At the altitude of Mexico City, 2,300 m; the reduction is in the order of 15 percent; at 4,000 m, it is about 30 percent (from 4.24 to 3.07 liters · min^{-1} in the study of Stenberg et al., 1966). The considerable scatter of the data observed when different studies are compared may be explained by several facts: (1) The effect of the reduced oxygen pressure on the maximal aerobic power is different in different individuals; (2) different techniques have been applied, especially concerning criteria for ascertaining that the maximal oxygen uptake has been reached; (3) conceivably, persons with a high maximal aerobic power are more affected than persons with a lower maximal aerobic power, in that the diffusing capacity may be more critical for the former. The progressive fall in arterial oxygen saturation as the exercise level is raised at high altitude, in spite of an increasing alveolar tension, and the resulting large alveolar-arterial oxygen differences, can be explained by diffusion limitations of the lung (West et al., 1962; Saltin, 1967; Grover and Reeves, 1967).

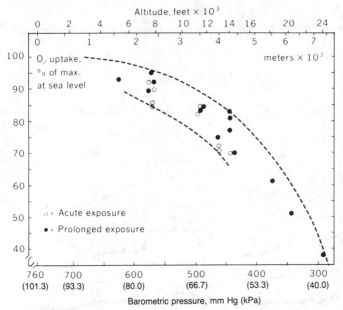

Figure 15-5
Reduction in maximal oxygen uptake in relation to altitude barometric pressure. Unfilled dots denote experiment in the acute hypoxic stage; filled dots, data obtained after various periods of acclimatization. In principle, the maximal aerobic power during acute exposure to reduced oxygen pressure falls at the lower part, within the dotted lines; during acclimatization, it is shifted toward the upper part of the field. *(Data from Balke, 1960; Pugh, 1964; Stenberg et al., 1966; Buskirk et al., 1967; Hansen et al., 1967b; Saltin, 1967; Roskamm et al., 1968.)*

Adaptation to High Altitude

We shall now discuss the effect of a prolonged stay at high altitude, i.e., *acclimatization* to reduced oxygen pressure in the inspired air. It is customary to distinguish between short-term adaptation when it is a matter of days, weeks, or a few months at high altitude, and long-term adaptation when the stay consists of years at high altitude.

1 The first few days' exposure to reduced oxygen pressure entails a further increase in the pulmonary ventilation at a given work rate. This hyperpnea will further raise the P_{O_2} and reduce the P_{CO_2} of the alveolar air. This ventilatory response is illustrated in Fig. 15-6. It shows data obtained at sea level, during acclimatization for 4 weeks to an altitude of 4,300 m (14,250 ft), and again at sea level during the reacclimatization. Four subjects were studied, but the figure presents the data on only one of the subjects. However, he represents the normal reaction reasonably well. Three comments should be made: (*a*) It is evident that within a week at high altitude, a new level for pulmonary ventilation is attained, exceeding the value noticed in acute exposure to the same degree of hypoxia. The prolonged exposure to this hypoxia caused a 40 to 100 percent

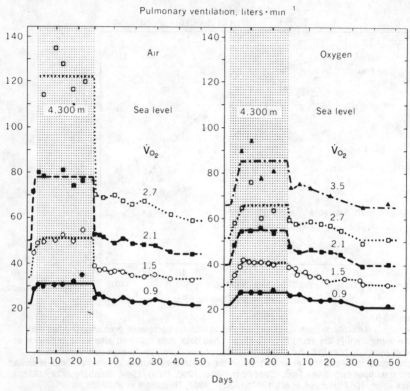

Figure 15-6
Pulmonary ventilation in one subject exercising on a cycle ergometer (1) at sea level (base line to the left), (2) during a 4-week sojourn at an altitude of 4,300 m (shaded area), and (3) again after return to sea level including almost 50 days' observation time. Oxygen uptake is indicated on respective line. *Left panel:* Subject breathing room air; *Right panel:* Subject breathing oxygen during exercise. *(Unpublished data obtained by I. Åstrand and P.-O. Åstrand at White Mountain Research Station, California, 1957.)*

increase in the pulmonary ventilation compared with the sea level controls, the increase being more pronounced the heavier the work rates. At the end of the stay, the ventilation at an oxygen uptake of 2.7 liters $\cdot$ min^{-1} (200 watts) was about 120 liters $\cdot$ min^{-1} when breathing air, compared with 60 liters $\cdot$ min^{-1} at sea level. The alveolar P_{CO_2} was 24 mm Hg (3.2 kPa) at 4,300 m, compared with 40 mm Hg (5.3 kPa) at sea level. (*b*) Even when oxygen is inhaled during exercise at high altitude, blocking the peripheral chemoreceptor drive, there is a gradual increase in pulmonary ventilation (see Fig. 15-6). In the example chosen, the ventilation in the control experiment at sea level was 50 liters $\cdot$ min^{-1}, but at the end of the 4-week sojourn at high altitude, it was raised to 65 liters $\cdot$ min^{-1} during the standard exercise. If oxygen is used during exercise at high altitude, it is clearly oxygen-saving if the individual is unacclimatized.

Since, on the other hand, oxygen can scarcely be supplied continuously for long periods of time, an acclimatization is desirable, especially at very high altitudes. Furthermore, the oxygen-providing equipment may fail. (c) When the subjects returned to sea level following exposure to altitude, it took several weeks before the control level was attained (see Buskirk et al., 1967; Forster et al., 1971). The return of the alveolar P_{CO_2} to control levels paralleled the shift in pulmonary ventilation.

A calculation of the pulmonary ventilation at STPD gives practically the same volumes at a given oxygen uptake in subjects acclimatized to various altitudes for about 4 weeks (Christensen, 1937; Dejours et al., 1963; Pugh et al., 1964). In other words, the number of oxygen molecules, which are inhaled per unit of time is constant at the various altitudes. It should be remembered, however, that the alveolar oxygen pressure will still not be the same at high altitudes as at sea level.

At high altitude, the energy cost for the respiratory muscles to move the greater volume of air may not be higher than at sea level because of the reduced density of the air. When sea level is again reached, the "abnormally" high ventilatory response at a given oxygen uptake must require extra respiratory work.

It has been suggested that the extra increase in ventilation in connection with exercise at high altitude must be attributed to a hypoxic drive via the peripheral chemoreceptors. This hypoxic drive is prevalent even during chronic hypoxia, at least for a considerable period of time (P.-O. Åstrand, 1954; Dejours et al., 1963). The induced alkalosis on sudden exposure to low P_{O_2} reducing the central chemoreceptor drive, becomes gradually compensated by a proportionate decrease in the blood bicarbonate, and restoration of a normal pH occurs in the acclimatized person. Prolonged alteration in arterial P_{CO_2} in either direction tends to bring about alterations in the renal acid-base excretion that slowly tend to return the arterial H^+ toward normalcy. As the alkalosis is reduced (pH is lowered), there is a further increase in pulmonary ventilation, as illustrated in Fig. 15-6. However, a restoration of CSF $[H^+]$ does not sufficiently explain the hyperventilation obtained upon sojourn to high altitude (Dempsey et al., 1972).

Evidence of a real change in the regulation of the body P_{CO_2} developing during the first week after ascent from sea level to high altitudes is presented in Fig. 15-7. The respiratory response to CO_2 was tested by breathing CO_2-O_2 mixtures during exercise with 100 watts (oxygen uptake, 1.5 liters · min^{-1}). There was a marked shift to the left so that, after acclimatization, a given pulmonary ventilation was attained at 15 to 20 mm Hg (2.0-2.7 kPa) lower alveolar P_{CO_2} compared with controls at sea level. The difference in the response to CO_2 can be illustrated by the following example: At sea level, a pulmonary ventilation of 35 liters · min^{-1} was obtained when the end-expiratory P_{CO_2} was 45 mm Hg (6.0 kPa); after 1 week at 4,300 m, the same P_{CO_2} was recorded when ventilation was as high as 83 liters · min^{-1}. (In this subject, the high CO_2 mixture had an almost narcotic effect, with a reduced ventilatory response as a consequence; see days 15 and 19 at high altitude, and day 1 at sea level.)

After the return to sea level, there was a gradual return of the CO_2 response curve to the control level. This slow reacclimatization shows a pattern parallel with the one illustrated in Fig. 15-6. When the hypoxic drive was reduced by the increased oxygen pressure in the inspired air, the pulmonary ventilation was reduced. This produces a reduced washout of CO_2 and an uncompensated acidosis with an increased central chemoreceptor drive as a

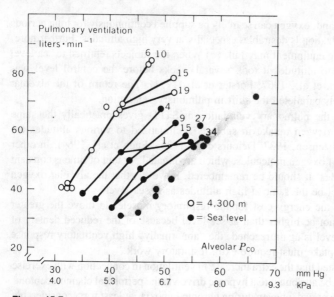

Figure 15-7
Ventilatory response during a standard work rate (100 watts) to inhalation of various CO_2-O_2 mixtures at sea level (shaded area), during a prolonged sojourn at altitude of 4,300 m (unfilled dots), and at various intervals during the reacclimatization to sea-level conditions (filled dots). Figures at top of line denote the day for the experiment after arrival at altitude and sea level respectively. (Same subject as in Fig. 17-6). *(Unpublished data by I. Åstrand and P.-O. Åstrand.)*

consequence. (For a more detailed discussion of the regulation of respiration during hypoxia, see Kellog, 1964.) Millhorn et al. (1980) suggest that the observed long-lasting facilitation of respiration, mediated endogenously by the central neurotransmitter serotonin, may explain the slow secondary increase in ventilation during acclimatization to high altitude hypoxia.

There is another effect of the *reduced alkaline reserve* in the blood of the person acclimatized to high altitude. The individual will have less ability to withstand acidosis from other acids which may arise in the course of metabolism (Roughton, 1964). During acute exposure of a few weeks' duration to high altitude, the person can apparently attain the same high blood lactate level as at sea level, but data by Edwards (1936) suggest a gradual decline in this maximal (Hansen et al., 1967a). The reduced alkaline reserve may be one factor behind this decrease in maximum anaerobic power.

Natives and long-term residents at high altitude have a lower pulmonary ventilation than the short-term acclimatized individual. However, Hackett et al. (1980) maintain, on the basis of their studies of the ventilatory control of Sherpas, that Sherpas differ from other high-altitude natives: Their hypoxic ventilatory response is not blunted, and they exhibit relative hyperventilation. Forster et al. (1971) found that the maximal ventilation increase in response to hypoxia usually occurred during the first 3 to 4 weeks at altitude, and then it began to decrease toward sea-level values. The highlanders were also less responsive to CO_2. With a lower exercise V_E the alveolar CO_2 tension is higher. The ventilatory response to a removal of hypoxemia is less in residents at high altitude than in individuals who have spent limited time at high altitude (see also Dempsey et al., 1972). There seems to be a gradual adaptation of the peripheral chemoreceptors (Severinghaus et al., 1966; Milledge and Lahiri, 1967).

2 With regard to the transport of O_2 from the alveolar air to the pulmonary capillary blood, the situation is somewhat controversial. West (1962) found no change in the diffusing capacity after a 6-month exposure to 5,800-m altitude. Natives and long-term residents at high altitude may have a greater diffusing capacity than comparable sea-level residents (Velasquez, 1959; DeGraff et al., 1965; Dempsey et al., 1971; Guleria et al., 1971; Cruz et al., 1975). Kreuzer et al. (1964) found that the values of the alveolar-arterial P_{O_2} gradient in the Andean natives was higher than in normal sea-level residents, which is in contrast to Hurtado's (1964) observation of a particularly small gradient in his native subjects in the same area. Similar findings of a relatively narrow alveolar-arterial O_2 pressure gradient in natives and long-term residents at high altitude are reported by Dempsey et al. (1971) and Cruz et al. (1975). Frisancho (1975) points out that the vital capacity and residual lung volume are larger in highland natives than in subjects from low altitudes. An exposure to hypoxia during the developmental period, i.e., during childhood, may be essential. A greater alveolar area and an increased capillary volume would facilitate gas diffusion in the lungs.

3 Pugh et al. (1964; Pugh, 1964) have done hemodynamic studies including maximal exercise at very high altitudes. They report that a prolonged sojourn at various altitudes brought the *cardiac output* at a given work rate down to the level typical for the same work rate performed at sea level. However, the maximal cardiac output was markedly reduced, and after several months' stay at 5,800 m, the values were 16 to 17 liters · min^{-1} compared with 22 to 25 liters · min^{-1} at sea level. This reduction of cardiac output was a combined effect of a lowered stroke volume and maximal heart rate (reduced from 192 down to 135 beats · min^{-1}). This study confirms the data by Christensen and Forbes (1937).

A number of measurements of cardiac output during exercise at altitudes between 3,000 and 4,300 m have been made with exposure up to a few weeks (Klausen, 1966; Alexander et al., 1967; Hartley et al., 1967; Vogel et al., 1967; Saltin et al., 1968; Vogel et al., 1974b). The results indicate that after a few days, the minute volume of the heart during submaximal exercise is already reduced, compared with the cardiac output during acute exposure to the hypoxic condition, and that it returns gradually to values typical for sea-level conditions or it may even become subnormal. During maximal exercise, the cardiac output is reduced. A reduced stroke volume appears to be the primary reasons for the reduced cardiac output; the lowering of the heart rate, in any case during maximal exercise, is a more inconsistent finding. (Figure 15-4 shows that the stroke volume during light exercise is reduced at acute exposure to the hypoxia.) Grover et al. (1976) rule out myocardial hypoxia as a basis for the decrease in stroke volume.

Horstman et al. (1980), studied the aerobic power at sea level and during a 2-week stay at 4,300 m in 9 subjects. After 2 weeks at this altitude, the maximal oxygen uptake and systemic oxygen transport increased by 10 percent, compared with the first day at altitude. The increased systemic oxygen transport resulted from a 19 percent increase in arterial oxygen content, despite a 9 percent decrease in maximal cardiac output. They concluded that the relative polycythemia was a major contributor to the increased maximal aerobic power.

Vogel et al. (1974a) studied eight subjects, who were native to an altitude of 4,350 m, at that altitude and also at sea level. At sea level, cardiac output was the same, heart rate was less, and stroke volume was greater at rest and during submaximal exercise than was observed at altitude. At sea level, the maximal oxygen uptake increased from an average of 2.97 to 3.25 liters $\cdot$ min^{-1} (9 percent), which was due to a greater maximal cardiac output (an 8 percent increase). (It is an open question as to why the a-$\bar{v}$ O$_2$ difference was not greater during maximal exercise at sea level than at high altitude. The oxygen content of arterial blood was namely 231 ml $\cdot$ l^{-1} at sea level and 200 ml at altitude.) In these subjects, the arterial blood pressure and peripheral resistance were higher at altitude than at sea level. This is in contrast to other studies indicating that the systemic blood pressure in adult highland natives is lower than in lowland natives at sea level (see Frisancho, 1975).

An example of the heart rate response to fixed work rates was presented by P.-O. Åstrand and I. Åstrand (1958); see Fig. 15-8. During acute exposure to a tracheal oxygen tension of about 85 mm Hg, i.e. 11.3 kPa (4,300 m), the heart rate was 15 to 30 beats higher per minute than at sea level. When the hypoxia was prolonged, there was a gradual reduction in heart rate at a given oxygen uptake. In the later stage of acclimatization, the heart rate attained during lower levels of exercise fell in the same range as those recorded at sea level. At the heavier loads, however, the heart rate was even lower than in experiments with high tracheal P_{O_2}. In this subject, the normal maximal heart rate was about 190 beats $\cdot$ min^{-1}. At the high altitude, it gradually declined to 135 (a decline in maximal heart rate at prolonged exposure to altitudes exceeding 3000 m has also been reported by Christensen and Forbes, 1937; Cerretelli and Margaria, 1961; Pugh et al., 1964; Cerretelli, 1976). When the subject was allowed to breathe 100 percent oxygen during almost maximal exercise, the heart rate increased within seconds by as much as 25 beats $\cdot$ min^{-1}. Pugh et al. (1964) noticed that a maximal heart rate of 130 to 150 was elevated to almost sea-level values when the subjects were breathing oxygen. Hartley et al. (1974) noted that intravenous atropine increased the maximal heart rate in subjects exposed to 4,600 m altitude (a mean decrease of 24 beats $\cdot$ min^{-1} was reduced to 13 beats $\cdot$ min^{-1} by atropine). An increased parasympathetic activity during prolonged exposure to hypoxia may partly explain the reduction in maximal exercise heart rate.

Within the first few days at high altitude, the *hemoglobin concentration* in the blood increases, but this increase is mainly due to a hemoconcentration secondary to a decrease in the plasma volume (Merino, 1950; Surks et al., 1966; Buskirk et al., 1967). Gradually the increased erythrocytopoiesis brings hemoglobin content to high levels so that the oxygen content per liter of arterial blood may be the same in the acclimatized person at 4,500 m as it is in a person at sea level (Christensen and Forbes, 1937; Hurtado et al., 1945; Reynafarje, 1967). At an altitude of 4,500 m at Morococha in Peru, the native residents had a hemoglobin concentration which averaged 208 g $\cdot$ l^{-1} blood (Hurtado et al., 1945).

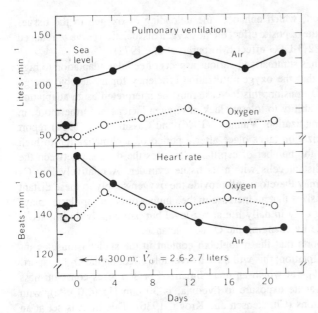

Figure 15-8
Pulmonary ventilation and heart rate during exercise on a cycle ergometer, work rate 200 watts at sea level, and during a 22-day sojourn at an altitude of 4,300 m. On some days, room air was inhaled during exercise, on other days pure oxygen was taken. (Same subject as in Figs. 17-6 and 17-7.) (Modified from P.-O. Åstrand and I. Åstrand, 1958.)

As already mentioned, Horstman et al. (1980), have emphasized that relative poly-cythemia is a major contributor to an increased maximal aerobic power during a prolonged high altitude sojourn. Robertson et al. (1982) conclude that induced ery-throcythemia increased hypoxia tolerance during physical exercise.

As a consequence of an increased hemocentration during acclimatization to high altitude, the oxygen offered to the tissues per liter of arterial blood may be the same in the individual living at high altitude as in the resident at sea level. However, these values of oxygen are really not comparable. The oxygen pressure gradient between blood and tissue is most important for the final transfer of oxygen to the mitochondria, and the gradient is reduced during hypoxia. However, the gradual decline in cardiac output during prolonged exposure to a low oxygen pressure in the ambient air can be partially explained by the concomitant rise in the oxygen-combining capacity of the blood. The increased viscosity of the blood with the elevated hematocrit must neces-sitate an increased cardiac work at a given cardiac output, but the net effect of the hematologic response to prolonged hypoxia in terms of work of the heart cannot be evaluated at present. In any case, the increase in hemoglobin concentration and the mentioned shift in the operational range to the steeper slope of the oxygen-dissociation curve provide major contributions to the gradually increased oxygen conductance within the body at altitude. During chronic hypoxia there is also an increase in the 2,3-DPG concentration in the blood, which enhances the unloading of oxygen to the tissues (see Chap. 5). However, the opposite effect on the oxygen dissociation curve of reduced P_{CO_2} may cancel out the 2,3-DPG effect (Morpurgo et al., 1972).

4 Barbashova (1964) has summarized various aspects of cellular adaptation to high altitude. It is concluded that the oxygen-utilization efficiency for an aerobic energy yield increases at a low O_2 tension; this increase must be interpreted as an adaptation at the enzyme level. The ability to tolerate lack of oxygen increases, furthermore, in connection with an acclimatization. Vannotti (1946) and Cassin et al. (1966) report that an increased capillarization takes place after a period of acclimatization to high altitudes. An increase in the number of capillaries reduces the distance between the capillary and the most distant cells within its tissue cylinder. A relatively low O_2 tension in the capillaries may therefore still provide the oxygen supply to these distant cells. Rahn (1966) emphasizes that an increase in the number of open capillaries plays a most important role not only in daily life at sea level but particularly during acclimatization to high altitude by changing the O_2 conductance.

Reynafarje (1962) reports that the myoglobin content in the skeletal muscles increases during altitude adaptation; this will have a favorable effect on the O_2 transport.

The critical alveolar P_{O_2}, at which an unacclimatized person loses consciousness within a few minutes in acute exposure to hypoxia, is 30 mm Hg (4.0 kPa), with minimal individual variations (Christensen and Krogh, 1936). This limit is set at an altitude of slightly more than 7,000 m. Down to this low P_{O_2}, the demand of the nerve cells for oxygen can apparently be maintained (Noell, 1944). The well-acclimatized individual can spend hours at an altitude above 8,000 m breathing the ambient air. This must be an example of adaptation on the cellular level. The comprehension of the last step in the oxygen transfer from air to tissue, i.e., from the capillary to the mitochondria, is, however, still very incomplete.

5 There is one more finding in high-altitude dwellers that should be mentioned. Residents at altitudes of around 3,500 m or above develop a pulmonary hypertension with increased pulmonary vascular resistance and hypertrophy of the right ventricle of the heart (Penaloza et al., 1963; Frisancho, 1975; Hultgren, 1982). This phenomenon does not change rapidly when returning to lower altitudes. Lockhart et al. (1976) noted that raising the arterial oxygen pressure to normal sea-level values had no effect on the pulmonary circulation at rest but prevented, to a large extent, the rise in pulmonary arterial pressure during exercise. This type of hypoxia modification of the cardiopulmonary system, as well as other changes in connection with adaptation to hypoxia, is believed to occur primarily in the fetal and infant period and is retained in adult life. The physiological consequences and the significance of these findings are not revealed. One effect might be a more even ventilation/perfusion ratio in the lungs, which reduces the difference in oxygen pressure between alveolar air and arterial blood (Bisgard et al., 1974; Cruz et al., 1975; Chap. 5).

6 The hormonal response to submaximal exercise under normoxic and hypoxic conditions were studied by Sutton (1977). Plasma glucose, free fatty acids, lactate, cortisol, and serum growth hormone concentrations all increased more during hypoxic exercise than under normoxic conditions. Serum insulin concentration decreased by 50 percent during hypoxic exercise, and was followed by marked rebound when exercise stopped. On the basis of these changes, it was suggested that energy substrate-hormone interrelationships are altered by hypoxic exercise, resulting in increased fat mobilization and increased gluconeogenesis.

Similarly, Young et al. (1982), studied muscle glycogen utilization during chronic high altitude exposure and observed an increased mobilization and utilization of FFA during exercise associated with a sparing of muscle glycogen.

Clench et al. (1982) studying the effects of chronic exposure to altitude hypoxia in an indigenous population, the Aymara in Chile, found an increase in 2, 3–diphospho–glycerate and a decrease in adenosine triphosphate, and suggest that pyruvate kinase plays a key role in the physiological adaptation to altitude hypoxia.

Summary It may be stated that with prolonged exposure to reduced oxygen pressure in the inspired air, the compensatory devices are more slowly acquired, such as (1) a further increase in pulmonary ventilation (in long-term dwellers at high altitude it is followed by a significant reduction in ventilation); (2) an increased hemoglobin concentration in the blood; and (3) morphological and functional changes in the tissues (increased capillarization, myoglobin content, modified enzyme activity). The initially observed increase in cardiac output during submaximal exercise is replaced by a gradual decline to, or even below, the values observed at sea level. During submaximal, as well as maximal exercise, the stroke volume becomes reduced. If the sojourn has been at very high altitude, 4,000 m or higher, the maximal heart rate may become reduced compared with sea-level values. All these adaptive changes are reversible, but it may take several weeks before the values return to sea-level values in the case of sea-level dwellers who have stayed at high altitudes for a month or more.

The net effect of this acclimatization to high altitude is a gradual improvement in the physical performance in endurance events or prolonged exercise. An increased oxygen availability to the active muscles is important for this improvement (see Pugh, 1965; Saltin, 1966; Maher et al., 1974). The maximal anaerobic power after a prolonged period of acclimatization has not been carefully analyzed.

It is also true that the oxygen content per liter of blood eventually increases, but the maximal cardiac output apparently is reduced at about the same rate. It should be pointed out that the well-trained individual is not acclimatized to high altitude any sooner or any more effectively than the untrained individual. The fact that acclimatization of highland natives does not depend on hyperventilation is perhaps due in part to their enlarged lung volume, which facilitates the receiving of an adequate oxygen supply at the alveolar level (Frisancho, 1975). The earlier the age when one becomes a sojourner at high altitudes, the greater the influence of the environment modifying the expression of inherited potential. The optimum seems to be during growth and development.

Saltin (1966) followed a top athlete during a 3-week period at the Mexico City altitude. His initial decrease in maximal oxygen uptake of 14 percent became less during the stay in Mexico City, but it was still 6 percent below the sea-level value after 19 days. The same trend was observed for the oxygen uptake measured during maximal efforts in a canoe. In a group of eight international top athletes, the reduction in maximal oxygen uptake at the altitude of 2,300 m averaged 16 percent (ranging from 9 to 22 percent); after 19 days at this altitude, the maximum was still 11 percent below the sea-level average, with a range of 6 to 16 percent (Saltin, 1967). This example illustrates the individual variations in response to hypoxic conditions.

Performance after Return to Sea Level

Opinions differ concerning the question of whether the performance capacity at sea level is improved following exposure to high altitude, or whether a certain amount of training at high altitude is more effective than the same amount of training conducted at sea level. The above-mentioned subject, studied by Saltin (1966), was the best adapted athlete in the Swedish test group sent to Mexico City in 1965. When he returned to Stockholm after the 3-week Mexican sojourn, his maximal oxygen uptake was no higher than before the trip. Buskirk et al. (1967), as well as Consolazio (1967), state that their subjects who stayed at altitudes up to about 4,000 m for 4 weeks or more did not attain any better results than usual when they returned to sea level. The measured maximal O_2 uptake was not improved. Buskirk and associates conclude that there is little evidence to indicate that performance on return from altitude is better than before going to high altitude, if training remains relatively constant. One should view with caution any statement about the physical superiority of the indigenous resident in any environment until that person can be compared with individuals of similar training and experience (and physical endowment).

Grover and Reeves (1967) also came to similar conclusions when studying the exercise performance at altitudes of 300 to 3,100 m in men native to those two altitudes. The maximal oxygen uptake was 26 percent less at the higher altitude for both groups; the sea-level athletes won all track competitions at both low and high altitudes; their performance after return to low altitude was not improved by their sojourn at medium altitude. (See also Hansen et al., 1967b; Kollias et al., 1968; Adams et al., 1975.)

Cerretelli (1976) reports studies on subjects acclimatized to high altitude (5,350 or higher) for about 4 months. When studied after their return to sea level (13 subjects), there was no significant increase in maximal oxygen uptake compared with the data obtained before departure. The hemoglobin concentration was still 12 percent higher than before the sojourn at high altitude. It is difficult to explain why an elevation of the hemoglobin content, after a prolonged sojourn at high altitude, does not effectively improve the maximal aerobic power. This is in contrast to the beneficial effect of an acute increase in the hematocrit by reinfusion of red cells (Chap. 4; see Fig. 15-9).

It is true that both training and prolonged exposure to hypoxia produce similar changes with respect to increased vascularization in the skeletal muscles, increased myoglobin content, and possibly similar changes in the oxidative transport system. On the other hand, training at sea level produces no increase in the Hb concentration. Exposure of more than a few days' duration to high altitudes evidently hinders the attainment of a maximal stroke volume, and the increased pulmonary ventilation following return to sea level represents no advantage. The same is true for the reduced alkaline reserve in the blood, with reduced buffer capacity for lactic acid as a consequence. The training intensity has to be reduced at high altitude. For these reasons,

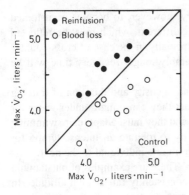

Figure 15-9
The effect of bloodletting (800 ml) and subsequent reinfusion one month later of the same red blood cells on maximal oxygen uptake in eight subjects (three in one study and five in a subsequent study). *(From Ekblom et al., 1976.)*

the improvements attained at high altitude cannot be transferred directly to sea-level conditions.

It should be emphasized that few records were broken in swimming and middle and long distance running when the athletes returned to sea level after the pre-Olympic Games and the actual Olympic Games in Mexico City. Apparently, performance at sea level is not enhanced by prolonged exposure to hypoxia.

Practical Applications

In order to attain top achievmement at altitudes of 2000 m or higher in activities requiring the engagement of the maximal aerobic power, an acclimatization period of 2 to 3 weeks seems necessary. At lower altitudes, the time required is probably less. A longer exposure to high altitude would probably be beneficial from a physiological point of view, but this advantage must be considered against possible psychological, social, and economic factors. After the initial acclimatization, the improvement per week is so small that it may easily be concealed by day-to-day variations in the physical fitness.

Athletes competing in events where the technique is of prime importance or in events primarily involving anaerobic metabolic processes may arrive at more or less the time of the competition if the altitude is not so high that mountain sickness may be expected.

A theoretical consideration shows that the performance capacity in certain events ought to be better in competitions at high altitude (e.g., 100- to 400-m, bicycling). Dickinson et al. (1966) state, on the basis of ballistic calculations, that at an altitude corresponding to Mexico City, one might expect an improvement of 6 cm in shot-putting, 53 cm in throwing the hammer, 69 cm in javelin throwing, and 162 cm in discus throwing. In view of the fact that throwers are used to mastering different wind conditions, the reduced density of the air at high altitude certainly does not require any special period of preparation.

There is no evidence to suggest that it is necessary to take it easy during the initial period of exposure to high altitude. It is necessary, however, to become accustomed to the fact that the subjective feeling of fatigue is different, which must be reflected in the choice of tactics. Experience has shown (especially in the case of cross-country skiers) that if the effort is too intense, a considerably longer recovery time will be required than is needed at sea level.

One is forced to accept a slower tempo, and the intensity and duration of training activities must be reduced. Swimmers discover that they can remain under water for a shorter time after turning than they normally do, and they must adapt their swimming strokes to a different breathing rhythm. The ability to tolerate an intense tempo for long periods of time at high altitudes is different from one individual to the next. This complicates the selection of the athletes for a team. There are examples of outstanding athletes in long-distance events at sea level who consistently fail at high altitude. In competitions at high altitude, collapse from unknown causes occurs more frequently than it does at sea level. This fact, however, does not appear to represent an increased health hazard. Ingestion of additional fluid and possibly also carbohydrates may be required during exposure to high altitude when it is combined with heavy physical exercise. There is no evidence to suggest that the preparation for competitions that place heavy demands on the aerobic power should include training at an altitude which is higher than that of the actual place of competition.

HIGH GAS PRESSURES

Although humans can become acclimatized to low air pressures, there is no way to become biologically acclimatized to high air pressures, such as those encountered in deep sea diving or when a submarine crew tries to escape from the inside of the craft, where the pressure is normal, to the surface through the sea where the air pressure is higher. (For a more comprehensive review of the subject, see Lambertsen, 1967; Behnke, 1971; Fagraeus, 1974; Behnke, 1978; Halsey, 1982).

Pressure Effects

For every 10 m (33 ft) of sea water the diver descends, an additional pressure of 1 atm is acting upon his or her body. Small changes in sea depth thus bring about great pressure changes. The effect of changing pressures on the blood P_{O_2} and P_{CO_2} and its consequence for underwater swimming and breath holding were discussed in Chap. 5. The body may tolerate high pressures as long as the pressure is the same inside and outside the body. When diving with a snorkel connected to the mouth, one maintains the atmospheric pressure in the lungs, while the surface of the thorax, in addition, is exposed to the pressure of the water. At a depth of about 1 m, the pressure difference becomes so large that the inspiratory muscles no longer have the strength to overcome the external pressure, and normal breathing becomes impossible. For this reason, a snorkel system does not permit diving to depths exceeding about 1 m. At greater depths, breathing apparatus has to be used in which the pressure in the system corresponds to that prevailing at the depth in question. If there is an overpressure in the system, the lung tissues may be damaged, with hemorrhaging as a consequence.

As the pressure increases, more gases can be taken up by the diver's body and dissolved in the various tissues. At a depth of 10 m, twice as much gas will be dissolved in the blood and tissues as at sea surface. This is apt to give the diver trouble, mainly because of the nitrogen.

Nitrogen

The problem with nitrogen is that is diffuses into various tissues of the body very slowly, and once dissolved, it also leaves the body very slowly when the pressure is once more reduced to the normal atmospheric pressure. This is especially bad when the pressure is suddenly reduced from several atmospheres, as may be the case during submarine escape or deep sea diving. Then the nitrogen is released from the tissues in the form of insoluble gas bubbles. These bubbles congregate in the small blood vessels, where they obstruct the flow of blood. This, then, gives rise to symptoms such as pains in the muscles and joints, and even paralysis may develop if the bubbles become trapped in the brain. These symptoms are known as the *bends*. Obviously, the severity of the symptoms depends on the magnitude of the pressure, which relates to the depth to which the person has descended under water, the length of time spent at that depth, and the speed of ascent to the surface.

The bends can be avoided to a large extent by a slow return to normal pressure so as to allow time for the tissues to get rid of their excess nitrogen without the formation of bubbles. Another way to avoid the bends is to prevent the formation of nitrogen bubbles by replacing atmospheric nitrogen by helium, which is less easily dissolved in the body. This is done by having the diver breathe a helium-oxygen gas mixture. Another advantage of this method is that it is more apt to prevent the so-called nitrogen narcosis, which occurs when air is breathed at 3 atm or more, when there is an onset of euphoria and impaired mental activity with lack of ability to concentrate. With increasing pressures, the individual is progressively handicapped and may be rendered helpless at 10 atm. Diving to depths exceeding 100 m while breathing ordinary atmospheric air may thus be fatal.

Pilots flying high-altitude aircraft may also suffer from bends if there is a sudden loss of pressure in the pressurized cabin, but the symptoms in these cases are usually not so severe as in the divers, and they usually do not occur at altitudes lower than 3,000 m in any case.

Oxygen

Prolonged breathing of 100 percent oxygen at 1 atm may be harmful (Lambertsen, 1965; Gilbert, 1981); irritation of the respiratory tract may occur after 12 hr and frank bronchopneumonia after 24 hr, and the peripheral blood flow (the flow through the brain) may be reduced. In most individuals, no harmful effects result from breathing mixtures with less than 60 percent oxygen, but newborn infants are particularly susceptible to oxygen poisoning and may suffer harmful effects with oxygen concentrations over 40 percent. The remarkable thing is that oxygen poisoning is apparently no problem when breathing 100 percent oxygen at altitudes over 6,000 m, no matter for how long. Oxygen poisoning, therefore, is not such a problem in aviation medicine, but it is indeed an important problem in deep sea diving where it may even affect the brain function when pure oxygen is breathed under increased pressure. This latter form of oxygen toxicity is apt to occur in divers at depths greater than 10 m, but there are great individual variations in sensitivity to 100 percent oxygen. The onset of symptoms may be hastened by vigorous physical activity at great depths; it starts with muscular twitchings and a jerking type of breathing and ends in unconsciousness and convulsions. The exact cause of this is unknown, but it is assumed that it is a matter of interference with certain enzyme systems in the tissues.

When a person breathes pure oxygen at a pressure of 3 atm or higher, the dissolved oxygen covers the oxygen need of the body at rest. No oxygen would be removed from HbO_2 during its passage through the capillary bed. Therefore, the hemoglobin of the venous blood would still be saturated with oxygen, which would interfere with the amount of H^+ ions taken up by Hb, a weaker acid than HbO_2 (Chap. 4). Thus, CO_2 entering the blood from metabolizing cells would raise the blood P_{CO_2} and the H^+ concentration would be higher than under normoxic conditions when the desaturation of HbO_2 simultaneously favors the removal of H^+ ions. The end result would be a CO_2 retention in the tissues and an acidosis.

When breathing air at a hyperbaric pressure during heavy exercise, the increased air density reduces the pulmonary ventilation. Depletion of energy stores in the inspiratory muscles may contribute to limiting pulmonary ventilation during maximal exercise at raised air pressure (Hesser et al. 1981). Secondarily, the maximal oxygen uptake may be reduced. By substituting nitrogen with helium, with a lower density, the pulmonary ventilation goes up, and the oxygen uptake may be higher than the maximum measured during normal atmospheric conditions (Fagraeus, 1974).

Carbon Dioxide

If the pressure of the carbon dioxide in the respiratory air is increased, as it is when the absorption of CO_2 fails in a closed breathing system, CO_2 pressure may be reached (about 75 mm Hg (10.0 kPa) or higher), which produces a narcotic effect.

One effect of *reduced* CO_2 pressure in the blood as a consequence of hyperventilation was discussed in Chap. 5: Breath holding during exercise can, after such a procedure, be prolonged until unconsciousness, which induces a potential danger when practiced during underwater swimming (see Craig. 1976).

Oxygen Inhalation in Sports

In athletic events, oxygen breathing is of very limited value. Whereas, with a lung volume of 5 liters, one normally has a supply of about 0.8 liter of oxygen, this oxygen volume may be increased to about 4.5 liters if 100 percent oxygen is inhaled. This volume may be consumed by the cells within less than a minute during heavy exercise. In most events the weight of the equipment for an adequate oxygen supply would be a definite handicap and, if used during competition, it would disqualify the athlete because the additional oxygen supply would be considered doping.

After a few breaths of atmospheric air, the oxygen is quickly diluted, since the nitrogen content of the air is as much as 79 percent. The elimination of anaerobic metabolites is not speeded up if pure oxygen is inhaled during recovery.

Oxygen inhalation may improve the performance capacity just prior to swimming under water and when there is a need to increase the supply of oxygen *during* heavy muscular exercise. (It increases the maximal oxygen uptake, reduces the anaerobic participation in the energy yield, and reduces the pulmonary ventilation, see Chap. 4; Chap. 7.) The "sniffing" of oxygen practiced by football players during the rest periods may perhaps have a psychological effect. From a physiological point of view, however, the extra oxygen is definitely wasted.

TOBACCO SMOKING

Larson et al. (1961) have prepared a very extensive summary on this subject. Only a few aspects of tobacco smoking and physical performance will be discussed here.

Circulatory Effects

Tobacco smoke contains up to 4 percent by volume of carbon monoxide. By inhalation, some of this carbon monoxide is absorbed. The affinity of the hemoglobin to carbon monoxide is 200 to 300 times greater than to oxygen. The presence of even small amounts of carbon monoxide may, therefore, noticeably reduce the oxygen-transporting capacity of the blood. Carbon monoxide also interferes in a negative way with the unloading of oxygen in the tissues by shifting the oxyhemoglobin dissociation curve to the left (Chap. 4). A study has shown that subjects who smoked 10 to 12 cigarettes a day had 4.9 percent carbon monoxide hemoglobin, those who smoked 15 to 25 cigarettes a day had 6.3 percent, and those who smoked 30 to 40 cigarettes per day had 9.3 percent. Other studies have confirmed these findings; it may take 1 day or more for the carbon monoxide content of the blood to return to normal (Larson et al., 1961). This amount of carbon monoxide in the blood gives no subjective symptoms at rest, as long as the concentration in the blood is below 10 percent. The adverse effect is noticeable only during physical exertion. A blocking of 5 percent of the hemoglobin by CO will reduce the maximal oxygen uptake and performance (see Ekblom and Huot, 1972). There is no way to compensate for a reduced oxygen content in arterial blood during maximal exercise.

In another study, normal subjects exercised on a cycle ergometer without having smoked 12 hr prior to the experiment and immediately after smoking one or two cigarettes (Juurup and Muido, 1946). Oxygen uptake, pulmonary ventilation, and respiratory rates were unaffected by smoking. A slight rise in arterial blood pressure was noted, but the difference was not significant. At a fixed oxygen uptake, the heart rate was 10 to 20 beats $\cdot$ min^{-1} higher when the work test was preceded by smoking. The difference in heart rate between smokers and nonsmokers was greater the higher the work rate. The effect of smoking could not be observed if the subject waited 10 to 45 min after smoking before the exercise test started (the exercise was at a sub-maximal level).

Respiratory Effects

In Chap. 5 (in the section "Airway Resistance"), it was pointed out that inhalation of smoke from a cigarette could, within seconds, cause a two- to threefold rise in the airway resistance. Da Silva and Hamosh (1973) also observed an increase in airway resistance (measured by body plethysmography) in subjects after smoking one cigarette. In addition to this acute effect, smoking also causes a more chronic swelling of the mucous membranes of the airways, leading to an increased airway resistance. At rest when the pulmonary ventilation is less than 10 liters $\cdot$ min^{-1}, the increased airway resistance is not noticeable, however. When the demand on respiration is elevated, the increased respiratory resistance caused by smoking may be noticeable. A reduced pulmonary ventilation capacity may cause a smaller volume of oxygen to reach the alveoli, resulting in an impaired gas exchange. The effect may be subjective symptoms of distress.

Metabolic Effects

Wahren et al. (1983) examined the metabolic response to acute tobacco smoking in 11 healthy, habitual smokers. They smoked three tobacco-free cigarettes, and then one hour later smoked three tobacco-containing cigarettes. Tobacco-free cigarettes stimulated O_2 uptake by 26 ± 6 percent and CO_2 production by 30 ± 8 percent of that observed after the smoking of tobacco-containing cigarettes. Cigarettes with tobacco elicited an 11 percent rise in O_2 uptake which was sustained for more than an hour. CO_2 output rose by 25 percent but decreased more rapidly. Venous concentrations of norepinephrine and epinephrine rose in some subjects by as much as 50 percent, while insulin and glucagon levels were unchanged. FFA concentrations increased during and after tobacco smoking by 50 to 125 percent, while glucose levels remained unchanged.

In one subject, who smoked 20 tobacco-containing cigarettes during 11 hours, 24-hr continuous respiratory exchange measurements were made, using a metabolic chamber (Wahren et al. 1983). His energy expenditure was found to be 14 percent greater than on the following day when he did not smoke, food intake and physical activity being the same. On this basis, Wahren et al. (1983) conclude that smoking significantly raises energy expenditure. This may possibly explain why smokers often gain weight when giving up smoking.

The reaction to exercise of different duration was studied in 16 healthy, habitual smokers twenty-two to forty years of age by Brundin (1980). He found that the exercise induced rise in blood temperature was considerably augmented after smoking. This effect was paralleled and probably caused by, an increased heat production reflected by an increased oxygen uptake. During long-term exercise, cardiac output and heart rate increased and the stroke volume decreased in response to smoking.

Smoking Habits among Athletes

A study of 285 top athletes in Great Britain (Report in *JAMA*, **170**:1106, 1959) showed that 16 percent of them smoked, but only 4 percent of them smoked more than 11 cigarettes per day. These findings are in agreement with the results of similar studies of top athletes elsewhere. Most of the smokers are among those engaged in athletic events requiring skill rather than endurance. None of the middle- and long-distance runners or swimmers were smokers. Among skiers, some of the ski jumpers and downhill skiers may smoke, but none of the cross-country skiers of the elite class are smokers. This may indicate that under conditions where the demand on the oxygen-transporting system is very great, smoking has been found to impair performance. If the current trends continue, even athletes participating in events primarily requiring skill will be required to undergo strenous physical training which taxes the oxygen-transporting system to a considerable degree. This may make it desirable for the athlete to refrain from smoking while training, even though tobacco smoking at the time of the competitive event may not affect the outcome.

ALCOHOL AND EXERCISE
Neuromuscular Function

It is well established that alcohol may temporarily cause impaired coordination. The performance of rather simple movements is used to test whether or not an individual is under the influence of alcohol (walking on a straight line, touching the tip of the nose with the index finger with the eyes closed, etc.). A precise assessment in borderline cases is impossible, however. The deleterious effect of alcohol may be particularly noticeable in tasks requiring divided attention (Brewer and Sandow, 1980). The tolerance of alcohol varies greatly from one individual to another. Ikai and Steinhaus (1961) noted that the maximal isometric muscle strength could actually be improved in some cases, especially with untrained subjects, after moderate alcohol consumption. They explain this on the basis of a depressing influence of alcohol on the central inhibition of the impulse traffic in the nerve fibers to the skeletal muscles during maximal effort. The result is an increased impulse activity and an increased strength (see Chap. 3). At times, individuals taking part in competitive events, such as target-shooting, maintain that they achieve better results following moderate alcohol consumption. They feel more relaxed. It is conceivable that the depression of the inhibiting effect of certain CNS centers may cause routine procedures to progress normally without the disturbing effect caused by the anxiety of the actual competition.

Aerobic and Anaerobic Power

Blomqvist et al. (1970) studied eight young male subjects during cycle exercise at two submaximal and one maximal work rate before and after peroral ethanol intake producing blood levels of 90 to 200 mg percent. Oxygen uptake and heart rate were determined. In four of the subjects, cardiac output, stroke volume, and intra-arterial pressure were also measured. Table 15-2 summarizes the most important changes. During maximal exercise with an O_2 uptake of about 3 liters $\cdot$ min^{-1} and a cardiac output of about 21 liters $\cdot$ min^{-1}, no difference was observed when the results before and after alcohol consumption were compared. Maximal heart rate, stroke volume, a-$\bar{v}$ O_2 difference, and calculated peripheral resistance were also unaffected. During

TABLE 15-2
HEMODYNAMIC EFFECTS OF ALCOHOL

	V_{O_2}	Heart rate	Cardiac output	Stroke volume	$a\text{-}\bar{v}O_2$ difference	Total peripheral resistance
Rest, sitting	↑*	↑	↑	0	↓	↓
Submaximal exercise	↑	↑	↑	0	↓	↓
Maximal exercise	0	0	0	0	0	0

*The arrows denote the observed changes, which are significant in all cases except the O_2 uptake.
Source: Blomqvist et al., 1970.

submaximal exercise, on the other hand, the heart rate was on an average 12 to 14 beats higher per minute in the alcohol experiments ($p < 0.01$). In the latter case, the cardiac output was greater while the stroke volume was unaffected. The O_2 uptake during submaximal exercise was slightly higher after alcohol, but the $a\text{-}\bar{v}$ O_2 difference was nevertheless reduced (i.e., the cardiac output was more elevated than would be expected from the increased oxygen uptake). At rest and during submaximal exercise, the calculated total peripheral resistance was reduced.

Asmussen and Böje (1948) studied the effect of alcohol on the ability to perform one activity of 9,400 Nm and another of 96,700 Nm as fast as possible. The first type of activity could be performed in 12 to 15 s, simulating a 100-m sprint in its effect on the organism, whereas the latter, lasting about 5 min, simulated a 1,500-m run. A blood concentration of alcohol of up to 100 mg percent did not significantly affect the ability to perform this kind of maximal exercise.

Metabolic Effects

In a study of the influence of alcohol on splanchnic and muscle metabolism during exercise, Jorfeldt and Juhlin-Dannfelt (1977), observed that most of the acetate released from the liver after ethanol ingestion is taken up by the skeletal muscle, but is of minor importance as an energy substrate during exercise. Even moderate doses of alcohol inhibit liver gluconeogenesis and reduce the splanchnic glucose release.

Summary These studies have revealed an effect on the circulatory response to submaximal exercise in individuals with elevated blood alcohol level. Measurements during maximal exercise, however, showed no effect on oxygen uptake, cardiac output, heart rate, stroke volume, or total peripheral resistance. Nor was the maximal time the subjects could tolerate a standard load affected by the alcohol intake.

It should be pointed out that the question of dosage in the case of alcohol consumption in connection with athletic performance is a most difficult one. The tolerance varies greatly from individual to individual, and probably also from time to time in the same individual. Furthermore, the use of alcohol is definitely to be considered as a form of doping. It should also be kept in mind that alcohol ingestion may, in some cases, be followed by a subsequent spell of hypoglycemia.

For a further discussion of the cardiovascular effects of alcohol, see Regan, 1982; Bing, 1982; B. M. Altura and B. T. Altura, 1982; and Rubin, 1982.

CAFFEINE

Caffeine has been claimed to increase fat metabolism and to spare muscle glycogen (Essig et al., 1980). Toner et al. (1982), studied the effect of caffeine on circulation and metabolism during maximal exercise on a cycle ergometer in 8 male subjects. They claim that caffeine increased the maximal oxygen uptake slightly, on the average by 140 ml $\cdot$ min^{-1}, and increased the average maximal heart rate by 5 beats $\cdot$ min^{-1}, but there were no other satistically significant changes in metabolic or cardiovascular variables during submaximal or maximal exercise.

Powers et al. (1983), observed that small doses of caffeine did not enhance performance during graded cycle ergometer exercise in trained subjects.

DOPING

The use of stimulants of many kinds to improve physical performance has been practiced since ancient times. The Romans tried to increase the speed of their chariot racing by giving the horses a mixture of honey and water. The Indians of South American chewed coca leaves on their long, strenuous mountain journeys in order to enhance endurance and suppress the feeling of fatigue. With the increasing commercialism and professionalism in human competitive sports, the problem of doping no longer is limited to racing animals, but relates to athletes as well. This is a matter of considerable concern, for it represents an unnecessary risk for the health of the user, while any real advantage can only be expected from it in exceptional cases (Ariëns, 1965). This is true both in the use of drugs for the purpose of depressing the feeling of fatigue or of stimulating the central nervous system to attain greater "pep," and in the use of hormones or other products to augment muscle strength or physical performance (amphetamine, cocaine, heroin, ephedrine, morphine, and anabolic steroids).

The most commonly used drugs for doping are the psychostimulants, or "pep pills," such as amphetamine (Latis and Weiss, 1981). In general, their main effect, at least as far as the athlete is concerned, is to suppress the feeling of fatigue and thus to permit individuals to exert themselves to complete exhaustion, and thereby to improve performance. Bucher and Smith (1965), in a double-blind study in highly trained swimmers, runners, and weight throwers, showed that amphetamine sulfate in a dose of 14 mg per 70 kg of body weight improved the measured performance significantly in about 75 percent of the cases. The weight throwers obtained the greatest improvement from amphetamine (3 to 4 percent); the runners obtained an improvement of approximately 1.5 percent; the swimmers showed varying degrees of improvement (0.6 to 1.2 percent). This does not mean, however, that amphetamine causes an improvement in performance which could not have been achieved without it had the individuals been properly motivated and able to mobilize all their capacities at their own will. It should be emphasized that maximal oxygen uptake (and cardiac output and arterio-venous oxygen difference) is attained at a submaximal effort. The maximal power of aerobic *and* anaerobic processes is difficult to measure in a reproducible manner. It is a general finding that the peak concentration of lactate in athletes' muscles and blood is lower in an "all out" experiment conducted in the laboratory as compared with a measurement in samples obtained after an important competition. Results from a laboratory test of the effect of various drugs cannot be directly applied to competitive situations.

Hanley (1972) reports observations from a world weight-lifting championship in 1970. In the first three classes, eight of the nine who finished in the first three places turned in urine specimens positive for amphetamines. (They were disqualified, but those who finished fourth, fifth, and sixth replaced them as winners without having dope control tests.) In the next six classes, there was only one specimen positive for amphetamines. It was pointed out that more records were actually broken in the classes where tests were negative than in the positive ones. The use of psychopharmaca is associated with the risk of addiction and of dangerous intoxications (Ariëns, 1965).

The use of anabolic steroids by athletes is based on the fact that these steroids, in controlled laboratory experiments, have been found to cause enhanced protein retention and increases in muscle mass and muscle strength. The reported results, however, are inconclusive. Thus, several studies have indicated such positive relationships, but other investigators have failed to find such a relationship (for references, see Johnson et al., 1975; Lamb, 1975; Ryan, 1981; Crist et al., 1983; Lamb, 1984). One problem is the difficulty involved in measuring muscle strength, especially in standardizing the subject's state of motivation. An anabolic effect may be achieved by doses of 10 to 20 mg per day. The effect is of short duration, however, since the ingested steroids may cause the endogenous production to be reduced. According to Lamb (1984), athletes have been known to use doses up to 2,000 mg per day. The results of such overdosage include not only an increase in muscle mass, but also an increased appetite leading to vastly increased food intake, amounting to as much as 25 to 33.5 MJ a day. This causes weight gain, and the added weight is localized primarily to the trunk, especially the neck, shoulders, chest, and arm regions, a result that may appear favorable for wrestlers, weight throwers, etc. This has led to a considerable abuse of anabolic steroids by athletes. The medication may give a sense of well-being and may also give the athlete more "appetite" for training. The side effects are serious, however. The endogenous hormone production is suppressed. The production of sperm cells may be reduced, sterility may result, and the hormonal balance may be upset, thus producing a variety of complications. In children, these steroids may cause a premature closure of the epiphysis of the long bones, leading to cessation of growth. In women, they may cause various degrees of masculinity, with changes in hair growth, voice, clitoris, and the production of the female sex hormones. These steroids may also cause liver changes and undesirable changes in the fat and carbohydrate metabolism. They also cause a stronger tendency to injuries of the ligaments and tendons due to a too rapid increase in muscle strength without a corresponding development of ligaments and tendons. All these possibilities clearly show that the use of anabolic steroids is dangerous (see Johnson, 1975).

Summary The use of large doses of anabolic steroids on a regular basis may or may not cause a modest increase in muscle mass and strength. This, however, is associated with many potentially hazardous effects (Lamb, 1984).

Another group of drugs, which have been used for doping purposes lately are the anticholinesterases: pyridostigmine and neostigmine (Mestinon). The effect is rather complicated, and overdosage may cause dangerous side effects. These drugs affect the motor end plate, causing a delay in the action of the cholinesterase, which normally would cause an instantaneous breakdown of acetylcholine. The general effect is claimed to be diminished fatigue and enhanced endurance. The danger is that large doses may have a paralyzing effect on the central nervous system leading to collapse.

In a study of eight subjects, Ekblom et al. (1976) have shown that bloodletting (800 ml blood) followed by reinfusion of the same red blood cells a month later caused a significant increase in maximal oxygen uptake (Fig. 15-9). The effect may be explained on the basis of the achieved increase in the transporting capacity of the blood as the result of an ultimate increase in the number of red blood cells. (See Chap. 4) (Williams et al., 1973, could not confirm that reinfusion of red cells improved physical performance. Unfortunately, they did not measure the oxygen content of arterial blood, and there is no proof that they were able to maintain the oxygen binding potential of the stored blood.)

In any case, a transfusion of blood or red cells to an athlete in connection with a competition is a violation of the rules: "Doping is defined as the administration or use of substances in any form alien to the body or of physiological substances in abnormal amounts and with abnormal methods by healthy persons with the exclusive aid of attaining an artificial and unfair increase in performance in competition. Furthermore, various psychological measures to increase performance in sports must be regarded as doping." (Statement by the International Olympic Committee.)

THE WILL TO WIN

Not everyone can win. Certainly the same person cannot win all the time. For every winner there is at least one loser, and to lose, at least at times, is part of life.

This is especially true in athletic competitions. Not all people are equally fit to win Olympic laurels or to rank among the best in any athletic competition. Some reach the top, but they are only a very small percentage of all those who try. The rest of them may still aspire to win for their own sake by striving to reach personal goals in performance. They can at least reach as high a level of performance as their own endowment will permit. Such an achievement is in itself a gratifying experience.

The attainment of athletic championship requires that the athlete meets a number of requirements. He or she must be endowed with the necessary talents to start with. But talent alone is not enough. Champion-quality athletes must master the proper techniques and have the suitable tools or equipment. And they must subject themselves to arduous training as well. But, above all, they must have the proper motivation, not only to apply all their resources in the final test during competition, but also to endure the hardship of their training. They need to prepare themselves, to gain the necessary competitive experience, and to master the art of competing. They have to nurse their health, and must be prepared to accept the day-to-day variations in their level of fitness. Finally, they need a certain amount of good luck. It is said that chance favors only the prepared mind. It is equally true to say that the athlete must be prepared to win when favored by chance.

The art of competing is the skillful application of all these factors with the single purpose of outperforming the competitors. This art includes the proper application of the stress hormones, aimed at preparing the organism for the supreme effort. The competitor must master the body's regulatory mechanisms, so as to allow the hormone-producing glands to be stimulated to the optimal level at the proper time. Some can, others cannot, do this; individuals differ also in this respect. It is in any case a matter of experience and adaptation to the stress of competing, which can be gained only through regular participation in competition.

So much depends on psychic factors. Some athletes are able to mobilize all their resources in an almost superhuman effort during an important competition. They can reveal an amazing ability to concentrate, to maintain a perfect control of the most intricate coordinated motions of the body and its various parts, to the almost complete exclusion of all extraneous disturbing factors, including the spectators. Others fail in this. They allow themselves to become tense; their muscles stiffen, their neuromuscular machinery fails at the crucial point. Such persons may win during training or at minor competitive events when they are naturally relaxed, but they fail when it matters, in the qualifying meet or during the major events. This possibility applies especially in events that require perfect coordination and timing, such as gymnastics. To some extent, it may also apply to runners, where the tense person tends to tighten the muscles, including the antagonists, so that both flexors and extensors of the same joint are contracted simultaneously, leading to clumsiness, waste of energy, and premature exhaustion. Examples of this are most clearly seen in events that require perfect coordination and exact timing, such as high jumping, diving, or ski jumping. If the jumper has only three trials and has failed twice, everything depends on that third and last chance. This crisis imposes a great strain on the performer. Some unique individuals will repeatedly succeed during the last trial, whereas others notoriously fail under such circumstances. Some may fail during the important qualifying test, and yet, only a week later at a less important competition, may attain a world record in the same event.

All these factors show the importance of participating regularly and frequently in competitions among top athletes where, by competing with those who are equal to or better than oneself, one learns by experience how to apply oneself to the fullest when it matters. The spectators' stimulation may have both positive and negative effects—negative in the sense that inexperienced competitors are encourage to exceed their capacity to exhaust themselves prematurely, positive in the sense that it may inspire more mature competitors to mobilize all their resources to the utmost in the supreme effort to reach their goal.

REFERENCES

Adams, W. C., E. M. Bernauer, D. B. Dill, and J. B. Bomar, Jr.: Effects of Equivalent Sea-level and Altitude Training on $\dot{V}_{O_2}$ max and Running Performance, *J. Appl. Physiol.*, **39**:262, 1975.

Alexander, J. K., L. H. Hartley, M. Modelski, and R. F. Grover: Reduction of Stroke Volume During Exercise in Man Following Ascent to 3,100 m Altitude, *J. Appl. Physiol.*, **23**:849, 1967.

Altura, B. M., and B. T. Altura: Microvascular and Vascular Smooth Muscle Actions of Ethanol, Acetaldehyde, and Acetate, *Fed. Proc.* **41**:2447, 1982.

Ariëns, E. J.: General and Pharmacological Aspects of Doping, in A. De Schaepdryver and M. Hebbelinck (eds.), "Doping," Pergamon Press, Oxford and New York, 1965.

Asmussen, E., and O. Böje: The Effect of Alcohol and Some Drugs on the Capacity for Work, *Acta Physiol. Scand.*, **15**:109, 1948.

Asmussen, E., and H. Chiodi: The Effect of Hypoxemia on Ventilation and Circulation in Man, *Am J. Physiol.*, **132**:426, 1941.

Asmussen, E., W. von Döbeln, and M. Nielsen: Blood Lactate and Oxygen Debt after Exhaustive Work at Different Oxygen Tensions, *Acta Physiol. Scand.*, **15**:57, 1948.

Asmussen, E., and M. Nielsen: Cardiac Output during Muscular Work and Its Regulation, *Physiol. Rev.*, **35**:778, 1955.

Åstrand, P.-O.: The Respiratory Activity in Man Exposed to Prolonged Hypoxia, *Acta Physiol. Scand.*, **30**:343, 1954.

Åstrand, P.-O., and I. Åstrand: Heart Rate during Muscular Work in Man Exposed to Prolonged Hypoxia, *J. Appl. Physiol.*, **13**:75, 1958.

Balke, B.: Work Capacity at Altitude, in W. R. Johnson (ed.), "Science and Medicine of Exercise and Sports," p. 339, Harper & Row Publishers, Inc., New York, 1960.

Barbashova, Z. I.: Cellular Level of Adaptation, in D. B. Dill (ed.), "Handbook of Physiology," sec. 4. Adaptation to the Environment, p. 37, American Physiological Society, Washington, D.C., 1964.

Behnke, A.: Decompression Sickness: Advances and Interpretations, The Harry G. Armstrong Lecture, *Aerospace Med.*, **42**(3):255, 1971.

Behnke, A. R.: Physiological Effects of Abnormal Atmospheric Pressures, chap. 9 in Patty's: Industrial Hygiene and Toxicology, 3rd rev. ed., vol. 1, General Principles. Edited by G. D. Clayton and F. E. Clayton, John Wiley and Sons Inc., New York, 1978.

Bert, P.: "La Pression Barométrique," Masson et Cie, Paris, 1878.

Bing, R. J.: Effect of Alcohol on the Heart and Cardiac Metabolism, *Fed. Proc.*, **41**:2443, 1982.

Bisgard, G. E., J. A. Will, I. B. Tyson, L. M. Dayton, R. R. Henderson, and R. F. Grover: Distribution of Regional Lung Function during Mild Exercise in Residents of 3100 m, *Resp. Physiol.*, **22**:369, 1974.

Blomqvist, G., and J. Stenberg: The ECG Response to Submaximal and Maximal Exercise during Acute Hypoxia, in G. Blomqvist: The Frank Lead Exercise Electrocardiogram, *Acta Med. Scand.*, **178**(Suppl. 440):82, 1965.

Blomqvist, G., B. Saltin, and J. H. Mitchell: Acute Effects of Ethanol Ingestion on the Response to Submaximal and Maximal Exercise in Man, *Circulation*, **42**:463, 1970.

Brewer, N., and B. Sandow: Alcohol Effects on Driver Performance under Conditions of Divided Attention, *Ergonomics*, **23**(3):185, 1980.

Brundin, T.: Effects of Tobacco Smoking on the Blood Temperature during Exercise, *Acta Physiol. Scand.* (Suppl. *479*):43, 1980.

Bucher, H. K., and G. M. Smith: Drugs and Athletic Performance, in A. De Schaepdryver and M. Hebbelinck (eds.), "Doping," Pergamon Press, Oxford and New York, 1965.

Buskirk, E. R., J. Kollias, R. F. Akers, B. K. Prokop, and E. Picón-Reátegui: Maximal Performance at Altitude and on Return from Altitude in Conditioned Runners, *J. Appl. Physiol.*, **23**:259, 1967.

Cassin, S., R. D. Gilbert, and E. M. Johnson: Capillary Development during Exposure to Chronic Hypoxia, Report SAM-TR-66-16, USAF School of Aviation Medicine, Randolph Field, Tex., 1966.

Cerretelli, P.: Limiting Factors to Oxygen Transport on Mount Everest, *J. Appl. Physiol.*, **40**:658, 1976.

Cerretelli, P., and R. Margaria: Maximum Oxygen Consumption at Altitude, *Intern. Z. Angew. Physiol.*, **18**:460, 1961.

Christensen, E. H.: Sauerstoffaufnahme und Respiratorische Funktionen in Grossen Höhen, *Skand. Arch. Physiol.*, **76**:88, 1937.

Christensen, E. H., and A. Krogh: Fliegerundersuchungen; die Wirkung niedriger O_2- Spannung auf Höhenflieger, *Skand. Arch. Physiol.*, **73**:145, 1936.

Christensen, E. H., and H. E. Nielsen: Die Leistungsfähigkeit der menschlichen Skelettmuskeln bei niedrigen Sauerstoffdruck, *Skand. Arch. Physiol.*, **74**:272, 1936.

Christensen, E. H., and W. H. Forbes: Der Kreislauf in grossen Höhen, *Skand. Arch. Physiol.*, **76**:75, 1937.

Clarke, C., M. Ward, and E. Williams (eds.): "Mountain Medicine and Physiology," Alpine Club, London, 1975.

Clench, J., R. E. Ferrell, and W. J. Schull: Effect of Chronic Altitude Hypoxia on Hematologic and Glycolytic Parameters, *Am. J. Physiol.*, **242**(5):R447, 1982.

Consolazio, C. F.: Submaximal and Maximal Performance at High Altitude, in R. F. Goddard (ed.), "The International Symposium on the Effects of Altitude on Physical Performance," p. 91, The Athletic Institute, Chicago, 1967.

Cotes, J. E.: Ventilatory Capacity at Altitude and Its Relation to Mask Design, *Proc. Roy. Soc. (London)*, ser. B, **143**:32, 1954.

Craig, A. B.: Olympics, 1968: A Post-mortem, *Med. Sci. Sports,* **1**:177, 1969.

Craig, A. B., Jr.: Summary of 58 Cases of Loss of Consciousness during Underwater Swimming and Diving, *Med. Sci. Sports,* **8**:171, 1976.

Crist, D. M., P. J. Stackpole, and G. T. Peake: Effects of Androgenic-anabolic Steroids on Neuromuscular Power and Body Composition, *J. Appl. Physiol.*, **54**(2):336, 1983.

Cruz, J. C., H. Hartley, and J. A. Vogel: Effect of Altitude Relocations upon $AaDO_2$ at Rest and during Exercise, *J. Appl. Physiol.*, **39**:469, 1975.

Cruz, J., C. R. Grover, J. T. Reeves, J. T. Maher, A. Cymerman, and J. C. Denniston: Sustained Venoconstriction in Man Supplemented with CO_2 at High Altitude, *J. Appl. Physiol.*, **40**:96, 1976.

Da Silva, A. M. T., and P. Hamosh: Effect of Smoking a Single Cigarette on the "Small Airways," *J. Appl. Physiol.*, **34**:361, 1973.

DeGraff, A. C., Jr., R. F. Grover, J. W. Hammond, Jr., J. M. Miller, and R. L. Johnson, Jr.: Pulmonary Diffusing Capacity in Persons Native to High Altitude, *Clin. Res.*, **13**:74, 1965.

Dejours, P., R. H. Kellogg, and N. Pace: Regulation of Respiration and Heart Rate Response in Exercise during Altitude Acclimatization, *J. Appl. Physiol.*, **18**:10, 1963.

Dempsey, J. A., W. G. Reddan, M. L. Birnbaum, H. V. Forster, J. S. Thoden, R. F. Grover, and J. Rankin: Effects of Acute through Life-long Hypoxic Exposure on Exercise Pulmonary Gas Exchange, *Resp. Physiol.*, **13**:62, 1971.

Dempsey, J. A., H. V. Forster, M. L. Birnbaum, W. G. Reddan, J. Thoden, R. F. Grover, and J. Rankin: Control of Exercise Hyperpnea under Varying Durations of Exposure to Moderate Hypoxia, *Resp. Physiol.*, **16:**213, 1972.

Dickinson, E. R., M. J. Piddington, and T. Brain: Project Olympics, *Schw. Zschr. Sportmed.*, **14:**305, 1966.

Dill, D. B. (ed.): "Handbook of Physiology," sec. 4, Adaptation to the Environment, American Physiological Society, Washington, D.C., 1964.

Edwards, H. T.: Lactic Acid in Rest and Work at High Altitude, *Am. J. Physiol.*, **116:**367, 1936.

Ekblom, B., and R. Huot: Response to Submaximal and Maximal Exercise at Different Levels of Carboxyhemoglobin, *Acta Physiol. Scand.*, **86:**474, 1972.

Ekblom, B., G. Wilson, and P.-O. Åstrand: Central Circulation during Exercise after Venesection and Reinfusion of Red Blood Cells, *J. Appl. Physiol.*, **40:**379, 1976.

Essig, D., D. L. Costill, and P. J. Van Handel: Effects of Caffeine Ingestion on Utilization of Muscle Glycogen and Lipid during Leg Ergometer Cycling, *Int. J. Sports Med.*, **1:**86, 1980.

Fagraeus, L.: Cardiorespiratory and Metabolic Functions during Exercise in Hyperbaric Environment, *Acta Physiol. Scand.*, (Suppl. 414), 1974.

Fenn, W. O.: The Pressure Volume Diagram of the Breathing Mechanism, in W. M. Boothby (ed.), "Handbook of Respiratory Physiology," USAF School of Aviation Medicine, Randolph Field, Tex., 1954.

Forster, H. V., J. A. Dempsey, M. L. Birnbaum, W. G. Reddan, J. Thoden, R. F. Grover, and J. Rankin: Effects of Chronic Exposure to Hypoxia in Ventilatory Response to CO_2 and Hypoxia, *J. Appl. Physiol.*, **31:**586, 1971.

Frisancho, A. R.: Functional Adaptation to High Altitude Hypoxia, *Science*, **187:**313, 1975.

Gilbert, D. L. (ed.): "Oxygen and Living Processes. An Interdisciplinary Approach," Springer-Verlag, New York, 1981.

Goddard, R. F., and C. B. Favour: United States Olympic Committee Swimming Team Performance in International Sports Week, Mexico City, Oct., 1965, in R. F. Goddard (ed.), "The International Symposium on the Effects of Altitude on Physical Performance," p. 135, The Athletic Institute, Chicago, 1967.

Grover, R. F., R. Lufschanowski, and J. K. Alexander: Alterations in the Coronary Circulation of Man Following Ascent to 3,100 m Altitude, *J. Appl. Physiol.*, **41:**832, 1976.

Grover, R. F., and J. T. Reeves: Exercise Performance of Athletes at Sea Level and 3,100 Meters Altitude, in R. F. Goddard (ed.), "The International Symposium on the Effects of Altitude on Physical Performance," p. 80, The Athletic Institute, Chicago, 1967.

Guleria, J. S., J. N. Parde, P. K. Sethi, and S. B. Roy: Pulmonary Diffusing Capacity at High Altitude, *J. Appl. Physiol.*, **31:**536, 1971.

Hackett, P. H., J. T. Reeves, C. D. Reeves, R. F. Grover, and D. Rennie: Control of Breathing in Sherpas at Low and High Altitude, *J. Appl. Physiol.*, **49**(3):373, 1980.

Hackett, P. H., D. Rennie, R. F. Grover, and J. T. Reeves: Acute Mountain Sickness and the Edemas of High Altitude: A Common Pathogenesis?, *Resp. Physiol.*, **46**(3):383, 1981.

Hackett, P. H., K. H. Maret, J. S. Milledge, R. M. Peters, C. J. Pizzo, J. B. West, and R. M. Winslow: Physiology of Man on the Summit of Mt. Everest, *J. Physiol.*, **334:**99, 1983.

Halsey, M. J.: Effects of High Pressure on the Central Nervous System, *Physiol. Rev.*, **62**(4):1341, 1982.

Hanley, D. F.: Pill Popping and Performance, *Modern Medicine*, **40:**81, 1972.

Hansen, J. E., G. P. Stelter, and J. A. Vogel: Arterial Pyruvate, Lactate, pH, and P_{CO_2} during Work at Sea Level and High Altitude, *J. Appl. Physiol.*, **23:**523, 1967a.

Hansen, J. E., J. A. Vogel, G. P. Stelter, and C. F. Consolazio: Oxygen Uptake in Man during Exhaustive Work at Sea Level and High Altitude, *J. Appl. Physiol,* **23:**511, 1967b.

Hartley, L. H., J. K. Alexander, M. Modelski, and R. F. Grover: Subnormal Cardiac Output at Rest and during Exercise in Residents at 3,100 m Altitude, *J. Appl. Physiol.,* **23:**839, 1967.

Hartley, L. H., J. A. Vogel, and M. Landowne: Central, Femoral, and Brachial Circulation during Exercise in Hypoxia, *J. Appl. Physiol.,* **34:**87, 1973.

Hartley, L. H., J. A. Vogel, and J. C. Cruz: Reduction of Maximal Heart Rate at Altitude and Its Reversal with Atropine, *J. Appl. Physiol.,* **36:**362, 1974.

Hermansen, L., and B. Saltin: Blood Lactate Concentration during Exercise at Acute Exposure to Altitude, in R. Margaria (ed.), "Exercise at Altitude," p. 48, Excerpta Medica Foundation, Amsterdam, 1967.

Hesser, C. M., D. Linnarsson, and L. Fagraeus: Pulmonary Mechanics and Work of Breathing at Maximal Ventilation and Raised Air Pressure, *J. Appl. Physiol.,* **50**(4):747, 1981.

Horstman, D., R. Weiskopf, and R. E. Jackson: Work Capacity during 3-week Sojourn at 4,300 m: Effects of Relative Polycythemia, *J. Appl. Physiol.,* **49**(2):311, 1980.

Hultgren, H. N.: Pulmonary Hypertension and Pulmonary Edema, in Loeppky and Riedesel (eds.): "Oxygen and Transport to Human Tissues," p. 243, Elsevier Biiomedical, North Holland, New York, 1982.

Hurtado, A.: Animals in High Altitudes: Resident Man, in D. B. Dill (ed.), "Handbook of Physiology," sec. 4, p. 843, Adaptation to the Environment, American Physiological Society, Washington, D.C., 1964.

Hurtado, A., C. Merino, and E. Delgado: Influence of Anoxemia on the Hemopoietic Activity, *Arch. Int. Med.,* **75:**284, 1945.

Ikai, M., and A. H. Steinhaus: Some Factors Modifying the Expression of Strength, *J. Appl. Physiol.,* **16:**157, 1961.

Johnson, L. F.: The Association of Oral Androgenic-anabolic Steroids and Life-threatening Disease, *Med. Sci. Sports,* **7:**284, 1975.

Johnson, L. C., E. S. Roundy, P. E. Allsen, A. G. Fisher, and L. J. Silvester: Effect of Anabolic Steroid Treatment on Endurance, *Med. Sci. Sports,* **7:**287, 1975.

Jokl, E., and P. Jokl (eds.): "Exercise and Altitude," S. Karger, New York, 1968.

Jokl, E., and P. Jokl: Heart and Sport, in D. Brumer and E. Jokl (eds.): "The Role of Exercise in Internal Medicine," Medicine and Sport, vol. 10, p. 36, S. Karger, Basel, 1977.

Jorfeldt, L., and A. Juhlin-Dannfelt: The Influence of Ethanol on Human Splanchnic and Skeletal Muscle Metabolism during Exercise, *Scand. J. Clin. Lab. Invest.,* **37:**609, 1977.

Juurup, A., and L. Muido: On Acute Effects of Cigarette Smoking on Oxygen Consumption, Pulse Rate, Breathing Rate and Blood Pressure in Working Organisms, *Acta Physiol. Scand.,* **11:**48, 1946.

Kellogg, R. H.: Central Chemical Regulation of Respiration, in W. O. Fenn and H. Rahn (eds.), "Handbook of Physiology," sec. 3, Respiration, vol. 1, p. 507, American Physiological Society, Washington, D.C., 1964.

Klausen, K.: Cardiac Output in Man in Rest and Work during and after Acclimatization to 3,800 m, *J. Appl. Physiol.,* **21:**609, 1966.

Kollias, J., E. R. Buskirk, R. F. Akers, E. K. Prokop, P. T. Baker, and E. Picón-Reátegui: Work Capacity of Long-time Residents and Newcomers to Altitude, *J. Appl. Physiol.,* **24:**792, 1968.

Kreuzer, F., S. M. Tenney, J. C. Mithoefer, and J. Remmers: Alveolar-arterial Oxygen Gradient in Andean Natives at High Altitude, *J. Appl. Physiol.,* **19:**13, 1964.

Lamb, D. R.: Androgens and Exercise, *Med. Sci. Sports,* **7:**1, 1975.

Lamb, D. R.: Anabolic Steroids and Athletic Performance, *Sports Med. Digest,* **6:**1, 1984.

Lambertsen, C. J.: Effects of Oxygen at High Partial Pressure, in W. O. Fenn and H. Rahn (eds.), "Handbook of Physiology," sec. 3, Respiration, vol. 2, p. 1027, American Physiological Society, Washington, D.C., 1965.

Lambertsen, C. J. (ed.): "Underwater Physiology," The Williams & Wilkins Company, Baltimore, 1967.

Larson, P. S., H. B. Haag, and H. Silvette: "Tobacco," The Williams & Wilkins Company, Baltimore, 1961.

Latis, V. G., and B. Weiss: The Amphetamine Margin in Sports, *Fed. Proc.*, **40:**2689, 1981.

Leary, W. P., and C. H. Wyndham: The Possible Effect on Athletic Performance of Mexico City's Altitude, *S. Afr. Med. J.* **40:**984, 1966.

Lockhart, A., M. Zelter, J. Mensch-Dechene, G. Antezana, M. Paz-Zamor, E. Vargas, and J. Coudert: Pressure-Flow-Volume Relationships in Pulmonary Circulation of Normal Highlanders, *J. Appl. Physiol.,* **41:**449, 1976.

Loeppky, J. A., and M. L. Riedesel (eds.): "Oxygen Transport to Human Tissues, " Sect. IV: Effects of Altitude on Oxygen Transport, p. 197, Elsevier Biomedical, North-Holland, New York, 1982.

Luft, U.C.: Laboratory Facilities for Adaptation Research: Low Pressures, in D. B. Dill (ed.), "Handbook of Physiology," sec. 4, p. 329, Adaptation to the Environment, American Physiological Society, Washington, D.C., 1964a.

Maher, J. T., L. G. Jones, and L. H. Hartley: Effects of High-altitude Exposure on Submaximal Endurance Capacity of Men, *J. Appl. Physiol.,* **37:**895, 1974.

Margaria, R. (ed.): "Exercise at Altitude," Excerpta Med. Found., Amsterdam, 1967.

McManus, B. M., S. M. Horvath, N. Bolduan, and J. C. Miller: Metabolic and Cardio-Respiratory Responses to Long-term Work under Hypoxic Conditions, *J. Appl. Physiol.,* **36:**177, 1974.

Merino, C.: Studies on Blood Formation and Destruction in the Polycythemia of High Altitude, *Blood,* **5:**1, 1950.

Miles, S.: The Effect of Changes in Barometric Pressure on Maximum Breathing Capacity, *J. Physiol.,* **137:**85P, 1957.

Milledge, J. S., and S. Lahiri: Respiratory Control in Lowlanders and Sherpa Highlanders at Altitude, *Respir. Physiol.,* **2:**310, 1967.

Millhorn, D. E., F. L. Eldridge, and T. G. Waldrop: Prolonged Stimulation of Respiration by Endogenous Central Serotonin, *Resp. Physiol.,* **42:**171, 1980.

Morpurgo, G., P. Battaglia, N. D. Carter, G. Modiano, and S. Passi: The Bohr Effect and the Red Cell 2, 3-DPG and HB Content in Sherpas and Europeans at Low and High Altitude, *Experientia,* **28:**1280, 1972.

Noell, W.: Über die Durchblutung und die Sauerstoffversorgung des Gehirns, VI, Einfluss der Hypoxämie und Anämie, *Arch. Ges. Physiol.,* **247:**553, 1944.

Norton, E. F.: "The Fight for Everest," Edward Arnold (Publishers) Ltd., London, 1925.

Penaloza, D., F. Sime, N. Banchero, R. Gamboa, J. Cruz, and E. Marticorena: Pulmonary Hypertension in Healthy Men Born and Living at High Altitudes, *Am. J. Cardiol.,* **11:**150, 1963.

Powers, S. K., R. J. Byrd, R. Tulley, and T. Callender: Effects of Caffeine Ingestion on Metabolism and Performance During Graded Exercise, *Eur. J. Appl. Physiol.,* **50:**301, 1983.

Pugh, L. G.: Animals in High Altitude: Man above 5,000 Meters-Mountain Exploration, in D. B. Dill (ed.), "Handbook of Physiology," sec. 4, p. 861, Adaptation to the Environment, American Physiological Society, Washington, D.C., 1964.

Pugh, L. G.: "Report of Medical Research Project into Effects of Altitude in Mexico City," report to the British Olympic Committee, 1965.

Pugh, L. G., M. B. Gill, S. Lahiri, J. S. Milledge, M. P. Ward, and J. B. West: Muscular Exercise at Great Altitudes, *J. Appl. Physiol.*, **19**:431, 1964.

Rahn, H.: Introduction to the Study of Man at High Altitudes: Conductance of O_2 from the Environment to the Tissues, in "Life at High Altitudes," Scientific Publ. 140, p. 2, Pan-American Health Organization, WHO, Washington, D.C., September 1966.

Raynaud, J., J. P. Martineaud, J. Bordachar, M. C. Tillous, and J. Durand: Oxygen Deficit and Debt in Submaximal Exercise at Sea Level and High Altitude, *J. Appl. Physiol.*, **37**:43, 1974.

Reeves, J. T., J. Halpin, J. E. Cohn, and F. Daoud: Increased Alveolar-arterial Oxygen Difference during Simulated High-altitude Exposure, *J. Appl. Physiol.*, **27**:658, 1969.

Regan, T. J.: Regional Circulatory Responses to Alcohol and Its Congeners, *Fed. Proc.*, **41**:2438, 1982.

Reynafarje, B.: Myoglobin Content and Enzymatic Activity of Muscle and Altitude Adaptation, *J. Appl. Physiol.*, **17**:301, 1962.

Reynafarje, C.: Humoral Control of Erythropoiesis at Altitude, in R. Margaria (ed.), "Exercise at Altitude," p. 165, Excerpta Medica Foundation, Amsterdam, 1967.

Robertson, R. J., R. Gilcher, K. F. Metz, G. S. Skrinar, T. G. Allison, H. T. Bahnson, R. A. Abbott, R. Becker, and J. E. Falkel: Effect of Induced Erythrocythemia on Hypoxia Tolerance during Physical Exercise, *J. Appl. Physiol.*, **53**(2):490, 1982.

Roskamm, H., L. Samek, H. Weidemann, and H. Reindell: "Leistung und Höhe," Knoll AG, Ludwigshafen am Rhein, 1968.

Roughton, F. J. W.: Transport of Oxygen and Carbon Dioxide, in W. O. Fenn and H. Rahn (eds.), "Handbook of Physiology," sec. 3, Respiration, vol. 1, p. 767, American Physiological Society, Washington, D.C., 1964.

Rubin, E.: Alcohol and the Heart: Theoretical Considerations, *Fed. Proc.*, **41**:2460, 1982.

Ryan, A. J.: Anabolic Steroids are Fool's Gold, *Fed. Proc.*, **40**:2682, 1981.

Saltin, B.: Aerobic and Anaerobic Work Capacity at 2,300 Meters, *Schw. Zschr. Sportmed.*, **14**:81, 1966.

Saltin, B.: Aerobic and Anaerobic Work Capacity at 2,300 Meters, *Schw. Zschr. Sportmed.*, **14**:81, 1966.

Saltin, B.: Aerobic and Anaerobic Work Capacity at an Altitude of 2,250 Meters, in R. F. Goddard (ed.), "The International Symposium on the Effects of Altitude on Physical Performance," p. 97, The Athletic Institute, Chicago, 1967.

Schoene, R. B.: Control of Ventilation in Climbers to Extreme Altitude, *J. Appl. Physiol,* **53**(4):886, 1982.

Severinghaus, J. W., C. R. Bainton, and A. Carcelen: Respiratory Insensitivity to Hypoxia in Chronically Hypoxic Man, *Respir. Physiol.,* **1**:308, 1966.

Shephard, R. J.: Athletic Performance at Moderate Altitudes, *Medicina Dello Sport,* **26**:36, 1973.

Somervell, T. H.: Note on the Composition of Alveolar Air at Extreme Heights, *J. Physiol.,* **60**:282, 1925.

Stenberg, J., B. Ekblom, and R. Messin: Hemodynamic Response to Work at Simulated Altitude, *J. Appl. Physiol.,* **21**:1589, 1966.

Surks, M. I., K. S. Chinn, and L. O. Matoush: Alterations in Body Composition in Man after Acute Exposure to High Altitude, *J. Appl. Physiol.,* **21**:1741, 1966.

Sutton, J. R.: Effect of Acute Hypoxia on the Hormonal Response to Exercise, *J. Appl. Physiol.* **42**:587, 1977.

Sutton, J. R., N. L. Jones, L. Griffith, and C. E. Pugh: Exercise at Altitude, *Ann. Rev. Physiol.,* **45**:427, 1983.

Toner, M. M., D. T. Kirkendall, D. J. Delio, J. M. Chase, P. A. Cleary, and E. L. Fox: Metabolic and Cardiovascular Responses to Exercise with Caffeine, *Ergonomics,* **25**(12):1175, 1982.

Ulvedal, F., T. E. Morgan, Jr., R. G. Cutler, and B. E. Welch: Ventilatory Capacity during Prolonged Exposure to Simulated Altitude without Hypoxia, *J. Appl. Physiol.,* **18**:904, 1963.

Vannotti, A.: The Adaptation of the Cell to Effort, Altitude and to Pathological Oxygen Deficiency, *Schweiz. Med. Wschr.,* **76**:899, 1946.

Velasquez, T.: Tolerance to Acute Anoxia in High Altitude Natives, *J. Appl. Physiol.,* **14**:357, 1959.

Vogel, J. A., J. E. Hansen, and C. W. Harris: Cardiovascular Responses in Man during Exhaustive Work at Sea Level and High Altitude, *J. Appl. Physiol.,* **23**:531, 1967.

Vogel, J. A., L. H. Hartley, and J. C. Cruz: Cardiac Output during Exercise in Altitude Natives at Sea Level and High Altitude, *J. Appl. Physiol.,* **36**:173, 1974a.

Vogel, J. A., L. H. Hartley, J. C. Cruz, and R. P. Hogan: Cardiac Output during Exercise in Sea-level Residents at Sea Level and High Altitude, *J. Appl. Physiol.,* **36**:169, 1974b.

Wahren, J., E. Jequier, K. Acheson, and Y. Schutz: Influence of Cigarette Smoking on Body Oxygen Comsumption, *Clin. Phys.,* **3**(1):91, 1983.

West, J. B.: Diffusing Capacity of the Lung for Carbon Monoxide at High Altitude, *J. Appl. Physiol.,* **17**:421, 1962.

West J. B. (ed.): "High Altitude Physiology," Benchmark Papers in Human Physiology, Vol. 15, Hutchinson Ross Publ. Co., Stroudsburg, Pa., 1981.

West, J. B.: Climbing Mt. Everest without Oxygen: An Analysis of Maximal Exercise during Extreme Hypoxia, *Resp. Physiol.,* **52**(3):265, 1983.

West, J. B., S. Lahiri, M. B. Gill, J. S. Milledge, L. G. Pugh, and M. P. Ward: Arterial Oxygen Saturation during Exercise at High Altitude, *J. Appl. Physiol.,* **17**:617, 1962.

West, J. B., and S. Lahiri (eds.): "Humans at High Altitude," Am. Physiol. Soc., Williams and Wilkins, Baltimore, 1984.

Williams, M. H., A. R. Goodwin, R. Perkins, and J. Bocrie: Effect of Blood Reinjection upon Endurance Capacity and Heart Rate, *Med. Sci. Sports,* **5**:181, 1973.

Young, A. J., W. J. Evans, A. Cymerman, K. B. Pandolf, J. J. Knapik, and J. T. Mayer: Sparing Effect of Chronic High-altitude Exposure on Muscle Glycogen Utilization. *J. Appl. Physiol.,* **52**(4):857, 1982.

APPENDIX

CONTENTS

DEFINITION OF TERMS AND UNITS
PREFIXES, SYMBOLS, UNIT ABBREVIATIONS
CONVERSION TABLES
LIST OF SYMBOLS

DEFINITION OF TERMS AND UNITS

Preferably the S. I. Units (Le Système International d'Unités) should be applied. In order to facilitate the transition from old units to this new system, both the "conventional" and new units are often used in the text.

Basic Units

Length: Meter (m)
Mass: Kilogram (kg)
Time: Second (s)
Quantity: Mole (mol)

The *meter* is the length equal to 1,650,763.73 wave lengths in vacuum of the radiation corresponding to the transition from the level 5 d_5 to 2 p_{10} of the krypton^{-86} atom.

The *kilogram* (kg., unit of mass), is the mass of a particular cylinder of platinum-iridium alloy, International Prototype Kg, which is preserved in a vault at Sévres, France, by the International Bureau of Weights and Measures.

The *second* is the duration of 9,192,631,770 periods of the radiation corresponding to the transition between the two hyperfine levels of the ground state of the cesium^{-133} atom.

The *mole* is the amount of substance of a system which contains as many elementary entities as there are atoms in 0.012 kg of carbon $-$ 12 (^{12}C) (When the mole is used, these entities must be specified and may be atoms, molecules, electrons, other particles, or specified groups of such particles). In biologic contexts it is often convenient to use the unit millimole, micromoles, or nanomoles.

From these basic units other units can be defined:

Force (F) can be defined as that which changes the state of rest or motion of a body, measured by the rate of change of momentum. The unit of force is the *newton* (N) which is the force producing an acceleration of 1 meter per second every second (1 m $\cdot$ s^{-2}) when it acts on 1 kg. Also F = m $\cdot$ a, where m = mass in kg and a = acceleration in meters per second every second.

In the old system, the commonly used unit for force was kilopond (kp). One kp is the force acting on the mass of 1 kg at normal acceleration of gravity: 1 kp = 9.80665 N, or approximately 10 N.

Work is done when a force acts against resistance to produce motion. It is therefore measured by the product of the force (F) and the distance moved (L) in the direction of the force: Work = F $\cdot$ L. The unit of work is the *joule* (J). One joule of work is done when the force of 1 N moves the point of application through a distance of 1 m, 1,000 J = 1 kJ; 1,000 kJ = 1 MJ; 1 calorie (cal) = 4.186 J; 1 kcal = 4.186 kJ.

Power is the rate of doing work and is measured in watts (W); 1 watt = 1 joule $\cdot$ s^{-1}. An old alternative is kpm $\cdot$ min^{-1}; 1 watt = 6.12 kpm $\cdot$ min^{-1}; 9.81 watts = 1 kpm $\cdot$ s^{-1}; 16.35 watts = 100 kpm $\cdot$ min^{-1}; 1 hp (horsepower) = 736 watts = 75 kpm $\cdot$ s^{-1} = 4,500 kpm $\cdot$ min^{-1}.

Pressure is the force per unit area and it is defined in newtons per square meter (N $\cdot$ m^{-2}). It is measured in *pascals* (Pa): 1 Pa = 1 N $\cdot$ m^{-2}. The practice of expressing pressures in terms of mm Hg or torr in medical contexts (blood pressure, osmotic pressure, gas pressures, etc.) will probably continue for some time. The pressure defined in mm Hg multiplied by 0.133 gives the answer in terms of kilopascals (kPa). Hence, 1 kPa = 7.5 mm Hg. (Torr is a provisional international standard term to replace mm Hg; 1 torr = 1 mm Hg.)

Concentrations may be expressed as moles per liter (mol $\cdot$ l^{-1}, as suggested in the S. I. system, or M; mmol $\cdot$ l^{-1} or mM, etc.). A one molar solution contains 1 mol or g mol weight of the solute in 1 liter of solution. Concentrations may also be expressed as moles per kg, e.g., lactate per kg of muscle tissue, wet weight or dry weight.

Frequency is the rate of oscillation; unit: 1 cycle $\cdot$ s^{-1} = 1 Hertz (Hz).

PREFIXES, SYMBOLS, UNIT ABBREVIATIONS
Prefixes for Multiples and Submultiples of Units

Multiple	Prefix	Symbol
10^{-1}	deci	d
10^{-2}	centi	c
10^{-3}	milli	m

Multiple	Prefix	Symbol
10^{-6}	micro	μ
10^{-9}	nano	η
10^{-12}	pico	p
10	deka	da
10^2	hecto	h
10^3	kilo	k
10^6	mega	M
10^9	giga	G
10^{12}	tera	T

Examples: m^{-3} = mm; m^{-6} = μm; m^{-9} = nm; g: gram; g^{-3} = mg; g^3 = kg; M^{-3} = mmolar concentration.

As mentioned above, the following abbreviations are used:

$$s = \text{second; min} = \text{minute;}$$
$$J = \text{joule; kJ} = \text{kilojoule; MJ} = \text{megajoule;}$$
$$V = \text{volt; mV} = \text{millivolt;}$$
$$W = \text{watt;}$$
$$kp = \text{kilopond}$$

SUMMARY

Force	Newton	N ($kg \cdot m \cdot s^{-2}$)
Work or energy	Joule	J ($kg \cdot m^2 \cdot s^{-2}$ = Nm)
Power	Watt	W ($kg \cdot m^2 \cdot s^{-3}$ = $J \cdot s^{-1}$)
Pressure	Pascal	Pa ($N \cdot m^{-2}$)

CONVERSION TABLES

Length and Weight

$$1 \text{ centimeter} = 0.39370 \text{ in.}$$
$$1 \text{ meter} = 39.37 \text{ in.}$$
$$1 \text{ kilometer} = 0.62137 \text{ mile}$$
$$1 \text{ inch} = 2.54 \text{ cm}$$
$$1 \text{ foot} = 30.480 \text{ cm}$$

$$1 \text{ milliliter} = 0.03381 \text{ fl oz}$$
$$1 \text{ liter} = 1.0567 \text{ U.S. qt}$$
$$1 \text{ kilogram} = 2.2046 \text{ lb}$$

Power

$$1 \text{ watt} = 0.001 \text{ kilowatt}$$
$$1 \text{ watt} = 0.73756 \text{ ft-lb} \cdot s^{-1}$$

1 watt = $1 \cdot 10^7$ ergs $\cdot$ s^{-1}
1 watt = 0.056884 BTU $\cdot$ min^{-1} = 3.41304 BTU $\cdot$ hr^{-1}
1 watt = 0.01433 kilocalories $\cdot$ min^{-1}
1 watt = $1.341 \cdot 10^{-3}$ hp (horsepower)
1 watt = 1 J $\cdot$ s^{-1}
1 watt = 6.12 kpm $\cdot$ min^{-1}

1 kilocalorie per minute = 69.767 watts
1 kilocalorie per minute = 51.457 ft-lb $\cdot$ s^{-1}
1 kilocalorie per minute = $6.9770 \cdot 10^8$ ergs $\cdot$ s^{-1}
1 kilocalorie per minute = 3.9685 BTU $\cdot$ min^{-1}
1 kilocalorie per minute = 0.093557 hp

1 horsepower = 745.7 watts
1 horsepower = 550 ft-lb $\cdot$ s^{-1}
1 horsepower = $7.457 \cdot 10^9$ ergs $\cdot$ sec^{-1}
1 horsepower = 42.4176 BTU $\cdot$ min^{-1}
1 horsepower = 10.688 kcal $\cdot$ min^{-1}
1 horsepower = 745.7 joules $\cdot$ s^{-1}
1 horsepower = 75 kpm $\cdot$ s^{-1}

Work and Energy

1 kilocalorie = $4.186 \cdot 10^{10}$ ergs
1 kilocalorie = 4186 joules = 4.186 kJ
1 kilocalorie = 3.9680 BTU
1 kilocalorie = 3087.4 ft-lb
1 kilocalorie = 426.85 kpm
1 kilocalorie = $1.5593 \cdot 10^{-3}$ hp $\cdot$ hr^{-1}

1 erg = $2.3889 \cdot 10^{-11}$ kcal
1 erg = $1 \cdot 10^{-7}$ joule
1 erg = $9.4805 \cdot 10^{-14}$ BTU
1 erg = $7.3756 \cdot 10^{-8}$ ft-lb
1 erg = $1.0197 \cdot 10^{-8}$ kpm
1 erg = $3.7251 \cdot 10^{-14}$ hp $\cdot$ hr^{-1}

1 joule = $2.3889 \cdot 10^{-4}$ kcal
1 joule = $1 \cdot 10^7$ ergs
1 joule = $9.4805 \cdot 10^{-4}$ BTU
1 joule = 0.73756 ft-lb
1 joule = 0.10197 kpm
1 joule = $3.7251 \cdot 10^{-7} \cdot$ hr^{-1}

1 BTU = 0.25198 kcal (BTU = British Thermal Unit)
1 BTU = 1.0548 · 10^{10} ergs
1 BTU = 1054.8 joules
1 BTU = 777.98 ft-lb
1 BTU = 107.56 kpm
1 BTU = 3.9292 · 10^{-4} hp · hr^{-1}

1 foot-pound = 3.2389 · 10^{-4} kcal
1 foot-pound = 1.35582 · 10^7 ergs
1 foot-pound = 1.3558 joules
1 foot-pound = 1.2854 · 10^{-3} BTU
1 foot-pound = 0.13825 kpm
1 foot-pound = 5.0505 · 10^{-7} hp · hr^{-1}

1 kilogram-meter* = 2.3427 · 10^{-3} kcal
1 kilogram-meter = 9.8066 · 10^7 ergs
1 kilogram-meter = 9.8066 joules
1 kilogram-meter = 9.2967 · 10^3 BTU
1 kilogram-meter = 7.2330 ft-lb
1 kilogram-meter = 3.6529 · 10^{-6} hp · hr^{-1}

1 watt = 6.12 kpm · min^{-1} (approx. = 6 kpm · min^{-1})
1 watt = 0.00134 hp
50 watts = approx. 300 kpm · min^{-1}
1 kpm · min^{-1} = 0.1635 watt
1 kp = 9.80665 newtons
1 kpm = 9.80665 joules

Pressure

1 atmosphere (ATA§) = 1013.25 mb (mb = millibar)
    1 atmosphere = 101325 pascal
    1 atmosphere = 101325 N · m^{-2}
    1 atmosphere = 1.03323 kg · cm^{-2}
    1 atmosphere = 1033.23 g · cm^{-2}
    1 atmosphere = 1033.26 cm H_2O at 4°C
    1 atmosphere = 760 torr
    1 atmosphere = 760 mm Hg at O°C
    1 atmosphere = 29.9213 inch Hg at 32°F

§The pressure of 760 mm Hg (101 kPa) at a density of 13.5951 g · cm^{-1} and acceleration of g = 980.655 cm · s^{-2}
  *or kilopand-meter

1 torr = $1.31579 \cdot 10^{-3}$ ATA

1 torr = 1.33322 mb

1 torr = 133.322 pascal

1 torr = $1.35951 \cdot 10^{-3}$ kg $\cdot$ cm^{-2}

1 torr = 1.35951 g $\cdot$ cm^{-2}

1 torr = 1.35955 cm H$_2$O at 4°C

1 torr = 1 mm Hg at O°C

1 torr = $3.93701 \cdot 10^{-2}$ inch Hg at 32°F

1 pascal = 0.0075 mm Hg

1 pascal = 0.001 kilopascal (kPa)

1 kPa = 0.0099 ATA

1 kPa = 0.01 bar

SPEED

km $\cdot$ hr^{-1}	mph	m $\cdot$ s^{-1}	km $\cdot$ hr^{-1}	mph	m $\cdot$ s^{-1}
10	6.22	2.78	200	124	55.6
20	12.4	5.56	220	137	61.2
30	18.7	8.34	240	149	66.7
40	24.9	11.1	260	162	72.3
50	31.1	13.9	280	174	77.8
60	37.4	16.7	300	187	83.4
70	43.6	19.4	320	199	88.9
80	49.8	22.2	340	211	94.5
90	56.0	25.0	360	224	100
100	62.2	27.8	380	236	106
120	74.7	33.3	400	249	111
140	87.1	38.9	420	261	117
160	99.5	44.5	440	274	122
180	112	50.0	460	286	128

ENERGY

J	kWhr	kpm	kcal	hp/hr	ft – lb	BTU
1	$0.27778 \cdot 10^{-6}$	0.10197	$0.23885 \cdot 10^{-3}$	$0.37251 \cdot 10^{-6}$	0.73756	$0.94782 \cdot 10^{-3}$
$3.6 \cdot 10^6$	1	$0.36710 \cdot 10^6$	859.85	1.3596	$2.6552 \cdot 10^6$	$3.4121 \cdot 10^3$
9.8066	$2.7241 \cdot 10^{-6}$	1	$2.3423 \cdot 10^{-3}$	$3.6529 \cdot 10^{-6}$	7.2330	$9.2949 \cdot 10^{-3}$
$4.1868 \cdot 10^3$	$1.163 \cdot 10^{-3}$	426.94	1	$1.5593 \cdot 10^{-3}$	$3.0880 \cdot 10^3$	3.9683
$2.6478 \cdot 10^6$	0.73550	$0.27 \cdot 10^6$	632.42	1	$1.9529 \cdot 10^6$	$2.5096 \cdot 10^3$
1.3558	$0.37662 \cdot 10^{-6}$	0.13826	$0.32383 \cdot 10^{-3}$	$0.50505 \cdot 10^{-6}$	1	$1.2851 \cdot 10^{-3}$
$1.0551 \cdot 10^3$	$0.29307 \cdot 10^{-3}$	107.59	0.25200	$0.39292 \cdot 10^{-3}$	778.17	1

POWER

W	kpm/s	kcal/s	kcal/hr	hp	ft · lb/s	BTU/hr
1	0.10197	$0.23885 \cdot 10^{-3}$	0.85985	$1.3410 \cdot 10^{-3}$	0.73756	3.4121
9.8066	1	$2.3423 \cdot 10^{-3}$	8.4322	$13.151 \cdot 10^{-3}$	7.2330	33.462
$4.1868 \cdot 10^3$	426.94	1	$3.6 \cdot 10^3$	5.6146	$3.0880 \cdot 10^3$	$14.286 \cdot 10^3$
1.163	0.11859	$0.27778 \cdot 10^{-3}$	1	$1.5596 \cdot 10^{-3}$	0.85779	3.9683
745.70	75	0.17811	641.19	1	550	$2.5444 \cdot 10^3$
1.3558	0.13826	$0.32383 \cdot 10^{-3}$	1.1658	$1.8182 \cdot 10^{-3}$	1	4.6262
0.29307	$29.885 \cdot 10^{-6}$	$69.999 \cdot 10^{-6}$	0.25200	$0.39302 \cdot 10^{-3}$	0.21616	1

PRESSURE

Pa	bar	kp/cm²	kp/mm²	torr	atm
1	$10 \cdot 10^{-6}$	$10.197 \cdot 10^{-6}$	$0.10197 \cdot 10^{-6}$	$7.5006 \cdot 10^{-3}$	$9.8692 \cdot 10^{-6}$
$100 \cdot 10^3$	1	1.0197	$10.197 \cdot 10^{-3}$	750.06	0.98692
$98.066 \cdot 10^3$	0.98066	1	$10 \cdot 10^{-3}$	735.56	0.96784
$9.8066 \cdot 10^6$	98.066	100	1	$73.556 \cdot 10^3$	96.784
133.32	$1.3332 \cdot 10^{-3}$	$1.3595 \cdot 10^{-3}$	$13.595 \cdot 10^{-6}$	1	$1.3158 \cdot 10^{-3}$
$101.32 \cdot 10^3$	1.0132	1.0332	$10.332 \cdot 10^{-3}$	760	1
$6.8948 \cdot 10^3$	$68.948 \cdot 10^{-3}$	$70.307 \cdot 10^{-3}$	$0.70307 \cdot 10^{-3}$	51.715	$68.046 \cdot 10^{-3}$

TEMPERATURE: CONVERSION OF DEGREES CENTIGRADE INTO DEGREES FAHRENHEIT AND VICE VERSA

$-50°C$ to $150°C$; $-58°F$ to $302°F$	$30°C$ to $45°C$; $86°F$ to $113°F$
1 division corresponds to $1°C$ or $1°F$	1 division corresponds to $\frac{1}{10}°C$ or $\frac{1}{10}°F$

°C	°F	°C	°F	°C	°F	°C	°F
0		50		100		150	
	+30		120		210		300
	+25		115		205		295
-5		45		95		145	
	+20		110		200		290
-10	+15	40	105	90	195	140	285
	+10		100		190		280
-15	+5	35	95	85	185 135		275
	0		90		180		270
-20	-5	30	85	80	175 130		265
	-10		80		170		260
-25	-15	25	75	75	165 125		255
	-20		70		160		250
-30	-25	20	65	70	155 120		245
-35	-30	15	60	65	150 115		240
	-35		55		145		235
-40	-40	10	50	60	140 110		230
	-45		45		135		225
-45	-50	5	40	55	130 105		220
	-55		35		125		215
-50		0		50		100	
°C	°F	°C	°F	°C	°F	°C	°F

°C	°F	°C	°F	°C	°F
35	-95	40	-104	45	-113
	-94		-103		-112
34		39		44	
	-93		-102		-111
	-92		-101		-110
33		38		43	
	-91		-100		-109
	-90		-99		-108
32		37		42	
	-89		-98		-107
	-88		-97		-106
31		36		41	
	-87		-96		-105
30	-86	35	-95	40	-104
°C	°F	°C	°F	°C	°F

Source: Documenta Geigy, "Scientific Tables," 5th ed., Geigy Pharmaceuticals, Ardsley, New York, 1956.

LIST OF SYMBOLS

$\bar{x}$	dash over any symbol indicates a mean value
$\dot{x}$	dot above any symbol indicates time derivate

Respiratory and Hemodynamic Notations

V	gas volume
$\dot{V}$	gas volume per unit time (usually liters $\cdot$ min^{-1})
R or RQ	respiratory exchange ratio (volume CO_2 $\cdot$ volume O_2^{-1})
I	inspired gas
E	expired gas
A	alveolar gas
F	fractional concentration in dry gas phase
f	respiratory frequency (breath $\cdot$ unit time^{-1})
TLC	total lung capacity
VC	vital capacity
FRC	functional residual capacity
RV	residual volume
T	tidal gas
D	dead space
FEV	forced expiratory volume
FEV$_{1.0}$	forced expiratory volume in 1 s
MVV	maximal voluntary ventilation
MVV$_{40}$	maximal voluntary ventilation at $f = 40$
D_L	diffusing capacity of the lungs (ml $\cdot$ min^{-1} $\cdot$ mm Hg^{-1})
P	gas presssure
B or Bar	barometric
STPD	0°C, 760 mm Hg (101.3 kPa), dry
BTPS	body temperature and pressure, saturated with water vapor
ATPD	ambient temperature and pressure, dry
ATPS	ambient temperature and pressure, saturated with water vapor
Q	blood flow or volume
$\dot{Q}$	blood flow $\cdot$ unit time^{-1} (without other notation, cardiac output; usually liters $\cdot$ min^{-1})
SV	stroke volume
HR	heart rate (usually beats $\cdot$ min^{-1})
BV	blood volume
THb or HB$_T$	total amount of hemoglobin in body
Hb	hemoglobin concentration (g $\cdot$ l^{-1})
Hct	hematocrit
BP	blood pressure
R	resistance
C	concentration in blood phase
S	percent saturation of Hb
a	arterial
c	capillary
v	venous

Temperature Notations

T or t	temperature
r or re	rectal

s	skin
e or oe	esophageal (oesophageal)
m	muscle
ty	tympanic
M	metabolic energy yield
C	convective heat exchange
R	radiation heat exchange
E	evaporative heat loss
S	storage of body heat
°C	temperature in degrees centigrade
°F	temperature in degrees Fahrenheit

Dimensions

W	weight
H	height
L	length
LBM	lean body mass
BSA	body surface area

Statistical Notations

M	arithmetic mean
SD or S.D.	standard deviation
SE or S.E.	standard error of the mean
n	number of observations
r	correlation coefficient
range	smallest and largest observed value
Σ	summation
D or d	difference
P	probability
*	denotes a (probably) significant difference; $0.05 \geqq P > 0.01$
**	denotes a significant difference; $0.01 \geqq P > 0.001$
***	denotes a (highly) significant difference: $P \leqq 0.001$

Examples

V_A	volume of alveolar gas
$\dot{V}_E$	expiratory gas volume $\cdot$ min^{-1}
$\dot{V}_{O_2}$	volume of oxygen $\cdot$ min^{-1} (oxygen uptake $\cdot$ min^{-1})
V_T	tidal volume
P_A	alveolar gas pressure
P_B	barometric pressure
FI_{O_2}	fractional concentration of O_2 in inspired gas
PA_{O_2}	alveolar oxygen pressure
pH$_a$	arterial pH
Ca_{O_2}	oxygen content in arterial blood
$Ca_{O_2} - C\bar{v}_{O_2}$	difference in oxygen content between arterial and mixed venous blood (often written a-$\bar{v}O_2$ diff.)
T_r	rectal temperature
$T_{\bar{s}}$	mean skin temperature

GLOSSARY

A number of key words are listed in the index, with reference to the appropriate sections in the text. In the following, definitions and explanations of some of the more common terms are listed.

A A prefix signifying without or not, as in Alactacid = not pertaining to lactic acid.

acceleration Increase in speed; to cause to move faster. (Basic S. I. units: m · s^{-2}).

acclimatization = acclimation The process of becoming accustomed to a new climate or new conditions.

acidaemia An increase in the hydrogen ion concentration in the blood above the normal value.

acidosis A situation in which the acid-base balance in the arterial blood shifts to the acid state (pH < 7.40).

adrenaline See epinephrine.

aerobic In the presence of oxygen.

aerobic power Used synonymously with the oxygen uptake per unit time.

afferent Conveying toward a center, centripetal.

alkalaemia A decrease in the hydrogen ion concentration in the blood below the normal value.

alkalosis A situation in which the acid-base balance in the arterial blood, shifts to the alkaline, or basic side (pH > 7.40).

amphetamine A synthetic drug that stimulates the central nervous system.

anabolic Constructive, building up, especially body protein from amino acids.

anabolic steroids Synthetic drugs related to the male hormone testosterone, in which the anabolic activity is enhanced at the expense of androgenic activity (for development and persistence of sexual characters).

anaerobic In the absence of oxygen.

androgenic Producing masculine characteristics.

anthropometry The study of body measurements.

apnea Cessation of breathing impulse.

735

apneustic Pertaining to apneusis, a condition marked by maintained inspiratory activity.

arteriovenous oxygen difference is usually expressing the difference in oxygen content between the blood entering and that leaving the pulmonary capillaries (a-$\bar{v}O_2$ diff.). Usually the oxygen content of the mixed venous blood ($C\bar{v}_{O2}$) is determined in blood withdrawn from a long thin tube (catheter) introduced into a cubital vein and then passed through the right atrium and ventricle into the pulmonary artery. The arterial oxygen content (Ca_{O_2}) is analyzed in blood samples taken from a systemic artery, usually the femoral, brachial, or radial artery.

atrophy Reduction in size and/or mass of cells and tissues.

autonomic Self-controlling; functionally independent.

baroreceptors Receptors sensing changes in pressure.

bends Pain occurring as a result of rapid reduction in air pressure (Decompression sickness).

biopsy The extraction of small pieces of tissues for chemical and/or histological analyses and studies.

bradycardia Very slow heart rate.

British thermal unit (BTU) The quantity of heat required to raise the temperature of one pound of water one degree Fahrenheit at or near its point of maximum density ($39.1°F$) = 0.252 kcal.

BTPS Body—Temperature—Pressure—Saturated.

calorie The unit of heat energy required to increase the temperature of 1 g of water 1°C (from 15 to 16°C). In the SI system, calorie is replaced by the unit joule: 1 cal = 4.186 joule.

capacity is often used to express the potential to do work, aerobically and/or anaerobically.

capacitance vessels Vessels capable of holding or storing a volume of blood.

cardiac output is the volume of blood ejected into the main artery by *each* ventricle, usually expressed as liters per minute ($\dot{Q}$). With small fluctuations, the cardiac outputs of the right and left ventricles are identical. The cardiac output divided by the estimated body surface area gives the "cardiac index," which relates the cardiac output to the body size.

catabolism The breaking down of complex substances into more simple compounds.

catalysis Change in the velocity of a chemical reaction produced by the presence of a substance which does not form part of the final product.

catecholamines The class of chemicals which includes epinephrine and norepinephrine.

central nervous system The denomination for the brain and spinal cord.

chemoreceptors Receptors sensing changes in their chemical environment.

ciliated Provided with cilia or a fringe of hairs.

circuit The course traversed by an electrical current.

concentric (con = together, centrum = center). Having a common center, towards the center.

concentric contraction A contraction of a muscle reducing its length.

conduction (thermal) The transfer of energy between objects of different temperatures in direct contact with each other.

convection (thermal) The transfer of energy by a heated particle (liquid or gas).

dehydration Excessive loss of body fluid.

diffusion The net movement of solute which occurs when a membrane or boundary separates fluids differing in their concentration of solutes.

dysmenorrhea Painful menstruation.

dyspnea Labored breathing.

eccentric Away from a center.

eccentric contraction A contraction of a muscle which is too weak to overcome the resistance imposed, so that the length of the muscle increases.

efferent Conveying away from a center.

electrolyte   A substance that ionizes in solution, and is capable of conducting an electrical current.

embolus   Blood clot.

endergonic   Accompanied by the absorption of energy.

endocrine gland   A gland that produces and releases hormones directly into the blood.

engram   A memorized motor pattern stored in the brain.

enzyme   An organic catalyst that accelerates the velocity of specific chemical reactions.

epinephrine   A chemical transmitter substance released from the adrenal medulla and from sympathetic nerve endings. Also called adrenalin.

ergogenic   Tending to increase work output.

ergometer   An instrument to measure work and power.

excitation   Response to a stimulus.

exergonic   Accompanied by the release of energy.

extrafusal fiber   An ordinary skeletal muscle cell.

fatigue   The inability to maintain a given force or power, due to one or a combination of several factors.

fibrillation   Spontaneous contraction of individual muscle fibers no longer under the control of a motor nerve.

Fick principle   Oxygen uptake $(\dot{V}_{O_2})$ = cardiac output $(\dot{Q})$ × arteriovenous O_2 difference $(a - \bar{v}\ O_2\ \text{diff.})$ or $\dot{V}_{O_2}$ = SV × HR × $(Ca_{O_2} \times C_{\bar{v}_{O_2}})$ (SV = stroke volume; HR = heart rate; C = content)

fulcrum   The axis of rotation for a lever.

functional residual capacity (FRC)   = the volume of air left in the lungs when the respiratory muscles are relaxed.

gamma motoneuron   Efferent nerve cell which innervates the ends of an intrafusal muscle fiber.

gluconeogenesis   The metabolic process by which glucose (and glycogen) is formed from non-carbohydrate precursors, which include lactate, pyruvate, glycerol, and amino acids. (Neos = new; genesis = formation).

glycogen   A polymer of glucose and the storage compound of carbohydrate in the body, predominantly in the liver and skeletal muscles.

glycogenolysis   The chemical breakdown of glucose to pyruvate and lactate (an anaerobic catabolism; see glycolysis).

glycolysis   The chemical breakdown of glycogen to pyruvate and lactate. The term is also used when glycogen is the substrate for the breakdown.

heart rate (HR)   is the number of ventricular beats per minute as counted from records of the electrocardiogram or blood pressure curves. The heart rate can also easily be determined by auscultation with a stethoscope or by palpation over the heart, both during rest and exercise.

homeostasis   A tendency to uniformity or stability in the normal body states (internal environment or fluid matrix) of the organism.

hyperglycemia   Concentration of glucose in the blood above the normal limit.

hyperplasia   Increase in the number of cells.

hyperpnea   Marked increase in pulmonary ventilation.

hypertrophy   Increase in the size or mass of body tissue.

hyperventilation   A pulmonary ventilation in excess of levels necessary to secure an adequate supply of oxygen to the lungs and elimination of the carbon dioxide produced by tissue metabolism.

hypoglycemia   Glucose concentration in the blood below the normal limit.

hypothermia   Abnormally low body temperature.

hypoventilation Reduced ventilation of the lungs.

hypoxia Low oxygen tension in the inspired air.

intrafusal Pertaining to the striated fibers within a muscle spindle.

isokinetic muscle contraction Contraction with the speed of movement maintained constant.

isometric muscle contraction Contraction in which tension is developed, but there is no change in the length of the muscle (static muscle contraction).

isotonic Having the same tension or concentration.

kinetic Pertaining to or producing motion.

menarche The onset of menstruation.

mole The gram-molecular weight or gram-formula weight of a substance. For example, one mole of glucose, $C_6H_{12}O_6$ weighs $(6 \times 12) + (12 \times 1) + (16 \times 6) = 72 + 12 + 96 = 180$ grams, where the atomic weight of carbon (C) $= 12$; hydrogen (H) $= 1$; and oxygen (O) $= 16$.

moment (moment arm) The perpendicular distance from the line of action of the force to the point of rotation.

motor unit An individual motor nerve and all the muscle fibers it innervates.

notochord (chorda dorsalis): The rod-shaped body below the primitive groove of the embryo, defining the primitive axis of the body.

orthostatic Pertaining to or caused by standing erect.

oxygen uptake is the volume of oxygen (at $0°C$, 760 mm Hg (101.3 kPa), dry = STPD = standard temperature and pressure, dry) extracted from the inspired air, usually expressed as liters per minute ($\dot{V}_{O_2}$). If the oxygen content of the body remains constant during the period of determination, the oxygen uptake equals the volume of oxygen utilized in the metabolic oxidation of foodstuffs. One liter of oxygen, then, corresponds to 19.7 to 21.1 kJ (4.7 to 5.05 kcal) of energy liberation. One liter at STPD $= 44.6$ mmol (one mol $= 22.414$ liters).

placebo An inert substance having the identical physical characteristics of a real drug.

polycythemia An increased number of red blood cells.

power is the rate of doing work; the rate of transfer of energy. It is defined in watts (W). 1 watt $= 1$ joule per second ($1 W = 1 J \cdot s^{-1}$).

proprioceptor Sensory organs found in muscles, joints and tendons, which give information concerning movements and position of the body.

pulse rate is the frequency of pressure waves (waves per minute) propagated along the peripheral arteries, such as the carotid or radial arteries. In normal, healthy individuals, pulse rate and heart rate are identical, but this is not necessarily so in patients with arrhythmias. In such cases, the output of blood by some beats may be too small to give rise to a detectable pulse wave.

radiation The transfer of heat between objects through electromagnetic waves.

receptor A sense organ that receives stimuli.

reciprocal Alternating; going backward and forward; mutual.

residual volume (RV) $=$ the volume of air left in the lungs after a forced maximal expiration.

respiratory exchange ratio (R, RQ) The ratio of the amount of carbon dioxide produced to the amount of oxygen consumed ($\dot{V}_{CO_2}/\dot{V}_{O_2}$).

saline A 0.9 per cent salt solution which is isotonic to the blood (in mammals).

second wind A phenomenon characterized by a sudden transition from an ill-defined feeling of distress or fatigue during the early portion of prolonged exercise to a more comfortable, less stressful feeling later in the exercise.

soma The body as distinguished from the mind.

somatic Pertaining to the body.

specific heat The heat required to change the temperature of a unit mass of a substance by one degree.

static Not in motion, not dynamic.

STPD Standard temperature, pressure, dry.

stroke volume (SV) is the volume of blood ejected into the main artery by each ventricular beat. The stroke volume is usually calculated by dividing the cardiac output by the heart rate ($\dot{Q}$/HR).

sudomotor Pertaining to activation of the sweat glands.

sympathetic Pertaining to the thoracolumbar portion of the autonomic nervous system

syncope Fainting. A temporary suspension of consciousness due to cerebral anemia.

tachycardia An increased or rapid heart rate.

temporal summation An increase in responsiveness of a nerve, resulting from the additive effect of frequently occurring stimuli.

testosterone The male sex hormone secreted by the testicles; it possesses masculinizing properties.

thermodynamics The science of the transformation of heat and energy.

thyroxin A hormone secreted by the thyroid gland.

tidal volume (V_T) = the volume of gas moved during each respiratory cycle.

tonus Resiliency and resistance to stretch in a relaxed resting muscle.

total lung capacity (TLC) = vital capacity + residual volume. The *diffusing capacity of the lung* (D_L) is defined as the number of milliliters of a gas at STPD (Standard Temperature, 0°C, Pressure 760 mm Hg (101.3 kPa), Dry) diffusing across the pulmonary membrane per minute and per millimeter of mercury (per 0.133 kPa) of partial pressure difference between the alveolar air and the pulmonary capillary blood. It should be emphasized that the diffusion path includes the blood.

trophic Pertaining to nutrition or nourishment.

Valsalva's maneuver Making an expiratory effort with the glottis closed.

vasoconstriction A decrease in the diameter of a blood vessel (usually an arteriole) resulting in a reduction of blood flow to the area supplied by the vessel.

vasodilation An increase in the diameter of a blood vessel (usually an arteriole) resulting in an increased blood flow to the area supplied by a vessel.

vasomotor Pertaining to vasoconstriction and vasodilation.

vital capacity (VC) = the maximum volume of gas that can be expelled from the lungs following a maximal inspiration.

WBGT (Wet Bulb Globe Temperature) index An index calculated from dry bulb, wet bulb, and black bulb temperatures. It indicates the severity of the environmental heat conditions.

work Application of a force through a distance; it is measureed in joules (J, kJ, MJ).

work rate Work performed per unit time-power.

INDEX

A band, 20, 22–23
Abbreviations, 726
Absolute refractory period, 63
Acceleration, 395
Acclimatization:
    to cold, 607–610
    to heat, 605–608, 621, 625
    to high altitude, 695–705
Acetylcholine (ACh), 16, 69, 100
Acetyl-CoA, 531
Acid-base balance, 134
Acidic phosphate, formation and excretion of, 139
Acidosis, 139
Actin, 21, 23, 26–27
Actin filaments, 28
Actin, structure of, 24
Actin-myosin interaction, 28
Action potential, 29, 62, 65–66, 69, 100, 101
Actomyosin, 23
Adaptation:
    to high altitude, 695–705
    to local cold exposure, 610
    to temperature, 588
Adenosine diphosphate (ADP), 528
Adenosine monophosphate (AMP), 528
    cyclic, 16
Adenosine triphosphate (ATP), 2, 27, 30–33, 303, 314, 324, 527–535
Adenyl cyclase, 16
Adipose tissue, 129, 541, 572
ADP (see adenosine diphosphate)
Aerobic energy output, measurement of, 298 (see also Oxygen uptake)

Aerobic energy yield, 2
Aerobic metabolic rate, 534
Aerobic mitochondrial oxidation, 532
Aerobic oxidation, 527
Aerobic power, 330 (see also Oxygen uptake)
    evaluation of, 384
    of cyclists, 654
    of elite runners, 653
Aerobic power, maximal, 330–342
    absolute values, 332–336
    age and sex, 330–336, 341–342, 401, 402
    evaluation of test results, 382
    in relation to body size, 336–339
    performance and, 338–341, 414, 464
    prediction of, 372
    termination of test, 381–382
    test procedure, 355–383
    tests in patients, 378
    training and, 334, 439
Aerobic processes, 299, 527
Afferent (receptor) neuron, 73
Afferent tracts, 88
After-hyperpolarization, 62
Afterdischarge, 75
Age: (chronological and biological), 330
    aerobic power, maximal, 401
    blood pressure, 197
    body dimensions, 408–410
    body proportions, 398
    cardiac output, 197, 402
    circulatory dimensions, 409
    correlation factor, 376
    effect on heat tolerance, 617

Age: (chronological and biological) (*Cont.*)
  heart rate, 189, 197, 409
  heart size, 198, 409
  lung volumes, 224
  maximal heart rate, 373
  maximal oxygen uptake, 332–337, 409
  muscle strength, 342, 410
  respiratory function, 224
  stroke volume, 197, 409
  total muscle mass, 333, 410
  training effect, 422
Agonist, 109
Air movement, 588, 628
Air sacs, 212
Air temperature, 587
Airways: 211
  "air conditioning", 218
  anatomical structures, 212–218
  compliance, 227
  general architecture of, 212–213
  resistance, 227
Alactic capacity, maximal, 316
Alactic power, 315
Alanine, 543, 550
Alkalosis, 139, 692, 697
Alcohol, effect on:
  aerobic and anaerobic power, 711–712
  exercise, 711
  neuromuscular function, 711
Alpha-receptors, 162
All-or-none response, 63
Alpine skiing, 666
  aerodynamic forces, 667
Altitude:
  adaptation to, 695
  air temperature, 685
  athletic competition, 687, 689
  blood lactate levels, 689
  cardiac output at, 692, 699–700
  cellular adaptation to, 702
  effect on Hb dissociation curve, 692
  effect on maximal oxygen uptake, 693, 695
  effect on physical performance, 685–688, 705
  energy cost of breathing, 697
  heart rate, 700
  hemoglobin concentration, 700
  hormonal response, 702
  limiting factors, 688
  myoglobin content, 702
  oxygen saturation, 692
  oxygen transport, 690
  partial pressure of O_2, 684
  physical aspects, 685–686
  pulmonary hypertension, 702
  pulmonary ventilation at, 691, 696, 698
  reduced alkaline reserve, 698
  solar radiation, 685
  stress, 689
Alveolar air, 242
Alveolar capillaries, 217
Alveolar ventilation, 233

Alveoli, 212
  microscopic structure of, 214–215
Ambient air temperature, 586
Amino acids, 542
Ammonia, 1
AMP (see Adenosine monophosphate)
Amphetamine, 713
Anabolic steroids, 434, 714
Anabolism, 527
Anaerobic energy yield, 10, 434, 527, 534
Anaerobic fermentation, 1
Anaerobic power:
  assessment of, 315, 317–319
  maximal, 315
  training, 434
Anaerobic processes, 314–330
Anaerobic threshold, 327–330, 383
Anaerobic glycogen breakdown, 316
Antagonist, 109
Anterior hypothalamus, 602
Anthropoid apes, 5
Antidiuretic hormone, 620
Antigravity muscles, 112
Aorta, pressure in, 143
Arm exercise, 192, 195, 312, 356, 496
Arterial blood pressure (see also Blood pressure):
  measurement of, 145
  vasomotor tone, 159
Arteries, 149
Arterioles, 149, 151–152
Arteriovenous anastomoses, 153
Arterio-venous oxygen difference, 179, 454
Athletic achievements, sex differences, 678–679
Athletic events, analysis of, 414–415, 647–678
Atmosphere, 1, 729
ATP, 2, 27, 30, 32–33
ATP-ADP system, 2
ATPase, 30
ATPase activity, 101
ATP Formation, 532
ATP hydrolysis, 32
Australopithecus, 6–7
Autonomic nervous system, 57
Axis cylinder, 70
Axolemma, 57
Axon, 56, 58, 70
Axon reflex, 168
Axonal terminals, 57
Axoplasm, 72

Backache, 286–288
Badminton, 676
Ball games, 674–678
Baroreceptors, 163
Basal ganglia, 91, 93, 95
Basal metabolic rate (BMR), 399, 567
Basic units:
  length, 725
  mass, 725

Basic units (*Cont.*)
  quantity, 725
  time, 725
Basket cell, 89
Bed rest, prolonged, effect of, 278, 438
Behavior, 73
Bends, 263, 707
Beta blockade, 190, 453
Beta receptors, 162, 190–191, 381, 453
Bicarbonate, 136, 325, 697
Bicycling, 654
  energy cost of, 654
Bile, 542
Biological clocks, 311, 515
Blood: 127–139
  buffer action of, 133–139
  buffer capacity of, 137
  clotting of, 132
  CO_2 transport of, 133–139
  coagulation, 132
  effective viscosity of, 133
  iso-electric point of, 134
  O_2 content of, 181
  O_2 transport of, 131
  oxygen pressure in, 180, 243–246, 247, 690
  pH of, 133–139
  specific heat of, 132
  transportation of materials by, 130
  viscosity of, 132–133
Blood cells, 131
Blood doping, 183–185, 600, 705, 715
Blood flow: 140–143, 147
  effect of muscle contraction, 177
  through heart muscle, 141
Blood glucose concentration, effect of exercise, 549
Blood lactate concentration, time course of, 320–321
Blood lactate, in relation to O_2 uptake, 300
Blood pH, metabolic effect on, 324–325
Blood plasma (see Plasma)
Blood pressure, 191–194
  effect of exercise, 193
Blood reinfusion, effect of, 183–185, 600, 705, 715
Blood vessels, 149
  classification of, 149
  pulmonary, 216–217
Blood volume: 130, 187
  determination of, 130
  effect of climate, 606
  effect of dehydration, 130, 621
  effect of training, 130, 417
  sex differences, 130
  variation in, 130
Body dimensions, 391–410, 466
Body fluids, 127–139
Body temperature, 585
  and oxygen uptake, 596–597
  limits of tolerance, 584, 613–618
Body weight, "ideal", 570
Bohr effect, 179

Bolus, 541
Bomb calorimeter, 529
Bone:
  absorption of, 275
  adaptation, 281
  calcium content, 276
  demineralization, 277
  effect of activity, inactivity, 277
  effect of bed rest, 278
  effect of pressure, 277
  effect of training, 277
  evolution of, 274
  formation of, 275
  function of, 275–279
  growth, 279
  main functions of, 271–274
  maintenance of, 275
  maximal metabolism of, 277
  remodeling, 280
  repair, 280
  structure of, 274–275
Bone matrix, 275
Bony levers, 42
Botsball thermometer, 614
Bradycardia, 161, 167
Bradykinin, 172
Brain, 9, 56, 68, 83–97
Brain stem, 88
Brain temperature, 586
Breath holding-diving, 264
Breathing:
  at rest, 249
  central chemoreceptors, 251
  during exercise, 255
  hypoxic drive, 259, 691–694
  neurogenic and chemical stimuli, 257
  peripheral chemoreceptors, 252
  regulation of, 249–261
  role of muscle spindles, 256
  mechanisms of, 219
  resistance to, 221–222
Breathlessness (dyspnea), 262
Bronchi, 216
Bronchial tree: nervous supply, 217
Bronchioles, 212, 216
Brown fat, 533, 590
Bruce's treadmill test, 362–363
BTU, 729
Buffer, 133, 137, 166
Buffer capacity of the blood, 137

CNS, 55–56, 70–83, 95–100, 111, 118
  metabolic rate of, 557
  CO_2 transport, 135–136
Caffeine, 713
  effect on circulation, 713
  effect on performance, 713
  glycogen sparing effect, 561, 713
Calcium ions (Ca^{2+}), 2, 23, 29, 30, 61, 69, 120, 532

Calcium ions, (Ca^{2+}) (*Cont.*)
  diffusion, 67
  influx, 67
Calcium receptor, 2
Calcium transport, mechanism of, 30
Callus, 280
Calmodulin, 2
Canadian home fitness test, 380
Canoeing, 667–670
  blood lactate levels, 669
  energy cost, 668
  physiological demands, 667–670
Capacitance vessels, 150, 156, 170
Capacity, definition of, 297
Capillaries, 151–154
  colloidosmotic pressure in, 132, 150
  fluid exchange in, 155
  hydrogen exchange in, 134–135
  osmotic pressure in, 132
Capillary bed, 150
Capillary density, 141, 157, 216, 430, 451, 458, 702
Capillary membrane, 153
  transport mechanism, 154
Capillary structure, 153
Capillary wall, ionic exchange, 130
  passage through, 130
Carbaminohemoglobin, 136
Carbohydrate, 541
Carbon dioxide, 1
Carbon dioxide tension:
  altitude, 687, 695–697, 708
  alveolar air, 238
  blood, 179, 238, 243–246
Carbon dioxide, narcotic effect of, 708
Carbon monoxide, 180–183, 709
Carbonic anhydrase, 244
Carboxyhemoglobin (HbCO), 180, 183, 247
Cardiac hypertrophy, 175
Cardiac muscle, 139, 405
Cardiac output, 174, 178, 194, 201
  afterload, 171
  altitude, 692, 699
  effect of exercise, 194–195
  effect of training, 198–201, 441
  maximal, in relation to body size, 399–402
  oxygen uptake and, 178–180, 195
  preload, 170
Cardiac patient, 175, 196, 471–474
Cardiovascular system, testing of, 378
Carotid sinus, 163, 193
Carrier protein, 16
Cartilage, 274
Catabolism, 130, 527
Catalyst, 527
Catecholamines, 162, 498, 516, 559, 591
Cell, 13
Cell body, 56
Cell membrane, 13–14, 17
  transport, 15

Cell-to-cell communication, 16
Cement (bone), 275
Centigrade, 732
Central chemoreceptors, 251
Central nervous system (CNS), 55–56, 70–83, 95–100, 111, 118
Central pattern generator, 95, 250
Central vasomotor integration, 164
Centrosome, 14
Cerebellar cortex, 89
Cerebellum, 86–87, 91–93, 95
Cerebral cortex, 83, 95
Cerebral motor cortex, 85
Cerebrum, 86, 88
Chemical potential, 525
Chemical transmitter substance, 58
Chemically gated, 59
Chemoreceptors, 55, 167, 249–255, 691, 697
Chin-ups, 394
Cholinesterase, 61, 69
Chromatin, 14
Chromosomes, 4
Chronotrophic action, 144
Chyme, 542
Ciliated cells, 219
Circadian rhythms, 515
Circuit training, 447
Circulation, 127–202
  altitude, 692, 699
  climatic effects, 605–609
  regulation of, 158–174
  training, 438–442, 449–456, 461
Circulatory strain, 496
Climate, effect on body temperature, 589–593
Climbing fiber, 89, 91
Clo unit, 632
Clothing, 630
  cold weather, 630–634
  Eskimo, 631
  requirement, 630–631
Coactivation, 77, 82, 85
Coated pit, 15
Coated vesicle, 15
Cohesive force, 147
Coinhibition, 85
Cold:
  combined with nutritional stress, 606
  effect on manual dexterity, 592
  exposure, effect on heart rate, 592
  injury, 610–612
  physical effects, 589–593
  stress, 608
Collateral, 58
Colloidosmotic pressure, 132, 150, 219
Compliance, 227
Concentration gradient, 13
Concentric muscle contraction, 40
Conductance (heat) of the tissue, 589
Conduction velocity, 70
Congestive heart failure, 178

Contractile properties, 33
Contraindications for physical training, 413
Control of body temperature, 585
Convective heat exchange, 584
Conversion tables, 727
Coordination, 49, 109
Coordinated movements, temperature effects, 618
Core, 585
Core temperature, 586, 591
Cori cycle, 536–537
Coronary artery flow, 141, 144
Cortisol, 558, 561, 702
Cramp, 122
Crank-length, 364
Creatinine, 410
Critical closing pressure, 148
Critical Reynolds number, 147
Cross-bridges, 24, 28, 32
Cross-country skiing, 661
Cycle ergometer, 46, 355, 359, 363–366
Cycle ergometer test, 363–366
    procedure, 363–366
    submaximal, 369
Cyclic AMP (3′,5′ -adenosine monophosphate),
    16, 186
Cytoplasm, 14, 17

Dallas study, 438
Darwinism, 4
Dead space, 232
    estimation of, 232–233
Decarboxylation, 532
Deep body temperature, 585–586
Deep sea diving, 263
Definition of terms and units, 725
Dehydration, 372, 565, 621
    effect on performance, 622–624
Dendrites, 56, 58
Depolarization, 16, 60, 63, 65–66
Detraining, 463
Diaphysis, 279
Diastole, 142–143
Diastolic filling, 143
Diet and exercise, 544–548, 552
Diffusion, 2, 3, 13, 59, 153, 240
Diffusing capacity of the lung, 244, 692, 699
Diffusion-Filtration, 156
Digestion, 541–542
    of carbohydrates, 542
    of fat, 542
    of protein, 542
Dimensions, 392, 406
    physical, 392, 406
    physiological, 393, 406
    secular increase in, 407
Dinosaurs, 4
Diphospho-glycerate (2-3 DPG), 247, 452, 703
Disaccharides, 542
Disinhibition, 67, 107, 431

Dissociation curve, 180
Disuse, effect on skeletal muscle, 430
Diving:
    effect of breath holding, 264–267
    effect of hyperventilation, 265–267
    snorkel, 706
Doping, 713–715
    effect on physical performance, 713
Double membrane, 14
Douglas bag, 228, 491
Douglas bag method, 298
Dry heat transfer, 633
Dynamic:
    exercise, 40, 433
    gamma axons, 78
    spirometry, 226
Dyspnea, 262

ECG (electrocardiogram), stress tests in patients,
    378
EMG (electromyogram), 81, 112–113, 118
Eardrum temperature, 586
Eccentric muscle contraction, 40, 433
Edema, 155, 166
Efferent (effector) neuron, 73
Efferent tracts, 88
Elastic element, 32, 649
Elastic energy, 46, 48
Electric transmission, 69
Electrocardiogram, 142, 386, 472
Electrochemical gradients, 59
Electromyogram (EMG), 43, 109, 497, 647
Electron carrier, 531
Electron transport chain, 19
Electroneurogram, 78
Endergonic reactions, 525
Endocrine system, 68, 467, 498, 557–562
Endocytosis, 16–17
Endomysium, 19
Endoplasmic reticulum, 17
Endorphins, 68, 467
Endplates, 20, 72
Endurance, 421, 430, 457, 462
    effect of blood glucose, 549
    effect of glucose ingestion, 550
    effect of muscle glycogen, 552
    effect of training, 442, 461–462
Energy, (conversion table), 728, 731
    kinetic, 398
Energy expenditure:
    assessment of, 491
    classification of, 491, 501
        maximal permissible limits, 504
    daily rates, 502–506
    different activities, 505–512
    effect of tools used, 510
    indirect assessment of, 491–497
        bicycling, 654
        building work, 510

Energy expenditure, indirect assessment of (*Cont.*)
  carrying a load, 509
  canoeing, 668
  climbing stairs, 510
  coal mining, 511
  commercial fishing, 492
  cycle ergometer exercise, 366, 491
  farming, 511
  fishing, 511
  housework, 507–508
  lifting a load, 509
  light industry, 507
  lumber work, 511
  manual labor, 509
  military activities, 512
  playing children, 512
  recreational activities, 511
  relation to maximal aerobic power, 501
  running, 46–48, 651–654
  sedentary work, 507
  skiing, 661–662, 666–667
  sleeping, 506
  swimming, 656
  treadmill running, 362
  walking, 649, 652
  walking versus running, 652
  washing clothes, 487, 501
Energy liberation, 524
  diagram, 530
Energy metabolism, 543–557
  regulation of, 557–562
Energy stores, 535–536
Energy supply, in relation to body size, 398
Energy transformation, 524
Energy yield, 534, 647
  aerobic and anaerobic, 325–327, 534
  in athletic events, 647
Energy-yielding fuels, 529
Environment, 128
Environmental temperature, preferred, 634
Enzymes, 14, 18
Epinephrine, 144, 452, 467, 558
Epimysium, 19
Epiphyses, 279
Erg, 728
Erythrocytes, 131
Erythrogenin, 131
Erythropoiesis, 131
Erythropoietin, 131
Esophagal temperature, 586
Eukaryote, 2
Evaporative heat loss, 584, 587, 632
Evolution of life, 1
Excitability, 73
Excitation, 29, 65, 74
Excitatory neurons, 66
Excitatory postsynaptic potential (EPSP), 63, 65
Exercise:
  cardiovascular responses to, 201
  circulatory response to, 173

Exercise (*Cont.*)
  continuous vs. intermittent, 305
  duration, 299
  effect of alcohol, 711
  effect on blood lactate, 316–324, 327–330,
    436, 556, 688
  effect on plasma FFA, 556
  electrocardiogram, 386
  intensity, 299
  intermittent, 304, 423–427
  muscular mass involved in, 311
  prolonged, 308–310
  prolonged, effect of, 121
  recovery, 303
  with arms, 312–313
  with arms + legs, 312–313, 316
Exergonic reactions, 525
Expiratory reserve volume, 223
Exteroceptors, 55
Extra- and intracellular fluid, 13, 129
Extracellular fluid, 129
  composition of, 130
Extrafusal muscle fibers, 76, 78
Extrapyramidal centers, 97
Extrapyramidal tracts, 91
Extrastitial fluid, 129

FFA (see Free fatty acids)
Facilitation, 63, 66
Facilitation-Excitation, 61
Fahrenheit, 732
Fainting, 177
Fast-twitch fiber, 33–34, 101, 104
Fat, 542
Fat utilization:
  effect of diet, 545–548
  effect of exercise, 544
  effect of training, 532, 545–546
Fat-free body weight (lean body mass), 129
Fatigue, 115, 119, 310, 512
  general, 512
  in joints, 122
  in prolonged exercise, 118, 308
  in static exercise, 115
  local, 356, 513
  muscular, 115
  objective symptoms, 513–514
  physical, 115
  respiratory muscles, 239
Fatty acids, (see Free fatty acids)
Fiber sprouting, 110, 469
Fiber typing, 33
Filaments, 21, 23, 32
Filtration, 153–155
Finnish sauna bath, 594
  effect on blood pressure, 594
  effect on cardiac output, 594
  effect on heart rate, 594
  effect on mean circulation time, 594

Flavin-adenine-dinucleotide (FAD), 533
Flow and resistance, 147
Flow meters, 229, 491
Fluid balance, 132, 566
Food for athletes, 562–566
Food intake:
  and body weight, 566–568
  and physical activity, 562, 566, 568
  regulation of, 569
Foot-pound, 729
Football (soccer), 674–676
Force (definition), 726
  dependence on dimensions, 393
  maximal, 102
Force-platform, 46, 315, 649
Force-velocity, 43–44
Force-length diagram, 41
Forced expiratory volume (FEV), 226
  determination of, 226
Free fatty acids (FFA), 542–548, 555–562
  metabolism:
    effect of diet, 545–548
    effect of exercise, 544
    effect of training, 458–461
    plasma levels and utilization, 559
Frequency code, 67
Frostbite, 611
Fructose, 542
Fuel utilization:
  effect of diet, 545
  effect of exercise intensity, 544–545, 556
  effect of fasting, 550
  effect of prolonged exercise, 547
  regulation of, 557–562
Fuel:
  for muscular exercise, 543
  selection of, 543
Functional residual capacity (FRC):
  determination of 223–224
Fåhraeus-Lindquist effect, 133

Galactose, 542
Gamma motor system, 76
  efferents, 78
  fibers, 76–77
Gas-dilution method, 223
Gas tension:
  in alveolar air, 241
  in blood, 241
  in expired air, 241
  in inspired air, 241
Gastric juice, 541
Generator potential, 63
Genetic material, 17
Glial cells, 68
Globin, 131
Glucagon, 550
Gluconeogenesis, 550
Glucose, 1, 467, 526, 542

Glucogenolysis, 532
Glucolysis, 1, 533
Gluconeogenesis, 533
  effect of glucagon, 550
Glucose-6-phosphatase, 537
Glycerides, 542
Glycerol, 542
Glycogen breakdown, 317, 531
  anaerobic, 316
Glycogen content, human muscle, 317, 446, 461,
  551, 555
Glycogen formation, 537, 556
Glycogen oxidation, 537
Golgi tendon Organ, 78–80
  function of, 79
Golgi apparatus, 14
Golgi cell, 89
Granule, 14
Granule cell, 89
Growth in height, 331
Growth hormone, 558, 561

Habituation, 99, 591
H-acceptor, 531
H zone, 20–23
Hb concentration, determination of,
  131–132
Heart, 139–144
  efficiency of, 174
  external work of, 145
  innervation of, 144
  mechanical efficiency of, 144–145
  pressure variation in, 142–143
Heart disease, ischemic, 471–474
Heart muscle, 19, 139–140
  capillary density, 141
  contraction, 140
  hypertrophy, 175
  mechanical work of, 144
  structure of, 139–140
Heart rate, 188–191, 369, 452
  age, 197
  altitude, 693, 699
  dimensions and, 404
  effect of β-blockers, 190
  heat exposure, 594, 605–606
  maximal, decline with age, 189
  nervous control of, 161
  recording in the field, 493–494
  reserve, 496
  tobacco smoking, 709
  training, 369, 417, 440
Heart volume, 195–197, 449
Heart weight, 404–405, 450
Heat, failure to tolerate, 612–613
Heat balance, 584–585
  assessment of, 585
Heat content of the body, 584–585, 587
  assessment of, 587

Heat exchange, 589
  at rest, 595
  conductive, 588
  convective, 588
  during exercise, 595, 598
  equation, 632
  evaporative, 588
  radiant, 588
Heat exposure:
  effect on body fluids, 621–625
  effect on cardiac output, 594
  effect on heart rate, 594
  effect on peripheral resistance, 594
  effect on skin temperature, 594
  effect on splanchnic blood flow, 594
  industrial, 616
  physiological effects, 593–595
Heat stress:
  index, 615
  practical solutions, 625–626
  response to, 612, 626
Heat stroke, 612
Heat tolerance:
  effect of age, 617
  effect of sex, 617
  effect of training, 618
Height:
  of athletes, 407
  maximal running speed and, 397–398
  muscular strength and, 393
  velocity, 330
Helium, 263
Hematin, 131
Hematocrit, 131, 181, 183
Hemodynamics, 144
  of circulation, 148
Hemoglobin, 131, 700
  affinity for CO, 131
  affinity for O_2, 131
  amount of, 131, 368, 450, 565
  determination of, 131
Hering-Breuer reflex, 250
Hexokinase, 549
High air pressures, 263
High altitude, 247–248, 684
High gas pressure, effect of, 706–709
High-density lipoprotein (HDL), 460, 466, 474
High-energy phosphates, 527
  breakdown, 315
  compounds, 537
Histochemical differentiation, 33
Homeothermy, 604
Hominid, 5
Homo erectus, 6
Homo habilis, 6
Homo sapiens neanderthalensis, 7
Homo sapiens, 17
Homo sapiens sapiens, 4–7
Horsepower, 728
Human skeletal muscle fibers:
  classification of, 36

Humidity of the air, 587
Hunger, 569
Hydrogen acceptor, 2, 60, 62, 64–66
Hydrops, 292
Hydrostatic pressure, 145
Hyperbaric pressure, 706, 708
Hypercapnia, 251
Hyperglycemia, 547, 549, 569
Hyperpnea, 691
Hyperpolarization, 64
Hyperthermia, 595
Hyperventilation, 236, 265
Hypocapnia, 254, 692
Hypoglycemia, 547, 549, 569
Hypohydration, 598, 620, 687
  due to heat, 598
  effect on endurance, 622–624
  effect on heart rate, 622
  effect on performance, 622–624
  effect on rectal temperature, 623
  effect on tilt-table tolerance, 623
  physiological effects, 598–599
Hypopolarized, 63
Hypothalamus, 92, 163, 602
Hypothermia, 591, 612
Hypoxemia, 264
Hypoxia, 183, 252, 689
Hypoxic drive during exercise, 259

I band, 20, 22
Ice Hockey, 677–678
Immobilization, effect on skeletal muscle, 430
Impulse, 57
Impulse traffic, 74, 98
Infrared thermography, 586
Inhibition, 57, 64–66
Inhibitory:
  interneuron, 64, 66, 76
  synaptic response, 65
Innervation, 100
  of the heart, 144
Insensible perspiration, 588
Inspiratory capacity, 223
Insulating requirements in the cold, 631
Insulating value, 631
  assessment of, 633–634
  of clothing, 633
  of different furs, 632
Insulin, 16, 558–559, 702
Integrated EMG, 45
Integrated electrical activity, 44
Integrating motor pneumotachygraph, 492
Intermittent exercise, 304, 423
  effect of rest periods, 305–306, 423
Interneuron, 57, 64
Interoceptors, 55
Interstitial fluid, composition of, 171
Intra-abdominal pressure, 290
Intracellular fluid, 129

Intracellular pH, 137
Intracerebellar nuclear cell, 89
Intradiscal pressure, 292
Intrafusal muscle fibers, 76, 78
Intravascular fluid, 129
Ionic flux, 65
Ionic permeability, 62
Iron, 16
Ischemic heart disease, 471
Isokinetic exercise, 429
Isokinetic muscle contraction, 43, 192, 433
Isometric contraction, 40, 101, 433
Isotropic, 22

Joint receptors, 80
Joint stability, 292
Joints, 281
  blood supply, 282
  cartilaginous, 282
  exercise, 282–284
  fibrous, 281
  innervation, 285–286
  movements, 282–284
  nerve endings in, 285–286
  nourishment of, 282
  synovia, 282
  synovial, 282
  synovial fluid, 282
  synovial membrane, 282
Joule, 727–728
Jumping, 396

K^+, 70
K^+ channels, 62
Katharometer, 223
Ketonbodies, 555
Kilocalorie, 728
Kilogram, 725
Kilogram-meter, 729
Kinematic analysis, 647
Kinetic energy, 109, 398
Krebs citric acid cycle, 531
Krebs cycle, 18, 530

Lactate (lactic acid, HLa), 10, 136–137, 320, 324, 660, 663, 665–673, 689
  concentration in blood and muscle, 318, 321
  fate of, 322–323
  free fatty acids and, 558–560
  removal of, 323
  threshold, 329
Lactate dehydrogenase (LDH), 322, 528
Lactic acid (HLa), 10, 136–137
Lactic oxygen debt, 303
Lactose, 542
Lacunas, 275
Laminar flow, 147
Laplace's law, 146

Lateral inhibition, 82
Lean body mass, 129, 336, 338, 570
Learning, 97
Leg exercise, 312
Length, 727
Leukocytes, 131
Lever arms, 43
Ligaments, 286
Ligament strength, 286
Limiting facts in exercise, 185, 236, 313
Lipase, 542
Lipoproteins, 460, 466, 474
Lipoproteins lipase, 460
Local fatigue, 356
Locomotion, 114, 395, 649
  energy cost of, 395
Locomotor movements, 95
  control of, 95
Low density proteins (LDL), 466
Low energy consumers, 574
Lumbago, 287
Lung:
  air condition, 218
  airway resistance, 221
  blood vessels, 216
  capacity, 222
  compliance, 227
  diffusion, 240
  gas exchange in, 240–244
  heat loss from, 588
  mechanics of breathing, 219
  nerve supply of, 217
  perfusion, 244
  regional gas tension, 242
  surface area, 211
Lung volumes:
  age and sex, 224–26
  dynamic volumes, 226–227
  tidal volume, 234
Lymph, 130
Lymphatic system, 157
Lysosomes, 15

M line, 22
Maltose, 542
Mammals, 4
Margaria staircase test, 384
Max Planck respirometer, 491
Maximal:
  aerobic power, 301
  anaerobic power, 315
  breathing capacity, 227
  force, 43
  oxygen power, assessment of, 302
  oxygen uptake, 301, 311, 312
    estimated vs measured, 377
  voluntary contraction (MVC), 115
  voluntary ventilation (MVV), 227
    determination of, 227
McArdle's syndrome, 329

Mean skin temperature, 586
  assessment of, 586
Mean transit time, 458
Mechanical efficiency (ME), 45–46, 49, 465, 590
Mechanoreceptors, 77, 165–166
Medullary peripheral chemoreceptors, 250
Medullary vasomotor area, 161
Membrane permeability, 59
Membrane potential, 62
Membrane proteins, 13
Memory, 97
Memory storage, 98
Mental performance:
  effect of temperature, 619
Meromyosin segments, 26
MET, 362
Metabolic rate, 587–590
Metaphysis, 279
Meter, 725
Mg.ATP, 33
Microcirculation, 152
Microclimate, 634–636
Micropauses, 307
Milieu interne, 128
Mitochondria, 3, 17–18, 72, 459, 527
Mobility, 10
Mole, 726
Monark cycle ergometer, 363
Monophosphate, (AMP), 528
Mossy fibers, 89, 91
Motivation, 103, 122
Motoneuron, 56–58, 61, 67, 69–70, 73–76, 82,
    85, 93, 100
  firing rate of, 104, 106
Motor cortex, 84, 95
Motor endplate, 16, 70–71, 100
Motor nerve, 101
Motor unit, 19, 70, 72, 100–102
  recruitment of, 104–105
Mountain sickness, 694
Movement, 73, 91–92
Multicellular organism, 4–5
Muscle, 10
  activation, 40
  blood flow, 117
  cell, 128
  fiber, 35, 100
  length, 40
  spindle, 76
  variation of metabolic rate, 128
Muscle contraction, 45
  concentric, 40
  dynamic, 40
  eccentric, 40
  effect on blood supply, 116
  isometric, 40
  isotonic, 40
  mechanical efficiency of, 45
  static, 40
Muscle engagement, assessed by EMG-activity,
    43, 109, 497–498

Muscle fatigue, 115, 137
  causes of, 120
Muscle fiber types:
  genetic aspects, 37
  individual variations of, 39
  metabolic profile, 37, 327
  training, 430, 461
  transformation of, 37
Muscle glycogen:
  effect of depletion, 554
  effect of exercise and diet, 553
  effect on endurance, 552
  effect on maintenance of running speed, 555
  stores, 551
Muscle pump, 156, 165, 176
Muscle spindle, 78, 87, 256
  role in regulation of breathing, 256–258
Muscle strength, 103–109, 342–346
  age and, 111, 342, 410
  CNS effect, 431
  in relation to cross sectional area, 344
  in relation to growth, 343
  measurement of, 106, 386
  regulation of, 104, 445
  sex and age, 342
  sex and, 343
  testing of, 103
  training effect, 108, 428–434
  variation in, 106
Muscle temperature, 597
Muscle tissue:
  glycogen content of, 551
  sampling of, 551
Muscular contraction, 100
Muscular fatigue, 308
Muscular soreness, 433, 470
Mutations, 4
Myelin, 58, 72
Myelin sheath, 57, 70
Myofibrils, 20–24
Myofibrillar:
  ATPase, 33–34
  fine structure, 23
  system, 24
Myoglobin, 248, 307, 327
Myoneural junction, 24
Myosin 21, 23–24, 26–27, 35
  ATPase, 35
  cross-bridges, 27
  filaments, 28
  head, 27
Myotatic stretch reflex, 112

N^+, 70
NAD-$NADH_2$, 528–534
Na^+ -K^+ pump, 62
Na^+ channels, 62
Natural endowment, 296
Natural selection, 4
Neanderthal, 6

Needle biopsy technique, 551
Nerve axon, 70
Nerve cell, 56, 59, 73
Nerve conduction velocity, effect of temperature, 592–593
Nerve fiber, 70
Nerve impulse, 62, 100
  transmission of, 56
Nerves:
  degeneration of, 110
  denervation of, 110
  regeneration of, 110
Neural control, 73
Neuromuscular function, 102, 647
Neuron circuits, 74
Neuronal activity, integration of, 92
Neurotransmission, 68
Neurotransmitters, 68–69
Newton, 727
Nicotinamide-adenine-dinucleotide (NAD), 531
Nitrogen, 707
Nitrogen excretion, 543
Nodes of Ranvier, 57
Nomogram, Åstrand & Åstrand, 364
Nonprotein respiratory quotient, 544
Nonshivering thermogenesis, 590
Norepinephrine, 144, 160, 452, 467, 499, 558
  effect on FFA, 558
  temperature regulation and, 607
Notochord, 274
Nuclear membrane, 14
Nuclear sap, 14
Nucleolus, 14
Nucleus, 14, 17
Nucleus pulposus, 289
Nutrients:
  optimal supply of, 574
  resorbtion of, 542
Nutrition:
  in general, 540, 574
  physical performance and, 538–540
Nutritional requirements:
  in general, 574
  of athletes, 563–567

Obesity, 571–573
OBLA (onset of blood lactate accumulation) test, 328, 383
Oral temperature, 586
Osmosis, 2, 154, 172
Osteoblasts, 275
Osteoclasts, 275
Osteocytes, 275
Osteoporosis, 277, 281
Oxaloacetate (OAA), 531
Oxidation, 527
  of fuel, 530
Oxidative decarboxylation, 531

Oxygen, 1, 707
  binding capacity of, 247
  breathing, 707–708
  effect of breathing 100 % O_2, 264
Oxygen debt, 299, 303
Oxygen deficiency, 300, 384
Oxygen deficit, 299, 300
Oxygen dissociation curve, 180, 692
Oxygen extraction, 181–182
Oxygen inhalation in sports, 708
Oxygen poisoning, 264
Oxygen pulse, 404
Oxygen radicals, 469
Oxygen tension:
  in alveolar air, 237
  in blood, 238
Oxygen transport, 174
  at altitude, 690
Oxygen uptake, 45, 178
  at start of exercise, 300
  estimated by heart rate, 495–496
  heart rate and, 372–373, 375
Oxygen uptake maximal:
  accuracy of prediction, 380
  bicycling vs. running, 357
  in relation to body size, 399–400
  in relation to performance, 338–341
  in various sports, 367–368
  measurement of, 359
  prediction from heart rate, 365–374
  prediction from submaximal tests, 372
  prediction of, 365–366
  related to cardiac output, 195
  test, 357
  test procedure, 359–367
  test results, 382
  type of exercise, 356
Oxyhemoglobin, 131, 179, 247
Ozone, 1–2

Pacemaker, 161
Pain receptors, 93
Parasympathetic nerve activity, 144
Partial refractory period, 63
Pascal, 727, 730
Peak height velocity, 330
Peak weight velocity, 330
Pedal frequency, 364
Pep pills, 713
Pepsin, 542
Peptidases, 542
Perceived exertion, subjective rating, 328, 371, 446
Performance capacity, 296
Performance, physical, 297, 487
  alcohol and, 711–712
  athletic, 646–679
  biochemical processes involved, 648
  doping and, 713
  factors affecting, 683–714

Performance, physical (*Cont.*)
  following exposure to high altitude, 704
  functions involved, 648
  high altitude and, 684–706
  motivation and, 715
  oxygen inhalation and, 708
  structural basis, 648
  tobacco smoking and, 709–711
Perimysium, 19
Periosteum, 279
Peripheral chemoreceptors, 249, 252
Peripheral resistance, 149
pH, 133–139, 179, 249–255, 324, 436, 691
  and kidney function, 138
  arterial blood, 137
Phagocytosis, 218
Phosphocreatine (PCr), 303, 314, 529
Phosphorylation, 531
Photosynthesis, 1, 525
Physical fitness, 296
Physical performance:
  capacity, 298
  effect of natural endowment, 296
  effect of training, 296
  factors affecting, 297, 488–489, 648
Physical work capacity, 378
Plasma, 131–132
  albumin, 132
  composition of, 132
  electrolytes, 132
  fibrinogen, 132
  globulin, 131–132
  proteins, 132
Plasma FFA level:
  and utilization, 559
  effect of epinephrine, 558
  effect of growth hormone, 560
  effect of hormones, 559
  effect of insulin, 559, 561
  effect of lactate, 560
  effect of norepinephrine, 558, 560
  response to exhausting exercise, 555
  skimming, 133
Pleurae, 219
Pneumothorax, 219
Poiseuille-Hagen formula, 148
Polycythermia, 700
Population code, 67
Postcapillary venules, 129
Posterior hypothalamus, 602
Postganglionic fibers, 57
Postsynaptic membrane, 68
Posture, 112, 114, 164, 245
  maintenance of, 112
Postural sway, 114
Potassium, 59, 64, 69, 129
Power (conversion table), 393, 727, 731
  definition of, 297, 726
Precapillary sphincters, 151, 153
Prediction of maximal oxygen uptake, 372

Prefixes, 726–727, 731
Pressure (conversion table), 729, 731
  definition of, 726
Presynaptic synapses, 67
Primates, 4
Prolonged exercise, 308
  blood lactate, 309
  body weight, 309
  heart rate, 309
  pulmonary ventilation, 309
  rectal temperature, 309
  respiratory rate, 309
Propranolol, 190
Proprioceptors, 55, 73
Protective clothing, 630–634
Protein, 540, 542–543, 564
Protein digestion, 542
Protein rods, 21, 23
Psychostimulants, 713
Pulmonary capillary, 243
Pulmonary perfusion, 244–247
Pulmonary ventilation, 228, 230, 232, 244–247
  at altitude, 691
  at rest, 228
  dead space, 232
  determination of, 228–229
  during exercise, 229, 255
  effect of exercise, 260
  maximal, 231
    in relation to body size, 399–400
  regulation of, 249–261
  respiratory frequency, 234
  sex difference, 232
  tidal volume, 234
Punctuated equilibrium, 4
Purkinje cell, 89–91, 140
Pyramidal cells, 84
Pyramidal motor centers, 97
Pyramidal tract, 73, 84, 91, 92
Pyruvate (pyruvic acid), 10, 531
Pyruvate dehydrogenase (PDH), 531

R (see Respiratory quotient)
Radiant heat, 628
  effect of shielding, 630
Radiant heat exchange, 584, 588, 628
Radiometer, 586
Ramapithecus, 5–6
Receptor potential, 63
Receptor-mediated endocystosis, 15
Receptors, 15–16, 55–56, 76, 163–168, 601
Reciprocal inhibition, 80, 109
Reciprocal innervation-inhibition, 81
Recovery, 320, 465
Rectal temperature, 585
Red cell volume, 130
Red nucleus, 91
Reduction, 527
Reflex activity, 73, 77

Regulation of food intake, 569
Regulation of fuel utilization, 557–562
    schematic diagram, 558
Reinfusion of red cells, 183–185
Relaxation, 30
Renshaw cells, 75–76
Residual volume (RV), 223
Respiration (see also pulmonary ventilation), 222
    "air conditioning", 218
    cost of breathing, 221, 236
    diffusion in lung tissues, 240
    filtration and cleaning mechanisms, 218–219
    main function of, 210–211
    regulation of breathing, 249
    respiratory frequency, 234
    respiratory work, 221, 236
    tidal volume, 234
    volume changes, 222–224
Respiratory gas exchange, 587
Respiratory muscles, 220–221
Respiratory quotient (R), 136
    effect of diet, 544–548
    effect of exercise, 543–547
    use in prediction of max. V_{O_2}, 380–381
Respiratory threshold, 329
Respiratory tree, 211
Respiratory work, 236, 685
Resting membrane potential, 59
Resting muscle, 28
Reticular formation, 91
Ribosomes, 17
Rowing, 356, 670–674
    blood lactate levels, 673
    energy demand, 671
    force acting on the oars, 672
    maximal O_2 uptake and performance, 671
    records, 670
    water resistance, 670
Running, 49, 653
    energy demand of, 653
Running speed, 396
    for children, 394–398
    in relation to body weight, 397
    maximal, 394

Salt intake, 566
    loss by sweating, 594
Sarcolemma, 19
Sarcomere, 21, 146
Sarcoplasm, 23, 72
Sarcoplasmic matrix, 23
Sarcoplasmic reticulum, 23, 186
Satiety, 569
Sauna bath, 594
Sciatica, 287
Scoliosis, 292
SDA (specific dynamic action), 506
Second wind, 262
Semistarvation, 375

Sensory feedback, 73
Serum, 132
Sex:
    differences, in anaerobic power, 341
    differences, in athletic achievements, 344,
        678–679
    differences, in maximal oxygen uptake, 332,
        402
    differences, in muscle strength, 343
    effect on heat tolerance, 617
    respiratory function, 224
Shell (the skin), 585
Sherrington's lengthening reaction, 122
Shift work, 515–517
Shivering, 591
Sinus node, 140, 161
SI system (Système International d'Unités),
    definition of, 725–726
Skeletal muscle, 19
    capillary density, 157–158
    composition of, 40
    vascularization of, 157
Skeletal system (see bone), 273
Skiing:
    alpine, 664–667
    blood lactate levels, 663, 665, 667
    cross-country, 356, 661–663
    energy cost of, 661–662, 666
Skin temperature, 586
    effect of cold, 591
Slimming diets, 573
Sling psychrometer, 587
Slow twitch fibers, 33–34, 101, 104
Smoking habits of athletes, 710
Smooth muscles, 19
Soccer, 674–676
"Soda loading", 325
Sodium, 129
Sodium bicarbonate, 325
Sodium chloride, 129, 566, 594, 621
Sodium pump, 59
Sodium-potassium ATP Pump, 59
Solar energy, 2
Soma, 56–57
Sore muscles, 433, 470
Spatial summation, 75
Speed of contraction, 40
Speed skating, 657–660
    effect of air resistance, 659
    effect of blood lactate levels, 660
    oxygen uptake, 658–659
        at different speeds, 659
        at different styles, 659
    versus running, 658
Speed, 730
Spinal column, 286–292
    calisthenics and, 292
    effect of extreme bending, 292
    effect of trunk muscles, 289–291
    lifting and, 291

Spinal column (*Cont.*)
  load carrying and, 291
  stability, 288
Spinal cord, 74
Spinocerebellar tracts, 88
Spirometer, 223
Splanchnic nerve, 160
Sports performance, sex differences, 346–347
Staircase test, 384
Starches, 542
Starling's heart law, 186
Static gamma axons, 78
Steady state, 299, 301, 308
Step test, 355, 366
Stimulants, 713
Stimulation, 101
STPD (Standard temperature and pressure dry), 733
Strength training:
  program, 432
  dynamic, 433
  hormonal response to, 498–499
  isometric, 433
  nervous response to, 500
  specificity of, 108
Stress reactions, 654
Stress tests in patients, 378
Stretch receptors, 163
Stretch reflex, 49, 93
Stretching, 468
Striated muscle, 22
Stride frequency, 653
Stride length, 650
Stroke volume, 185–188, 201, 441, 598, 622
  altitude, 692, 699
  heat, 598–600
  training, 440
Subjective rating, of perceived exertion, 371
Subliminal, 63
Sublingual temperature, 586
Submarine escape, 263
Submaximal, cycle ergometer test, 369
Sucrose, 542
Summation, 100
Supermaximal intensity, 302
Supraspinal control of motoneurons, 82
Swallowing reflex, 541
Sweat, 594, 602, 605
  loss, 565–566, 587
  nitrogen content of, 543
  production, individual differences, 593
Swimming, 356, 464, 591, 654, 706
  average speed, 657
  energy cost of, 656
  flume, 657
Symbols, 726–734
Sympathetic nerve activity, 160, 169, 172, 500
Synapses, 74–75
Synaptic:
  cleft, 57–58
  functions, 75

Synaptic (*Cont.*)
  knobs, 57–58, 61
  potentials, 63
  vesicles, 61

T system, 23
Table tennis, 675
Tachycardia, 161, 163
Temperature, 732
  acclimatization, 605–610
  conversion table, 732
  core, 585
  eardrum, 585
  effect on coordinated movements, 618
  effect on mental performance, 619
  effect on muscle performance, 592
  esophagus, 585, 596
  rectal, 585, 596
  regulation, 585, 600–604
  regulatory center, 600
  shell, 585
  skin, 586
  tolerance, upper limit of, 613–617
Temporal summation, 63, 75
Tendons, 286
Tension, 146
Terminal cisternae, 23
Terminals, 56
Termistors, 586
Terms, definition of, 725
Tetanus, 100–102
Thalamus, 87–88, 92
Thermal:
  comfort, 588
  neutrality, 588
  physiology, practical application, 625–626
  properties of clothing systems, 633
  regulation, effects of fluid-electrolyte changes, 601
Thermocouple, 586
Thermogenesis, 568, 590, 608
Thermoreceptors, 588
Thermoregulatory center, 586, 602–603
Thirst, 620
Thorax, 220
Threshold, 66
  anaerobic, 327–330
  firing (nerve cell), 67
  respiratory, 329
Thrombocytes, 131
Thyroid hormones, 469, 591
Tidal volume (VT), 223, 234
Tidal volume-respiratory frequency, 234
Tilting table, 164
Time, 394
Tissue pH, metabolic effect on, 324–325
Tobacco smoking, 707–711
  circulatory effects, 709
  metabolic effects, 710
  respiratory effects, 710

Tolerance to climate, limits of, 610
Torr, 730
Total lung capacity (TLC), 223
Tracheal air, 242
Training, 37, 296, 412–476
  assessment of effects, 412–416, 449
  contraindications, 474–476
  effect of anabolic steroids, 434
  effect of beta-adrenergic blockers, 453
  effects of continuous vs intermittent, 423–427
  effect of hypoxia, 468
  effects of ski pole walking, 424
  effects of steroids, 434
  effect on:
    aerobic power, 442–447
    anaerobic power, 436–438
    anteriovenous oxygen differences, 440, 454
    blood lipids, 464
    blood lipoproteins, 464
    blood pressure, 455
    capillary density, 430
    cardiac output, 440, 454
    central circulation, 454
    children, 454
    heart rate, 369, 440, 452
    heart volume, 449
    heat tolerance, 468
    locomotive organs, 427–428
    maximal oxygen uptake, 421, 439
    mechanical efficiency, 465
    mitochondria, 459
    muscle fiber composition, 461
    muscle mass, 430
    muscle strength, 428–429
    nerve sprouting, 469
    organ functions, 416–419
    oxygen radicals, 469
    oxygen-transporting system, 449
    peripheral circulation, 461
    plasma catecholamines, 453
    plasma hormones, 467
    pulmonary function, 455
    skeletal muscle, 456–461
    stroke volume, 440
    technique, 465
    total hemoglobin, 450
    vascularization of muscle, 450–451
  following bed rest, 438
  long-term effects, 427–468
  muscle strength, 432–434
  psychological effects, 469
Training, specificity of, 463
Transfer effect, 98
Transferrin, 16
Transmission of impulses, 70
Transmitter mechanism, 62
Transmitter substance, 58, 61, 65, 68, 72
Transverse tubular system, 23–24
Treadmill, 355, 359, 360–361
Treadmill running, 361–363
Treadmill test, 360–363

Tricarboxylic acid cycle, 531
Tropomyosin, 23–24, 26, 28, 186
Troponin, 23–24, 28, 186
Troponin C, 29
Troponin-tropomyosin-actin complex, 29
Trypsin, 542
Turbulent flow, 147
Twitch, 101–102
Type I fibers, 33, 35, 445, 461
Type II fibers, 33, 35, 445, 461
Type IIb fibers, 36
Type IIc fibers, 36

Ultraviolet radiation, 1
Unicellular organism, 2
Units, definition of, 725
Upright position, 112, 164
Urine, 498

Vacuole, 14
Valsalva maneuver, 266, 310
Vascular bed, 150
Vascular system, 129
Vascular tension, 146
Vasoconstriction, 160–165, 168–174, 589–590, 602
Vasoconstrictor center, 161
Vasodepressor area, 161
Vasomotor tone, 160–162
Vasovagal syncope, 160
Veins, 156
Venous blood return, 176, 178
Vibration, 515
Visceroceptors, 55
Vital capacity (VC), 223, 234
Voltage-gated, 59, 62

Walking, 49, 649–652
  effect of body weight, 650
  effect of speed, 650
  energy cost of, 652
Warm-up, 627–628
  active versus passive, 627
  effect on muscle temperature, 629
  effect on physical performance, 627–628
Water:
  balance, 619
  deficit, 621
  loss, 619
  pool, 129
  space, compartments of, 129
Watt, 727–729
Weight, 727
Weight velocity, 330
Wet Bulb Globe Temperature index (WBGT), 614
Wet Globe thermometer, 614
Will to win, 715–716
Windkessel vessel, 143, 150

Wingate test, 384
Work (conversion table), 728
  capacity (aerobic power), assessment of, 490
  classification of, 490, 502
  definition of terms, 487–488, 726
  energy and, 728
  factor, 604
  load, assessment of, 491–495
  relative, 490
  versus work capacity, 487–488

Work, external, in relation to body size, 398
Wrestling, 624

Year-round training, 447

Z line, 20, 22–23